GUIDELINES FOR

Safe Handling of Powders and Bulk Solids

GUIDELINES FOR

Safe Handling of Powders and Bulk Solids

CCPS
CENTER FOR CHEMICAL PROCESS SAFETY

An AIChE Industry Technology Alliance

CENTER FOR CHEMICAL PROCESS SAFETY
of the
American Institute of Chemical Engineers
Three Park Avenue, New York, New York 10016

American Institute of Chemical Engineers
3 Park Avenue
New York, New York 10016-5991

Library of Congress Cataloging-in-Publication Data:

CIP data applied for

ISBN 978-0-8169-0951-3

CCPS Publication G-95

CONTENTS

4
Assessing Particulate Hazards

5
Equipment Hazards and Preventive/Protective Measures

7
Plant Operation and Maintenance

8
Occupational Health and Environmental Considerations

ACKNOWLEDGMENTS

The American Institute of Chemical Engineers wishes to thank the Center for Chemical Process Safety (CCPS) and those involved in its operation, including its many sponsors whose funding made this project possible; the members of its Technical Steering Committee who conceived of and supported this Guidelines project, and the members of the Bulk Solids Handling subcommittee.

It is the subcommittee's belief that if the information contained in this book happens to prevent one bulk solids fire or dust explosion, the efforts of all those involved in preparing this work will be justified and rewarded.

Principle authors:

Stanley S. Grossel (Process Safety & Design, Inc., Clifton, NJ)
Robert G. Zalosh (Worcester Polytechnic Institute, Worcester, MA)

The members of the CCPS Bulk Solids Handling subcommittee were:

Russell Kahn (Chair, Syngenta Crop Protection)
Dan Sliva (CCPS Staff Liason)
John Bresland (US Chemical Safety and Hazard Investigation Board)
Larry Britton (consultant)
Warren Greenfield (International Specialty Products)
Dave Herrmann (DuPont)
Dave Kirby (Baker Engineering & Risk Consultants, Inc.)
Melvin Nelson (Syngenta Crop Protection)
Al Ness (Rohm & Haas)
Jeff Philiph (Monsanto)
Gary Pilkington (Abbott Laboratories)
Monica Stiglich (3M)

The following individuals also made contributions to the subcommittee:

Walt Frank (ABSG Consulting, Inc.)
Carla Hardy (Eastman Kodak)
Ronald Kersten (TNO)
Roy Winkler (Sunoco)
Paul Wood (Eli Lilly)

Finally, these individuals are thanked for their contributions to our principal authors:

Richard E. Bassett (Gustafson, LLC)
Lyndon Bates (Ajax Equipment Limited)
Robert P. Benedetti (National Fire Protection Association)
Timothy J. Carney (Palamatic Handling USA, Inc.)
Cristoph Cesana (Kuhner)
C. A. Crouch (Consultant)
Vahid Ebadat (Chilworth Technology)
Henry L. Febo (FM Global Technologies LLC)
John W. Fields (LCI Corporation)
Julie Janicke (B.A.G. Corporation)
Trevor A. Kletz (Process Safety Consultant)
R. Glenn Lunger (Fuller Bulk Handling Corporation)
Bill Mahoney (The Young Industries, Inc.)
Peter E. Moore (Kidde International)
Matthew Paine (Chilworth Technology)
Andre Petric (Glatt Air Techniques, Inc.)
Richard D. Pickup (formerly with Chilworth Technology)
John L. Roberts (Steri Technologies, Inc.)
Martyn Ryder (Extract Technology)
Robert W. Schoeff (Kansas State University)
Jim Shell (Niro Incorporated)
Albert J. Shohet (Processall, Inc.)
Amy Spencer (NFPA)
Bruce Teeling (Key International, Inc.)
William A. Thornberg (formerly with GE Global Asset Protection Services)
Alan Tyldesley (U.K. Health and Safety Executive)
Edward B. Weisselberg (Wyssmont Company, Inc.)
Jack Zoppa (Vac-U-Max)

Prior to publication, all CCPS books are subjected to a thorough peer review process. For such effort, CCPS also gratefully acknowledges the thoughtful comments and suggestions of

John Alderman (RRS)
Sheila Beattie (Syngenta Crop Protection)
John V. Birtwistle
Larry Bowler
Reinhard E. Bruderer (PRED)
Wayne Cannell (Gowan Milling)
William Chandler (OSHA)
Don Connolley (Akzo)

Tony Downes (FMC)
Bob Gravell (DuPont)
Tom Hoppe (CIBA Specialty Chemicals)
Pete Lodal (Eastman Chemical Co.)
Brian Dean Moore (Eli Lilly)
Samuel J. November (Rohm & Haas)
Tony Powell (3M)
Sam Rodgers (Honeywell)
Joe Senecal (Kidde-Fenwall)
Bob Stankovich (Eli Lilly)
Kenan Stevick (Dow)
Tony Thompson (Monsanto)
Alan Tyldesley (Health & Safety Executive, U.K)
Erdem A. Ural (Loss Prevention Science & Technologies, Inc.)

Chapter 1

INTRODUCTION AND OVERVIEW

1.1 PURPOSE OF BOOK

This book is intended to be a resource for process design and plant engineers who are responsible for designing and running processes handling powders and bulk solids in the chemical, pharmaceutical and related manufacturing industries. The book can also be an aid for process hazard analysis (PHA) teams and leaders, and for people operating small plants and toll operations. It may also be useful to insurance and regulatory personnel with assignments at industrial facilities that process, store, or transport large quantities of solid particulates.

The main focus of the book is on the instability, reactivity and combustibility hazards of particulate solids manufactured or handled in the chemical and pharmaceutical industries. Toxicity hazards are also discussed, but to a lesser extent than the other hazards. Much of the material presented may also apply to the food processing, grain handling and coal mining industries. The book does not cover the hazards of Explosives (UN-DOT Class 1 Materials) but does include UN-DOT Class 4 material (flammable solids, spontaneously combustible materials and materials that are dangerous when wet) Class 5 materials (oxidizers and organic peroxides), and Class 6.1 toxic materials, as well as the testing to distinguish explosives from the other UN-DOT categories.

Definitions and examples of these hazards and some key national and international standards covering them are presented in Section 1.2. All four generic hazards depend on particle size and various other particulate properties. Descriptions of these properties and their measurement are provided in Chapter 2 of this book. Accident scenarios and case histories are discussed briefly in Section 1.3, and in much more detail in Chapter 3. Particulate hazard assessment, via laboratory testing and other methods, is described in Chapter 4, with Appendix A being a listing of laboratories that conduct these tests. The types of particulate storage and handling equipment, are described in Appendix B. Chapter 5 is a discussion of the hazards and corresponding protection methods for the various equipment and operations in Appendix B. General protection measures applicable to particulate handling/processing equipment and facilities are described in Chapter 6. Chapter 7 discusses how plant operation and maintenance practices can influence particulate hazards. The final chapter, Chapter 8, describes occupational health and environmental concerns and regulations pertinent to potentially hazardous particulate material processing.

1.2 PARTICULATE HAZARDS

1.2.1 Combustibility Hazards

Combustibility hazards refer to the fire and explosion hazards of particulates in either bulk form, layer form, or in the form of a suspended dust cloud. NFPA 704 (2001) has a five-category flammability rating that provides an indication of the general combustibility hazard. The criteria for placing a particulate material in one of the five categories are shown in Table 1-1.

TABLE 1-1
NFPA 704 Flammability Categories for Particulates

NFPA 704 Flammability Hazard Category	Criteria for Particulate Materials
0	Materials will not burn in air when exposed to a temperature of 815.5°C (1500°F) for 5 minutes.
1	Combustible pellets with a representative diameter greater than 2 mm (# 10 mesh).
2	Solid materials in the form of powders or coarse dusts of representative diameter between 420 microns (# 40 mesh) and 2 mm (# 10 mesh) that burn rapidly but that generally do not form explosive mixtures with air; or Solid materials in a fibrous or shredded form that burn rapidly such as cotton and hemp.
3	Flammable or combustible dusts of representative diameter less than 420 microns (# 40 mesh); or Materials that burn with extreme rapidity, usually by reason of self-contained oxygen e.g., many organic peroxides; or Materials that on account of their physical form can form explosive mixtures with air.
4	Materials that ignite spontaneously in air.

One common particulate fire scenario that is applicable to many materials that are in flammability categories 1, 2, or 3 is the smoldering fire that develops in silos, bunkers, and hoppers. There have been numerous incidents of this type in grain silos, coal bunkers, and plastics manufacturing and processing facilities, and many of these fires have been very difficult to extinguish. Another common fire scenario is the overheating of particulates in various types of dryers. Both the drier fire scenario and the bulk storage smoldering are usually examples of particulate self-heating and spontaneous combustion. Many agricultural products are prone to self-heating due initially to microbiological activity, and later to oxidation during bulk storage. Examples include bagasse, compost, hay, pecans, soya beans, and walnuts. Activated carbon, hafnium and zirconium powder are examples of materials that can undergo oxidative self-heating when they are stored as fine particles.

A dust explosion hazard exists when flammability category 3 particulates are suspended in air at a concentration above the Minimum Explosible Concentration (MEC). As documented in Section 1.3.1, prevalent sites for particulate explosion scenarios include blenders, pulverizers, hoppers, conveyor/elevator transfer stations, and dust collectors. Important features of these locations are frequent dust clouds, moving mechanical parts representing potential ignition sources, and confinement to allow potentially damaging pressures to develop as a result of an accidental ignition. Descriptions of these and other particulate processing and transport equipment are provided in Chapter 5 along with a discussion of specific hazards associated with the equipment. Generic dust explosion hazard scenarios are described in Section 3.7.

Particulate fire and explosion prevention measures for general processing and handling facilities are described in NFPA 654. Preventive measures for electrical and electrostatic ignition sources are contained in additional standards such as NFPA 499, NFPA 77, and IEC 61241. Particulate explosion prevention systems and deflagration venting systems are presented in NFPA 69 and NFPA 68, respectively. There are also fire protection standards for specific particulate materials such as pesticides (NFPA 434) and organic coatings (NFPA 35).

1.2.2 Instability Hazards

Particulate instability is the tendency of certain bulk solids to vigorously decompose, polymerize, become self-reactive, or oxidize at the temperatures and other conditions they are subjected to during physical processing, transport and storage. These exothermic reactions can generate potentially dangerous temperatures, pressures, or hazardous gases, or otherwise become violent.

NFPA 704 (2001) defines five hazard categories for unstable materials, with the lowest (zero) category for materials that do not have an exotherm at temperatures at or below 500°C. The four higher categories are defined qualitatively in terms of their instability initiation requirements, and quantitatively in terms of their instantaneous power density (heat of reaction multiplied by reaction rate) at 250°C. The instability category of a material is one of three factors that must be prominently displayed in industrial and commercial facilities manufacturing, processing, storing, or using hazardous materials. The U.S. Department of Transportation and the United Nations regulations for shipping of hazardous materials have generic classifications for self-reactive solids (UN 3224 and 3234), and specify packaging and testing requirements for these materials (49CFR Parts 172-173). One other source of instability hazard ratings is the Hazardous Materials Identification System promulgated by the National Paint and Coatings Association (NPCA).

Particulate materials that have either high NFPA 704 reactivity ratings, or are designated by criteria as U.N. self-reactives, or have been involved in noteworthy incidents include ammonium perchlorate, azodicarbonamide, methyl parathion, potassium nitrate, and sodium azide. The latter, which is designated as a U.N. Class 6.1 toxic material, has been involved in several explosion incidents at airbag propellant manufacturing facilities. Hydroxylamine is a self-reactive particulate material that is so prone to violent self-decomposition that it is always stored/transported in aqueous solutions, and has been involved in several explosions when the solution became too concentrated. Other decomposition incidents are described in Section 1.3.3.

Instability hazard scenarios involving external heating, self-heating, and other initiation modes are discussed in Section 3.1. Laboratory tests to assess particulate instability hazards are described in Section 4.3. In addition to the federal and U.N. standards mentioned above and various NFPA standards for different types of potentially unstable materials, there are general protection recommendations for unstable materials in the CCPS Guidelines (1995), and in VDI Guideline 2263 for powders and dusts.

1.2.3 *Reactivity Hazards*

Particulate reactivity is the tendency of certain bulk solids to react with other materials that they may contact during bulk storage, transport, or physical processing. These materials can be the container material itself, contamination from previous loads or batches, or, in the case of water-reactive materials, water leakage into the container or process vessel. NFPA 704 has a provision to designate water-reactive materials so that emergency responders will be aware of the reactivity hazard when they determine appropriate response measures. Four different NFPA 704 categories of water reactivity are defined in terms of the heat of reaction. Some examples of particulate materials with high water reactivity ratings are calcium carbide and calcium hypochlorite. The National Paint and Coatings Association's Hazardous Materials Identification System® has a similar provision for alerting plant personnel to the reactivity hazard of chemicals used in paint and coatings.

One well-known example of a reactive incident occurred when water inadvertently entered a blender containing water reactive materials, and caused the blender to explode because of an inadequately sized emergency vent (EPA/OSHA 1997). Another water reaction occurred in 1998 when steam was deliberately used in an attempt to clear an aluminum and aluminum chloride sludge blockage at the bottom of a linear alkylbenzene reactor. There has also been a series of fire incidents initiated from inadvertent wetting of the chlorinated swimming pool chemicals, calcium hypochlorite and trichloroisocyanuric acid, while stored in warehouses and building supply stores.

More complete descriptions of some of these water reactivity hazard incidents and scenarios are provided in Sections 3.3. Reactivity hazard scenarios involving contamination of particulates during transport and storage, and container/packaging reactivity are also presented in Section 3.3. Updated information on U.S. government activities on chemical reactivity hazards can be found in the OSHA Reactivity Web site, http://www.osha.gov/dep/reactivechemicals/index.html.

1.2.4 Toxicity Hazards

The most common toxicity hazard associated with particulates is the respirable hazard associated with particles in the size range 0.2 to 7 μm. Particles in this size range can flow through the bronchi and penetrate into the alveoli, where some particles can remain for decades (King, 1990). Submicron particles are more readily exhaled and therefore represent a lower hazard level than those in the 1–7 μm range. Once being lodged in the lungs, the chronic and acute effects of these particles depend on their biological activity and their solubility. Some examples of dust materials that are particularly hazardous in this regard are silica, coal dust, aluminum, and many heavy metals, such as beryllium, chromium, and plutonium (Kerfooot et al., 1995).

NFPA 704-2001 has five health hazard categories in its classification scheme for potentially hazardous materials. The criteria for placing a powder or dust in one of these categories are based in part on the LC_{50} concentration for acute inhalation toxicity. The specific criteria are given in Table 1-2.

Besides inhalation, the other pathways for small particles to enter the body include accidental ingestion, dermal contact, and eye entry. Toxicity hazards that can be manifested after entry into the body include systemic toxicity, allergic reaction, mutagenic effects, and carcinogenic reactions

TABLE 1-2
NFPA 704 Health Hazard Categories for Particulate Material Inhalation Toxicity

Health Hazard Category	LC_{50} (mg/L)
0	> 200
1	< 200 and > 10
2	< 10 and > 2
3	< 2 and > 0.5
4	< 0.5

(Kerfoot et al., 1995). The NFPA 704 health hazard categorization scheme includes criteria based on the LD50 values for acute dermal toxicity and for acute oral toxicity. Specific scenarios associated with both chronic exposures and acute exposures are discussed in Section 3.8. Asphyxia scenario examples are presented in Section 1.3.4.

1.3 ACCIDENT DATA AND CASE HISTORIES

As an introduction to the numerous case histories and other incident accounts described throughout this book, a statistical overview is presented here along with some representative examples of how the various particulate hazards have been manifested in accidents at industrial facilities.

1.3.1 Dust Explosion Data and Case Histories

Tabulations of materials and equipment involved in dust explosions have been compiled by various organizations. Representative data from organizations in the United States, Germany, and the United Kingdom are presented here. The data used to represent U.S. dust explosions are taken from insurance company loss history (Febo and Thornberg, 2001) because the losses were obtained from a broad cross-section of industrial facilities handling combustible particulates. The data from the U.K. were obtained from the Health and Safety Executive (HSE) and include particulate fires as well as explosions in U.K. facilities. The data for Germany were compiled by the German Institute for Safety at Work of the Trade Unions, as presented by Eckhoff (1997).

The data cited in Tables 1-3 and 1-4 represent only a small fraction of all the dust explosion incidents in the U.S., U.K., and Germany. In the U.S., there

TABLE 1-3
Particulate Materials Involved in Reported Dust Explosions

Material	U.S. (1985–1995) (FM Global, Febo, 2001)		U.K. (1979–1988)[a] (HSE)		Germany (1965-1980) (Eckhoff, 1997)	
	Number Incidents	%	Number Incidents	%	Number Incidents	%
Wood/Paper	56	37	69	23	120	34
Coal	27	18	24	8	33	9
Metals	19	13	55	18	47	13
Plastics	8	5	10	3	46	13
Food/Grain	?[b]	?[b]	94	31	88	25
Pharmaceuticals/Organic	?[b]	?[b]	27	9	?[b]	?[b]
Other/Unknown	4	27	24	8	23	6
Total	150	100	303	100	357	100

[a]The U.K. data include particulate fires as well as 140 reported explosions.

[b]This material category was not explicitly identified in the cited reference.

is no centralized national database and no requirement to report all explosion incidents. In the U.K., the HSE maintains a centralized national database, but receives reports on only a small fraction of all the incidents. The British Materials Handling Board (BMHB) conducted a voluntary survey in 1984 to assess the frequency of dust fires and explosions (Abbott, 1988). For the years 1979–1984, 84 incidents were reported in the BMHB survey, but only 3 of these were reported to the HSE. Furthermore, the data sources do not necessarily contain proportionate representation from the various industries and facilities handling combustible particulates. Therefore, the following tabulations are merely indicative of the types of materials and equipment that have been involved in dust explosions, and are not a reflection of the relative risks of specific materials and equipment.

Both the U.S. (FM) data and the German data in Table 1-3 indicate that the material most frequently involved in reported dust explosions is some form of wood or paper dust. In the U.K., food/grain particulate matter has the highest frequency of reported explosions. Food/grain is the second most frequently involved material in German dust explosions, and is also involved in a large percentage of U.S. dust explosions despite its absence

TABLE 1-4
Equipment Involved in Dust Explosions

Material	U.S. (1985–1995) (FM Global, Febo, 2001)		U.K. (1979–1988) (HSE)		Germany (1965–1980) (Eckhoff, 1997)	
	Number Incidents	%	Number Incidents	%	Number Incidents	%
Dust Collectors	156	42	55	18	73	17
Grinders/Pulverizers	35	9	51	17	56	13
Silos/Bunkers	27	7	19	6	86	20
Conveying System[a]	32	9	33	11	43	10
Dryer/Oven	22	6	43	14	34	8
Mixers/Blenders	>12	>3	7	2	20	5
Other or Unknown	84	23	95	31	114	27
Total	372	100	303	100	426	100

[a] Conveying systems include conveyors, ducts, and elevators.

from the FM tabulation in Table 1-3. A tabulation reported by Schoeff (2001) indicates that there have been 122 U.S. grain dust explosions in the 10-year period 1991 to 2000.

Metal powders/dusts have been involved in 13–18% of reported dust explosions in the three compilations shown in Table 1-3. The combined category of plastics and pharmaceuticals has been responsible for 37 U.K. explosions (12%) in the 10-year reporting period, and at least 46 explosions (13%) in Germany. Similar percentages of plastic and pharmaceutical dust explosions are contained in the 222 dust explosion losses reported by Industrial Risk Insurers (IRI) for the years 1975–2001 (Thornberg, 2001).

Process equipment frequently involved in dust explosions can be ascertained from the compilations in Table 1-4. In both the U.S. and the U.K., dust collectors have been most frequently involved. Three possible reasons for the high occurrence of dust collector explosions are (1) they are almost omnipresent in particulate handling facilities, (2) they inherently concentrate the smaller particles which are easier to ignite than the mostly larger particles in other equipment, and (3) dust collectors are often structurally weaker than other process equipment, and therefore more prone to explosion damage. In Germany, silos and bunkers have been most frequently involved, whereas they have only been involved in 6% to 7% of the reported dust explosions in the U.S. and the U.K. In all three compilations, grinders/mills and pulverizers have been involved in between 9% and 17% of all the reported

TABLE 1-5
Fires Following Dust Explosions (Thornberg, 2001)

Fires after dust explosions?	Number	%
Yes	156	70.3
No	56	25.2
Unknown	10	4.5

incidents. Particulate conveying systems have been involved to 9 to 11% of the reported explosions, and dryers/ovens have been involved in 6 to 14% of the tabulations in Table 1-4. Many of the larger explosions involved multiple types of equipment, with conveying systems and dust collectors often receiving damage from explosions initiated in other process equipment.

Most dust explosions are followed by fires as evidenced by the statistics in Table 1-5 from the IRI database (Thornberg, 2001). The fires are presumably caused by burning particles landing on nearby combustible materials.

The dust explosions reported to the various national safety authorities have caused numerous injuries and fatalities. For example, there were 103 fatalities and 492 injuries in the 357 dust explosions reported to the German Institute for Safety at Work of the Trade Unions, as presented by Eckhoff (1997). There were 100 injuries and 5 fatalities in the 140 dust explosions reported to the HSE for the period 1979-1988. More recent (1988–1993) HSE data reported by Owens and Hazeldean (1995) reveal that there were 827 injuries and 30 deaths in the 1273 dust explosions. There were 16 fatalities and 147 injuries in the 122 U.S. grain dust explosion reports compiled by Schoeff (2001) and the U.S. Department of Agriculture. The ratio of injuries per reported dust explosion in these data compilations ranges from 0.65 to 1.38, and the ratio of deaths per dust explosion ranges from 0.024 to 0.289.

A few brief case studies can best illustrate how and why some dust explosions are relatively inconsequential, while others involve tragic losses of life, numerous injuries, and major facility destruction.

Yowell (1968) described three minor dust explosions that occurred in a polycarbonate manufacturing plant in 1966-1967. The first two explosions occurred during loading of a phenolic intermediate called bisphenol-A into a storage silo. In both silo explosions, the bisphenol-A was being transferred from hopper trucks via positive pressure blowers in the trucks. The most probable ignition source in both incidents was reported to be an electrostatic discharge in the silo. Apparently, electrostatic charging of the powder occurred at it was transferred at a relatively high flow rate through an unbonded rubber hose connection from the truck to the transfer piping, and then directly into the silos. The transfer system was subsequently changed to a vacuum transfer from the hopper car by means of a vacuum blower downstream of the filters on top of the silos. The powder enters the silos by first passing through a rotary air lock valve below the filter.

Both silo explosions caused the explosion venting silo covers to lift and relieve the deflagration pressure as intended. There was some minor damage to the covers and piping on top of the silo, but no damage to the silo itself, and no personnel injuries. After the phenolic transfer system was changed, Yowell reports there were no further silo explosions but there was one minor explosion caused by an employee trying to free a plugged transfer line with a compressed air hose. Compressed air pressure caused the transfer line to separate and a cloud of bisphenol-A formed and was ignited, perhaps again by an electrostatic discharge. Although the employee was injured, he managed to extinguish the fire before seeking first aid. Fortunately, the explosion did not propagate away from the vicinity of the site where the transfer line was blown off.

On February 25, 1999, a devastating dust explosion occurred involving a phenol-formaldehyde resin being used along with sand to make foundry casting molds. After blending, the sand–resin mixture was conveyed to eight shell mold fabrication booths. A central dust collection system served all eight booths, and over a period of time resin dust accumulated in the ducting and on the various equipment and structural surfaces in and around the mold fabrication booths. Each booth had gas-fired ovens for curing the molds. On the day of the explosion, the oven burner flame ignited either a gas-air mixture formed following a temporary flameout, or a dust cloud formed from the shaking/striking of a flexible hose dust collection line (often called an elephant trunk). The initiating event caused flame and a pressure wave to enter the main dust collection ducting network and propagate the explosion to all the other mold booths in the building. The secondary dust explosion that occurred in the building caused extensive burn injuries to twelve employees, three of whom subsequently died. One entire masonry wall and portions of two other walls collapsed from the deflagration pressure (Joint Foundry Explosion Investigation Team Report, 2000). This was one of two similar multifatality secondary dust explosions that occurred that month (Zalosh, 2000).

The primary difference between the phenolic intermediate dust explosions at the foundry and the phenolic intermediate explosions described by Yowell (1968) was the propagation of the dust explosion away from the initiating site, and the eventual involvement of dust/powder that had accumulated in the ducting and on structural surfaces. The occurrence of secondary dust explosions is due in large part to the extended accumulation of dust layers throughout a large portion of either interconnected process equipment or building surfaces. These secondary explosions can be prevented by (1) designing and maintaining equipment to prevent particulate accumulations, (2) frequent and thorough cleaning of ducting and surfaces on which accumulated dust layers have developed, and (3) installing explosion isolation systems of the type described in NFPA 69 and in Section 6.5.5 of this book.

1.3.2. Other Particulate Incident Databases

Many organizations maintain accident databases that can be searched for listings of incidents involving particulates. However, the authors of this book are not aware of any published general surveys of particulate incidents besides the dust fire and explosion incident compilations described in Section 1.3.1. Moreover, the authors and most readers do not have access to proprietary databases maintained by insurance companies and other private organizations. On the other hand, there are several public organizations and professional associations that maintain relevant databases. Table 1-6 is a listing of the salient features of these potentially accessible databases.

Since most process industry and hazardous material incidents often involve gases and liquids rather than solid particulates, most of the incidents in each of the Table 1-6 databases do not involve particulates. However, most of these databases can be either computer-searched or visually perused to focus on particulate incidents. One example is the OSHA online database, which contains both powder and dust as keywords for online searching. Web sites for the various databases are listed in Table 1-6.

1.3.3 Sample Case Histories for Particulate Instability, and Reactivity Incidents

Thermal decompositions have caused several incidents including the May 8, 1997 fire and subsequent explosion at an agricultural chemical packaging facility in Arkansas. The facility received bulk shipments of pesticides, insecticides, etc. and repacked them into smaller containers. On the day of the incident, the facility received a shipment of Flexible Intermediate Bulk Containers (FIBCs) of a pesticide called Azinphos methyl (AZM 50W). The FIBCs were loaded into the northwest corner of an approximately 7800 ft^2 warehouse. The AZM FIBCs were placed next to (and probably in contact with) a 15-ft-long hot compressor discharge pipe. Tests conducted by the EPA accident investigation team (EPA/OSHA 1999) indicated that the discharge pipe temperature was probably in the range 124°C to 149°C (255°F to 301°F) depending on how much of the FIBC was actually in contact with the pipe. Thermal stability testing of AZM indicates that it begins decomposing at a temperature of about 100°C (at least 24°C below the discharge pipe temperature), with an intense exothermic reaction beginning to occur at 170°C.

A few hours after storing the 26 AZM FIBCs, each containing about 1600 pounds of AZM, several plant employees noticed a large cloud of yellow smoke and a strong sulfurous odor of decomposing AZM emanating from the northwest corner of the warehouse. The plant employees evacuated and

TABLE 1-6
Accessible Incident Databases

Source	Years Covered	Types of Incidents	Locations of Incidents	Access Restrictions	Web Site	Comments
U.S. Chemical Safety Board Chemical Incident Report Center (CIRC)	1998 – Present	Nontransport incidents involving chemical fires, explosions, releases to environment, and asphyxiations.	No limits, but most are in U.S.	None for online access.	www.chemsafety.gov/circ/	Database can be searched online. Approximately 1500 incidents recorded through March 2002.
AIChE Center for Chemical Process Safety (CCPS)	Unknown	Process safety incidents with potentially important lessons-to-be-learned.	Unknown	Accessible only to companies that have contributed to database with their own incident accounts.	www.aiche.org/ccps/lldb.htm	Data do not include the name of the company involved, or the location of the incident. 24 Companies currently participate in database.
NFPA	1970 – Present	Explosions and fires in various properties.	U.S.	Summaries of large-loss fires published annually in NFPA Journal.	www.nfpa.org	The NFPA Research Division provides a service to customers that want to sort through NFPA databases for incidents involving particular materials.[b]

OSHA	1972[a]–Present	Incidents involving employee injuries and resulting OSHA investigations.	U.S.	None in principle.[a]	155.103.6.10/cgi-bin/inv/inv1	Searchable online database with brief narratives and results of regulatory investigations.
ISPRA[b] Major Accident Reporting System (MARS)	1980 to date	Major industrial accidents involving hazardous materials.	Europe	Public access only to short reports with plant names and locations deleted.[d]	mahbsrv.jrc.it/mars/Default.html	Short report database (< 10% of the reports) can be searched and sorted over Web site. Only abbreviated listings available online.
UK Chemical Reactions Hazards Forum	Not Specified	Mostly unintended or runaway reactions.	Europe	None.	www.crhf.org.uk/	Forum members meet twice a year to review new incidents and update the database.
United Nations Environmental Program (UNEP) Awareness and Preparedness on a Local Level (APELL)	1970 – 1998	Hazardous materials incidents with >24 dead or >124 injured or >10,000 evacuated or >10,000 people deprived of water.	International	None	www.unepie.org/pc/apell/disasters/lists/disastercat.html	About 14 incidents per year from 1979 to 1997, and fewer in other years Many incidents are taken from press reports, and are often not accurate. Listings only without any narrative.

[a] Many recent OSHA accident reports have not been reviewed yet, and are not available online. Particulate/dust incidents occurring after1996 were not accessible in March 2002.

[b] NFPA also processes data in the National Fire Incident Reporting System (NFIRS) maintained by the U.S. Fire Administration.

[c] ISPRA is a European Community Joint Research Center in Italy.

[d] The designated U.S. organization for MARS liaison is the EPA Chemical Emergency Preparedness Office.

called the local fire department, and firefighters arrived at the facility 13 minutes later. Firefighters remained outside the warehouse while deciding on a plan of attack. Approximately 30 minutes after the smoldering fire was first observed, the warehouse automatic sprinkler system actuated.

Unfortunately, the water spray discharge from the sprinklers wet some pallet loads of Maneb (polymeric manganese ethylenebisdithiocarbamate) stored near the AZM. Maneb reacts with water, releasing a heat of hydration and volatile decomposition products including carbon disulfide. Several minutes after the sprinkler system activated, while an electrical utility service employee started disconnecting the electrical power feed to the warehouse, an explosion occurred and blew out a cinder block wall. The collapsing wall struck four firefighters; three were killed and the fourth was seriously injured. The EPA/OSHA accident investigation team concluded that the explosion was probably due to an arc (generated at power disconnect) ignition of the gases and vapors generated by the decomposing AZM and Maneb.

Shortly after the explosion, a shifting plume of toxic combustion and decomposition products caused local authorities to initiate a temporary three-mile radius evacuation. The warehouse materials continued to burn unabated because firefighters did not want to apply water to the Maneb. On May 14th (6 days after the start of the fire), the firefighters implemented a recommendation to spread the Maneb into thin layers and apply a water fog. This technique was successful in extinguishing the fire. Accounts of other warehouse storage fire scenarios and firefighting experiences are discussed in Section 3.5.3.

Although the 1997 Arkansas warehouse fire and explosion was tragic, costly and disruptive, the explosion itself was far less energetic than several other bulk particulate explosions. Two of the most energetic explosion events, as measured in terms of calculated blast wave energy, were the 1988 ammonium perchlorate explosion in Henderson, Nevada (described in Chapter 3), and the September 21, 2001 ammonium nitrate explosion in Toulouse, France.

The September 21, 2001 ammonium nitrate explosion at the Grande Paroisse Toulouse Factory in Toulouse, France caused 30 fatalities, approximately 2500 injuries, and about $2 billion in damage (Financial Times, February 6, 2002). Figure 1-1 shows the destruction in the vicinity of the explosion: the remains of buildings in the area surrounding a crater approximately 40 m in diameter and 7 m deep. Windows were blown out in the center of Toulouse, about 3 km from the explosion site. The estimated blast wave energy required to produce this devastation is equivalent to 20–40 tons of TNT (Barthelemy et al., 2001).

The explosion occurred in a warehouse in which granular ammonium nitrate was stored flat, separated by partitions. Between 200 and 400 tonnes of ammonium nitrate, used for fertilizers and industrial chemical supplies, were stored in the warehouse. The ammonium nitrate stored in the warehouse consisted of industrial nitrates that did not meet commercial specifica-

Figure 1-1. Aftermath of September 21, 2001 ammonium nitrate explosion in Toulouse (from UNEP APELL web site).

tions in terms of particle size and possibly composition. The day before the explosion, 15 to 20 tonnes of product containing a new additive/coating at the qualification stage were placed in the building (Barthelemy et al., 2001). On the morning of the explosion, other off-specification product was brought into the building. Approximately 15 minutes prior to the explosion, a bin of disputed contents was dropped off in the airlock at the entrance to the warehouse. The worker who transported the bin said it contained recyclable bags labeled "nitrate," but French government investigators found bags of different chlorine-based products and a leaky bag of a fine white powder consisting of sodium dichloroisocyanuarate (DCCNa) (http://www.saunalahti.fi/ility/AZF.htm#ExMag), which was also manufactured in the plant.

Although the cause of the Toulouse explosion is still in dispute, the government inquiry reached the following preliminary conclusion (Kersten et al. 2002). Numerous contaminants (oils, organic debris, iron oxides, asphalt, etc.) had accumulated on the concrete floor of the warehouse, and contaminated the ammonium nitrate such that it would decompose and react energetically. The DCCNa, which may have been released just before the explosion, reacts with ammonium nitrate to produce nitrogen chloride (NCl_3), a particularly unstable gas that will explode at ambient temperature. This reaction is enhanced by high humidity, such as existed on the day of the explosion. Grand Paroisse argues that this contamination/reaction scenario is less credible than the explosion being triggered by large electrical faults that occurred shortly before the explosion.

1.3.4 Sample Case Histories for Asphyxia Incidents

The following accounts are taken from summaries of the OSHA accident investigations of fatal accidents involving asphyxia due to immersion in particulate piles.

On January 11, 1992, Employee #1, the yard foreman, went inside a cement silo to unclog the pouring spout from the inside. Employee #1 was tied off to a rung of a 16 ft ladder. While he was inside the silo, cement was being discharged. Employee #2 was outside the silo, checking on Employee #1, and saw him stuck in the cement powder. He went down the ladder to try and pull him out. Employee #2 could not pull Employee #1 out and also became stuck in the cement. Rescue was called and two fire fighters, who had climbed down into the silo, became stuck. The discharge pipe was enlarged by firemen cutting the rubber boot, which was part of the discharge pipe, allowing a free flow of cement from the tank's center, but the cement falling from the sides of the tank covered the men. Employee #1 died of suffocation and Employee #2 was hospitalized (OSHA Accident 000740761).

On October 13, 1990, employee #1 was one of two workers hired to assist in the installation of two baghouse (dust collector) clogging indicator devices. Prior to the installation, Employee #1 and a coworker entered the baghouse through a 19-in. hatch, stood on an 18-in. diameter auger which had been locked out, and used a pitchfork to loosen a buildup of nuisance dust. The co-worker in the first baghouse stated that the dust flowed around him up to his chest when it let loose, but he did not mention this to Employee #1, who entered the second baghouse to release the clog. Several minutes later, after not responding to a call, Employee #1 was found lying dead under several feet of the dust, asphyxiated by dust aspiration. It is possible that he tripped on the auger as he backed away from the dust mass as it was released (OSHA Accident 000785931).

Another important asphyxia hazard is associated with nitrogen inerting of vessels and silos containing certain particulates. Following is one account of a fatality associated with nitrogen inerting of particulate containers/vessels.

At approximately 12:55 P.M. on March 15, 1995, Employee #1, a chemical operator was found slumped in the manway of reactor XR30. According to the batch sheet, the employee had been dry charging bromoketone powder into the nitrogen-inerted reactor. The medical examiner determined that the employee died of cerebral anoxia due to inhaling nitrogen gas (OSHA Accident 170022818).

1.4 PARTICULATE HANDLING AND STORAGE EQUIPMENT HAZARD OVERVIEW

Large quantities of bulk particulate at industrial facilities are usually stored in bins, hoppers, and silos, as described in Section 5.3.15 of this book. Since silos are larger and more expensive than bins and hoppers, they are usually used for longer term storage, and are often grouped together with a common conveying system for loading and unloading. The common conveying system is often an avenue for dust explosion propagation between silos, such as occurred in the damaged grain elevator complex shown in Figure 1-2. Another important hazard consideration in silo/hopper design is whether to use a mass flow or core flow design with differences illustrated in Figure 1-3. There is a greater chance of particulate material being inadvertently retained

Figure 1-2. Silos destroyed in grain elevator explosion.

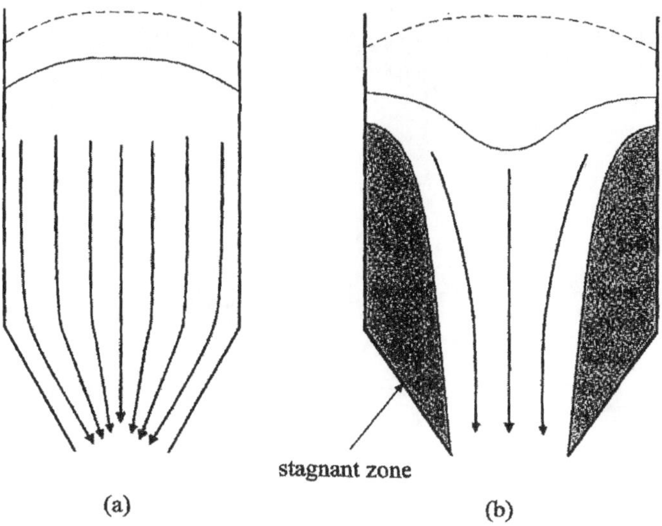

Figure 1-3. (a) Mass flow hopper and (b) core flow hopper (from Fan and Zhu, 1998).

near the silo/hopper walls for a longer duration, and possibly undergoing spontaneous heating, in the core flow design than in the mass flow design. Practical problems and solutions associated with silo/bin/hopper design and operation are discussed in the Silos, Hoppers, & Bins forum on the Bulk/Online forum Web site: http://www.bulk-online.com/Forum/.

Smaller quantities of particulate are stored in bags, drums, and Flexible Intermediate Bulk Containers. These smaller, portable storage containers are described in Sections 5.3.10 to 5.12 of this book. Dust explosion hazards associated with these portable containers arise during loading and unloading because the suspended dust concentration is often between the minimum and maximum explosible concentrations. Other hazards associated with these containers include container damage causing product leakage, and contamination with incompatible materials because of either storage of incompatible materials, or container recycling/mislabeling, and/or container breaches. These hazards were apparently manifested in the Toulouse ammonium nitrate explosion described previously. They were also manifested in the 1992 Allied Colloids Ltd. warehouse fire, which started when two or three drums of combustible powder ruptured, and the released combustible reacted with an oxidizing powder that had been stored in bags under the drums (HSE, 1993). Figure 1-4 is a photograph of the resulting pyrotechnics caused by the burning of the combustible powder while in intimate contact with an oxidizer.

Bulk particulate transport at industrial facilities occurs either in various types of mechanical conveyors and bucket elevators (described in Section

Figure 1-4. Allied Colloids warehouse fire due to storage of incompatible partiulates (from Gary Pilkington).

5.3.8), or in pneumatic conveying systems (see Section 5.3.9). Negative pneumatic conveying systems are operated at negative gage pressures by locating the exhauster fan or blower at the downstream end of the system. Positive pneumatic conveying systems are operated at positive gauge pressures by placing the blower at the upstream end of the system. Negative conveying systems have an inherent advantage for toxic and combustible particulates in that minor leakages will not produce releases of material.

Several reported dust explosions have been ignited in the boot or head of bucket elevators because of the normal presence of explosible dust concentrations together with frictional-heating ignition sources associated with misaligned moving parts and worn out bearings. Five of the fourteen grain dust explosions investigated by Kauffman through 1982 were ignited in the bucket elevator (Eckhoff, 1997, p. 172). Figure 1-5 is a photograph of a bucket elevator damaged from a corn dust explosion that was ignited by a hot spot due to welding on the elevator casing. Mechanical conveyors usually present less of an explosion hazard than bucket elevators and pneumatic conveying systems, but the case history summarized in Section 5.3.9 involved three

Figure 1-5. Bucket elevator damaged by grain dust explosion (from C.W. Kauffman with permission).

fatalities due to an explosion in a screw conveyor. The more common hazard in mechanical conveyors is a fire ignited by frictional heating at a damaged roller or bearing.

The particulate handling/processing equipment most often involved in dust explosions as indicated in Table 1-4 are dust collectors. The breakdown of the 98 dust collector explosions in the IRI/Thornberg database is as follows: 60 involved bag type collectors, 13 involved cyclone collectors, and 25 were either other or unspecified collector type. The large bag type collectors are usually referred to as baghouses, and they are often situated either on the roof or adjacent to the process building as shown in Figure 1-6, and as recommended in NFPA 654. The outdoor location of the baghouse has a mitigating effect in that it usually prevents the triggering of a secondary dust explosion in the process building, providing there is some type of isolation system for the collector ducting. Personnel entry into the baghouse does present a confined entry and associated asphyxiation hazard as illustrated by one of the preceding case histories. The various types of dust collectors and their associated hazards are discussed in detail in Section 5.3.4 of this book.

Dryers and ovens have been responsible for numerous fires due to overheating of combustible or unstable particulate materials. Some of the cited reasons for the overheating are given in Section 5.3.3 along with a detailed description of the various types of dryers. The recent Chemical Safety Board (CSB) investigation into the February 2003 dust explosion at the CTA Acoustics plant has indicated that a resin fire in a continuous web oven with malfunctioning combustion controls preceded the explosion. According to the

Figure 1-6. Baghouse dust collector (courtesy U.S. Air Filtration, Inc.).

CSB preliminary findings, flames escaping from an open oven door probably ignited a dust cloud in the area adjacent to the oven. The preliminary findings in the CSB investigation of the January, 2003 West Pharmaceuticals dust explosion also indicate that a drying operation may have been involved, but the West Pharmaceuticals drying process apparently was not enclosed and allowed polyethylene powder to be entrained into the air flow above a suspended ceiling. Other dryer/oven fire scenarios are discussed in Chapter 3 of this book.

As indicated in Table 1-4, various types of size reduction equipment, i.e. grinders, pulverizers, and mills, have been involved in a large number of dust explosions. Section 5.3.17 provides descriptions of the various types of particulate size reduction equipment and their associated hazards. The ignition sources for the two grinder/mill explosion case histories in Section 5.3.17 were frictional hot spots caused by tramp metal rubbing against the grinder/mill wall. This has also occurred in numerous other mill/pulverizer fires and explosions. Sometimes the tramp metal is due to the breaking of a mill hammer, ball, or other moving object.

Blenders have also been involved in numerous dust explosions and fires. Often the blending generates electrostatic charges on the combustible particulates, and there is a subsequent electrostatic discharge. Besides blenders and the previously cited equipment, other particulate handling and processing equipment discussed in Chapter 5 include feeders, samplers, screens, and granulators.

1.5 HISTORICAL AND REGULATORY PERSPECTIVE

The evolution of particulate handling and processing equipment and facilities has been accompanied by an evolution of consensus guidelines and government safety regulations. U.S. federal government regulations have been promulgated by the U.S. Occupational Safety and Health Administrartion (OSHA), the Environmental Protection Agency (EPA), the Food and Drug Administration (FDA), and the Department of Transportation (DOT). DOT hazardous material regulations have evolved to incorporate the material/packaging categorization scheme recommended in the U.N. Model Regulations (1999). However, there are indications that additional regulations may be forthcoming. For example, the Chemical Safety Board has recommended that the OSHA Process Safety Mangement regulation and possibly EPA regulations be expanded to include coverage of chemical reactivity hazards, including reactive particulate materials. The three agencies have started meeting to discuss possible approaches to deal with reactivity hazards. OSHA Administrator John Henshaw, in a September 2003 speech at the CCPS Conference, said OSHA prefers a collegial, cooperative approach to the reactive chemicals issue rather than expanded PSM regulations.

The CSB is also concerned about the possible need for additional safety standards for dust explosions, beyond the existing OSHA regulations for grain elevators (CSB July 8, 2003 announcement). Hence, it is entirely possible that future editions of this book may describe either new government regulations for particulate hazards or new joint government-industry safety initiatives.

Professional organizations and trade associations have also played an important role in the evolution of particulate hazard control and safety practices. Many consensus guidelines and standards have been developed under the aegis of safety organizations such as the National Fire Protection Association and the American Conference of Governmental Industrial Hygienists. Representative professional, trade, and safety organizations are listed in Tables 1-7 and 1-8, along with particular programs and resources they provide for safety research, publications, conferences and training. Readers are urged to stay abreast of current and future developments in this field by contacting the organizations most relevant to their facilities and particulate materials.

Europe also has several government regulations and professional and industry initiatives pertinent to particulate material safety. For example, the European Union Seveso I and II Directives govern siting of hazardous materials processing and storage facilities, including explosive and toxic materials. High-risk facilities such as the Toulouse ammonium nitrate facility have to submit safety reports describing accident scenarios potentially involving the release of large quantities of toxic materials. However, since the ammonium nitrate explosion scenario had not been envisaged as part of the Seveso directive requirements, Kersten et al. (2002) and others suggest there may be a need for new requirements that include analyses of "off-spec" materials. There may also be a need for new interpretations of the calculated risks in these safety studies, with more attention being paid to injuries as well as projected fatalities.

One of the pertinent new European Union regulations is the ATEX Directive, which is intended to provide uniform technical and legal requirements for commercial products designed for use in potentially explosive atmospheres, including those containing combustible dusts. Products covered include electrical and mechanical equipment and explosion protection systems. As of July 1, 2003, covered products sold in EU member states need to comply with the Essential Health and Safety Requirements of the Directive, and be marked accordingly. Explosion protective systems such as vent panels, suppression systems and explosion barrier devices will need third party certification, by a test house based in the EU. The requirements for other equipment depend on the zone in which it will be installed. All will be marked with the symbol of explosion protection, CE in a hexagon. Help for manufacturers on understanding the requirements is set out on

TABLE 1-7
Professional Organizations with Activities and Resources in Particulate Safety

Organization	Pertinent Activities/Resources	Web Site
AIChE/Center for Chemical Process Safety	Guidelines Series of Publications, CCPS and AIChE Loss Prevention Conferences, Continuing Education Courses, Reactivity Mangement Roundtable	http://www.aiche.org/ccps/
American Chemical Society	Chemical Health & Safety Publications, Conferences, Newsletters	http://www.chemistry.org/portal/a/c/s/1/home.html
American Filtration and Separations Society	Conferences, Exhibits, Short Courses	http://www.afssociety.org/
American Society of Agricultural Engineers	Journal, Conferences, Online Discussion Forums	http://www.asae.org/
Canadian Centre for Occupational Health and Safety	IPCS INCHEM	http://www.ccohs.ca/
International Society of Explosive Engineers	Journal of Explosives Engineering, Courses, Symposia	http://www.isee.org/
International Society for Pharmaceutical Engineering	ISPE Pharmaceutical Engineering Baseline® Guides Series	http://www.ispe.org/
National Fire Protection Association	Standards, Conferences, Continuing Education Courses, NFPA Handbook	www.nfpa.org
Society of Plastics Engineers	Conferences, Continuing Education Courses	http://www.4spe.org/
Society of Fire Protection Engineers	SFPE Handbook for Fire Protection Engineering, Various Symposia and Professional Development Courses	http://www.sfpe.org
Society of Toxicology	Conferences, Continuing Education Courses	http://www.toxicology.org/

TABLE 1-8
Trade Associations with Activities and Resources in Particulate Safety

Association	Pertinent Activity/Resource	Web Site
American Chemistry Council	Chemtrec®, Responsible Care Program	http://www.cmahq.com/
American Crop Protection Association	Stewardship Program for the Crop Protection Industry, Fire and Spill Emergency Pre-Plan for Handling Agricultural Chemicals, Other publications on safe handling and storage of pesticides	http://www.acpa.org/
American Flock Association	Publications on Flock Industry Health and Safety	
American Fibers Manufacturers Association	Publications, Conferences	http://www.fibersource.com/
American Textile Manufacturers Institute	Safety and Health Committee, Best Practices Publications for Textile Dusts	http://www.atmi.org/
The Association of Powder Process Industry and Engineering	Technical Center of Powder Technology	http://www.iijnet.or.jp/APPIE/index_e.html
Bulk-Online, The Powder/Bulk Portal	Online forums on the handling of powders and bulk solids	http://www.bulk-online.com/
European Chemical Industry Council	Responsible Care Programme	http://www.cefic.org/
INDA: Association of the Nonwovens Fabrics Industry	Conferences and Publications on Nonwovens Industry Health and Safety, Standardized Test Methods	http://www.inda.org/
Institute of Makers of Explosives	IME Guidelines and Recommended Practices	http://www.ime.org/
National Paint and Coatings Association	HMIS®—Hazardous Materials Identification System Special Safety Studies such as respirable silica study	http://www.paint.org/
Pharmaceutical Research and Manufacturers Association	Publications, Conferences	http://www.phrma.org/index.phtml?mode=web
Synthetic Organic Chemical Manufacturers Association	Conferences, Workshops, Training, Responsible Care Program	http://www.socma.org/

Association	Pertinent Activity/Resource	Web Site
Society for the Plastics Industry	Safety Statistics (Members Only)	http://www.socplas.org/
The Fertilizer Institute	Health and Safety Testing, Publications, and Conferences	http://www.tfi.org
Powder Coating Institute	Training Manual, Test Methods, Health and Safety Technical Briefs	www.powdercoating.org/ membership_roster/a1list.htm

http://europa.eu.int/comm/enterprise/atex/guide.htm, and detailed standards for mechanical equipment are gradually being produced.

European and Asian professional and trade organizations analogous to the predominantly U.S. organizations listed in Tables 1-7 and 1-8 also provide guidance and assistance in safe handling of hazardous particulates. There are also similar organizations in other continents and regions, and readers are encouraged to seek the most applicable organizations for their industry and location. Bulk-Online, which is listed in Table 1-8, is a particularly pertinent source of worldwide guidance and assistance devoted exclusively to powders and bulk solids.

REFERENCES

Abbot, J. A., 1988, "Dust Explosion Prevention and Protection," IBC Symposium, London.

Barthelemy, F., Hornus, H., Roussot, J., Hufschmitt, J-P., and Raffoux, J-F., 2001, "Accident on the 21st of September 2001 at a Factory Belonging to the Grande Paroisse Company in Toulouse," French Ministry for Regional Development and the Environment, 24 October 2001.

Bartknecht, W., 1989, *Dust Explosions: Course, Prevention, Protection.* New York: Springer-Verlag.

CCPS, 1995, *Guidelines for Safe Storage and Handling of Reactive Materials,* New York: AIChE.

Eckhoff, R., 1997, *Dust Explosions in the Process Industries,* 2nd ed. Burlington, MA: Butterworth-Heinemann.

"EPA/OSHA Joint Chemical Accident Investigation Report," EPA 550-R-99-03, April 1999.

Fan, L-S. and Zhu, C., 1998, *Principles of Gas-Solid Flows,* Cambridge: Cambridge U. Press.

Febo, H., 2001, FM Global, personal communication to S.S. Grossel.

HSE, 1993, "The Fire at Allied Colloids Limited: A Report of the HSE's Investigation into the Fire at Allied Colloids Ltd, Low Moor, Bradford, on 21 July, 1992." Sudbury, Suffolk, UK: Health & Safety Executive.

Johnson, R. W., Rudy, S. W., and Unwin, S.D., 2003, *Essential Practices in Managing Chemical Reactivity Hazards,* New York: AIChE/Center for Chemical Process Safety.

"Joint Foundry Explosion Investigation Team Report," 2000, U.S. Occupational Safety and Health Administration, Massachusetts Office of the State Fire Marshall, and the Springfield Massachusetts Arson and Bomb Squad.

Kerfoot, E., Steinmetz, D., and Ozimek, T., 1995, "Industrial Toxicology." In S. Grossel and D. Crowl, eds., *Handbook of Highly Toxic Materials Handling and Management,* New York: Marcel Dekker.

Kersten, R., van der Steen, A., Creemers, A., and Opschoor, G., 2002, "The Ammonium Nitrate Explosion in Toulouse, France—The Incident and its Consequences for Industrial Activities." Paper presented at the 17th Annual CCPS Conference.

King, E., 1990, "Health Risks of Fine Powders." In M. J. Rhodes, ed. *Principles of Powder Technology,* New York: John Wiley & Sons.

NFPA 35, 1999, "Manufacture of Organic Coatings," Quincy, MA: National Fire Protection Association.

NFPA 68, 2002, "Deflagration Venting Guide,"Quincy, MA: National Fire Protection Association.

NFPA 69, 2001, "Explosion Prevention," Quincy, MA: National Fire Protection Association.

NFPA 77, 2000, "Static Electricity," Quincy, MA: National Fire Protection Association.

NFPA 434, 1998, "Storage of Pesticides," Quincy, MA: National Fire Protection Association.

NFPA 499, 1997, "Classification of Combustible Dusts and of Hazardous (Classified) Locations for Electrical Installations in Chemical Process Areas," Quincy, MA: National Fire Protection Association.

NFPA 654, 1997, "Standard for the Prevention of Fire and Dust Explosions from the Manufacturing, Processing, and Handling of Combustible Particulates Solids," Quincy, MA: National Fire Protection Association.

NFPA 704, 2001, "Identification of the Hazards of Materials for Emergency Response," Quincy, MA: National Fire Protection Association.

National Institutes of Health, "Eighth Report on Carcinogens," http-server.niehs.nih.gov/htdocs/8_RoC/.

National Paint and Coatings Association, 2003, *HMIS®—Hazardous Materials Identification System,"* 3rd ed.,Washington, DC: NPCA.

Owens, K., and Hazeldean, J., 1995, Fires, explosions, and related incidents at work, *J. Loss Prevention,* 8:291–297.

United Nations, 1999, *Recommendations on the Transport of Dangerous Goods, Model Regulations,* 11th rev. ed., New York: United Nations.

Rhodes, M., 1990, "Pneumatic Conveying," In M. J. Rhodes, ed., *Principles of Powder Technology,* New York: John Wiley & Sons,.

Schoeff, R.W., 2001, Personal communication from Professor Schoeff of Kansas State University to S.S. Grossel, August.

Thornberg, W., 2001, "Personal communication from Mr. Thornberg of IRI to S.S. Grossel,.

Williams, J., 1990, "The Storage and Flow of Powders," In M. J. Rhodes, ed. *Principles of Powder Technology,* New York: John Wiley & Sons.

Yowell, R.L., 1968, "Bisphenol-A Dust Explosions," AIChE Loss Prevention Symposium, volume 2. New York: American Institute of Chemical Engineers.

VDI 2263, 1992, *Dust Fires and Dust Explosions: Hazards Assessment—Protective Measures,* Berlin and Koln: BeuthVerlag.

Chapter 2

PARTICULATE CHARACTERISTICS AND PROPERTIES

2.1 HOW PARTICULATE CHARACTERISTICS AND PROPERTIES AFFECT HAZARDS

This chapter discusses particulate physical, chemical, and toxicological properties that can affect the hazards associated with particulate material storage, transport, and processing. All four generic hazards (combustibility, instability, reactivity, and toxicity) generally increase with decreasing particulate size. Therefore it is important to have an appropriate measurement and characterization of particle sizes in a heterogeneous sample. Measurement and representation of particle size distributions for different shaped particles, including evolving advances in pertinent instrumentation, are discussed in Sections 2.2.1 to 2.2.4 of this chapter. Changes in particle size distribution during handling and transport, and the influences of hardness, friability, agglomeration, and abrasiveness, are discussed in Sections 2.2.5 through 2.2.8.

Several hazards are also dependent on other fundamental physical and chemical properties discussed in this chapter. For example, both the dust explosion hazard and the inhalation hazard depend upon the suspended dust concentration in air. Dust cloud concentration measurement and reporting techniques are described in Section 2.2.10. Spontaneous combustion hazards and instability and reactivity hazards depend on particulate bulk density, which is described in Section 2.2.9. The ignitability and explosibility of combustible particulate decreases sharply with increasing moisture content, as summarized in Chapter 4. Moisture measurement methods are described in Section 2.2.11. The propensity to form hazardous dust clouds depends to a great extent on the fluidity and dispersability of particles, as described in Section 2.2.12.

Chemical composition has a strong bearing on all four generic hazards. Section 2.3 provides a summary of how chemical composition influences particulate flammability, explosibility, instability, reactivity, and corrosivity. Toxicological properties of particulate are summarized in Section 2.4 with regard to respiratory hazards, allergenic hazards, and carcinogenic hazards for acute single exposures and for systemic repeating exposures.

2.2 PARTICULATE PHYSICAL CHARACTERISTICS

2.2.1 *Size Measurement Methods*

There is a wide assortment of methods for measuring particle size distribution. These methods can be divided into two categories: laboratory tests for sampled materials, and test methods that can also be used for in-situ measurements during particulate transport or processing. In the case of the sampled material test methods, sampling methods can sometimes produce biased size distributions. Techniques and consensus standards to minimize biasing for laboratory testing of particulate material are described in Chapter 4. Specific recommendations for sampling bulk materials for particle size measurements are described in NIST SP 960 (Jillavenskatesa et al., 2001), and in ISO/WD 14888 (2000). In the case of the in-situ measurements, there are often limitations on particulate concentrations and flow rates associated with particular instrumentation as discussed later for some of the optical (diffraction) devices. Therefore, the instrumentation is sometimes installed on a small sample line connected to the processing equipment or conveying line (Bumiller, 2001).

The traditional laboratory test method used for size measurements on sampled material is sieve analysis. Woven wire sieves are nested together progressing downward from larger openings to smaller openings, with a pan on the bottom. After weighing the empty sieves, ASTM D 1921–96, for example, specifies that 50 grams of particulate plastic sample be placed into the top sieve (a different size sample may be needed for significantly smaller or larger particles), and that the sieve stack is covered and placed into a mechanical shaker. After shaking the sieve stack for 10 minutes, each sieve is weighed and the net weight retained is determined. Particle size distribution is reported in terms of weight-percent retained on each sieve.

The relationship between U.S. standard sieve number and nominal sieve opening is given in ASTM D 1921. Table 2-1 is an abbreviated listing covering the more commonly used sieve sizes. Particles are segregated such that their second largest dimension lies between the sieve openings of the sieve that captures them, and the largest sieve through which they passed.

The particle size corresponding to a particular sieve opening is called the sieve diameter, defined as the width of the minimum square aperture through which the particle will pass (Fan and Zhu, 1998). As shown in Figure 2-1, the sieve diameter for an irregular shaped particle is larger than the min-

TABLE 2-1
ASTM Standard Sieve Sizes (abbreviated listing)

Sieve Number	Nominal Sieve Opening (μm)
40	425
60	250
80	180
100	150
200	75
230	63
325	45
400	38

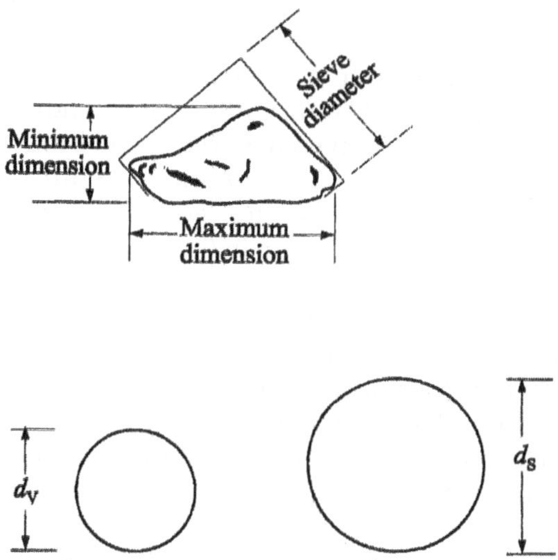

Figure 2-1 Relationship between sieve diameter, volume diameter, and surface diameter (from Fan and Zhu, 1998, Figure 1.2)

imum dimension and smaller than the maximum dimension of the particle. The particle surface diameter, which is defined as the diameter of a sphere having the same surface area as the particle, is generally larger than the sieve diameter of an irregular shaped particle. The volume diameter, which is defined as the diameter of a sphere having the same volume of the particle, is smaller than the sieve diameter, as illustrated in Figure 2-1.

ASTM D 1921 states that the lower size limit for sieving plastic materials with wire mesh sieves is about 38 μm (corresponding to a No. 400 sieve), providing the wire cloth is maintained so that it is in accord with ASTM E 11. The increased uniformity and distortion resistance of electroformed sieves can extend the range of particle diameters down to 20 μm (No. 635), as recommended in ASTM Standard D 4513 for catalytic powders, or even down to 5 μm, as indicated in Table 2-2a.

Sieving can be difficult with particulate materials that are susceptible to electrostatic charging and charge retention because of high resistivity. The electrostatic attractive forces tend to form agglomerates that are not readily dispersed during sieve shaking. Some of the techniques that have been used to sieve these materials include brushing them or adding dispersants. Suitable dispersants include silica, aluminum oxide, titanium dioxide, and commercial dispersants such as Daxal 19 (from Hampshire Chemicals). Osman et al. (2001) report that the optimum addition for breaking up agglomerates of ceramic powders was about 0.50% added Daxal 19. ASTM D 1921 suggests adding 1% of an antistat agent to plastic particulate samples, and that the specific antistatic agent used be reported with the test data.

Good reproducibility can be achieved among laboratories sieving the same material, providing the laboratories use consistent particulate handling, conditioning (primarily drying), and testing procedures. For example, round-robin results reported in ASTM D4513-97 for a catalytic material with an average measured median particle diameter of 64.3 μm, produced a seven-laboratory standard deviation, of 1.9 μm (3.0%), with a corresponding 95% expectation that the median diameter for a given material tested by different laboratories should not differ by more than 2.77(1.9 μm) = 5.3 μm, or ±8.2%. However, there is a larger uncertainty associated with the gap in mesh sizes; there was nothing between 44 μm and 74 μm in the ASTM D4513 tests. Jillavenkatesa et al. (2001) report a larger standard deviation of 4.9 μm (7.4%) for a silicon nitride with an average measured median diameter of 66.4 μm, but the wider scatter in this case is partly due to one laboratory reporting skewed data compared to the other six laboratories. ASTM D1921-96 round-robin test results showed interlaboratory standard deviations of 4.8 μm, and 1.8 μm, for polyethylene and polyvinyl chloride particulates with mass-mean diameters of 291 μm, and 137 μm, respectively. However, the 196 μm mean diameter acrylic powder used in the ASTM D1921-96 round-robin testing produced a standard deviation of 13 μm, corresponding to 6.6%. The larger standard deviation of the acrylic powder compared to the other powders may be due to the fact that the acrylic particle size distribution was bimodal (two peaks in the density distribution), or perhaps the acrylic powder was less spherical than the other powders. Sieving data are inherently less accurate for powders with unusual shapes, with multi-modal size distributions, and with large fractions of small particles.

Other particle size measurement techniques are needed for smaller particle sizes, for in-situ measurements during processing and transport, or when additional data, such as particle velocity, shape, or concentration, are

TABLE 2-2a
Particle Size Measurement Methods for Sampled Materials or Batch Mode

Method	Size Range[a] (μm)	Advantages of Method	Disadvantages of Method	Type of Size Distribution	Representative Manufacturers[b]
Sieve Analysis: Woven Wire Sieve	38–4000	Relatively inexpensive and simple.	Low resolution produces uncertainty in average diameter. Can erode particles and reduce sizes.	By weight based on sieve diameter	Farleygreene, Hosokawa Micron, Laval Lab, Rotex
Electroformed Sieve	5–120	Same as woven wire sieve	Same as woven wire sieve		Gilson Co., Topas-Gmbh,Christison Particle Technologies Ltd.
Optical Microscope	0.8–150	Two-dimensional images.	Manual sizing is labor intensive and somewhat subjective.	By number	American Optical, Bausch & Lomb, Leica, Leitz, Nikon, Olympus, Omicron, Reichert, WILD, Zeiss
Scanning Electron Microscope	0.001–5	Three-dimensional images. Applicable to sub-micron particles.	Expensive equipment. Manual sizing is labor intensive.	By number	Amray, Carl Zeiss, Coulter Electronics, Electroscan, Hitachi, Jeol, LEO,
Electrozone (Electronic Sensing Zone)	1.4–1,000 depending on aperture size	Relatively inexpensive. Electronic signal processing.	Requires suspending particles in liquid electrolyte.	Volume equivalent diameter by number	Coulter Electronics, Micromeritics Instrument
Cascade Impactor	0.3–50	Can be integral to sampling system.	Cannot be used with particles larger than 50 μm.	By weight based on aerodynamic equivalent diameter	California Measurements, Dekati, KNJ Engineering, MSP, Thermo Andersen, TSI
Gravitational Sedimentation	5–100	Traditional method for paint and ceramics industry.	Requires suspending particles in liquid.	By weight based on Stokes diameter	Laval Lab, Micromeritics Instrument

TABLE 2.2a (*continued*)

Method	Size Range[a] (μm)	Advantages of Method	Disadvantages of Method	Type of Size Distribution	Representative Manufacturers[b]
Ultrasonic Spectroscopy	0.1–1,000	Broad size range. Provides velocity also.	Usually requires suspending particles in liquid.	By number	Malvern, Sympatec
Time-of-Flight	0.6–200	High resolution. Some versions can be used in field.	Cannot be used with particles larger than 200 μm.	Aerodynamic diameter by number	TSI

[a]Data primarily from Table 3.6 of Svarovsky (1990) and Table 1.2 of Fan and Zhu (1998). See also Table 1.1. of NIST SP 960 (Jillavenskatesa et al., 2001).

[b] Individual manufacturers equipment may cover size ranges that extend beyond, or fall within, those listed in this table, and readers are urged to consult individual manufacturer specifications. This list is only representative of availabilities in 2002.

needed. Tables 2-2a and 2-2b list the particle size ranges for the various measurement techniques, advantages and disadvantages of each method, and representative instrument manufacturers. In some cases, the sizing data is collected as number distributions, whereas other techniques inherently provide weight (mass) distributions. Since the conversion from particle diameter to particle mass involves cubing the diameter, errors are magnified accordingly. For example, a 20% error in particle diameter measurement becomes an error of 73% $[(1.2^3 - 1) * 100]$ in particle mass calculation, even if the particle density is known accurately.

Microscopic observations of particle samples allow for direct visualization of particle morphology and size. Optical microscopes produce a two-dimensional image, while scanning electron microscopes provide three-dimensional information. The lower limit of resolution for an optical microscope depends on the ratio of the wavelength of light to the numerical aperture of the objective, and is typically in the range 0.2 μm to 1.0 μm (Fan and Zhu, 1998). Maximum useful magnifications range from 80 to 1250, with the higher magnifications being achieved at the expense of a very limited depth of focus (0.4 μm at a magnification of 1250).

The scanning electron microscope (SEM) can be used for far smaller particles than any other method listed in Table 2-2. It is particularly useful for viewing the particle surface morphology and observing details such as pores and dendrites. Sizing analysis can be conducted either by manual counting or by computer interface and image analysis software. Some of the SEM manufacturers listed in Table 2-2 provide image analysis software options,

TABLE 2-2b

Particle Sizing Methods Compatible with Continuous In-Situ Measurements

Method	Size Range* (μm)	Advantages of Method	Disadvantages of Method	Type of Size Distribution	Representative Manufacturers[b]
Laser Light Scattering: Fraunhofer and Mie Theory Diffraction	0.1–1000	Broad size range, including sub-micron particles.	Need to know particle refractive index. Errors can occur at large concentrations.	By number Software produces mass and other distributions.	CILAS, Horiba, Laval Lab, Malvern, Microtrac, Micromeritics Instrument, Particle Sizing Systems, Sci-Tec,
Laser Light Scattering: Phase Doppler Diffraction	1–10,000 depending on lens size	Broad size range. Produces velocity data also.	Expensive equipment. Need to know particle refractive index. Errors can occur at large concentrations.	By number Software produces mass and other distributions.	Dantec, TSI, Sympatec
Laser Light Obscuration	1– 9000	Broad size range, some versions produce two-dimensional images.	Errors can occur at large concentrations.	By number based on projected surface area	Particle Sizing Systems
Video Imaging Analysis	10– 100,000	Can produce two-dimensional or three-dimensional images.	Errors can occur at large concentrations. Accuracy is dependent on software algorithms used.	By number based on software defined diameters	Horiba, Sci-Tec, Malvern
Impact Sensor	50–500	Works at large particle concentrations (about 500 g/m^3)	Limited experience with this method.	By number based on volumetric diameter	CSIRO (Australia)

[a]Data primarily from Table 3.6 of Svarovsky (1990) and Table 1.2 of Fan and Zhu (1998). See also Table 1.1. of NIST SP 960 (Jillavenskatesa et al., 2001).

[b] Individual manufacturers equipment may cover size ranges that extend beyond, or fall within, those listed in this table, and readers are urged to consult individual manufacturer specifications. This list is only representative of availabilities in 2001.

and some provide laboratory measurements for submitted samples. The electrozone method, also known as the Coulter counter, involves suspending the particles in an electrolyte with immersed electrodes on both sides of a small orifice. As the particles flow through the orifice, the displacement of electrolyte changes the conductivity and induces a voltage pulse with amplitude that is proportional to the particle volume. Although a wide range of particle diameters is listed for the electrozone method in Table 2-2, the typical size range is 1 to 50 μm (Fan and Zhu). According to ASTM F577, which describes the use of the electrozone method for sizing toner particles, the particle size range corresponds to the interval from 2% to 40% of the aperture diameter. Four different orifice apertures are prescribed in ASTM F577 to span the particle size range of 1.4 μm to 80 μm.

Laser light scattering methods have experienced a dramatic increase in availability and applicability in recent years, as is evident from the large number of instrument manufacturers. The primary reason for their popularity is that they are noninvasive and in many processing applications can be used for in-situ measurements. A stream of particles passes through the laser beam and the scattered light signal is sensed on one or more photo-detectors situated at the desired scattering angle(s). If measurements are made on a previously collected particulate sample, a required accessory is a device for channeling a stream of particles through the laser beam. The particular type of light scattering phenomenon depends on the particle size in relation to the laser wavelength. Fraunhofer diffraction is applicable to particle diameters that are large (at least five times as large) as the wavelength. Mie scattering is applicable to particle diameters that are comparable to the wavelength. Thus, the Mie scattering instruments can be used for smaller particles (0.2 μm) than the Fraunhofer diffraction instruments (which have about 3 μm lower limit).

One limitation of laser light scattering methods is the occurrence of coincidence errors at large particle concentrations. The optimum sample size for dry powders is 4–10 g, so as to provide a statistically significant sample without incurring coincidence errors.

Phase Doppler laser diffraction instruments use two intersecting laser beams, and a detector that measures both the phase shift and the frequency shift caused by particle diffraction at the beam intersection. The phase shift is used to calculate the particle diameter, and the frequency shift is used to determine the particle velocity. The combination allows for the determination of particle concentrations. Coincidence error concerns usually limit the maximum allowable concentration to approximately 1000 particles per mm^3. After testing several laser diffraction systems, Neumann and Kramer (2001) report that they obtain consistently accurate results if the solids concentration is no greater than about 1.5 vol%, and if they use similar signal processing algorithms. Particle index of refraction must be input to the signal processing software to use both phase Doppler diffraction and Mie scattering sizing instruments.

Laser light obscuration depends on particle shadowing of laser light projected onto a photodetector. Rapid electronics counts the number and intensity of momentary reductions in received light as different size particles pass through the beam. This technique can be used for online real time measurements.

Cascade impactors consist of a series of plates interconnected with aerodynamic flow channels of progressively decreasing width such that particles are collected on the plates according to their aerodynamic diameters. The amount of particulate collected on each plate depends on the air flow rate passing through the impactor. The particle diameter range collected depends on the number of plates (4 to 12) and their design. Traditional cascade impactors, which are used for particle diameters in the range 0.5 μm to 50 μm, require manual weighing of the various stages. Some of the newer designs allow for online data processing using either real time weight measurements or measurements of electrical current carried by charged particles. The latter design is applicable to the size range 0.03 μm to 10 μm.

Gravitational sedimentation instruments measure the terminal settling velocities, which are related to the particle aerodynamic diameter through the Stokes equation (see Fan and Zhu, or NIST SP 960). The settling medium is usually a liquid, and the minimum particle diameter measured by gravitational sedimentation is usually 5 μm, although some instruments use centrifugal action to allow measurements with smaller particles. X-ray absorption measurements of settling velocities of heavier materials, such as metals, can also extend the minimum particle size down to 0.1 μm, as described in ASTM B761. Light absorption measurements provide an alternative to x-ray absorption for measuring the velocities and concentrations of the settling particles.

The last two particle size measurement techniques listed in both Table 2-2a and Table 2-2b are new developments associated with emerging technologies. Ultrasonic spectroscopy entails passing a high frequency acoustic wave through a suspended aerosol cloud, and analyzing the spectral distribution of the signal received. In order to have measurable attenuation/modification of the transmitted signal, the particulate volume fraction usually needs to be greater than 1% (Jillavenskatesa et al., 2001). If these conditions exist during normal particulate transport, ultrasonic instrumentation can be used for continuous online measurements. Time-of-flight instruments rapidly accelerate a particulate sample such that particle aerodynamic diameters and density determine the arrival times of the individual particles at the receiving light absorption station. Automated imaging analysis involves the generation of digitized video or still images, and allows various rules to be implemented to recognize individual particles of various shapes, and with small levels of overlapping. The impact sensor has been developed and patented by Coghill (2001) to determine particle sizes at high particle loadings (tested up to 500 g/m^3) in pneumatic transport systems.

2.2.2 *Particle Size Distribution*

The particle size distribution is critical in determining particulate behavior during transport and handling (Kaye et al., 2001) as well as determining the type and extent of particulate hazards. What is the best way to characterize the distribution of measured particle diameters in a sample? The answer depends on the particular hazard in question and the form of the hazard test and classification data, as discussed in Chapter 4. For example, the rate of burning and associated maximum-rate-of-pressure-rise is dependent on the particulate surface area, whereas the total energy released and the associated maximum pressure in a dust explosion are dependent on the mass, that is, the particle diameter weight distribution. Here, diameter distribution functions and various characteristic diameters are defined.

The distribution of particle diameters, x, can be expressed either in terms of a cumulative distribution $F(x)$, or in terms of the distribution density $f(d)$. The relationship between these two functions is

$$f(x) = \frac{dF}{dx} \qquad [2\text{-}1]$$

The cumulative number distribution, $F_n(x)$, represents the fraction of particles having a diameter less than or equal to x. The number density distribution, $f_n(x)$, is defined such that $f_n(x)dx$ is the fraction of particles having a diameter between x and $x + dx$. $F_n(x)$ is dimensionless, whereas $f_n(x)$ has dimensions of inverse length (usually μm^{-1}). In the case of size distributions based on mass measurements (e.g., sieve analysis), the corresponding mass distributions are denoted by $F_m(x)$ and $f_m(x)$. The number density and mass density distributions are related by

$$f_m(x) = \frac{N m_p(x)}{M} f_n(x) \qquad [2\text{-}2]$$

where N is the total number of particles, M is the total sample mass, and $m_p(x)$ is the mass of a particle of diameter x.

Simplification often motivates using a mean diameter to characterize a particle size distribution. Several different types of mean diameter can be defined. The number mean, $\bar{d}_n$ (also called the arithmetic mean) is defined as

$$\bar{d}_n = \int_0^1 x\, dF_n = \int_0^\infty x f_n(x)\, dx \qquad [2\text{-}3]$$

The volume mean diameter, $\bar{d}_v$ (also called the cubic mean) is defined as

$$\bar{d}_v^3 = \int_0^1 x^3\, dF_n = \int_0^\infty x^3 f_n(x)\, dx \qquad [2\text{-}4]$$

Other characteristic diameters in a particle size distribution are the mode, defined as the peak in the density distribution, and the surface mean diameter (also called the quadratic mean), defined as in Equation 2-4 but with x^2 in the integral instead of x^3, and with $\bar{d}_v^3$ becoming $\bar{d}_s^2$. The relationship between these characteristic diameters is illustrated in Figure 2-2, which shows two different particle size distributions having the same number (arithmetic) mean diameter. The volume (cubic) mean and the quadratic (surface) mean are always larger than the number (arithmetic) mean. In the top distribution in Figure 2-2, the volume mean is more than twice as large as the number mean, while in the bottom distribution it is about 30% larger. In a broad distribution (top distribution in Figure 2-2), the mode is significantly smaller than the mean diameter (about one-third of the mean in the top distribution), while in a narrow distribution the mode is closer to the mean (about 70% of the mean in the bottom distribution). Analogous definitions of mean diameters are applicable to the particulate mass distribution function, $f_m(x)$, so that $\bar{d}_m$ is the mass mean particle diameter. In the case of highly nonspherical particle shapes, there are several different definitions of equivalent diameter, as described by Fan and Zhu (1998).

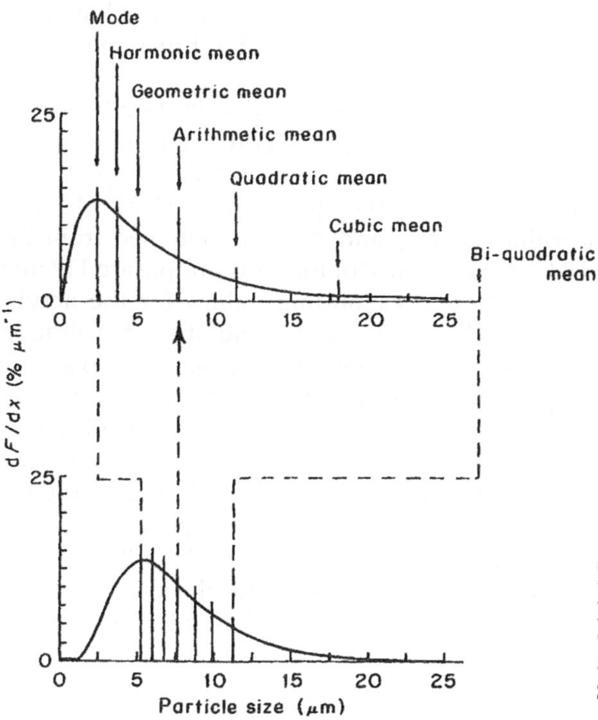

Figure 2-2 Comparison of particle size parameters for distibutions with the same arithmetic mean (from Svarovsky, 1990, Figure 3.5).

One more type of representative particle size is the Sauter mean diameter, d_{32}, defined as

$$d_{32} = \frac{\int_0^\infty x^3 f_n(x)\,dx}{\int_0^\infty x^2 f_n(x)\,dx} = \frac{\bar{d}_v^3}{\bar{d}_s^2} \qquad [2\text{-}5]$$

The Sauter mean diameter is conceptually the ratio of the representative particle volume to the representative surface area, and is a very appropriate parameter for characterizing the combustion of a suspended dust cloud because both surface area effects and volumetric effects influence the rate of burning. One example is the series of correlations for the time for vapor generation in dust cloud flame propagation (Eckhoff, 1997, p. 291).

In many applications, the particle size distribution has a long tail corresponding to a small number of relatively large particles. This is the case in the top distribution in Figure 2-2. A lognormal distribution function often fits the data for this type of distribution. The lognormal density distribution is given mathematically by

$$f_n(x) = \frac{1}{\sqrt{2\pi}\sigma_{dl} x} \exp\left[-\frac{1}{2}\left(\frac{\ln x - \ln d_{med}}{\sigma_{dl}}\right)^2\right] \qquad [2\text{-}6]$$

where d_{med} is the median particle diameter, and σ_{dl} is the natural log of the ratio of the diameter at which the cumulative distribution function is equal to 0.84 to the median diameter.

Figures 2-3a and 2-3b show the cumulative mass distribution functions $F_m(x)$ for an acrylic powder sample and a sample of polyvinyl chloride particulate with carbon black antistatic agent. The data for both samples have been obtained from the sieve size data examples given in ASTM D1921–1996. After plotting the data, the values of d_{med} and σ_{dl} were obtained from the plot. The 50% (median) and 84% $F_m(x)$ values for the PVC sample are 117 mm and 152 mm, respectively, and the corresponding values for the acrylic sample are 130 mm and 395 mm. The lognormal cumulative distribution function values corresponding to the particular values of d_{med} and σ_{dl} were obtained from the MS Excel spreadsheet used to develop the plots shown in Figures 2-3a and b. The acrylic powder size distribution is much broader than the PVC sample. However, in both cases the lognormal distribution provides an excellent fit to the data. Thus lognormal distribution curve fits provide an attractive method for interpolating or even extrapolating particle size data in many, but not all, applications. According to the NIST Recommended Practice for Particle Size Characterization (2001), the lognormal distribution is applicable to granulated powders produced by spray-drying and milling of fine-grained materials.

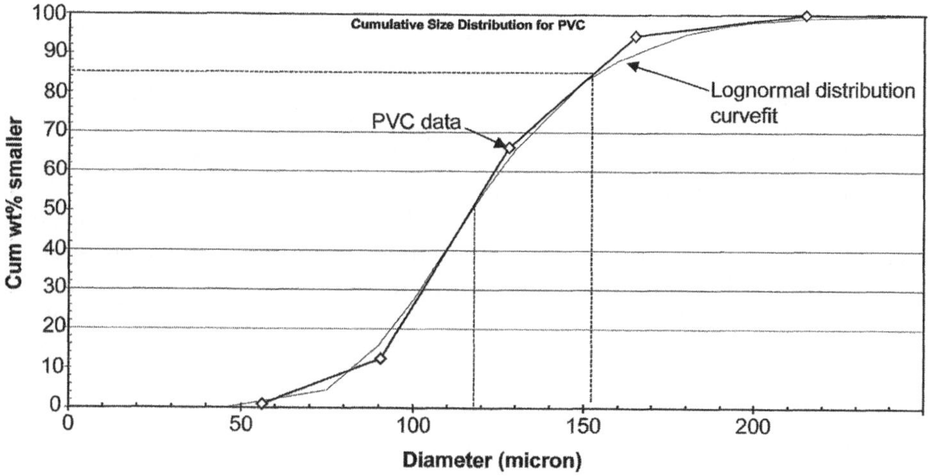

(a)

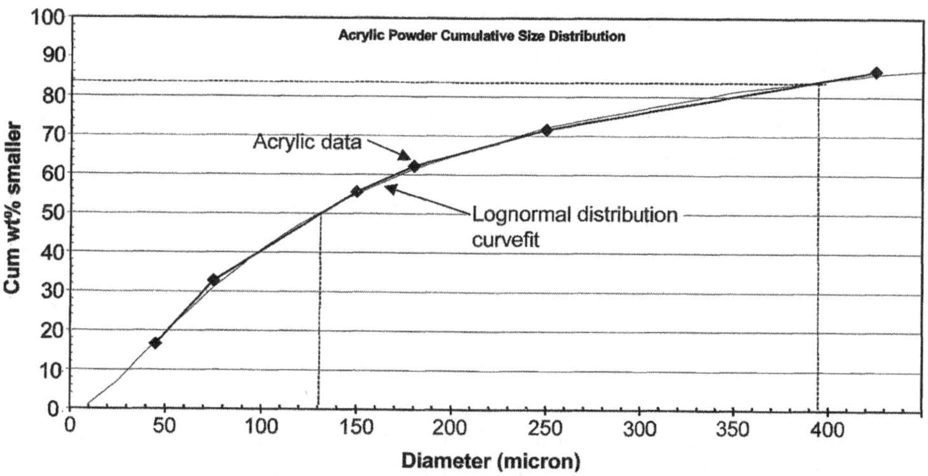

(b)

Figure 2-3 (a) Lognormal curvefit to PVC particle size distribution. (b) Lognormal curvefit to acrylic powder particle size distribution.

Another distribution function that is sometimes used for particulate samples is the Rosin-Rammler distribution, which has the following density function:

$$f_m(x) = \alpha\beta x^{\alpha-1} \exp(-\beta x^{\alpha}) \qquad [2\text{-}7]$$

where α and β are the two parameters that characterize the distribution. The Rosin-Rammler cumulative distribution function is:

$$F_m(x) = - \exp(-\beta x^{\alpha}) \qquad [2\text{-}8]$$

Large values of β denote small characteristic particle size, and vice versa. For example, Fan and Zhu (1998) report that a finely ground coal sample had a β value of 0.021 μm^{-1}, whereas two different coarsely ground coals had β values of 6.7×10^{-5} μm^{-1} and 15×10^{-5} μm^{-1}. Large values of α also imply small particle sizes, since from Equation 2-8 we know that 63.2% of the particles are smaller than $(1/\beta)^{1/\alpha}$. In many cases, α is in the range 0.70 to 1.20. By knowing the fraction of particles smaller than a certain critical diameter, one can determine the fraction of particles that represent certain hazards, such as being respirable or being capable of producing a dust explosion when suspended in air. The NIST SP 960 Recommended Practice (Jullavenkatesa et al., 2001) states that the Rosin-Rammler distribution is applicable to "some milled, coarse-grained materials."

Commonly used distribution functions such as the lognormal distribution are included in most spreadsheet software, such as Microsoft Excel. These and the less commonly used distribution functions are also available in the statistical software that comes packaged with many of the size distribution instruments listed in Table 2-2b, and in the software that has been developed for use with the microscopy methods for particulate sizing listed in Table 2-2a.

ASTM E1617 prescribes three different levels of reporting particle size distribution data, with different levels of detail on the instrumentation and data analysis provided at each level. All three levels require reporting the modal diameter and median diameter. The Level II and III reporting requirements include detailed statistical information on the particle size distribution and software used. ISO 9276 – Part 1 is an international standard for particle distribution graphs and nomenclature, both for the density function, $f(x)$, and the cumulative distribution, $F(x)$. Other parts of the standard, which are still in draft stages, describe the calculations of the various average diameters, and the use of the lognormal particle size distribution function.

Highly asymmetric shaped particles cannot readily be characterized in terms of any of the diameters defined in Section 2.2.2. Two examples are fibers and flakes, as discussed in Sections 2.2.3 and 2.2.4, respectively.

2.2.3 *Fiber Characteristics*

Fibers are defined as particles with length-to-diameter ratio of 3 or greater. In many cases, the fiber length is at least an order of magnitude larger than its diameter. Fibers can either be straight as shown in Figure 2-4, or curved as shown in Figure 2-5.

Airborne clouds of particulate fibers are often generated during fiber cutting and handling/conveying operations, particularly when friable materials such as asbestosis are being handled. The fiber dimensions that render them potentially respirable are discussed below following a brief description of particulate fiber size measurement methods.

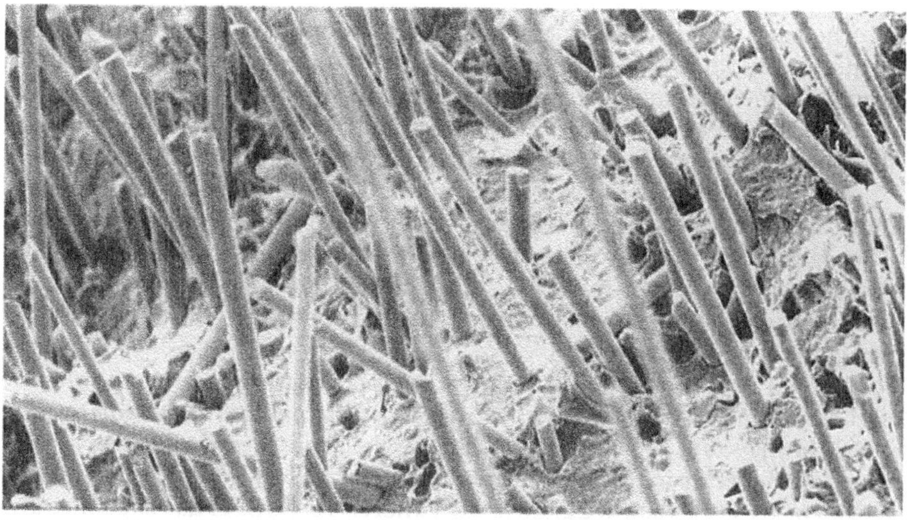

Figure 2-4 Glass fibers in fiber reinforced polypropylene composite (with permission by Dr. Fabrizio Parodi, www.fpchem.com/fap_5a-en.html).

Figure 2-5 Friable asbestos fibers (from Fuller.com).

The high length-to-diameter (*L/d*) ratio renders many particle-sizing methods inapplicable to fibers. For example, standard sieves can be used to provide some qualitative sizing segregation of fibers, but there is no clear relationship between the sieve opening and the fiber diameter. Straight fibers are more amenable to sieving than curved fibers.

The most widely used fiber sizing method is optical or SEM microscopy. There are automated digital imaging systems that are specifically designed to recognize and measure the sizes of individual fibers. In some cases, the fiber recognition is entirely in the software, but in other cases the fibers are dispersed in a flowing liquid carrier so that the fibers will tend to be oriented parallel to the flow direction.

NIOSH defines respirable fibers as having a mass median aerodynamic diameter of 3.5 μm or less. The aerodynamic diameter is the diameter of a sphere of the same particle density having the same terminal velocity in air or some other relevant fluid. (The terminal velocity of a 3.5-μm sphere with a density of 1 g/cm^3 is about 2 cm/s in air.) Therefore, NIOSH has exposure airborne concentration limits for many materials if they have a median diameter of 3.5 μm or less and a length of 10 μm or greater.

The World Health Organization, as reported by Sagehorn et al. (2001) classifies fibers as hazardous if they have lengths greater than 5 μm, diameters less than 3 μm, and *L/d* ratios greater than 3:1. These fibers are not only respirable, but are capable of being retained deep in the respiratory tract. Sagehorn et al. describe the development of a light scattering probe intended to automatically detect and monitor the concentration of fibers in this size and *L/d* range.

Certain types of asbestos fibers are notorious for easily forming respirable debris with diameters on the order of 1 μm and lengths in the range 10 to 20 μm. These are the fibers that seem to be associated with many occurrences of asbestosis and other dangerous pulmonary ailments. Airborne silica fibers also pose a significant respiratory hazard potentially leading to the development of silicosis.

In the textile industry, a parameter called the denier is used to characterize the effective diameter or fineness of a fiber. The denier is defined as the fiber weight in grams per 9000 m length. Thus, the diameter in microns is equal to $12(\text{denier/sg})^{1/2}$, where sg is the fiber specific gravity. In Switzerland and Germany, the equivalent parameter is the titer measured in dtex and defined as weight in grams per 10 km fiber length. The diameter in microns is equal to $11.3(\text{dtex/sg})^{1/2}$.

Flock fibers are cut to a length that allows them to be attached at one end to a textile substrate such that the fibers provide a desired surface texture. The typical flock fiber length is on the order of 1 mm, and the typical range of deniers is 0.5 to 10 (corresponding to a diameter range of about 7 to 30 μm for a representative sg of 1.4. Flock fiber explosibility tests reported by Bartknecht (1989) suggest that the data for a variety of materials can be correlated in terms of the product of dtex and fiber length. The smaller the value of [dtex · length], the easier it was to ignite the flock, and the more rapid was the rate of combustion. Since the dtex (or denier) varies as the square of the diameter, it seems that the rate of combustion is dependent on [(diameter)^2 · length], that is, to the volume of the flock fiber. Details of this relationship and data are presented in Section 4.3.6.

2.2.4 *Flake Characteristics and Specific Surface Area*

Flake shaped particles are produced commercially for a variety of applications including additives for thin films, paints, lubricants, and adhesives. Their inherently high surface area per unit mass also makes them attractive for use as catalysts and as vapor/liquid adsorbents. Polyethylene flakes are often produced in the first stage of the polymerization of ethylene monomer. Many grinding and milling operations also produce flakes even if they are not the desired end product.

Figure 2-6 is a scanning electron micrograph of commercially produced copper flakes. According to the manufacturer's specifications, the volume median diameter of the flakes in Figure 2-6 is 4.5 μm, and the 90% volume diameter is 9.7 μm, and their specific surface area is 1.1 m^2/g.

Figure 2-6 Copper flakes SEM (from Umicore, Cu Flake U9-240a, November 2001).

The specific surface area of a collection of particles is defined as the total particle surface area divided by the total mass of particles. Flakes have a specific surface area, S_w, given by

$$S_w = \frac{2A_f}{\rho_p A_f t} = \frac{2}{\rho_p t} \qquad [2\text{-}9]$$

where t is the flake thickness, and A_f is the one-side surface area of a flake, and ρ_p is the particle density. Since copper has a particulate density, ρ_p, of 8.95 g/cm^3, the average thickness of the flakes in Figure 2-6 must be

$$\frac{2}{(8.95\ \text{g/cm}^3)(1.1\times10^4\ \text{cm}^2/\text{g})} = 0.2\times10^{-4}\ \text{cm} = 0.2\ \mu\text{m}.$$

Since the specific surface area of uniformly sized spherical particles is equal to $6/\rho_p D$ (Arakawa, 1999), spherical copper particles would require a diameter of 0.6 μm to have the same specific surface area as the 0.2-μm-thick flakes.

Internal surfaces can also contribute to the total surface area of flakes and other shaped particles. These internal surfaces arise from cracks, holes, cavities, and pores in the particle. They can add appreciably to the external surface areas calculated on the basis of the external surface area alone.

Eckhoff (2003) has shown that the maximum rate-of-pressure-rise in a closed vessel dust explosion is linearly proportional to the specific surface area for a given combustible material. This relationship is shown in Figure 2-7a for potato/corn starch dusts and fish protein dusts, and in Figure 2-7b for aluminum particles. Since the data in Figure 2-7a were obtained in a 1.2-liter Hartmann cylinder, and the data in Figure 2-7b were obtained in a 1-m^3 sphere, the results are not compatible. The rates-of-pressure-rise for metal flakes with specific surface areas in excess of 1 m^2/g are often sufficiently high to create difficulties in establishing effective explosion protection measures. This is discussed further in Chapter 4 and 6.

Measurements of specific surface area are usually conducted using gas adsorption instruments. The most commonly used method involves measuring the decrease in gas volume due to adsorption by a sample of powder in a tube. Commercial devices using nitrogen or helium or other adsorbate gases are available (Arakawa, 1991). The specific surface area is linearly proportional to the adsorbed molar volume multiplied by the cross-sectional area of adsorbate gas, which is a function of temperature. Pertinent equations and representative values of cross-sectional area for different gases are reported by Arakawa (1991).

2.2.5 *Abrasiveness*

Many particulate solids are abrasive in nature and as they travel through piping and process equipment can erode the inner surfaces. If the erosion

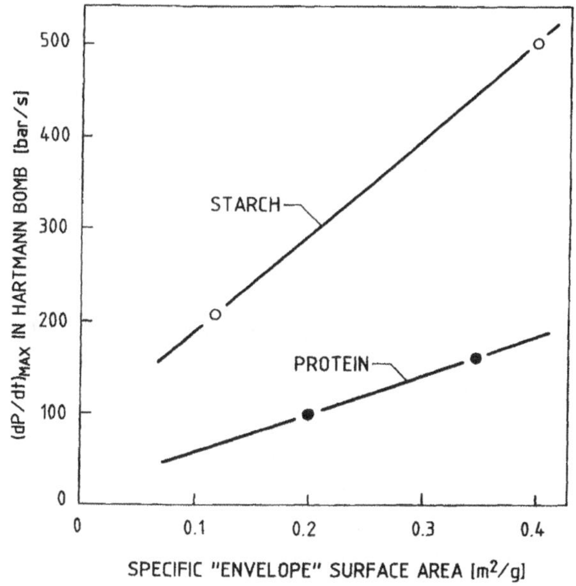

(a)

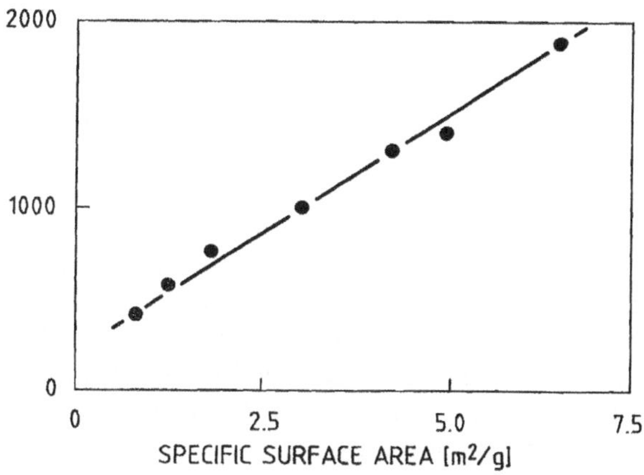

(b)

Figure 2-7 (a) Rates-of-pressure-rise in starch and protein dust explosions versions versus specific surface area. (b) Rates-of-pressure-rise in aluminum dust explosions versus specific surface area. (From Eckhoff, R., *Dust Explosions in the Process Industries,* 3rd ed. Amsterdam: Elsevier, 2003.)

becomes severe enough a hole can be produced in the piping or process equipment and the particulate solids will be discharged into the atmosphere. This should be avoided for the following reasons:

1. The emitted solids can deposit as a dust layer and, if they are combustible, they could be ignited, resulting in a fire or a secondary dust explosion.
2. If the emitted solids are toxic, or respirable or eye irritants, they could pose a serious hazard to personnel in the area.

3. The loss of product from the processing stream can pose a significant business/monetary loss.

Particulate abrasiveness is the combination of the particle's physical characteristics that enables it to erode surfaces with which it comes in contact. The physical properties that affect abrasiveness include:

- particle hardness
- particle shape
- particle size

Particle hardness is discussed in Section 2.2.6. Berns and Koch (1999) have shown that the room temperature wear resistance of a surface increases as the ratio of the surface hardness, H_s, to the particle hardness, H_p, increases, with particularly good wear resistance occurring at ratios greater than 1.2. The situation is more complicated at high temperatures because the abrasive particles can become imbedded in the opposing surface.

Particle shape abrasiveness effects include the overall length/width aspect ratio, and various measures of particle surface irregularity and angularity. One measure of surface irregularity is the ratio perimeter^2/area. Stachowiak and Stachowiak (2001) note that erosion rates increase with increasing values of both the length/width ratio and the perimeter^2/area ratio. Various measures of particle angularity have been proposed by researchers based on image processing techniques (Leavers, 2000; Stachowiak, 2000). These techniques quantify the extent and sharpness of protrusions from the particle surface. Stachowiak and Stachowiak (2001) have shown that wear rates in many cases are linearly proportional to an angularity parameter called the spike parameter quadratic fit.

The effect of particle size on abrasiveness is more complicated and less apparent than particle hardness and angularity. For example, Kelly and Hutchings (2001) studied abrasivity using a microscale abrasive wear test, and found no significant effect of particle size. They explain this observation by the breaking of larger particles into small fragments. Additional work on particle fragmentation is described in Section 2.2.8.

A much less scientific measure of particle relative abrasiveness was developed by the Conveyor Equipment Manufacturers Association (CEMA) for the purpose of selecting abrasive resistant conveying equipment. The method presented in CEMA Book 550 (1980) assigns a factor to each of the above three characteristics and, together with particulate bulk density, these factors are multiplied together to give the CEMA Abrasive Index. If the calculated Abrasive Index indicates that the particulate solid is highly abrasive, then extra heavy wall piping is recommended, as is process equipment with extra thick walls or a wear plate at the entry point. Fayed and Skocir (1997) also present recommendations on how to minimize abrasion problems for different types of mechanical conveyors.

2.2.6 *Hardness and Friability*

Material hardness is loosely defined as resistance to deformation while being subjected to an applied force (Okuyama and Kousaka, 1991). There are several different types of hardness ratings and measures. The Mohs Hardness rating is a relative rating in which talc has a rating of 1 and diamond has a rating of 10. Eight other materials have defined integer ratings of 2 through 9. Ratings of 1 through 4 are considered soft, while ratings of 5 (glass) through 10 represent increasing degrees of hardness for nominally hard materials.

Vickers hardness is a quantitative measure of hardness based on a materials resistance to being indented by a pyramid shaped diamond device applied with a known force, typically corresponding to a weight in the range 10 kgf to 120 kgf. The Vickers hardness value is proportional to the applied force divided by the indented area (actually square of the diagonal) produced in the test material. Talc has a Vickers hardness in the range 40–56, glass has a Vickers hardness of 500, tantalum carbide has a Vickers hardness of 1800 (Okuyama and Kousaka, 1991), and silicon carbide has a Vickers hardness of 2350. Diamond has a Vickers Hardness that often ranges from 6000 to 9400, with the latter value being representative of bulk natural diamond. ASTM E384 describes the test procedures for measuring Vickers hardness.

There are several other quantitative measures of hardness, including Knoop hardness and Rockwell hardness. ASTM E140 provides conversion tables and equations for correlating the various hardness values.

Friable particulate solids are those that may be easily broken mechanically and reduced in size as a result of impact, agitation, or attrition. The resulting material will therefore have a different size distribution from the original material, as discussed in Section 2.2.8. The reduced particle size resulting from the handling or processing of friable particulate solids may cause operating problems and represent increased hazards of inhalation and possibly dust explosions.

Friabilility is usually measured in terms of fracture toughness. ASTM E1820 describes the standardized test procedure for measuring various fracture toughness parameters.

If friable particulates are being handled, process equipment that gently handles friable particulate solids should be used, for example, dense-phase (low velocity) conveyors versus dilute-phase (high velocity) conveyors.

2.2.7 *Agglomeration*

Agglomeration is the adherence of particles to each other such that they form a substantially larger particle, called an agglomerate. Agglomerations of individual particles can form from a variety of interparticle attraction forces including van der Waals' force (important primarily for particles smaller 0.1 mm), surface adhesion, electrostatic attraction, and liquid surface tension. An example of surface adhesion is the smooth rubbery surface of ethylene vinyl acetate copolymer (Kendall and Stainton, 2001). Stirring beads of the copolymer in a beaker results in the formation of a large aggregate held together via tacky surface adhesion. If the particles do not have naturally soft sticky surfaces, adhesives can be used to achieve surface adhesion. Electrostatic attractive forces are important for many plastics with low electrical conductivities. Liquid surface tension forces are important when moisture levels are sufficiently high to form liquid bridges as shown in Figure 2-8a. The other diagrams in Figure 2-8 illustrate increasing moisture contents, such that in Figure 2-8d the particles become a suspension rather than a cohesive aggregate.

The liquid bridge agglomeration regime occurs when the liquid filled fraction of powder void space is in the range of about 0.02 to 0.25. At a void space fraction of 0.4 and a particle density of 1 g/cm^3, a liquid fraction of 0.25 corresponds to moisture content of 14%. In this regime, the interparticle attractive forces increase with increasing moisture content. Sample data are provided by Eckhoff (2003). This regime is applicable to particles that adsorb moisture rather than absorb it. The liquid does not have to be water, and the

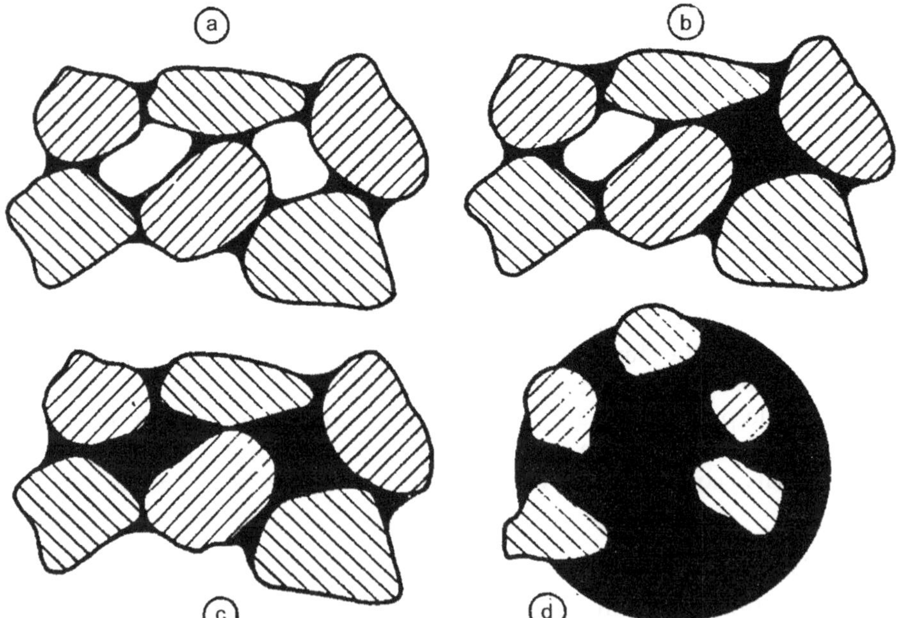

Figure 2-8 Liquid in particulate interstitial spaces (from Schubert, 1973).

grain industry sometimes uses soybean oil as dust control measure based on this phenomenon.

Dry adhesive forces between spherical particles can be represented by the following equation derived by Kendall and Stainton (2001).

$$\sigma = 15.6\varphi^4 \frac{R}{d} \quad [2\text{-}10]$$

where σ is the theoretical tensile strength of the agglomerate, φ is the particulate packing fraction (1 – void fraction), R is the nonequilibrium adhesion energy per unit surface contact area, and d is the particle diameter. Equation 2-10 implies that stable agglomerates of a given material are more likely for small particles with large packing fractions (small void spaces). Figure 2-9 is a SEM micrograph of such agglomerates formed from 10 mm spherical corn starch particles. Eckhoff notes that the presence of such agglomerates during dust explosibility testing can cause misleadingly low explosibility values because larger particles ignite and burn more slowly than small particles.

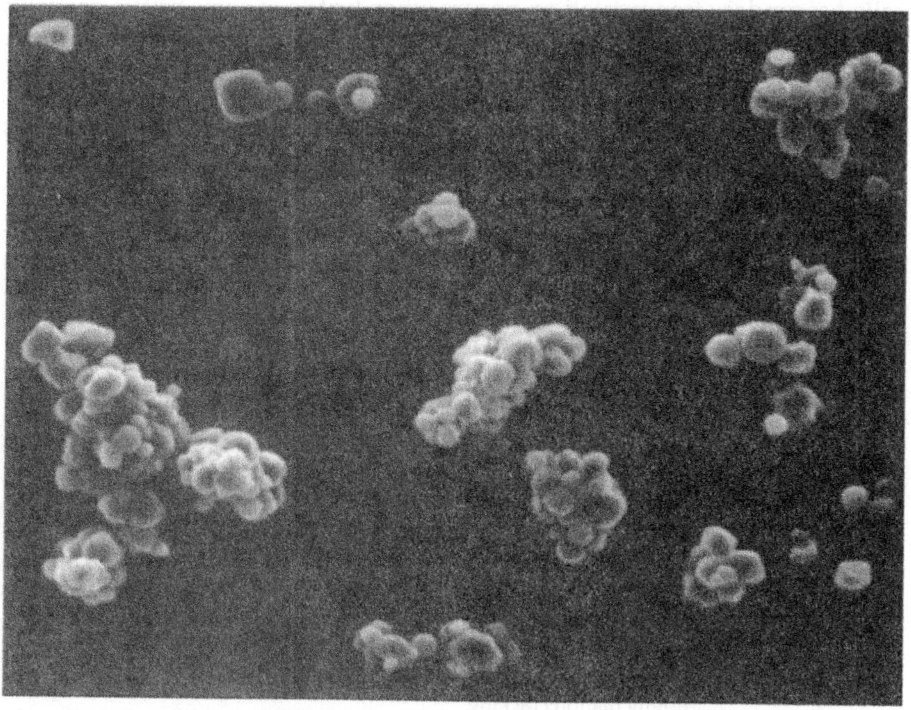

Figure 2-9 Corn starch agglomerates. (From Eckhoff, R., *Dust Explosions in the Process Industries,* 3rd ed. Amsterdam: Elsevier, 2003.)

2.2.8 Particle Size Changes due to Friability and Agglomeration

As particulates are processed and transported, there are competing forces promoting size reduction by attrition, and size increase via agglomeration. Although some unit operations are specifically designed for size reduction or size enlargement, the types of attrition and agglomeration phenomena discussed here are those that are an unintended result of solids processing and transport. Operations that incur unintended attrition include pneumatic transport, cyclone dust collection, flow through narrow clearances such as in star valves, and catalyzed reactors, particularly in fluidized bed reactors. Operations that promote agglomeration include compaction, moisture addition (often from condensation), and heating to incipient melting or polymerization.

Particulate attrition occurs by four different mechanisms as listed in Table 2-3. Thermal attrition occurs when certain materials are heated rapidly or become sufficiently hot to produce devolatization, or phase change, or chemical decomposition. The most common example of particulate devolatization occurs with coal heating. Abrasion is the gradual surface erosion due to the relative tangential motion of particles with each other and with the walls of the conveying or confining equipment. Chipping, also called secondary fragmentation, occurs when brittle or semi-brittle particles collide at a sufficiently high velocity perpendicular to the particle surface. Primary fragmentation occurs when the impact velocity exceeds a threshold that depends on the particle's mechanical properties.

Figure 2-10 shows SEM micrographs obtained by Scala et al. (2000) for limestone particles after impact fragmentation at a velocity of 15 m/s, that is, at a velocity sufficiently high to cause primary fragmentation. The four micrographs on the left show fresh (F) limestone particles with a characteristic size in the range 600 to 850 μm, and limestone subjected to calcination (C)

TABLE 2-3
Attrition Mechanisms

Mechanism	Description	Particle Relative Velocity
Thermal Attrition	Particulate mass loss or breakup due to thermal stresses, devolatization, phase change, or chemical reaction.	Not applicable
Abrasion	Wearing and rounding of particle surface due to frictional contact with walls or other particles.	Low
Chipping	Material removal at corners and edges via surface (secondary) fragmentation	Moderate
Fragmentation	Breaking and/or disintegration due to (primary) fragmentation cracking	High

Figure 2-10 Original and fragmented limestone particles. (From Scala et al., "Attrition of Sorbents during Fluidized Bed Calcination and Sulphation," *Powder Technology*, 107:153–167, 2000.)

and/or sulfatization (S) in fluidized beds. In most cases, the particles are angular with rounded edges (the exception being the CS particles which have sharper edges and some surface cracks). After impact, in all cases the fragmented debris is significantly smaller than the original limestone. The fresh limestone debris shows evidence of brittle disintegration such that many of the fines are smaller than 100 mm. The calcined limestone and sulfated limestone debris consists of larger fragments with the sulfated fragments having sharp edges.

Figure 2-11 is a pair of SEM micrographs obtained by Repenhagen and Werther (2000) showing the result of abrasion (a) and of chipping (b) of catalyst particles after flowing through a cyclone at inlet velocities of 20 m/s and 24 m/s, respectively. The abrasion regime produces a more spherical shape than the original particle shape, whereas the chipping regime produces distinctive surface fragmentation leaving the particle with a rougher, more irregular surface than the original particle. The fines produced from the abrasive attrition of these catalyst particles are shown in Figure 2-12. All of the fines are on the order of 1 μm or smaller, whereas the original catalyst particles were in the range 30μm to 200 μm.

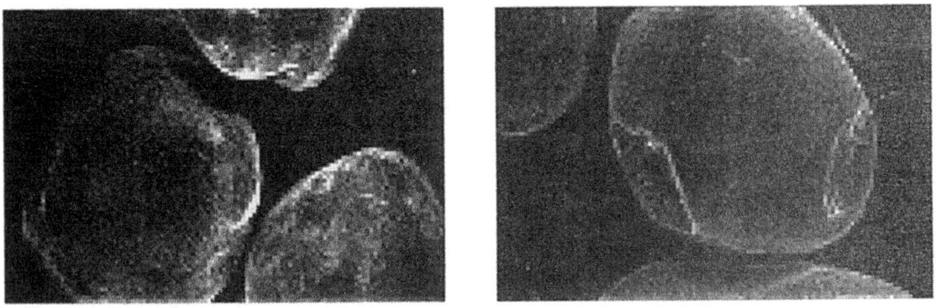

(a) (b)

Figure 2-11 Catalyst particles after abrasion (a) and chipping (b). (From Repenhagen and Werther, "Catalyst Attrition in Cyclones," *Powder Technology*, 113:55–69, 2000.)

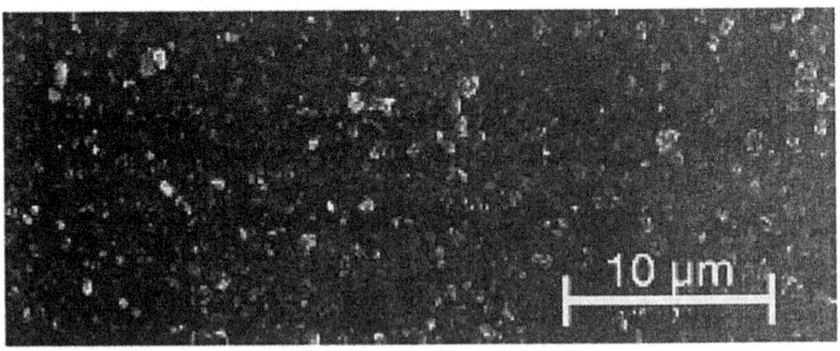

Figure 2-12 Abrasion-generated debris particles (From Repenhagen and Werther, "Catalyst Attrition in Cyclones," *Powder Technology*, 113:55–69, 2000.)

Single particle impact tests have shown that the fractional loss of material by chipping can be represented as (Taylor, 1998)

$$\zeta = a\frac{\rho_p v_p^2 (d/H)}{K_c K_f^2} \qquad [2\text{-}11]$$

where ζ is the fractional weight loss upon impact of a particle of density ρ_p (kg/m^3), diameter d (m), hardness H (Pa), and fracture toughness, K_c (Nm$^{-3/2}$), impacting at a velocity, v_p (m/s). K_f is a constraint factor (often taken as 1), and a is proportionality constant, approximately equal to 0.87. Calculations by Taylor (1998) for a particle with a density of 1300 kg/m^3, a diameter of 300 μm, a hardness of 2×10^8 Pa, and a fracture toughness of 4×10^5 N/m$^{3/2}$, indicate that the chipping attrition is negligible at impact velocities less than 5 m/s, but increases to about 10% at an impact velocity of 10 m/s. This is roughly consistent with Taylor's observation that there is 5–15% increase in small particles (under 180 μm) at a pneumatic conveying velocity of about 20 m/s, if we assume that the particle impact velocity is approximately half the conveying velocity.

Taylor also reports that attrition via fragmentation begins to occur at impact velocities greater than a critical velocity, v_{pc}, given by

$$v_{pc} = \left(\frac{K_c}{H}\right)^4 \frac{E}{\rho H^{1/2} d^2} \qquad [2\text{-}12]$$

where E is Young's modulus of elasticity (Pa). Based on Equation 2-12, Taylor suggests that particles with the characteristics cited above start experiencing fragmentation attrition at impact velocities of about 30 m/s. Attrition tests for individual particles and for bulk particulates are described by Bemrose and Bridgewater (1987).

Repenhagen and Werther (2000) have conducted extensive testing of the abrasive attrition rates during flow through cyclone dust collectors. The cyclones are intended to separate and remove large particulates from the stream, and allow most of the smaller particles to flow out with the exiting airflow. However, the efficiency of cyclone particulate removal is degraded by the attrition of particles via wall impact. Repenhagen and Werther have determined that the fractional rate of attrition, r_c, can be correlated as

$$r_c = \frac{\dot{m}_{\text{fines}}}{\dot{m}_p} C_c d_s \frac{u_{c,\text{in}}}{\sqrt{\dot{m}_p/\dot{m}_a}} \qquad [2\text{-}13]$$

where

$\dot{m}_{\text{fines}}$ = mass generation rate of fines by attrition (kg/s),
$\dot{m}_p$ = mass flow of particulate entering the cyclone (kg/s),
C_c = material dependent attrition rate constant (s^{2}m^{-2}),
d_s = particle surface mean diameter (m),
$u_{c,\text{in}}$ = cyclone inlet velocity (m/s),
$\dot{m}_a$ = mass airflow through cyclone (kg/s).

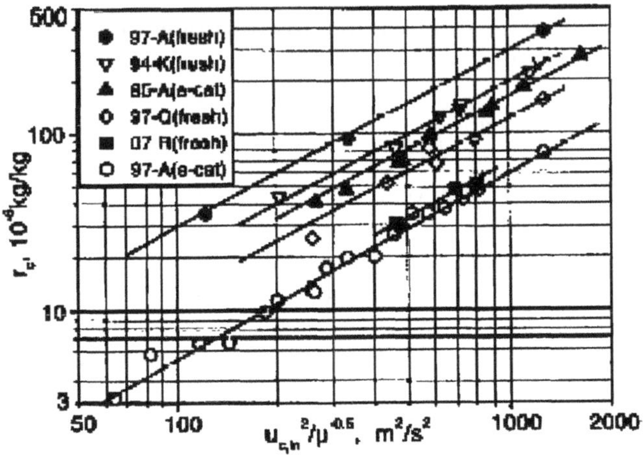

Figure 2-13 Attrition rate for various catalyst particles versus cyclone inlet velocity/load ratio$^{1/2}$ (From Repenhagen and Werther, "Catalyst Attrition in Cyclones," *Powder Technology*, 113:55–69, 2000.)

Figure 2-13 is a plot of r_c versus $u_{c,in} / (\dot{m}_p / \dot{m}_a)^{1/2}$, that is, the inlet velocity divided by the square root of the solids load fraction. As suggested by Equation 2-13, the data for each catalyst material is linear in these coordinates. The fact that the attrition rate varies inversely as the square root of the load ratio suggests that the chipping is due to impact against the cyclone wall, as opposed to interparticle collisions. This was confirmed by Repenhagen and Werther's observations of significant erosion of the cyclone inlet wall surface.

The material-specific attrition rate constant, C_c, depends on the particulate morphology as well as its mechanical strength. Angular, irregular shaped particles were found to have a value of C_c twice as high as the value for spherical shaped particles of the same size and material. Thus, it appears that the steps to minimize particle attrition during processing and transport include using small rounded particles at a large solids load ratio and a relatively small velocity.

Agglomeration without adhesives is most likely to alter particulate size distribution in applications involving small particles at a high packing fraction (small void fraction). It occurs most frequently with submicron particles flowing at concentrations of at least 1 g/m^3 (Koizumi et al., 1995). Larger particles (up to 100 mm) also experience electrostatic induced agglomeration in pharmaceutical coating operations because of the high binding strength of the coating materials (Jono et al., 2000). Chemical reaction induced agglomeration occurs in fluidized bed olefin polymerization reactors (Yiannoulakis et al., 2001) at temperatures above the polymer softening or sticking temperature, which have been measured for several materials by Tardos and Pfeffer (1995). Agglomeration caused by the addition of liquid binders is designed to occur in granulators, which are described in Chapter 5. Agglomeration rates in high shear and low shear mixers/granulators are dependent on the binder viscosity as well as particle size (Mills et al., 2000). Agglomeration can also affect particulate flow characteristics, bulk density measurements, and dust cloud concentration measurements as explained in the following sections.

2.2.9 Bulk Density Measurements and Characterizations

The bulk density, ρ_b, of a powder–air mixture is defined as (Fan and Zhu, 1998)

$$\rho_b = \varphi\rho_p + (1 - \varphi)\rho_a = (1 - \alpha_v)\rho_p + \alpha_v\rho_a \qquad [2\text{-}14]$$

where ρ_a is the air density (1.2 kg/m^3 at standard temperature and pressure), ρ_p is the particle density, and α_v is the void fraction in the mixture. In the overwhelming majority of cases, the second term is negligible compared to the first term in Equation 2-14,that is, the particle mass fraction is virtually equal to one. From the standpoint of practical bulk density measurements,

$$\rho_b = M/V_b \qquad [2\text{-}15]$$

where M is the mixture mass and V_b is its bulk volume. Thus, the determination of bulk density merely requires the collection of a mixture sample in a known volume, and weighing the sample after taring the container volume.

The complication associated with this seemingly simple determination is that the filling of a container can sometimes entail an arbitrary degree of packing the particulates. As the particles packing fraction increases, the bulk density increases proportionally, as indicated by Equation 2-14. Furthermore, as the particles settle toward the bottom of the container, the particulate fraction also increases. Hence, there is a need for a standardized measurement method to determine bulk density.

ASTM D6393 describes methods for determining the loose bulk density and the packed bulk density. Since the methods described in ASTM D6393 were developed by Carr (1965), they are called Carr indices. (Other Carr indices described in ASTM D6393 pertain to cohesion, angle of repose, angle of spatula, compressibility and dispersability.) The Carr loose bulk density measurement involves passing the particulate through a sieve with a 710-mm opening into a 100-cm^3 cup. The Carr packed bulk density measurement involves placing the filled cup onto a tapping device and subjecting it to 180 seconds of tapping, while adding more powder to the cup so that the final level will coincide with the top of the cup. The 180-second tapping duration is presumably sufficient to allow the settled particles to reach a limiting equilibrium packing fraction. Both the loose bulk density and the packed bulk density depend to a great extent on the nature of the electrostatic forces that can either attract or repel adjacent particles. Tapping can increase the electrostatic charging of the particles.

Once the bulk density has been determined, particulate volume fraction and/or the void fraction can be calculated from Equation 2-14, as

$$\varphi = \frac{(\rho_b/\rho_a)-1}{(\rho_p/\rho_a)-1} \qquad [2\text{-}16]$$

Similarly, the particulate mass fraction, Y_p, can be calculated from,

$$Y_p = \frac{\varphi\rho_b}{\rho_p} = \frac{\varphi(\rho_p/\rho_a)}{1+\alpha_p\left[(\rho_p/\rho_a)-1\right]} \qquad [2\text{-}17]$$

For example, suppose powder with a particle density of 2000 kg/m^3 has a bulk density of 400 kg/m^3. The particulate volume fraction is equal to (332/1666) = 0.1995 (void fraction = 0.80), and the particulate mass fraction is equal to (0.1995 * 2000)/400 = 0.998. Although the simplifying approximation of assuming the mass fraction to be unity should be perfectly acceptable for powder deposits and piles, the full equations should be used in the case of flowing or air-suspended powders; in the flowing situation the bulk density becomes the mixture density.

2.2.10 Dust Cloud Concentration Measurements

The mass concentration, C, of airborne particulates is related to the particulate volume fraction, φ, by

$$C = \varphi\rho_p \qquad [2\text{-}18]$$

and is usually easier to measure than the volume fraction. The concentration is a critical parameter for explosibility hazards in particulate process and transport equipment. Concentrations in occupied areas are of great concern for toxicology considerations. There is also a need to measure and control concentrations in stacks and other emission paths to satisfy environmental regulations.

Traditional measurements of airborne concentrations entail air sampling at a known flow rate on filters and weighing the collected samples. If the sample is obtained from within a duct, pipe, or stack, it is necessary to do isokinetic sampling (so that the collected sample will be representative of the suspended stream) and measurements of the air/gas flow rate as well as the sampling duration. This usually entails a sampling system or probe with air flow instrumentation and possibly air drying provisions. Air sampling in open areas entails using a sampling probe equipped with a calibrated air pump for sample collection at a known flow rate. The samples can be collected manually or automatically at periodic intervals, and there are many commercial devices for doing both the sampling and concentration determinations. Standards on gravimetric concentration measurements and sampling include ASTM D6331 and ISO 9096. Air sampling with gravimetric concentration determinations are not suitable for reactive, unstable, or very volatile materials.

Modern instrumentation now makes continuous in-situ concentration measurements possible in process equipment and in pneumatic transport systems, as well as in laboratory and large-scale testing. Table 2-4 lists different methods that have been used for in-situ concentration online measurements. In most cases, commercial instruments are available based on the indicated measuring methods, and one or more commercial manufacturers or developers are indicated.

Light attenuation, the first listed method, is based on the Beer-Lambert law, which for a cloud of monosized spheres of diameter d, can be expressed as (Louge, 1996)

$$I/I_0 = \exp[-(\kappa\varphi^x/d)] \qquad [2\text{-}19]$$

where I is the light intensity transmitted through a cloud of length x and particulate volume fraction, φ. I_0 is the light intensity entering the cloud, and κ is a material dependent constant approximately equal to 2 for many materials. In one version of the light attenuation concentration probe, the light source and sensor are situated on opposite walls of a duct (Figure 2-14), and the attenuation path length, x, is equal to the duct diameter. The other version of the light attenuation probe has the emitter and sensor immersed in the flow, and the path length, x, is approximately 3 cm, as shown in Figure 2-15. Both versions use an air purge to prevent deposits on the light- receiving window. Calibration of these probes entails generating a semilog scale plot of I/I_0 as a function of φ or C for a given particulate sample. Eckhoff (2003) shows calibration curves for both types of light attenuation curves over the ranges of concentrations shown in Table 2-4.

One drawback of the light attenuation probes reported by Eckhoff is that each configuration has a limited concentration range. The local concentration probe shown in Figure 2-15 has a second drawback in that it is suffi-

Figure 2-14 Light attenuation dust concentration probe in duct. (From Eckhoff, R., *Dust Explosions in the Process Industries,* 3rd ed. Amsterdam: Elsevier, 2003.)

TABLE 2-4
In-Situ Dust Cloud Concentration Measurement Devices

Method	Sensing Length	Concentration (g/m^3) or φ Range	Reference or Commercial Manufacturer
Light attenuation	15 cm	C = 10–90	Eckhoff (1994)
	3 cm	C = 100–600	Eckhoff (1994)
	20–360 cm	<10	Afriso
Optical fiber backscattering	≈1 cm	—	Zhang et al. (1998)
		$0.07 < \varphi < 0.4$	Bellino et al. (2001)
	0.2 – 0.8	$C = 10^3–10^6$	Meili et al. (1995, 1998) MSE Meili
	1 –1.5 cm	C = 0.1–10,000	
Digital video imaging	—	$\varphi \approx 1$	Grasa and Abanades (2001)
		$0.05 < \varphi < 0.3$	Lasentec
Laser Doppler scattering	≈ 1 cm	C = 1–1000	Dantec, TSI
Capacitance probe	~15 cm	?	Louge (1996) Yan and Reed (1999)
Ultrasonic attenuation	~5 cm	$\varphi > 0.01$	Harker et al. (1991)
		Capacitance probe	Malvern Ultrasizer
		Ultrasonic and acoustic attenuation	Moss et al. (1999)
Triboelectric (electrodynamic effect)	—	$C = 10^{-3}–1000$	Yokogawa
Beta radiation attenuation through particulate layers on filters	NA	$C = 5 \times 10^{-6}–2 \times 10^{-2}$	Andersen Thermo Environmental
Oscillating microbalance	NA	$C = 10^{-6}–5$	Ruprecht and Pataschnick Co.

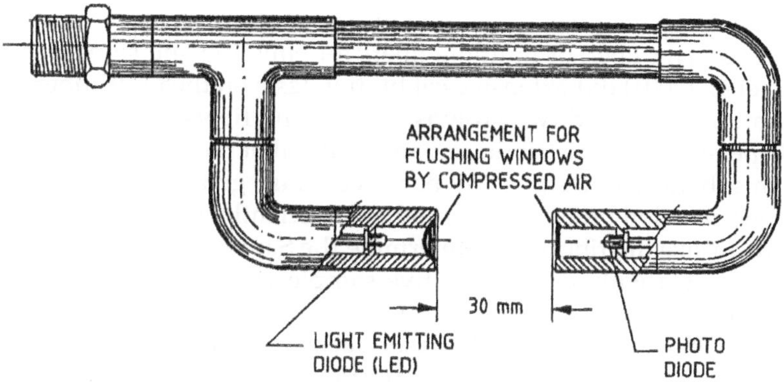

Figure 2-15 Local dust concentration probe. (From Eckhoff, R., *Dust Explosions in the Process Industries*, 3rd ed. Amsterdam: Elsevier, 2003.)

ciently large that it can disturb the particulate flow and possibly alter the local concentration.

The optical fiber back scattering probe is currently the most widely used concentration probe for laboratory testing, and at least one rugged design has recently been commercialized for use in industrial facilities as shown in Figure 2-16. The single fiber version of an optical fiber probe measures the back-scattered light through the same fiber as the source. Louge (1994) has shown that the return signal should be a function of $\varphi \kappa D_f / d$, where D_f is the

Figure 2-16 Fiber optic measuring instrument: "Labasys® Control Ex" for inline monitoring of dust concentration, velocity and flow rate. (From MSE Meili, Zurich, www.msemeili.ch, with permission.)

fiber diameter. Other versions of the probe employ either one emitter fiber and one receiver fiber, or multiple fibers for emission and signal reception. Bellino et al. (2001) did extensive calibration tests on the multi-fiber version (one emitting fiber and two crowns of receiving fibers) of the probe using polydisperse particle distributions, and found that the signal varied as φ/d_{32} for both spherical and aspherical particles. Thus, general use of the optical fiber concentration probe requires either sufficiently accurate knowledge of the particle size distribution to determine the Sauter mean diameter, d_{32}, or calibration data with a representative sample of the specific dust/powder being used.

Meili (1998) has developed a laser backscatter system with pneumatic cleaning that can measure both concentrations and velocities in spray driers and other particulate processing equipment. By using different optical configurations, his commercial instrument is reported to cover a remarkably wide concentration range as indicated in Table 2-4. The other fiber optic backscatter probe listed in Table 2-4 is the two-fiber in a small diameter tube device developed by Zhang et al. (1998), which also measures particle velocities as well as mass concentrations.

Video digital imaging systems measure concentrations either by associating the pixel fraction of an image with the particulate volume fraction, or by calibrating the grayscale of the image. At least one company (Lasentec) has developed an explosion-proof system consisting of a purged probe, cable in conduit connection to power supply/controller, and fiber optic communication to a computer with video image analysis software. Figure 2-17 is a schematic diagram of this system. Grasa and Abanedes (2001) have developed and tested another video imaging system for determining mixing fractions of blended bulk powder.

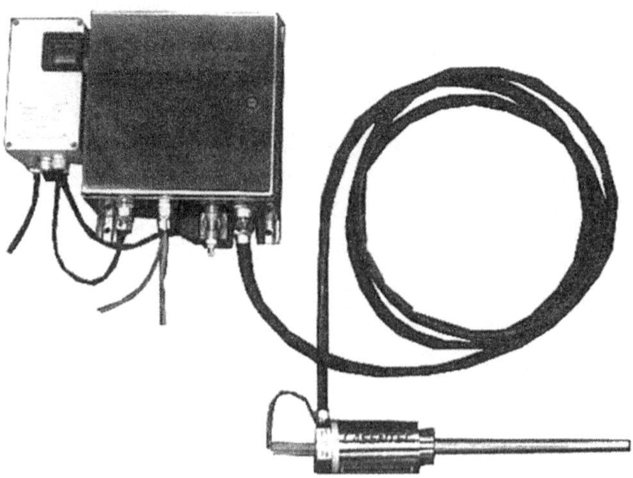

Figure 2-17 Digital video image dust concentration system (from Lasentec).

The laser Doppler scattering systems listed in Table 2-4 are virtually identical to the systems mentioned earlier for particulate size measurement. They are accurate sophisticated systems, but require precise focusing and near spherical shaped particles. Therefore, they may be difficult to adapt for general process applications.

The capacitance probe listed in Table 2-4 has one or more sensor electrodes that can measure the local capacitance of the airborne particle cloud and relate it to concentration. The ultrasonic probe measures the sound speed in a particulate cloud or flow, and uses a mathematical relationship between sound speed and particle volume fraction in the solids-air mixture. The acoustic frequency version of this technique utilizes the attenuation of sound waves by the airborne particles. The triboelectric effect probe utilizes the electrostatic charging associated with inter-particle collisions and collisions of the charged particles with the electrostatic probe. According to the manufacturer (Yokogawa), it can sense a wide range of particulate concentrations flowing in an air or gas stream with a velocity in the range 4 m/s to 30 m/s.

The last two methods listed in Table 2.4 are intended for significantly lower particulate concentrations than the other methods. The beta radiation attenuation method uses a low flow rate sampling pump to capture particles on a continuously moving filter tape, and pass low-energy beta rays through the exposed and unexposed portions of the tape. The beam attenuation is a calibrated function of the accumulated concentration of particulate. The oscillating microbalance method involves the capturing of particles on a filter cartridge mounted on the tip of a tapered hollow glass element that oscillates in an applied electric field. A patented microbalance is used to monitor the rate of accumulation of particulate. The EPA has approved several of these beta ray attenuation instruments and one oscillating microbalance instrument for the monitoring of ambient air quality for concentrations of particulate with a aerodynamic diameters less than 10 μm (EPA, 1999). These instruments are well suited for monitoring for respiratory hazards, but their upper limits of measurable concentrations are far too small for them to be used to monitor for Minimum Explosible Concentrations of combustible dusts.

In the case of steady flow of particulate through a duct or pipeline of cross-sectional area, A, the solids mass flow rate, $\dot{M}_s$, is related to the average concentration, $\overline{C}$, by

$$\dot{M}_s = A\overline{v}_s\overline{C} \qquad [2\text{-}20]$$

where $\overline{v}_s$ is the average velocity of the solid particulates. Therefore, another method of determining the particulate concentration in these applications is to measure the mass flow rate and velocity, and use Equation 2-20 to calculate $\overline{C}$.

The U.K. Institute of Management and Control (2001) has recently published a guide to measuring particulate flow rates, velocities, and concentrations. Besides the measurement methods described above and listed in Table 2-4, the IMC guide included several other methods. These include measuring flow rates using office plates, turbine rotors, and impact plates, all of which are considered restrictive in that they cause pressure drops and reduced flow rates, and they can also cause abrasion of the flowing particles. One of the nonrestrictive measurement methods not previously discussed is the use of microwave transmitters and receivers that either measure the back scattering from the moving particulates, or else measure the microwave absorption resonant frequency associated with the flowing particulates. Another noninvasive technique is the recording and analysis of particle generated sound waves (due to wall collisions and aerodynamic turbulence) by attaching a microphone or piezoelectric transducer to the duct/pipe wall. Table 2-5, reproduced from the IMC guide, lists various manufacturers of particulate flow measurement instrumentation. The tabulation includes manufacturers of both invasive and noninvasive instrumentation.

2.2.11 Bulk Powder Moisture Measurements

Many powders may contain moisture, the amount depending on the presence of moisture from previous processing steps, the hydrophilic nature of the powder (hygroscopicity), and the relative humidity of the surrounding atmosphere. The moisture content of a particulate layer or aggregate can significantly affect its dispersability, its combustibility, and in certain cases, its thermal stability, and reactivity. The presence of moisture may be beneficial, as it tends to decrease both the dispersability and combustibility of a dust.

As the moisture content increases, the dust particles generally become more cohesive and form agglomerates (as described in Section 2.2.7) that are more difficult to disperse in the air. Second, any heat applied to a suspension of moist dust will first be used to vaporize the moisture (water and/or solvent) and will delay or even possibly prevent the dust from being heated to its ignition temperature.

When a powder becomes too moist it may become sticky, and this increases the propensity for the particulate solids to adhere to the inner surface of piping and process equipment. As time progresses, the piping or process equipment may become plugged, and an overpressure situation may occur.

Moisture effects on chemical reactivity and thermal instability are more complicated, and depend on the particular material. For example, moisture is detrimental to the thermal stability of many water reactive particulate materials (for example, calcium hypochlorite) and to materials that are subject to microbiological heating when wet (for example, agricultural feedstocks).

TABLE 2-5
Particle Flow Instrumentation: Mass Flow Rate and Concentration

Company	Country	Principle	Parameters	Mode of Sensing
ABB Automation	UK	Electrostatic	Velocity Concentration	Nonrestrictive
Acoustica	Norway	Ultrasonic	Concentration?	Nonrestrictive
CalmpOn	Finland	Ultrasonic	Concentration?	Nonrestrictive
CSIRO Minerals	Australia	Acoustic	Concentration Velocity	Nonrestrictive
Flow Force Technology	Australia	Impact plate	Mass flow rate	Restrictive
Flumesys	?	Orifice plate?	Mass flow rate?	Restrictive
GE (EER)	USA	Rotorprobe	Mass flow rate	Restrictive
Inerco	Spain	Rotorprobe	Mass flow rate	Restrictive
M&W	Denmark	Rotorprobe	Mass flow rate	Restrictive
Milltronics	UK	Impact plate	Mass flow rate	Restrictive
Mission Instruments	USA	Rotorprobe	Mass flow rate	Restrictive
PCME	UK	Electrostatic Optical	Mass flow rate Velocity Concentration	Restrictive and nonrestrictive
Oxford Instruments	UK	Capacitance? Electrostatic?	Mass flow rate Concentration Velocity	Nonrestrictive
Physel	Finland	Radiological	Concentration Velocity	Nonrestrictive
Promecon	Germany	Microwave	Mass flow rate Concentration Velocity	Nonrestrictive
Ramsey	USA	Capacitance Microwave	Mass flow rate Concentration Velocity	Nonrestrictive
Rospen Industries	UK	Mechanical	Mass flow rate	Restrictive
Schench	?	Orifice plate?	Mass flow rate?	Restrictive
S-E-G	UK	Coriolis	Mass flow rate	Restrictive
SWR Engineering	Germany	Microwave	Concentration	Nonrestrictive
TR-Tech Int. Oy	Finland	Electrostatic	Mass flow rate	Restrictive
Truscott	UK	Impact plate	Mass flow rate	Restrictive

Therefore, it is important to have an accurate and convenient method to determine moisture content. Oven heating with sample weighing before and after the heating to drive off the water does represent an accurate method, but it lacks convenience and speed, and usually requires sampling and laboratory submittal. Semiautomatic drying and weighing ovens are now available to minimize the labor and time involved. Although drying via heating is a suitable technique for most materials, it is problematic for materials that either pyrolyze or oxidize at temperatures around 100°C.

Commercially available moisture meters are based on a variety of physiochemical phenomena as listed in Table 2-6. Electrical conductivity based moisture probes have been developed for materials that have a direct correlation between resistance and moisture content. These materials include many textiles, wood, grain, and paper. For most of these materials, the correlation between resistance and moisture content is linear up to the saturation point, which varies from 12% to 25% moisture, depending on the materials (Eckhoff, 2003). Commercial devices are available with both surface probes, and with long needle probes for measurements in the interior of bulk materials (e.g., Strandberg Engineering Laboratories, Inc.).

In many materials, the addition of moisture sharply increases its dielectric constant. Capacitance type moisture probes typically use a radio frequency power supply in a capacitance bridge circuit containing a sample of the moist material. Either a bridge imbalance or a frequency change is measured and correlated against moisture content. The microwave attenuation meters are reportedly suitable for the full range of moisture contents from dry to fully saturated.

Karl Fischer titration methods for moisture determination are based on the chemical reaction between an iodine bearing reagent and water, such that the determination of the iodine content of the product is tantamount to

TABLE 2-6
Types of Moisture Meters (from Iinoya et al., 1988)

Physiochemical Principle	Measurement Method
Electrical resistance decrease	Conductivity meter
Electrical capacitance increase	Capacitance meter
microwave attenuation	Microwave moisture meter
Infrared reflection and absorption	three color infrared moisture meter
Nuclear magnetic resonance	NMR moisture meter
Neutron reflection and absorption	High-energy neutron absorption
Cobalt chloride reaction	Color change due to reaction
Equilibrium relative humidity	Saturated temperature difference
Karl Fischer titration	pH meter or coulometer

the determination of the water content. Various reagents have been used since the original pyridine based reagent discovered by Karl Fischer. The volumetric method (often involving a pH meter) is used for high moisture contents, while the coulometric measurement is used for samples with lower moisture levels.

Description of most of these and the other types of moisture probes listed in Table 2-6 are provided by Iinoya et al. (1988).

2.2.12 Fluidity and Dispersibility

Particulate fluidity is the property that represents the particles tendency to flow when subjected to sufficiently large stresses, that is, when the forces acting on the powder/dust produce stresses in excess of the solid strength. The appropriate measure of fluidity depends on the flow context and application. For example, in the context of hopper flow, particulate fluidity is best characterized by the particulate unconfined yield strength, consolidating strength, angle of internal friction, and angle of wall friction. These and other pertinent properties for hopper flow can be measured by laboratory tests developed by Jenike, Johanson (1981), and others. Fan and Zhu (1998) provide a good summary of the use of those tests and properties for hopper flow.

In the context of unconfined solid piles, the angle of repose is the most common measure of particulate fluidity. It is the angle formed between a horizontal plane and the slope line extending along the face of a pile of material. The angle of repose for a given material may vary, however, depending on how the pile is created and the density, particle shape, moisture, and particle size distribution of the material. The angle of repose can be measured by several methods (Woodcock and Mason, 1987, pp. 31–32), perhaps the one most commonly used yields a value of "poured" angle of repose, which is the angle between the horizontal and the sloping side of the material poured gently from a funnel onto a flat surface. The angle of repose test method involving the use of a vibrating sieve is described in ASTM D6393, and is called the Carr angle of repose test.

The angle of repose may be used as a rough guide to the flow behavior of particulate solids, as shown below (Woodcock and Mason 1987, pp. 31–32):

Angle of Repose	Qualitative Fluidity
25–30°	Very Free-Flowing
30–38°	Free-flowing
38–45°	Fair-flowing
45–55°	Cohesive
>55°	Very Cohesive

The fluidity of powders in rotating process equipment such as mixers can be characterized using a new instrument developed by TSI called an AeroFlow. The test entails rotating a sample of powder in a disc with transparent end walls so that powder displacement and avalanche formation can be recorded by a light source and an array of photocells. Time intervals between avalanche flows and the time-averaged centroid of the powder image are used to characterize the powder fluidity. Kaye et al (2001) reported preliminary results using the AeroFlow to determine the affect of flow enhancing agents (silica and magnesium stearate) on the fluidity of lactose powders.

Dispersibility is the tendency of a powder or dust to form a suspended dust cloud during either routine handling or accidental upset conditions. The dispersibility test developed for routine discharge of particulates from a hopper or bin or conveyor transfer station is called the Carr dispersibility test (ASTM D6393, Test J). It involves discharging 10 g of powder from a 5-cm diameter cylinder with a bottom shutter. When the shutter is opened, the powder falls through a 10-cm diameter, 33-cm-high open cylindrical tube (Item A) shown in Figure 2-18. A 10-cm diameter watch glass (Item B) placed under the tube collects the powder that has not been dispersed during the discharge. The powder dispersibility in this test is the percentage of the 10-g powder charge that is not recovered in the watch glass.

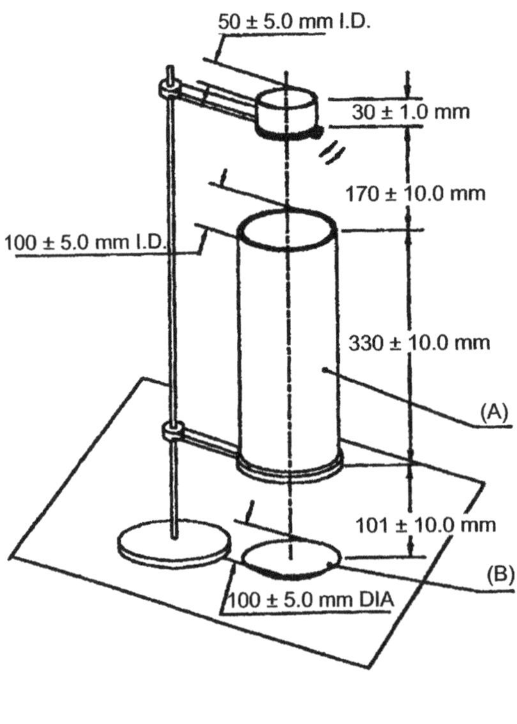

Figure 2-18 Carr dispersability test apparatus (reproduced with permission from ASTM D6393).

Eckhoff (1997) and Ural (1989) have described test methods to determine the dispersibility of dust in the context of dust cloud formation by an aerodynamic disturbance to a deposited dust layer. This property is particularly important in the occurrence of secondary dust explosions. The most commonly reported parameter for this application is the minimum air velocity required to entrain a specified fraction of dust from the dust layer. In the case of 50% entrainment from corn starch dust layers, Ural measured minimum air velocities in the range 10 to 35 m/s, depending on the deposited dust layer mass density (in g/m^2) and its method of deposition. The minimum required air velocity for entrainment decreased slightly as the dust layer mass density increased. More recently, Scherpa (2002) has found that the minimum required air velocity for dispersal is lower for transient air blasts than for steady air flows.

2.2.13 *Electrical Resistivity*

Dust layer electrical resistivity is pertinent to the accumulation of electrostatic charges generated during particulate transport and processing. It is also pertinent to the accidental occurrence of electrical shorts in electrical equipment that may contain deposited dust layers. Both phenomena can lead to dust layer or dust cloud ignition. It is therefore important to have standardized tests to measure dust layer resistivity.

IEC 61241-2-2 describes a test apparatus and method for measuring dust layer resistivity. The standard defines layer resistivity as the minimum value of electrical resistance measured between electrodes spaced a unit distance apart, and with each electrode having a unit area in contact with the dust layer. Figure 2-19 shows the apparatus consisting of two stainless steel electrodes of height, *H*, of 1 cm, and Length, *W*, of 10 cm, and separated by a gap of 1 cm in which the dust layer is placed. The dust layer resistivity, ρ_R, is calculated from:

$$\rho_R = 0.001 R_s (HW/L) = 0.01 R_s \qquad [2\text{-}21]$$

where R_s is the resistance of the dust-filled test cell. Dusts with resistivity values under 10^3 Ω-m are considered conductive, and dusts with resistivity values greater than 10^3 Ω-m are considered nonconductive. Resistivities above 10^8 Ω-m are indicative of the potential for significant electrostatic charging (IEC 61241-2-2).

NFPA 77 (2000) has the following three-category classification scheme for powder volume resistivity:

Low-resistivity powders have resistivities of up to 10^8 Ω-m. These powders can become charged during flow, but the charge rapidly dissipates when the powder is collected in a grounded container. Medium-resistivity powders have resistivities in the range 10^8 to 10^{10} Ω-m. When medium resistivity powders are collected in a grounded container, the charge retention

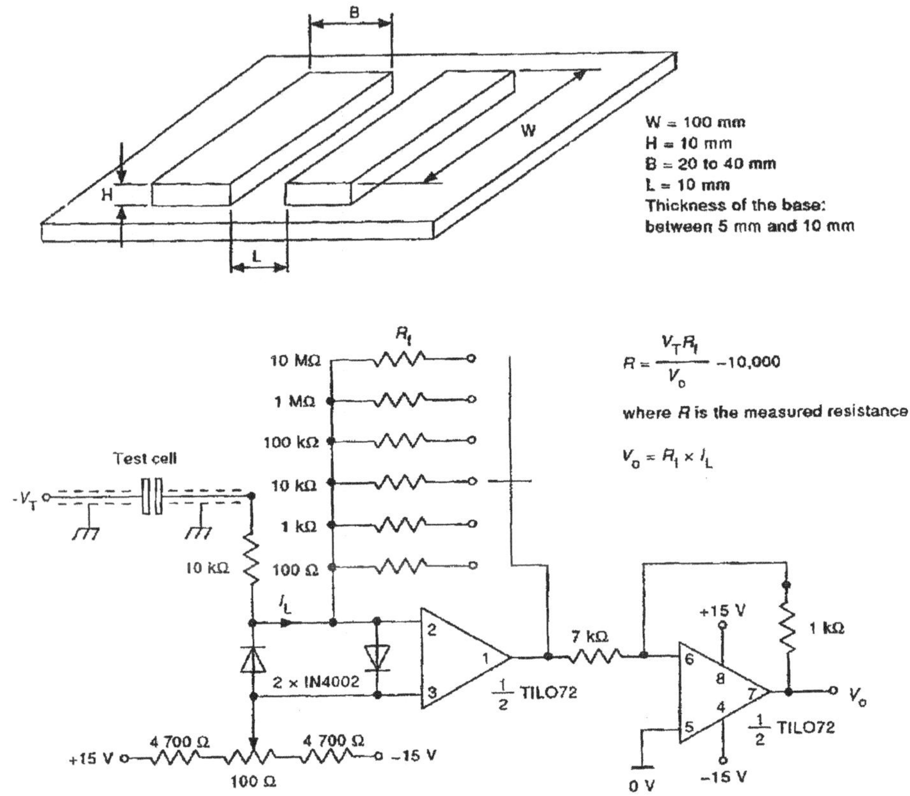

Figure 2-19 IEC apparatus for measuring dust resistivity.

time of the bulk powder is governed by the powder resistivity multiplied by its dielectric constant. High-resistivity powders have resistivities greater than 10^{10} W-m. According to NFPA 77-2000 (paragraph 8.4.2.2), high resistivity powders can produce corona, brush, bulking brush, and propagating brush discharges.

Resistivity tests conducted by Probst and Grivei (2002) on various types of carbon black powder show that the particulate resistivity decreases with increasing specific surface area and with increasing layer bulk density. The lowest resistivities (about 5×10^{-3} Ω-m) were measured with activated carbon, which had a specific surface area of 1600 m^2/g, that is, at least twice the value for the next smallest form (next largest specific surface area) of carbon black.

Additional descriptions of particulate layer resistivity testing are provided in Chapter 4.

2.3 OVERVIEW OF PARTICULATE CHEMICAL CHARACTERISTICS

2.3.1 *Flammability and Explosibility*

Flammability and explosibility refer to a material's ease of ignition, and to its rate of burning and associated energy release rate after ignition. Increasing chemical reactivity of particulate solids, similar to gases and vapors, leads to increasing flammability and explosibility. Examples of highly reactive powders are metals (e.g., al., Mg, Ti, Zr) that possess very high heats of oxidation, and correspondingly high flame temperatures. The maximum premixed flame temperature of some small diameter metal powders may approach or exceed 3000°K. For example, Eckhoff (1997) reports that the flame temperature for 6 mm aluminum particles at a concentration of 300 g/m3 is about 2900°K, and it is greater than 3000°K at a concentration of 500 g/m3. The corresponding flame temperature for most organic powders will usually be 2000°K to 2700°K (about the same as a gas explosion). Likewise, the maximum closed vessel explosion pressures listed by Eckhoff (1997) for aluminum powder (11–12 bar(g)) and magnesium powder (17 bar(g)) are substantially higher than those for most organic dusts (typically 7-10 bar(g)).

The presence of specific chemical groups in an organic molecule can give an indication of the relative flammability and explosibility. For example, Abbott (1990) provides the following list of chemical groups as being indicative of potentially deflagrating or detonating particulate explosives: $-NO_2$ or $-ONO_2$; N=N or N–N; NX_2 (e.g. NCl_2); C=N; $OClO_2$; O–O or O–O–O (e.g,. peroxides); C≡C; and a metal atom connected by an unstable bond to C or certain organic radicals. According to Field (1982), COOH, OH, NH_2, C=N, C≡N, and N=N tend to increase the explosion hazard, while molecules which have the halogens Cl, Br, and F generally exhibit reduced flammability and explosibility. The CCPS book on chemical reactivity hazards (Johnson et al., 2003) and the NOAA Chemical Reactivity Worksheet provide additional guidance on identifying potent oxidizing agents and other chemical groups that can exacerbate flammability and explosibility hazards.

The most widely used measure of a material's fire severity potential is the heat release rate per unit fuel surface area. Tewarson (1995) has shown that the chemical heat release rate per unit surface area, $\dot{Q}''_{ch}$, is proportional to the ratio of the heat of combustion, ΔH_c, to the heat of vaporization, ΔH_v. Representative values of heats of combustion, heats of vaporization, and the ratio are shown in Table 2-7. Based on the ratio values, it is clear that the polyolefins are significantly more flammable than polyamides, PMMA, and partcularly PVC.

Data of the type shown in Table 2-7 can be sensitive to the presence of additives and fillers in these polymers. If the fillers used are nonreactive,

TABLE 2-7
Representative Chemical Flammability Properties[a]

Granular Material	ΔH_c (kJ/g)	ΔH_v (kJ/g)	$\Delta H_c/\Delta H_v$
Polyethylene (high density)	43.6	2.32	18.8
Polyethylene (low density)	43.6	1.75	24.9
Polypropylene	43.4	2.03	21.3
Polystyrene	39.2	1.70	23.1
Nylon 6/6	30.8	2.35	13.1
Polymethylmethacrylate	25.2	1.63	15.5
PVC (rigid)	16.4	2.47	6.6

[a]Based on data reported by Tewarson (1995).

such as silica, or a fire retardant, the flammability hazard is usually reduced. In the case of PVC, plasticizers are often added to provide flexibility, and these often increase the effective heat of combustion and flammability hazard.

Chapter 4 contains a comprehensive discussion of particulate flammability and explosibility tests. In the case of dust layers, ignitability tests include the IEC 61241-1 hot plate ignition test, the DIN glow coil ignition temperature test, and dust layer flame spread time. Data from the glow coil ignition temperature test are listed by Eckhoff (1997). However, the test is not useful for thermoplastic materials because melting occurs before the material is ignited. The U.N. dust pile flame spread test is used to determine the combustibility classification of particulate materials within Category 4.1 (U.N., 1999).

In the case of combustible dust clouds, relevant explosibility tests include dust cloud Minimum Ignition Temperature, Minimum Ignition Energy, Minimum Explosive Concentration, Maximum Explosion Pressure and maximum rate-of-pressure-rise (ASTM E1226). Results of these tests, which are described in Chapter 4, depend on chemical composition as well as particle size, concentration, moisture level, and the details of the test apparatus. As for composition, copolymer test results are directly related to the explosibility of the component monomers.

One other important chemical consideration in flammability and explosibility testing is the oxidant composition and concentration. Limiting oxygen concentration (LOC) test data are presented in NFPA 69 for numerous particulate materials inerted with nitrogen and with carbon dioxide. In most cases, the LOC test data are in the range 12–17 volume%, but can be significantly lower for certain organic materials. LOC values of 2% and lower are listed for following metal dusts: aluminum, magnesium, thorium, titanium, and zirconium. Data presented in a wide assortment of publications have shown that combustible dust flammability and explosibility increase substantially at oxygen concentrations above 21 volume%. The LOC varies with initial pressure such that it increases with pressure for some materials, and decreases with pressure for other materials (Siwek, 1996).

2.3.2 *Thermal Degradation and Instability*

Particulate thermal decomposition and instability properties can be described either in terms of fundamental chemical and physical properties, or in terms of the laboratory self-heating test parameters that are intended to determine safe storage and handling temperatures for actual particle size, composition, oxygen availability, and storage pile or container size. Two fundamental thermochemical equilibrium properties that govern thermal instability and self-reactivity are the thermochemical heat of self-reaction, ΔH_r, and the adiabatic reaction temperature, T_{ar}. Murphy et al (2003) have recently shown how these parameters can be calculated from chemical heats of formation, ΔH_f^0, and how the results can be used to obtain relative rankings and categorizations of the instability hazard. The calculations entail using readily available thermochemical software such as the ASTM CHETAH code (2002). Example calculations for nitrate particulate compounds given by Murphy et al., are ammonium nitrate ($\Delta H_r = -2.44$ kJ/g, and $T_{ar} = 1723$°K), and TNT ($\Delta H_r = -5.87$ kJ/g, and $T_{ar} = 2090$°K). Based on these values, Murphy et al. (2003) put TNT in a higher hazard category than ammonium nitrate.

Chemical kinetics parameters that account for the expected reaction rate are also useful for a more comprehensive analysis of a material's instability propensity. Probably the most important chemical kinetics parameter is the activation energy, E_a, for each significant decomposition reaction. The most unstable chemicals from the standpoint of self-reactivity are those with a combination of a high ΔH_r, and low value of E_a, as described in CCPS *Guidelines for Storage and Handling of Reactive Chemicals* (1995). Activation energies, heats of formation, and adiabatic reaction temperatures for unstable materials can be determined experimentally using laboratory reactivity testing described in Section 4.3.2 of this book.

In the case of particulate materials prone to self-heating and possible spontaneous ignition, tests listed in Table 2-8 have been used for parameter and hazard evaluations. Most of the tests entail placing the sample in an instrumented wire mesh basket or other container, and then placing the sample (in some cases alongside an inert sample) in an oven, furnace, or special apparatus to slowly heat the sample while its temperature remains near-uniform. The tests continue for either a designated duration, or until the exotherm is initiated. Further elaboration is provided in Section 4.3.4 of Chapter 4.

TABLE 2-8
Standard Tests for Spontaneous Heating/Ignition

Test Designation (Reference)	Test Name	Sample Size	Heating Apparatus	Test Duration (hr)
ASTM E771 (withdrawn 2001)	Spontaneous heating tendency	≥ 10 ml	Insulated vessel with air supplied to sample well	24
ASTM D3523	Differential Mackey test	10 g	Open top double chamber for test sample and inert sample	4–72
BAM SADT	Self-accelerating decomposition temperature	400 ml	0.5 liter dewar in oven	168
VDI 2263	Relative self-ignition temperature	8 ml	Grewer oven with preheated air flow	Up to 5.5 hr (1°C/min to 350°C)
Bureau Mines Adiabatic Heating Oven (RI 8473)	Coal dust adiabatic heating	100 g	Oven with preheated air flow	Varies
JRIIS SIT (Kotoyori, 1989)	—	0.5 – 3 g	Open or closed cell adiabatic heating apparatus	Varies
UN SIT	Self-ignition test	400 – 3000 ml	Oven with heated air flow	72 hr

Data from the self-ignition test in varying size mesh baskets can be used with Bowes-Cameron self-ignition theory (Bowes, 1984) to scale up the data to industrial size storage containers. The theory is summarized in Section 4.3.4. Application of the theory for hazard evaluation entails either determining fundamental chemical and thermal property data, or analysis of the self-heating data to provide a basis for scaling.

Some of the particulate materials that have been evaluated for self-heating hazards are listed in Table 2-9 along with some of the corresponding references describing the evaluations.

2.3.3 Chemical Reactivity: Incompatible Chemical Groups

Chemical Groupings

Several reactivity hazard evaluations are presented in the form of chemical groupings rather than individual materials. The premise is that there are

TABLE 2-9
Examples of Spontaneous Heating Evaluations for Particulate Materials

Material	Reference
ABS and MBS Powders	Babrauskas (2003, p. 909)
Activated Carbon	Bowes (1984)
Benzoyl Peroxide	Bowes (1984)
Coal	Babrauskas (2003)
Cotton, Cottonseed	Babrauskas (2003), Gray (2002)
Milk Powder	Babrauskas (2003), Beever (1995), Gray (2002)
Nylon	Babrauskas (2003)
Sawdust (with oil contamination)	Bowes (1984)
Sodium Dithionite (sodium hydrosulfite)	Babrauskas (2003)

generic hazards inherent with certain groups, and generic hazards caused by mixing of incompatible reactive groups. These generic hazards include generation of heat and gaseous reaction products that can lead to fires and/or pressurize inadequately vented vessels.

One of the limitations inherent in classifying hazards in terms of chemical compatibility is that it is very difficult to quantify the degree of hazard. Therefore, the tabulations of incompatible chemical groupings often do not indicate the severity of the hazards.

EPA/NOAA Reactivity Worksheet Reactivity Groupings

The EPA/NOAA Reactivity Worksheet defines 47 chemical groupings for reactivity hazard considerations. Representative groupings and combinations of groupings are shown in Table 2-10. Some of the generic reactivity hazards associated with mixing chemicals from different groupings are also shown in Table 2-10. Many other combinations of groupings can result in temperature and pressure increase, but not necessarily at a dangerous rate. For example, the Reactivity Worksheet has numerous examples of acid-base reactions, most of which probably occur sufficiently slowly that they would not usually represent a serious hazard. Table 2-10 is meant to be illustrative rather than comprehensive.

Coast Guard Chemical Compatibility Chart

The Coast Guard (2001) chemical grouping compatibility chart has 22 primary reactive groupings, and another 14 groups that may potentially react with some of the primary groups. An X in a particular cell of the chart indicates that the corresponding combination of groupings should be avoided because of reactivity problems. The Coast Guard regulations include provisions to prevent contamination of bulk cargo containers from incompatible cargo carried either in connected containers or on previous voyages.

Chemical-Specific Compatibility Charts

Chemical-specific compatibility charts are matrices that provide summary descriptions or hazard ratings for pairs of specific chemicals, as opposed to chemical groupings. The hazards of the individual materials are indicated in the cells along the diagonal, while the hazards of binary combinations are indicated in the cells below the diagonal. These charts are very useful for applications in which there are a manageable number of individual chemicals to be included in the chart. Gay and Leggett (1993) have described an approach for developing compatibility charts with a mixing hazard rating from 0 to 4 that is analogous to the NFPA rating scheme for self-reacting chemicals.

TABLE 2-10
Representative Reactive Groupings in NOAA/EPA Reactivity Worksheet

Chemical Group[a]	Generic Reactivity Hazards
Acid Halides	Water reactive, often violently.
Inorganic Acids	React with metals to form hydrogen; catalyze polymerization.
Alcohols and polyols	React with alkali metals and reducing agents to form flammable/toxic gases.
Aldehydes	Subject to polymerization when in contact with acid catalyst. Phenol-aldehyde reactions are highly exothermic, and the cause of several incidents.
Anhydrides	React exothermally with water, sometimes violently.
Azo, Diazo, and Azido Compounds	Can detonate when sensitized by metal salts or acids.
Bases	Can initiate polymerization; react with aluminum or zinc to form hydrogen
Chlorosilanes	Exothermic reactions with water, acids and bases.
Epoxides	Violent polymerization reactions in presence of catalysts.
Esters	Vigorous reactions with oxidizers can ignite reaction products.
Ethers	Form unstable peroxides when exposed to oxygen.
Halogenated Organic Compounds	Low molecular weight haloalkenes react violently with aluminum and are subject to violent polymerization and peroxide formation.
Hydrocarbons, Aliphatic Saturated	Although usually unreactive, they are incompatible with strong oxidizing agents.

Chemical Group[a]	Generic Reactivity Hazards
Hydrocarbons, Aliphatic Unsaturated	In the presence of certain acids, vigorous polymerization reactions can occur. React exothermally with reducing agents to form hydrogen.
Hydrocarbons, Aromatic	Explosive reactions with strong oxidizing agents.
Inorganic Oxidizing Agents	Can undergo explosive reaction with reducing agents if initiated by a spark, heat, or catalyst.
Inorganic Reducing Agents	May undergo explosive reactions with oxidizing agents.
Metals, Alkali	React vigorously with alcohols, acids, and water.
Metals, Powdered	These are reducing agents that react vigorously with oxidizing agents.
Metal Hydrides	These are reducing agents that react dangerously with oxygen.
Nitrates and Nitrites	Range from slight to strong oxidizing agents; some reactions with hydrides and sulfides result in a detonation.
Organometallics	Strongly reactive with most other groups and reacts violently with water.
Organic Peroxides	Can ignite on contact with various organics and reducing agents.
Phosphates	React with reducing agents such as hydrides to form phosphine gas.
Salts	Although not usually reactive, acidic salts can catalyze organic reactions.
Sulfides	React vigorously with oxidizing agents and with acids.

[a]Some groups have been combined, and others omitted for brevity.

2.3.4 Corrosivity

Corrosion is an electrochemical process in which metal atoms are oxidized to form positive ions while other chemical species are reduced. If severe corrosion occurs in piping or process equipment, it can cause failure and result in release of potentially hazardous particulates. Therefore, it is advisable to determine by tests if a particulate material, which one might think is not normally corrosive, can become corrosive due to contact with water, acids, or alkaline solutions.

For this electrochemical process to occur there must be an anode (site where oxidation occurs), a cathode (site where reduction occurs), and an electrolyte (fluid which allows the movement of electrical charge from the anode to the cathode). There are many factors which influence the rate of corrosion including the type of metal involved, the presence of other metals having a different Redox Potential, the availability of an oxidizer (usually oxygen), the pH (acidity or alkalinity), the ion concentration of the electrolyte, the localized concentration of ions, the system temperature, and the ability of the metal to maintain a protective film. The presence of particulate solids in contact with metallic surfaces can affect the corrosion process in a variety of ways. In many cases particulate solids can promote corrosion, but in some cases a coating of solids can actually inhibit corrosion.

Often, particulate solids promote corrosion because of their ability to attract and hold moisture against the metallic surface. The moisture, along with any ions present, acts as an electrolyte connecting anodic and cathodic areas of the metal surface. Neutral salts (e.g. sodium chloride, calcium chloride, or potassium nitrate) combine with water to form strong electrolytes which allow the flow of electrical charge from the anodic to the cathodic areas of the metal surface. Sometimes granular or powdery materials may contain residual acid solutions. In addition to providing a strong electrolyte, the presence of acids contribute excess H+ ions which are reduced to hydrogen gas in the cathodic areas of the metal surface resulting in a corresponding increase in the oxidation of the metal in the anodic areas.

Biological action within accumulations of damp particulate solids on metal surfaces can sometimes contribute to corrosion. Under certain conditions, bacteria and other organisms can grow in accumulations of damp materials producing acidic compounds. These acidic compounds can concentrate on the metal surface and cause severe localized corrosion.

In some cases an accumulation of powder on a metal surface can inhibit the corrosion process by preventing oxygen or moisture from reaching the surface. In fact, the way some metals naturally resist corrosion is by the formation of a protective oxide film which inhibits any further corrosion.

There are many ways of preventing or minimizing the effects of corrosion. These methods include the use of compatible materials, the use of protective coatings, the minimization of cracks and crevices, the exclusion of moisture or oxygen, the control of pH, and the use of cathodic protection devices.

Some particulate solids are corrosive to metallic materials of construction because of their acidity or alkalinity (pH). For example, acid fertilizers, caustic soda, sodium chloride, and many other chemical salts may be corrosive to some metals. Many "inert" granular or powdery materials that are wet with solutions of various acids (e.g., sulfuric, hydrochloric, nitric, etc.) and alkalis (e.g., ammonium, potassium, and sodium hydroxides) become corrosive. Also, some powders that contain chlorine in the molecule become corrosive when the dry powder is contacted with atmospheric air and the chlorine reacts with the water in air to form hydrochloric acid.

Although there are apparently no specific corrosivity tests for particulate chemicals, there are several ASTM tests for characterizing soil corrosivity and soil-induced corrosion rates that may also be applicable to some other particulate materials. Chaker and Palmer (1989)have provided a good review of soil corrosivity including the ASTM tests and the applicable corrosivity mechanisms. An example of the latter is the particulate deposit

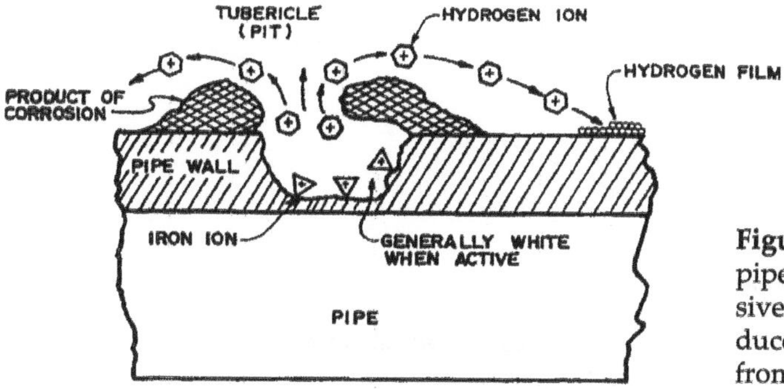

Figure 2-20 Pitting on pipe wall due to corrosive particulate. (Reproduced with permission from ASTM STP 1013.)

induced pitting corrosion on steel pipes as illustrated in Figure 2-20. This pitting occurs when the soil has relatively high concentrations of chloride, sulfate, or nitrate. These ionic components can set up an electrochemical cell in which the pit is the anode. A localized high current density at the pit polarizes the surrounding metal surface. Ferrous ions at the soil-steel interface are converted to hydrous ferrous oxide, which acts as an adhesive attaching the particulate to the surface of the pipe. The pit grows rapidly following this adhesion and the corrosion process is accelerated to eventually form a hole in the pipe. If this phenomena occurs from deposits on the inside of a pipe carrying hazardous particulates, the particulates will be emitted into the surrounding atmosphere. Chaker and Palmer (1989) list various corrosion prevention and control standards developed by the National Association of Corrosion Engineers to deal with this and other corrosive mechanisms.

2.4 OVERVIEW OF PARTICULATE TOXICITY

2.4.1 Particulate Properties Pertinent to Respiratory Hazards

The depth of penetration of airborne particles into the respiratory tract is dependent primarily on particle size shape, density, and its electrostatic charge. The variation of percent deposition with particle size for high-density particles is shown in Figure 2-21 for three respiratory tract locations. The majority of particles with aerodynamic diameters larger than 4 μm are deposited in the nose. Less than 10% of the particles are deposited in the trachea and bronchi. Deep penetration into the lung peaks at about 35% for 2-μm-diameter particles, and remains above about 20% for diameters between 0.5 μm and 4 μm.

Since fiber particles tend to orient themselves parallel to the airways, their depth of penetration is dependent primarily on diameter, rather than length (Craighead, 1993). Therefore, Figure 2-21 should represent a first approximation to fiber penetration based on fiber diameter. However, fiber curvature and surface configuration can also be important. For example,

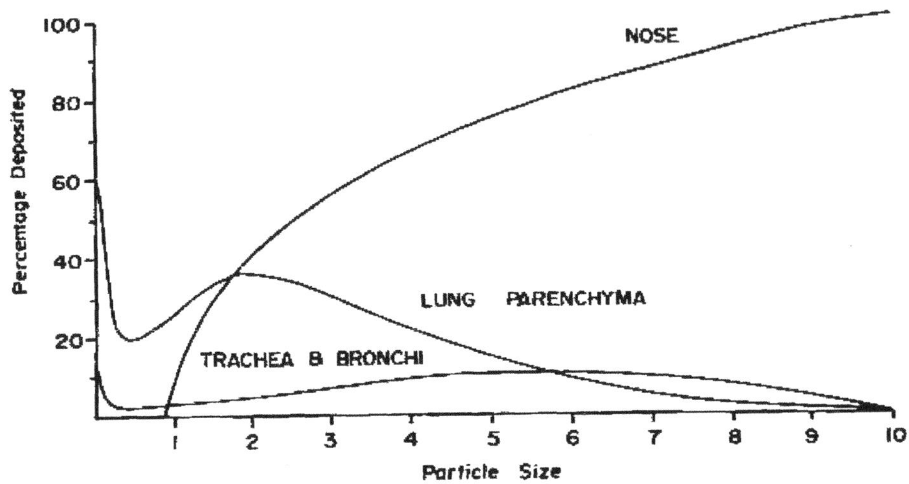

Figure 2-21 Deposition of particulates in respiratory tract. (From W. Keith et al., *Occupational Lung Diseases,* W.B. Saunders, 1984.)

flexible irregular shaped chrysotile asbestos fibers are deposited predominantly in the larger respiratory passages, while sleek rigid amphibole asbestos fibers penetrate deeper at lower airflow rates (Craighead, 1993). Fiber length plays a more important role in its resistance to being expelled by normal physiological functions in the respiratory tract, such that is very difficult to expel durable fibers longer than 10 μm.

Although airborne dust concentrations are usually sufficiently low to allow the lung to expel most nonfibrous low pathogenic particulate, there is a threshold concentration above which the normal physiological clearance capacity is overwhelmed. Craighead (1993) notes that this threshold concentration must have been exceeded at Chinese clay and talc handling facilities because workers in these facilities had accumulated massive amounts of particulate material leading to pneumoconiosis, i.e. lung disease resulting from inhalation and retention of organic or inorganic dust. The primary manifestation of pneumoconiosis (other than breathing difficulty) is fibrosis, which is the production of increased collagen in the lymph glands adjacent to the alveoli.

Frequent and prolonged exposures to some particulates can also lead to emphysema. Textile industry workers have experienced a form of emphysema called bysinosis.

2.4.2 Allergenic and Irritant Materials

Normally innocuous particulates can produce severe allergenic or irritant reactions in sensitized people. Common allergenic and irritant reactions include hay fever, asthma, dermatitis, and eczema. A broad array of organic materials including some pesticides and powdered enzymes are known allergens. Several metal powders are also known to produce allergenic and irritant reactions. Some allergic reactions, such as anaphylactic shock, can be fatal.

2.4.3 *Systemic and Single Exposure Toxicity*

Maximum allowable airborne concentrations have been established for many toxic and other particulates that are known health hazards. The exposure concentration limits depend in part on the expected exposure duration, and in part of the organizational objectives and criteria in establishing these limits. Commonly used terms are defined here.

- Threshold Limit Value (TLV®) is the maximum exposure concentration recommended by the American Conference of Government Industrial Hygienists (ACGIH) for long term exposures.
- Permissible Exposure Limit (PEL) is the maximum permissible exposure limit for systemic workplace 8-hour time-weighted average exposures as established by OSHA.
- Immediately Dangerous to Life and Health (IDLH) is the concentration that poses a threat of either death, adverse health effects, or inability to escape danger without respiratory protective equipment. IDLH values are determined from animal toxicity data either as the lowest lethal concentration, or as one-tenth the median lethal concentration.
- Short Term Exposure Limit (STEL) is the limiting exposure concentration for exposure durations up to 15 minutes, as developed by the ACGIH.

Some sample TLV®, PEL, and STEL values for particulate materials are listed in Table 2-11.

TABLE 2-11
Examples of TLV®, PEL, and STEL Values

Substance	TLV®	PEL	STEL	Source of information
Acrylamide	0.03 mg/m³	0.3 mg/m³		2
Arsenic	0.01 mg/m³	0.5 mg/m³		1,2
Asbestos (all forms)		0.1 fiber/cc	1.0 fiber/cc	3
Calcium Chromate	0.001 mg/m³	0.1 mg/m³ (C)		1
Calcium Silicate	10 mg/m³ total	15 mg/m³ total 5 mg/m³ respirable		1

Table continues on next page

Substance	TLV®	PEL	STEL	Source of information
Cellulose		15 mg/m^3 total 5 mg/m^3 respirable		2
Chlorinated Camphene	0.5 mg/m^3	0.5 mg/m^3	1.0 mg/m^3	1
Chromium II	0.5 mg/m^3	0.5 mg/m^3		1
Coal Dust (> 5% SiO_2)	0.1 mg/m^3	= (10 mg/m^3) / (% SiO_2 + 2)		1
Coal Dust (< 5% SiO_2)	2 mg/m^3	2.4 mg/m^3		1
Cyanamide	2 mg/m^3			1
2,4-D (Dichlorophenoxyacetic acid)	10 mg/m^3	10 mg/m^3		1
p-Dichlorobenzene	60 mg/m^3	450 mg/m^3		1
Dinitrobenzene (all isomers)	1 mg/m^3	1 mg/m^3		1
Fiberglass	10 mg/m^3	15 mg/m^3 5 mg/m^3 respirable		2,4
Heptachlor	0.5 mg/m^3	0.5 mg/m^3		1
Hexachloroethane	9.7 mg/m^3	10 mg/m^3		1
Hexachloronaphthalene	0.2 mg/m^3	0.2 mg/m^3		1
Lead	0.05 mg/m^3	0.05 mg/m^3		5
Methyl Acrylate	2 ppm	35 mg/m^3		2
4,4′-Methylene bis (2-chloroaniline)	0.11 mg/m^3			1
4-Methoxyphenol	5 mg/m^3			1
Nickel soluble compounds	0.1 mg/m^3	1 mg/m^3		1
Phenol	19 mg/m^3	19 mg/m^3		1
Silica (Crystalline Cristobalite)	0.05 mg/m^3 respirable			2
Silicon	10 mg/m^3 total			2
Temephos	10 mg/m^3	15 mg/m^3 total 5 mg/m^3 respirable		1
Tetrasodium pyrophosphate	5 mg/m^3			1
Tetryl	1.5 mg/m^3	1.5 mg/m^3		1
Thiram	1 mg/m^3	5 mg/m^3		1
Trichloroacetic acid	6.7 mg/m^3			1

Substance	TLV®	PEL	STEL	Source of information
Trimellitic anhydrid	0.04 mg/m³ (C)			1
Triphenyl amine	5 mg/m³			1
Vanadium pentoxide dust	0.05 mg/m³	0.5 mg/m³ (C) Respirable		1
Vinyl chloride	1 ppm	1 ppm	5 ppm	2
Warfarin	0.1 mg/m³	0.1 mg/m³	0.3 mg/m³	1
Wood Dust, All Soft & Hardwoods Except Western Red Cedar	1 mg/m³ for certain hardwoods, such as beech 1and oak	15 mg/m³ total	10 mg/m³ for soft wood	1
	5 mg/m³ for soft wood	5 mg/m³ respirable		
Western red cedar dust	5 mg/m³	15 mg/m³ total	10 mg/m³	1
		5 mg respirable		
Zinc chromate	0.01 mg/m³	0.1 mg/m³ (C)		1
Zinc oxide	10 mg/m³ total	15 mg/m³ total		1
		5 mg/m³ respirable		
Zinc stearate	10 total	15 mg/m³ total		1
		5 mg/m³ respirable		

(C) : Ceiling Limit

Sources of information:

1. OSHA/NIOSH/EPA Health Guidelines (OSHA web)
2. Chemical Sampling Information (OSHA web)
3. ASBESTOS IN AIR - (Inorganic Method #160) (OSHA web)
4. 11/19/1991—Fiberglass and the HCS Standard (OSHA web)
5. Safety and Health Regulations for Construction (OSHA web)

2.4.4. Carcinogenic Classifications

Various organizations have established carcinogenic classifications based on laboratory test data with animals and, where available, epidemiological studies with humans. Table 2-12 shows the category definitions of the International Agency for Research on Cancer (IARC) and the American Conference of Government Industrial Hygienists (ACGIH). Examples of particulate materials in each category are also listed.

TABLE 2-12
***Carcinogenic Categories*[a]**

Criterion	IARC Category
Confirmed Human Carcinogen	Group 1
Probable/Suspected Human Carcinogen	Group 2A
Possible Human Carcinogen (ACGIH: via unlikely or uncommon routes)	Group 2B
Not Classifiable as Human Carcinogen	Group 3
Probably not Carcinogenic in Humans	Group 4

[a] Source: www.hc-sc.gc.ca/whmis

REFERENCES

Abbot, J, 1990, *Prevention of Fires and Explosions in Dryers—A User Guide,* 2nd ed., Rugby, U.K.: The Institution of Chemical Engineers.

Arakawa, M. 1991. "Specific surface area," in K. Iinoya, K. Gotoh, K Higashitani, eds., *Powder Technology Handbook,* New York: Marcel Dekker.

ASTM, 2002, "CHETAH," Version 7.3, West Conshohocken, PA: American Society for Testing and Materials.

ASTM B761, "1998, Standard Test Method for Particle Size Distribution of Powders and Related Compounds by X-Ray Monitoring of Gravity Sedimentation," West Conshohocken, PA: American Society for Testing and Materials.

ASTM D6331, 1998, "Standard Test Method for Determination of Mass Concentration of Particulate Matter from Stationary Sources at Low Concentrations, West Conshohocken, PA: American Society for Testing and Materials.

ASTM D6393, 1999, "Standard Test Method for Bulk Solids Characterization by Carr Indices," West Conshohocken, PA: American Society for Testing and Materials.

ASTM D1921, 2001, "Standard Test Methods for Particle Size (Sieve Analysis) of Plastic Materials," West Conshohocken, PA: American Society for Testing and Materials.

ASTM D3523-92, 1997, Standard Test Method for Spontaneous Heating Values of Liquids and Solids (Differential Mackey Test), West Conshohocken, PA: American Society for Testing and Materials.

ASTM E 11-01, 2001, "Specification for Wire-Cloth Sieves for Testing Purposes," West Conshohocken, PA: American Society for Testing and Materials.

ASTM E140, 2002. "Standard Hardness Conversion Tables for Metals, Relationship Among Brinell Hardness, Vickers Hardness, Rockwell Hardness, Superficial Hardness, Knoop Hardness, and Scleroscope Hardness," West Conshohocken, PA: American Society for Testing and Materials.

ASTM E384, 1999, "Standard Test Method for Microindentation of Materials," West Conshohocken, PA: American Society for Testing and Materials.

ASTM E771-90, 1996, Standard Test Method for Spontaneous Heating Tendency of Materials (Withdrawn 2001), West Conshohocken, PA: American Society for Testing and Materials.

ASTM E1617, 1997, "Standard Practice for Reporting Particle Size Characterization Data," West Conshohocken, PA: American Society for Testing and Materials.

ASTM E1820, 2001, "Standard Test Method for Measurement of Fracture Toughness," West Conshohocken, PA: American Society for Testing and Materials.

ASTM F577, 1997, "Standard Test Method for Particle Size Measurement of Dry Toners," West Conshohocken, PA: American Society for Testing and Materials.

Babrauskas, V., 2003, *Ignition Handbook,* Fire Science Publishers, Society of Fire Protection Engineers. Quincy, MA: National Fire Protection Association.

Bartknecht, W., 1989, *Dust Explosions, Course, Prevention, and Protection,* Berlin: Springer-Verlag.

Beever, P., 1995, "Self-heating and spontaneous combustion," in *SFPE Handbook of Fire Protection Engineering,* 2nd ed., Section 2, Chap 12, Quincy, MA: National Fire Protection Association.

Bellino, I., Bergougnoux, L., 2001, Misguich-Ripault, J., and Firpo, J., 2001, "Particle size effects on an optical fiber sensor response," *Powder Technology,* 115: 68–74.

Bemrose C. R., and Bridgewater, J., 1987. "A Review of Attrition and Attrition Test Methods," *Powder Technology,* 49: 97–126.

Berns, H. and Koch, S., 1999. "Influence of abrasive particles on wear mechanism and wear resistance in sliding abrasion tests at elevated temperatures," *Wear,* 233–235: 424–430.

Bose, A., 1995, *Advances in Particulate Materials,* Burlington, MA: Butterworth-Heinemann.

Bowes, P. C., *Self-Heating: Evaluating and Controlling the Hazards,* Amsterdam: Elsevier, 1984.

Bumiller, M., 2001, "Analyzing Powder Particle Size," www.ChemicalProcessing.com pp. 24-26, December 2001.

CEMA. 1980. *Classification and Definitions of Bulk Materials,* Book 550, Rockville, MD: Conveyor Equipment Manufacturers Association.

Chaker, V., and Palmer, J. D., 1989, "Effects of Soil Characteristics on Corrosion," ASTM STP 1013, West Conshohocken, PA: American Society for Testing and Materials.

Coast Guard, 2001, "Chemical hazard response information system (chris) manual," Compatibility Chart URL: www.chrismanual.com/Intro/compatib.htm.

Coghill, P., 2001, "Particle size determination by impact measurement in pneumatically conveyed solids," *Particle System Characteristics,* 18: 114–119.

Craighead, J. 1993. "Inorganic mineral particulates in the lung," *Handbook of Hazardous Materials,* San Diego: Academic Press.

Eckhoff, R., 1997, *Dust Explosions in the Process Industries,* 2nd ed. Burlington, MA: Butterworth-Heinemann.

Eckhoff, R., 2003, *Dust Explosions in the Process Industries,* 3rd ed. Amsterdam: Elsevier.

EPA, 1999. "Continuous measurement of PM_{10} suspended particulate matter (SPM) in ambient air," Chapter 10-1 in *Compendium of Methods for the Determination of Inorganic Compounds in Ambient Air,* Washington, DC: EPA/625/R-96/010a.

Fan, L-S., and Zhu, C., 1998. *Principles of Gas-Solid Flows,* Cambridge, U.K.: Cambridge Univ. Press.

Fayed, M. E. and Skocir, T. S. 1997. *Mechanical Conveyors: Selection and Operation.* Lancaster, PA: Technomic Publishing Company.

Field, P., 1982. "Dust Explosions" in *Handbook of Powder Technology,* vol. 4, Amsterdam: Elsevier.

Gay, D.M. and Leggett, D.J., 1993, "Enhancing thermal hazard awareness with compatibility charts," *J. Testing and Evaluation,* 21: 477–480.

Grasa, G. and Abandres, J., 2001, "A calibration procedure to obtain solid concentrations from digital images of bulk powders," *Powder Technology,* 114: 125–128.

Gray, B., 2002, "Spontaneous combustion and self-heating," in *SFPE Handbook of Fire Protection Engineering,* 3rd ed., Quincy, MA: National Fire Protection Association.

Harker, A., Schofield, B., Stimpson, R., Taylor, R., and Temple, J., 1991, *Ultrasonics,* 29: 427.

IEC 61241-2-2, 1993. "Electrical Apparatus for Use in the Presence of Combustible Dust, Part 2, Test Methods, Section 2: Method for Determining the Electrical Resistivity of Dust in Layers," Geneva: International Electrotechnical Commission.

Iinoya, K., Masuda, H., and Watanabe, K., 1988. *Powder and Bulk Solids Handling Processes,* New York: Marcel Dekker.

IMC, 2001, *Guide for the Flow Measurement of Particulate Solids in Pipelines,* London: Institute of Measurement and Control (IMC).

Institute for Measurement and Control, 2001, *Guide to the Flow Measurement of Particulate Solids in Pipelines,* ISBN 0 904457 34 6.

ISO WD14888, 2000, *Sample Splitting of Powders for Determination of Particle Size Characterization,* Geneva: International Organization for Standardization.

ISO 9096, 2003, *Stationary source emissions. Manual Determination of Mass Concentration of Particulate Matter,* Geneva: International Organization for Standardization.

ISO 9276-1, 1998, *Representation of Results of Particle Size Analysis—Part 1: Graphical Representation,* Geneva: International Organization for Standardization.

Jillavenkatesa, A., Dapkunas, S., and Lum, L-S., 2001, "Particle Size Characterization," Gaithersburg, MD: NIST Recommended Practice Guide, SP 960-1.

Johanson, J.R., 1981. "How to predict and use the properties of bulk solids," in *Solids Handling,* New York: McGraw Hill.

Johnson, R., Rudy, S., and Unwin, S. 2003, *Essential Practices for Managing Chemical Reactivity Hazards,* New York: AIChE/Center for Chemical Process Safety.

Jono, K., Ichikawa, H., Miyamoto, M., and Fukumori, Y., 2000. "A review of particulate design for pharmaceutical powders and their production by spouted bed coating," *Powder Technology,* 113: 269–277.

Kaye, B., Hollinger, M., and Tobyn, M., 2001, "The effect of shape and particle size distribution on the flow of powders," *Powder/Bulk Solids,* pp. 19–26, November.

Kelly, D.A., and Hutchings, I.M., 2001, "A New Method for Measurement of Particle Abrasivity," *Wear*, 250: 76–80.

Kendall, K. and Stainton, C., 2001, "Adhesion and aggregation of fine particles," *Powder Technology*, 121: 223-229.

Kotoyori, T., 1989. "Critical ignition temperatures of chemical substances," *Journal of Loss Prevention in the Process Industries*, 2.

Leavers, V.F. 2000. "An active angularity factor for the characterization of abrasive particles," *Wear*, 239: 102–110.

Louge, M., "The Measurement of Particle Concentration with Optical Fiber Probes." Paper presented at the AIChE Winter Annual Meeting, 1994.

Louge, M., "Guarded Capacitance Probes for Measuring Particle Concentration and Flow," US patent 5,546,006, 1996.

Masuda, H. 1991. "Electrical Properties," in K. Iinoya, K. Gotoh, K Higashitani, eds., *Powder Technology Handbook*, New York: Marcel Dekker.

Meili, R., Nicolai, R., Shi, D., Hugi, E., and Reh, L., "An Enhanced Optical Measuring System for In-Line Two-Phase Flow Monitoring." Paper presented at the AIChE Annual Meeting, 1995.

Meili, R., "Adaptation of a Laser Backscatter Photometry System for Concentration and Velocity Measurements in Spray Drier.," Paper presented at PARTEC 98 – 1st European Symposium on Process Technology in Pharmaceutical and Nutritional Sciences, 1998.

Mills, P., Seville, J., Knight, P., and Adams, M., 2000. "The effect of binder viscosity on particle agglomeration in a low shear mixer/agglomerator," *Powder Technology*, 113: 140–147.

Moss, S. Attenborough, K., and Woodhead, S., 1999. "Measured Dependence of the Attenuation of Audio-Frequency Sound on Concentration in Flowing Particulate Suspensions," *Proceedings of the Institute of Mechanical Engineers*, 213: 45–58.

Murphy, M., Singh, S., and Shanley, E., 2003, "Computationally evaluate self-reactivity hazards," *Chem. Eng. Progress*, 99(2): 54–61.

Neumann, A., and Krmaer, H., 2002. "A comparative study of various size distribution measurement systems," *Part. Part. Syst. Charact.* 19: 17–27.

NFPA 77, 2000, *Recommended Practice on Static Electricity*, Quincy, MA: National Fire Protection Association.

NOAA 2002. "Chemical Reactivity Worksheet," Version 1.5. U.S. National Oceanic and Atmospheric Administration. [http://response.restoration.noaa.gov/chemaids/react.html].

Okuyama, K., and Kousaka, Y., 1991, in K. Iinoya, K. Gotoh, K Higashitani, eds., *Powder Technology Handbook*, New York: Marcel Dekker.

Osman, H., Muchtar, A., and Tasirin, S., 2001, "Effects of dispersant on the rheological properties and microstructure of ceramic powder compacts," Proceedings NSF Workshop, Kuala Lumpur.

Probst, N. and Grivei, E., 2002. "Structure and electrical properties of carbon black," *Carbon*, 40: 201–205.

Repenhagen, J. and Werther, J., 2000, "Catalyst attrition in cyclones," *Powder Technology,* 113: 55-69.

Sagehorn, H., List, J., Wiegrand, T., Weichert, R., and Wriedt, T., 2001, "characterization of airborne fibers via fraunhofer theory: Examination of the validity of of fraunhofer theory using the exact scattering theory MMP," *Particle System Characteristics,* 18: 55–63.

Scala, F., Salatino, P., Boerefijn, R., and Ghadiri, M., 2000, "Attrition of sorbents during fluidized bed calcination and sulphation," *Powder Technology,* 107: 153-167.

Scherpa, T., 2002. WPI M.S. Thesis, **Title to be added.**

Schubert, H., 1973, "Kapillardruck und Zugfestigkeit von feuchten Haufwerken aus körnigen Stoffen." *Chem.-Ing. Techn.* 45(6): 396–401.

Siwek, R., 1996. "Determination of technical safety indices and factors influencing hazard evaluation of dusts," *J. Loss Prevention in the Process Industries,* 9: 21–31.

Stachowiak, G.W. 2000. "Particle angularity and its relationship to abrasive and erosive wear," *Wear,* 241: 214–219.

Stachowiak, G.B. and Stachowiak, G.W. 2001. "The effects of particle characteristics on three-body erosive wear," *Wear,* 249: 201–207.

Svarovsky, L., 1990. "Characterization of powders," in M.J. Rhodes, ed., *Principles of Powder Technology,* New York: John Wiley & Sons.

Tardos, G. and Pfeffer, R., 1995. "Chemical reaction induced aglomeration and defluidization of fluidized beds," *Powder Technology,* 85: 29–35.

Taylor, T., 1998, "Specific energy consumption and particle attrition in pneumatic conveying," *Powder Technology,* 95: 1–6.

Tewarson, A, 1995. "Generation of Heat and Chemical Compounds in Fire," in *SFPE Handbook of Fire Protection Engineering,* Quincy, MA: National Fire Protection Association.

U.N. 1999. "Recommendations on the Transport of Dangerous Goods, Manual of Tests and Crtieria," UN ST/SG/AC.10/11/Rev 3.

VDI 2263—Part 1, 1990. *Dust Fires and Explosions,* Verein Deutcher Ingenieure, VDI-Handbuch Reinhaltung der Luft, Band 6.

Yan, Y. 1996. "Mass flow measurement of bulk solids in pneumatic pipelines," *Measurement Science and Technology,* 7: 1687–1706.

Yan, Y., and Reed, A., 1999. "Experimental evaluation of capacitance transducers for volumetric concentration measurement of particulate solids," *Flow Measurement and Instrumentation,* 10: 45–49.

Yiannoulakis, H., Yiagopoulos, A., and Kiparissides, C., 2001. "Recent Developments in the particle size distribution modeling of fluidized-bed olefin polymeriztion reactors," *Chemical Engineering Science,* 56: 917–925.

Woodcock, C. R. and Mason, J. S. 1987. *Bulk Solids Handling.* New York: Chapman and Hall.

Zhang, H., Johnston, P., Zhu, J., de Lasa, H., and Bergougnou, M., 1998, "A novel calibration procedure for a fiber optic solids concentration probe" *Powder Technology,* 100: 260–272.

Chapter 3

PARTICULATE HAZARD SCENARIOS AND EXAMPLES

This chapter describes accident scenarios and examples pertinent to the various particulate hazards described in Chapter 1. The discussion is divided into the hazard categories: instability (thermal and shock), chemical reactivity, particulate fires, dust explosions, and toxic material exposures.

3.1 THERMAL AND SHOCK INSTABILITY SCENARIOS

3.1.1 Exothermic Decomposition Explosions

Certain chemical groups are either unstable or metastable with exothermic decomposition energies. Examples include azides, azo compounds, epoxides, nitrates, perchlorates, and peroxides. These and other materials that are prone to exothermic self-reactive decomposition can explode when heated to temperatures above their autodecomposition temperature. Laboratory tests to determine a material's nominal decomposition temperature and its related self-accelerating decomposition temperature (SADT) or its self-ignition temperaure (SIT) (Kotoyori, 1989), which are applicable to packaged materials, and combustible materials, respectively, are described in Chapter 4. However, the results can be sensitive to particle size and the presence of minor additives and contaminants (Kotoyori, 2003). Furthermore, the applicability of an effective SADT or SIT can be ambiguous when applied to large-scale, nonuniform heating scenarios such as exposure fires.

Laboratory tests to determine the speeds of exothermic decompositions have shown that the decomposition rate and corresponding hazard level depends on the presence of stable and unstable intermediates during the decomposition. Mohan and Tang (1983) found that the lifetimes of the intermediates determined whether explosions or controlled decomposition would occur in five different self-decomposing powders, including picryl azide (an azido trinitrobenzene). In practice, the lifetimes of these intermediates depend on the presence of contaminants, substrates, and material packaging. For example, Haberman and Castorina (1972) have shown that various adsorbates can significantly affect the initiation times for the thermal decomposition of lead azide, and that these times were influenced by the fraction of the azide surface covered by these adsorbates and associated intermediates. Thus, the nominal SADT or SIT or critical temperature, T_c, for exothermic decomposition should serve as a guide rather than an absolute determinant of exothermic decomposition hazards.

The consequences of maintaining an unstable material at a temperature above its SADT or SIT depend on whether or not the reaction is autocatalytic, i.e. whether a reaction product also acts as a catalyst. Autocatalytic reactions often are explosive because the decomposition occurs very rapidly after a delay in which decomposition product accumulates. Bou-Diab and Fierz (2002), Fierz et al. (1994), Keller et al. (1997), and Kotoyori (1989) provide guidance on how the results of laboratory thermal testing can be used to determine whether the decomposition is autocatalytic. Kotoyori (1999) further differentiates between solids that are truly autocatalytic and those that are quasi-autocatalytic. The latter have exothermic decomposition reactions that occur almost simultaneously with endothermic melting.

In transportation applications, the UN-DOT SADT value is determined via tests with packaged material, and these tests account for reaction initiation times up to seven days at a given ambient temperature. If the UN-DOT test data indicate that the SADT value of a 50 kg package is less than 75°C, the material is classified as self-reactive (Division 4.1) for transportation purposes (U.N., 1999). Depending on the SADT value, the U.N. regulations require that self-reactives be maintained at a temperature of at least 10°C to 20°C below its SADT. Kotoyori (2003) recommends that materials should be stored and maintained at a temperature at least 30°C below their SADT or SIT values. The 30°C safety margin is suggested to account for possible SADT or SIT or T_c value lowering due to contaminants and freshness, i.e. newly produced materials sometimes are more reactive because passivating oxidation layers have not yet formed.

One scenario that has led to particulate thermal decomposition incidents is prolonged overheating in a dryer or oven. In one such incident (CSB CIRC Incident Number 2000-4968), an organic peroxide was inadvertently left in a hot oven overnight. Noxious gaseous decomposition products emitted from the oven filled the building and the surrounding neighborhood. Sixteen people, including ten firefighters, were treated for eye and throat irritations at a nearby hospital.

The CCPS *Guidelines for Chemical Reactivity Evaluation and Application to Process Design* (1995a, p. 153) cites an example of the self-accelerating exothermic decomposition of 3,5-dinitro-*o*-toluamide (dinitrolmide) remaining in dryer at a temperature of about 125°C for 27 hours after the drying process was completed. This decomposition, which caused a detonation in the dryer, was subsequently found to be autocatalytic and therefore to occur at temperatures much lower than had been indicated previously from Differential Scanning Calorimetry (DSC) test data. The use and interpretation of different modes of DSC testing and other thermal stability screening tests is discussed in Chapter 4.

Another overheating scenario is hot-work on a pipe or container with a residue of the unstable material. This occurred in an air bag factory when a cutting torch was applied to pipes that had been used to transport sodium azide (NaN_3). The resulting explosion blew out walls and injured three workers (CSB CIRC Incident Number 2001-5086). Laboratory testing reviewed by Pegg et al. (1997) indicate that the SIT for sodium azide is 450°C and its heat of decomposition is 21.3 kJ/mol.

The most frequently reported scenario for thermal decomposition explosions has been stored material exposed to a fire. Considerations of additives, packaging etc. have played an important role in the evolution of the current understanding of fire exposure scenarios that may or may not lead to the explosive decomposition of common oxidizers such as ammonium nitrate and ammonium perchlorate. This understanding is summarized here with references to key explosion incidents initiated by fire exposures.

Ammonium Nitrate Explosions Due to Fire Exposure

The combination of a strong reducing group (NH_4^+) and a powerful oxidant group (NO_3^-) in the same molecule causes ammonium nitrate to be metastable. It decomposes spontaneously in the following oxidation-reduction reaction (Guiochon, 2002):

$$NH_4NO_3 \rightarrow N_2O + 2H_2O$$

This reaction probably takes place in the molten salt, above its melting point of 169.6°C (for the anhydrous product). The exothermic heat of this reaction is 36 kJ/mole. SADT values have been reported to be between 160°C and 200°C.

The other decomposition reaction, which occurs simultaneously with the preceding reaction, is

$$NH_4NO_3 \rightarrow NH_3 + HNO_3$$

which has an endothermic reaction energy of 176 kJ/mole. This endothermic dissociation reaction can stabilize the exothermic oxidation–reduction reaction providing that the nitric acid dissociation product does not itself decompose into a mixture of nitrogen oxides. The possible presence of chloride ions (from a contaminant) near the reaction zone catalyzes a vapor phase ammonia reaction with nitric acid such that the preferred ammonium nitrate decomposition reaction is

$$5NH_4NO_3 \rightarrow 4N_2 + 2HNO_3 + 9H_2O$$

which is the most energetic of the preceding decomposition reactions, but is less energetic than the following detonative decomposition reaction:

$$NH_4NO_3 \rightarrow N_2 + 2H_2O + \frac{1}{2}O_2$$

According to Guiochon (2002), the detonation decomposition is more likely to occur in the intimate presence of a molten or particulate reducing agent, such as hydrocarbon.

The following account of the 1947 Texas City ammonium nitrate explosion is based primarily on the information from the *Houston Chronicle* Web site: www.chron.com/content/chronicle/metropolitan/txcity/main.html, and the Société Française de Chemie Web site: www.sfc.fr/Guiochon%20VO/exincendieVO.htm.

A fire started in a hold of the *Grandcamp* ship shortly after 8 A.M. April 16, 1947. There has been much speculation over the years as to what caused the initial fire on the *Grandcamp* but over 50 years later there has been no definitive answer. According to the insurance claim report, a discarded cigarette ignited its cargo of 2200 tons of ammonium nitrate fertilizer. This particular fertilizer had 4% mineral additives and 1% wax coating as an anticaking agent.

When the small fire inside the *Grandcamp* could not be doused with jugs of drinking water and a portable extinguisher, an order was given to batten down the ship's hatches and cover them with tarpaulins. The *Grandcamp*'s fire-smothering steam system was activated to keep the cargo from being damaged by water. But instead of killing the fire, the heat and pressure accelerated decomposition of the ammonium nitrate fertilizer and produced sufficient gas to blow off the ship's hatch covers, sending the red smoke (from the N_2O) skyward. Burning continued for a few minutes in the open hold and then a devastating explosion occurred at 9:12 A.M.

Pieces of the *Grandcamp* were hurled several thousand feet in the air. Some of the heavy debris landed on pipelines and storage tanks, igniting secondary explosions. Other pieces landed on houses and people. The *Grandcamp*'s 1.5-ton anchor was flung 2 miles and was embedded 10 feet into the ground at the Pan American refinery.

A second explosion, 16 hours after the *Grandcamp* blew up, came from another ship loaded with the same fertilizer bags as were on the *Grandcamp*. The *High Flyer* had been loaded with 961 tons of ammonium nitrate, slightly more than what exploded on the *Grandcamp*. The vessel had been torn from its moorings by the first explosion and had drifted across its slip and come to rest in the ship channel.

The *Grandcamp* explosion had blown the hatches off the *High Flyer*, but no fire aboard the vessel had been detected. The only damage appeared to be a hole in the deck the size of a dinner plate. Despite the absence of fire, the ship's crew was ordered off the vessel after fumes of sulfur from its cargo proved too powerful. When flames were later spotted aboard the *High Flyer*, an alarm was sounded and the waterfront was cleared just before the ship blew up. The *High Flyer* ignited a string of secondary explosions equal to the *Grandcamp* catastrophe. Crude oil tanks were ignited, a grain elevator was destroyed, warehouses on the piers were set ablaze. Although the shockwave caused a displacement of the water in the harbor and created a small tidal wave that washed inland over 150 feet, it did little to save the people from the fires.

The Texas City disaster killed at least 581 and injured about 3500, making it the most catastrophic industrial accident in U.S. history. Figure 3-1 is an aerial view of the physical devastation.

Coincidently, another multifatality, fire-induced ammonium nitrate decomposition explosion occurred in a cargo ship in Brest, France, 3 months after the Texas City disaster. The ship was carrying 3300 tons of ammonium nitrate and an abundant cargo of mixed, combustible goods (fuels, paints, lubricants, polystyrene, tires). When a fire started for unknown reasons about 12:30 P.M., the ship captain ordered the hold to be closed and high-pressure steam to be released in the hold. As the fire continued burning for about 90 minutes, the ship was towed into the outside harbor. Abundant billows of red and black smoke were flowing from the hold and the fire turned

Figure 3-1 Aerial view of Texas City explosion aftermath, copyright Tobin International (reproduced with permission).

violent around 5:00 P.M.; witnesses reported that barrels of petroleum products or solvents caught fire at that time. The cargo exploded at 5:25 P.M., causing 29 deaths and extensive damage to the city of Brest.

There have also been numerous documented accounts of large ammonium nitrate storage fires that did not result in any explosion. The explanations usually provided for detonations occurring in some ammonium nitrate fires and not in other fires are: (1) the amount of combustible material in intimate contact with the nitrate, and (2) confinement of the burning nitrate. In the case of the Texas City disaster, the pertinent combustible is said to be the 1% wax coating, and the cargo hold provided confinement. In the case of the Brest explosion, the combustibles were the fuels, paints, and lubricants. The scarcity of combustibles and the unconfined burning in the other fires prevented the development of explosions.

Based on the cited fire exposure incidents and associated laboratory testing, ammonium nitrate containing less than 0.2 percent combustible is classified as an oxidizer, whereas a higher combustible content brings ammonium nitrate into the UN-DOT explosive category. NPFA 490 requires that ammonium nitrate storage be separated from combustible materials by either approved fire partitions or a spacing of at least 30 ft (9.1 m).

Ammonium nitrate explosions have also occurred in several manufacturing processes described in the 1997 EPA alert on ammonium nitrate explosion hazards. EPA guidance for preventing ammonium nitrate explosions include avoiding specified contaminants that can catalyze the decomposition reaction.

Ammonium Perchlorate Incidents

Ammonium perchlorate (AP) is another powerful oxidizer subject to detonations during some fire exposures. It undergoes exothermic decomposition at a minimum temperature of 440°C (834°F), probably via the following reaction

$$2NH_4ClO_4 \rightarrow 2NH_3 + Cl_2 + H_2O + 3.5\ O_2$$

If the decomposition occurs in the presence of certain combustible materials, an intense fire or explosion will result. This happened in the May 4, 1988 fire/explosion at the Pepcon manufacturing facility in Henderson, Nevada. The Pepcon AP had a characteristic diameter in the range 100 μm to 200 μm, and was sold for use as a solid rocket propellant. According to Mniszewski's account (1994), a small fire started in a polyethylene drum of contaminated ammonium perchlorate. The fire spread to other drums and to fiberglass wall panels in the building. Some of the drums rocketed and spread the fire to other areas of the plant and to asphalt ground covering that melted from the flame radiation. Two major detonations occurred after about 25 minutes of fire development. Two people were killed, and 372 people were injured by the detonations. A video news crew situated on a nearby mountaintop recorded the spreading fire and detonations on videotape. The detonations probably involved large aluminum tote bins, each containing approximately 5000 lb of ammonium perchlorate. Blast damage analyses and calculations of shock velocity recorded on videotape indicate that the blast wave energy was equivalent to approximately 250 tons of TNT (Mniszewski, 1994). However, later tests with fires of ammonium perchlorate filled aluminum shipping containers were unable to replicate these detonations (Rockett, 2001).

Other ammonium perchlorate explosion incidents have been triggered by friction and impacts as described in Section 3.1.2.

The gaseous decomposition products shown in the reactions for both ammonium nitrate and ammonium perchlorate can significantly increase the pressure in a closed container, such as the drums of ammonium perchlorate. The more general pressurization hazard associated with self-reactive chemicals is illustrated in the flow chart in Figure 3.2. The right side of the flow chart pertains to multiple chemical reaction hazards discussed in Sections 3.3 and 3.4.

3.1.2 Shock/Friction Sensitive Instability Scenarios

Frictional heating, impact loads, or shock waves can sometimes trigger explosive reactions in sensitive particulate materials. Numerous explosions of this type have occurred at fireworks manufacturing plants and at some chemical processing facilities, as summarized in the following four representative accounts.

Mniszewski's (1994) analysis of the Pepcon explosion states that bins, drums, and bags of ammonium perchlorate were involved in sympathetic detonations initiated by shock waves from the first major explosion, which was caused by prolonged fire exposure as described above. The detonations produced row craters where the bins and drums were stored. Mniszewski estimates the critical distance is 8.5 ft for the propagation of sympathetic detonations in 550 lb drums of ammonium perchlorate. This critical distance increases to about 17 ft for 4500 lb bins of ammonium perchlorate.

A fireworks plant in Oklahoma suffered a devastating explosion triggered by employees dragging a galvanized metal container of explosive powder over a surface contaminated with some previously spilled powder. According to the NFPA incident report (Kyte, 1986), the powder was a mixture of potassium perchlorate and sulfur that has an ignition temperature of 560°C but can explode when subjected to friction generated sparks. The explosion and resulting fire killed 21 people and injured five others.

According to OSHA Accident Report 170150924, a chemist at a pyrotechnics and ordinance manufacturing facility accidentally exploded two spoonfuls of zirconium potassium perchlorate in 1973 when he started to sieve the

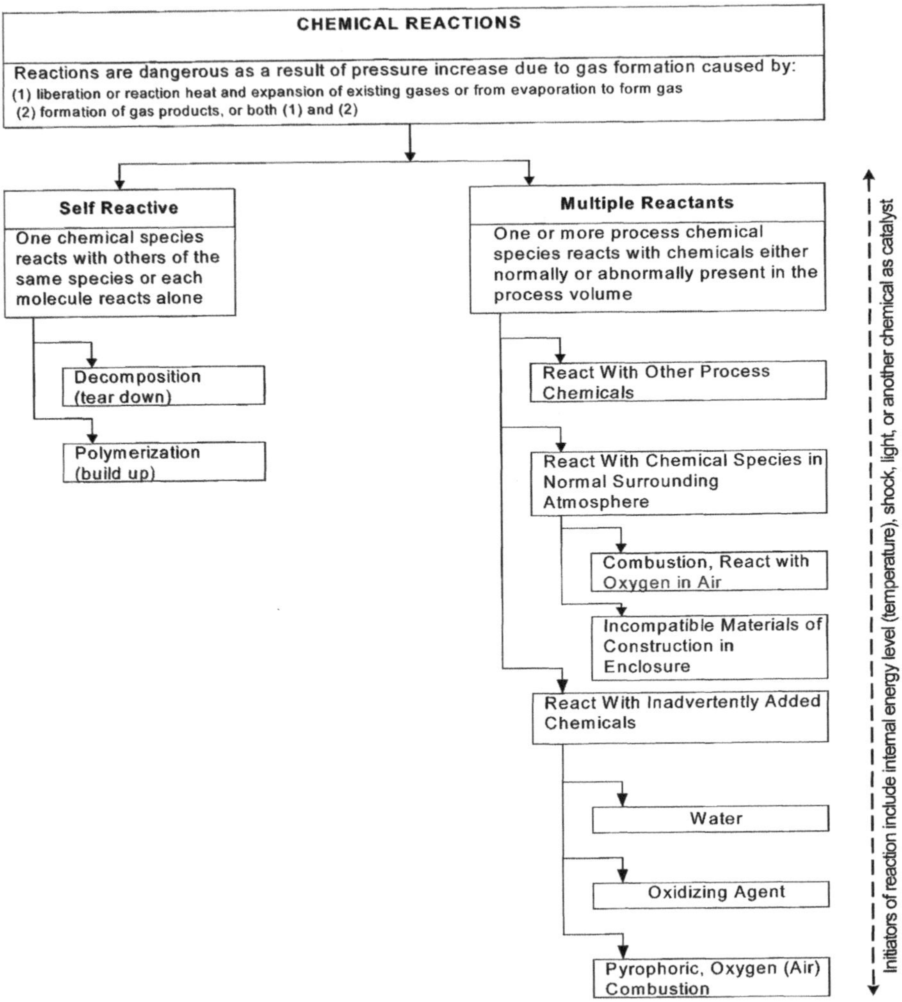

Figure 3-2 Chemical reactivity hazard flow chart. (From Syngenta.)

powder. The chemist was doing this by inserting his hands through openings in a blast resistant steel wall. The chemist's injured hand/arm required amputation.

Sodium azide has been involved in several fatal accidents caused by frictional or impact ignitions. One accident occurred when employees were trying to clear a plugged dust collector. A more recent accident (CSB CIRC Incident number 2002-5549) occurred at the same facility as an employee was cleaning a filter drum used in the azide production process. Both accidents resulted in single fatalities, with four other employees also being injured in the first incident.

CSB Incident number 2001-5271 also describes an ammonium perchlorate incident that started from a friction spark. Apparently, a saw blade was inadvertently started and hit a metal brush used for cleaning the cutting machine. Sparks ignited the ammonium perchlorate particles in the area of the saw. The resulting fire activated the building sprinkler system, and water apparently penetrated into a barrel of magnesium powder. This incident, which killed one employee and severely injured two others, is discussed further under water reactive material accident scenarios (Section 3.3.2).

Dibenzoyl peroxide is a notoriously shock and friction sensitive powder. The CCPS (1995b) *Guidelines for Safe Storage and Handling of Reactive Materials* includes an account of a serious fire ignited by sweeping a dibenzoyl peroxide floor deposit or spill with a broom. One specific preventive measure given is to avoid using containers with screw covers or other tight fitting closures because the frictional heat developed in closing the container may initiate decomposition of any powder residue in the closure.

MARS Incident DE/1998/005-[02] describes the explosive decomposition of about 2 kg of diazonium salt particulate upon falling to the ground during inspection of a clarification press. The explosion resulted in one fatality, six injuries, and almost $3 million in damages. Following the accident, a new hazard analysis was conducted of all azo compounds and clarification filtration of diazonium salts and similar shock sensitive materials was eliminated.

The UK Chemical Reaction Hazards Forum collection of incidents (CRHF, 2003) includes a decomposition explosion triggered by friction and/or impact of a wetted residue of tribromphenyl diazonium bromide. Plant operators were not aware of this particular hazard because the diazonium bromide was usually mixed with a diazonium hydrogen sulfate such that the mixture was stable and invulnerable to mechanically induced decomposition.

According to the Societe Francaise de Chemie Web site, pure ammonium nitrate and high concentration ammonium nitrates are not sensitive to mechanical shocks. They have never been reported to detonate merely because they had been exposed to any shock or vibrations. When these fertilizers have turned into large solid masses because they are hygroscopic, they can be broken down into pieces sufficiently small to be handled and reprocessed by bulldozers. Under the impact of a 200-kg hammer, ammonium nitrates do not detonate. They do not detonate when struck by high velocity bullets. The only reported exception is for low-density pellets of pure ammonium nitrate at 140°C, when hit by 1200-m/s bullets. Sensitivity of ammonium nitrates to mechanical and explosive shocks increases with increasing temperature and with decreasing density, and with chloride and hydrocarbon contamination as discussed in Section 3.3.1.

One possible mechanical impact scenario for very sensitive unstable materials is the collapse of a cavity formed at the bottom of a storage pile. During storage of a granulated material, an internal cavity can develop when some material is withdrawn from the bottom of the storage. This happens frequently in silo storage, and can happen in other storage configurations with bottom withdrawal or large drain openings. Although the authors are not aware of any confirmed, documented instability incidents of this nature, it has been suggested as one of the least plausible initiating events for the contaminated ammonium nitrate storage explosion at Toulouse.

3.1.3 Self-Heating Hazard Scenarios

Particulate self-heating occurs when heat generation from an exothermic chemical or biochemical reaction is not adequately balanced and dissipated by heat transfer through the particulate bed and from the surface of the particulate bed to the surroundings. In some cases, the chemical/biochemical reactions are initiated at slightly elevated temperatures caused by physical processes such as process heating, friction, or moisture absorption. If the temperature in the heated region within the interior of the particulate bed exceeds some critical temperature, a thermal runaway reaction or spontaneous combustion can result.

Following Bowes (1984), with later elaboration by Gray (2002), the various types of self-heating processes can be identified as shown in Table 3-1. Sometimes, there can be multiple processes such that it is not clear which process is dominant.

TABLE 3-1
Self-Heating Processes

Self-Heating Category	Initiating Process	Examples
Oxidative self-heating of organic materials	Oxidation	Activated carbon, coal, polyolefin pellets, various food products
Biologically initiated self-heating	Microbiological activity	Haystacks, compost and mulch piles
Moisture absorption/reaction initiated self-heating	Heat of water vapor condensation or exothermic hydrolysis	Freshly manufactured/ dried wood chips, anhydrous calcium hypochlorite
Chemically reactive/unstable materials	Exothermic reaction, often a decomposition reaction	Hydrated calcium hypochlorite

Self-heating fire scenarios tend to fall into the following four categories: (1) excessively large piles of particulates stored at ambient temperature for extended durations; (2) heated particulate stored or stacked without adequate cooling; (3) particulate remaining in heated process equipment for unusually long periods of time, and (4) oil-saturated particulate. Examples follow.

Excessive Storage Pile Spontaneous Combustion Fires

Large coal piles are often involved in spontaneous heating fire incidents. These incidents occur both in large outdoor (often wet) piles, and in coal bunker storage. High volatility, low-rank bituminous coals are more prone to self-heating than higher rank, low volatility coals The DOE *Primer on Spontaneous Heating and Pyrophoricity* describes one coal bunker fire in which the hot spot was discovered while it was still quite small, but fire grew over a period of 21 hours to eventually involve most of the bunker. The fire spread was due to a reluctance to apply water because of fear that a steam explosion might occur. When water was applied, the fire was extinguished. Moisture either from rain or condensed from humid air exacerbates the spontaneous heating propensity for most coals, but application of massive amounts of water is often effective in extinguishing these fires.

Moisture content is known to be a critical factor in the spontaneous heating tendency of storage piles of certain materials, such as bagasse (sugar cane residue after sugar extraction). Gray (2002) has offered an explanation for the complicated effect of moisture in terms of an exothermic reaction that occurs at a relatively rapid rate at a temperature of 55–60°C. Similar reactions are suspected in outdoor storage piles of other materials prone to spontaneous heating.

Self-heating of activated carbon storage has also been responsible for many fires such as the six reported (Bowes, 1984, p. 316) shipboard fires of bagged carbon that occurred within one year. Bowes described the use of isothermal self-heating tests to derive the following equation between the critical storage pile radius (half-thickness), r (in mm), and the ambient temperature, Ta (°K), of storage:

$$\ln \frac{\delta_c T_a^2}{r^2} = 35.9 - \frac{11{,}670}{T_a} \qquad [3\text{-}1]$$

where δ_c is a nondimensional parameter that is geometry dependent, and has the value of 0.88 for a large slab, and 2.52 for a cube. Bowes suggests that average temperature for an extended voyage could be as high as 38°C = 311°K. Using that temperature and values of δ_c between 1 and 2 in Equation 3-1, the calculated values of r are in the range 700 mm to 988 mm, suggesting that the maximum allowable storage pile size ($2r$) for activated carbon should be between 1.4 m and 2 m. This is significantly smaller than many cargo holds on large ships, and therefore explains the occurrence of the many reported fires.

Other particulate materials that have been involved in spontaneous heating initiated storage pile fires include fertilizer pellets, nitrocellulose filter media, cottonseed meal, and a wide variety of agricultural products including grains, nuts, and animal feed.

Heated Particulate Storage Self-Heating Fires

Inadequate cooling and stacking of heated product has caused numerous fires due to spontaneous combustion. Black (1981/82) and Bowes (1984) describe how this occurs in ligno-cellulosic materials such as wood fiberboard and wood shavings. Black notes that there have been many spontaneous ignitions of dry wood shavings placed in storage bins soon after being dried in particle-board plants. After some discussion and analysis, he concludes that the spontaneous ignition is due in large part from the heat of condensation of water absorbed from the humid atmosphere onto the dry wood chips and shavings. Other authors have reached similar conclusions about spontaneous ignitions in the storage of dried wool and other hygroscopic textile fabrics. Black recommends remoistening these materials to a moisture content of 8–10% before storage.

Composting facilities are also prone to spontaneous heating fires. Microbiological activity in the normal compost process raises the compost pile temperature to about 60°C. However, large piles can sometimes (depending on composition, moisture content, and porosity) reach temperatures of 80°C before all biological activity ceases. At this temperature oxidation can generate sufficient heat to continue to heat the center of a large pile. If the pile is left intact for a sufficiently long period of time, fires can result. A fire of this nature destroyed a $27 million composting facility in Hartford (Block and Rynk, 2000). The Hartford facility had cubical composting cells 26 ft on a side, and tried to limit storage in the composting vessels to 28 days. However, product delivery problems significantly delayed the normal turnover at the time the fire broke out. Similar problems have been responsible for spontaneous combustion silo fires at other composting facilities.

Bulk storage of direct-reduced iron has also caused some self-heating and smoldering pile fires. Eckhoff (1997) has a brief review of some studies that suggest the critical pile size is sensitive to the ambient humidity level because the iron oxidation reaction is moisture dependent, as is well known from iron corrosion studies.

Sometimes the self-heating can occur even when the reactive material is mixed with a large quantity of inert material. For example, CRHF (2003) Incident 5 involves pharmaceutical tablets containing a small amount of a self-heating-prone active ingredient. The tablets were dried at a temperature of 90°C, which is only about 10°C below the self-heating initiation temperature of the active ingredient. On one occasion, fumes filled a room containing fresh kegs of tablets that had been dried several hours earlier. Later examination revealed that tablets in the center of at least one keg had decomposed. This incident demonstrates the importance both of testing the particulate formulation actually being used in the process and allowing an adequate margin of safety between dryer temperatures and the measured critical temperature for self-heating.

Prolonged Process Heating Spontaneous Combustion Fires

Prolonged heating of materials that have accumulated in driers has led to spontaneous combustion fires at various processing facilities. The following incidents provide examples.

Beever (1982) described a fire caused by the accumulation of animal feed in a rotary disc drier operating at a temperature of 174°C. Laboratory tests showed that the critical thickness of a feedstock layer at that temperature is about 2 cm; that is, layers thicker than 2 cm could undergo spontaneous combustion if allowed to remain at 174°C. In order to prevent future accumulations, the drier outlet had to be redesigned.

Beever (1995) also investigated a spontaneous combustion fire resulting from the accumulation of milk powder in a spray drier. A test data correlation similar to Equation 3-1 showed that milk powder layers thicker than 1.7 cm could undergo spontaneous combustion at the drier operating temperature of 200°C. Accumulations were more likely in a region of the drier at a temperature of 80°C, and these accumulations were found to undergo spontaneous combustion at a layer thickness greater than 4 cm.

Accumulations of pecan husks and pecan dust on the 165°C steam coils of a pecan drier over a period of several weeks led to a spontaneous combustion fire. The drier and the surrounding area were supposed to be cleaned on a weekly basis, with accumulations limited to a ⅛in. thickness, but apparently these instructions were not always implemented. The fire spread to accumulations on beams and ledges and opened 40 sprinkler heads before it was contained.

Spontaneous combustion of carbon black deposits caused a fire in the preheater section of a carbon fiber composite production plant. The deposits had formed on screens that were inaccessible because of an obstruction in the preheater. The fire spread to deposits on the ductwork leading from the preheater to the heated rolls on which the composite was formed.

Oil/Vapor Adsorption Self-Heating Fires

Carbon bed adsorbers have had a long history of spontaneous combustion fires caused by the cumulative heat of adsorption. One such fire occurred at a vapor adsorber in a plastic resin manufacturing facility. An investigation revealed that carbon deposits had formed on the inside wall of the adsorber vessel, and these deposits did not go through the vapor desorption cycle that was conducted periodically by removing the carbon baskets from the adsorber and placing them in a special desorber. Many other carbon bed adsorber fires have occurred in vapor recovery systems at flammable liquid tank farms, and at solvent recovery facilities.

Wood shavings with deposits of lacquer and other liquid sealants have been responsible for spontaneous combustion fires at furniture manufacturing plants. Similar fires have developed on discarded particulate debris resulting from sanding painted surfaces.

3.2 DECISION TREES FOR ASSESSING THERMAL INSTABILITY HAZARD SCENARIOS

A variety of flow charts and decision trees have been developed for assessing instability hazards, classifying self-reactive materials, and making appropriate protection determinations. For example, Figure 3-3a is a flow chart that begins by screening powders for explosive behavior, and if the results are negative, going on to test for thermal instability and ignitability and dust explosibility characteristics. Figure 3-3b indicates that the type of thermal instability hazard evaluation depends on the anticipated storage or residence time, the scale (size) of the storage container or material handling equipment, air access to the powder, and any suspected contaminants. Depending on these conditions, one or more of the five possible laboratory tests are suggested. For example, a heated air over powder deposits test is suggested for drier hazard evaluations, whereas either a small basket test series or a larger bulk storage simulation is suggested. Test methods and equipment for most of these tests are described in Chapter 4. Other thermal instability test methods described in Chapter 4 include fundamental thermal analysis testing, the Reactive System Screening Test, and various versions of the Vent Sizing Package (VSP). Preliminary evaluations that do not require testing, are also discussed in Section 4.3.

Figure 3-4 is a material characterization logic diagram developed by Markowski and Mujumdar (1995) for dealing with materials that are to be thermally dried during processing. Besides considering explosiveness based on chemical composition, their characterization includes considerations of thermal decomposition and self-heating, dust combustibility, and possible dust-vapor hybrid explosibility. The chart includes the specific test laboratory test parameters needed to evaluate these hazards for a particular powder or dust.

The CCPS *Guidelines for Safe Storage and Handling of Reactive Materials* also has a reactivity evaluation flow chart. The first consideration in the CCPS (1995b) chart is a review of prior experience, and this is an essential step for materials that have been evaluated or used previously. There is a provision for theoretical evaluations via unstable chemical groups, oxygen balance, and thermodynamic calculations as described in the CCPS *Guidelines for Safe Storage and Handling of Reactive Materials,* which includes a complete tabulation of unstable chemical bonds and groups. There is a recommendation for expert evaluation in organizations that have ready access to such reactive chemical experts. The last step on screening tests includes con-

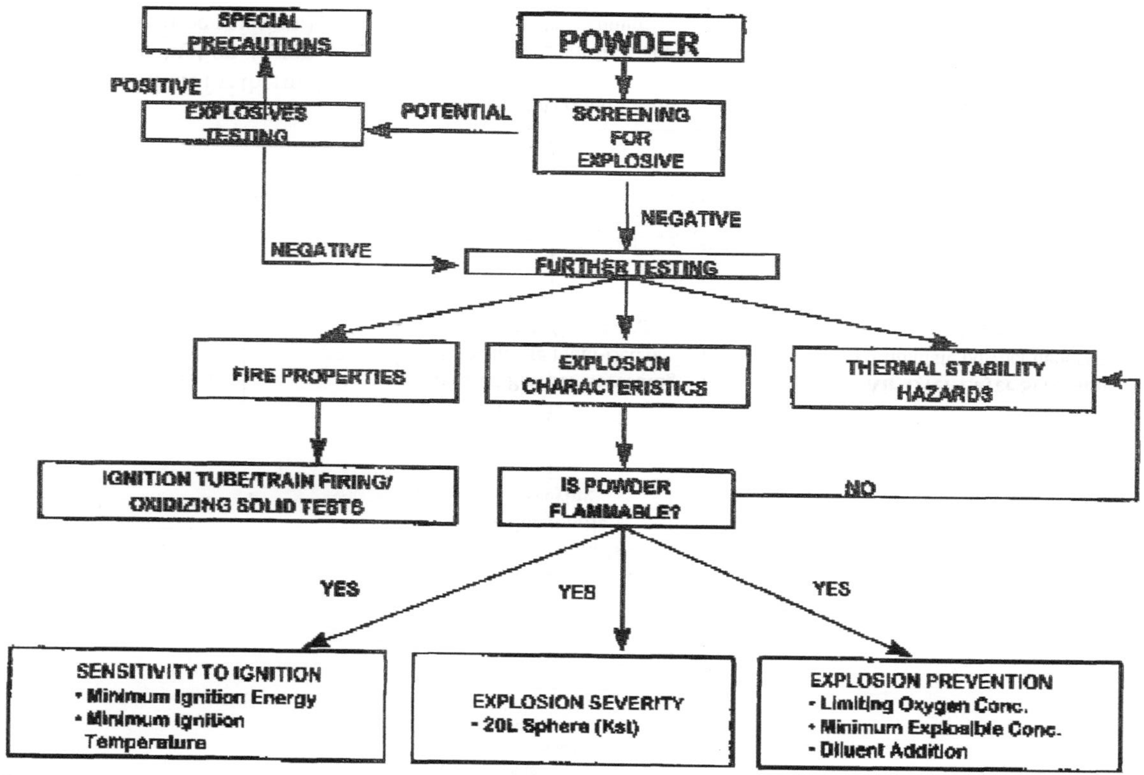

Figure 3-3a Flow chart for powder instability and combustibility evaluations. (From Chilworth, Inc., with permission.)

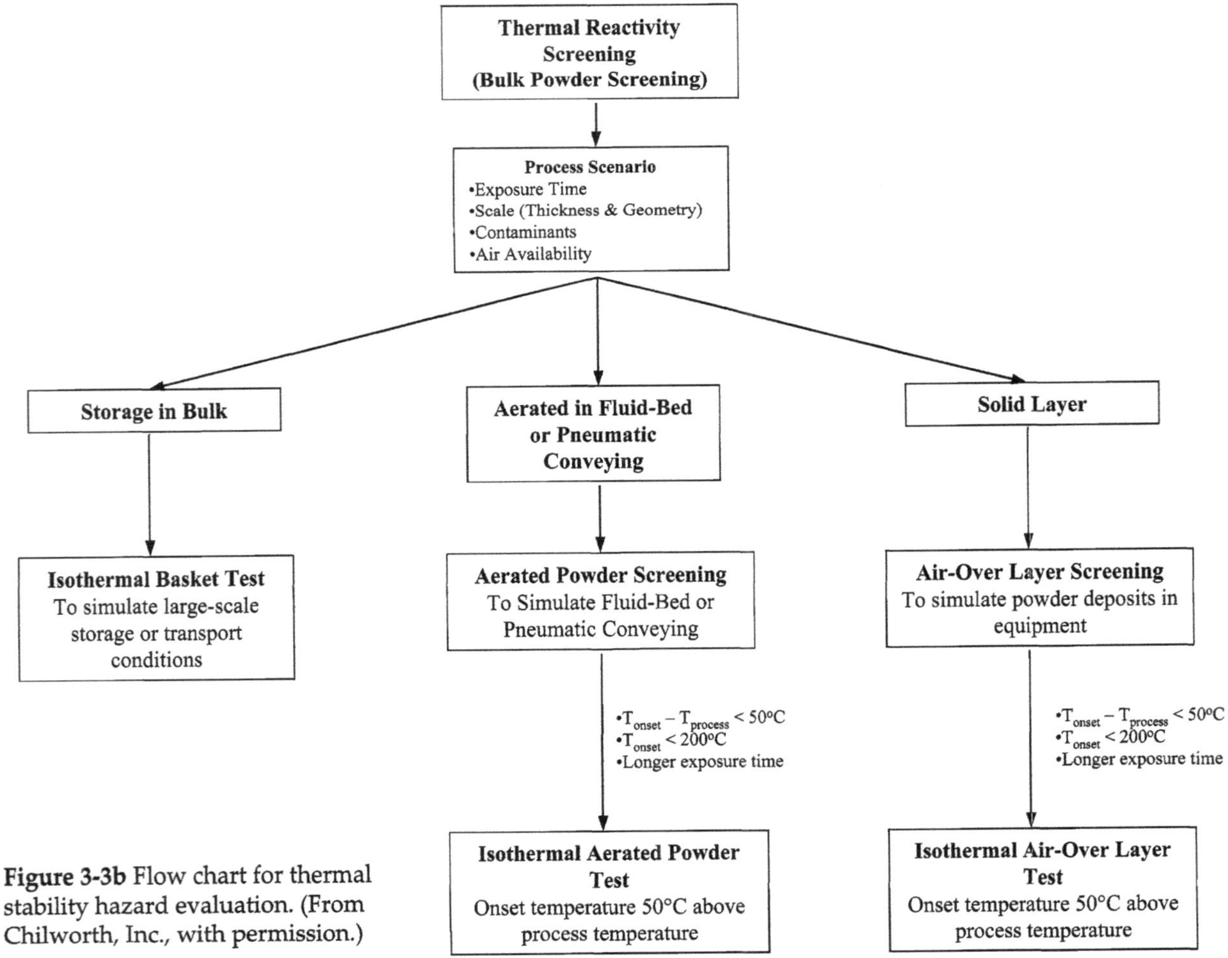

Figure 3-3b Flow chart for thermal stability hazard evaluation. (From Chilworth, Inc., with permission.)

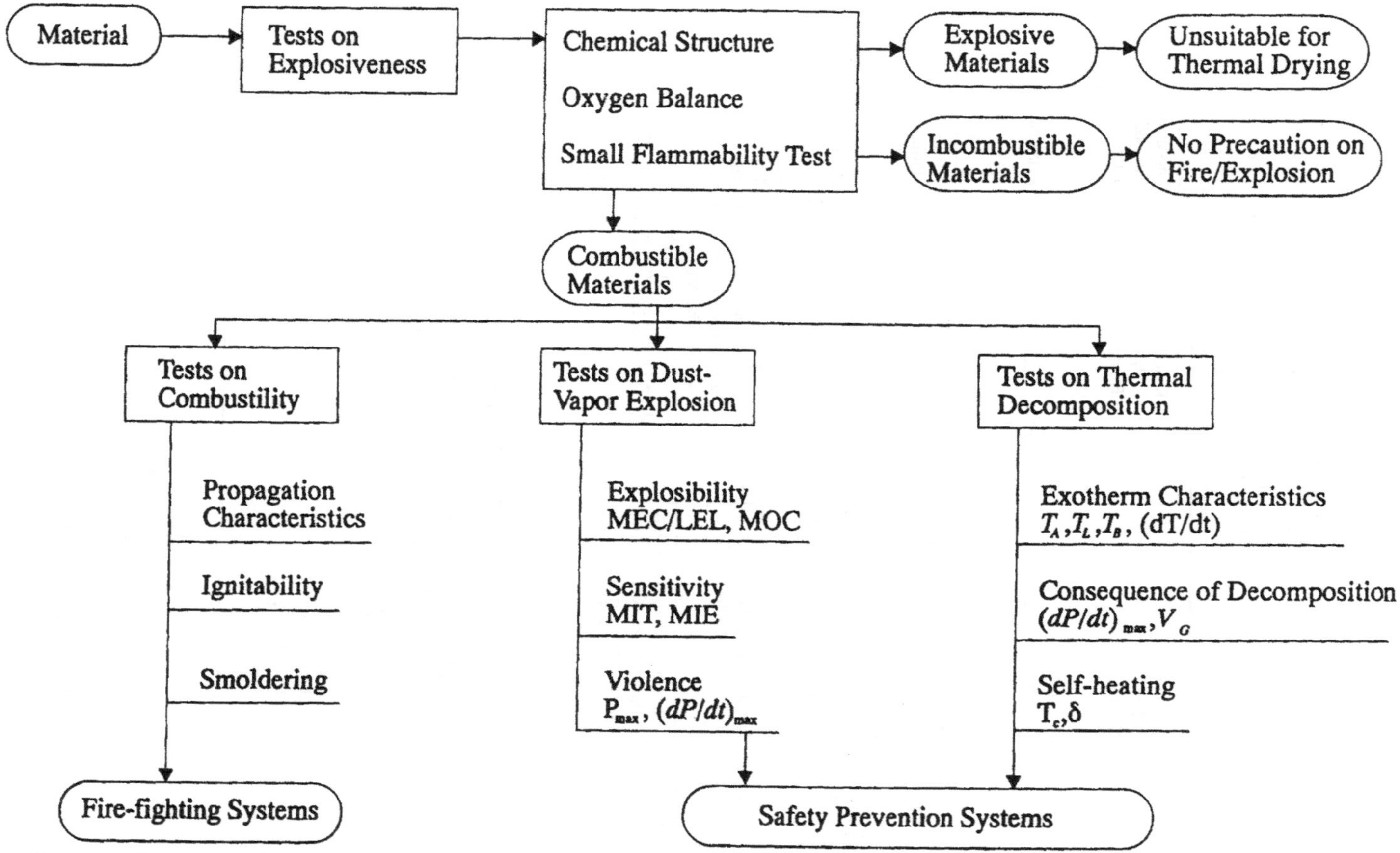

Figure 3-4 Reactivity characterization scheme for dryers. (From Markowski, A. and Mujumdar, A., 1995. "Safety aspects of industrial dryers," in *Handbook of Industrial Drying*, 2nd ed, New York: Marcel Dekker.

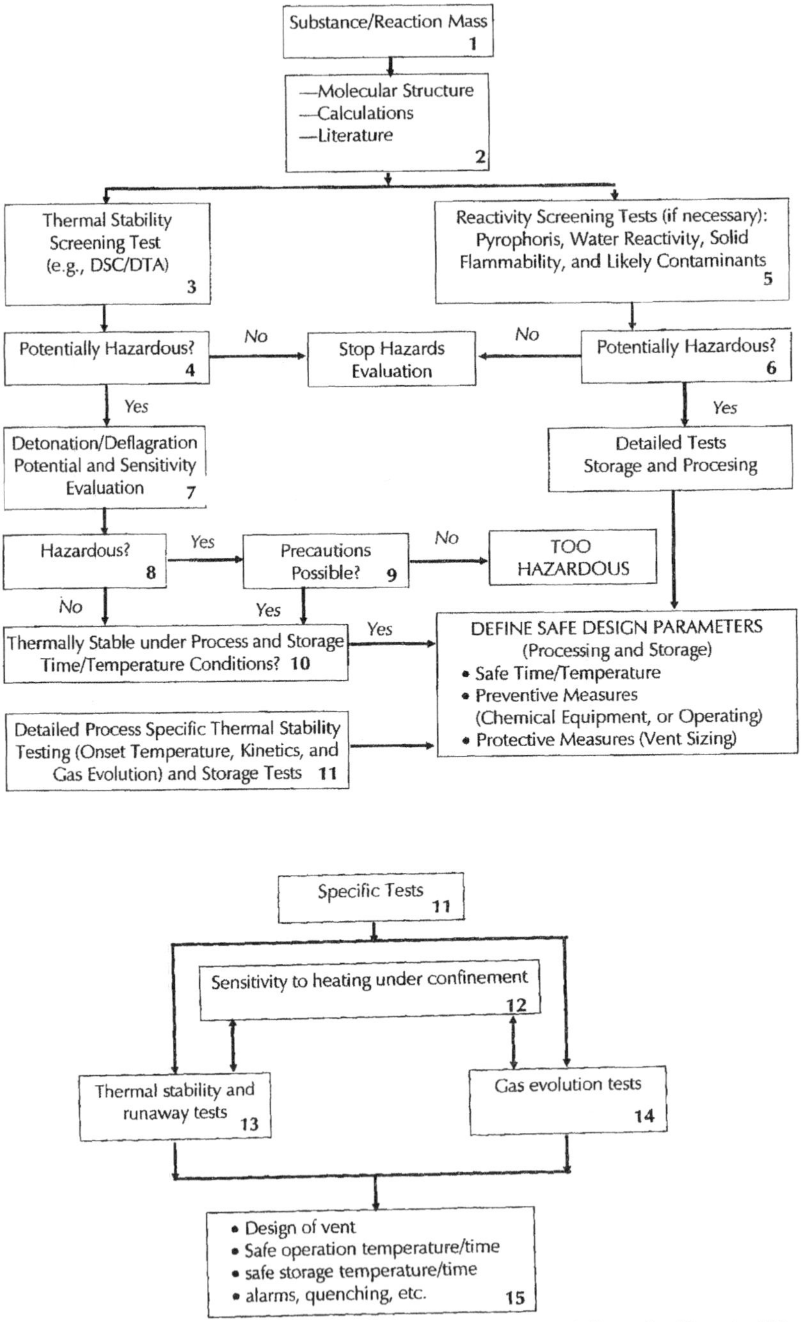

Figure 3-5 Thermal stability flow charts. (From CCPS *Guidelines for Chemical Reactivity Evaluation*.)

siderations of pyrophoricity, water reactivity, peroxide formation, and chemical compatibility, as well as thermal stability.

Figure 3-5 is a pair of reactivity evaluation flow charts taken from the CCPS (1995a) *Guidelines for Chemical Reactivity Evaluation and Application to Process Design*. It starts with a consideration of chemical structure, thermodynamic calculations and a literature search. It then divides into laboratory screening tests for thermal stability and reactivity. It specifically lists runaway reactivity and gas evolution tests to determine sensitivity of heating under confinement. This is an important consideration because accident histories summarized above have demonstrated that materials like ammonium nitrate are much more prone to explosive decomposition when heated under confinement such as storage in a ship cargo hold.

Figures 3-6a and 3-6b are flow charts to determine suitable package sizes and labels for transporting self-reactive materials per the UN-DOT regulations for hazardous materials. The decision points in the flow charts involve whether or not the instability can result in detonation propagation or deflagration propagation, and the effects of heating the material/package under confinement. Similar charts and corresponding test methods are shown in Chapter 4 for the determination of packaging and shipping requirements that may be sufficiently energetic and sensitive to be classified as an explosive for transportation purposes.

3.3 CHEMICAL INCOMPATIBILITY HAZARD SCENARIOS

A variety of scenarios can cause incompatible materials to come into contact with particulates and thereby initiate chemical reactions that result in the development of a fire or explosion. One obvious scenario not discussed here is the use of incompatible chemicals in a product formulation. The scenarios that are reviewed here are grouped into the categories of contamination, inadvertent water entry, container/packaging incompatibilities, and air access to pyrophoric particulates.

3.3.1 *Contamination Hazard Scenarios*

The most frequently occurring contamination scenarios involve inadvertent mixing of a strong oxidizer with a combustible or unstable material. Oxidizers involved in these incidents include peroxides, nitrates, and chlorinated oxidizers of the type used in swimming pool chemical biocides and disinfectants. Contamination scenarios include: (1) incomplete cleaning of an incompatible material previously used in either process equipment or a storage container; (2) use of an incompatible cleaner for processing equipment or storage containers; (3) storing incompatible reactive chemicals near the particulate material; (4) collecting and/or disposing of particulate in a manner

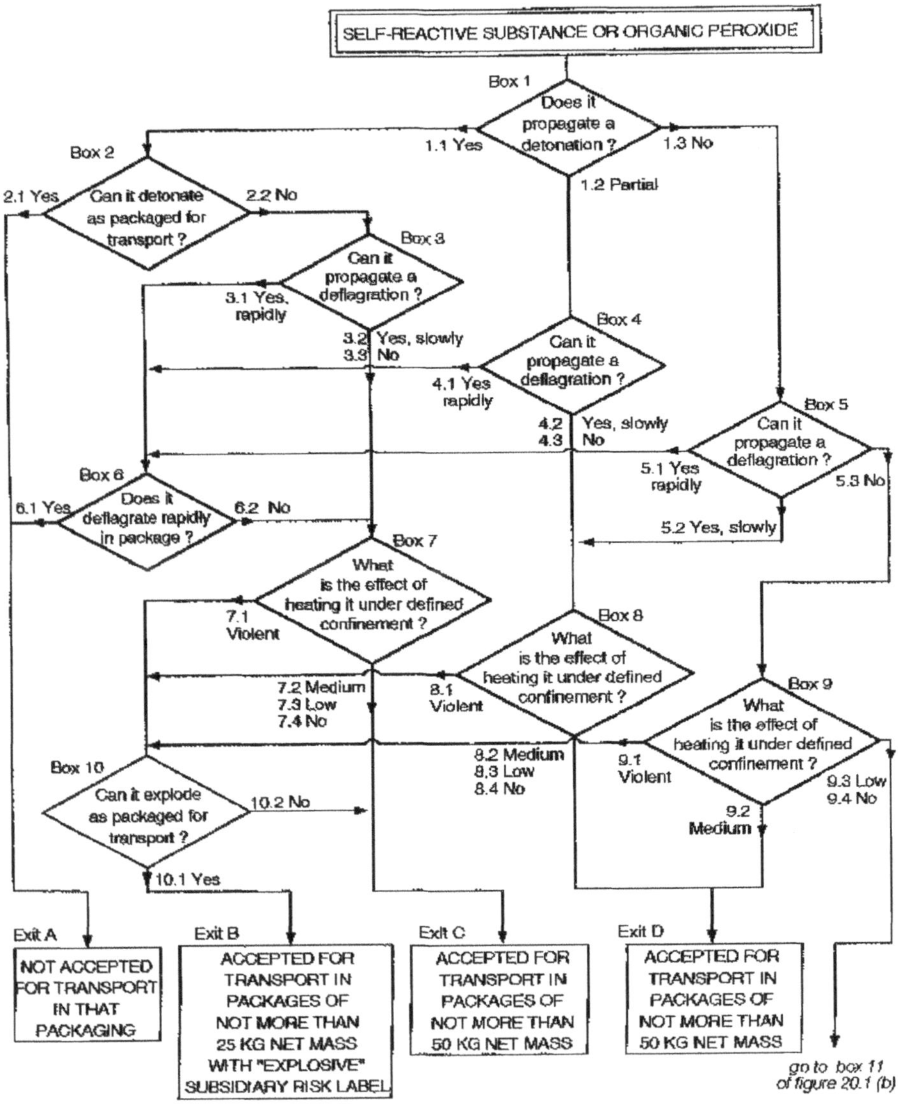

Figure 3-6a UN-DOT decision tree for transportation of self-reactive materials and organic peroxides.

that allows it to become contaminated with an incompatible material; and (5) transportation accidents involving both the oxidant and a reducing agent.

The chlorinated swimming pool chemicals (primarily calcium hypochlorite [CaClOH] and trichloroisocyanuric acid [$C_3Cl_3N_3O_3$]) have been involved in a number of storage incompatibility incidents. These materials are distributed as either powders or tablets in either burlap bags or high

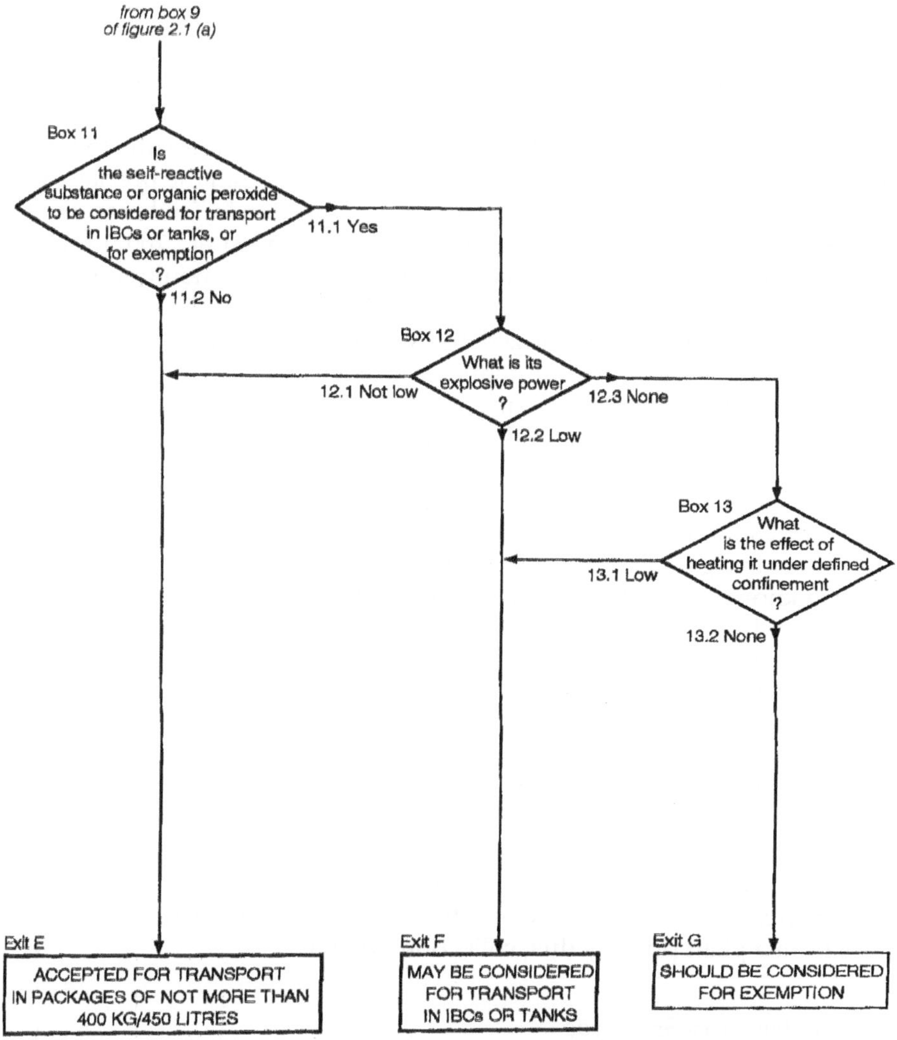

Figure 3-6b UN-DOT decision tree for transportation of self-reactive materials and organic peroxides.

density polyethylene containers. Retail warehouse stores sometimes store containers of the powdered or tablet CaClOH or $C_3Cl_3N_3O_3$ near flammable liquids such as lighter fluids, or near charcoal bags, and the proximity enhances the chances of contamination. Several large fires in these stores have started because of this contamination. Waste disposal fires due to contamination of these products have also occurred.

Ammonium nitrate (AN) has been involved in numerous fires and explosions associated with contamination. Heather (2002) described several of these incidents in his review of ammonium nitrate accidents. They include: a 1972 fire in the UK involving the storage of wooden furniture near AN fertilizer; a 1997 Brazilian highway accident in which a gasoline tank truck caught fire while attempting to pass a burning truck of AN, causing both the tank truck and the AN to explode; a 2000 Florida fire due to a collision between a gasoline tanker and a AN truck. Although the precise cause of the September 21, 2001 ammonium nitrate explosion in Toulouse, France has not been established publicly, it is clear (Kersten et al., 2002) that the explosion involved about 300 tons of off-specification ammonium nitrate in terms of particle size and possibly composition. Furthermore, the French government findings summarized by Kersten et al., indicate that "substances of all sorts (oils, organic debris, iron oxides, sulfur, bitumen, etc) had accumulated on the ground mixing with the ammonium nitrates, contaminating them as they decomposed, predisposing them to react with an energetic tendency. Just before the explosion, a chlorine compound (sodium dichloroisocyanurate) may have been tipped in the storage. This compound reacts with ammonium nitrate to produce nitrogen chloride (NCl_3), a particularly unstable gas that will explode at ambient temperature." Chlorine ions are also known to catalyze the decomposition of ammonium nitrate. However, Grand Paroisse denies that the ammonium nitrate was mixed with sodium dichloroisocyanurate, and believes the explosion was initiated by an energetic electrical fault. Both hypothesized initiation scenarios are still being studied.

Contamination incidents often arise when filters are used to collect incompatible materials without replacing the filter cartridge or bags and thoroughly cleaning the filter discharge container before the second material is processed. The CRHF (2003) incident descriptions (Incident # 2) included a near-miss incident in which there was charring of incompatible materials in a filter discharge container, and another incident caused by the recycling of fines from a plant primary dust collection system.

3.3.2 Water Entry Scenarios

Many particulate materials are water reactive and have been involved in fires and explosions caused by inadvertent water entry. Table 3.3-2 is a tabulation of some chemical groups that are known to be water reactive. A few examples are listed for each chemical group. Most of the example materials are frequently encountered particulate materials.

Since the water reactivity of these and other materials is well known, it may be helpful to describe pathways and scenarios for inadvertent water entry. The following three incident summaries obtained from the DOE *Primer on Spontaneous Heating* (1994) illustrate incidents that have occurred with water reactive metal particulates.

TABLE 3-2
Some Chemical Categories Susceptible to Water Reactivity (CCPS 1995b)

Category	Examples
Alkali and alkaline-earth metals	Calcium, potassium, sodium, lithium
Anhydrous metal halides	Aluminum tribromide, germanium tetrachloride, titanium tetrachloride
Anhydrous metal oxdes	Calcium oxide
Grignard reagents	Ethylmagnesium chloride, methylmagnesium bromide
Metal alkyls	Aluminum alkyls, lithium alkyls
Metal amides	Lead amide, potassium amide, silver amide, sodium amide
Metal hydrides	Calcium hydride, lithium aluminum hydride, sodium borohydride, sodium hydride
Nonmetal hydrides	Boron trifluoride, phosphorus trichloride, sodium tetrachloride
Nonmetal halide oxides (inorganic acid halides)	Phosphoryl chloride, sulfuryl chloride, chlorosulfuric acid
Nonmetal oxides	Phosphorus pentoxide, sulfur trioxide
Low-molecular-weight organic acid halides and anhydrides	Acetic anhydride, acetyl chloride
Other	Calcium carbide

Prior to 1955, zirconium shavings were stored outside in open-top bins. Several days after a heavy rain, fires erupted in a few of those bins, with flames reaching heights of more than 100 ft. The fire spread to the other bins and eventually burned 159,000 pounds of zirconium.

In January 1956, water contaminated zirconium powder was repackaged in steel drums. Four months later, one of those drums exploded and produced 100-ft-long streaks of red flame and black smoke. Two employees were killed and a third had an arm amputated as a result of the explosion.

When some ¼-in.-thick pieces of a magnesium–zirconium alloy were submerged in water for washing, there was a slight generation of gas (presumably hydrogen). The next day, several more pieces of the alloy were washed, and shortly later a violent explosion occurred. Residual pieces of the alloy were burning on the floor.

A devastating water–magnesium powder explosion also occurred in the previously reported ammonium perchlorate fire that was initiated by a friction spark (Section 3.1.2). Besides the water flowing from the opened several sprinkler heads, responding firefighters tried to cool the barrel/drum of magnesium powder with hose stream application. When the barrel lid became dislodged, water contacted the magnesium powder and caused a devastating explosion. The municipal fire department and the company are disputing how the lid opened and whether the water penetration came from the hose stream or the sprinkler heads.

Water contamination of water-reactive powders has also occurred from leaky water coolant lines on process equipment. One such 1995 incident involved the contamination of a blend of aluminum powder and sodium hydrosulfite, triggering a self-accelerating decomposition of the sodium hydrosulfite and a reaction with aluminum. The resulting explosion and fire killed five employees (EPA/OSHA Joint Chemical Accident Investigation Report, 1997).

Discarded pellets of an aluminum phosphide pesticide mixed with rain water in a trash container and the resulting chemical reaction caused a series of explosions (CSB Incident Number 2002-5492). One police officer was injured, and a vapor cloud disrupted activities in neighboring buildings.

Rain water penetration through a leaky roof wet fiberboard drums of trichloroisocyanuric acid stored in a Springfield, Massachusetts, warehouse in 1988, and the resulting exothermic reaction started a persistent fire that eventually resulted in 60,000 people being evacuated from the warehouse neighborhood (Zalosh, 2003). Similar reactions of chlorinated pool treatment particulates with water have occurred in many other facilities, including the 2001 explosion and fire at a Manchester, Connecticut, pool supply company (CSB Incident Number 2001-5178).

3.3.3 Container/Packaging Incompatibility Scenarios

Chemically incompatible packaging/containers for particulate products have caused several fire and toxic fume incidents. One example occurred with the same type of chlorinated water treatment product described in the water entry and contamination scenarios. In this case (CSB Incident Number 1999-2223), the pellets spilled from a drum in a government warehouse. The spilled product was cleaned up and placed in a container with an incompatible packing material. Reaction with the packing material started a small fire in the warehouse. Incidents such as the Henderson, Nevada, ammonium perchlorate (AP) fire and explosions involved strong oxidizers being stored in polyethylene containers.

Other incidents have occurred as a result of metal container corrosion and rust formation. For example, CRHF (2003) Incident Number 48 notes that incidents have been caused by rust contamination induced decomposition of 4-nitrophenyl chloroformate and other nitro compounds.

3.3.4 Air Access to Pyrophoric Particulates

Pyrophoric particulates involved in fire incidents include the following materials: iron sulfide, metal hydrides, platinum catalysts, plutonium, white phosphorus, Rainey nickel, uranium, finely divided sodium and potassium, and possibly thorium, titanium, tungsten, and zirconium. The latter four materials were involved in fires without an apparent ignition source, but their chemical treatments (acid baths) may have been a factor in their ignitability. Likewise, very fine barium, cobalt and magnesium are sometimes listed as pyrophoric.

According to the DOE *Primer on Spontaneous Heating and Pyrophoricity*, white (or yellow) phosphorus ignites spontaneously on contact with air at a minimum temperature of 30°C (86°F). The primary combustion product is phosphorus pentoxide, such that dense, white, irritating fumes are produced. When moisture is present, H_3PO_4 is also generated from phosphorus fires. The recommended preventive measure is to store phosphorus under water in either drums or hermetically sealed cans. However, there has been at least one reported fire (MARS Incident Number GB/1990/003) in which a drum seal failed due to an extreme diurnal temperature variation and associated pressure changes causing drum "breathing." The accident report states that the air entering the drum oxidized the white phosphorus and produced phosphoric acid, which then corroded the drum and allowed the white phosphorus to be released. The released white phosphorus ignited and a large fire erupted in the storage building.

The iron sulfide ignition incidents primarily occur in refineries and in hydrocarbon liquid tanks. The iron sulfide is a reaction product of the tank wall or roof with the sulfur components in the hydrocarbon. One scenario is the formation of iron sulfide under a layer of iron oxide or some other covering, and then some disturbance exposes the iron sulfide to the air in the tank. Burning iron sulfide then ignites the flammable vapor–air mixture in the tank and causes an explosion. In the case of finely divided iron particles, only a small percentage as iron sulfides may be sufficient to trigger the ignition of the iron itself.

One reason for the ambiguity about the pyrophoricity of many particulate materials is that they may ignite long after they are first exposed to air. This has occurred with some of the transuranic metals. The 1969 fire at the DOE Rocky Flats plant started with the ignition of pyrophoric plutonium scrap in a metal can in a glovebox. The fire spread through several interconnected gloveboxes, and then to rows of machining boxes and inspection boxes. Radioactive contamination was extensive, and property damage was in the range $26 million to $50 million (DOE Primer).

CRHF (2003) Incident Number 8 is a fire in a charge pot used to mix catalyst with the feed to a batch hydrogenator. Three days prior to the incident some partially spent catalyst was inadvertently added to the charge pot along with partially reacted materials. Later investigation revealed that the finely divided, partially oxidized catalyst is pyrophoric even though the purchased fresh catalyst is not pyrophoric.

Another CRHF incident (Number 41) involved spontaneous combustion of sodium hydride when it was loaded from a 5-kg bag into a supposedly nitrogen inerted vessel. The discharge of some type of powdered fire extinguishing agent into the vessel put the fire out temporarily, but apparently additional air entered the vessel during extinguishment and the fire reignited. The secondary fires were eventually extinguished with dry sand.

3.4 CHEMICAL COMPATIBILITY CHARTS FOR ASSESSING HAZARDS

One method for anticipating possible adverse interactions between pairs of incompatible materials is the development of compatibility charts, also called interaction matrices. These charts/matrices list all the pertinent materials in a process or storage facility on both the horizontal and vertical axes. At a cell representing the combination of two particular materials, there is either some summarizing description of the type of interaction anticipated, or some symbolic designation of the hazard associated with this particular interaction. ASTM E 2012 provides guidance on the construction of chemical compatibility charts. The CCPS publication on managing chemical reactivity hazards (CCPS, 2003) provides guidance on the use of these charts in the overall evaluation of reactivity hazards.

The NOAA Chemical Reactivity Worksheet (NOAA, 2002) software package is a convenient tool for constructing compatibility charts. The NOAA Worksheet groups chemicals into 40 groups based on their chemical composition and chemical reaction propensity. Charts can be drawn to indicate reactions between binary combinations of these chemical groupings, or between pairs of individual chemicals. For example, Table 3-3, which was constructed using the NOAA Reactivity Worksheet, shows the interactions between six groups of chemicals. Three of the groups in Table 3-3 are inorganic materials, and the other three groups are various types of metals. The reaction hazard designations in each cell of the matrix are explained in the notes below the matrix. As indicated in the matrix, explosive interactions can occur between inorganic oxidizing agents and reactive alkali and powdered metals, and between oxidizing agents and reducing agents. The other reactions are less hazardous.

Table 3-4, which was also produced using the NOAA Reactivity Worksheet, is an example of a compatibility chart showing interactions between binary combinations of three particular generic chemicals: soda lime (sodium hydroxide mixed with calcium hydroxide), a solid organic peroxide, and coated aluminum powder. Reactions between soda lime and aluminum powder, and between organic peroxide and aluminum powder are considered sufficiently violent to possibly result in a detonation, that is,. the reaction front propagates at a supersonic speed in the reacting medium. Although the other reactions are less violent, they can be sufficiently

TABLE 3-3

Chemical Groupings Compatibility Chart

Chemical Reactivity Worksheet

COMPATIBILITY CHART for metals and inorganic compounds

Chemical Name	1) Inorganic Compounds/Neither Reducing nor Oxidizing	2) Inorganic Oxidizing Agents	3) Inorganic Reducing Agents	4) Metals, Alkali, Very Active	5) Metals, Elemental & Powder, Active	6) Metals, Less Reactive
1) Inorganic Compounds/Neither Reducing nor Oxidizing	—	C1	C1	No reaction	No	
2) Inorganic Oxidizing Agents	C1	—	A9,B1,C1	A8,A9,B1,C1	A7	
3) Inorganic Reducing Agents	C1	A9,B1,C1	—	C1	No	
4) Metals, Alkali, Very Active	No reaction	A8,A9,B1,C1	C1	—	No	
5) Metals, Elemental & Powder, Active	No reaction	A7,B4,C1	No reaction	No reaction	—	
6) Metals, Less Reactive	No reaction	B1,C1	No reaction	No reaction	No	

HAZARD STATEMENTS

A7 - Explosive when mixed with oxidizing substances

A8 - Explosive when mixed with combustible material

A9 - Heat generated from chemical reaction may initiate explosion

B1 - May cause fire

B2 - Contact with combustible material may cause fire

B4 - Fire from exothermic reaction-ignition of products or reactants

C1 - Heat generation by chemical reaction, may cause pressurization

TABLE 3-4
Sample Chemical Compatibility Chart for Individual Chemicals

Chemical Reactivity Worksheet Ver. 1.5 4/4/2003

SAMPLE COMPATIBILITY CHART

Chemical Name	1) SODA LIME, WITH MORE THAN 4% SODIUM HYDROXIDE	2) ORGANIC PEROXIDE TYPE B, SOLID	3) ALUMINUM POWDER, COATED
1) SODA LIME, WITH MORE THAN 4% SODIUM HYDROXIDE	—	C1,D3	A6,B5,C1
2) ORGANIC PEROXIDE TYPE B, SOLID	C1,D3	101,102,104	A6,C1,D3
3) ALUMINUM POWDER, COATED	A6,B5,C1	A6,C1,D3	101,105,107,108

HAZARD STATEMENTS

A6 - Explosive due to vigorous reaction or reaction products may produce detonation
B5 - Flammable gas generation
C1 - Heat generation by chemical reaction, may cause pressurization
D3 - Contact with substance liberates toxic gas; causes pressurization

101 - Highly Flammable
102 - Explosive
104 - Strong Oxidizing Agent
105 - Strong Reducing Agent
107 - Water-Reactive
108 - Air-Reactive

exothermic to cause pressurization of closed containers, or they can generate toxic gaseous reaction products.

Frurip et al. (1997) have described the ASTM E27 Committee's approach and guidance in generating compatibility charts. They emphasize a scenario-based definition of incompatibility, and a numerical hazard rating scheme to delineate different hazard levels associated with various incompatibilities. They also suggest that the chart be utilized in conjunction with laboratory test data such that possible needs for additional data are identified. The CHEMPAT software package developed by Dow Chemical and distributed by AIChE is designed to generate compatibility charts along the lines recommended by Frurip et al.

One limitation of these compatibility/interaction matrices is that they only account for binary interactions. In constructing such charts, it would be prudent to somehow account for possible three-way interactions or for catalytic effects of a third material. For example, the combination of iron and sulfur particulates can result in the formation of iron sulfides, and the subsequent exposure to oxygen/air can result in a fire. Similarly, the presence of traces of iron sulfide or of a peroxide can ignite a cloud of combustible particulate in air.

The main value of compatibility charts is their use in hazard identification exercises such as a HAZOP. The hazard identification method serves to determine if/how the combination of materials occurs, and the compatibility chart facilitates the determination of the result of that combination. If the combination is hazardous, the HAZOP can be used to determine potential preventive/control measures.

3.5 PARTICULATE FIRE SCENARIOS

3.5.1 Smoldering Fires in Storage Piles and Dust Collectors

The most frequent cause of smoldering particulate storage pile fires is self-heating—spontaneous combustion. As discussed in Section 3.1.3, the self-heating is a consequence of either excessively large piles or being allowed to remain without turnover for an excessively long period of time. Critical pile sizes and expected times to ignition can be calculated using the data from laboratory tests described in Section 4.3.2.3, together with the scaling methods described by Bowes (1984), Beever (1995), and Gray (2002).

A frequent cause of smoldering fires in dust collectors is the transport of a burning ember generated during particulate processing on some equipment serviced by the dust collector. Some examples include grinders and pulverizers in which either some tramp metal or some misaligned moving part can heat a few particles to a temperature at which they oxidize rapidly enough to begin glowing or burning. The burning ember, which sometimes forms a larger smoldering nest, then travels through the ducting on its way to the dust collector, where it is then covered by a large quantity of combustible particulate. The heat generated by the smoldering nest can produce a smolder region propagating slowly outward toward the exposed surface of the collected dust. If the collected dust is then discharged into a bin or other container, the smoldering particles can come in contact with enough air to suddenly flare up and rapidly enlarge the burning region.

Eckhoff (1997) has reviewed and summarized the results of studies of the conditions required for the smoldering nest to continue burning as it travels through ductwork. The tests indicated that a 10-g smoldering nest would rapidly self-extinguish at a dust loading concentration in the duct greater than or equal to 1 kg/m^3. Smaller dust loading concentrations allowed the smoldering nest to travel appreciable distances without self-extinguishing, with smaller particle nests burning over significantly longer distances than larger particles. Rather than rely on self-extinguishment it is prudent to use certified spark/ember detection and extinguishing systems as described in Appendix C of NFPA 654, as well as in Chapter 6 of this book.

It is often possible to intervene in the smoldering process while the smoldering is still limited to a small portion of the pile interior. Intervention requires first recognizing that the smoldering is occurring, and then either breaking the pile up to get at the burning region, or applying a fire suppression agent to an inaccessible pile as in a large silo or bunker. Automatic detection of smoldering within outdoor stockpiles entails use of thermal monitoring devices. There are two options: monitoring the pile surface with an infrared system or installing some type of temperature monitor (usually thermistors) within the pile interior. Interior monitoring can detect smoldering at a significantly earlier stage than infrared monitoring, but requires an extensive preinstalled array of thermal sensors.

Automatic detection of smoldering fires in bunkers, silos, and enclosed dust collectors can be achieved with either a CO or hydrocarbon vapor detector, or with infrared monitoring of the silo/collector wall temperature. In the case of coal bunkers, NPFA 850 recommends alarming at a CO concentration of 1.25%.

The most success in extinguishing silo/bunker fires has been achieved with carbon dioxide, and to a lesser extent with nitrogen. For example, Tuomisaari et al. (1998) conducted approximately 50 laboratory tests with carbon dioxide or nitrogen applied to barrels of smoldering wood chips or peat. The gaseous agents were applied to the top of the barrel in some tests and to the bottom in other tests. Better results (i.e., earlier suppressions), were achieved with bottom application because the gases flowed up through the smoldering material and gradually extinguished the fire.

3.5.2 Dust Layer Fires

Dust layer fires most often occur on heated surfaces such as exist in dryers and on space heaters, bearings, and motors. Prolonged heating on these surfaces (often due to upset conditions or inadequate cleaning) can initiate either combustion or exothermic decomposition. Standardized tests to determine the hot surface ignition temperature of combustible dust layers are described in Section 4.3. Unstable material decomposition on hot surfaces has been discussed in Section 3.1.1.

If not detected and suppressed rapidly, the fire can spread throughout the dust layer, and possibly ignite the dried product exiting the dryer. One such incident (CRHF Incident Number 2, Case 1) occurred in a dryer that developed a blockage in the discharge of dried material from the bottom of the dryer. A 5-hour delay in the arrival of the maintenance crew allowed the particulate layer in the dryer to be heated for a much longer time than in normal operation. Furthermore, a small leak in the slide valves at the bottom of the dryer allowed air entry into the normally nitrogen inerted dryer. The combination of air access and prolonged heating caused a charring of the particulate material, which was eventually discharged into a drum under the dryer. In this case, the charring was observed in time to prevent a large fire.

There have been several reported fires in which the fire destroyed the dryer and continued to spread to accumulated dust layers in the surrounding area. Some of the particulate materials involved in these fires include ammonium nitrate based fertilizers, various pharmaceutical products, and various polymers and copolymers. One particular incident in a plastics and rubber manufacturing facility involved overheating during the drying of a copolymer of acrylonitrile and vinyl acetate, downstream of the polymerization reactor. The dryer and adjacent equipment were destroyed such that there was a complete loss of production capability as a result of this 1997 fire (MARS Incident IT/1997/001).

Prevention of dust layer fires primarily entails minimizing dust layer accumulations via equipment design and cleaning up any accumulations soon after they occur. The example of the pecan dryer fire cited in Section 3.1 occurred because the weekly cleaning of the steam coils was put off to the extent that dust layer grew far deeper than the $\frac{1}{8}$-in. limit established by plant management.

Another dust layer fire scenario that can be readily prevented is cutting or welding hot work on ducting or piping containing internal accumulations of particulates. Prevention of this fire scenario entails verifying that the ducting or piping is free of combustible particulate accumulations before issuing the hot work permit.

Dust layer fires are often easily extinguished with either automatic sprinklers or fixed water spray nozzles, unless the particulate material is water reactive. Metal dust fires require use of certified Class D extinguishing agents. Manual application of a hose stream or a portable extinguisher to non water reactive materials should proceed very cautiously so that the force of the discharge stream will not generate a dust cloud and thereby cause either a large flash fire or even a dust explosion.

3.5.3 Warehouse Storage Fires

There have been a wide variety of ignition scenarios for particulate warehouse storage fires. The 1997 fire at the Bartlo Packaging Incorporated (BPS) warehouse in West Helena, Arkansas was reportedly started when a bulk sack of a thermally unstable pesticide (Azinphos methyl 50W) was stored too close to a hot compressor discharge pipe (EPA/OSHA Joint Report, 1999). The storage arrangement and proximity to the compressor room are shown in Figure 3-7. Flammable decomposition products (possibly a hybrid dust/vapor mixture) accumulated and later exploded killing three firefighters.

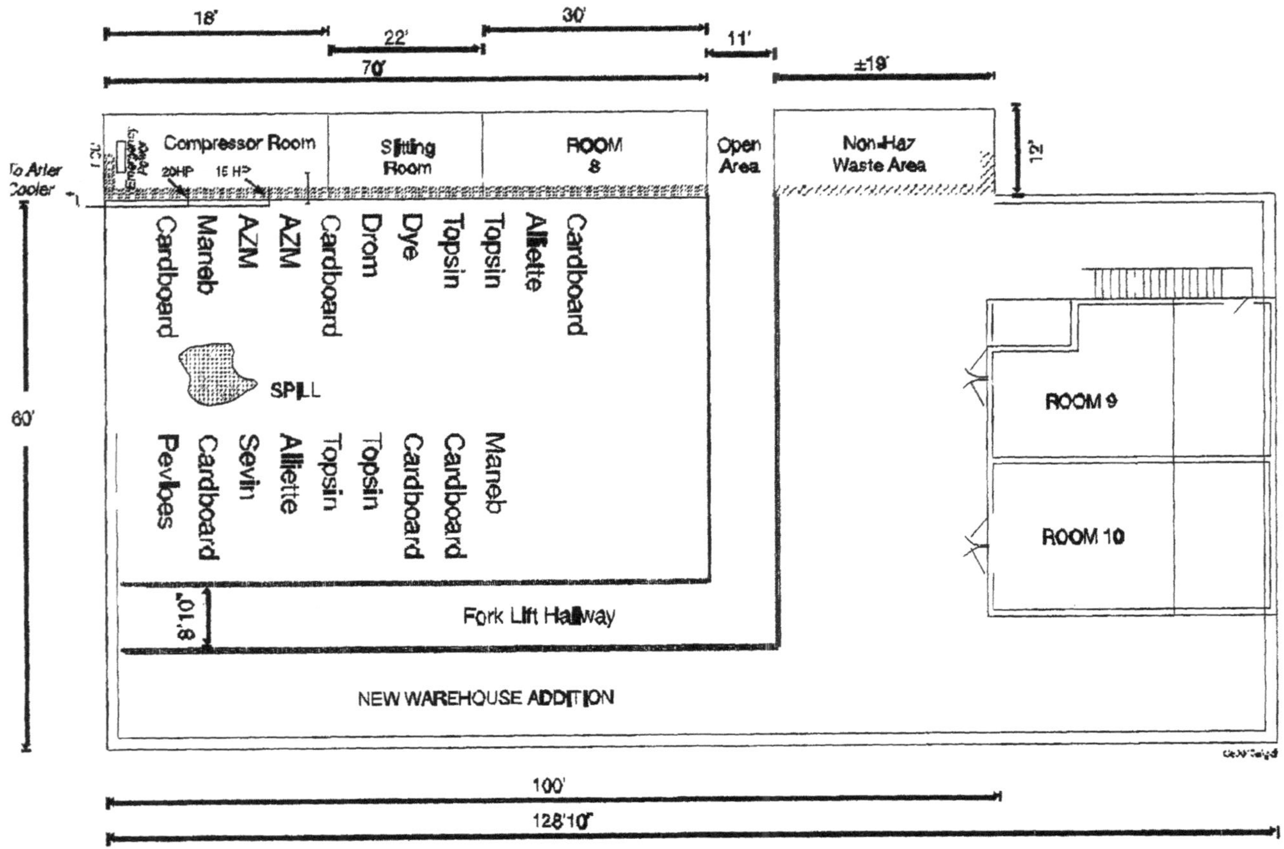

Figure 3-7 Location of chemicals in 1997 BPS warehouse fire. (From EPA/OSHA Joint Accident Investigation Report.)

The 1986 fire in the Sandoz warehouse in Basel, Switzerland reportedly started when a sack of a thermally unstable pigment was shrink wrapped in a manner such that the heat source initiated the subsequent thermal decomposition of the pigment (Zalosh, 2003). The 1995 chemical warehouse fire in Tonawanda, New York was reportedly started because a production problem initiated exothermic decomposition of ammonium persulfate, and the decomposing persulfate was packaged and stored without noticing the incipient decomposition (Isner, 1995). Closely stacked bags of the ammonium persulfate and the adjacent drums and FIBCs containing sodium persulfate and potassium persulfate, shown in Figure 3-8, allowed the fire to overwhelm the sprinkler system and destroy the warehouse.

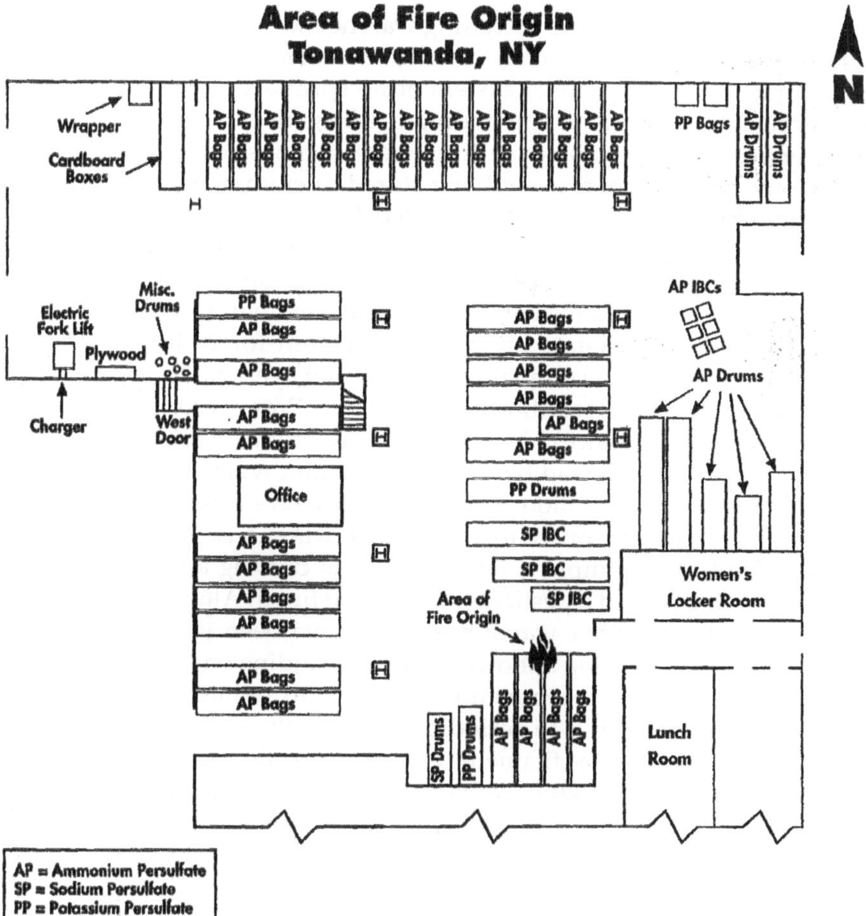

Figure 3-8 Chemical storage layout and location of fire origin in 1995 Tonawanda, NY, warehouse fire. (Reprinted with permission from *Persulfate Warehouse Fire, Tonawanda, NY, August 18, 1995,* Copyright © 1996, National Fire Protection Association.)

The preceding fire initiations notwithstanding, many warehouse storage fires are started by malicious deliberate ignition. An intruder started the 1992 fire at the B&R Haulers chemical warehouse in Salford, England (Health and Safety Executive report). The fire rapidly spread through the warehouse and initiated an explosion in 25 metric tons of sodium chlorate. Another intruder in 1989 started a fire in an agrochemical warehouse in Denmark (MARS Accident Report DK/1989/001). Materials stored in this warehouse included starch glue, Methiocarb (CAS code 2032-65-7), Thiram (CAS code 137-26-8), Hymexazol (CAS code 10004-44-1), Iprodione (CAS code 36734-19-1), Carbofuran (CAS Code 1563-66-2), calcium peroxide (CAS Code 1305-79-9), and saw dust. As the local fire brigade was extinguishing the fire, a dust explosion occurred.

Major factors in all of the preceding warehouse fires were the inadequate isolation of the special chemical hazards (unstable and/or toxic materials), either an inadequate sprinkler system or no sprinkler system at all, and an uninformed or ill informed responding fire brigade. The challenges of special chemical hazard warehouse fires were demonstrated in the May 25, 2004, Conyers, Georgia, fire involving stored calcium hypochlorite product for swimming pool treatment (Associated Press, May 25, 2004). The large smoke plume containing chlorinated decomposition and combustion products produced a major disruption to neighboring properties and highway traffic. Several thousand residents were evacuated within a 1.5-mile radius of the fire and an interstate highway was closed for 20 hours. About 40 people went to hospitals complaining of burning eyes and lungs. Most were treated and released. According to Associated Press accounts (May 25 to May 28, 2004), the local fire chief said the fire overwhelmed the warehouse's sprinkler system and spread beyond fire-resistant walls. A company statement said that about 15 million pounds of product were destroyed in the fire.

One other ignition source for particulate warehouse storage fires is an electrical fault. For example, a lightning strike causing a power surge and sustained arcing started a pesticide warehouse fire in Alliance, Ohio in 1974 (Diefenbach, 1982). Other warehouse fires have been ignited by halogen lighting failures producing a shower of sparks landing on combustible packaging. One company's recommended practice to reduce the frequency of halogen lighting failures is to periodically (typically weekly) cycle the lighting off and on, to stress the filaments and cause a bulb to fail on restart while it is still cold. This reduces the likelihood of a hot bulb failing and releasing a shower of sparks on combustibles below.

3.5.4 Particulate Flash Fires

Flash fires occur when a dust cloud at a concentration above the minimum explosive concentration is exposed to an ignition source, but there is not sufficient confinement to cause a dust explosion. Some examples follow.

A company routinely filled FIBCs with a powdered food additive. After a tramp metal detector was installed in the filling nozzle, there was a series of incidents where there was an ignition and flash of the powder in the FIBC as it was filling. Although there was no major damage, the incidents charred the material and frightened the workers. Investigation of the incidents revealed that the body of the tramp metal detector was a nonconductive plastic tube which was graphite coated on the outside surface only. This allowed a static charge to build up on the inside of the plastic tube until it discharged to an adjacent conductive surface and ignited the powder flowing past it. Replacing the nonconductive plastic tube with a conductive plastic tube and bonding it to the rest of the filling apparatus resolved the problem.

Another facility had a pneumatic conveying system, which included a section that ran over a warehouse. In the middle of the night, the conveying piping came apart, dumping combustible powder onto the warehouse roof. The problem went unnoticed until several tons of the material had accumulated on the roof and caused the warehouse roof to collapse. The collapsing roof broke some electrical wiring and the electric arc ignited the material as it fell. Because the collapsing roof also damaged the warehouse sprinkler system, the ensuing fire destroyed the warehouse.

Flash fires can sometimes occur when a fire engulfs stacked bags of combustible powder. In this scenario, the fire starts at the base of a pile of bags made from combustible packaging and grows up the outside of a stack, causing a series of bags at higher level to fail and allow powdered material to flow into the fire. Some of this powder burns as a suspended cloud before it reaches the floor. Tyldesley (2004) reports that the UK Health and Safety Executive has investigated a fatal incident involving 25-kg paper bags of rubber crumb stored on pallets. A pallet loaded with 46 bags in a stack 2 m high was ignited by a flame gun used for shrink-wrapping, but the operator did not immediately notice the fire. As the pallet was being moved using a forklift truck into a storage area, an extremely rapid fire growth then occurred.

Subsequent HSE tests with crumb rubber and a variety of other bagged powders sometimes produced rapid fire growth that by burning of powder that is either falling from a burst bag, or being carried upward on convection currents from the fire. Lower fire growth rates sometimes occurred because the powder softened and would not flow, because it charred and did not flow, or because the inherent cohesive forces within the powder did not permit free flow. Previous fire tests conducted elsewhere have shown that larger free-flowing particulates can also form large heaps that restrict air access to the lower level combustion zone, and thereby reduce the fire intensity.

Unfortunately, there are no quantitative criteria yet to distinguish between potential flash fire formation and the reduced fire growth rates in many of these bagged powder tests. Therefore, for a large range of powdered products it must be assumed that a tall stack of bags, or a release from tall single container caused by fire at low level can create a fire capable of growing very suddenly, with a corresponding risk to people in the vicinity. Since normal provision of means of escape might not be adequate, multiple egress paths should be available, and personnel should be forewarned not to attempt manual firefighting with portable extinguishers in these situations.

3.6 DECISION TREES FOR ASSESSING PARTICULATE FIRE SCENARIOS

Figure 3-9 is a decision tree that can help determine suitable prevention and protection measures for smoldering fires in a storage pile or in a dust collector. The tree starts with a consideration of whether the particulate material is prone to self-heating or exothermic decomposition. If so, there are considerations of the pile size compared to the critical pile size for spontaneous combustion (from thermal explosion theory), and the particulate residence time in the pile compared to the expected time-to-ignition. If the material is not prone to self-heating or thermal instability, the primary considerations are the combustibility rating of the particulate (perhaps a flame spread rating from a dust layer fire test), and whether or not the pile is enclosed. Depending on these factors, the tree leads to the selection of alternative fire detection and firefighting measures. If the material is water compatible (including reactivity and spoilage considerations), a water-based suppression agent is suggested. The suppression agent selection does not specify whether or not the agent should be applied from a preinstalled fixed suppression system, or from manual nozzles.

Figure 3-10 is a decision tree for warehouse storage fire scenarios. The primary factors in this tree are whether or not the particulate is prone to self-heating or thermal decomposition, and whether the material is water compatible. If the material is either self-heating or thermally unstable, as determined by thermal stability tests described in Section 4.3, the tree leads to storage in an isolated area. If the material is water compatible, the tree leads to a selection of an automatic sprinkler system. The premise for this recommendation is that the total loss of the warehouse and product would not be an acceptable risk. There are also exceptions for warehouses that are amenable to some type of special detection/suppression system. If NFPA 13 is the jurisdictional standard for sprinkler system design, the system should probably be designed for a free-flowing particulate commodity. In the case of NFPA 13, a free-flowing plastic material, would be classified as a Class IV commodity, which is less of a challenge in terms of sprinkler system design than Group A plastic commodities.

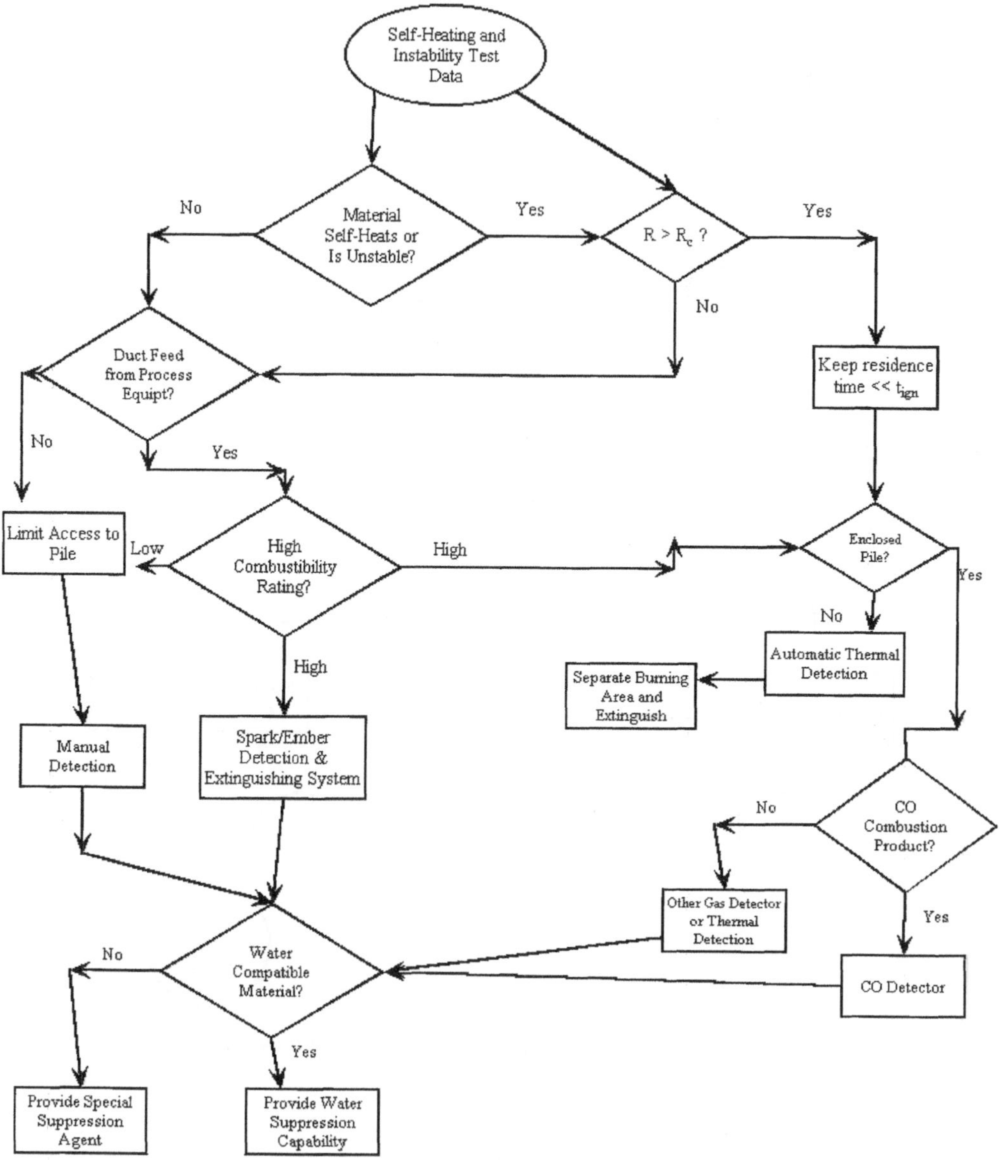

Figure 3-9 Decision tree for assessing smoldering fire scenarios.

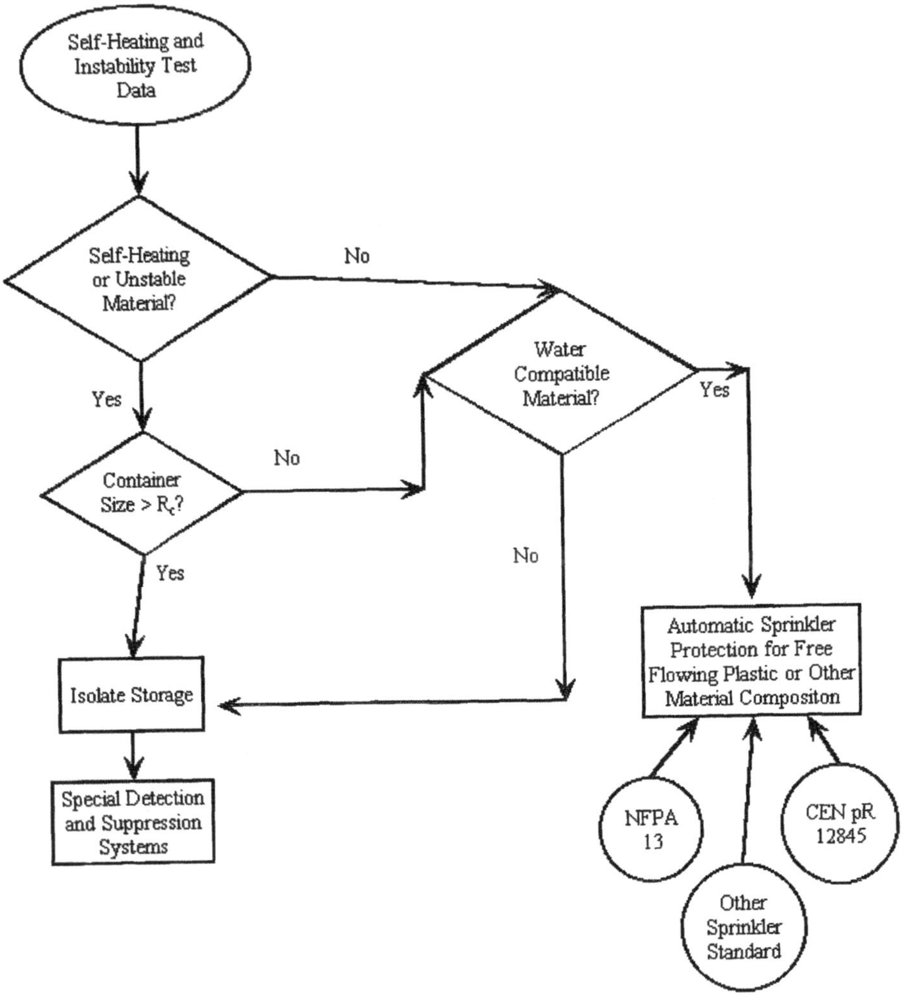

Figure 3-10 Decision tree for particulate warehouse fires.

3.7 DUST EXPLOSION SCENARIOS

3.7.1 Primary Dust Explosions in Process Equipment

Dust explosions occur in process equipment when there is a particulate concentration between the minimum explosible concentration (MEC) and the upper explosive concentration (both of which depend on the oxidant present), and then an ignition source develops or reaches the combustible cloud.

The types of process equipment that routinely have combustible dust concentrations in at least a portion of the equipment volume include blenders, dryers, dust collectors, and grinders/pulverizers. Examples of primary dust explosion scenarios follow for each of these types of equipment. Additional examples are provided in Chapter 5.

Blender Explosion Scenarios

The blending of particulates of two or more different compositions inherently involves the kind of interparticulate friction and particulate-wall friction that causes electrostatic charge generation. If the particulate resistivity is sufficiently high, the electrostatic charge can continue to accumulate with correspondingly increasing voltage differences. Furthermore, if the blender wall is not well grounded, charge and associated high voltages can accumulate on the blender wall. If the particulate minimum ignition energy (MIE) is sufficiently low, and if the eventual electrostatic discharge occurs in a location where combustible concentrations exist, the result is a dust explosion. This has occurred in a plastics manufacturing plant in a blender used to mix the primary polymer with various additives. The primary polymer had a resistivity of 2×10^{16} ohm-cm, an MIE of about 7 mJ, and an MEC of 20 g/m^3. The latter two values are lower than those of most organic powders. However, even with significantly larger MEC values, concentrations above the MEC should be anticipated toward the top of the blender during normal operation and throughout most of the blender volume during batch loading and unloading.

Another potential ignition source in many blenders is the overheating associated with either a failed bearing or a misaligned ribbon or screw scraping against the blender wall. This overheating can cause a clump of particulate to be heated to an incipient ignition temperature. Still another potential ignition source is any tramp metal that may enter the blender. See the discussion in Chapter 6 on methods for dealing with tramp metal in process equipment.

German statistics on dust explosion ignition sources in blenders are as follows (Eckhoff, 1997):

Ignition Source	Percent of Explosions in Blenders/Mixers
Electrostatic discharges	45
Mechanical heating, friction	25
Mechanical sparks	15
Other/unknown	15

Eckhoff (1997) has a detailed description of a 1973 aluminum dust explosion in a screw blender/mixer. A rubber-lined steel tube surrounded the center screw of the mixer. Eckhoff states that the explosion was probably ignited by a propagating brush discharge due to electrostatic charge generation on the rubber lining within the grounded steel tube. The aluminum flakes had a specific surface area of 7.5 m^2/g, and an MIE of only 1 mJ. Even though there was an attempt to nitrogen-inert the mixer, Eckhoff suggests that the volume within the steel tube was not inerted and oxygen concentrations were sufficiently high to support combustion. Thus, the explosion was initiated at the 3.3-m-long enclosed screw, and then propagated as a flame jet into the 5.2-m^3 mixer.

According to preliminary findings in the investigation of the January 29, 2003 devastating explosion at the West Pharmaceuticals Services plant in Kinston, North Carolina, a dust explosion was initiated in or below a mixing area where blenders had previously experienced fires (Chemical Safety Board February 3, 2003, news release and information posted February 10, 2003, in the Chemical Incident Report Center on the CSB Web site). One previous blender "fire was strong enough to blow off the mixer's door." The blenders in the West Pharmaceuticals automated compounding system are used to mix bulk rubber materials into formulations for molding into medical device components, such as rubber stoppers for syringes.

Dryer Explosion Scenarios

Overheating of particulate by a hot surface is by far the most likely ignition source in dryers. In some cases, the particulate accumulates on the hot surface and forms a smoldering nest, while in other cases the hot surface temperature is sufficiently high to directly ignite the suspended dust cloud. Although particulates near the dryer inlet may be too wet to be readily ignited, particulates exiting the dryer are both dry and often suspended in concentrations above the MEC.

FM Data Sheet 7-76 contains an account of a nondairy creamer powder explosion in a large spray dryer. The explosion occurred during normal dryer operation, without any indication of overheating. However, the size of the dryer (15.2 m high and 3.4 m in diameter) suggests that it would be very difficult to achieve uniform heating without some overheating of pockets of accumulated powder. The explosion deformed the dryer sufficiently to render it inoperable. Explosion propagation to adjacent equipment also did extensive damage.

Acrylonitrile-butadiene-styrene (ABS) copolymer resin particulate exploded in a 13.7-m-long, 4.6-m-diameter rotary dryer. A new type of steam coil in the dryer normally operated at 182°C, and may have been hotter on the day of the explosion. Resin on the coils was found sintered to a tarlike consistency, even though the dryer had been designed to prevent resin contact with the steam coils. Steel panels at the heater end of the dryer were distorted, and there was considerable damage to adjacent equipment, ducting, and surrounding structures.

A starch dust explosion occurred in a flash dryer soon after the dryer feed was restarted after being shut down for a minor repair. The dryer had continued in operation during the feed shutdown, and caked starch on the dryer wall overheated and ignited. The small fire in the dryer was not detected, and when the dryer feed re-started, the moisture content of the entering starch powder was too low to prevent the starch stream from being ignited by the fire. The explosion was vented out through both the dryer inlet and outlet, and damaged adjacent equipment and the building walls.

The Chemical Safety Board June 10, 2003, presentation on the West Pharmaceuticals explosion attributes the primary explosion to fugitive polyethylene powder from an air-blown dryer designed to evaporate water from an aqueous polyethylene solution applied to the rubber. The ignition source and precise location has not been determined as of this writing.

Dust Collector Explosion Scenarios

As indicated by the following data on German dust explosions (Eckhoff, 1997), the most prevalent ignition source in dust collector explosions has been a mechanical spark probably due to tramp metal being conveyed into the collectors. Smoldering nests, the second most frequent ignition source, probably originated in upstream process equipment such as dryers. Electrostatic discharges are also important and may contribute to a higher percentage of dust explosions in certain industries.

Ignition Source	Percent of Explosions in Dust Collectors
Mechanical sparks	41
Smoldering nests	11
Electrostatic discharges	10
Mechanical heating, friction	7
Other/unknown	31

FM Data Sheet 7-76 describes one wood-dust collector explosion apparently caused by mechanical sparks or overheating at a sander at the upstream end of the duct feeding the cyclone collector. Although the collector was outside, ducting ruptured in the building and emitted burning dust that formed a large fireball over most of the 930-m^2 ceiling. Deflagration vents on the dust collector prevented any major damage to it.

A disintegrated nylon bushing on a bearing for a discharge screw conveyor on a dust collector caused the conveyor shaft to make a small hole in the conveyor housing and burn methyl-methacrylate-butadiene-styrene (MBS) particles carried in the conveyor. The fire apparently propagated back into the dust collector when the product flow was stopped. Since the air flow through the dust collector had continued during the process interruption, combustible dust concentrations existed in the collector when the flame reached it. The explosion destroyed the dust collector and propagated into upstream equipment.

Matsuda and Yamaguma (2000) describe a tantalum dust collector explosion that they attribute to an electrostatic discharge in the collector. The small, corral shaped tantalum particles have a high resistivity, and become electrostatically charged by rubbing against the collector wall. Apparently an electrostatic discharge from the charged particles triggered the explosion in the 5-m-high, 1.5-m-diameter bag type collector.

Chapter 5 of this book provides a discussion of generic operating conditions associated with dust collectors and peripheral equipment that can eventually lead to dust collector explosions.

Grinder/Pulverizer Scenarios

Grinders, pulverizers, and other size-reduction equipment inherently dissipate large energy inputs required to break up the particles. This energy dissipation inevitably causes heating of the particles and metal surfaces. Particles accumulating in the grinder can easily overheat, smolder, and ignite a dust explosion during grinder loading or unloading. Zalosh (1984) reviewed reports of numerous coal pulverizer fires and explosions, and found that most of those were due to spontaneous combustion of overheated pulverized coal.

Phenol formaldehyde resin was ignited in a roller mill due to a loose blade on a spinner assembly striking the mill wall and causing either mechanical sparks or frictional heating of resin on the wall. The explosion propagated into a downstream dust collector, and a secondary dust explosion occurred in the building.

Pneumatic Conveying Equipment

Dust explosions can occur in enclosed conveying equipment when the ratio of the transfer rate to the air flow rates is equivalent to a concentration above the MEC. They can also occur at transfer points on unenclosed conveyors due to dust cloud formation and the confinement represented by surrounding structures and equipment. Upset conditions also temporarily cause suspended dust concentrations to exceed the MEC. Ignition sources have included electrical equipment not rated for a Class II hazardous location, electrostatic discharge (often due to faulty or incomplete grounding), and hot work on or near the conveyors.

3.7.2 *Hybrid Explosion Scenarios*

Hybrid vapor–dust explosion scenarios involve the combination of a vapor concentration and combustible particulate concentration that renders the vapor–dust–air mixture flammable, that is, capable of supporting flame propagation away from an ignition site. Hybrid explosions can sometimes occur even when the flammable vapor concentration is below the vapor lower flammable limit *and* the dust concentration is below the material's MEC. Dryers for flammable solvents on particulate products often operate at conditions that allow these concentrations to develop. The separation of some polymers from their volatile monomers (such as vinyl chloride and propylene) also often produces hybrid mixtures with concentrations exceeding the mixture lower flammable limit, which can usually be approximated via a Le Chatelier's law calculation (Cardillo and Anthony, 1978).

Ebadat (1999) has provided an account of a hybrid mixture explosion that occurred during railcar loading of a "fine chemical powder discharged from a dryer." The powder contained flammable solvent such that the MIE of the powder was reduced from about 325 mJ to 25–50 mJ. Ebadat determined that the powder probably became highly charged during transfer to the railcar, and that the powder in turn electrostatically charged a sanitary coating on the railcar walls. Since the sanitary coating could retain surface potentials greater than 10,000 volts, the energy of the subsequent electrostatic discharge most likely exceeded the hybrid mixture MIE. Flames emitted through the railcar's inspection port badly burned an employee near the opening.

3.7.3 *Explosion Propagation to Connected Equipment*

Many of the explosion incidents described in this chapter and in Chapter 5 resulted in flame propagation through interconnected process equipment. The resulting explosion damage extended far beyond the site of the originating explosion. The path for the explosion propagation is usually ducting used for pneumatic transport of particulate. In facilities such as grain elevators, the path is often enclosed or underground conveyor galleries/tunnels, usually leading to large, vulnerable silos.

Any approach to preventing explosion propagation needs to distinguish between the propagation of the ignition source and the propagation of the deflagration itself. If the ignition source is a smoldering nest or burning ember traveling through the ducting, properly designed/installed spark detection and extinguishing systems have been effective in preventing this scenario from escalating into a downstream deflagration. On the other hand, if the particulate loading in the ducting corresponds to a concentration above the MEC, a deflagration, rather than a mere smoldering ember/nest, can propagate through the duct. According to the FM Data Sheet 7-76 description of the wood-sander-initiated dust collector explosion incident, this occurred in a duct equipped with a spark detection and extinguishing system, and the deflagration overwhelmed the extinguishing system.

Isolation of interconnected equipment to prevent full deflagration propagation requires the more robust types of isolation systems described in Chapter 6 and in NFPA 69. Use of these systems also requires that the ducting be sufficiently strong to withstand the design flame speed and pressure associated with the isolation system certification.

3.7.4 *Secondary Dust Explosions in Buildings*

Perhaps the most devastating dust explosion scenario is the generation of a secondary dust explosion in the building surrounding the equipment in which some primary explosion occurs. The secondary explosion occurs when the blast wave emanating from the ruptured equipment/conveyor lifts the accumulated dust into suspension, and the flame from the primary explosion subsequently ignites the suspended dust cloud. The resulting devastation and casualties are associated both with the burning of building occupants and with the structural damage to the building.

One classic example of a devastating secondary dust explosion is the magnesium stearate explosion that occurred in 1976 in a plant manufacturing chewing gum. The magnesium stearate is applied as a lubricant on the freshly produced gum before it is cut into chewable pieces. The primary explosion occurred in the cutting machine several minutes after the machine started vibrating with sufficient intensity to generate a magnesium stearate combustible dust cloud. The rupture of the cutting machine generated a blast wave that displaced and lifted fugitive magnesium stearate powder from beams, ledges, and light fixtures, and the emerging fireball ignited the suspended cloud of powder. According to FM Data Sheet 7-76, the secondary dust explosion blew out windows on two sides of the building, demolished a cinder block wall about 3 m away from the cutting machine, and destroyed the equipment in the vicinity. The fireball and subsequent fires opened 166 sprinkler heads in the building. According to the New York City Fire Department account of this incident, 6 people were killed, and 50 other people suffered burn and blast wave injuries.

The extensive destruction and casualties that occurred in the January 29, 2003, explosion at the West Pharmaceuticals plant, as indicated by the aerial view photographs shown here as Figures 3-11 and 3-12 indicate that there was probably a secondary dust explosion. The Chemical Safety Board June 10, 2003 presentation on the West Pharmaceuticals explosion attributes the primary explosion to fugitive polyethylene powder from an air-blown dryer designed to evaporate water from an aqueous polyethylene solution applied to the rubber. The ignition source and precise location of the primary explosion has not been determined as of this writing. The CSB investigators determined that polyethylene powder accumulations on the upper surface of the ceiling tiles on a suspended ceiling were dispersed either by the primary

Figure 3-11 Aerial view of West Pharmaceuticals explosion residual fire. (Courtesy of WRAL.)

Figure 3-12 Aerial view of West Pharmaceuticals explosion destruction. (Courtesy of WRAL.)

explosion or by some unidentified disturbance, and that the polyethylene dust cloud burned in the space between the ceiling tiles and the concrete floor above, such that "a rapidly expanding chain of explosions moved through the ceiling space and literally tore the building apart" (transcript of June 2003 CSB Hearing in Kinston, NC).

A primary gas/vapor explosion can also initiate a secondary dust explosion. Zalosh (2000) described two such explosions that started in gas-fired equipment. The secondary explosions were due to accumulated coal dust in one case, and to accumulated phenol formaldehyde resin in the other case. According to the Chemical Safety Board, phenolic resin dust accumulations in the CTA Acoustics Corbin, Kentucky plant in February 2003 was also responsible for devastating secondary explosions following ignition of a dust cloud in an oven (CSB July 8, 2003 press release).

One issue that often arises in assessing the threat and potential consequences of a secondary dust explosion is whether it is feasible to design and install building deflagration vents for such an event. Usually, the initiating explosion will create a dust cloud that occupies only a small fraction of the building volume. A new method to determine the deflagration vent area needed to cope with such partial volume secondary dust explosion is described in the 2002 edition of NFPA 68. The new method was developed from the treatment of partial volume deflagrations in the Dust Calc computer program used by FM engineers in conjunction with Data Sheet 7-76.

3.8 DUST EXPLOSION DECISION TREES AND PROTECTION FLOW CHARTS

Figure 3-13 is a flow chart representing the approach offered by Eckhoff (1997) for evaluating potential dust explosion scenarios and determining suitable dust explosion protection measures. The top of the flow chart entails making a yes/no determination of whether the dust is explosible, and if so, which explosibility class it falls into. This is much more than a trivial determination because of the variations in particle size and the presence of additives and potential contaminants in many particulate processing/transport facilities. Eckhoff references an example in which fine calcium stearate particles are added to a coarse polypropylene powder. The nominal percentage of calcium stearate (<1%) is sufficiently small that would not significantly affect the inherent explosibility of the polypropylene, but segregating during processing and transport can create areas with a much greater explosion hazard. A similar situation occurred in the iron foundry explosion described by Zalosh (2000), because the phenol formaldehyde resin particles sometimes became airborne and separated from the sand with which it was mixed, and the resin accumulated in dust collection ducting, and on building surfaces.

The protection measures represented in Figure 3-13 include both preventive measures and damage mitigation measures. Preventive measures encompass limiting process and surface temperatures to eliminate one type of ignition source, possibly inerting process equipment, and emphasizing good housekeeping to reduce the chances of a devastating secondary dust explosion. Damage control measures in the flow chart include explosion containment, deflagration venting, explosion suppression, and deflagration isolation.

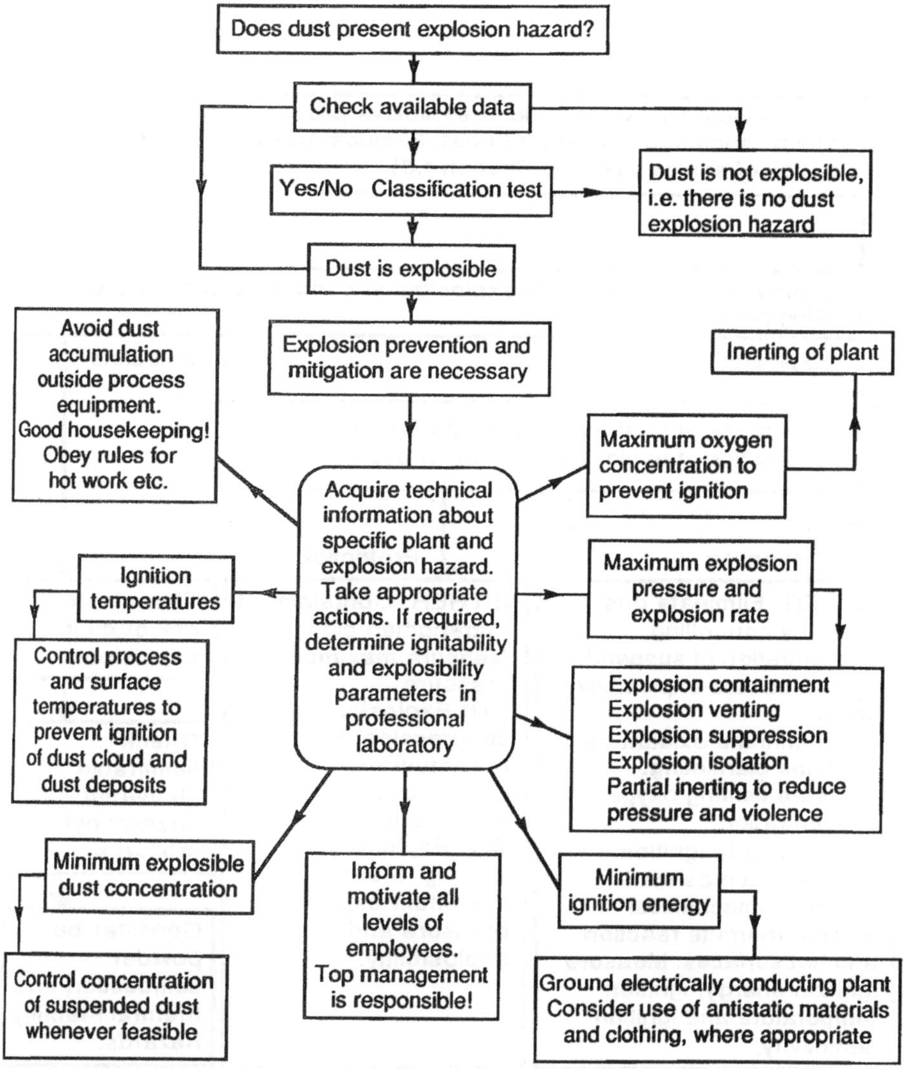

Figure 3-13 Eckhoff's approach to dust explosion protection. (From Eckhoff, R., *Dust Explosions in the Process Industries,* 3rd ed. Amsterdam: Elsevier, 2003.)

Figure 3-14 is a similar flow chart developed by Barton (2002). It begins with specific screening tests to establish any pyrophoric or explosive propensity of the dust, as well as its sensitivity to impact, friction, and heat. If the screening tests indicate the dust is explosible at conditions for which it is handled, Barton's flow chart offers several options for prevention and for

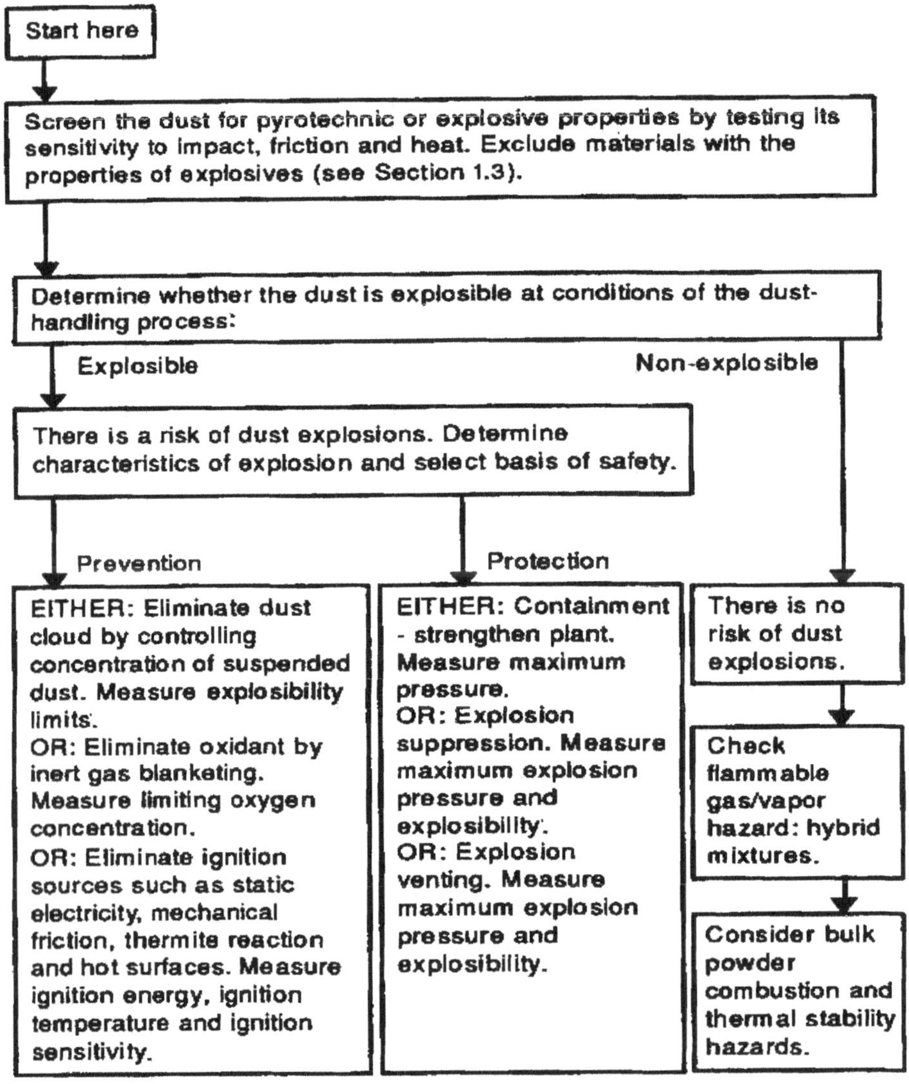

Figure 3-14 Flow chart for determining dust explosion protection. (From Barton, 2002.)

postignition protection. Barton provides a long list of ignition sources that need to be eliminated if that is the chosen prevention measure.

Figure 3-15 provides a comprehensive methodology recommended by one testing organization (Chilworth Technology Inc.) for determining the potential for electrostatic ignitions of combustible dust clouds. The evaluation begins with a dust explosibility screening test (described in Section 4.3.7) to determine a particular material's ability to produce a dust explosion

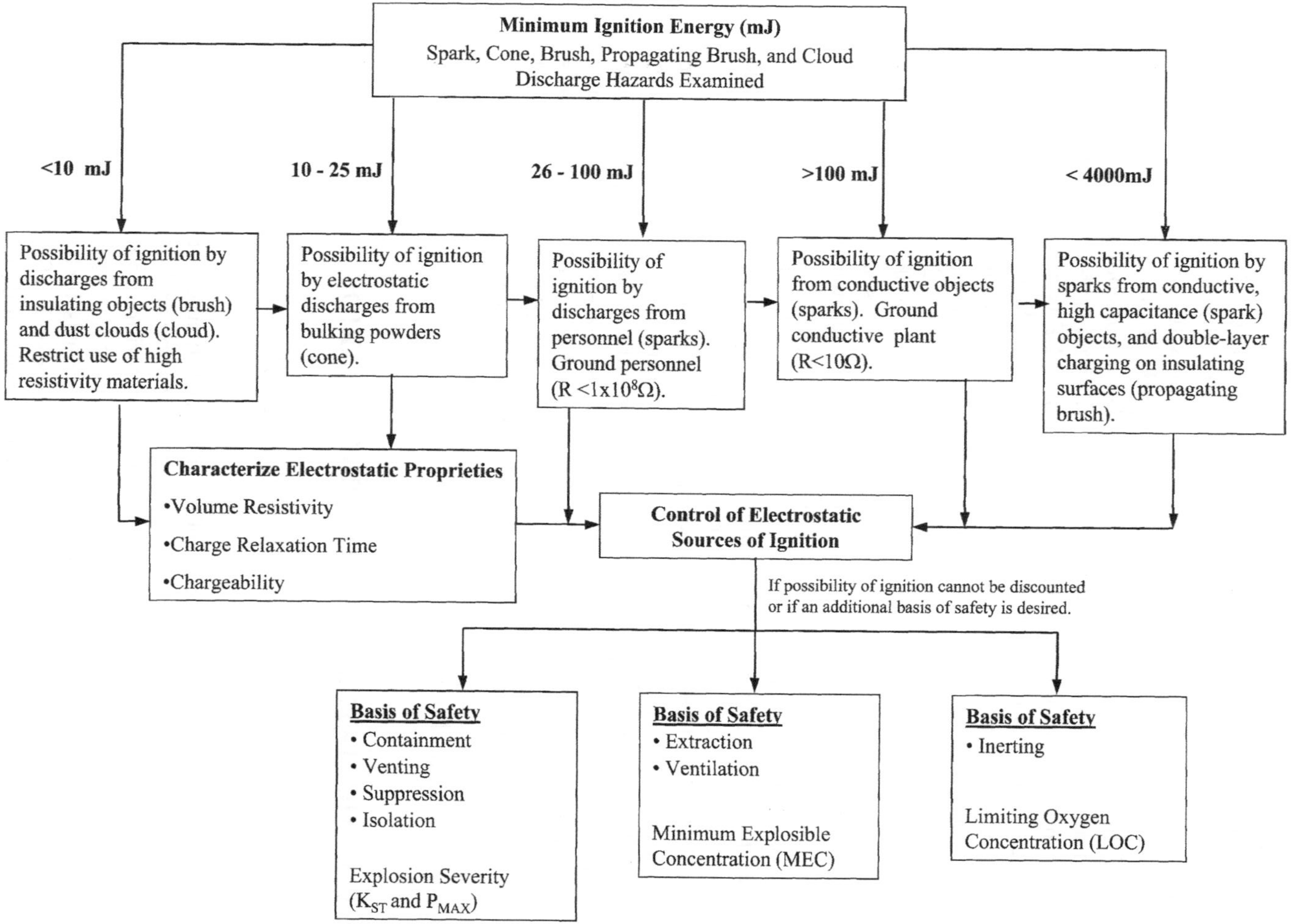

Figure 3-15 Evaluation of electrostatic hazards. **(**From Chilworth Technology Inc., with permission.)

when a concentrated suspension is produced in the presence of a strong ignition source. Assuming the material can produce a dust explosion, the next step is to determine its ignition sensitivity in terms of its MIE and minimum ignition temperature (MIT) for a dust layer (see Section 4.3.5) and for a dust cloud (see 4.3.7). According to Figure 3-15, if the MIE is in the range 100 to 1000 mJ, there is an ignition hazard due to sparks from conductive objects, and if the MIT is less than 500°C, there is also a possibility of ignition by heated surfaces and sparks caused by friction/impact of steel objects (such as in ball mills or misaligned hammermills). If the MIE is in the range 25–100 mJ, there is a further possibility of ignition due to electrostatic discharges from personnel, and therefore personnel should be grounded when handling or working near this material. If the MIE is less than 25 mJ, Figure 3-15 states that there is an additional possibility of ignition due to electrostatic discharge from insulating objects and from the surface of the bulk powder, and therefore additional electrostatic property testing is needed (see Section 4.3.6). Besides grounding, other explosion protection measures indicated in Figure 3-15 include deflagration containment, venting, and suppression (as discussed in Chapter 6), ventilation to maintain the dust concentration below the MEC, and inerting to maintain the oxygen concentration below the limiting oxygen concentration for dust cloud ignition (see Section 4.3.7).

CENELEC TR 50404 (2003) also has detailed recommendations on the assessment and control of electrostatic hazards during combustible powder and bulk solid handling. The CENELEC document says that no special measures are needed if the powder material's MIE is greater than 10 J and there are no flammable vapors present. Recommended precautions for powders with lower MIE values and for hybrid mixtures depend on both the MIE and volume resistivity values and the resistivity of the particulate container. Detailed guidance is provided for FIBCs. Similar guidance is summarized in Chapter 5 of this book.

3.9 TOXIC MATERIAL EXPOSURE SCENARIOS

3.9.1 Chronic Exposure Scenarios during Processing and Material Handling

Chronic exposure to toxic particulates can occur during opening and filling of normally closed containers and during processing and transport in either open systems or in closed systems that have deteriorated to the extent that there are frequent releases of particulates. Container opening and filling operations that may pose chronic exposure hazards include bag opening (see Section 5.3.1), portable container filling described in Sections 5.3.10 and 5.3.12, material sampling (see Section 5.3.13), silo and hopper filling and emptying (5.3.15), and railcar and hopper truck filling and emptying (see

5.4). Exposures during processing in open systems can include spray painting and powder coating operations. Open system exposures can also entail processing of materials with bound particulates that are released during operations such as cutting or trimming. If these particulate releasing operations proceed without suitable worker personal protective equipment, serious or fatal illnesses can develop, sometimes after long latency periods (Levin et al., 1998).

Toxic particulate chronic exposures can also occur as a result of the material degrading over an extended time period. For example, some of the asbestos exposures at the Tyler Pipe Industries of Texas plant were allegedly due to long-term degradation of asbestos thermal insulation (http://www.riskworld.com/PressRel/2000/00q3/PR00a064.htm).

Exposures due to closed system deterioration can occur from a variety of causes including: severe erosion or corrosion, tearing of filters and gaskets, operating the equipment beyond its intended limits of pressure, temperature, or vibration, or operating the equipment well beyond its normal life expectancy.

Repetitive exposures can also occur during maintenance, repair, and cleanup activities. One example of repetitive exposure during cleanup is the following account from the OSHA database (Report ID 0522300, 07/11/1991).

The baghouse for a silica dust collector system was dumping directly on the floor inside of the shop. The employee had to shovel and sweep the accumulated silica sand into a hopper and then dump it into an outside dumpster. He did this while wearing a nonapproved disposable dust mask. The employee died of severe, accelerated pulmonary silicosis. The cleanup process was done on a regular basis. An air hose was also used to blow silica dust off parts and off the employee's clothing. The company did not perform any medical or air monitoring and the personal protective equipment was inadequate. This is an example of a combination of an unnecessarily hazardous operation (repeated shoveling, sweeping, and blowing of silica) together with use of ineffective personal protective gear.

There have been numerous claims of debilitating and sometime fatal illnesses due to repeated exposures to asbestos during facility repair and renovation; these facilities include chemical processing facilities. Various heavy metals, insecticides, crystalline silica, biological agents, and radioactive material particulates have also been responsible for fatalities due to prolonged repeated exposures. Equipment and procedures to contain and control the release of these and other toxic particulates are discussed in Section 6.11 of this book.

3.9.2 *Acute Exposure Accident Scenarios*

Figure 3-16 is a fault tree diagram illustrating the combinations of events that can lead to acute inhalation of toxic particulates. The acute inhalation

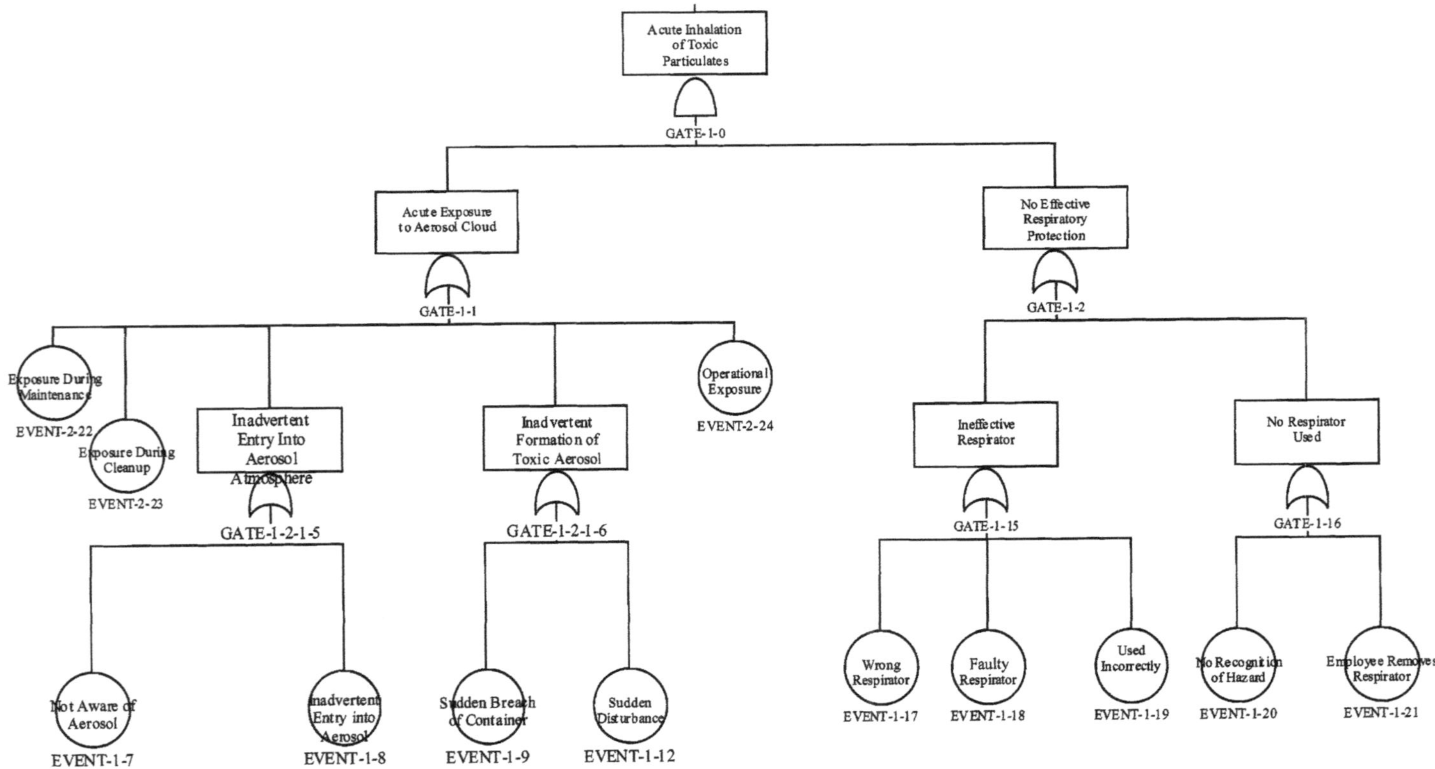

Figure 3-16 Acute inhalation fault tree.

requires both an acute exposure incident and the absence of effective respiratory protection. Acute exposure incidents can be either inadvertent or during routine operations involving personnel unaware of the particulate material's health hazards. One extreme example of dangerous particulate exposures to uninformed workers is the following description of one facility's pipe insulation manufacturing operations.

> Unibestos pipecovering is manufactured in the following fashion: the amosite asbestos bag is slit open and the contents are dumped into a machine that "fluffs" the fiber, separating all the clumps and bundles of fiber so that the resultant material is, well, fluffy. The asbestos is dry-mixed with other dry ingredients, such as diatomaceous earth, and a machine spreads the mixture out evenly on a conveyor belt. As it went down the conveyor belt, it was sprayed by overhead nozzles with liquid sodium silicate, the binder that holds everything together. As it went further down the line it was rolled up on a mandrel. A mandrel is a cylindrical male mold that functions somewhat akin to a rolling pin. After the material was rolled up on the mandrel to the desired thickness the mandrel was removed and the piece—now a hollow tube a little longer than three feet—went onto another conveyor through an oven for drying. Once dry, the ends of the product were trimmed off with a bandsaw and then the tube was sawed down the middle to make two halves for pipecovering." (http://www.mesothelioma.net/PittsburghCorning.html)

Perhaps a less extreme example is the unprotected exposure of workers doing manual powder coating using some type of spray application in an unenclosed area. Polyester powders containing triglycidyl isocyanurate (TGIC) as a cross-linking agent, are often applied via an electrostatic spray. Since TGIC can cause severe eye and skin irritations and genetic changes to sperm, precautions are needed to prevent exposures (http://www.hse.gov.uk/pubns/eis15.pdf).

The acute inadvertent exposure might occur from the sudden breach of a container or bin, such as occurred in the following incidents from the OHSA accident database.

> **Accident 014442271 — Report ID: 0355122 — Event Date: 11/16/1984**
> At 5 p.m.on 11/16/84 an elevated container full of dry materials used in the manufacturing of plastic containers was being dumped into a larger vat used in mixing-blending the compound. The elevated container slipped away from the edge of the vat spilling approximately 80% of the contents to the floor below; the result was 5 exposures to the generated dust. Symptoms were dizziness and throat–upper respiratory tract irritation. Employee #1 was the primary exposure. Employees #2– #5 were exposed as a result of the clean-up operation. All were hospitalized (between 8:00 p.m. and 1:30 a.m.) over night for observation and released.

> **Accident 014518161 — Report ID: 0419700 — Event Date: 05/12/1986**
> A drum of *p*-nitroaniline was accidentally struck by the tine of a forklift truck. The ensuing spill was not cleaned up; therefore, as forklift trucks

would pass through the spilled material, the substance became airborne. Subsequently, 13 employees were exposed to the p-nitroaniline and were taken to a hospital for treatment. All but one was treated and released. Employee #1 remained in the hospital for 1½ days for observation. He complained of weakness and experienced frequent urination.

Acute inhalation accidents during maintenance exposure activities often result from either a lack of hazard recognition (Event 1-20 on the fault tree in Figure 3.9-1) or the use of the wrong respirator (Event 1-17). The following two summaries from the OSHA database illustrate these scenarios.

Accident 014198956 — Report ID: 0352450 — Event Date: 05/24/1985
Employee #1 helped in the removal and replacement of an evaporator used in the manufacture of sodium bichromate. He was provided, but not required to wear, a half-face dust respirator. From two days' exposure the employee received chrome sores on his shoulder, chest, and face, an ulcerated nasal cavity, and lung damage. The employer provided no training and education as to what he would be exposed to or why the respirator was provided. The employer had no respirator fit testing program.

Accident 014534168 — Report ID: 0931300 — Event Date: 07/18/1990
Employee #1 was changing the polishing compound in a bowling ball cleaner. When he started the machine, the compound block jumped off the axle center, releasing a cloud of dust from the bowling ball compound. Soon thereafter, Employee #1, who was a chronic asthmatic, suffered an acute attack. During transport to the hospital, he suffered heart failure and subsequent brain damage and pneumonia. Employee #1 died 24 days later.

Another type of inadvertent acute exposure has occurred during aerial spraying of insecticides and other particulate materials. Farm workers working in nearby fields are particularly vulnerable to this type of exposure.

3.9.3 *Fire and Explosion Exposure Scenarios*

Exposures to toxic particulates can occur during fires and explosions. The toxic particulate can either be a preexisting material that is dispersed by the fire/explosion, or it can be generated by the combustion event. Fires causing the dispersal of preexisting toxics have occurred in several pesticide manufacturing and storage facilities, as described by Diefenbach (1982). Toxic plumes from some of these fires have required evacuations in the surrounding community.

Fire plumes from the burning of nominally nontoxic organic chemicals may or may not contain toxic particulates depending on the particular chemicals and the completeness of combustion, but the emergency responders and the media often treat the plume as toxic and advise the public accordingly. The following excerpts from Chemical Safety Board accounts of recent fires/explosions provide examples.

CSB Incident Number2002-5792
A massive fire at a plastics recycling building in Alexandra (Prince Edward Island, Canada) sent choking, black smoke across a large portion of eastern Queens County Monday (7/15/2002) night. Island Plastics manufactures and distributes finished lumber and other value-added products made from recycled waste plastic. The building on fire was filled with plastics, which prompted fire officials to fear that the smoke being produced was toxic. They started evacuating houses in the area and put out a warning to homeowners as far north as Mermaid and Mount Herbert to leave their homes to escape the smoke. By late Monday evening no one had been injured in the fire, but six residents of the rural community of Tea Hill had left their homes. Officials from the province's Environment Department were also at the factory investigating the possibility that toxic fumes may have been created by the burning plastics in the warehouses. But by late Monday night a spokesperson for the department was advising the public the fumes weren't toxic and no large-scale evacuation would be necessary.

CSB Incident Number 2002-5727
More than 2,000 residents of Kingsville (Ontario, Canada) were allowed to return to their homes yesterday after a stubborn fire in a plastics plant was brought under control. The residents were forced to abandon their homes Saturday afternoon when fire officials moved in on a huge plastics fire at the Horti-Pak Inc. industrial plant that had been burning since Thursday (06/20/2002) night. One family told of sore throats and breathing problems prior to leaving. And they saw the jet black clouds and thick black smoke near their home. Area medical officer of health Dr. Allen Heimann said there would be no "significant health risk" as a result of three days of billowing toxic smoke that placed Kingsville under a black cloud throughout a heat wave and provincial smog advisory. While assuring the public that the ash and soot particles covering large areas of Kingsville "does not pose a significant health risk," Dr. Allen Heimann urged local residents not to eat from their gardens until the province has the results of vegetation testing. The Ministry of the Environment is also conducting tests of water samples from Lake Erie. As of Sunday afternoon, he said no injuries associated with either fighting the fire or from the smoke and gases it produced had been reported. Heimann said "a strong plastic smell" will continue in some areas for the next few days but described it as more of an irritant than a health risk.

CSB Incident Number 2000-4991
Black plumes of toxic smoke rose into the sky above Richmond (California) on October 26, 2000, forcing 12 schools to shut down, businesses to evacuate and shelter-in-place warnings for residents and workers. One man was killed in the 2 a.m. explosion at MBA Polymers, a 90,000-square-foot plastics recycling facility. More than 200 people, including factory workers and firefighters, crowded area hospitals with complaints of irritated throats and

eyes, headaches and other ailments. Investigators for the California Division of Occupational Safety and Health said that the explosion and fire that claimed one life was related to an accumulation of toner-cartridge dust. They said an electrostatic charge in a grinder at the company may have ignited the explosive toner dust used in copy machines, causing the fire that killed a 26-year-old fork lift operator. Cal/OSHA fined the company more than $221,000 for, among other items, failing to prevent the dust from accumulating and failing to warn the employees of its fire hazards.

9.3.4 Incident Cleanup Exposure Scenarios

Sometimes a major spill or breach does not cause a direct exposure to toxic particulates, but the exposure occurs later during cleanup without adequate precautions and personal protective equipment. This is illustrated in the following OSHA accident investigation summary.

Accident 000570010 – Report ID: 0522000 – Event Date: 02/17/1986
Two bags of sodium meta-bisulfite broke open in the back of a truck. The five employees assigned to clean up the spill inhaled the dust and became ill. The fire department was called. A rescue unit transported the five employees to the hospital, where they were treated and released.

Another type of cleanup toxic exposure can occur during the changing of filter bags or cartridges in clogged dust collectors. This has occurred with heavy metals and with polymer particulates such as TGIC.

REFERENCES

Abbot, J, 1990, *Prevention of Fires and Explosions in Dryers—A User Guide,* 2nd ed., Rugby, U.K.: The Institution of Chemical Engineers.

Barton, J., 2002, *Dust Explosion Prevention and Protection: A Practical Guide,* Rugby, U.K.: The Institution of Chemical Engineers.

Beever, P., 1982, "Characteristics of spontaneous fires," *Journal of Occupational Accidents,* 3: 299–305.

Beever, P., 1995, "Self-heating and spontaneous combustion," in *SFPE Handbook of Fire Protection Engineering,* 2nd ed., Section 2, Chap 12, Quincy, MA: National Fire Protection Association.

Black, E., 1981/82, "autoignition in hygroscopic, organic materials—especially forest products—as initiated by moisture absorption from the ambient atmosphere," *Fire Safety Journal,* 4: 185–196.

Block D., and Rynk, R., 2000, "Fire destroys hartford in-vessel composting facilities," *Bio Cycle,* 41: 51–52.

Bou-Diab, L. and Fierz, H., 2002, "Autocatalytic decomposition reactions, hazards, and detection," *Journal of Hazardous Materials,* 93: 137–146.

Bowes, P. 1984, *Self-Heating: Evaluating and Controlling the Hazards,* Amsterdam: Elsevier.

CCPS, 1995a, *Guidelines for Chemical Reactivity Evaluation and Application to Process Design,* New York: AIChE/Center for Chemical Process Safety.

CCPS, 1995b, *Guidelines for Safe Storage and Handling of Reactive Materials,* New York: AIChE/Center for Chemical Process Safety.

Cardillo, P., and Anthony, E., 1978, "The flammability limits of hybrid gas and dust systems," *La Revista dei Combustibiliti,* 32: 390–395.

CENELEC, 2003, "Electrostatics—Code of practice for the avoidance of hazards due to static electricity," CLC/prTR 50404, January.

CRHF, 2003. Chemical Reactivity Hazards Forum, http://www.crhf.org.uk/

Diefenbach, R. E. 1982, "Pesticide fires," in G. Bennett, F. Feates, and I. Wilder, eds., *Hazardous Materials Spills Handbook,* New York: McGraw Hill.

DOE, 1994, *Primer on Spontaneous Heating and Pyrophoricity,* DOE-HDBK-1081-94, Washington, DC: Department of Energy.

Ebadat, V., 1999. "Investigation: how dust exploded during railcar loading," *Powder and Bulk Engineering.*

Eckhoff, R., 1997, *Dust Explosions in the Process Industries,* 2nd ed. Burlington, MA: Butterworth-Heinemann.

Eckhoff, R., 2003, *Dust Explosions in the Process Industries,* 3rd ed. Amsterdam: Elsevier.

EPA/OSHA, 1997, Joint Chemical Accident Investigation Report, Napp Technologies, Lodi, New Jersey. October. www.osha.gov

EPA/OSHA, 1999, Joint Chemical Accident Investigation Report, BPS, Inc. West Helena, Arkansas. http://yosemite.epa.gov/oswer/ceppoweb.nsf/vwResourcesByFilename/bps_full.pdf/$File/bps_full.pdf

EPA/CEPP,"Explosion Hazard from Ammonium Nitrate," 1997. EPA 550-F-97-002d. http://yosemite.epa.gov/oswer/ceppoweb.nsf/vwResourcesByFilename/ammonitr.pdf/$file/ammonitr.pdf?OpenElement

Factory Mutual Loss Prevention Data Sheet 7-76, 1996. "Prevention and Mitigation of Combustible Dust Explosions and Fires."Johnston R.I.: FM Global.

Fierz, H., Dien, J.-M., Stoessel, F., and Kille, G., 1994. "The thermal risk of autocatalytic decompositions: a kinetic study," *CHIMIA,* 48(12): 542–550.

Frurip, D., Hofelich, T., Leggett, D., Kurland, J., and Niemeir, J., 1997. "A Review of Chemical Compatibility Issues," *Proceedings AIChE Loss Prevention Symposium.* New York American Institute of Chemical Engineers.

Gray, B., 2002. "Spontaneous combustion and self-heating," *SFPE Handbook of Fire Protection Engineering,* 3rd ed, pp. 2-211 – 2-228, Quincy, MA: Society of Fire Protection Engineers and National Fire Protection Association.

Guiochon, G. 2002. "What might and might not have happened in Toulouse (France) on September 21, 2001," http://www.sfc.fr:8000/index.html.

Haberman, J., and Castorina, T., 1972. "Effect of adsorbates on time of thermal initiation of lead azide," *Thermochimica Acta,* 5: 153–164.

Heather, D., 2002, "A Review of Past Ammonium Nitrate Accidents and Lessons Learned," Paper presented at the European Commission Joint Research Centre Workshop on Ammonium Nitrate.

Isner, M., 1995. "Persulfate Warehouse Fire," NFPA Fire Investigation Report.Quincy, MA: National Fire Protection Association.

Johnson, R., Rudy, S., and Unwin, S. 2003, *Essential Practices for Managing Chemical Reactivity Hazards,* New York: AIChE/Center for Chemical Process Safety.

Keller, A, Stark, D., Fierz, H., Heinzle, E. & Hungerbuehler, K. Jan. 1997, "Estimation of the time to maximum rate using dynamic DSC experiments," *Journal of Loss Prevention in the Process Industries,* 10(1): 31–41.

Kersten, R., van der Steen, A.C., Creemers, A., and Opschoor, G., "The Ammonium Nitrate Explosion in Toulouse France—The Incident and Its Consequences for Industrial Activities." Paper presented at Center for Chemical Process Safety 17th Annual International Conference and Workshop, 2002. New York: AIChE/Center for chemical Process Safety

Kotoyori, T. 1989. "Critical ignition temperatures of chemical substances," *Journal of Loss Prevention in the Process Industries,* 2: 16–21.

Kotoyori, T., 1999. "The self-accelerating decomposition temperature (SADT) of solids of the quasi-autocatalytic decomposition type," *Journal of Hazardous Materials,* 64: 1–19.

Kotoyori, T., 2003. "Critical Teperatures for the Thermal Explosion of Chemicals", Draft manuscript.

Kyte, G., 1986. "Aerlex Fireworks plant explosion, Jennings, Oklahoma, June 25, 1985" Summary Fire Investigations Report, NFPA. Quincy, MA: National Fire Protection Association.

Levin, J. L., McLarty, J. W., Hurst, G. A., Smith, A. N., Frank, A. L., 1998. "Tyler asbestos workers: mortality experience in a cohort exposed to amosite," *Occupational and Environmental Medicine,* 55: 155–160.

Matsuda, T., and Yamaguma, M. 2000. "Tantalum dust deflagration in a bag filter," *Journal of Hazardous Materials,* A77: 33–42.

Markowski, A. and Mujumdar, A., 1995. "Safety aspects of industrial dryers," in *Handbook of Industrial Drying,* 2nd ed., New York: Marcel Dekker.

Matsuzawa, T., Wada, Y., Yabashi, H., Kaneko, Y., Ikeda, Y., Tamura, M., and Yoshida, T., 1991. "Evaluation of fire and explosion hazards of sodium azide according to flow chart," *Explosion and Explosives* (English translation of *Kogyo Kayaku*), 52(3): 200–205

Mniszewski, K., 1994. "The Pepcon plant fire/explosion a rare opportunity in fire/explosion investigation," *Journal of Fire Protection Engineering,* 6: 63–78.

Mohan, K., and Tang, T., 1983. "Slow and explosive decomposition of thermally unstable solids: a study with time-resolved mass spectrometry," *Journal of Chemical Physics,* 79: 4271–4278.

NFPA 13, 2000 Edition. "Installation of Sprinkler Systems,"Quincy, MA: National Fire Protection Association.

NFPA 68, 2002 Edition, "Guide to Deflagration Venting," Quincy, MA: National Fire Protection Association.

NFPA 430, 2000 Edition. "Code for the Storage of Liquid and Solid Oxidizers," Quincy, MA: National Fire Protection Association.

NFPA 490, 1998. "Code for the Storage of Ammonium Nitrate," National Fire Protection Association.

NFPA 654 – 2000 Edition. "Standard for the Prevention of Fire and Dust Explosions from the Manufacturing, Processing, and Handling of Combustible Particulate Solids,"Quincy, MA: National Fire Protection Association.

NFPA 704, 2000. "Identification of the Hazards of Materials for Emergency Response," Quincy, MA: National Fire Protection Association.

NFPA 850, 1996. "Recommended Practice for Fire Protection for Electric Generating Plants and High Voltage Direct Current Converter Stations," Quincy, MA: National Fire Protection Association.

NOAA 2002. Chemical Reactivity Worksheet, Version 1.5. U.S. National Oceanic and Atmospheric Administration. [http://response.restoration.noaa.gov/chemaids/react.html].

Pegg, M., Amyotte, P., Lightfoot, P., and Lee, M., 1997. "Dust explosibility characteristics of azide-based gas generants," *Journal of Loss Prevention in the Process Industries,* 10: 101–111.

Primer on Spontaneous Heating and Pyrophoricity, 1994. DOE-HDBK-1081-94, Washington, DC: U.S. Department of Energy.

Rockett, J., 2002. "The 1998 Ammonium Perchlorate Explosion in Henderson, NV," Howard Emmons Distinguished Lecture Summary, WPI.

Rodgers, S., 2004. Personal communication.

Tuomisaari, M., Baroudi, D., and Latva, R., 1998. "Extinguishing Smouldering Fires in Silos," Brandforsk Project 745-961, VTT Publication 339, Technical Research Centre of Finland.

Tyldesley, A., 2004. Personal communication.

UN, 1999. "Recommendations on the Transport of Dangerous Goods, Manual of Tests and Criteria," Third revised edition, New York and Geneva: United Nations.

Zalosh, R., 1986. "Review of Coal Pulverizer Fire and Explosion Incidents," Proceedings of the ASTM/NFPA Symposium on Industrial Dust Explosions, available from ASTM.

Zalosh, R., 2000. "A Tale of Two Explosions," Proceedings 34th AIChE Loss Prevention Symposium.

Zalosh, R. 2003. *Industrial Fire Protection Engineering,* New York: John Wiley & Sons.

Chapter 4

ASSESSING PARTICULATE HAZARDS

The first step in assessing the hazards of particulate materials is to review the information for those materials in Material Safety Data Sheets, handbooks, industry and consensus guidelines, and various applicable codes and standards. In some cases the review yields pertinent property and hazards test data that are directly applicable to the material in question. These data, when combined with the information on processing, handling, and storage hazards and protection (as described, for example, in Chapters 5 and 6 of this book) may be sufficient to determine appropriate protection measures. In many other cases, there will be a need to acquire material/plant-specific laboratory test data. In fact, several companies insist on acquiring such data, irrespective of the results of the preliminary review. This chapter explains how this two-step hazard assessment can be conducted for each generic hazard: instability, chemical reactivity, combustibility/explosibility, and toxicity. It also offers some guidance on the possible use of large-scale testing and/or theoretical analysis to assess hazards for which laboratory test data may not suffice or for which the existing guidelines, codes, and standards may not be suitable.

4.1 PRELIMINARY ASSESSMENTS VIA MATERIAL SAFETY DATA SHEETS, HANDBOOKS, GUIDELINES, CODES, AND STANDARDS

4.1.1 Preliminary Assessment of Instability Hazards

NFPA 704 Instability Ratings

NFPA 704 (2001 Edition) defines criteria to be used in assigning an Instability Rating for all materials in the context of emergency response to a fire exposure. The criteria refer to a material's susceptibility to release energy through either decomposition or polymerization. The ratings range from zero to four, depending on the temperature and pressure required to trigger the instability and the rate of energy release. Materials that are normally stable, but can become unstable under fire exposure conditions would probably have a rating of 1 or 2, depending on the energy release rate. Materials that can detonate during decomposition or polymerization would have an instability rating of 3 or 4 depending on the detonation initiation requirements. The instability rating is placed in the right diamond or right portion of the diamond quadrilateral in the NFPA 704 hazards identification scheme, as shown in Figure 4-1.

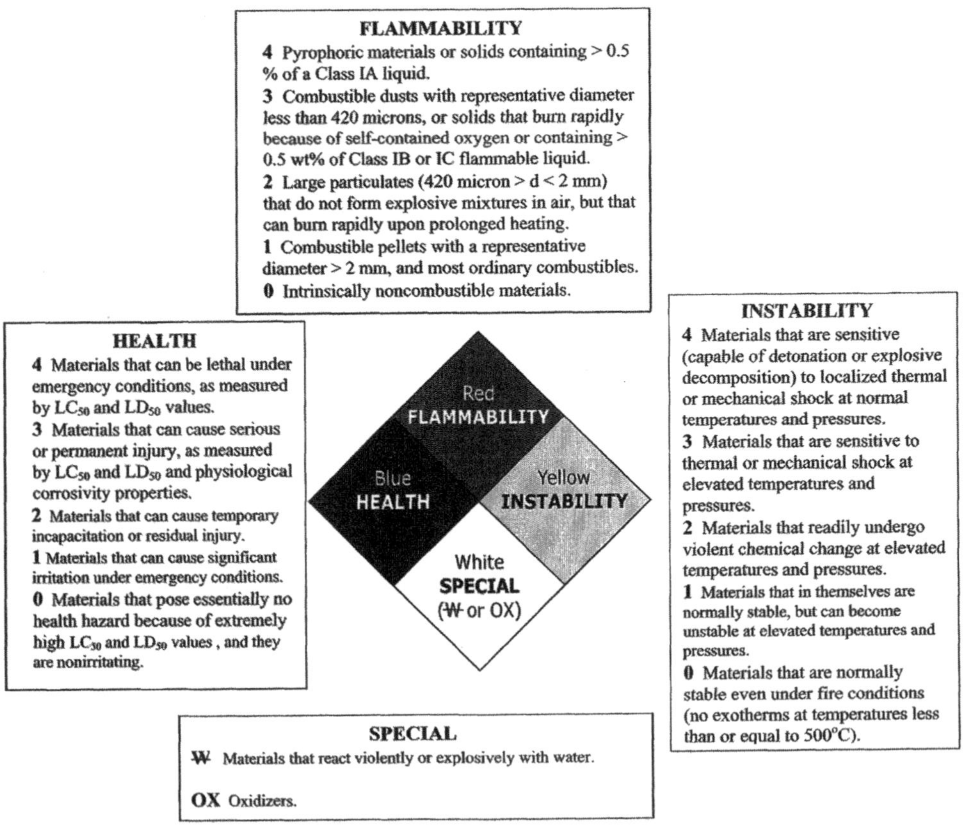

Figure 4-1 NFPA 704 Diamond for Solid Particulate Material Hazard Communication, the diamond is Copyright © 1991, National Fire Protection Association. This warning system is intended to be interpreted and applied only by properly trained individuals to identify fire, health, and stability hazards of chemicals. See NFPA 704.

UN Hazardous Materials Instability Designations

The United Nations Recommendations for the Transport of Dangerous Goods (U.N., 1999a,b) includes a category Division 4.1 that is pertinent to potentially unstable materials. One of the criteria for a material to fall into the Class 4.1 category is that it is self-reactive and likely to undergo a strongly exothermic reaction. The quantitative criteria are a heat of decomposition of at least 300 J/g and a self-accelerating decomposition temperature (SADT) of 75°C or less in a 50 kg package. Within this category of self-reactives, there are seven subcategories designated as type A through type G, with the designation depending on the packaging as well as the material. Type A self-reactors are not accepted for transport as packaged because they can detonate. Types B through D have deflagrative rather than detonative self-reactions, and have limitations on the amount of material that can be placed in one package. Types E and F neither detonate nor deflagrate under laboratory testing, with the distinction between them depending on their heat release rate. Type F self-reactive materials can be shipped in intermediate bulk containers.

Materials that are prone to spontaneous heating under normal transport conditions are categorized in the U.N. scheme as Division 4.2 materials. The laboratory test to determine whether a particulate material should be classified as a Division 4.2 substance is described in Section 4.3.4. Oxidizing materials are designated as U.N. Division 5.1 materials unless they are organic peroxides, which are designated as Division 5.2 materials.

Material Safety Data Sheets

Section 10 of the ACC/ANSI (1998) standard format Material Safety Data Sheet (MSDS) is entitled Stability and Reactivity. It should state clearly whether there is an instability hazard and, if so, should provide some guidance on instability onset conditions. For example, one company's MSDS for an organic peroxide powder product states that this material is chemically unstable and refers the reader to the Handling and Storage Section of the MSDS for guidance on the maximum safe storage temperature of 38°C (100°F). Section 9 (Physical and Chemical Properties) lists the SADT as being greater than 50°C. Section 10 of the MSDS should also contain a brief listing of hazardous decomposition products.

Products of non–U.S.-based companies often do not follow the ACC/ANSI format and are less likely to have NFPA 704 Hazardous Material Information System ratings than products of North American companies. On the other hand, non-U.S. companies are more likely to include the U.N. hazard class on the MSDS. For example, the organic peroxide powder MSDS cited previously included a U.N. Hazard Class designation of 5.2.

Section 4.1.5 of this book describes important precautions needed when using generic MSDS data for instability and other particulate material hazards.

Theoretical Considerations: Chemical Structure and Thermochemical Equilibrium Calculations

If the chemical structure and thermodynamic properties of a particulate material are known, several different theoretical considerations may be used to provide a preliminary indication of a material's tendency to be thermochemically unstable. One simple criterion used in the ASTM CHETAH program (ASTM, 2002) is the oxygen balance. The oxygen balance is the difference between the available oxidizing atoms and the available reducing atoms in a particular chemical compound. Conceptually, a near-zero oxygen balance suggests that the compound is potentially self-reactive because it contains just about enough oxygen to oxidize the reducing atoms without resorting to an externally supplied oxidizer. Conversely, a large negative calculated oxygen balance suggests a deficiency of oxidizing atoms, and a large positive calculated balance suggests the compound has an excess of oxidizing atoms. The CCPS (1995a) *Guidelines for Safe Storage and Handling of Reactive Materials* cautions that there are several important exceptions and limitations to the use of an oxygen balance as an indication of self-reactivity, and notes that many industrial high explosives have a significantly large negative oxygen balance. One particularly important limitation of the oxygen balance calculation is that it does not account for the type of oxygen bond in the molecule.

Thermochemical equilibrium calculations in CHETAH and other computer programs can provide the heat of reaction for a compound self-reacting to produce its most stable reaction products. The CCPS 1995a and 1995b guidelines and the paper by Murphy et al. (2003) contain a summary of the thermochemical principles and computational techniques. The CHETAH code offers a second instability criterion based on the heat of reaction being greater than 3.0 kJ/g, that is, a factor of 10 larger than the U.N. heat of decomposition criterion for potential classification (subject to SADT data) as a Division 4.1 self-reactive material (CCPS,1995a, p. 76). The CHETAH code uses another criterion, called the "plosive density" (CCPS, 1995a, p. 82), which accounts for chemical groups that contribute to self-reactivity.

Murphy et al. (2003) have proposed that the thermochemical heat of reaction should be supplemented by a calculation of the adiabatic reaction temperature (CART) using a method based on the NASA thermochemical equilibrium computer code. They provide CART values for various compounds in five different chemical groups in order to suggest benchmark CART values that denote potential instability for each group. Based on empirical classifications to distinguish compounds that are nonexplosive from those that are capable of being explosive when unconfined, the CART values reported by Murphy et al. indicate that the CART explosive thresholds apply to the following chemical groups:

- Nitro compounds and nitrates = 1737 ± 14°K
- Other nitrogen-bearing compounds = 1196 ± 142°K
- Organic Peroxides = 900 ± 110°K.

The large differences in threshold adiabatic reaction temperatures between these and other groups suggest that the primary value of this criterion is for chemicals that are closely related in chemical structure to other materials for which there is laboratory test data and experience on chemical instability tendencies.

4.1.2 Preliminary Assessments of Reactivity Hazards

Material Safety Data Sheets

Section 10, Stability and Reactivity, of an ACC/ANSI (1998) formatted MSDS should list incompatible materials that pose a reactivity hazard with the subject material. In the case of a combustible particulate MSDS, there is usually a statement in Section 10 stating that the subject material can react dangerously with strong oxidizing agents, such as chlorates, nitrates, peroxides, etc.

Besides reacting themselves, potential incompatibles include chemicals that can trigger a violent decomposition or polymerization reaction. For example, in the case of the previously mentioned MSDS for a powdered organic peroxide, the MSDS includes the admonition that strong acids, bases, oxidizers, amines, reducing agents, and other potential promoters/accelerators can cause a violent decomposition reaction.

If a material is water reactive, it should be so indicated in MSDS Sections 10 and 5 (Fire Fighting Measures). It should also be denoted in MSDS Section 3 (Hazards Identification) and on the NFPA 704 placard system for identifying hazards of materials by the symbol ₩ on the bottom of the placard.

Incompatibility Charts

Chemical incompatibility charts can provide a preliminary indication of potential reactivity hazards associated with binary combinations of chemicals or chemical families. Some examples are presented in Section 3.4 of this book; other examples can be found in the references cited in that section. The NOAA/EPA Chemical Reactivity Worksheet (NOAA, 2002) is a particularly useful and convenient software tool for the preparation of material specific incompatibility charts.

Reactivity Listings in NFPA Standards and in Other References

NFPA 430 provides a reactivity categorization scheme for oxidizers. Four classes of oxidizers are defined, with the hazard increasing in the progression from Class 1 through Class 4. Classes 1 and 2 are distinguished by the relative increase in the burning rate of a combustible material that a particular oxidizer will cause. Class 3 oxidizers are those that either cause "a severe increase in the burning rate of combustible materials with which [they] come in contact or that will undergo vigorous self-sustained decomposition due to contamination or exposure to heat." A Class 4 oxidizer is one that "can undergo an explosive reaction due to contamination." Although the nature of the contamination is not described in the standard, a combustible particulate material would seem like a logical candidate. The NFPA Committee has provided classification listings of particular oxidizers in the Appendix of NFPA 430.

NFPA 432 provides a classification scheme for organic peroxides based on their anticipated reactions to fire exposure. Reactivity hazard is a key factor in the classifications. For example, Class II organic peroxides present a severe reactivity hazard, whereas Class III organic peroxides present a moderate reactivity hazard, and Class IV materials present a minimal reactivity hazard. Class V peroxides present no reactivity hazard in the NFPA 432 classification. The Appendix of NFPA 432 has a two-page listing of the classifications of specific organic peroxides, as well as tabulations of typical organic peroxides in the various categories. The concentration of the peroxide and the diluent material are included in the tabulations, and several concentrations/diluents are listed for some peroxides such as dibenzoyl peroxide, diisopropyl peroxydicarbonate, and methyl ethyl ketone peroxide.

Perhaps the most widely utilized and comprehensive handbook for preliminary evaluations of chemical reactivity hazards is *Bretherick's Handbook* (Urben, 1999). Other very useful references for this purpose include *Sax's Handbook* (Lewis and Irving, 2000), Grewer (1994), Pohanish and Green (2003), and the CCPS *Guidelines* (1995, 2003) on reactivity hazard evaluations. Although less comprehensive, recent papers by Frurip et al. (1997) and Leggett (2002) provide excellent guidance on good current practices being followed by organizations experienced in this type of hazard evaluation. In the specific case of water-reactive and pyrophoric materials, the Gibson and Weber (1969) handbook contains property data for about 425 such materials.

Theoretical Considerations

Combinations of chemical compounds with known thermochemical properties are amenable to calculations of heat of reaction and of adiabatic reaction temperature as described previously for potentially self-reacting chemicals. Conceptually, it is possible to use these calculated values to provide a preliminary indication of the hazard of these compounds reacting adiabatically. However, the thermochemical equilibrium calculations do not provide any indication of the ease of reaction initiation and the rate of reaction. Therefore, these theoretical calculations are of far less value than preliminary reactivity indications based on reported experience and testing.

4.1.3 Preliminary Assessments of Combustibility and Explosibility Hazards

Material Safety Data Sheets

Unfortunately, most MSDSs for particulate products contain only very limited information on particulate combustibility and explosibility hazards. MSDS Section 5 on Fire Fighting Measures indicates whether or not the material is combustible and poses some sort of fire hazard. If the material is combustible there is often a corresponding warning that the generation of concentrated dust clouds creates a dust explosion hazard. Likewise, MSDS Section 7 on Handling and Storage Hazards sometimes has a precaution that airborne concentrations above the Minimum Explosible Concentration create a dust explosion hazard, with the appropriate MEC value listed. Section 7 often also has an admonition to avoid electrostatic charging during handling of particulate materials with a high resistivity, and usually suggests using bonding and grounding during transfer of such materials.

Although MSDS Section 9 has information on Physical and Chemical properties, dust explosibility is often not well represented. It has been common to find minimal dust flammability data, which is insufficient to give a full appreciation of the degree of hazard that the material presents. Because there has been no explicit requirement for reporting these data, if the data are available on the MSDS, the user should exercise care with their use. If these data are included, the particular test method also needs to be cited. In addition, the use of the material in the users' processes may change the physical properties of the material, and hence its flammability characteristics.

Usually, the MSDS includes the NFPA 704 Flammability Rating for the material, with different companies putting the rating in different sections of the MSDS. Until the clarification in the recent edition of NFPA 704, there were different interpretations of the ratings for powdered materials. Most combustible powders should have an NFPA 704 (2001 Edition) Flammability Rating of 3, which refers to flammable or combustible dusts with representative diameter less than 420 μm (40 mesh). A rating of 3 also refers to particulate materials of any size that burn with extreme rapidity, usually by reason of self-contained oxygen (e.g., dry nitrocellulose and many organic peroxides). Pyrophoric materials should have a combustibility rating of 4. In the case of materials that are only marginally combustible, NFPA 704 specifies that the material be assigned a Flammability Rating of either 0 or 1 depending on whether it will burn in air when exposed to a temperature of 1500°F (815.5°C) as determined using the test procedure described in ASTM D 6668-01.

Handbook and Textbook Dust Combustibility/Explosibility Databases

Combustibility and explosibility data for many commonly used combustible powders and for dusts from other commonly used combustible materials (wood, paper, grain, coal, etc.) are tabulated in several handbooks and textbooks. Perhaps the most widely used textbook is Eckhoff's *Dust Explosions in the Process Industries.* The appendix in Eckhoff's book provides 12 pages of data for 375 dusts categorized by type of material. It represents a subset of the much larger data compilation in German by the Berufsgenossenschaftliches Institut für Arbeitssicherheit (BIA, 1987). Eckhoff's appendix provides values for the combustibility and explosibility parameters for these dusts, as shown in Table 4.1, although not every parameter value is available for every dust.

Besides providing an abbreviated listing of the data in Eckhoff (2003), Schwab's chapter on dust explosions in the NFPA *Fire Protection Handbook* (2003) contains data tabulations for roughly 25 dusts from various U.S. Bureau of Mines reports cited by Schwab. The explosibility parameters are similar to those in Table 4.1, but the test apparatus and methods used by the Bureau of Mines to obtain the data in those reports are significantly different from the apparatus/methods specified in many contemporary consensus standards.

Besides listing the individual dust explosibility parameters, Schwab (2003) also lists values of the Ignition Sensitivity Index, the Explosion Severity Index, and the overall Explosibility Index, originally developed by the Bureau of Mines, and defined as follows.

$$\text{Ignition Sensitivity} = \frac{((\text{MAIT})(\text{MIE})(\text{MEC}))_{\text{Pittsburgh coal dust}}}{((\text{MAIT})(\text{MIE})(\text{MEC}))_{\text{Sample dust}}} \qquad [4.1]$$

$$\text{Explosion Severity} = \frac{(P_{\text{max}}(dP/dt)_{\text{max}})_{\text{Sample dust}}}{(P_{\text{max}}(dP/dt)_{\text{max}})_{\text{Pittsburgh coal dust}}} \qquad [4.2]$$

The multiplicative product of Ignition Sensitivity and Explosion Severity is called the overall Explosibility Index. As Schwab points out, the use of these three indices for measuring the relative explosion hazard of combusti-

TABLE 4.1
Combustitity/Explosibility Parameters in Eckhoff (2003) Appendix

Parameter	Source or Test Method
Particle Size Distribution	Sieve Analysis
MEC	1 m^3 or 20-liter vessel
P_{max}	1 m^3 or 20-liter vessel
K_{St}	1 m^3 or 20-liter vessel
Dust Cloud Minimum Ignition Temperature	Godbert-Greenwald Furnace, BAM Furnace
Minimum Ignition Energy, MIE	VDI
Dust Layer Ignition (Glow) Temperature	DIN
Dust Layer Flammability Class	VDI/UN

ble dusts (compared to Pittsburgh seam coal dust) has been criticized extensively. Hertzberg (1987) has documented the reasons for this criticism, which entail both the test methodology, the concept of relying on one or two numbers to represent explosibility, and the mathematical implications of multiplying the values of the individual parameters to determine the indices. Nevertheless, these indices are still being used in the NFPA 499 (1997) standard and by OSHA to determine if the explosibility of a dust material is sufficiently great to warrant hazardous area classifications for installing electrical equipment. Table 3-8A of Fire *Hazards of Materials* has a 6+ page listing of ignition sensitivity and explosion severity values, as well as other combustibility and explosibility data, for a wide variety of materials.

Bartknecht's 1981 and 1989 books and Field's book (1982) contain laboratory dust explosibility data scattered throughout. Babrauskas (2003) has a 4-page table of dust cloud ignition data as well as good descriptions of both dust cloud and particulate layer ignition test methods and fundamental combustion property data for pure chemical substances.

Babrauskas (2003) has published a handbook with an extensive compilation of ignition and combustibility data for a wide variety of materials, including several pages with tables of data for various types of dusts and powders. One interesting addition to the Babrauskas tabulations is a table listing dust materials that have been found to be nonexplosible by virtue of screening tests conducted by Field.

Explosibility Data Listings in NFPA and Other Standards

NFPA 68 (2002) has an appendix with tabulated explosibility parameters for 65 dust samples. The parameters in this tabulation are: Median Diameter, MEC, P_{max}, and K_{St}. There is also a limited amount of MEC and P_{max} data for some hybrid dust–vapor mixtures. NFPA 69 provides data on limiting oxygen concentrations to prevent dust cloud ignition when inerting with nitrogen and with carbon dioxide. NFPA 499 (1997) provides an extensive (4-page) tabulation of data on minimum cloud/layer ignition temperatures for various materials. NFPA 484 (2000) has a tabulation of combustibility and explosibility properties for metal powders/dusts.

4.1.4 Preliminary Assessments of Toxicity

Material Safety Data Sheets

Section 11 of the ACC/ANSI (1998) standard format MSDS contains pertinent toxicological information. Often the information consists only of qualitative descriptions of the materials effects on the eye (usually irritation), skin (either irritation or possible allergic reaction), oral/gastrointestinal functions, and inhalation/pulmonary functions (usually pulmonary edema). The latter effects are quite dependent on particle size, as discussed in Chapter 2.

If there are any tumorigenic or carcinogenic effects, they would be reported here also. Most MSDS toxicological descriptions are reports of animal effect studies, sometimes with multiple animal species. Sometimes there are no toxicity data for the particular material, but data/effects are reported for the constituent components. In other cases, data/effects are reported for a chemically similar material.

The MSDS should also list available quantitative data on toxicity threshold concentrations, such as Short Term Exposure Limit (STEL), Permissible Exposure Limit (PEL), Threshold Limit Value (TLV), and/or LC_{50} levels or LD_{50} levels for the various toxicological effects. Particulate materials that do not cause chronic diseases and are not biologically active fall into the category of nuisance dusts with no known toxicity effects other than respiratory impairments at high concentrations. The ACGIH TLV concentrations for these materials are 3 mg/m^3 for the respirable fraction and 10 mg/m^3 for total dust.

The NFPA 704 Health Hazard Rating (symbol on the left in the diamond quadrilateral) refers to the relative exposure hazard to emergency responders. The rating categories of 0 through 4 are defined in terms of the LC_{50} and LD_{50} values of the material. Materials with an LC_{50} for acute inhalation toxicity greater than 200 mg/m^3 are given a Health Hazard Rating of 0. Most common polyolefins fall into this category.

Section 8 of the ACC/ANSI standard MSDS describes appropriate exposure controls and personal protection for the subject material. Factors usually covered in this section are eye/face protection, ventilation recommendations (often prescribed to keep concentrations below TLV values), skin protection (protective clothing and gloves), and respiratory protection.

Handbook and Textbook Toxicity Listings

Several textbooks and handbooks used by industrial hygienists and toxicologists contain listings of particulate materials, properties, concentrations, and dosages pertinent to health risks. Perhaps the most comprehensive set of toxicity handbooks is the eight-volume *Patty's Toxicology* (Bingham et al., 2001). Each volume is devoted to one or more categories of materials. For example, inorganic particulates and dusts are included in Volume 1, metals and metal compounds are treated in Volumes 2 and 3, and hydrocarbons and organic nitrogen compounds are covered in Volume 4.

Other widely used handbooks with toxicity listings that include particulate materials are edited by Sax and Lewis (1987), Sittig (1997), and Carson and Mumford (2002). With regard to textbooks, Gad (1995) provides an overview of toxicology regulations and associated data, including the NOAEL, LOAEL, and other important concentrations and dosages of agricultural chemicals. A good example of a textbook focusing on inhalation properties and particle size and concentration effects is the Hatch and Gross 1964 AIHA publication, and some examples of handbook chapters are Kerfoot et al. (1995) and King (1990). In other cases there are generic guidelines based on the molecular composition. For example, the Institute of Chemical Engineers Guidelines for Plant Safety No. 28 states that a special toxicity warning is needed for materials that contain more than two of the following elements per benzene ring: halogens, N, P, or S. Two examples cited are polychlorodibenzofurans and mercaptobenzothiazole. The Guidelines do suggest this rule of thumb should be "confirmed where necessary by laboratory testing."

Government and Other Toxicity Databases and Listings

The Environmental Protection Agency (EPA) maintains perhaps the most comprehensive and extensive database for health effects of chemicals. Their database is called IRIS, which is an acronym for Integrated Risk Information System. It is available online at http://www.epa.gov/iris/subst/index.html. According to the EPA, "the information in IRIS is intended for those without extensive training in toxicology, but with some knowledge of health sciences." The type of data covered for individual chemical includes both descriptive and quantitative information on:

- Oral reference doses and inhalation reference concentrations (RfDs and RfCs, respectively) for chronic noncarcinogenic health effects, and
- Hazard identification, oral slope factors, and oral and inhalation unit risks for carcinogenic effects.

OSHA regulations and publications include Permissible Exposure Limit (PEL) values for both short-term exposures and 8-hour exposures to numerous materials. OSHA Web site searches for specific materials can be conducted at http://www.osha.gov/pls/oshaweb/owasrch.full_site_search. Where not otherwise listed, the OSHA general 8-hour exposure PEL requirement, as given in 29CFR1910.1000 Table Z-1, for particulates is 15 mg/m^3 for total particulates and 5 mg/m^3 for respirable particulates.

The National Institute for Occupational Safety and Health (NIOSH) developed a comprehensive database called the Registry for Toxic Effects of Chemical Substances (RTECS). The RTECS database includes toxicity data and summaries of pertinent journal articles, government reports, and EPA test submissions. Since December 2001, responsibility for maintaining RTECS has been transferred from NIOSH to various private and foreign organizations listed at http://www.cdc.gov/niosh/rtecs.html. These individual organizations update RTECS and make it available for purchase or lease along with software for searching and retrieving specific records. According to one organization (STN), there were 153,120 records in RTECS as of October 2002.

The International Agency for Research on Cancer maintains an online database (http://monographs.iarc.fr/monoeval/grlist.html) listing materials that have been subjected to scientific evaluations of potential carcinogenicity. The database currently contains 885 materials and exposure conditions, with many listings referring to particulate materials such as carbon black, heavy metals, and various fibrous particulates.

Threshold Limit Values for more than 700 chemical substances and physical agents are contained in the latest ACGIH (2003) listing. The TLV® values are determined by a ACGIH committee review of pertinent scientific literature. Proposed changes and new listings can be found on the ACGIH Web site.

The Canadian government provides a useful online resource for toxic material occupational exposure information called the Workplace Hazardous Materials Information System (WHMIS). The WHMIS database (www.hc-sc.gr.ca/whmis) for carcinogenic materials includes listings and classifications from ACGIH, the California EPA, the European Union, and IARC.

4.1.5 Special Considerations and Cautions in Using MSDS and Generic Databases

Generic data and hazard classifications may not be applicable to particulate materials that have even minor additives, contaminants, etc., as well as slightly different formulations. In addition, since particle size, shape, and moisture significantly affect most particulate hazards, the use of MSDS and generic data may not be directly applicable to particulate material that may have been altered as a result of physical transport and processing. Furthermore, many of the hazard classifications and categorizations described above are subjective judgments and are therefore subject to review and possible reclassification by a different authority or committee. Therefore, it is important to have updated and accurate MSDS data and hazard assessments that are specifically applicable to the particulate material being handled or stored at a particular facility. In view of the limitations of MSDS data, readers are advised to treat the data as a preliminary indication of a material's hazardous properties. It is often necessary to obtain more applicable data using the test methods described in Section 4.3.

4.1.6 Publicly Available Computer Databases

There are numerous publicly available computer codes and online and offline databases for hazardous materials. In addition to the previously cited toxicity databases, there are general chemical property databases (for example, the NIST Chemistry Webbook, http://webbook.nist.gov/chemistry/), and there are a few online databases specifically for hazardous materials. One example for assessing chemical reactivity hazards is the Brethericks

Reactive Chemical Hazards database, which can be accessed by opening an account on the Elsevier collection of chemistry databases at http://www.chemweb.com/databases. This Web site also provides access to a database of about 56,000 MSDSs.

There are far fewer publicly accessible/acquirable computational databases for particulate materials. One exception is The Bulk/Bulk Portal Web site (http://www.bulk-online.com/Forum/), which is devoted to particulate materials, and has several forums that are pertinent to particulate properties, handling and hazards. There is an explanation for the lack of an online particulate properties database in one of the forums. The explanation offered is that these properties are so sensitive to the specifics of particle size distribution, additives, moisture level, etc. that it is easier to conduct site/material-specific tests than to perhaps misuse inapplicable data from a complex database. Readers of this book can form their own impression of that explanation upon reading the descriptions of various laboratory test methods in Section 4.3.

4.1.7 Company and Consortium Databases

Several companies maintain proprietary databases for assessing hazards either of their materials and products or materials and products of companies with which they have a business relationship. In addition, testing laboratories inevitably maintain a database for rapid retrieval of test data generated for their clients.

Often the in-house database for particulate materials will focus on one hazard, such as explosibility. For example, the following items are contained in the explosibility database maintained by one company (Herrmann, 2003):

- Material name
- Job number
- Log number
- Particle size distribution ($D_{10}/D_{50}/D_{90}$)
- Moisture level (%)
- MEC (g/m^3)
- LOC ($\%O_2$)
- MIE (mJ)
- K_{St}
- P_{max}
- Cloud AIT
- Layer AIT
- Impact
- Client
- Site
- Lot #
- CAS #

- MSDS #
- Assay Purity

Another company database and computer code for dust explosion considerations is the FM Global code Dust Calc. It is used by FM field engineers and engineers at their insured facilities to determine appropriate dust explosion protection in accord with FM Loss Prevention Data Sheet 7-76.

4.2 WHEN ARE MORE DETAILED PARTICULATE HAZARD DATA NEEDED?

The flow charts in Chapter 3 and in Chapter 7 indicate some of the conditions that may require a need for either site-specific material property data or for more detailed data than can be obtained from the preliminary assessments discussed in Section 4.1. The discussion in 4.1.5 also implies that the generic data and classifications are often not directly applicable for accurate hazard evaluations of a specific particulate material.

In the case of dust combustibility and explosibility considerations, Eckhoff (2003, appendix) offers the following admonition to encourage use of material specific test data. "As a general rule, the tabulated data should only be used as indications, and not as the ultimate basis for design of actual safety measures in industry. . . . It is generally advisable to have the actual dust of interest tested in a professional laboratory." This admonition should also be extended to considerations of instability and reactivity hazards.

Most particulate material manufacturers generate some basic hazard data for all their materials, and acquire other data depending on the specific type of processing used in the material's production and/or intended use. Figure 4-2 shows one example of a company's policy for site-specific or material-specific hazard testing. The basic data obtained by this company for all particulate materials include particle size distribution, powder resistivity, Minimum Autoignition Temperature, and Minimum Ignition Energy.

4.3 LABORATORY TEST METHODS FOR DETAILED ASSESSMENTS OF PARTICULATE HAZARDS

4.3.1 Particulate Sampling and Conditioning for Testing

One of the first considerations in doing laboratory hazard assessment tests with particulate samples is how to acquire the field/site samples, and how many different samples to submit. The fundamental issue is whether to attempt to acquire representative samples from the different stages of processing and handling, or whether to obtain "worst-case" samples in terms of

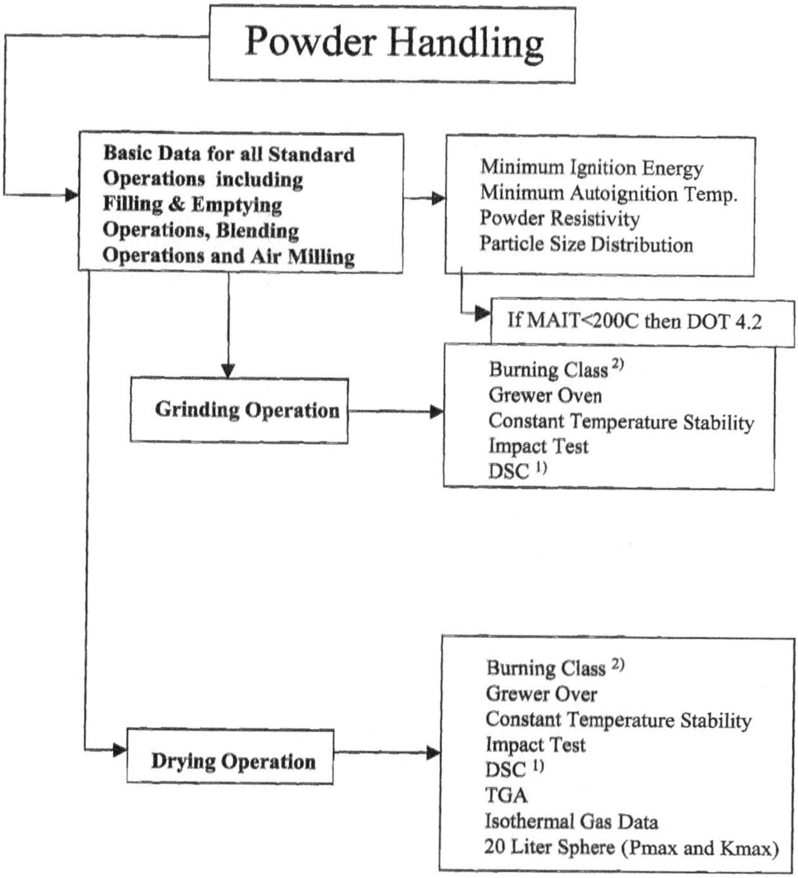

1) If $\Delta H > 250$ J/g request determination about possible deflagration potential
> 300 J/g run SADT

2) If Burning Class = 5, request Rate of Burning Determination in accordance to DOT 4.1

Figure 4-2 Example of Basic Data Needs and Process Dependent Particulate Testing, Copyright Syngenta 2004, with permission.

particle size, shape, composition, moisture content, and the like. The same issue is also applicable to the preparation or conditioning of the sample in the laboratory prior to testing.

In most cases, a prudent resolution of this issue would entail taking "worst-case" samples by sampling, for example, from a dust collector or collector duct so as to get a high proportion of fines. However, if the main purpose of the tests is to characterize a product or feedstock, then representative sampling might be warranted along with the measurement and reporting of the size distribution of the sample. A cautionary statement should be reported about the incremental hazard anticipated for smaller particles. The ASTM dust explosibility tests described in Section 4.3.7 recommend using a sieved test sample such that at least 95% has passed through a 200 mesh sieve, but as-received samples can be tested if the test report includes the cautionary statement about particle size effects.

Let us assume that a decision is made to submit representative field/site samples. If the sample is to be obtained from a moving air stream, an isokinetic sampling system is needed to provide a representative sample of the particulate being carried by the stream in order to avoid preferential deposition or deflection of particulates on their way to the collection device. Several commercially available sampling devices and filters are available for this purpose. If the sample is being obtained from a conveyor belt or bucket, it is necessary to peruse the belt/bucket and probably take representative small samples from a few different locations.

In the case of sampling particulates from a drum or similar container, there is an ASTM standard practice (ASTM D5680-95a) for obtaining representative particulate samples. The latter includes descriptions and references to standard practices for use of scoop samples (ASTM D5451) and Trier (elongated scoop) samplers for extracting a plug sample from some known depth beneath the surface (ASTM D5633). The basic principle emphasized in ASTM D5680 is that the sampling procedure should provide some element of randomness in selecting sample locations (to prevent biases associated with particulate segregation) and should minimize biases and contamination associated with the characteristics of the sampling device and material.

Eckhoff (2003) has a brief discussion of various sampling methods for producing laboratory test size samples from a large submittal. He mentions various techniques, including coning and quartering, but concludes that the most homogeneous subsamples are obtained using a spinning riffler. As shown in Figure 4-3, the spinning riffler consists of a conical bottom bin or funnel and rotating turntable on which contiguous subsample containers are situated. The primary sample is poured into the open bottom bin/funnel, which continuously fills the rotating small containers as they pass under the funnel opening. A rigorous comparison of the accuracy of various sampling techniques (see NIST, 2001, p. 15) confirms Eckhoff's conclusion, as is evident in Table 4-2, which shows the particle size standard deviations measured on multiple samples obtained from a known source using different sampling techniques.

Although the spinning riffler provides the most representative sample for free flowing powders, it has some difficulties with non-flowing particulates. NIST (2001) compares the advantages and disadvantages of the various sampling techniques shown in Table 4-3.

In the case of dust layer sampling, there are likely to be different size particulates in layers located in different parts of the facility. For example, dust layers on elevated surfaces and ledges are likely to contain significantly

Figure 4-3 Spinning riffler for particulate sampling. (Reprinted with permission from Laval Lab.)

smaller particles than floor dust layers. Thus, it is important to submit at least one adequate size sample from these elevated surface deposits. The question of what is an adequate size depends on the test method, such that the laboratory personnel provide instructions on the sample size needed.

4.3.2 Laboratory Testing for Instability Hazards

There are a wide variety of laboratory test methods to determine the chemical stability of powders. The more commonly used thermal stability tests and decomposition sensitivity tests are summarized here, beginning with a brief overview of instability screening tests for powders.

Instability Screening Tests

The objective of one or more instability screening tests is to obtain an indication of whether the material exhibits an instability, and to get a rough idea of the approximate temperature at which the unstable/runaway reaction occurs and possibly whether the instability can be triggered by a mechanical impetus. In some cases, screening tests are also conducted to determine if there is the potential for significant gas generation during the instability.

TABLE 4-2
Sampling Method Accuracy/Reliability in terms of Homogeneous Particle Size (from NIST, 2001)

Sampling Method	Relative Standard Deviation (%)
Cone and quartering	6.81
Scoop sampling	5.14
Table sampling	2.09
Chute riffling	1.01
Spin riffling	0.125

TABLE 4-3
Advantages and Disadvantages of Different Sampling Methods (from NIST, 2001)

Sampling Device	Advantages	Disadvantages
Cone and quartering	Good for powders with poor flow characteristics	Operator-dependent
Scoop	Reliable for homogeneous and nonflowing powder	Particle segregation
Table sampler	Ability to separate large quantity of material	Dependent on the initial feed
Chute riffler	Ability to reduce powder samples in half after one pass	Operator bias
Spinning riffler	Reliable for free flowing powder	Inability to do large quantity of powder efficiently

Most organizations start screening with differential scanning calorimetry and differential thermal analysis tests as described below. Some companies also conduct a Carius (or ICI) sealed tube test (Johnson et al., 2003) with continuous monitoring of the temperature and pressure outputs from a sample tube in the oven. The Carius sealed tube pressure data can be used to plot ln P against $1/T$ and thereby infer whether there is noncondensable gas generation, since the plot should be essentially a straight line if the pressure increase is due solely to the vapor pressure.

According to the 1995 CCPS *Guidelines on Chemical Reactivity*, some organizations also use a hot plate screening test for powders. A layer of the powder is deposited in a circle on the hot plate, and a thermocouple is inserted into the center of the powder. The powder temperature is monitored as the hot plate temperature slowly increases. Besides seeking an indication of self-heating or decomposition, observations are also recorded about whether there is any evidence of gas generation, smoldering, or spontaneous ignition. If this type of hot plate test is conducted, test operators need to be particularly careful about the possibility of a violent reaction causing an explosion or ejecting flaming or molten material.

Differential Scanning Calorimetry and Differential Thermal Analysis Testing

Differential scanning calorimetry (DSC) and differential thermal analysis (DTA) are by far the most widely used traditional test methods for chemical stability and reactivity screening. Their popularity stems from the virtual omnipresence of these devices and trained users in chemical laboratories, and from the relatively small quantity of reactants needed for each test (typically 1–50 mg, depending on the expected energy release). Thermal analysis—that is, the combined use of DTA and DSC—provides the temperatures at which potentially hazardous exothermic reactions occur and the changes in enthalpy associated with the reactions. One drawback of these methods is that they do not provide the same level of accuracy, quantification, and scale up methodology as some other test methods described here. Furthermore, for combustible powders, they often yield higher reaction onset temperatures than are experienced with test methods that utilize near-adiabatic conditions and continuous airflow.

As illustrated in Figure 4-4, the traditional DTA instrument consists of an oven, two sample capsules (one for the inert reference sample), a controller to produce a linear temperature rise, thermocouples to measure the sample and oven temperatures, and recorders or a computer to record the reference material temperature and the sample-minus-reference temperature for the duration of the test. The DSC instrumentation is similar, with the heat flow being measured instead of the temperature difference. The heat flow is traditionally measured by keeping the test sample and reference ther-

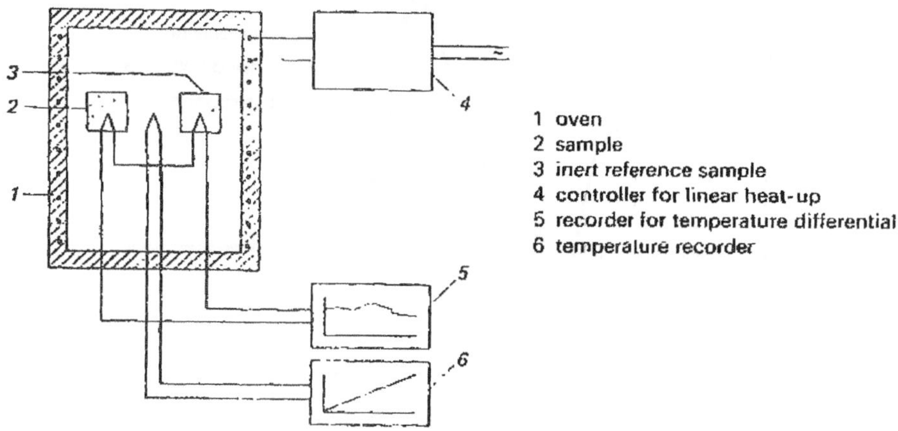

Figure 4-4 DTA apparatus schematic (from Grewer, 1994).

mally balanced by adjusting the current to the heaters under the two capsules (Grewer, 1994, p. 63). The difference in power is recorded as a function of time in the DSC. When the test sample undergoes some endothermic or exothermic reaction, there is a sample-reference temperature difference in the DTA and a deviation in the heat flow to the sample and reference in the DSC. Details of typical instrumentation and sample container construction and instrument operating procedures have been presented in Grewer (1994) and the CCPS *Reactivity Evaluation Guidelines* (1995b) as well as several other good references.

Figure 4-5 shows a generic DTA or DSC recorded output for a sample undergoing an exothermic reaction. Characteristic exotherm temperatures are indicated on the drawing using the notation and definitions in ASTM E 537. The peak temperature is the maximum temperature recorded during the exotherm. The exotherm onset temperature, T_0, is the temperature at the first noticeable deviation from the baseline temperature curve. The baseline temperature represents the programmed increase in sample/reference temperature as they are heated at a constant rate somewhere in the range 2 to 30°C per minute (ASTM E 537). The extrapolated onset temperature, T_s, is the temperature obtained at the intersection of the extrapolated pre-peak baseline temperature and the maximum rate of temperature rise as indicted in Figure 4-5.

Often an exothermic reaction is preceded by an endothermic reaction as illustrated in Figure 4-6. As indicated on the diagram, the endotherm is

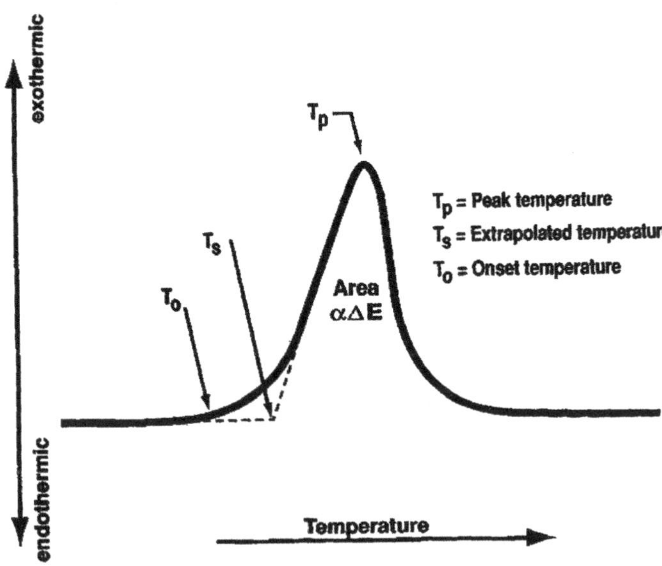

Figure 4-5 Characteristic exotherm temperatures obtained from a DTA or DSC curve. (From ASTM E 537. Copyright ASTM International, reprinted with permission.)

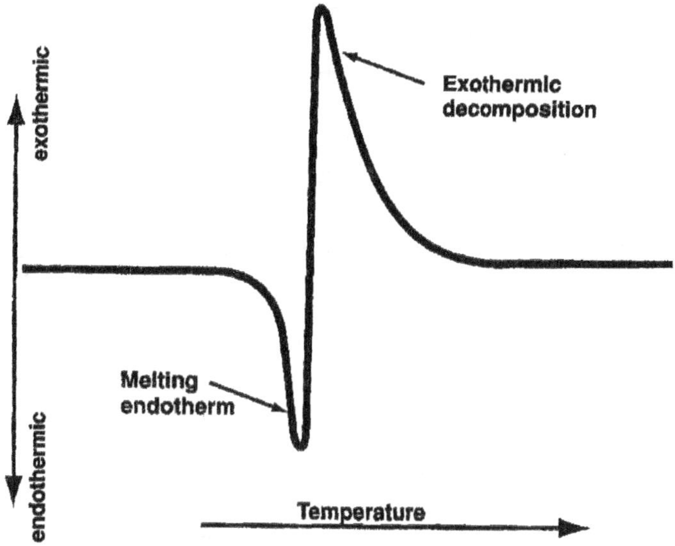

Figure 4-6 Exothermic decomposition following melting endotherm. (From ASTM E 537. Copyright ASTM International, reprinted with permission.)

sometimes associated with the melting of the test sample. Figure 4-7 shows an actual DSC output for a test sample exhibiting a 187 J/g endothermic reaction at 85°C and a 1082 J/g exothermic reaction at 203°C. Grewer (1994) points out that the indicated temperature interval between an endothermic and exothermic reaction can be influenced by the DSC/DTA heating rate, i.e. rapid heating rates tend to separate the reactions even if they both occur within a relatively narrow temperature interval. However, a rapid heating rate may indicate that the reaction occurs at a higher temperature than would be indicated with a slower heating rate. Thus, experience and judgment are often needed in providing an accurate interpretation of DSC/DTA data. Grewer's (1994) tabulation of decomposition energies and onset temperatures for 177 compounds includes heating rates from 0.2°C/min to 20°C/min, but most were in the narrow range 2 to 5°C/min.

Other potentially important aspects of DTA/DSC testing are the sample atmosphere and pressure, and the sample container material. The test cell atmosphere can be inerted with nitrogen to eliminate oxidation reactions or it can be run in an oxygen atmosphere to emphasize oxidation reactions. The DSC test result shown in Figure 4-8 was obtained in a nitrogen atmosphere at 150 psig. Querol Aragon et al. (2002) recommend the use of oxygen atmosphere thermal analysis in order to categorize the various particulate materials that are prone to spontaneous combustion. Figure 4-8 shows a comparison of thermogravimetric analysis (TG) records they obtained with an air atmosphere and with an all oxygen atmosphere. The output in a TG test is

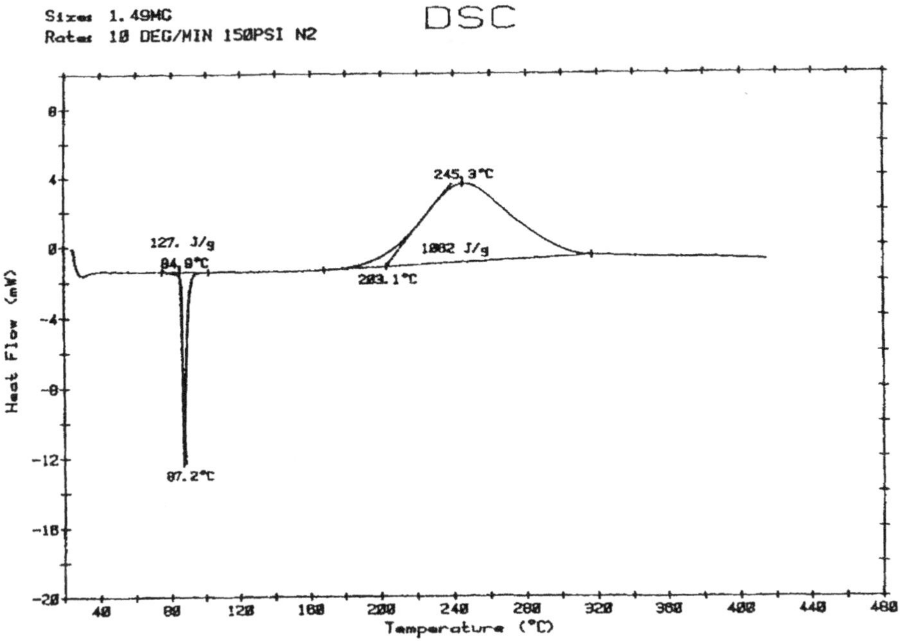

Figure 4-7 Typical DSC data for exothermic reaction (from CCPS, 1995b).

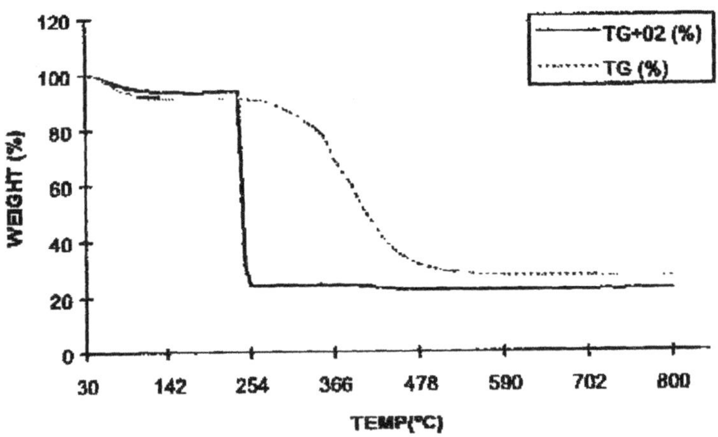

Figure 4-8 TG record for coal tested in air and in oxygen atmospheres (from Querol Aragon et al., 2002).

the weight loss change of the sample during the programmed temperature. The temperature at which the oxidation reaction weight loss occurs is much sharper in the test with an oxygen atmosphere. Querol Aragon et al. recommend that materials with clearly defined oxidation reactions occurring at a temperature less than 250°C should be considered very high risk.

Accelerating Rate Calorimeter Tests

The accelerating rate calorimeter (ARC) developed by Townsend (1981) is intended to achieve and maintain near-adiabatic test conditions as an exothermic reaction generates pressures as well as high temperatures. The closer a test vessel approaches adiabatic conditions, the more accurate is the measurement of reaction energy. The ARC design shown in Figure 4-9 seeks adiabatic conditions by minimizing heat losses from the test vessel. This is achieved by monitoring the test vessel temperature and increasing the temperature of the ARC jacket/housing so that it remains approximately equal to the vessel temperature. Three separate heaters and temperature controllers are installed in the top, side, and base of the ARC jacket so that that a near uniform temperature can be maintained with nonuniform heating of the various jacket surfaces.

The time lag between sample temperature increase and test vessel temperature increase depends on the relative thermal inertia of the sample and test vessel. This thermal inertia factor is expressed as the ratio φ, defined as

$$\varphi = \frac{m_s c_{ps} + m_v c_v}{m_s c_{ps}} \qquad [4\text{-}3]$$

where m_s = sample mass
c_{ps} = sample specific heat
m_v = vessel mass
c_v = vessel specific heat

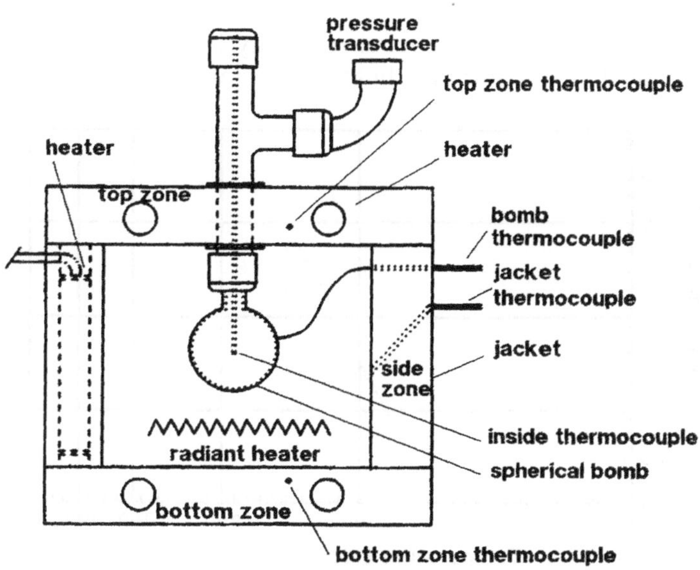

Figure 4-9 ARC test apparatus (from CCPS, 1995b).

Heavy test vessels have large φ values, while lightweight test vessels have values much closer to 1. The ARC has an assortment of sample vessels with varying design pressures and associated thickness and mass. Stainless steel test vessels have φ values of about 8 for a typical sample mass of about 2 g. The newer and more expensive titanium sample vessels have φ values of about 2 (Grewer, 1994, p. 119). Computer corrections to the measured vessel temperatures are made to account for these elevated values of φ.

Figure 4-10 is an example of a self-heating rate curve measured in an ARC test. The rate of sample temperature rise is plotted as a function of sample temperature. Pressure measurements are also plotted as a function of temperature for reactions with gaseous reaction products. The titanium sample vessel has a maximum pressure strength of 300 bar, and several energetic materials (such as mono-nitro compounds) can generate pressures in excess of 300 bar when the sample vessel is nearly filled (Grewer, 1994, p. 122). Thus, it is sometimes necessary to use a very small reactant mass and provide corrections for the corresponding high φ values. According to Grewer, these corrections are effective, but can be particularly difficult for self-accelerating reactions.

Although the ARC apparatus provides low φ value conditions, like the DSC/DTA test methods, it also produces results that may be limited by the lack of free air access to the reactants. Thus, it can also yield a higher exotherm onset temperature than results from tests with unlimited air access.

Isothermal Storage Tests

Isothermal storage tests for chemical reactivity/instability are intended to provide kinetic data to determine heat release rates and/or reaction induc-

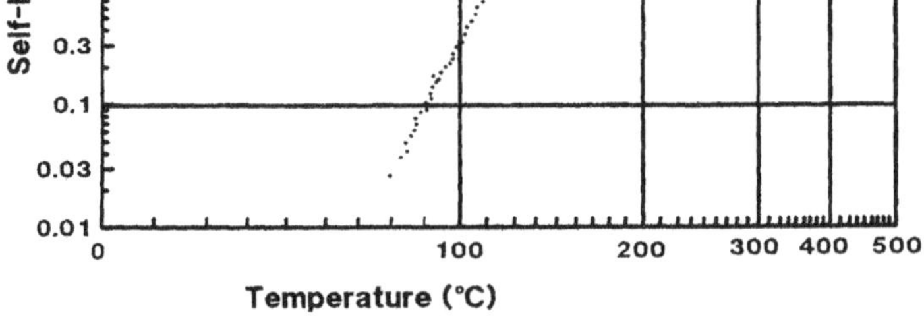

Figure 4-10 Sample output from ARC test (from CCPS, 1995b).

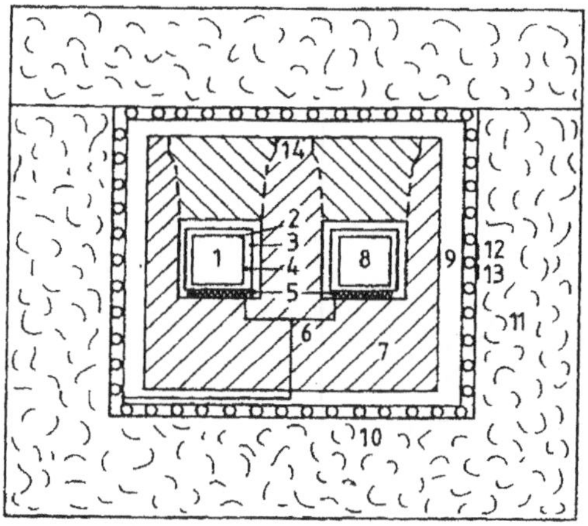

1. Sample
2. Sample vessel
3. Cylindrical holder
4. Air spaces
5. Peltier elements
6. Electrical circuit
7. Aluminum block
8. Inert material
9. Insulation wool
10. Heating wires
11. Glass wool
12. Platinum resistance sensor temperature control
13. Platinum resistance sensor for safety control
14. Platinum resistance thermometer

Figure 4-11 Example of an isothermal storage test setup (from CCPS, 1995b)

tion times. Figure 4-11 shows an example of an isothermal storage test setup in which both a test sample and a reference material are monitored during isothermal storage. The Peltier elements measure and control the temperature of the sample and reference material and pump heat from the test sample when necessary to maintain its temperature. The rate of sample cooling is effectively the rate of heat generation at a particular temperature. Tests are usually run at several different temperatures. Other types of isothermal storage test designs are described by Grewer (1994).

Figure 4-12 shows the type of heat release rate data as a function of temperature that can be obtained from an isothermal test chamber of the type shown in Figure 4-11. The linear correlation versus 1/T indicates an Arrhenius reaction rate variation with temperature, and the reaction activation energy, E, can be obtained from the slope of the curve fits. Since each isothermal storage test can last several hours or days, and since many tests are required to obtain the type of data shown in Figure 4-12, these tests often take significantly longer to obtain results than the thermal analysis tests described previously. However, they do provide valuable data on reaction rate and heat release rate that are not easily obtained with other methods. In order to make efficient use of staff time and equipment, several laboratories

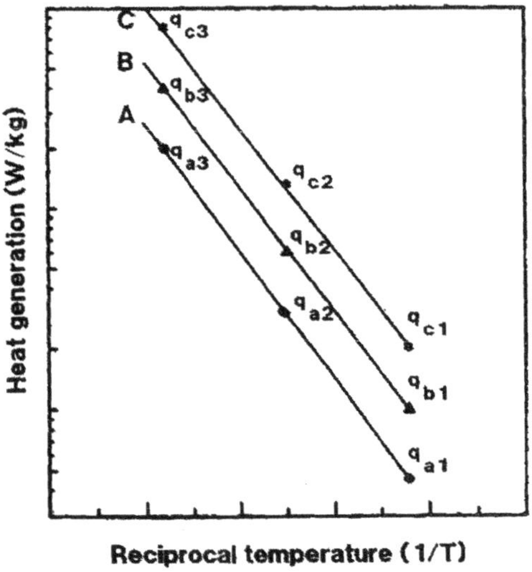

Figure 4-12 Heat generation rate data obtained from an isothermal storage test (from CCPS, 1995b).

use slow, programmed rates of temperature rise in the oven. Gibson et al. (1985) report using a rate of 0.5°C/min, but caution that it is important for the powder to remain at a temperature at or above the intended process equipment temperature for at least the same duration as the process/storage itself.

Gibson et al. (1985) presented a histogram distribution of the measured exotherm onset temperatures obtained with 150-g samples of 200 organic powders tested in the ICI glass cylinder-in-an oven apparatus sometimes referred to as a Diffusion Cell to distinguish it from the aerated cell. Approximately 50% of the powders had onset temperatures in the range 125°C to 175°C, about 7% had onset temperatures below 125°C, and the rest had onset temperatures above 175°C.

Heated Air Flow Tests: Grewer Oven and Aerated Cell

The Grewer oven test differs from the previous isothermal oven tests in that there is a flow of heated air over the test sample and reference sample. The airflow is intended to simulate conditions in heated process equipment such as a spray dryer or fluidized bed dryer. The test apparatus, which is shown schematically in Figure 4-13, has the air entering at the bottom and flowing over heated packings on its way past the test and reference samples. The oven temperature is increased until there is an indication of oxidation that leads to sample self-heating and eventually to autoignition. The typical sample size is about 100 g placed in an 8-ml wire mesh basket. Kuhner (1991) specifies that the sample particle size should be less than 250 μm; thus it should be sieved to pass through a No. 60 ASTM sieve.

The Kuhner (1991) test guidelines for the Grewer oven test suggest that the test be conducted at a slow temperature rise (1°C/min) and continue to

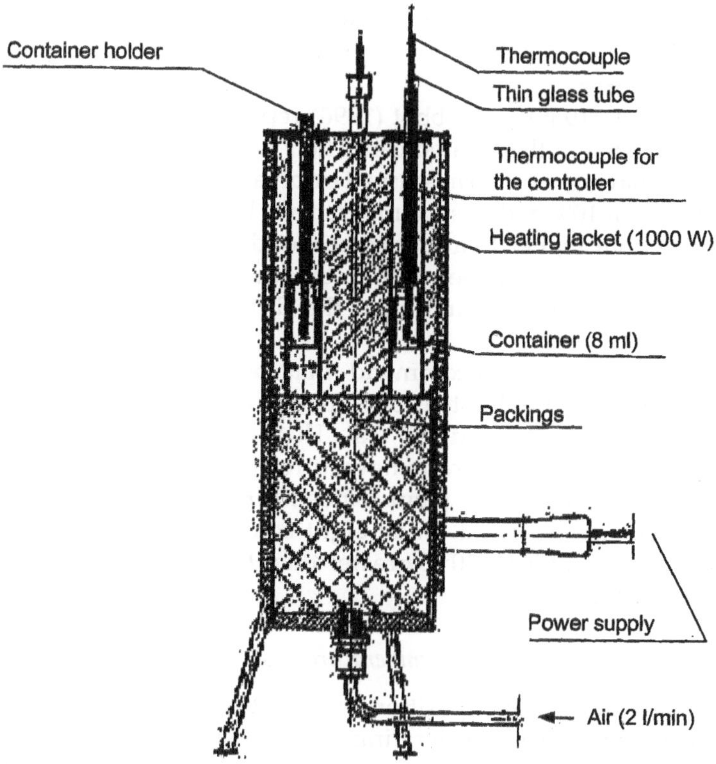

Figure 4-13 Grewer oven test apparatus (from Ciba Lab Brochure).

determine whether there is any self-heating or autoignition at temperatures up to 50°C above the maximum drier temperature. If any self-heating exotherm is observed at temperatures of 200°C or less, Kuhner recommends that an isothermal storage test be conducted for 24 hours with a much larger sample (at least 400 ml) at a lower temperature than the exotherm temperature.

A modified version of the Grewer oven test can be conducted with a nitrogen flow instead of an air flow. This configuration allows for the determination of exothermic decomposition without any complications due to oxidation. Results can be used to determine if the heated process equipment should be inerted to prevent oxidation self-heating and possibly autoignition.

The aerated cell test uses the test apparatus developed by Gibson et al. (1985) to determine the thermal stability of powders in driers with large air flow rates. It utilizes a heated air flow down through a 8.5 cm high by 5 cm diameter glass cell. The cell is situated in an oven with a temperature range up to 420°C. Thermocouple data for samples exposed to programmed oven temperature ramps can indicate approximate exothermic decomposition onset temperatures.

Gibson et al. (1985) and Abbot (1990) reported a histogram of data for 200 samples tested in the aerated cell. Their histogram shows about 60% of the samples having exotherm onset temperatures in the range 125°C to 175°C, and about 10% had onset temperatures below 125°C. Since these percentages are both higher than the corresponding percentages for samples tested in the diffusion cell, one can infer that the aerated cell produces somewhat lower onset temperatures than the diffusion cell for at least some powders.

Gibson et al. (1985) recommend that powder drying temperatures should be maintained at least 30–50°C below the exotherm onset temperature measured in the aerated cell test. They compared the recommended maximum allowable exposure temperatures obtained with Diffusion Cell and aerated cell tests to the plant exposure temperatures at which eight exotherm incidents occurred. In all eight comparisons, the incident onset temperatures were at least 10°C greater than the recommended maximum based on test data.

Self-Accelerating Decomposition Temperature Test

The self-accelerating decomposition temperature (SADT) is the minimum temperature at which self-accelerating decomposition occurs in a material as packaged or in transport. Since the SADT is influenced by both chemical decomposition kinetics and heat transfer through the packaging (as well as the criterion to discern self-accelerating decomposition), there have been various attempts to determine SADT values from a limited amount of testing, and then use theoretical modeling to scale the results to other packages. The U.N. *Manual of Tests and Crite*ria (U.N., 1999b), lists four specific tests used to measure SADT, and allows three types of theoretical models to "assist in interpreting results." The test procedures are applicable to both particulate materials and liquids.

The most direct test to determine SADT uses the test material in the actual full size packaging. This test, which is designated as the United States test in the U.N. Manual (U.N., 1999b), entails placing the loaded package in an insulated test chamber equipped with a heater and a thermostatically controlled air circulation capability. Different size test chambers can be used depending on the size of the package. Figure 4-14a shows a small package (up to 25 liters) test chamber constructed from a 55-gallon (220-liter) steel drum. Figure 4-14b shows a large package test chamber constructed as a wood frame, insulated plywood clad, cubic box, 1.2 m on each side.

Per the U.N. Manual (U.N., 1999b), the US test requires that the test package be heated continuously and the test package center temperature be

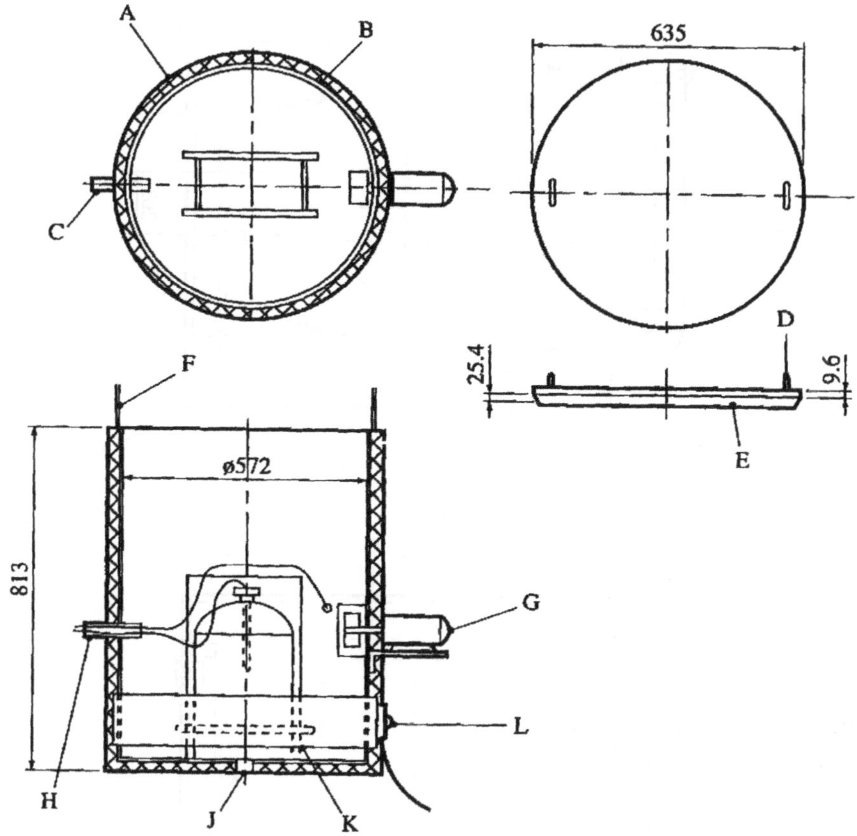

(A)	Insulation 25 mm thick	(B)	220 litres open top drum
(C)	19 mm pipe	(D)	9.6 mm eye bolt in steel cover
(E)	Insulation on steel cover	(F)	3 mm control cable
(G)	Fan	(H)	Thermocouples and controls
(J)	Drain	(K)	25 mm angle stand
(L)	2 kW drum heater		

Figure 4-14a Small package test chamber for SADT tests (from U.N., 1999b, Figure 28.4.1.1).

monitored and recorded starting from a temperature 2°C below the test chamber temperature, and until the earlier of either seven days or the time at which the test sample temperature increases at least 6°C above the chamber temperature. The SADT determination via this test is the lowest chamber temperature at which the sample temperature exceeds the chamber temperature by 6°C or more. If the sample temperature never exceeds the chamber temperature by 6°C, the SADT is deemed to be the highest chamber temperature tested.

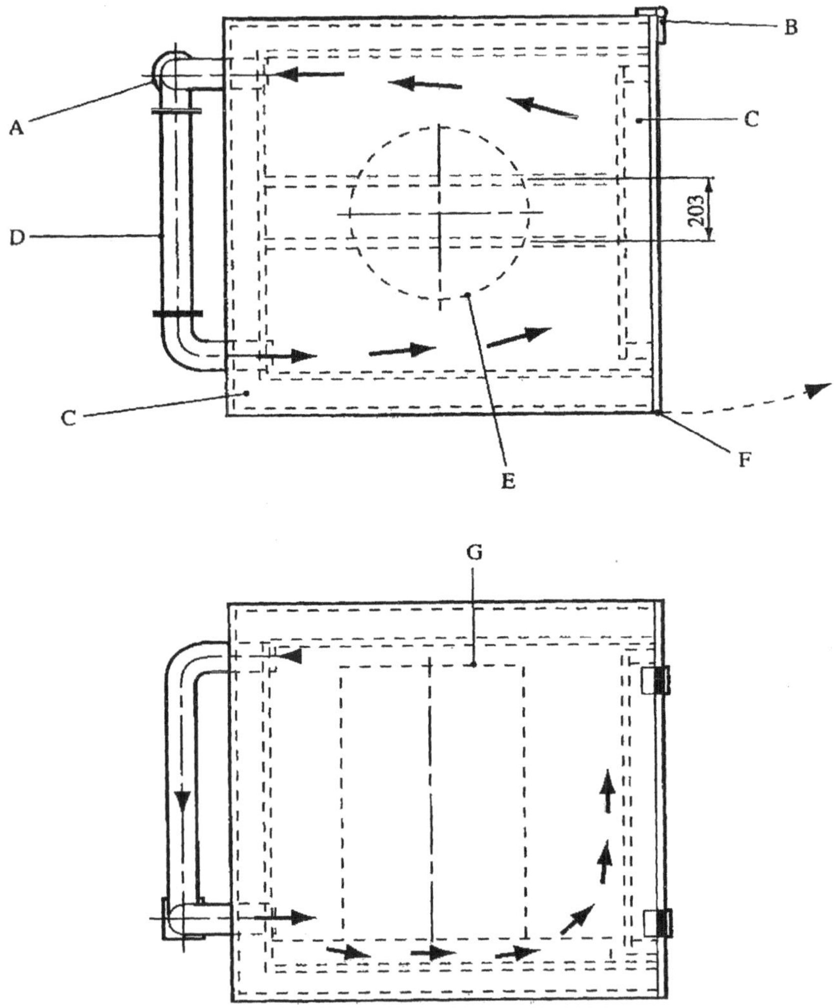

(A) Fan
(C) Insulation
(E) Drum
(G) Drum (e.g. 0.58 m x 0.89 m)
(B) Hinges (2)
(D) Heater
(F) Latch

Figure 4-14b Large package test chamber for SADT tests (from U.N., 1999b, Figure 28.4.1.2)

Another U.N. Manual accepted SADT test is the previously described isothermal storage test using the apparatus shown in Figure 4-16. When used for SADT testing, the isothermal storage test sample should include "a representative quantity of packaging material (if metal)." The SADT determination using the Isothermal Storage Test is based on a comparison of the measured heat generation rate per unit sample mass (mW/kg), and the calculated heat loss rate from the test sample, using an inert substance heat loss calibration test and a calculation procedure described in the U.N. Manual (1999b). Both the heat generation rate and a linear curve fit to the calculated heat loss rate are plotted as a function of oven temperature as indicated in Figure 4-15. The shape of the curves in this type of plot follows the theoretical exposition from thermal explosion theory as described, for example, by Grewer (1994), Bowes (1984), and Gray (2002). The line shown representing

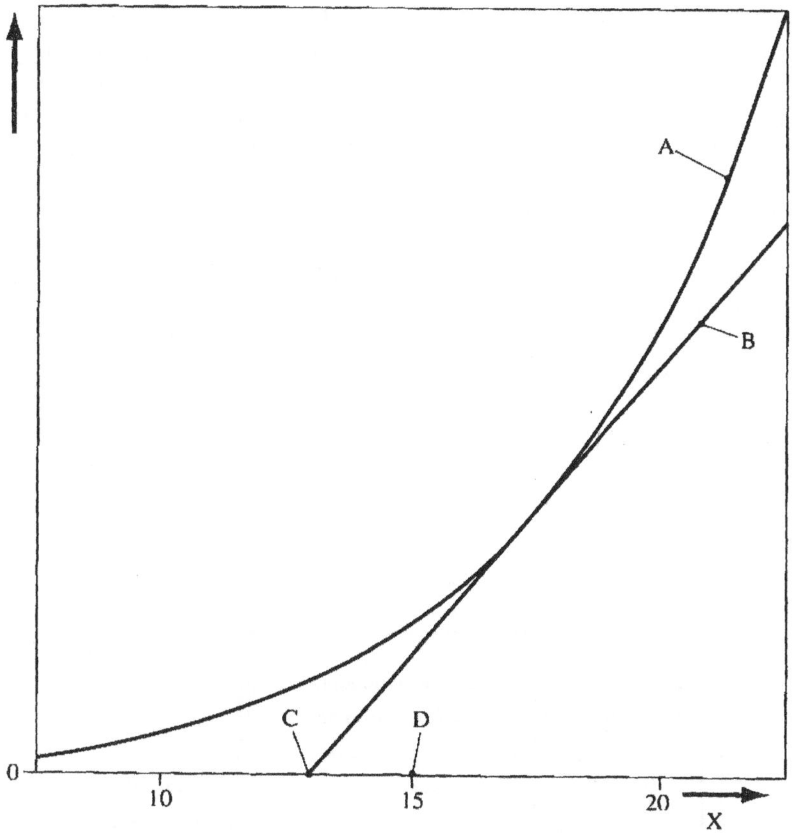

(A) Heat generation curve
(B) Line with gradient equal to the rate of heat loss and tangential to the heat generation curve
(C) Critical ambient temperature (intercept of heat loss line with the abscissa)
(D) Self-accelerating decomposition temperature (SADT) - critical ambient temperature rounded up to next higher multiple of 5 °C
(X) Temperature
(Y) Heat flow (generation or loss) per unit mass

Figure 4-15 SADT determination graph for U.N. isothermal test and adiabatic storage test (from U.N., 1999b).

the heat loss rate is actually a line with a slope equal to that of the heat loss rate, but displaced such that it is just tangent to the heat generation rate curve. The SADT determined from this procedure is the critical test chamber temperature corresponding to a zero value of heat generation rate (point C in Figure 4-15) rounded up to the next higher multiple of 5°C (U.N., 1999b).

The SADT Adiabatic Storage test (also known as the dewar flask test) is conducted with the sample placed in a dewar of volume 1.0 to 1.5 liter, that is in turn placed in an oven of the type shown in Figure 4-16. The use of a glass test vessel instead of the metal pressure vessels used for adiabaticity testing in the ARC allows for a relatively low φ, which is typically in the range 1.2 to 1.6 (Grewer, 1994, p. 114) as measured by monitoring the oven cooling of a pre-heated liquid in the Dewar. Although it is desirable to use a φ value representative of the transport packaging, the U.N. Manual (U.N., 1999b) does not require this, and merely specifies that "a representative quantity of packaging material (if metal)" also be placed in the Dewar. The test vessel heat loss rate as a function of dewar sample temperature is determined by a procedure similar to that for the isothermal storage SADT test. After determining test sample heat generation rates and heat loss rates, the SADT is determined as indicated in Figure 4-15 and described above for the isothermal test. Because the SADT data are determined with representative packaging materials, they should not be used for applications in which there is no packaging.

The SADT is a factor in the U.N. Transport (1999a) classification of particulate materials as self-reactive solids (Division 4.1). A material is deemed self-reactive (i.e., a Division 4.1 material) if the SADT is less than 75°C for a 50 kg package, and if the heat of decomposition is greater than 300 J/g. One

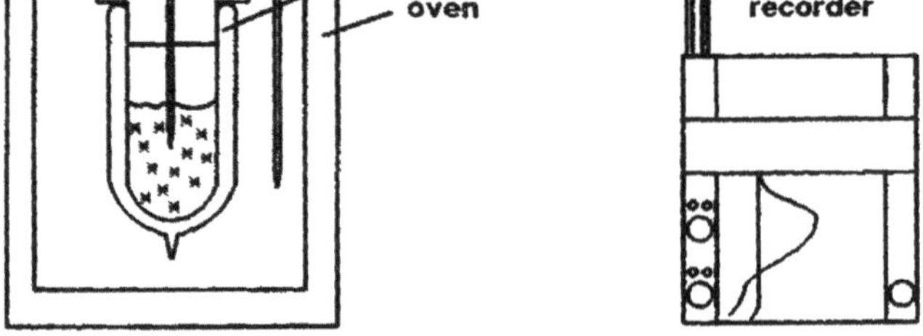

Figure 4-16 Dewar flask test to determine SADT (from CCPS, 1995b).

example is dibenzoyl peroxide (CAS # 94-36-0), which had a SADT of 70°C when a 0.45-kg sample was tested in U.N. Type 1G (fiberboard drum) package (UN 1999b, p. 284). Larger packages normally have lower SADT values because of the decreasing package surface area/volume ratio compared to smaller packages of the same material and configuration. One example of a self-reactive particulate material that is reported to have a SADT above 75°C for a 30-kg sample in a Type 1G package is azodicarbonamide (UN 1999b, p. 289).

Several researchers have shown how laboratory-scale testing using kinetic data obtained from the ARC, or some similar test apparatus can also be used to calculate the SADT (designated as T_{SADT} in the following) from the application of thermal explosion theory. One simple approach suggested by Fauske (2000b) for use with his Reactive System Screening Tool (RSST) is the following Semenov theory steady-state heat balance:

$$h\left(\frac{T_{SADT}R}{E}\right)A = V\rho c_p Z \exp\left(1-\frac{E}{T_{SADT}R}\right) \quad [4\text{-}4]$$

where h (W/m-K) is the surface heat transfer coefficient, A is its surface area, V is its volume, ρ is the reactant density, and Z is the preexponential factor in the Arrhenius reaction rate. Another approach is to use the following relationship between the adiabatic induction time (time to thermal runaway), τ_{ad}, and the initial temperature, T, preceding the runaway (Grewer, 1994, Eqn (2.37)):

$$\tau_{ad} = \frac{emc_p}{kA} \quad [4\text{-}5]$$

and

$$k = Z\exp\left(-\frac{E}{RT}\right) \quad [4\text{-}6]$$

where m is the mass of the reacting material. Although the use of a constant Z factor in Equation [4-6] implies this method is inapplicable to autocatalytic reactions, Grewer's comparison (p. 224) of SADT values determined from ARC test data and larger scale tests for various materials is very encouraging. Fauske (2000b) also compared calculated SADT values obtained for various organic peroxides from ARC and RSST data and Semenov explosion theory with those measured using the US tests for a 25-kg package. The values differed by at most 10°C, and in some cases were within 2°C. He also pointed out some cases (including the powder dicumyl peroxide) in which different organizations obtained very different kinetic parameters that produced correspondingly different SADT values (from 56°C to 85°C).

Kyotori's (1999, 2003) research suggests that the Semenov theory is applicable to liquids and to materials that are powders at room temperature but melt before they undergo significant decomposition. He finds that the

SADT values for particulate materials that thermally decompose while in the solid state are better predicted by the Frank-Kamenetskii spontaneous ignition theory described in Section 4.3.4. Materials prone to autocatalytic decomposition are not readily represented by either theory, and require some type of isothermal storage test to determine the SADT.

It is important to emphasize that the SADT values, whether determined by calculation or by measurement, are only applicable for the storage of isolated packages for a period of no more than 7 days. Gray (2002) points out that the stacking of packages decreases the effective heat loss rates for each package because there are fewer exposed surfaces for cooling. Thus, the effective safe storage temperature may be significantly less than the SADT, especially if the storage is stacked and expected to be of a much longer duration than 7 days. In conclusion, the SADT should be recognized and treated as a configuration and storage time dependent parameter, rather than a fundamental material property.

Gas Evolution and Other Test Methods

The Lutolf oven test shown schematically in Figure 4-17 is very useful when there is a need to determine the gas evolution rate, as well as the onset temperature, for decomposition reactions. Siwek (1996) calls the resulting temperature from this test the relative decomposition temperature because it is based on a comparison to the temperature of an inert reference material. He uses a sample size of about 2 g, and heats the material up to about 350°C at a rate of 2.5°C/min. An 8-hour period is used to determine if the test sample shows any increased temperature compared to the reference material. He

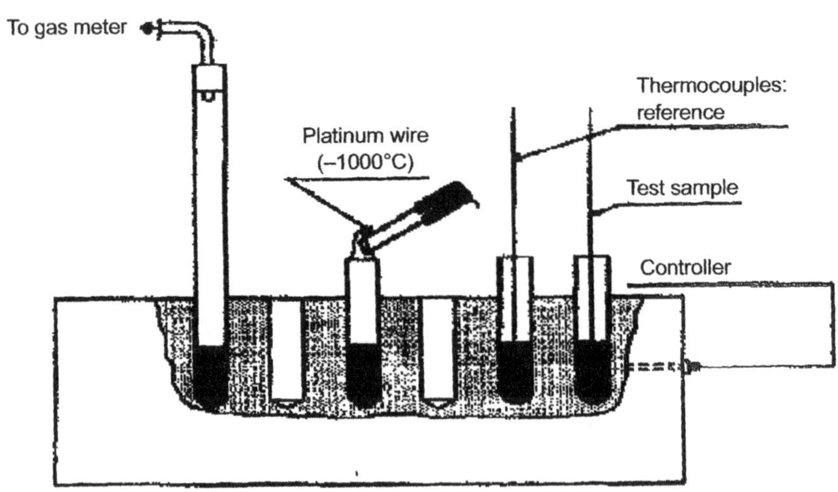

Figure 4-17 Lutolf oven test setup to measure gas evolution (from Ciba brochure).

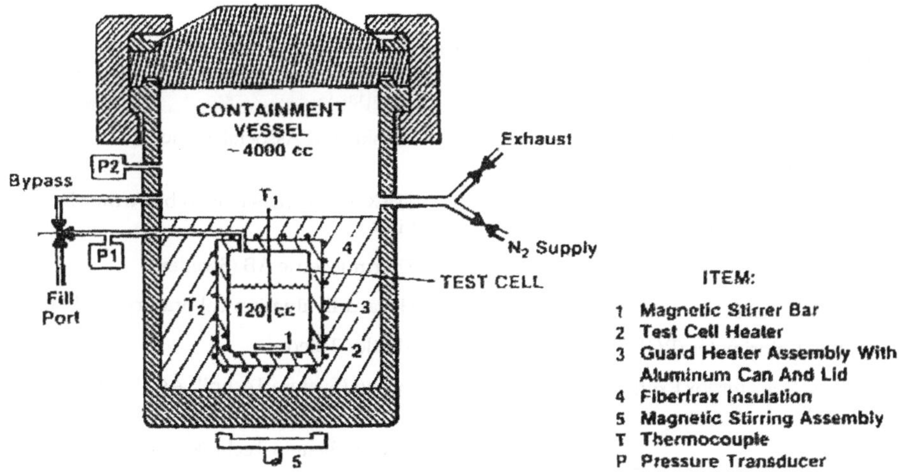

Figure 4-18 Vent Sizing Package test apparatus (from Grewer, 1994).

recommends using a 50°C safety factor when applying the results of the Lutolf Oven test to process equipment such as vacuum ovens and dryers.

One application in which it is crucial to know the gas production rate during a runaway is the design of emergency vents. The AIChE DIERS program and Fauske and Associates, Inc. developed the Vent Sizing Package (VSP) test apparatus shown in Figure 4-18 to determine reaction parameters that are need for emergency vent design. The VSP uses a thin walled metal sample container of 120 cm^3, which corresponds to a φ of about 1.07 (Grewer, 1994, p. 135), that is, much lower than most other reaction test equipment. Since the thin-walled sample container cannot withstand much of a pressure difference across its wall, the VSP has a provision to pressurize the containment vessel to match the pressure developed in the sample container. The containment vessel maximum pressure is 100 bar. Besides developing thermal stability data to characterize the runaway reaction hazard, the VSP, and its successor known as VSP2, are designed to generate data on possible effluent two-phase flow regimes. The flow regime data are needed for the emergency vent design, as described by Fauske (2000a).

During the past approximately ten years, other laboratory test equipment for reactivity/instability determinations have been developed and commercialized. Table 4-4 is a listing of much of the newer equipment and the associated companies from which they can be purchased. Descriptions and comparisons of the equipment are available, for example, in CCPS Guidelines (1995b, 1998). Suggested approaches to using the equipment for the systematic evaluation of chemical reactivity hazards are also available (see, for example, CCPS, 1995b, Johnson et al., 2003, and Melhem, 2003).

TABLE 4-4
Other Reactivity/Instability Laboratory Test Equipment

Test Equipment	Company
Advanced Reactive System Screening Tool (ARSST™)	Fauske and Associates, Inc.
Automatic Pressure Tracking Adiabatic Calorimeter (APTAC™)	TIAXX, LLC (Developed by Arthur D. Little)
CPA	ThermoMetric AB (Sweden)
PHI-TEC II	Hazard Evaluation Laboratory (HEL, UK)
Reaction Calorimeter (RC1™)	Mettler Toledo
Sensitive Detector for Exothermic Processes (SEDEX)	Systag (System Technik AG, Switzerland)
SIKAREX	Systag (System Technik AG, Switzerland)

Although most of the materials tested in this equipment are liquids, powders can be tested in at least some of the devices. For example, the ARSST can be used with a slightly larger fill tube to allow solid particulates to flow into the test chamber without excessive compaction and pressurization. Equipment manufacturers should be consulted about special operational considerations recommended for testing powders.

Impact and Friction Sensitivity Tests

Certain particulate materials can decompose violently when subjected to high impact or locally severe friction. ASTM E 680 is a standard test method for impact sensitivity. Figure 4-19 is a diagram of the drop hammer test apparatus. A drop hammer with a mass in the range 1.0 to 3.5 kg is dropped from various heights in an attempt to determine the drop height, H_{50}, that will produce a 50% probability of triggering a reaction in the test sample. The sample volume per unit area is standardized at 31.5 mm^3/cm^2, corresponding to a thickness of 0.315 mm. Usually the sample is placed into a confinement cup to allow it to be aligned properly under the raised weight. Drop hammer heights can vary from 5 cm to 320 cm. The standard provides a data analysis method for determining the H_{50} height based on the heights tested and the intervals between test heights.

Siwek (1996) describes an impact sensitivity test in which the test sample is placed in a die, which in turn is placed on an anvil and subjected to dropped weights. The test is conducted in a dark chamber to detect any flame or other luminous indication of a reaction. Based on the results of this test, decisions are made about processing the material in a grinder or hammermill.

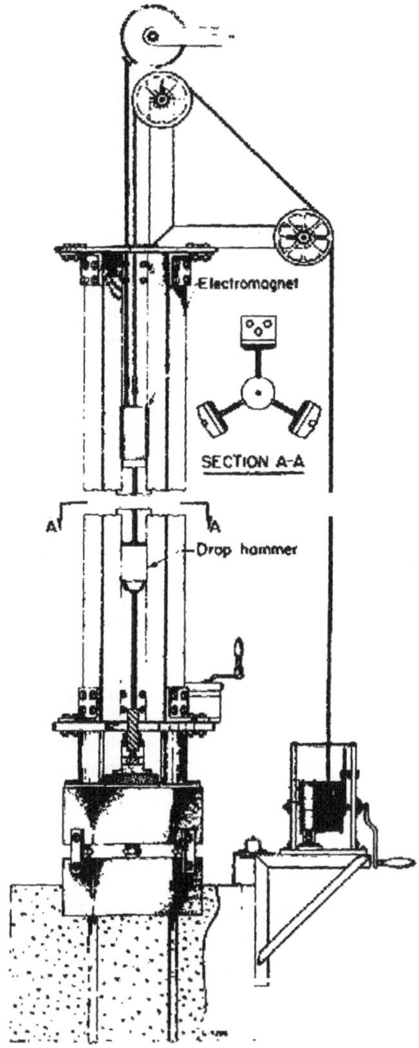

Figure 4-19 ASTM E 680 drop hammer test for impact sensitivity. (Reproduced with permission from ASTM.)

The BAM friction sensitivity test (CCPS, 1995b) determines the smallest frictional force that can trigger a decomposition reaction or ignition of particulate material. Although the test does not simulate actual processing conditions, it does provide a relative measure of a material's sensitivity to a friction triggered hazardous reaction.

Summary of Instability Tests

How does a process safety engineer or analyst decide which of these many instability tests to use for a specific particulate material or application? It

TABLE 4-5
Summary of Thermal Instability Hazard Tests

Test	Standard or Reference	Applications	Advantages	Limitations
DSC/DTA	ASTM E537, E968, VDI 2263	Preliminary screening for exothermic/ endothermic reactions	Small sample size; Low cost. Readily available.	Accuracy depends on heating rate and other factors. Onset temperatures often underestimated. Possible oxygen limitation in combustion reactions.
ARC	Townsend (1981)	Exotherm energy and kinetic data	Widely used and well documented.	Not frequently used with powders. Assumptions in Towsend and Tou original paper not applicable to consecutive and autocatalytic reactions.
SADT	UN Test Manual (1999)	Determine safe storage or transport temperature for packaged powders.	UN/DOT categorization for shipping	Four different test versions. Only applicable to 7- day hold period. Questionable basis for scaling up test data to larger packages.
Isothermal Storage	ASTM E487 Abbot (1990)	Processing or storage at elevated temperatures	Representative of high temperature process/ storage conditions	Can require long test durations. Results for some materials may be influenced by restricted air access in the oven.
Grewer Oven and Aerated Test Cell	Grewer (1994) Gibson et al. (1985) VDI 2263	Powders in process equipment with heated air flows	Accounts for Heated Air Flow Effects	Heat losses in test apparatus may differ from those in actual equipment; therefore results are only approximate.
VSP and VSP2	CCPS (1995) Fauske (2000a)	Exotherm data and relief vent design data.	Near-adiabatic test conditions; high pressure capability Well suited for 2-phase vent flow.	Need DIERS technology know-how to use data.
ARSST	CCPS (1995b)	Exotherm screening SADT approximation	Smaller sample size than VSP	Primarily a screening tool and for use with DIERS technology.
APTAC, PHI-TEC et al.	CCPS (1995b)	Exotherm data and relief vent design data.	Near-adiabatic test conditions; high pressure capability	Relatively new developments with less experience base than other methods.

TABLE 4-6
Summary of Mechanical Instability Tests

Test	Standard or Reference	Applications	Advantages	Limitations
Impact Sensitivity	ASTM E680 UN (1999) VDI 2263	hammermill processing, UN explosives class	UN/DOT categorization for shipping	Primarily used for explosives. Limited scientific/scaling background.
Friction Sensitivity	UN (1999)	Grinder, ball mills UN explosives class	UN/DOT categorization for shipping	Primarily used for explosives. Limited scientific/scaling background.

depends on the experience level of the organization and engineer/analyst with this and similar materials, the amount of material available for testing, and the intended processing, storage, and handling applications. Tables 4-5 and 4-6 provide simplified summaries comparing the types of applications, advantages, and limitations of the instability tests described above.

4.3.3 Laboratory Test Methods for Chemical Incompatibility Hazards

Several of the test methods described in Section 4.3.2 can also be used to assess chemical incompatibility reactions. For example, DSC tests are an excellent way to determine at what temperature two materials might react exothermically, and what the corresponding reaction energy is. Similarly, ARC tests and VSP tests are used routinely to determine fundamental kinetic data on chemical reactions that can escalate into runaways. Frurip et al. (1997) reviewed some early use of DTA tests for compatibility evaluations. However, since the mixture samples for these methods are prepared and inserted into the apparatus before any measurements can be made, these methods cannot assess the hazards associated with the act of mixing per se.

Frurip et al. (1997) caution that ad-hoc laboratory mixing tests with new material combinations should be conducted with small quantities of materials (<<1 g) in order to minimize dangers to test personnel. One small-scale test method developed specifically for determining chemical incompatibility hazards, including the hazards of mixing, is the two drop mixing calorimeter developed by Dow (Hofelich, 1997, and Frurip et al., 1997). This calorimeter is intended to measure the heat of mixing, and to collect any gases produced from the mixing reaction. The apparatus consists of a twin-cell microcalorimeter that allows one material (usually a liquid) to be injected from a small syringe (0.050 to 0.10 ml) into a small sealed vial containing about 50 mg of the second material. The temperature difference between the test sample and an inert reference vial is measured and integrated, with a small electrical resistor used for calibrating heat generation calculations. In order to collect and measure the volume of gas generated, the test vial is pierced with a small syringe leading to a gas burette. A similar test protocol could be setup for particulate material reactivity testing (Stevick, 2003).

NFPA 704 (2001) has a water reactivity hazard classification scheme based on the heat of reaction or heat of mixing determined via this method. The demarcations between the four hazard classes are at reaction energies of 30 cal/g-mixture, 100 cal/g, and 600 cal/g using a 1:1 weight ratio of chemical to water. Adjustments to some of these classifications are made based on the generation of gases during the reaction. Examples of particulate materials in Water Reactivity Hazard Categories 1 and 2 are: ~~W~~1—sodium hydrosulfite, ~~W~~2—calcium carbide (which produces calcium hydroxide plus acetylene from an exothermic reaction with water).

DOT regulations for hazardous materials and U.N. (1999a) *Recommendations for Transport of Dangerous Goods* include a Division 4.3 category for Dangerous When Wet Materials. The test methods for making this determination, as described in U.N. 1999b, involve first doing a preliminary determination of spontaneous ignition by mixing a very small quantity of material with water, and then doing a quantitative measurement with a larger quantity of material that has not reacted violently or produced spontaneous ignition in the preliminary test. The criterion for assigning a material to this category involves flowing water into a flask containing up to 25 g of the material, and observing whether there is either spontaneous ignition or the production of flammable gas at a rate greater than 1 liter per kilogram of material per hour (0.0167 liter/kg-min). Materials in this category are assigned to one of the three different packaging groups depending on the test results as delineated in Table 4-7.

The U.N. Recommendations specify that particulate or friable materials first be ground to a powder before undergoing the water reactivity test. The test itself is required to continue for 7 hours, with gas generation rates calculated at 1-hour intervals.

Another pertinent concern with regard to contact with water is the possible deterioration of the particulate material packaging or container. In par-

TABLE 4-7
DOT/UN Packing Groups for Dangerous When Wet Materials

Division 4.3 Packing Group	Flammable Gas Generation Rate (liter/kg-min)	Spontaneous Ignition?
I	≥10 for any one minute	Yes[a]
II	<10 but ≥0.33	No
III	<0.33 but >0.0167	No
Not Division 4.3 material	≤0.0167	No

[a]A material is assigned to Packing Group I if it is spontaneous ignition *or* it generates > 10 liter/kg-min.

ticular, there is concern about the effects of water on flexible intermediate bulk containers (FIBCs). The DOT requires that paper wall materials used in the construction of FIBCs be subjected to a water immersion test (49CFR Part 178.710), and retain at least 85% of its dry tensile strength.

Some ad hoc chemical incompatibility test methods and test equipment have been reported to address issues such as particulate material reactivity with multiphase reactants. One of these methods was used by Reza et al. (2002) to simulate the exothermic runaway reactions that led to the October 1998 explosion at the Condea Nast plant in Baltimore. The explosion occurred when steam was injected into a plugged reactor vessel outlet in an ill-advised attempt to clear the plug. The plug was composed of a sludge-like mixture of coarse aluminum powder, aluminum chloride, and various hydrocarbon liquids. The test vessel used by Reza et al. to determine the reactivity of various combinations of reactants is shown in Figure 4-20. The agitator at the vessel bottom allows particulate materials to be mixed together and added to liquid phase reactants. The vent line allows for the continuous injection of steam or some other vapor. Reza et al. used this vessel to determine that the runaway reaction was initiated by steam reacting with aluminum chloride to produce hydrochloric acid, which in turn

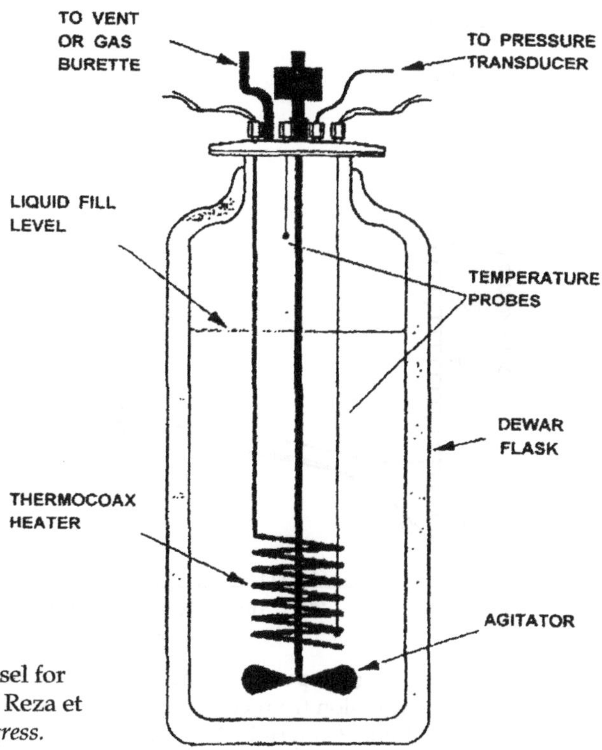

Figure 4-20 Ad-hoc test vessel for multiphase reactivity, from Reza et al. (2002) *Process Safety Progress.*

stripped the oxide coating off the aluminum powder. The aluminum then reacted uncontrollably with the hydrocarbon liquids in the sludge. Pressures in excess of 45 bar g (660 psig) generated during these reactions ruptured the test vessels.

4.3.4 *Self-Heating, Spontaneous Combustion, and Pyrophoric Solids Test Methods*

Constant Temperature Oven Tests

Most particulate self-heating tests involve filling wire mesh baskets with the particulate material and placing the filled baskets in a preheated oven at some temperature, T_{a1}. One or more thermocouples inserted in the particulate sample are used to monitor the sample temperature as a function of time. If no self-heating occurs, the oven temperature is increased to a higher value, T_{a2}, and a new instrumented sample-filled basket is inserted. The procedure is repeated until there is a sharp rise in sample temperature above the oven temperature as shown in Figure 4-21. Once a positive self-heating test has been observed, the tests are usually repeated to achieve a small temperature

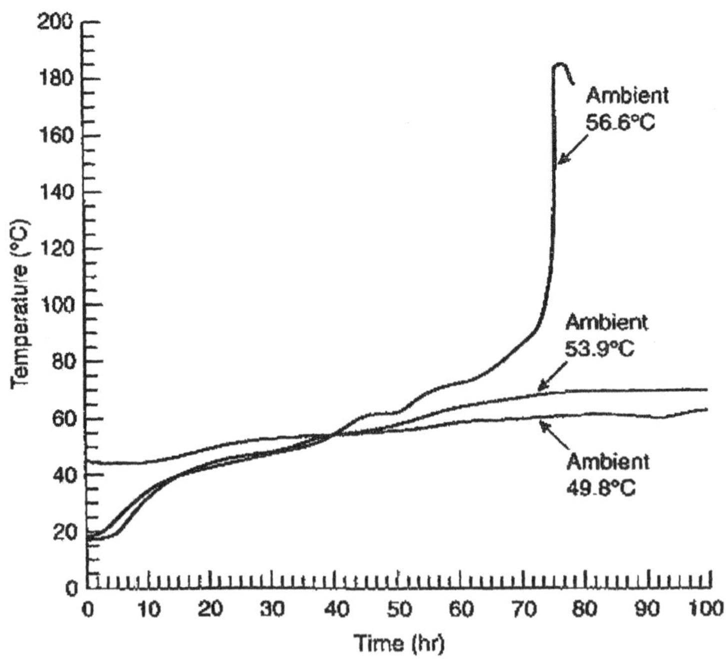

Figure 4-21 Self-heating temperature traces for anhydrous calcium hypochlorite. (Reprinted with permission from Gray (2002) in the *SFPE Handbook of Fire Protection Engineering*, 3rd Edition, Copyright 2002 the Society of Fire Protection Engineers.)

difference between a positive self-heating reaction and a test with only a small gradual increase above the oven temperature, as in Figure 4-21 for hydrated calcium hypochlorite. The average of these two temperatures is called the critical temperature, T_{ac}, corresponding to the particular sample basket size used. The value of T_{ac} for the 35-cm-diameter basket data in Figure 4-21 is 55.2°C (Gray, 2002).

Since the critical temperature is a function of the sample basket size, the preceding sequence of tests is repeated with the same material in a different size basket. Typically baskets are used with characteristic dimensions (diameter of cylinder or side of cube) varying from about 5 cm to about 50 cm. After doing several sets of these experiments, the data are correlated and extrapolated as explained in the following description of Frank-Kamenetskii theory in order to estimate either the maximum safe storage temperature for a given size pile of material, or the maximum safe pile/silo size for a given storage temperature or storage time.

Theoretical Basis for Data Correlations

Classical spontaneous ignition theories (Frank-Kamenetskii theory and Semenov theory) are based on the simplifying assumption that there is one exothermal chemical reaction responsible for the material self-heating. This reaction is assumed to generate heat at a rate governed by an Arrhenius reaction temperature dependence, and usually the reactant consumption is neglected. The Frank-Kamenetskii approach is to first model a steady-state heat transfer process in which the heat generated by this reaction is balanced by the heat transferred at the surface of the pile, and then to use the model to determine the conditions under which a steady-state solution is not possible because the surface heat transfer rate is not adequate. More comprehensive descriptions of this approach for different boundary conditions are provided by Bowes (1984), Beever (1995), and Gray (2002); the brief summary presented here is an abbreviated version of the discussion in Chapter 6 of Zalosh (2003).

Using the approximation that the temperature rise above ambient is small in comparison to the ratio of the activation energy to the universal gas constant, the differential equation governing the steady-state heat transfer with Arrhenius rate internal heat generation is (Gray, 2002, p. 2-219 and Bowes, 1982, p. 27):

$$\nabla^2\theta = -\delta e^{\theta} \qquad [4\text{-}7]$$

where θ is a nondimensional temperature rise,

$$\theta = \frac{E(T - T_a)}{RT_a^2}$$

and δ is the Frank-Kamenetskii parameter,

$$\delta = \frac{E\rho QAr^2}{RT_a^2\lambda}\exp\left(-\frac{E}{RT_a}\right)$$

which represents the ratio of the heat generation rate at T_a to the heat conduction rate from the center of the material.

The other parameters in Eq. [4-7] are as follows:

E is the reaction activation energy (J/mol),

A is the preexponential factor in the Arrhenius reaction rate equation, $f(c_0)$, (s^{-1})

R is the universal gas constant = 8.314 J/mol-K,

ρ is the material bulk density (kg/m^3),

Q is the heat of reaction (J/kg),

r is characteristic length of the storage pile (m),

λ is the material thermal conductivity (W/m-K),

T_a is the material storage temperature (K).

The boundary conditions associated with Eq. [4-7] are:

$$\frac{d\theta}{dz} = 0 \qquad \text{at } z = 0 \text{ (center of storage pile)}$$

$$-\frac{d\theta}{dz} = \alpha\theta_s \qquad \text{at } z = r \text{ (center of storage pile)} \tag{4-8}$$

where the Biot number, α is given by $\alpha = h_r/\lambda$, h is the sum of the natural convection and radiation heat transfer coefficient at the pile surface. The radiation contribution, h_r, is (Bowes, 1984, p. 190)

$$h_r = \frac{4\varepsilon_1\varepsilon_2\sigma T_a^3}{\varepsilon_1 + \varepsilon_2 - \varepsilon_1\varepsilon_2}$$

where ε_1 and ε_2 are the emissivities of the particulate surface and surrounding environment, and σ is the Stefan-Boltzman constant.

Unstable storage conditions, as predicted by this theory, correspond to situations in which the value of δ is larger than a critical value denoted by δ_c. The value of δ_c depends on the geometry of the storage pile and the value of α. In most spontaneous heating applications with pile characteristic dimensions of at least several m, the Biot number is sufficiently large (greater than about 20) for the asymptotic limit, $\alpha \to \infty$ to be applicable. Values of δ_c corresponding to this limit for different geometry storage piles are listed in Table 4-8. Adjustments for smaller values of α are given by Bowes (1984). Relationships for α_c in the case of an initially heated material in a cooler environment are given by Bowes (1984), with updates by Beever (1995), and some experimental results by Anthony and Greaney (1979).

By taking logarithms of the defining equation above for the Frank-Kamenetskii parameter, we can obtain

TABLE 4-8
Critical Values of Frank-Kamenetskii Parameter for Different Shape Storage Piles (compiled from values in Bowes, 1984)

Pile Geometry	Dimension	δ_c
Plane slab	Height $2r$ << width and length	0.88
Rectangular box	Height $2r$, width $2w$, length $2l$	
Cube	$2r \times 2r \times 2r$	2.52
Cylinder	Diameter << Height	2.00
Cylinder	Diameter/Height = $2r/2l$	$\delta_c(r) = 2.0 + 0.84(r/l)^2$
Cone[a]	$D/H = 1$	9.1 ± 0.1
(Diameter = D	$D/H = 2$	3.2
Height = H)	$D/H = 4$	1.7

[a] δ_c for cones is based on the characteristic dimension $H/2$.

$$\ln\frac{\delta_c T_a^2}{r^2} = M - \frac{P}{T_a} \quad [4\text{-}9]$$

where

$$M = \ln\frac{EQ\rho A}{R\lambda}$$

and $P = E/R$.

Writing the equation in this form provides a framework for correlating and extrapolating the results of the spontaneous heating initiation tests in which the critical values of storage temperature, T_a are measured for a laboratory sample of dimension r. If the tests are repeated with other size samples and results are plotted in the form $\ln(\delta_c . T_a^2/r^2)$ versus $1/T_a$, the data should be linear as shown in Figure 4-22 for anhydrous calcium hypochlorite. The slope of the best-fit line through the data should correspond to the value of material parameter P, that is, E/R, and the extrapolated intercept at $1/T_a = 0$ should correspond to the material parameter M. Results will depend on the units for T_a and r used to plot the data and obtain the best-fit. According to Gray (2002), many particulate spontaneous combustion activation energies determined in this manner are on the order of 100 kJ/mol.

Using the tabulated values of M and P, the preceding equation can be used to calculate the value of r for a maximum safe storage pile at an ambient temperature T_a, or the maximum safe material curing temperature T_a for a given value of r. Solving Eq. [4-9] for the critical radius, r_c, for spontaneous ignition,

$$r_c = T_a\sqrt{\delta_c e^{-(M-P/T_a)}} \quad [4\text{-}10]$$

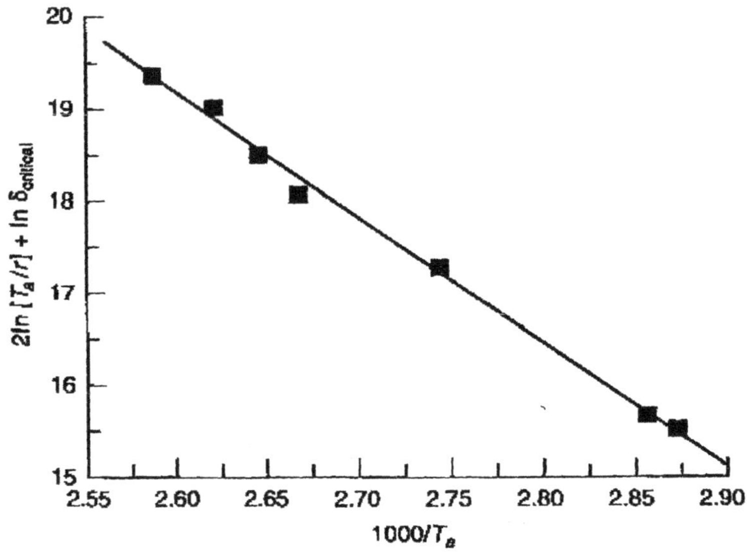

Figure 4-22 Frank-Kamenetski parameter plot for anhydrous calcium hypochlorite. (Reprinted with permission from Gray (2002) in the *SFPE Handbook of Fire Protection Engineering*, 3rd Edition, Copyright 2002 the Society of Fire Protection Engineers.)

As an example, consider the case of a chemically activated carbon with *M* and *E* values of 11,670 (for *r* in mm) and 97 kJ/mol, respectively, as given by Bowes (1984). If the activated carbon is to be stored in a silo with a height/diameter ratio of 1, such that $\delta_c = 2.84$, the critical silo radius, r_c, for a storage temperature of 20°C is 3.5 m. Jones (1998a,b, 1999) reported on results for 10 other activated carbon samples with activation energies ranging from 88 kJ/mol to 113 kJ/mol. Since the relationship between r_c and T_a is very sensitive to the value of *E*, the critical storage pile size calculations should be based on process-specific and site-specific test data.

There are some variations of this approach for data analysis. Rather than use the Frank-Kamenetskii equations, VDI 2263 states that the results from several basket tests be plotted using $1/T_a$ versus ln(*V*/surface area). The VDI standard allows the critical temperature for several sizes of cones, cylinders, cubes, spheres, and tetrahedrons, all to be plotted on the same straight line graph. Time to ignition can also be estimated with this approach, i.e. by plotting induction time as a function of *V* with constant T_a or vice versa (VDI 2263, Section 1.4.2).

The data analysis method and the microcalorimeter test (for *Q*) used by Jones (1998a, 1999) allowed him to run only one set of oven tests (using different basket size baskets at one oven temperature) for each sample of activated carbon. This allows for a considerable savings in time over the more tedious multiple test series. Gray (2002) points out that these shortcut methods are attractive when there is only one chemical reaction, but there are many applications for which the more extensive set of oven tests produce correlations that reveal the occurrence of two different chemical reactions, with each reaction being dominant over a limited range of temperatures. Gray (2002) concludes, "measurements at a single temperature or over a limited temperature range can give dangerously flawed results."

Oven Testing for U.N. Dangerous Goods Transport Classification of Self-Heating Materials

The U.N. classification scheme for transport of hazardous materials has a category in Division 4.2 for self-heating substances. An oven testing procedure (U.N., 1999b) is used to determine if a particular material should be so classified and, if so, what Packaging Group it should be assigned. The particular tests and criteria for classification are shown in flow chart form in Figure 4-23.

The U.N. classification tests indicated in Figure 4-23 are conducted in a hot air recirculation oven with a minimum volume of 9 liters. Stainless steel mesh sample baskets of various sizes are filled with particulate material, instrumented with 0.3 mm diameter chromel-alumel thermocouples in the sample center, and then hung in the oven. The first test is made with a 100 mm sample cube (1000 ml) at an oven temperature of 140°C. If the sample temperature does not reach 200°C after 24 hours, the material is not a self-heating material of U.N. Division 4.2. On the other hand, if the material does self-heat to at least 200°C, it is subjected to additional tests as indicated to determine if it is a Packaging Group II or Packaging Group III material, and whether it is eligible for the exemptions corresponding to the indicated small package sizes. In each case, a temperature rise of 60°C above the oven temperature is the criterion. According to the U.N. (1995) report, the criteria are based on 50°C self-ignition temperature for charcoal in a 27 m^3 container. Materials with extrapolated self-ignition temperatures higher than 50°C in a 27 m^3 container are considered Division 4.2 self-heating materials.

Other Types of Spontaneous Combustion Laboratory Tests

Siwek (1996) and Eckhoff (1995) described a constant oven temperature test to determine the so-called Self-Heating Temperature or the Self-Ignition Temperature (SIT) of particulate materials. The test is conducted as shown in Figure 4-24, and is almost the same as the U.N. self-heating classification test described above, but the oven heating time (72 hours) and the threshold self-ignition temperature criterion (400°C) are greater than those used in the U.N. tests. If the temperature increases above the oven temperature, but does not reach 400°C, the result is termed self-heating rather than self-ignition. The cylindrical wire mesh baskets in these Bartknecht/Siwek tests range in size from 400 ml to 3000 ml, and the variation of measured self-ignition tempera-

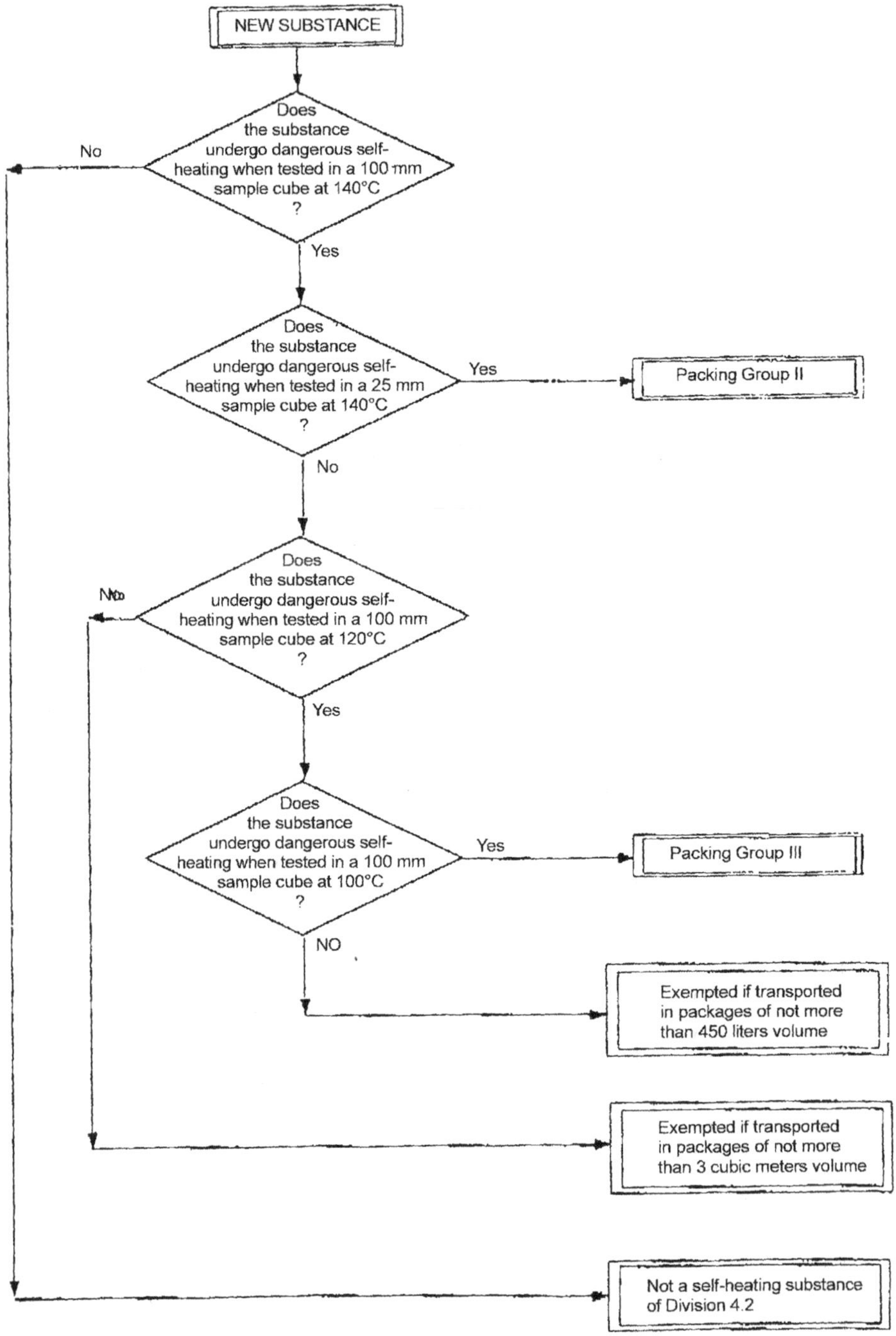

Figure 4-23 UN Classification testing scheme for self-heaters, from UN (1995).

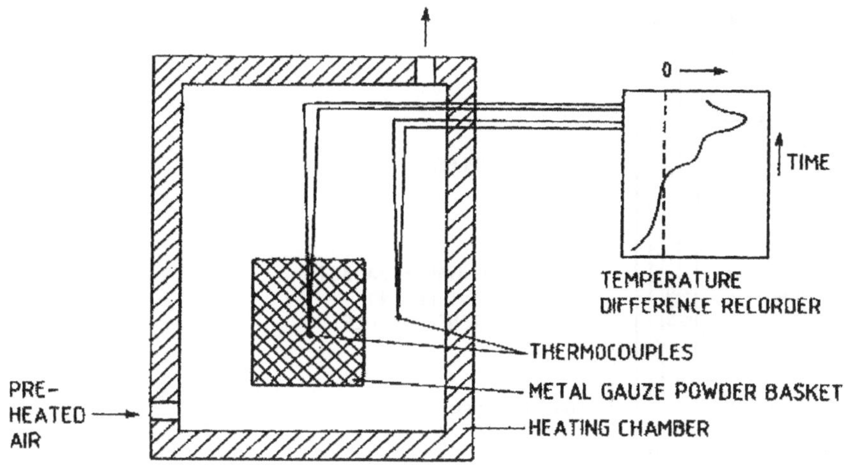

Figure 4-24 CEN apparatus for self-heating and spontaneous ignition temperature tests (from Eckhoff, 2003).

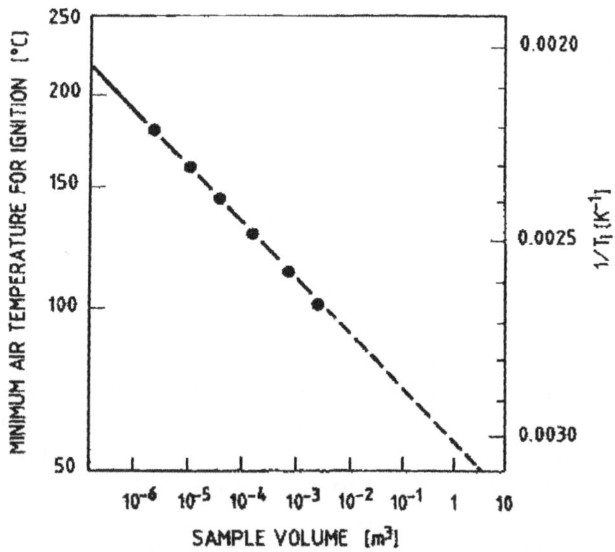

Figure 4-25 Variation of self-ignition temperature with test sample volume in Bartknecht/Siwek test apparatus (from Eckhoff, 2003).

ture with sample volume for a representative material is shown in Figure 4-25. These SIT tests are now codified in the annex to EU Directive 92/69.

The Grewer oven test described in 4.3.2 is also used sometimes to determine a so-called relative self-ignition temperature, RSIT (Siwek, 1996). Samples of the test material and of graphite powder (the inert reference material) are placed into 8-ml mesh baskets and heated up to 350°C at a rate of 1°C per

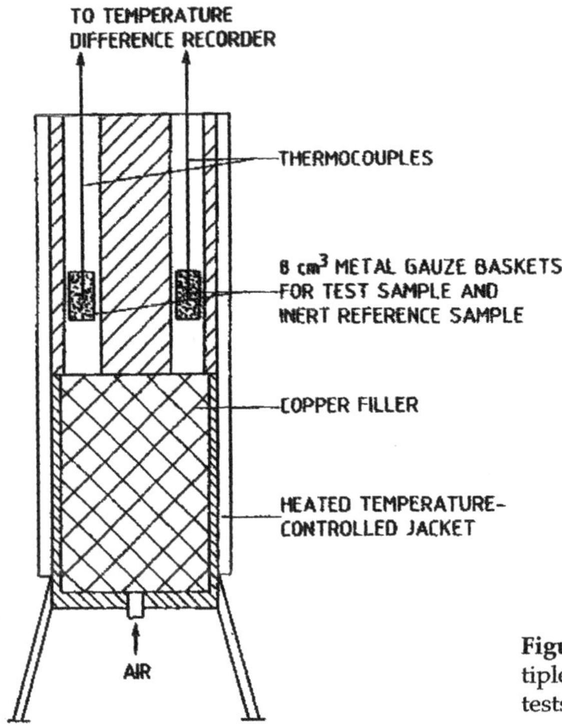

Figure 4-26 Grewer oven for multiple sample spontaneous ignition tests (from Eckhoff, 2003).

minute in the Grewer oven as illustrated in Figure 4-26. The RSIT is the oven temperature that causes the test material to self-heat up to a minimum temperature of 400°C. Siwek suggests that materials should not be processed (in equipment such as spray or pneumatic conveyor dryers) at temperatures within 50°C of their measured RSIT value.

The aerated cell, shown in Figure 4-27, can be used in a constant temperature mode as well as in the slow, programmed temperature ramp used for exothermic decomposition temperature measurements. Data at a constant oven temperature are more likely to indicate any tendency toward self-heating in a drier with a large air flow.

ASTM D 1929-96 (2001) is another oven test method to determine the spontaneous ignition temperature, in this case for plastic pellets or powders. The test sample consists of 3 g of particulates (or other solid forms) in a small (4-cm diameter) specimen pan within the furnace shown in Figure 4-28. There is a provision for heated air to flow up and around the sample pan at a velocity of 2.5 cm/s. After the furnace test temperature is established, the sample pan is raised to the opening at the top of the furnace to allow the test sample to be inserted. The standard calls for a maximum 10-minute observation period after the pilot flame is ignited to determine whether piloted (flash) ignition has occurred. After the piloted/flash ignition temperature is

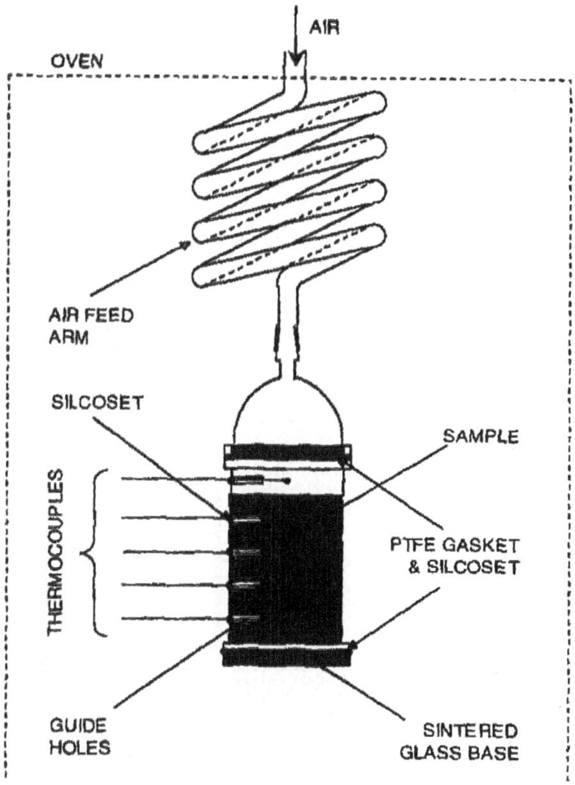

Figure 4-27 Aerated cell thermal stability test, copyright Syngenta 2004, used with permission.

determined, the test is repeated at higher temperatures to determine the spontaneous ignition temperature in the absence of any pilot flame. The sample form (pellets, sheets, fibers, etc.) is required to be included in the test report, but not necessarily the actual particulate size.

Table 4-9 shows some flash ignition temperatures and spontaneous ignition temperatures reported (ASTM D 1929, 2000) for polystyrene and for polyamide 6 (Nylon 6) samples. The nominal (average for round robin tests) flash ignition temperature is 31°C lower than the spontaneous ignition for polyamide, and 88°C lower for polystyrene. The reproducibility among the six laboratories doing the round robin tests is also indicated in the table. In addition, the last column shows the reported (Eckhoff, 1997) dust cloud ignition temperature measured using the Godbert-Greenwald furnace test described in Section 4.3.7. The dust cloud ignition temperature is approximately equal to the ASTM D 1929 spontaneous ignition temperature for these two polymers.

ASTM E 659-78 (2000) is yet another ignition temperature test. It is intended to determine the autoignition temperature for either liquids or for particulates that melt and vaporize at temperatures below their autoignition

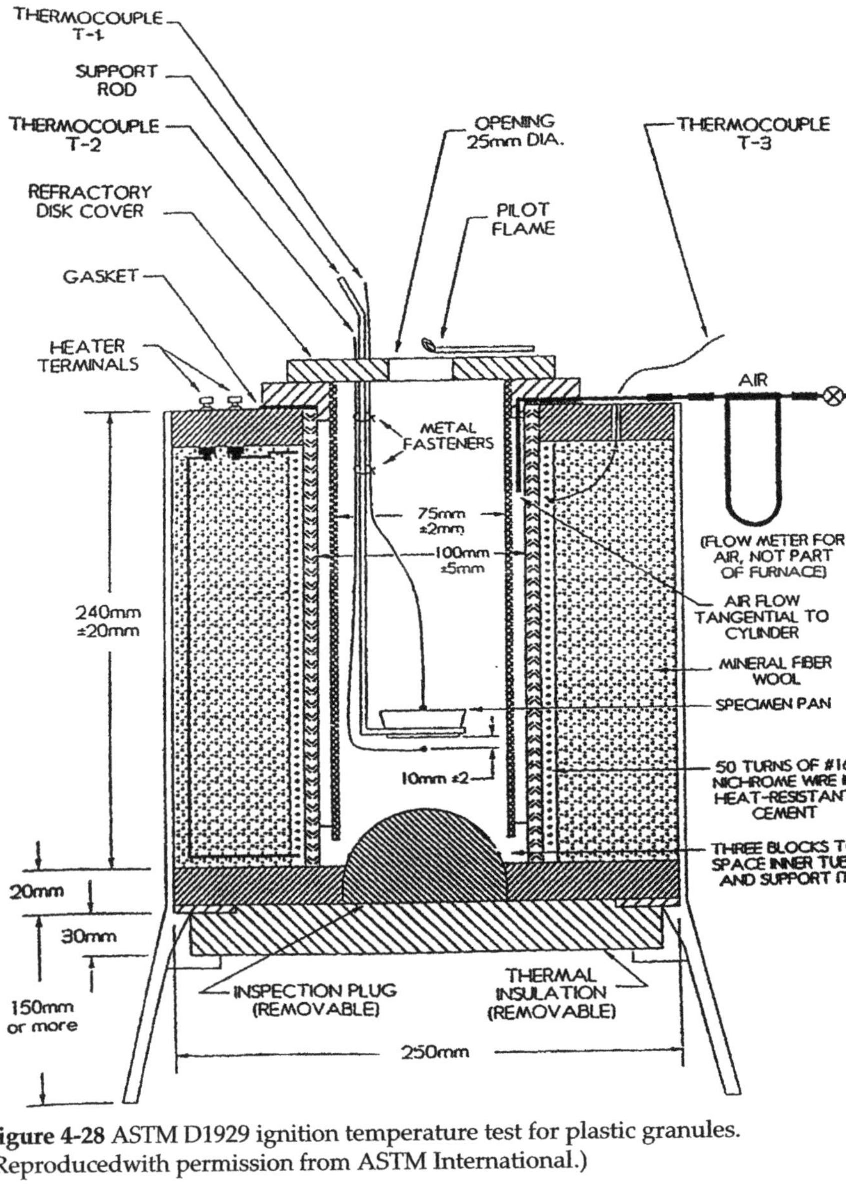

Figure 4-28 ASTM D1929 ignition temperature test for plastic granules. (Reproducedwith permission from ASTM International.)

temperatures. Figure 4-29 shows the ASTM E 659 test apparatus, which consists of a 500 ml borosilicate flask suspended inside a cylindrical furnace. After the furnace/flask test temperature is reached, a powder sample is inserted into the flask via a filling funnel, and a timer is started. The possible occurrence of flaming ignition within ten minutes is noted by viewing the flask through a mirror above the top opening in the furnace.

TABLE 4-9
ASTM D 1929 Granule Ignition Temperatures Compared to Dust Cloud Ignition Temperatures

Polymer	Flash Ignition Temperature for Granules (°C)	Spontaneous Ignition Temperature for Granules (°C)	Godbert-Greenwald Dust Cloud Ignition Temperature (°C)
Polyamide 6	413 ± 38	439 ± 56	450
Polystyrene	370 ± 52	458 ± 59	450

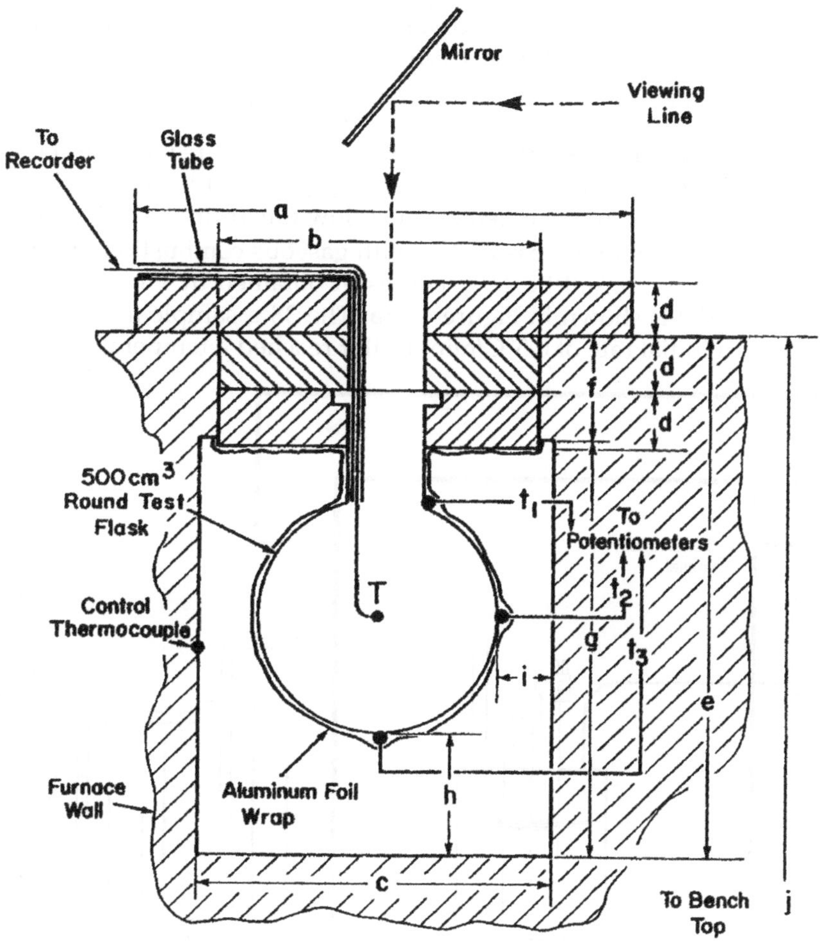

Figure 4-29 ASTM E 659 ignition temperature test. (Reproducedwith permission from ASTM International.)

According to ASTM E 659, the first sample tested should be 100 mg. After determining the autoignition temperature for the 100-mg sample, the procedure is repeated for a 150-mg sample. If autoignition occurs at a lower temperature with the larger sample, the procedure is repeated again for 200 mg and possibly larger samples of the same material in order to find the minimum autoignition temperature. The standard also suggests that the entire test procedure be repeated in larger test flasks (as large as 5 liters), since the autoignition temperature often decreases with increasing test volume.

Greenfield (2003) describes the use of the VSP2 test equipment for screening for self-heating and spontaneous combustion tests. The test is run with the VSP2 programmed heater providing known rates of sample temperature rise, and with a controlled purge air flow into and out of the test cell as shown in Figure 4-30. Air flow rates of 350 cc/min to 400 cc/min (corresponding to approximately 10 test vessel volume changes per minute) at room temperature are typically used. The onset of self-heating can be detected by sharp increases in sample rate of temperature increase above the imposed self heat rate, reductions in heater power input, and in some cases by the occurrence of an increased exhaust flow rate from the test vessel. The increased vent flow rate is associated with gaseous combustion products.

An example of a self-heating test result obtained by Greenfield (2003) using the VSP2 is shown in Figure 4-31a. The 38 g sample in this test was heated at an initial rate of 0.5°C/min until it reached a temperature of 80°C,

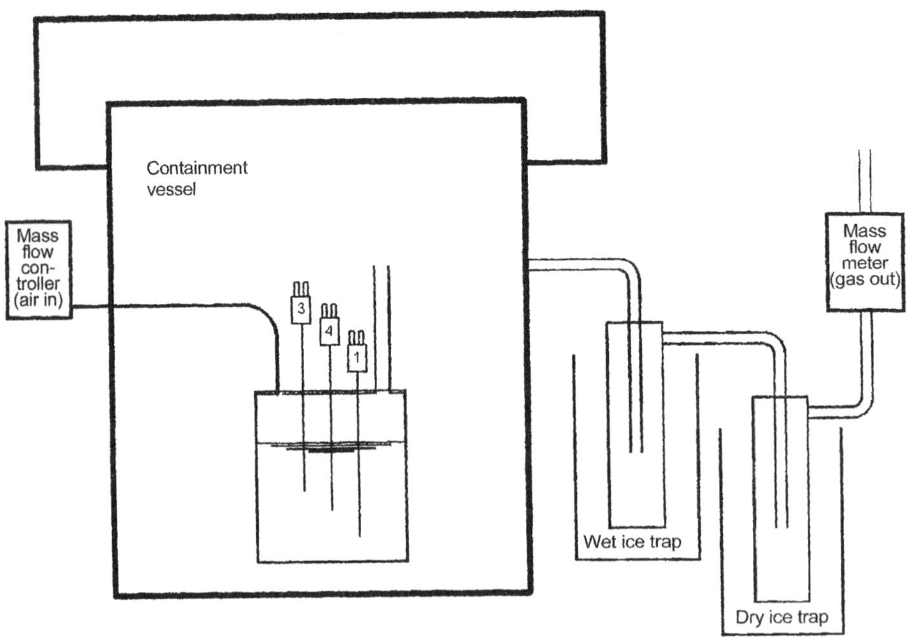

Figure 4-30 VSP2 spontaneous combustion test (from Greenfield, 2003)

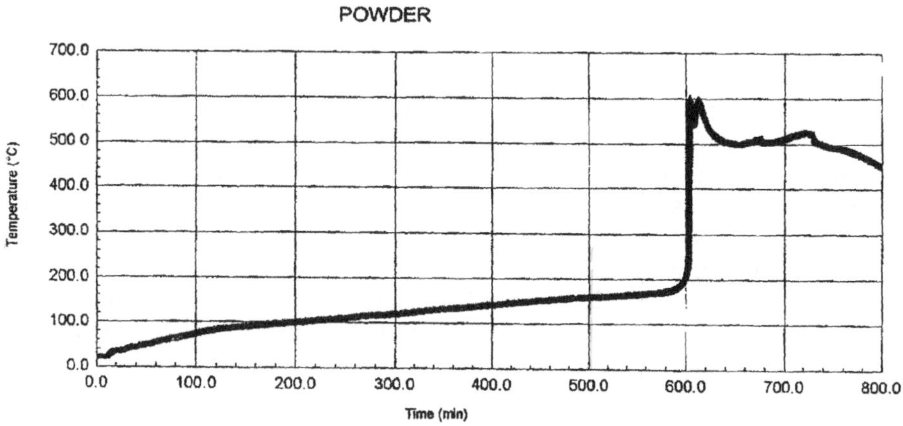

Figure 4-31a Temperature versus time plot from VSP2 spontaneous combustion test (from Greenfield, 2003).

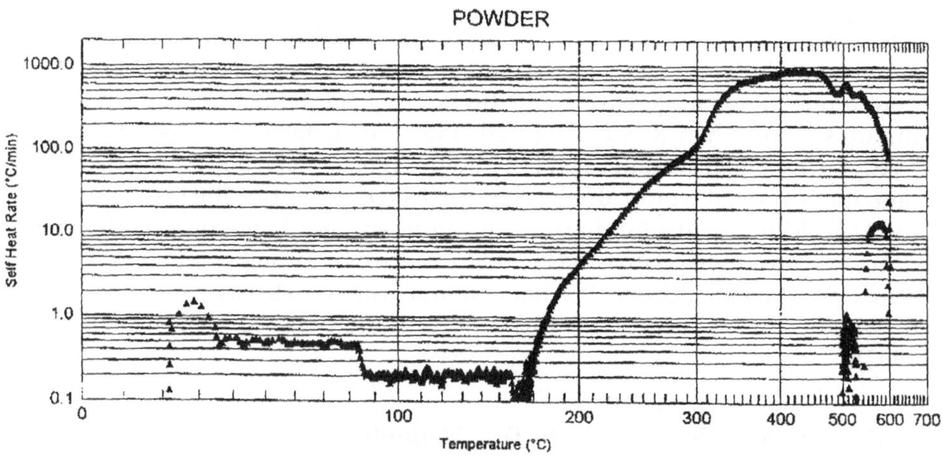

Figure 4-31b Self-heating rate measured by Greenfield for VSP2 test with temperature variation shown in Figure 4-31a.

and then was heated at decreasing rates until it reached its apparent self-heating temperature of 163°C, as evidenced by the steep rise in self heat rate and sample temperature up to about 600°C. The recording of sample heating rates shown in Figure 4-31b shows the self-heating rate increasing sharply starting at 163°C and reaching a peak at about 400°C. Observations of the sample after the test indicated that it had decomposed such that only a small fraction of the original sample remained in the pan. Greenfield notes that this type of test is a good indicator of the potential for self-heating and possible spontaneous combustion for particulate materials being processed in drying and dust collection systems.

ASTM D3523 describes the Differential Mackey Test to determine the spontaneous heating value (SHV) of solids and liquids on cellulosic surfaces. The standard defines the SHV as the maximum amount by which the temperature of the sample exceeds the temperature of a reference material (water) when exposed at a given temperature in the standard apparatus. The apparatus consists of a hot plate heated vessel containing two cylindrical chambers, one for water and the other for the test material. The test material sample is 10 g of solid in 20 g of fresh cotton gauze, and packed into a mesh basket containing a thermocouple. ASTM D3523 notes that the SHV measured by this method varies with the particulate size of the sample. Gray (2002) warns that this type of completely empirical test is "not reliable and cannot be properly related to the basic principles of spontaneous ignition theory."

Pyrophoric Solid Test Methods

The U.N. test N.2. for Division 4.2 pyrophoric solids (U.N., 1999b, p. 328) simply entails dropping one or two ml of powder from a height of about 1 m onto a noncombustible surface. If the powder ignites within a period of 5 minutes from when it was dropped onto the surface, it is classified as a pyrophoric solid. Presumably if the powder ignites even before it is dropped onto the surface, it would also be classified as pyrophoric. The test method in the U.N. Manual (1999b) does not provide any guidance on powder handling procedures to control the level of surface oxidation prior to testing. The thickness of the oxide layer has been demonstrated to be an important factor in whether or not powders exhibit pyrophoric behavior (Glassman et al. 1992).

4.3.5 *Dust Layer Combustibility Test Methods*

The ignition temperature for a dust layer immersed in a heated air environment for an indefinitely long period of time can be determined from the previously described ASTM oven tests, for example, ASTM D 1929 or ASTM E 659, or from the 5 mm depth heated layer test denoted as the minimum ignition temperature test MIT-1 by Querol Aragón et al. (2002). The tests described in this section pertain to the ignition and subsequent combustion of a dust layer in a near room temperature air environment.

Dust Layer Hot Surface Ignition Temperatures

The standard test method for determining the minimum heated surface temperature that can ignite a dust layer deposit of a given thickness is described in ASTM E 2021 and in IEC1241-2-1. ASTM E 2021 describes the test method

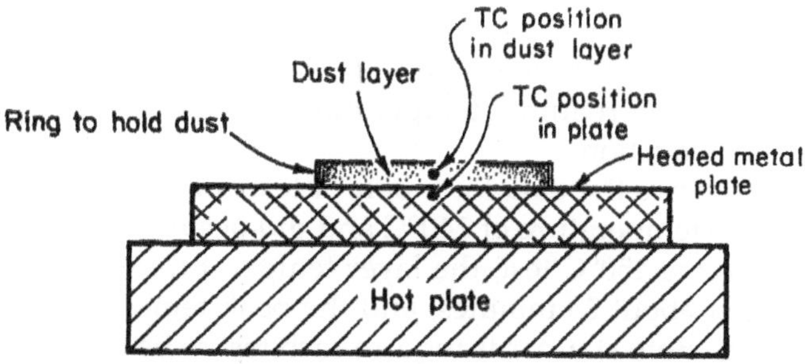

Figure 4-32 ASTM E 2021 and IEC 1241 hot surface test. (Reproducedwith permission from ASTM.)

and ways of reporting and interpreting the data but does not prescribe maximum safe surface temperatures based on test results. IEC 1241-2 describes the same test method but with different ignition criteria and sample particle size.

Both ASTM E 2021 and the IEC 1241-2 test utilize a 10-cm diameter disk-shaped dust sample placed in a metal ring on a 20-cm diameter heated plate as shown in Figure 4-32. The plate temperature is set at some predetermined value for each test, and a thermocouple located in the middle of the dust sample monitors the dust temperature as the plate heats it. The loosely packed dust sample is exposed to the heated surface for a period of about 30 minutes unless there is a positive indication of ignition earlier than that. Tests are repeated until there is only a 10°C difference between a test that results in ignition, and a no ignition test.

The dust sample preparation and the criteria for ignition differ in the two standards as indicated in Table 4-10. The dust sample used in the ASTM E 2021 test (and in virtually all the dust combustibility tests) is supposed to

TABLE 4-10
Comparison of ASTM E 2021 and IEC 1241-2

	ASTM E 2021-00	IEC 1241-2-1
Particle size	At least 90% <75 μm	100% <200 μm
Layer thickness	12.7 mm (½-inch) Other depths may also be used.	5.0 mm (Optional: second depth of 12.5 mm or 15.0 mm)
Ignition criteria	Glowing or flaming, or a temperature rise ≥50°C above surface temperature.	Glowing or flaming, or a temperature of 450°C or a temperature rise ≥250°C above surface temperature.
Extrapolation to other layer depths	Obtain data for at least three depths (emphasizing thicker layers) and plot log d versus $1/T_{si}$	Preferable to test with deeper layer, but allowable to plot log d versus $1/T_{si}$

be 90% by weight through a 75-μm sieve size. IEC 1241-2-1 requires only that the test sample passes through a 200-μm sieve. Ignition criteria differ in terms of the minimum required temperature rise for ignition if there is no visible sign of combustion.

Figure 4-33 shows the data for a pair of tests in which ignition did not occur at a surface temperature of 240°C and did occur at a surface temperature of 250°C. In the 250°C test, the layer temperature reached the surface temperature after about 25 minutes, and the peak temperature occurred at about 35 minutes. Both standards require each test to be run at least 30 minutes if there is no indication of self-heating above the surface temperature.

Dust layer thickness can significantly affect test results, such that the resulting hot surface ignition temperature decreases with increasing dust layer thickness as shown in Figure 4-34. The Bureau of Mines data in Figure 4-34 (Miron and Lazzara, 1988) suggest the ignition temperatures are approaching an asymptote beyond a depth of 25 mm, but more data would be needed with deeper layers and other materials to confirm such an asymptote. Both standards allow extrapolation by plotting the log of the depth versus the inverse surface temperature for ignition. This is tantamount to plotting depth versus $1/T_{si}$ on a semi-log plot as shown in Figure 4-35. The data for oil shale dust in Figure 4-35 satisfy a linear correlation in a semi-log plot, but the data for the two other materials (brass and coal dust) exhibit a nonlinear correlation. Miron and Lazzara (1988) state that the curvilinear correlation does not lend itself to extrapolation to much thicker layers, and they recommend conducting larger scale tests to get accurate/realistic surface ignition temperatures.

The self-heating theoretical equations described in Section 4.3.4 can also be extended to the surface heating application discussed here. The different boundary condition represented by the hot surface requires use of a more complicated equation for the Frank-Kamenetskii parameter, δ_c, in this case (Bowes, 1984, p. 99). Plotting data in a manner consistent with Eq. [4-10] has been shown by Bowes (1984, p. 216) to provide an excellent correlation for sawdust. Although it would be desirable to plot data for other dusts in a similar manner, the values of δ_c may differ because of the different chemical reactions and thermal properties. Thus, neither ASTM E 2021 nor IEC 1241-2 requires or even encourages use of other correlation methods for data extrapolation.

Once the minimum surface temperature for igniting a 5-mm-thick layer, $T_{5\,mm}$, has been determined by testing, the maximum safe surface temperature per IEC 61241-10 is given by Figures 4-36 and 4-37. In applications where the layer thickness can be controlled (typically on exposed outer surfaces of equipment), the maximum allowable surface temperature for a thin deposit of thickness ≤5 mm, is given by $T_{5\,mm} - 75°C$ (Rule 1). When the thickness exceeds 5 mm but is less than 50 mm, the curves shown in Figure 4-37

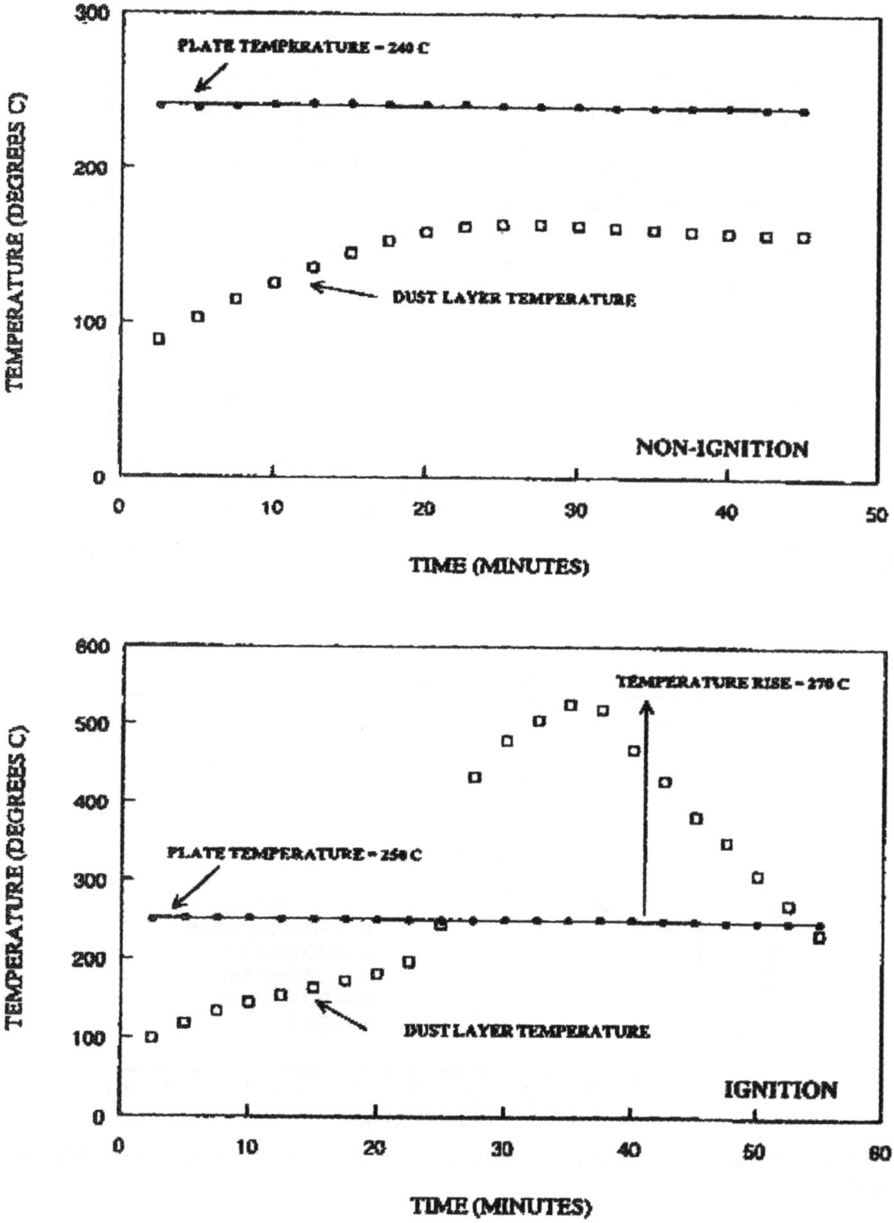

Figure 4-33 Dust layer temperature measurements for tests with and without ignition. (Reproduced with permission from ASTM International.)

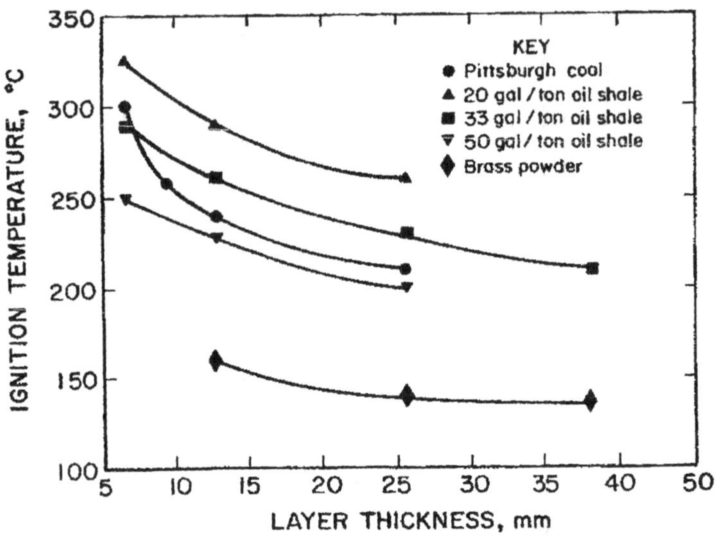

Figure 4-34 Surface ignition temperature versus layer thickness (from Miron and Lazzara, 1988).

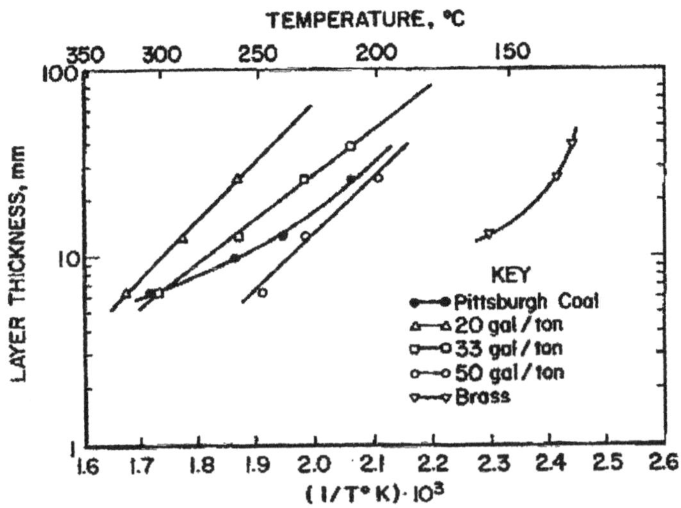

Figure 4-35 Dust layer thickness versus inverse ignition temperature (from Miron and Lazzara, 1988).

show how the maximum allowable surface temperature decreases correspondingly (Rule 2). Recent deep layer tests reported by Lunn et al. (2001) and Bennett et al. (2002) have confirmed that the maximum allowable surface temperature curves in Figure 4-37 are indeed conservative in establishing a large safety margin between actual surface temperatures causing igni-

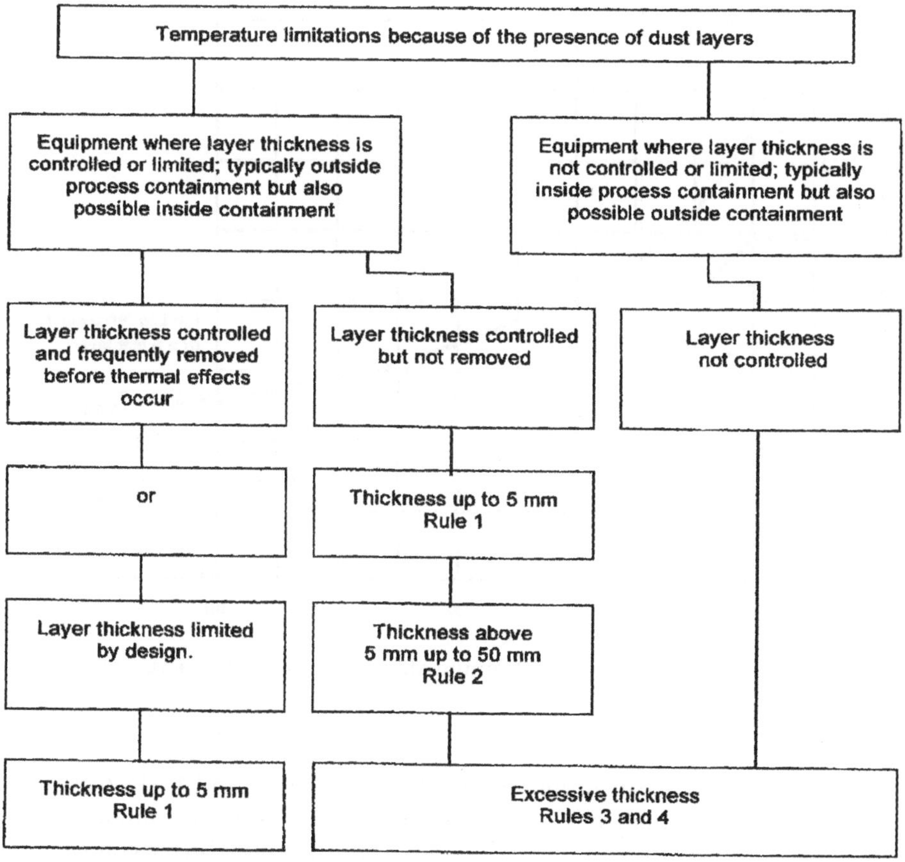

Figure 4-36 IEC 61241-10 flow chart to determine maximum allowable equipment surface temperatures for dust layers.

tion and the maximum allowable surface temperatures. For example, Bennett et al. report in the case of a dust with a $T_{5\ \mathrm{mm}}$ of 305°C, Figure 4-37 would dictate that a 30-mm-thick layer would have a maximum allowable surface temperature of 95°C, whereas their measured surface ignition temperature was 270°C.

In applications where the dust layer thickness cannot be controlled (usually because the surface and layer are not exposed or accessible), the IEC standard requires that the following Rules 3 and 4 be applicable.

Rule 3 states that the electrical power to the equipment must be limited to a value determined by either experiment or "recognized calculation methods." Rule 4 states that laboratory tests and/or calculations must be conducted to determine the maximum allowable surface temperature.

Lunn et al. (2001) also developed a test apparatus that can be used to determine maximum allowable surface temperature for dust piles deeper

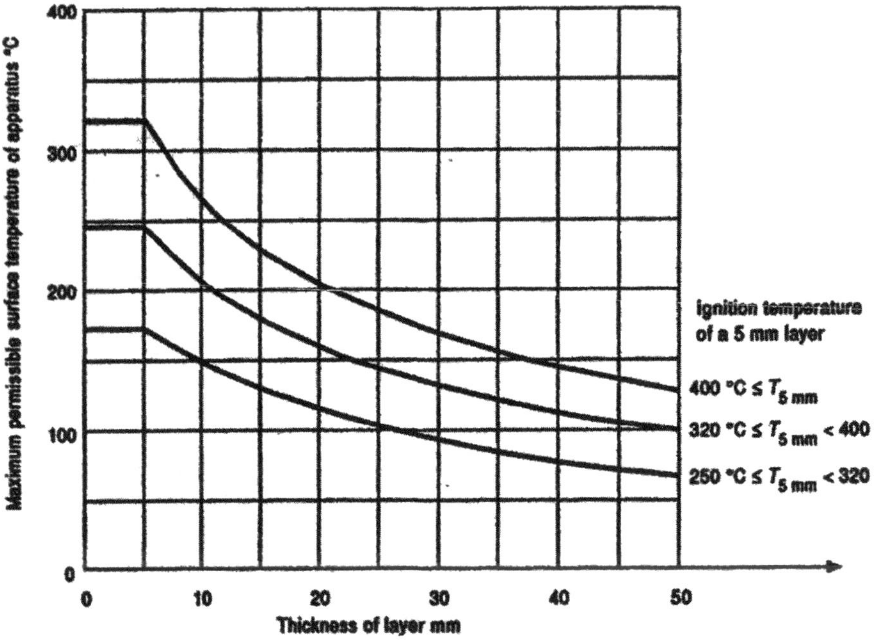

Figure 4-37 IEC 1241-10 maximum allowable surface temperature.

than 50 mm, i.e. to use in applying IEC 61241-10 Rules 3 and 4. The test apparatus is shown schematically in Figure 4-38. It consists of a rectangular heating block of dimensions 20 cm × 10 cm × 5 cm high, onto which a conical dust pile is deposited via a funnel. Four liters of test powder are deposited onto the block, and a constant power is applied to the heating element. Figure 4-39 shows a section of a sawdust pile used in these tests. Four thermocouples measure the surface and dust deposit temperature as it is heated for a period of 8 hours if no ignition is observed prior to that time. The test is then repeated with a new power application depending on whether or not the last

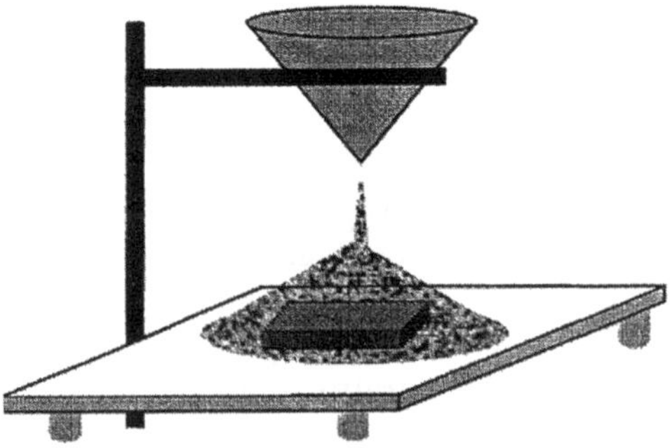

Figure 4-38 Apparatus for producing particulate pile on heating block (from Lunn et al., 2001).

Figure 4-39 Section of sawdust pile on heating block (from Lunn et al., 2001).

test resulted in ignition. This procedure continues until there is only a 10°C difference between the surface temperature for an ignition and a no ignition result. As an example, the 25-μm sawdust particles that had a 5-mm-deep layer surface-ignition temperature of 340°C, had a deep deposit surface ignition temperature of 230°C as determined via this test procedure.

Dust Layer Hot Spot, Impact, and Friction Ignition Tests

Another dust layer ignition scenario involves a small hot spot (representing perhaps a frictional heating hot spot) within or under the dust layer. Various laboratory tests have been conducted to examine this situation. For example, Siwek and Cesana (1995) describe tests with small hot spots of 3 cm^2 area that ignited dust layers when the spot/surface temperature was 1100°C. They also showed that smoldering lumps (also called nests) of powder would ignite many dust layers when the surface area of the smoldering lumps/nests was 96 cm^2 and its temperature was 900°C. Lunn (2002) reviewed more recent and extensive test data that showed that 75 g heated metal cylinders would ignite deep dust layers at temperatures between 450°C and 700°C depending on the dust material. Other tests with smoldering nests at a temperature of about 400°C deposited in the deep layers would ignite the layers if the nests were sufficiently large. The minimum size nest required to ignite the layers depended on the dust material. Since results are material dependent, it would be useful to develop a standardized test protocol to test various materials. Lunn also reviewed test data on the ignition of coal dust deposits by embedded electrical wires of various diameters and electrical currents. Here too, standardized tests would help establish safe current levels for intrinsically safe electrical equipment.

Figure 4-40 shows a test apparatus developed by Gibson and Harper (1981) to determine whether frictional heating from a rotating shaft would ignite dust deposits at the base of the shaft. There are also a variety of tests to

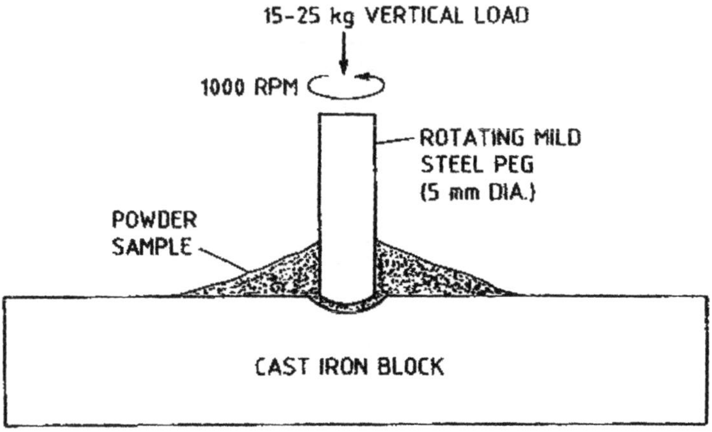

Figure 4-40 Gibson and Harper (1981) test apparatus for frictional heating of dust layers (from Eckhoff, 2003).

determine the reaction of a combustible dust layer to an impact from a dropped hammer. Eckhoff (2003) describes several versions of these impact tests and cautions that it is often difficult to determine whether or not the impact did produce a localized burning.

Dust Layer Combustion Tests

After a dust layer has been ignited at one location, a variety of combustion phenomena can ensue. Several similar dust layer combustion tests have been developed to observe and classify the behavior of particulate material when subjected to a small flame or heat source. ICI developed a test configuration to represent the conditions occurring in a spray dryer (Abbot, 1990). The particulate test sample was 20 cm long by 2 cm wide by 1 cm deep, and was heated at one end by a small gas flame. Some materials just burned/decomposed at the site of the flame but the reaction front did not propagate. Other materials developed propagating smoldering fronts, while still others reacted in the form of flame propagation.

The German/Swiss version of this test is shown in Figure 4-41. A 4-cm-long triangular ridge of dust with a 2-cm-wide base is exposed to a glowing platinum wire at one end. The test is usually performed at room temperature with the test sample sitting on a ceramic plate, but it can also be conducted at some elevated process temperature by placing the sample in a heated glass tube. An air flow velocity of 0.2 m/s is imposed over the sample. Observations of the result of the heated wire exposure determine into which of the six classes given in Table 4-11 the particulate material should be categorized (Eckhoff, 1997 and Lunn, 2002).

TABLE 4-11
European Combustibility Categories for Dust Layers

Combustibility Class	Result of Hot Wire Exposure	Example Material[a]
1	No Ignition	Stearic Acid (1300 μm)
2	Short duration localized combustion followed by self-extinguishment	Polypropylene (162 μm)
3	Local sustained combustion without any propagation	Polypropylene (25 μm)
4	Propagation of smoldering combustion	Phenol formaldehyde resin (60 μm)
5	Flaming combustion propagation	Methyl cellulose (29 μm)
6	Explosive combustion	Calcium/aluminum (22 μm)

[a]Example materials from listings in Eckhoff (2003) Appendix.

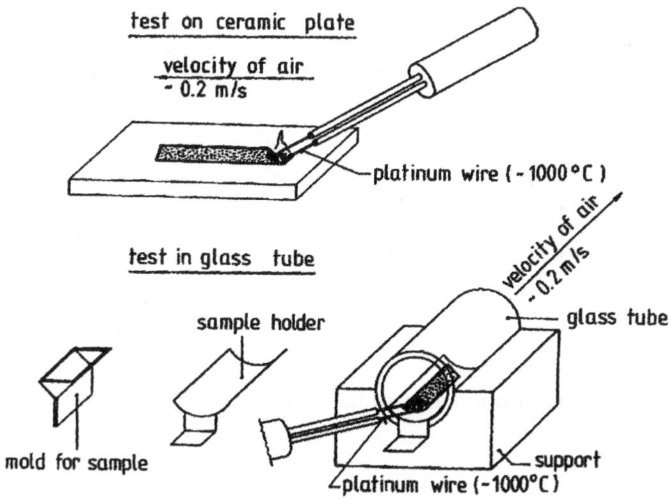

Figure 4-41 Dust layer combustibility test (from Siwek, 1996).

As indicated by the examples in the table, small particle size layers tend to have more vigorous combustion and correspondingly higher combustibility classes than larger particles of the same material. Higher temperatures also can increase the combustibility class. For example, Dextrin is Class 2 when tested at 22°C, but increases to Class 5 when tested at 100°C (Lunn, 2002). Likewise, an increase in the air velocity above the test sample can also significantly alter the combustion behavior and associated categorization.

United Nations classification testing (U.N., 1999b; Siwek, 1996) utilizes a 25-cm-length ridge as opposed to the 4-cm-length ridge. The powder sample is prepared by loosely filling the triangular cross-section mold shown in

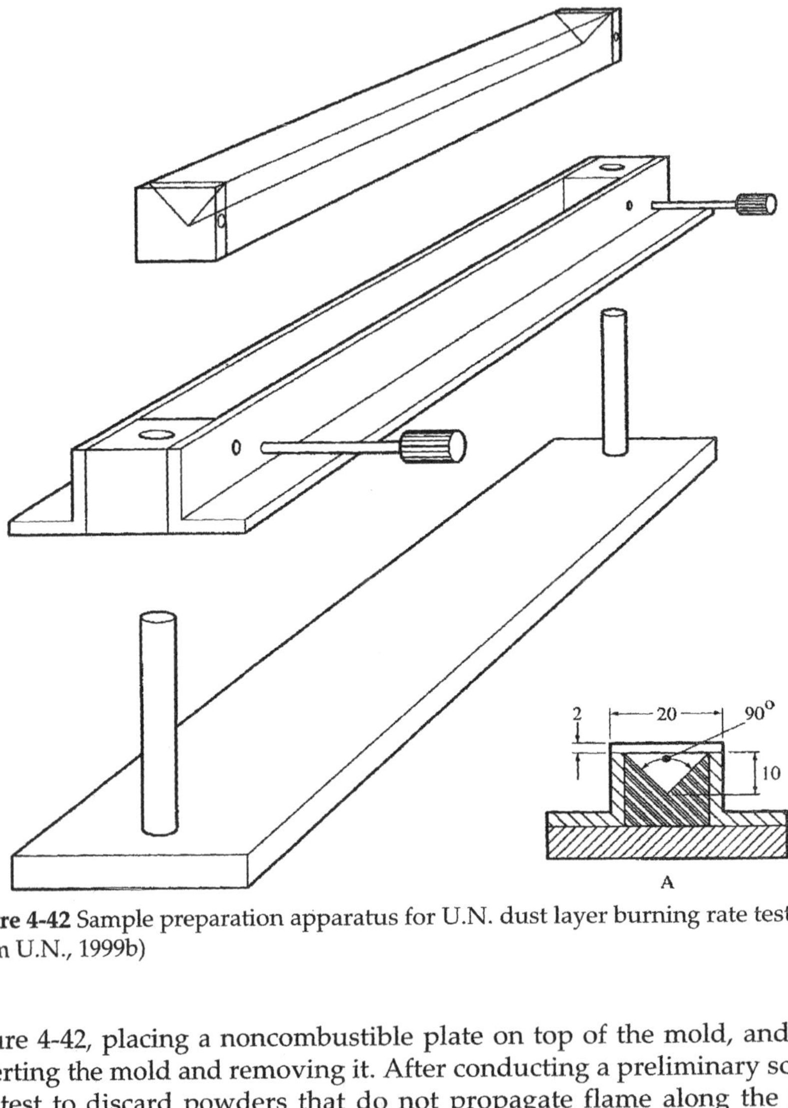

Figure 4-42 Sample preparation apparatus for U.N. dust layer burning rate test (from U.N., 1999b)

Figure 4-42, placing a noncombustible plate on top of the mold, and then inverting the mold and removing it. After conducting a preliminary screening test to discard powders that do not propagate flame along the ridge within 2 minutes, the flame propagation time over a 10-cm length of the ridge is measured, and the material is classified according to Table 4-12. A wetted region of powder (formed by applying water with the possible addition of wetting agent to the powder, if necessary) beyond the 10-cm end of the zone is used to determine the appropriate Packing Group for the Readily Combustible (Division 4.1) Solid.

Sample test data given in the U.N. Manual (1999b) shows how the addition of a zinc salt to a manganese ethylene bis (dithiocarbamate) complex

TABLE 4-12
U.N. Flammability Categories (Siwek, 1996 and U.N., 1999b)

Flame Propagation Time (rate)	Reaction at Moistened Region of Ridge	U.N. Category
≥45 s (<2.2 mm/s)	—	Not a 4.1 material
<45 s (>2.2 mm/s)	Flame propagates through wetted zone	Division 4.1, Packing Group II
<45 s (>2.2 mm/s)	Flame is stopped for at least 4 minutes at wetted zone	Division 4.1, Packing Group III

(Mancozeb) increases the flame propagation time and prevents the combustible powder from being categorized as a Division 4.1 material.

Metal powders have a different set of criteria as follows.

Propagation time > 10 min: Not a 4.1 material.
5 min < Propagation time < 10 min: Division 4.1, Packing Group III
Propagation Time < 5 min: Division 4.1, Packing Group II

Although the rationale for the different criteria for metal powders is not explained in the regulations, it is probably associated with the higher flame temperature of metal particulate and the difficulty in extinguishing metal powder fires using ordinary suppression agents.

The European and U.N. layer burning rate tests measure the rate of lateral flame spread, but they do not quantify the mass burning rate and the associated fire heat release rate. The most commonly used laboratory test apparatus for measuring burning rates and fire heat release rates is the cone calorimeter. The cone calorimeter uses a test sample with a 10-cm by 10-cm area exposed to the radiant heat flux from a cone shaped radiant heater as shown in Figure 4-43. Instrumentation includes a load cell for the mass burning rate, and gas analysis sampling in the exhaust duct in order to determine heat release rates as explained by Babrauskas (2002). Although the cone calorimeter is normally used for testing slabs of combustible materials, it has also been used to characterize powder layer fires.

Varshney et al. (1990) and Sharma et al. (1993) describe another type of powder burning rate test involving ignition in an indented cavity in the center of various size deposits of powder on various substrates. Thermocouples imbedded at various levels below the surface of the powder deposit provide a measure of the rate of downward propagation of the combustion front. Their data showed that the type of substrate material under the powder deposit plays an important role in the rate of burning, such that the burning rates are substantially larger and of longer duration on thermally insulating substrates than on metal substrates. The tests also show that the burning rate and surface temperatures increase with increasing amount of

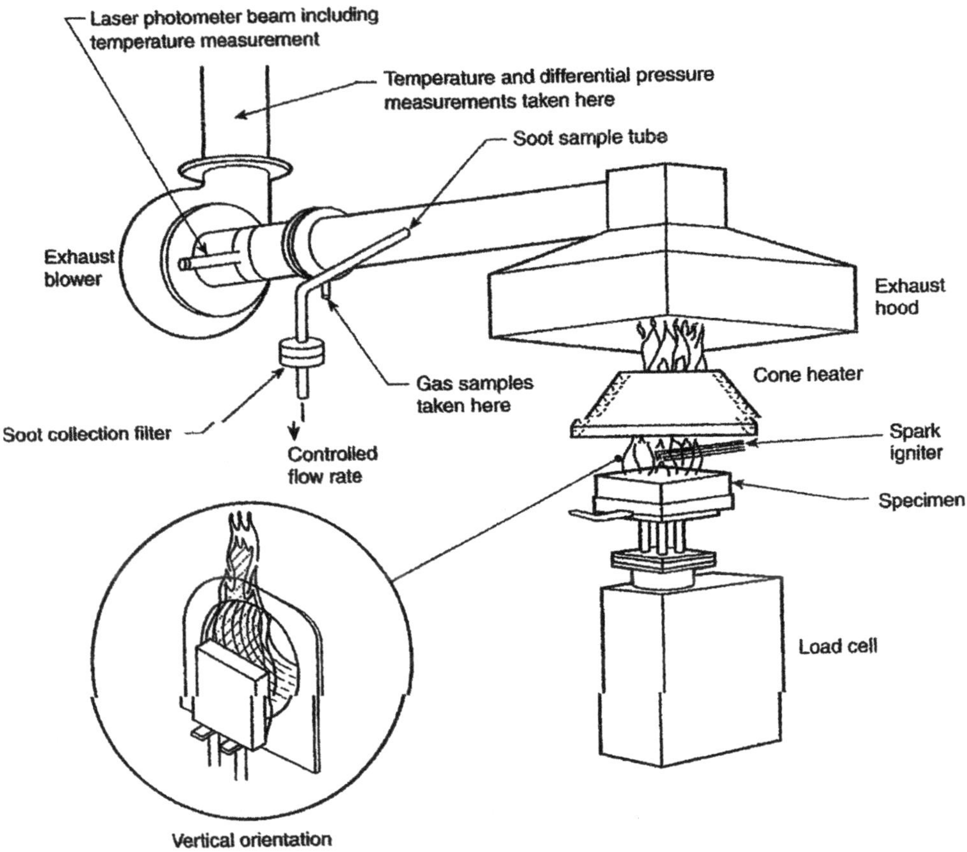

Figure 4-43 Cone calorimeter fire test apparatus. (From Babrauskas, 2003, in the *SFPE Handbook of Fire Protection Engineering*, 3rd Edition. Copyright 2002 the Society of Fire Protection Engineers.)

powder deposited over the range from 25 g to 800 g. Thus tests with small samples of powder can underestimate the dust layer fire hazard.

Table 4-13 provides a comparison and summary of the various dust layer ignition and combustion tests.

4.3.6 Electrostatic Charging and Discharge Testing for Particulates

The electrostatic charging and charge accumulation propensity of particulate material is primarily dependent on the material's electrical resistivity and charge retention characteristics. Commercial instrumentation to measure particulate material volume resistivity is available in a variety of configurations such as a cylindrical cell with electrodes on both end faces (Britton, 1999), and equipped with a sensitive ammeter.

TABLE 4-13
Summary of Dust Layer Combustibility Test Methods

Test	Standard or Reference	Applications	Advantages	Limitations
Hot spot/nest ignition	Siwek and Cesana (1995) Lunn (2002) Eckhoff (1997)	Dust ignition scenarios involving frictional heating or cigarette, or hot electrical wire.	Results can be used to establish well-founded dust fire prevention measures.	No standardized test protocol yet.
Hot surface ignition temperature	ASTM E 2021 IEC 1241-2-1	Maximum allowable surface temperature to avoid igniting combustible dust layer.	Can be used to establish safe surface temperature limits for process equipment and electrical equipment.	Maximum safe surface temperature decreases with increasing layer depth, and no depth-scaling law has been developed yet. Exception: isothermal oven tests.
Lateral flame spread rate and extent	UN (1999b)	UN/DOT Burning Class categorization for shipping.	Accounts for possible flame spread over areas of wetted dust.	Describes behavior with limited ignition source and one substrate. For large ignition source, the burning behavior usually is a more robust and more rapid and extensive combustion.
	Abbot (1990)	Simulates flame spread for a dust layer in a dryer	Gas flame ignition source	As above
	Eckhoff (1997)	European categorization of dust materials into one of six combustibility classes.	Test is conducted with an air flow to promote flame spread, and can be conducted at elevated dryer temperatures.	As above
Mass burning rate and heat release rate	Babrauskas (2002, 2003)	Will determine the threat of fire spreading to other nearby combustible materials.	Fire heat release rate is the most commonly used parameter in fire protection engineering.	Small (10 cm × 10 cm) sample size used in Cone Calorimeter.
	Sharma et al. (1993)	Deep layer deposits.	Accounts for downward flame propagation into a deep layer.	Results are sample size dependent, and there is no standard sample size.

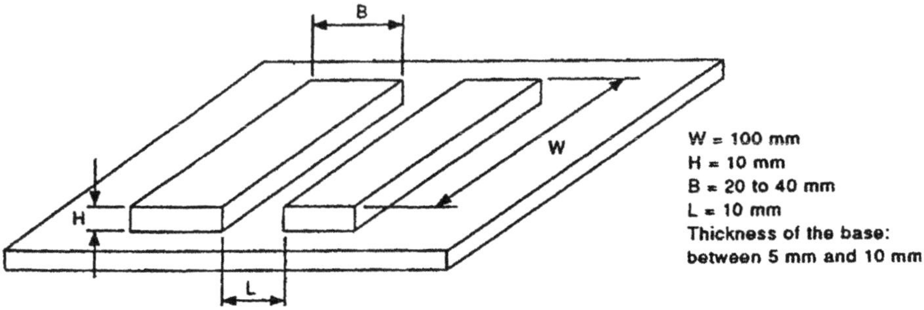

Figure 4-44 IEC 1241 dust layer resistvity test apparatus.

The resistivity test configuration in IEC 1241-2-2 uses a rectangular channel of powder formed between two metal bars as illustrated in Figure 4-44. After two glass bars are placed across the ends of the metal bars, powder is poured into the 10 mm high by 10 mm wide channel, and the excess is wiped away. The standard specifies that the powder should have a particle size less than 71 μm, and should have its moisture content measured, but does not specify any maximum allowable moisture level. An electrical circuit shown in the standard is used to apply various dc voltages in the range 110 V to 2000 V across the powder channel, and the resistance, R_s, of the channel is measured at each voltage. The powder volumetric resistivity, ρ, is then calculated from the following equation (providing the resistance of the empty test cell is at least 10 times the resistance measured with the test sample):

$$\rho = 0.001R_s[H(W/L)] \quad [4\text{-}11]$$

where *H, W,* and *L* are the height, length, and spacing, respectively, of the metal bars in the test cell.

If the measured resistivity is less than or equal to10^3 Ω-m, the powder is classified as a conductive dust per IEC 1241-2-2. If the measured resistivity is greater than 10^3 Ω-m, the powder is classified as a nonconductive dust. These classifications have important implications regarding electrical equipment used in areas containing combustible powders. Other resistivity criteria are used for particulate handling applications as discussed in Chapters 6 and 7.

Charge Generation Tests

Charge generation on flowing powder samples can be measured by collecting the powder in a Faraday cage, sometimes called a Faraday pail. A Faraday cage or pail consists of two concentric conducting enclosures with the outer enclosure being grounded and insulated from the inner enclosure. The powder sample is collected in the inner enclosure, which is electrically connected to an electrometer. The construction suggested in ASTM D4470 consists of two cups, with a removable lid on the outer cup and a small opening on top of the inner cup to allow entry of the powder sample. Britton (1999) suggests that the two metal enclosures can be constructed of either sheet metal or metal mesh, with PTFE or some other highly resistive plastic as the insulator.

The measurement of charge is often facilitated by using a shunt capacitor to reduce the voltage on the electrometer to a value that can be measured readily. The charge generation, Q, on the powder sample is then given by

$$Q = CV \qquad [4\text{-}12]$$

where C is the shunt or electrometer input capacitance, and V is the voltage reading. The charge density is given by Q/m, where m is the mass of powder collected in the Faraday cage. ASTM D4470 calls for charge density determinations on five or more identical samples, and the report should include both the average charge density and the standard deviation.

The charge generation mechanism for these laboratory tests can consist of the powder sample flowing down a trough or tube into the Faraday cage. Measurement of electrostatic charges generated during powder processing are more difficult, but possible. For example, Glor and Schwenzfeuer (1997) used the test setup shown in Figure 4-45 to measure charge generation and conical discharges during silo filling. Ring shaped capacitors were electrically connected to an RC-circuit and computer outside the silo as illustrated in the figure. The instrumentation and computer recorded charge transfers of at least 30 nC from powders of resistivities greater than 5×10^{12} Ω-m. The measurements indicated that the conical discharges were sufficiently energetic to ignite powders with low minimum ignition energies.

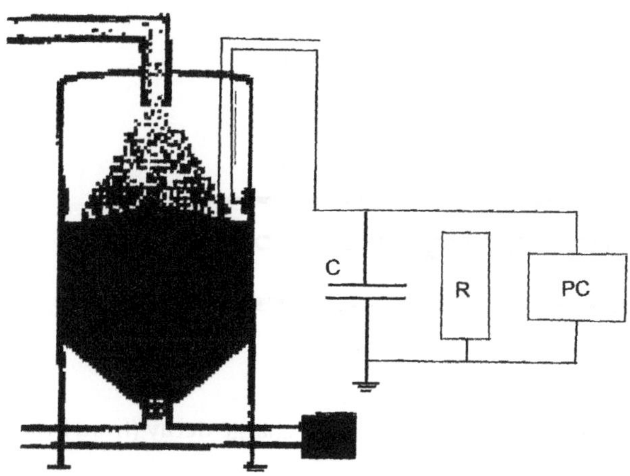

Figure 4-45 Test setup used by Glor and Schwenfeuer (1997) for measuring charges generated during silo filling.

Charge Decay Test

British Standard 7506 (1996) describes a method for measuring the decay time of an electrostatic charge on the surface of a powder deposit. A layer of powder at least 1 cm deep is deposited in a grounded container and clamped by a grounded annular ring to a baseplate as shown in Figure 4-46. The clamping ring diameter is 5 cm, and its width is at least 5 mm. A corona discharge is deposited on the powder surface by temporarily exposing it to a high voltage (at least 7 to 8 kV) electrode plate as shown in the figure. The electrode plate is then rapidly removed and the fieldmeter is used to measure the dissipation of the surface charge. The decay time is recorded on either an oscilloscope or computer or an electronic timing circuit. The nominal decay time is the measured time for the surface voltage to decrease to $1/e$ of its initial hazard threshold value. In the case of a combustible powder, the hazard threshold voltage is presumably the voltage at which a propagating brush discharge can occur. The longer the decay time, the longer is the duration of the hazard. Another version of the test entails replacing the grounded baseplate with an insulated plate, and allowing the surface charge to decay via contact with a grounded electrode.

The electrical resistivity, capacitance, and breakdown strength of processing equipment are other important parameters in evaluating the hazards of electrostatic charging and discharges. Britton (1999) provides tabulations of representative values of resistivity, dielectric constant, and breakdown strength for a number of solid dielectric materials. However, these values are easily influenced by material additives, contaminants, etc, and should be measured on-site if/when a serious electrostatic hazard is suspected.

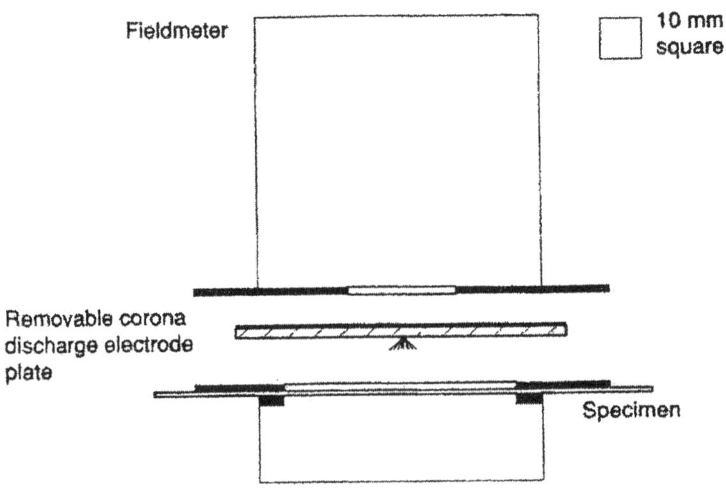

Figure 4-46 Test setup for measuring surface electrostatic charge decay (from British standard 7506).

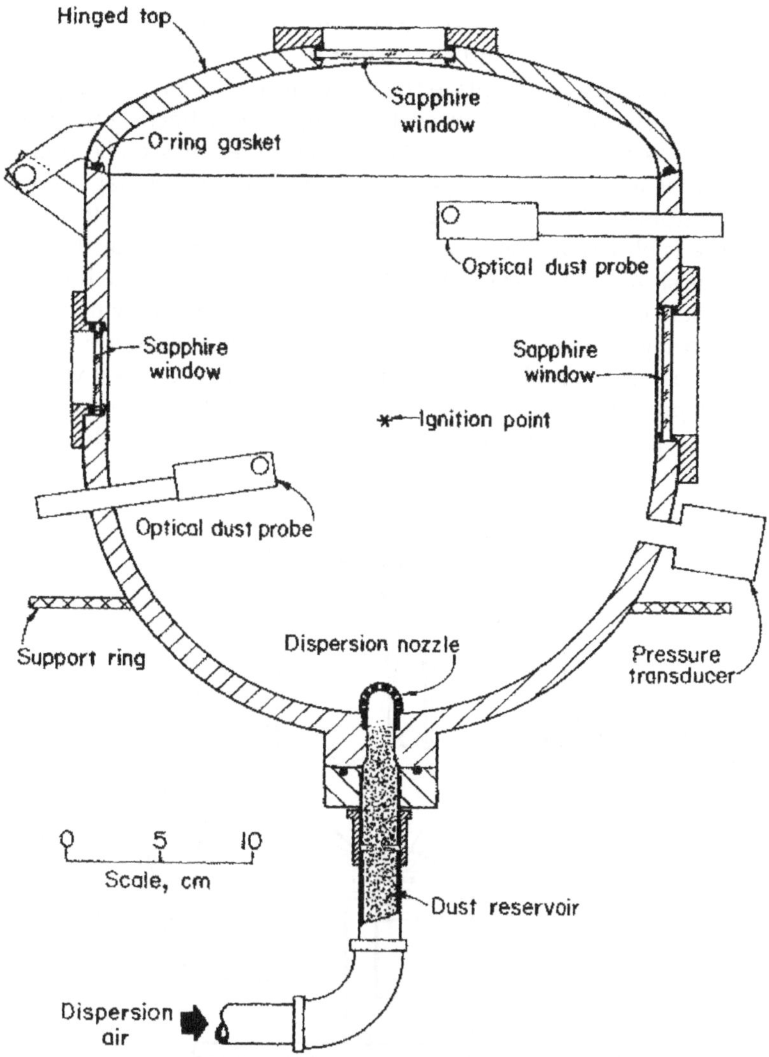

Figure 4-47 Twenty-liter test vessel with dust dispersion nozzle. (From ASTM E1515, with permission from ASTM International.)

4.3.7 Dust Cloud Explosibility Test Methods

General Considerations in Dust Cloud Explosibility Testing

Combustible dust cloud explosibility testing inevitably entails first generating a suspended cloud of particulate of some nominal concentration in a laboratory test vessel, and then attempting to ignite the suspended cloud. Two critical aspects of this type of test are (1) how to generate the dust cloud suspension, and (2) what size and shape test vessel should be used. The most widely used methods today place the dust sample in an auxiliary chamber, pressurize the chamber with air, and then open a fast-acting valve to allow the dust and air to enter the partially evacuated test vessel. Sometimes, the dust enters the test vessel via a perforated hemispherical dispersion nozzle near the bottom of the vessel as illustrated in Figure 4-47. More often, the dust charge is blown against a so-called rebound nozzle near the bottom of the vessel as shown in Figure 4-48. Still another technique is to inject the dust through a perforated tube shaped as a C around the inside wall of the vessel as shown in Figure 4-49. In all three techniques, the objective is to rapidly produce a near-uniform dust cloud throughout the test vessel volume.

The air pressure used to disperse the dust and the time delay between dust injection and ignition are critical factors in determining the level of tur-

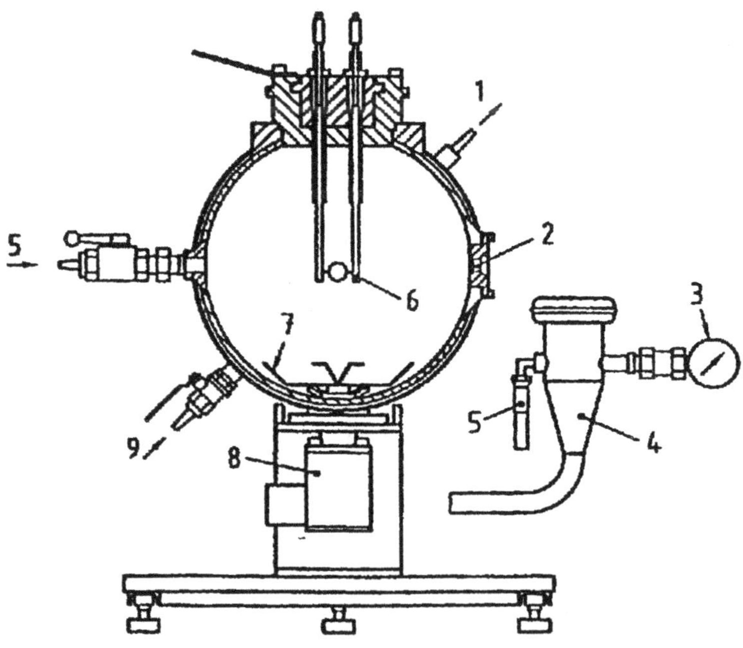

Legend

1	water outlet	6	ignition source
2	oxygen analyzer, pressure sensors	7	rebound nozzle
3	manometer	8	fast acting valve
4	dust container (0,6 l)	9	water inlet
5	air/inert gas inlet		

Figure 4-48 Twenty-liter dust explosion test vessel with rebound nozzle (from Draft EN 14034-1)

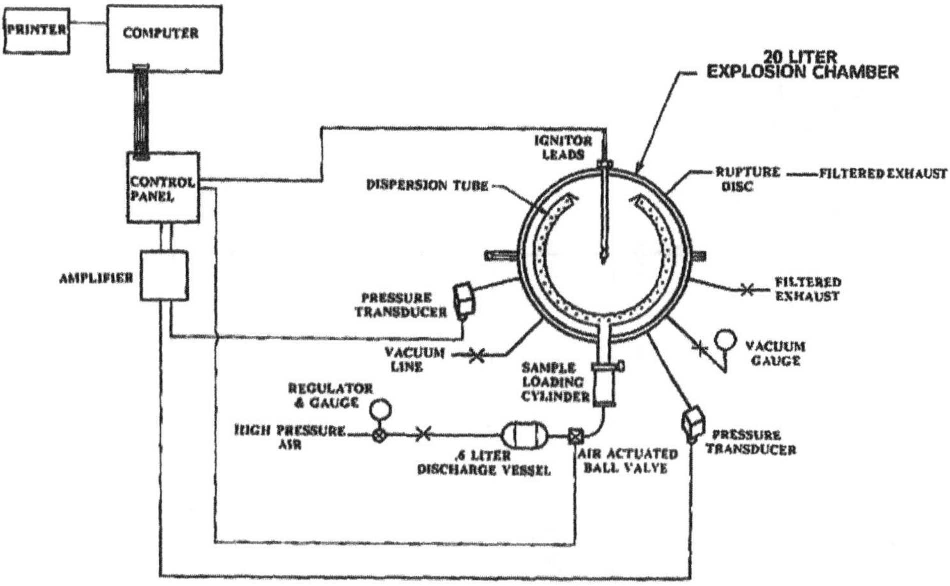

Figure 4-49 Twenty-liter vessel with perforated C-ring.

bulence in the cloud at the instant the igniter is fired. The turbulence level has a major bearing on whether or not the dust cloud will be ignited, and how rapidly the cloud will burn if it is ignited. Figure 4-50 is one example of the pressure decay in the auxiliary dust chamber following actuation of the fast-acting valve, the delay time between injection and ignition, and the corresponding pressure versus time curve following ignition. Since the air pressure, dust injection method, ignition source, and delay time between dust injection and ignition are critical factors influencing the test results, round-robin tests are often conducted with several standardized dust samples to establish benchmarks and verify reproducibility among different laboratories.

Most of the dust explosibility laboratory tests (other than preliminary screening tests) are now conducted in a 20-liter test vessel of the type shown schematically in Figures 4-47 through 4-49. The spherical shape enables a near-uniform dust cloud. The 20-liter test vessel volume has been shown by Bartknecht (1989) to be scaleable to larger test vessels, when comparing the K_{St} values defined as

$$K_{St} = \left(\frac{dP}{dt}\right)_{max} V^{1/3} \qquad [4\text{-}12]$$

where $(dP/dt)_{max}$ is the maximum rate-of-pressure-rise measured in a closed vessel of volume, V. Hence most of the testing today is conducted in 20-liter test vessels.

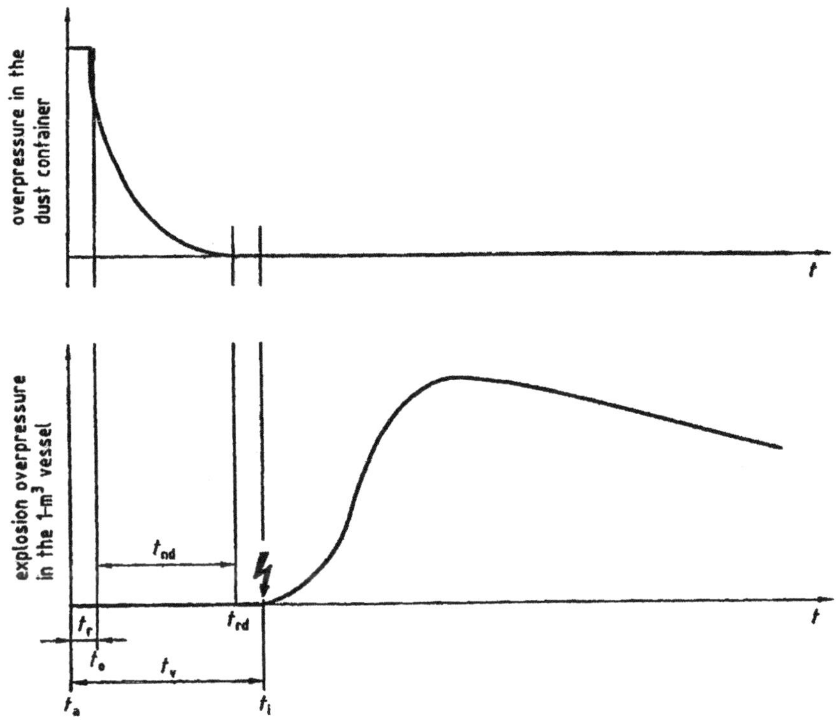

t_a initiation of the fast acting valve

t_o start of the dust dispersion

t_r reaction time of the fast acting valve

t_{nd} release time of the dust container into the 1 m^3 vessel without dust

t_i moment of the compensation of the pressure between the dust container and the 1 m^3 vessel

t_{rd} activation of the ignition source

t_v ignition delay (= 0,6 s)

Figure 4-50 Time delay between dust injection and ignition (from Draft EN 14034-1).

Dust Cloud Screening Tests

Explosibility screening tests for dusts and powders are conducted primarily to determine whether or not a particular dust or powder will be an explosive hazard. The process includes a preliminary analysis of the chemistry and particle size of the material, followed by a dust cloud flammability/explosibility test. Since there is no international standard developed for this process, the specifics of the preliminary analysis and type of screening test depend on the testing organization and the intended application and jurisdiction. These tests may often be referred to as yes/no or A/B tests, this is due to the simplicity of their results (Yes = explosive = Group A and No = non-explosive = Group B). Eckhoff (1997) indicates that for a screening test to be conducted the following parameters must be present:

- Reasonably sized dust cloud
- Dust concentration in the most explosive range (will vary depending on dust or powder)
- Sufficient ignition source

Material Chemistry and Particle Size Distribution

The objective of the preliminary chemistry tests is to determine if the material is indeed combustible and, if so, the approximate heat of combustion. These tests are often conducted via DSC or some similar apparatus requiring only a small test sample. If material has a moderate to high heat of combustion (for comparison, many polyolefins have a heat of combustion of about 40 kJ/g), and if the mass median particle size is sufficiently small (see criterion under United States/NFPA), then the material should be considered to pose a dust explosion threat. Conversely, if low or negligible amounts of heat are produced via oxidation, or if the characteristic particle size is sufficiently large, then the material may not be considered to pose an explosion hazard. Of course, there are many intermediary or ambiguous test results that warrant continuing on with one of the explosibility screening tests described below.

UK—Modified Hartmann Apparatus

The explosibility screening test often used in the United Kingdom (Barton, 2000) consists of a modified Hartmann 1.2-liter glass cylinder apparatus. The dust or powder is dispersed in the cylinder and is ignited by either an electric spark or by an electrically heated coil. If the material is ignited and the dust flame is observed to propagate away from the igniter, then the material is considered explosive or Group A. However, if the material fails to ignite, a series of additional steps are conducted prior to deeming the material to be nonexplosive or a Group B material. The dust sample is dried at 105°C for 1 hour before it is retested. Once retested if the material still does not exhibit Group A qualities, it is sieved down to 25 μm and the individually sieved samples are tested (based on particle size). If any portion of the sieved sample produces an explosion, as determined from flame propagation, the material is considered explosible.

Continental Europe—1-m³ and 20-Liter Pressure Vessels and Modified Hartmann Apparatus

According to the overview in Eckhoff's book (1997), there are a wide variety of methods currently being used in continental Europe. Some use 1-m^3 or 20-liter pressure vessels with a large ignition source (approximately 10 kJ), while others follow a similar methodology to that of the United Kingdom. For those who use the 1-m^3 and 20-liter pressure vessels, a positive or

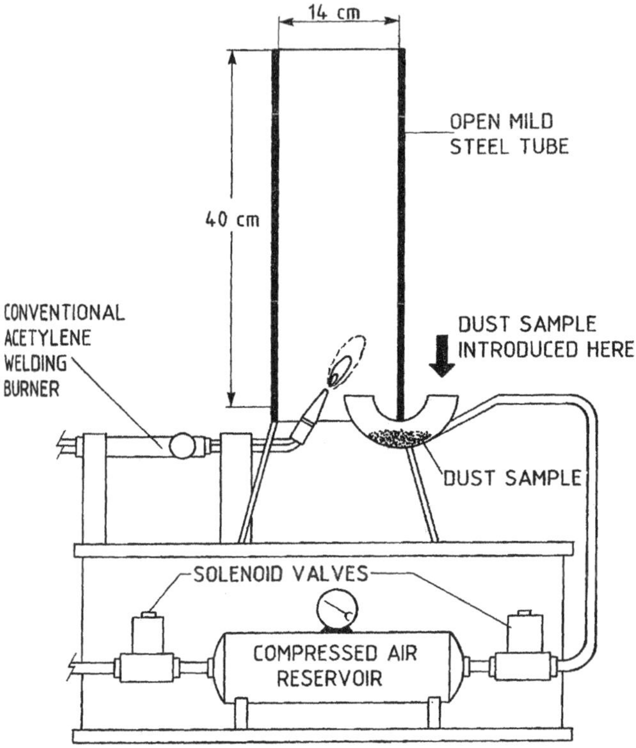

Figure 4-51 Dust Explosibility screening test apparatus used in Norway. (From Eckhoff, R., *Dust Explosions in the Process Industries,* 3rd ed. Amsterdam: Elsevier, 2003.)

explosible material is determined by the pressure increase measured within the vessel. The other methodology uses a modified Hartmann apparatus similar to that used in the United Kingdom; however, the spark used for ignition is somewhat weaker.

After deciding that the strongest ignition source that would be present either deliberately or accidentally in most industrial processes would be from a welding torch, the Norwegian dust explosion researchers at the Christian Michelson Institute (Eckhoff, 2003) developed the apparatus shown in Figure 4-51. The apparatus consists of an open top vertical steel tube (14 cm diameter), a port connected to a compressed air reservoir to inject the dust sample into the tube, and a conventional acetylene welding burner at the base. The material is dispersed in the apparatus and ignited via the welding torch. If the material ignites produces a vented flame as shown in Figure 4-52, the material is considered to represent a dust explosion hazard.

United States/NFPA

Both NFPA 654 and NFPA 68 define a combustible dust as finely divided solid of 420 μm or smaller (U.S. No. 40 Standard Sieve) diameter, and ignit-

Figure 5-52 Silicon dust explosion vented from Norwegian screening test apparatus. (From Eckhoff, R., *Dust Explosions in the Process Industries,* 3rd ed. Amsterdam: Elsevier, 2003.)

able when dispersed in air. Appendix B of NFPA 68 provides a brief description of the screening test recommended for determining whether a dust/powder is combustible/ignitable when suspended in air. The screening test entails repeated attempts to ignite the dust cloud using a strong (10 kJ) chemical ignition source and a range of dust concentrations. No specific test vessel is specified, but there is a cautionary statement that the pressure generated by the 10 kJ igniter in a small vessel (≤20 liter) can cause difficulty in discerning any pressure rise due to dust combustion. A practical alternative mentioned in NFPA 68 Appendix B is to use only sub-200 mesh (≤75 μm) particles in a small vessel with a weaker ignition source, such as an electric match.

NFPA 654 also states that combustible dust deflagration hazards should be determined by actual test data, with the type of test selected based on the application/situation. Thirteen tests are listed in NFPA 654 as factors sometimes used in determining the deflagration hazard.

In the absence of a standard screening test, individual companies and laboratories use different procedures for dust explosibility screening. Some laboratories use a version of one of the European or NFPA 68 screening tests described above. For example, one large chemical company conducts screening tests in a 1.3-liter Hartmann cylinder (see Figure 4-53) equipped with

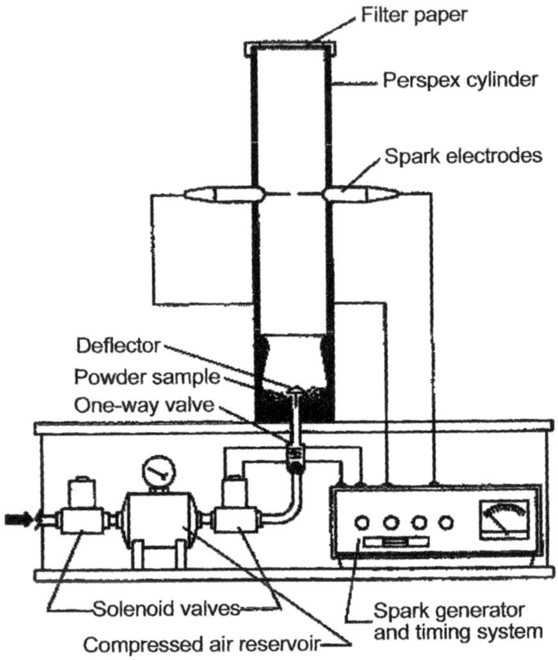

Figure 4-53 Original Bureau of Mines apparatus for MIE. (From Eckhoff, R., *Dust Explosions in the Process Industries,* 3rd ed. Amsterdam: Elsevier, 2003.)

spark electrodes and a 10 kV continuous AC power supply. If there is either a flame or a pressure rise sufficient to break the paper rupture disc at the top of the cylinder, the material is considered to be a combustible dust. Other laboratories/companies focus on one of the standard tests described below for their screening. For example, one company uses the results of minimum ignition energy (MIE) tests and DSC tests during pilot studies and then may proceed to maximum rate-of-pressure rise tests (K_{St}), minimum cloud ignition temperature, minimum oxygen concentration testing, etc. depending on a material/process specific hazard analysis. A similar screening procedure often used by another company is based on MIE test results, with a MIE value under 10 mJ indicating a need for K_{St} data and volume resistivity. Another company bases its decisions on particle size distribution testing and the possible presence of hybrid mixtures before deciding to proceed and select from tests such as minimum ignition energy, maximum explosion pressure and rate-of-pressure rise, minimum explosible concentration, and limiting oxygen concentration. Still another company focuses on the maximum rate-of-pressure rise as the basis for determining what further testing, if any, is needed.

Minimum Explosible Concentration Tests

ASTM E 1515 (2000) is the U.S. standard test method for determining minimum explosible concentrations (MEC) of combustible dusts, that is, the minimum concentration that will propagate a deflagration in a near-spherical closed vessel with a volume of at least 20 liters. The European Union has a similar standard in draft form (prEN 14034-3) using the term lower explosion limit (LEL). Values of the MEC or LEL obtained by this or any other method are not intrinsic material constants; they are dependent on the test method used as well as the moisture level and size distribution of the sample. The standard recommends using a test sample with at least 95% through a 200 mesh sieve (75 μm sieve diameter), but does allow using the as-received sample as long as the laboratory notes that smaller particles may have a lower MEC. Eckhoff (1997) reports that MEC values are not very sensitive to particle diameter for diameters less than about 60 μm, but increase significantly with increasing diameter above this approximate threshold. Since high sample moisture content can also affect MEC data, it should be a maximum of 5% per the ASTM standard.

ASTM E 1515 calls for the use of either a 2500 J or a 5000 J pyrotechnic igniter because those are the largest ignition energies that will not mask the pressures generated by the burning of the dust in a 20-liter vessel. The bar graph shown in Figure 4-53 shows the influence of these alternative ignition energies on the measured MEC for the five dusts which the standard suggests using as benchmark materials. In the 20-liter vessel, the 5000 J igniter produced lower MEC values than the 2500 J igniter for all five dust materials shown. However, the tests with the 10 kJ igniter in the 1-m^3 vessel resulted in MEC data that were actually closer to the 2500 J igniter data in the 20-liter vessel. Therefore, the 5000 J igniter may be having an undue influence (so-called overdriven deflagration) on the 20-liter sphere data for these dusts.

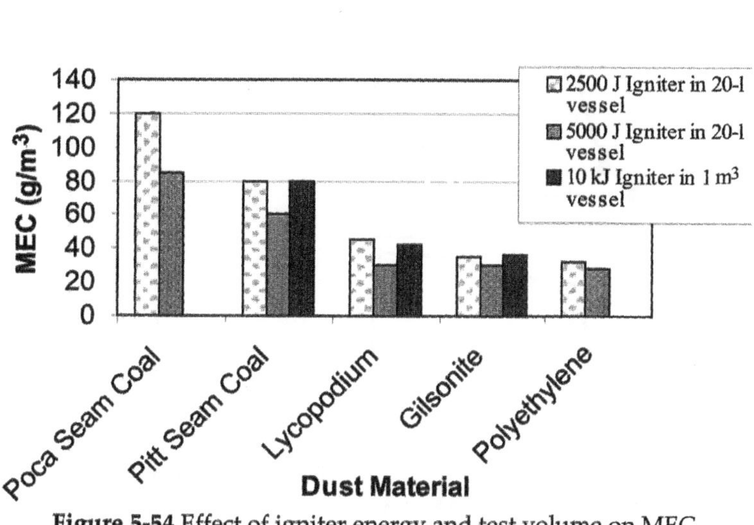

Figure 5-54 Effect of igniter energy and test volume on MEC.

The criterion recommended in ASTM E 1515 to delineate the MEC value is the concentration that produces a deflagration pressure ratio, *PR*, equal to 2. *PR* is defined in the standard as:

$$PR = \frac{P_{\text{ex,a}} - \Delta P_{\text{ignitor}}}{P_{\text{ignition}}} \qquad [4\text{-}13]$$

where, using ASTM E 1515 nomenclature, $P_{ex,a}$ is the maximum explosion pressure (absolute) in a test at a given dust concentration, $\Delta P_{ignitor}$ is the pressure increase caused by the igniter, $P_{ignition}$ is the vessel pressure at the time of ignition. Although the tests are nominally conducted at an initial (ignition) pressure of one atmosphere, the actual pressure, $P_{ignition}$, depends on the amount of preevacuation, the dust chamber charge pressure, and the ignition time delay.

The MEC corresponding to a *PR* value of 2 is determined from a plot of PR versus tested dust concentration as shown in Figure 4-55. In this case, the MEC would be approximately 120 g/m^3. The top graph in Figure 4-55 is a plot of $(dP/dt)_{max}V^{1/3}$ versus dust concentration. The appendix of ASTM E 1515 recommends using a second criterion that $(dP/dt)_{max}V^{1/3}$ be greater than or equal to 1.5 bar-m/s at the MEC in order to verify that the pressure rise is sufficiently rapid to represent an actual deflagration. The data in Figure 4-55 indicate that both criteria yield roughly the same MEC value in that example.

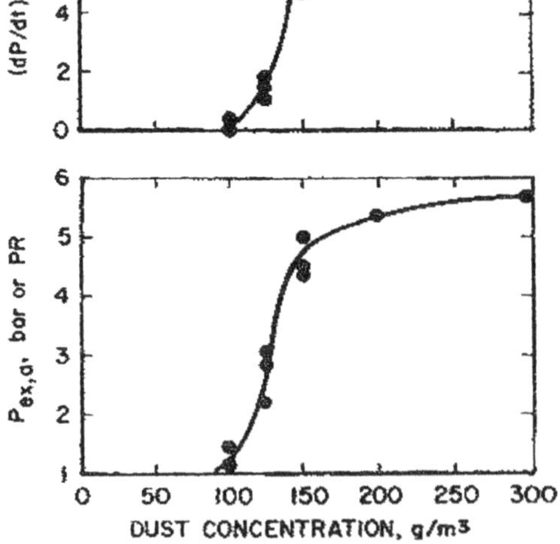

Figure 4-55 Explosion pressure ratio and rate-of-rise. (Reprinted from ASTM E1515 with permission.)

Since many applications involve handling or storing particulates in enclosures that cannot withstand a deflagration pressure ratio of 2, readers should consider using a significant safety margin between the measured MEC value and the maximum allowable dust cloud concentration during dust transport and container/silo loading. NFPA 69 specifies that the concentration be no greater than 25% of the MEC unless automatic concentration measurements and safety interlocks are provided. This safety margin is based primarily on expected concentration inhomogeneities rather than MEC test criteria. Eckhoff (1997) suggests that although limiting dust concentrations to values below a fraction of the MEC is difficult in many types of process equipment, two examples where it should be feasible are dust collector ducting and electrostatic powder paint spray booths.

Minimum Dust Cloud Ignition Temperature

Two types of furnaces with different orientations are used extensively to measure the minimum surface or air temperature that will ignite a dust cloud. The vertical oriented furnace shown schematically in Figure 4-56 is called the Godbert-Greenwald furnace. The horizontal oriented furnace shown schematically in Figure 4-57 is called the BAM furnace.

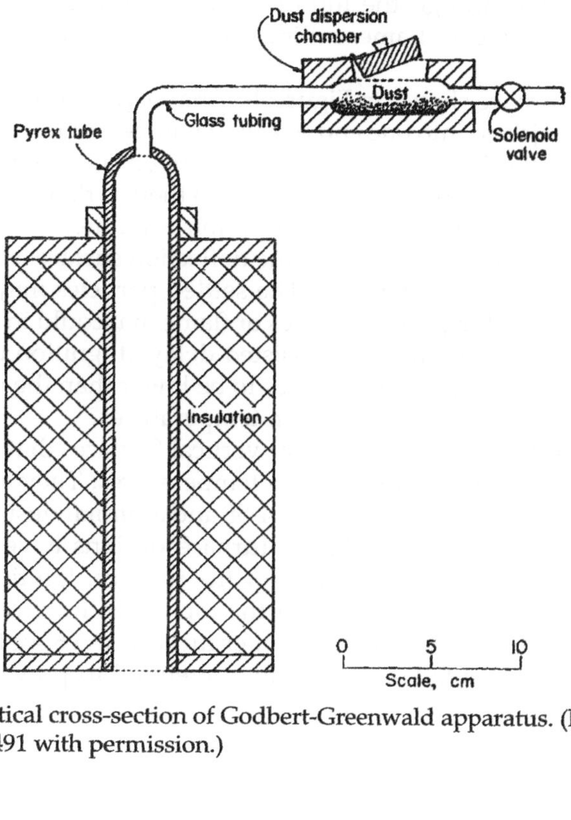

Figure 4-56 Vertical cross-section of Godbert-Greenwald apparatus. (Reprinted from ASTM E1491 with permission.)

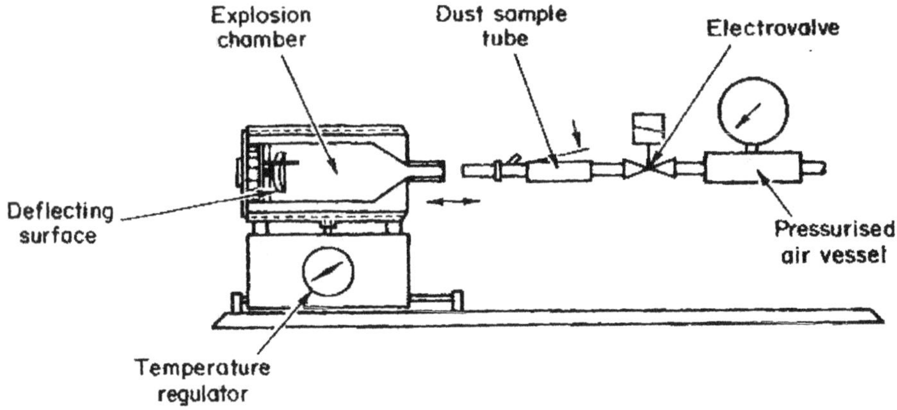

Figure 4-57 BAM automated furnace for dust cloud ignition. (Reprinted from ASTM E1491 with permission.)

The Godbert-Greenwald furnace has an internal diameter of 3.9 cm, and a height of 23 cm (ASTM E 1491). Its internal volume is 0.27 liters. After the furnace is pre-heated to a measured temperature, the preweighed dust sample is injected through the top of the furnace. Dust cloud ignition is observed when there is flame exiting through the open bottom of the furnace.

The BAM furnace shown in Figure 4-57 has a volume of 0.35 liters. The dust sample is injected into one end of a test chamber within the furnace and it strikes a hot deflecting surface in front of a hinged end plate at the opposite end of the chamber. Ignition is observed when there is flame exiting from the hinged end plate. One important complication associated with the BAM furnace is that particles can rebound off the deflection surface and settle on the hot lower wall of the test chamber. The settled particles may smolder and generate smoke and hot pyrolysis gases or flame that could in turn ignite the dust cloud. This smoldering layer assisted delayed ignition usually causes the BAM furnace ignition temperatures to be lower than the corresponding Godbert-Greenwald minimum ignition temperatures.

ASTM E 1491 also shows two other vertical furnaces that can be used to measure dust cloud minimum ignition temperatures. They are the 6.8-liter furnace shown schematically in Figure 4-58 and a similar design 1.2-liter furnace. Both of these Bureau of Mines furnace designs have dust samples injected through the bottom opening of the furnace. Ignition is observed when flame emerges from the vent at the top of the furnace within a prescribed period of time (3 sec for the 1.2-liter furnace and 6 sec for the 6.8-liter furnace). The time limits are presumably intended to prevent smoldering induced ignitions.

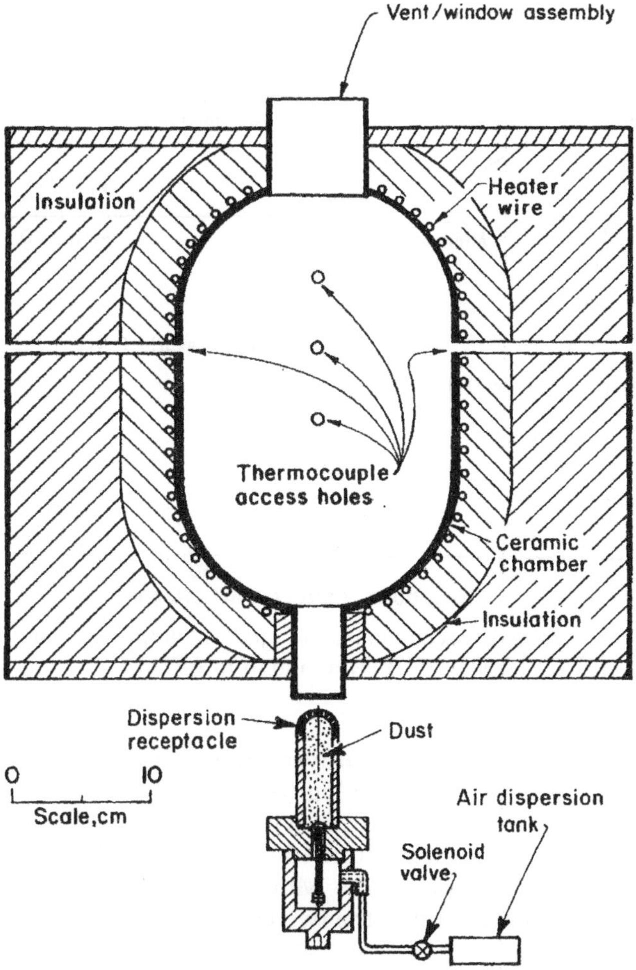

Figure 4-58 Bureau of Mines 6.8-liter furnace. (Reprinted from ASTM E1491 with permission.)

ASTM E 1491 specifies that the "minimum autoignition temperature (MAIT) of the dust cloud is the average of the lowest temperature at which ignition was reproducibly observed and the highest temperature at which ignition was not observed at any concentration. If there is an intermediate temperature at which ignition is observed for about half the tests, report this as the MAIT. Round the reported MAIT to the nearest 10°C." The data examples shown in the standard indicate that the autoignition temperature does not vary appreciably with concentration over the concentration range from 150 g/m^3 to 650 g/m^3, which is the range corresponding to the MAIT for many dusts. Figure 4-59 shows the variation of the AIT measured in the

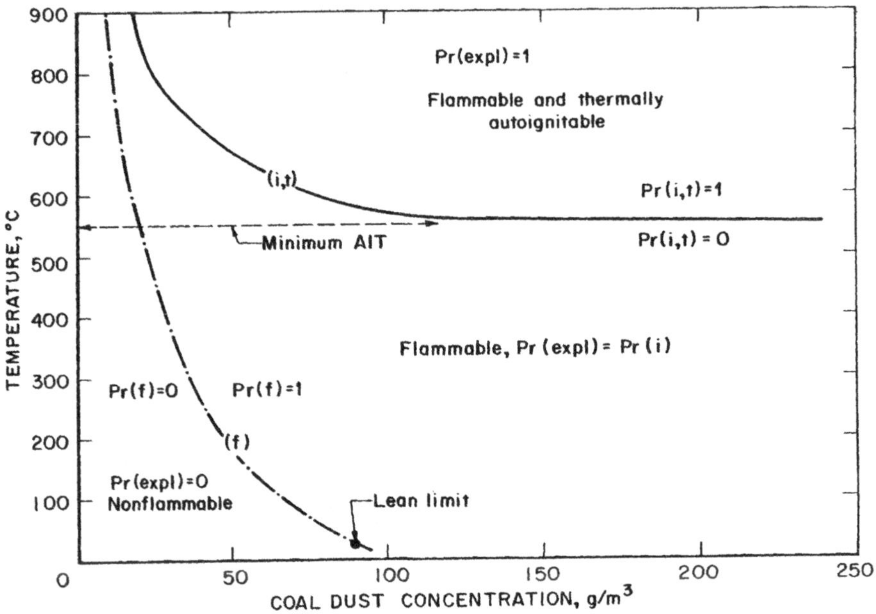

Figure 4-59 Relationship between dust cloud ignition temperature and MEC. (From Hertzberg, in ASTM STP 958, with permission from ASTM.)

Bureau of Mines 1.2-liter furnace for lycopodium as a function of cloud concentration. The variation of AIT in Figure 4-59 is minimal over a large concentration range, but increases sharply with decreasing concentrations when the concentration is less than about 650 g/m^3. The other curve shown in Figure 4-59 is the variation of the MEC with initial cloud temperature. Hertzberg (1987) points out that the two curves coincide at temperatures of 700°C and less.

Table 4-14 shows the minimum dust cloud ignition temperatures obtained using the four different furnace designs in ASTM E 1491. Minimum Ignition Temperatures measured in the BAM furnace are lower than those measured in the other three furnaces for four of the six dusts, and are lowest in the 6.8-liter furnace for the other two dusts. Maximum differences in cloud ignition temperature among the four test vessels range from 50°C to 80°C for five of the six materials, but extend to possibly 300°C for the anthracite coal, which has the highest ignition temperature of the various materials in Table 4-14.

The IEC 1241-2-1 (1994) standard for MAIT tests uses a furnace very similar to the Godbert-Greenwald furnace. The open bottom tube diameter is 4.4 cm, and its length is 21.6 cm, corresponding to a tube volume of 0.33 l. The recommended dust sample to be used with this furnace is a sample passing

TABLE 4-14
Examples of MAIT Data, °C (from ASTM E 1491)

	G G Furnace	BAM Oven	1.2-liter Furnace	6.8-liter Furnace
Anthracite coal	>900	>600	740	730
Anthraquinone	670	>500	620	680
Pocahontas (Poc) bituminous coal	640	580	610	600
Pittsburgh (Pgh) bituminous coal	600	570	540	530
Lycopodium	460	410	440	380
Sulfur	260	240	290	260

through a 71-μm sieve opening, that is, just slightly smaller than the 75-μm sieve opening specified for the ASTM dust tests. The minimum ignition temperature in the IEC standard is defined as the lowest furnace temperature at which an ignition occurred (for any dust concentration) minus either 10°C or 20°C depending on whether the furnace temperature is above or below 300°C. The rationale for the two temperature corrections to the furnace temperature is not explained in the standard.

In addition to the furnace orientation, the residence time of the dust cloud in the furnace has been shown to be an important factor influencing the results of these types of cloud ignition temperature tests. Eckhoff (1997) shows a plot of cloud minimum ignition temperature versus residence time that indicates that an increase in residence time from 0.12 s to 0.32 s can decrease the ignition temperature of coal dust by as much as 200°C. ASTM E 1491 has the following caveat about the effect of residence time and test vessel size. "Because of the short duration of the test, the data obtained are most applicable to industrial equipment where dust is present as a cloud for short time. Because of the small scale of the test and the possible variation of the MAIT value with scale, the data obtained by this test method may not be directly applicable to all industrial conditions."

Finally, readers should also realize that the surface temperatures required for dust cloud ignition increase sharply with decreasing area of the heated surface. Data reviewed by Lunn (2002) show that dusts with a BAM measured MIT of about 400°C require surface temperatures of about 1000°C for ignition if the surface area is only about 20 mm^2.

Minimum Ignition Energy and Other Spark Ignition Tests

The original Bureau of Mines test apparatus for measuring the electrical spark energy required to ignite a combustible dust cloud is illustrated schematically in Figure 4-53. A preweighed dust sample was placed in the dispersion cup situated at the bottom of the 1.2-liter plastic Hartmann tube, and a short blast of air generated a transient dust cloud. After a short predetermined time delay, a high-voltage capacitive discharge caused a spark across the electrodes in the center of the tube, and visual observation determined whether or not the spark caused flame propagation through at least part of the dust cloud. The spark energy was taken as $1/2CV^2$, where C is capacitance and V is the voltage to which the capacitor is charged.

Although the basic approach still exists today, numerous modifications to both the test vessel and the spark discharge circuit have been made by many experimenters in several laboratories. ASTM E 2019 allows use of either the transparent Hartmann tube or a 20-liter steel spherical vessel. The standard suggests use of either one of three different capacitive discharge spark generation circuits; a trickle charge circuit, a three-electrode circuit, and a compressed air driven moving electrode circuit. Since there can be some residual voltage, V_f, in the capacitor, and some additional capacitance in the wiring, ASTM E 2019 specifies that the spark energy, E, be calculated from:

$$E = \tfrac{1}{2} C(V_i^2 - V_f^2) \qquad [4\text{-}14]$$

where C is the total circuit capacitance, and V_i is the capacitor voltage prior to discharge.

The ASTM E 2019 minimum ignition energy (MIE) is determined from a curve drawn through each pair of go and no-go data points in a plot of spark energy versus dust sample weight or cloud concentration. The MIE is the minimum in this curve. Ten repeat tests are required at the most ignitable concentration since spark ignition tests near the MIE inevitably have an element of randomness associated with the value of the instantaneous dust concentration near the spark gap at the precise time of spark occurrence. Calibration tests are required with at least three different benchmark dust samples, and measured MIE values should fall within the range specified for each dust.

All three of the suggested circuits in ASTM E 2019 have a provision to place an inductor in the circuit. The circuit inductance often has an appreciable effect on the measured minimum ignition energy as can be seen from Figure 4-60. In many cases, the MIE with an inductance, L, on the order of 1 mH is an order-of-magnitude lower than the MIE measured without any significant inductance, that is, L less than 0.025 mH. Inductance causes a substantial increase in the spark duration, which allows the dust particles to be heated for a longer period of time. The relationship between inductance, L, and spark duration, t_d, for a circuit with $L \geq 1$ mH is (Eckhoff, 1997):

$$t_d = 9.2L/R \qquad [4\text{-}15]$$

where R is the circuit resistance. Large values of R in a R,C circuit without inductance can also substantially increase spark duration, but the resistance dissipates the capacitor discharge energy such that only a small fraction of

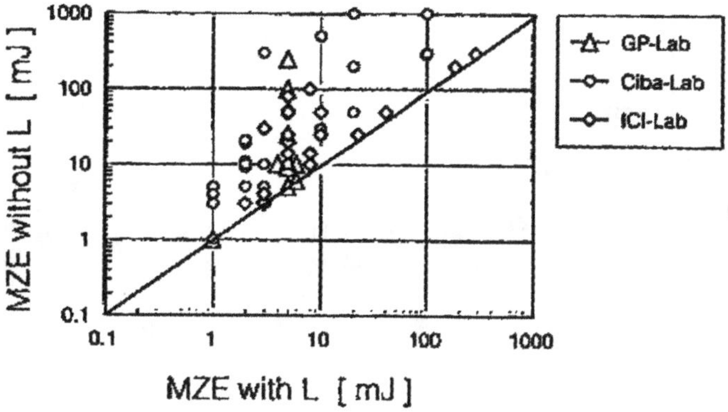

Figure 4-60 Comparison of MIE with and without inductance (from Siwek and Cesana, 1995).

that energy is transmitted in the spark itself (Bennett et al., 2002). According to data reviewed by Eckhoff (1997) the optimum spark duration corresponding to the lowest MIE value for a given dust is often in the range 0.1 ms to 1.0 ms, but can be significantly smaller for very reactive materials with MIE values below 1 mJ. Furthermore, the MIE value does not seem to be sensitive to the spark duration for many of these very reactive materials.

In view of the sensitivity of MIE measurements to inductance, standardized tests need to be conducted either with or without a substantial inductance. ASTM E 2019 notes that since electrostatic discharges from plant equipment are pure capacitive discharges, assessments of electrostatic ignition hazards should be made without using inductance in the circuit. The benchmark MIE values given in ASTM E 2019 for calibration purposes are values measured without inductance. However, there is also a note in the standard stating that the strict MIE value for general use should be based on the optimized spark duration obtained using inductance. If one adopts the position that the MIE test should utilize an inductance comparable to the inductance measured for typical process equipment, one could utilize the measured inductances recently reported by Bailey et al. (2002), which are up to about 3 mH with one exception of an unusually high inductance (220 mH) for a metal coil in a flexible rubber connection. MIE test data obtained with various levels of inductance indicate that the measured MIE values above 1 to 2 mJ are not very sensitive to the value of inductance beyond an inductance of about 1 mH.

In practice, many dust-testing laboratories have been utilizing the MIKE 3 apparatus manufactured by Kuhner AG, and described in the Kuhner MIKE 3 manual (1996). Kuhner has coordinated annual round-robin calibration test exercises to help ensure that these laboratories use proper standardized test procedures and properly maintained equipment. They distributed a standardized dust sample to any laboratory wishing to participate. The measured MIE results from 27 participating laboratories are shown in Figure 4-61. The nominal MIE value for the benchmark dust sample (pyridine-3-

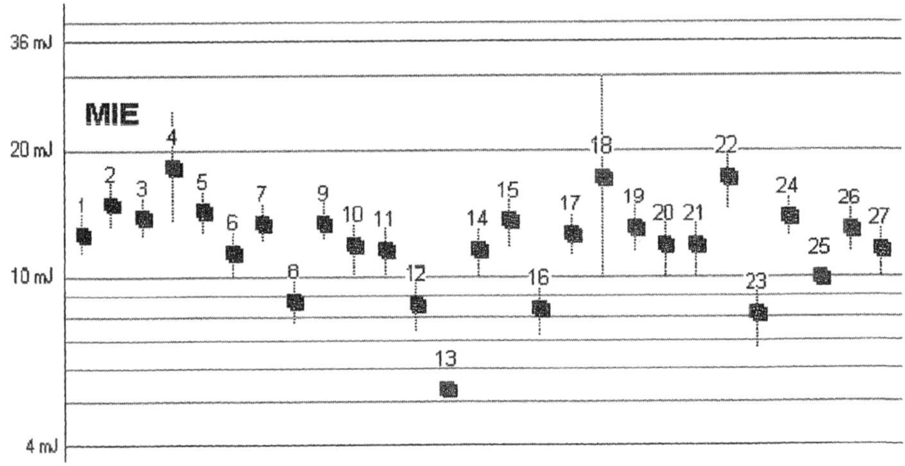

Figure 4-61 Round robin test data for MIE (from Kuhner with permission).

carboxamide with a 10%–90% particle size range of 14 μm–90 μm) is 12 mJ. As indicated in the figure, all 27 laboratories successfully obtained MIE values falling within a factor of three above or below the nominal MIE value, as required to establish conformity per prEN 13821. In fact, all but one of the laboratories obtained values falling in the range 8 to 18 mJ, that is, within 67% to 150% of the nominal value.

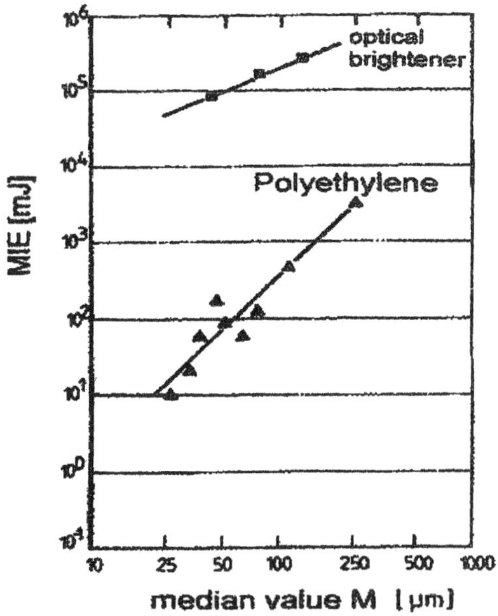

Figure 4-62 Dust cloud MIE versus particle diameter. (From ASTM E2019 with permission.)

MIE values increase sharply with increasing particle diameter as shown in Figure 4-62 for two different materials. The variation of MIE with polyethylene median particle diameter is consistent with the following empirical correlation recommended by Siwek (1999) and by Kuhner (1997):

$$\mathrm{MIE}_2 = \mathrm{MIE}_1\left(\frac{d_2}{d_1}\right)^{2.5} \qquad [4\text{-}16]$$

where d_1 and d_2 are two different median diameters for the same dust material. Although this correlation may fit the data for a number of materials, it substantially exaggerates the effect of particle size for the optical brightener dust data shown in Figure 4-62. Therefore, it should only be used for materials that have been checked to verify the applicability of this correlation.

MIE values also increase substantially with increasing dust cloud velocity and turbulence levels. The variation of MIE with air velocity through the spark gap is shown in Figure 4-63. For example, the MIE for polypropylene dust (with $d < 75$ μm) increases from about 5 mJ to about 15 mJ as the cloud velocity increases from < 1 m/s to 20 m/s.

FM Global Loss Prevention Data Sheet 7-76 uses a MIE value of 10 mJ as the threshold below which a combustible dust is deemed susceptible to electrostatic discharge ignitions and therefore should be processed in an inert atmosphere to prevent ignition. The data sheet also states that the "vast majority of dusts have MIE values above 10 mJ." As described in Chapter 7 of this book, other organizations (e.g., Ciba, 2000) have developed protection criteria that combine the material electrical resistivity along with the MIE and the type of process equipment and its operating characteristics.

MIE values decrease with increasing temperature. Siwek (1996) has correlated data at temperatures up to almost 1000°C to obtain the following empirical relation:

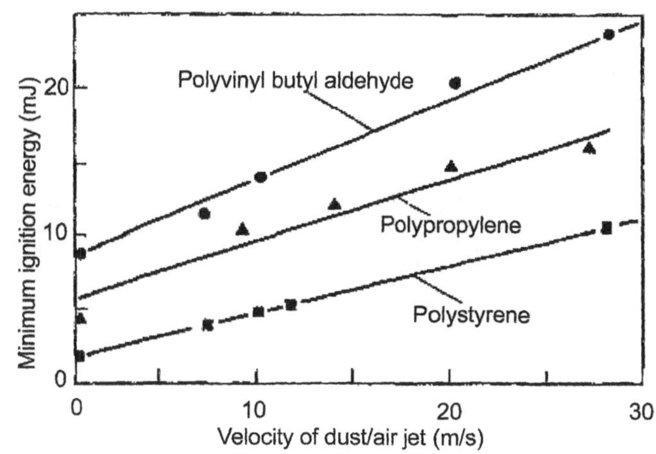

Figure 4-63 Dust cloud MIE versus air velocity. (From Eckhoff, R., *Dust Explosions in the Process Industries*, 3rd ed. Amsterdam: Elsevier, 2003.)

$$\text{MIE}(T) = 10 \exp[-4.056 + (1.873 - 0.624 \log T)(\log \text{MIE}(T_0) + 4.056)] \quad [4\text{-}17]$$

MIE values obtained from capacitive discharge sparks are not applicable to the hazards of electrical equipment and wiring sparks caused by the making and breaking of an electrical circuit. The spark energy associated with the breaking of an energized circuit is given by

$$E = \tfrac{1}{2}Li^2 \quad [4\text{-}18]$$

where i is the circuit current, and L is the inductance. These inductive sparks are inherently longer duration than the pure capacitive discharge sparks.

Rather than vary the spark duration to test dust cloud susceptibility to this circuit breaking type of spark, Bennett et al. (2002) have developed a test apparatus in which numerous sparks are generated and repetitive dust clouds are generated. Their European spark test apparatus shown in Figure 4-64 is a modified version of the spark test apparatus used in British Standard EN 50 020 for certification testing of intrinsically safe electrical equipment intended for use in a flammable gas atmosphere.

The current in the electrical circuit of the European spark test apparatus is repeatedly triggered and broken by a motor controlled switch that is designed to turn over 400 times per test if no ignition occurs. The electrical circuit can be either coupled to a particular piece of electrical equipment or a separate circuit set up specifically to study dust cloud ignitability criteria. The tests described by Bennett et al. (2002) used a fixed voltage supply and a variable resistance to obtain the desired current for each test. Multiple air blasts directed into the dust dispersion cup at the bottom of the 1.2-literiter Hartmann tube are used to generate the repetitive dust clouds. The concept is that eventually a spark will be imposed on a local dust concentration that is most readily ignitable for a given dust material. For each dust sample and current-voltage combination, multiple tests are conducted with varying amounts of dust in the dispersion cup. The tested dusts were all sub 63-μm and predried.

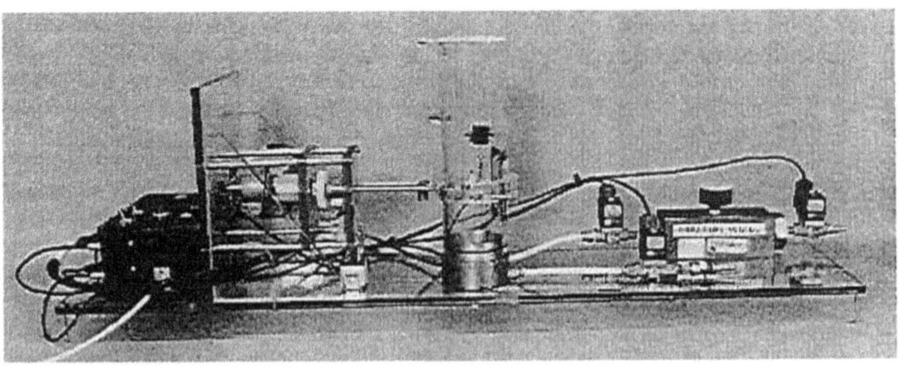

Figure 4-64 European spark test apparatus for electrical equipment sparks (from Bennett et al., 2002).

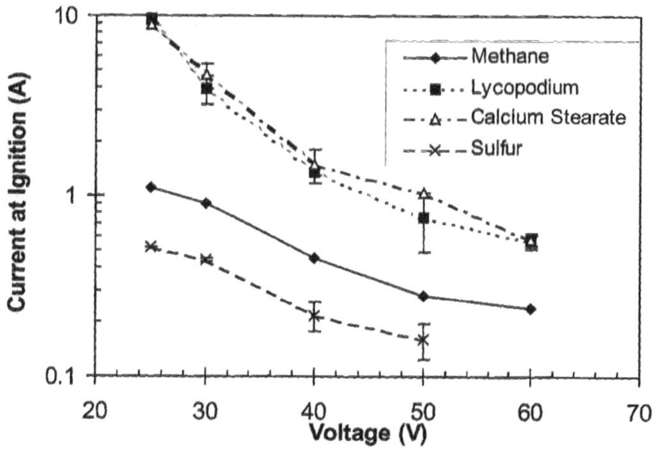

Figure 4-65 Dust cloud spark ignitability criteria obtained from European spark test apparatus.

Bennett et al. describe a series of round robin tests with three European laboratories using the same set of three dusts in a Spark test apparatus. The dusts were lycopodium, calcium stearate, and sulfur. Methane was also used for comparison. The Bennett et al. reported results have been plotted as shown in Figure 4-65. Each data point represents the minimum current needed at a given voltage to ignite the material. The data points are the average of the three laboratories, and the error bars denote the data variation among the laboratories. Results for calcium stearate and lycopodium are almost identical, whereas methane and sulfur were ignitable at significantly lower currents.

One of the interesting issues raised by the Bennett et al. results is whether or not separate dust ignitability categories should be established for hazardous location certification of electrical equipment. Certification testing laboratories in the United States have adopted the practice of certifying equipment for dusty locations if they pass the spark ignition certification testing for flammable gas atmospheres. The rationale for that practice is that the MIE values for most dusts are greater than the MIE for methane (0.2 mJ), which is the least ignitable of the various flammable gases used in certification tests. Bennett et al. suggest that separate tests using their European spark test apparatus should be established for dusts, and that dusts should be divided into various categories to account for the fact that sulfur is more readily ignitable than methane and the other two dusts tested, and that many other dusts are so difficult to ignite that they should not be subjected to the same requirements as methane. IEC committee deliberations will eventually lead to a decision about the adoption of an IEC standard for such combustible dust cloud equipment certification testing.

Tests for Maximum Explosion Pressure and Rate-of-Pressure Rise

The U.S. and international standards for measurement of maximum explosion pressure and maximum rate-of-pressure rise are ASTM E1226 and ISO 6184, respectively. Both standards emphasize the use of a 20-liter spherical

Figure 4-66 Kuhner 20-liter sphere (from Kuhner with permission).

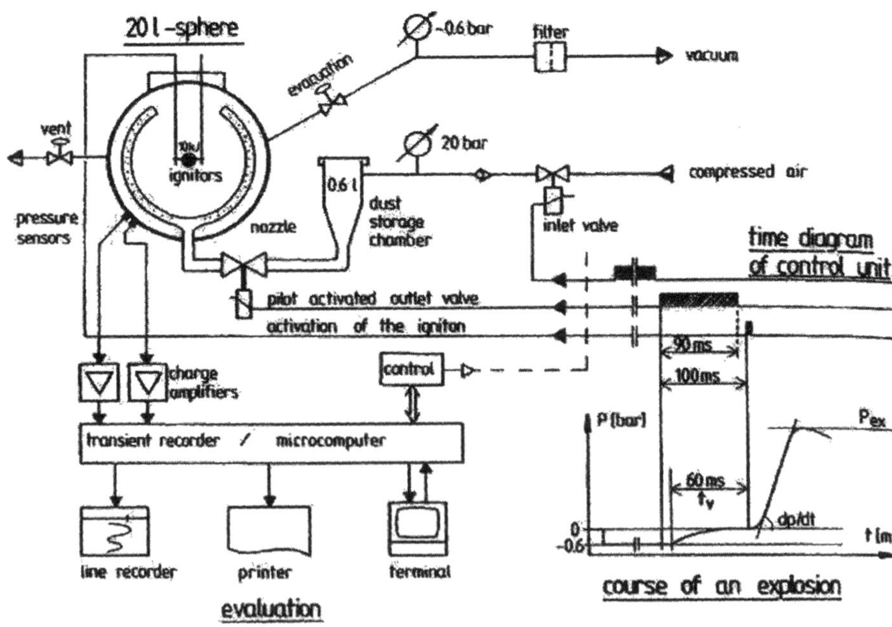

Figure 4-67 Test apparatus and data recording for 20-liter sphere tests. (From ASTM E 1226, with permission.)

test vessel for such tests, although larger test vessels up to 1-m^3 are sometimes used. Figure 4-66 is a photograph of the most widely used 20-liter spherical test vessel. Figure 4-67 is a schematic drawing of the test vessel, the associated components used for injecting dust into it, and the electronic control and data acquisition equipment. A sketch of a representative pressure versus time recording is also shown. The maximum-rate-of-pressure-rise, $(dp/dt)_{max}$ is determined from the maximum slope of the pressure rise curve as indicated in the top sketch of Figure 4-68. The reported values are the peaks in the plots versus tested dust concentration, as indicated in the bottom two plots in Figure 4-68.

The discussion at the beginning of Section 4.3.7 about the importance of the ignition time delay and the dust injection method is especially pertinent to the measurement of explosion pressures and rates-of-pressure-rise. Both standards emphasize the need for calibrating the test vessel and test methodology by checking data for benchmark dust samples that have been previously tested in a standardized 1-m^3 test vessel. The benchmark dusts specified in ASTM E 1226 are lycopodium clavatum (28 μm mean diameter), for which $P_{max} = 7.0$ barg and K_{St} (K_{max}) is 151 bar-m/s, and Pittsburgh seam coal (80% through a 200 mesh), for which $P_{max} = 7.0$ bar and K_{St} is 117 bar-m/s. The specified procedure for determining P_{max} and K_{St} in ASTM E 1226 is to run three tests at each concentration, and then take the highest value of the three-test-average over the entire range of concentrations.

Kuhner AG, the manufacturer of the most commonly used test vessel for these tests, coordinates round robin testing each year to verify that participating laboratories are maintaining and calibrating the equipment and methods in a way that can produce reproducible test data. The dust sample used for the 2001 round robin tests was the same dust sample described previously for the MIE round robin tests. Results for the P_{max} and K_{max} round robin tests are plotted in Figure 4-69. All but three of the 42 laboratories that participated achieved results for P_{max} that were within ±0% of the nominal value of 8.4 bar. Although the scatter was wider in the measured values of K_{max}, all but two of the 42 laboratories were able to obtain results that were within ±20% of the nominal value $K_{max} = 220$ bar-m/s, and 30 of the laboratories achieved results within 10% of this reference value.

Since the deflagration vent area is directly proportional to K_{St} in the new NFPA 68 guidelines, a 10% or 20% scatter/uncertainty in the value of K_{St} corresponds to a 10% or 20% uncertainty in the required deflagration vent area. There are similar implications in the design of deflagration suppression systems using manufacturer proprietary design formulas.

A more fundamental and difficult issue associated with the use of K_{St} values from 20-liter sphere tests for deflagration vent and suppression system design is that there is no known relationship between the turbulence level in the 20-liter sphere tests and the turbulence level during normal par-

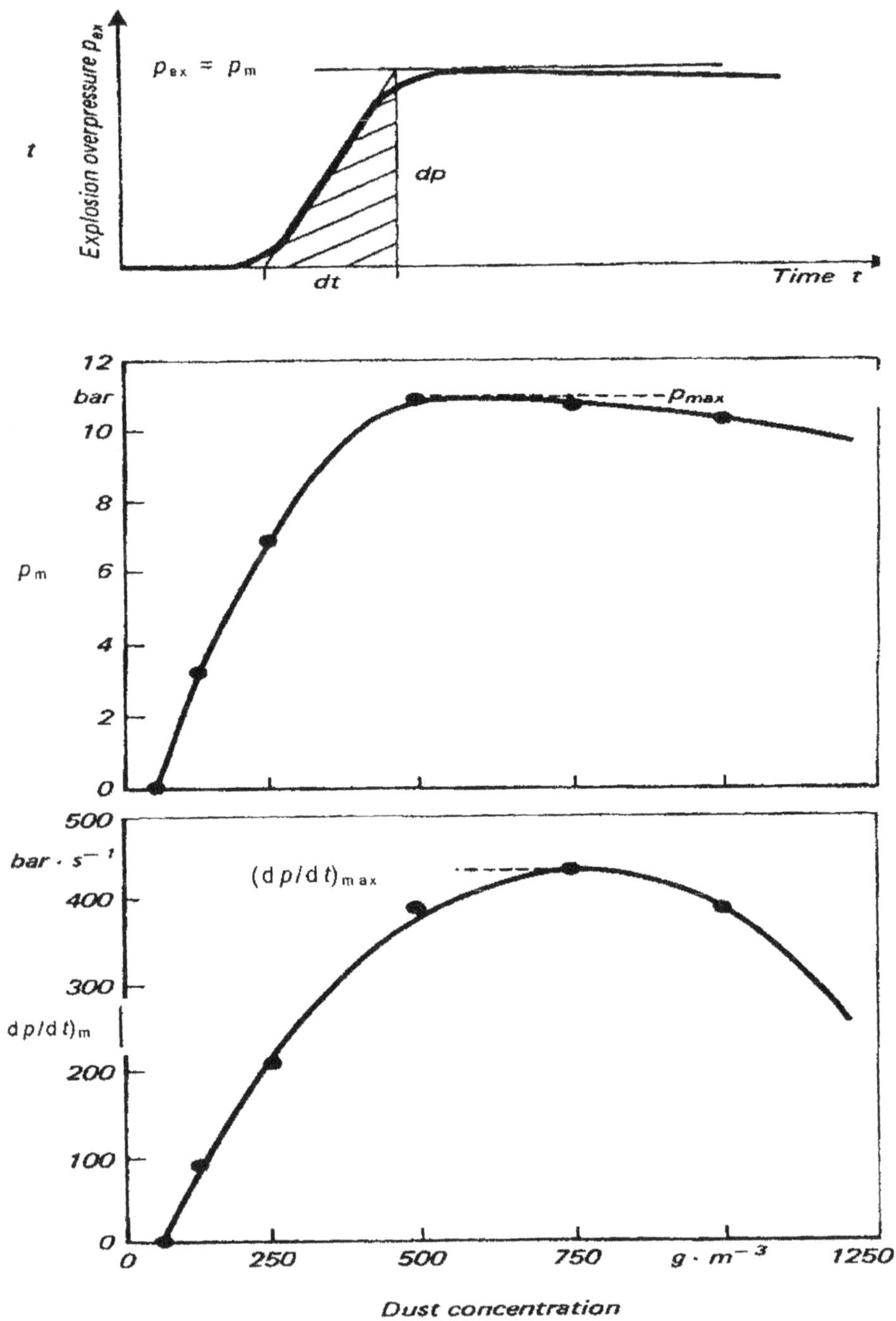

Figure 4-68 Determination of P_{max} and dP/dt_{max} from 20-liter sphere test data (from Draft EN 14034-1).

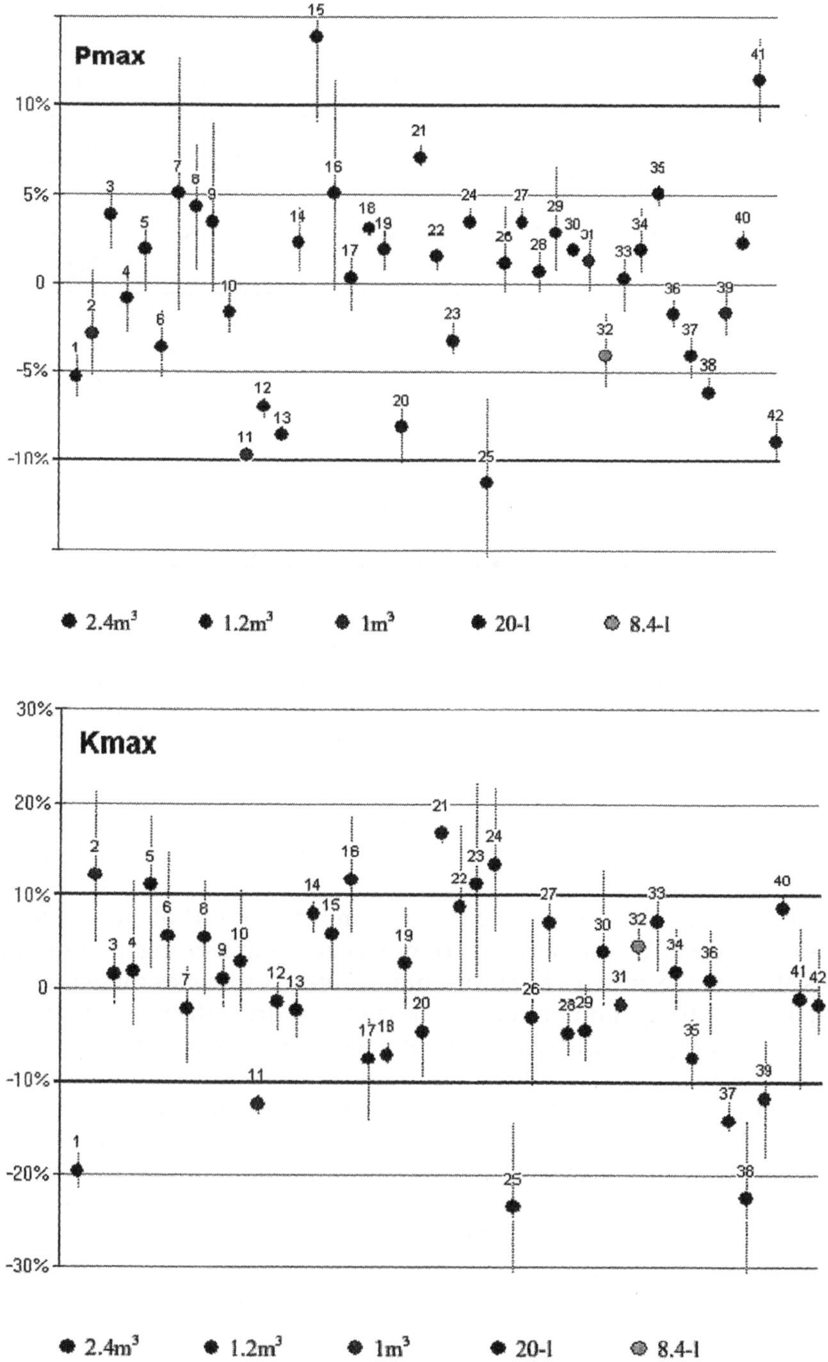

Figure 4-69 Round Robin test data for P_{max} and K_{max} (from Kuhner with permission).

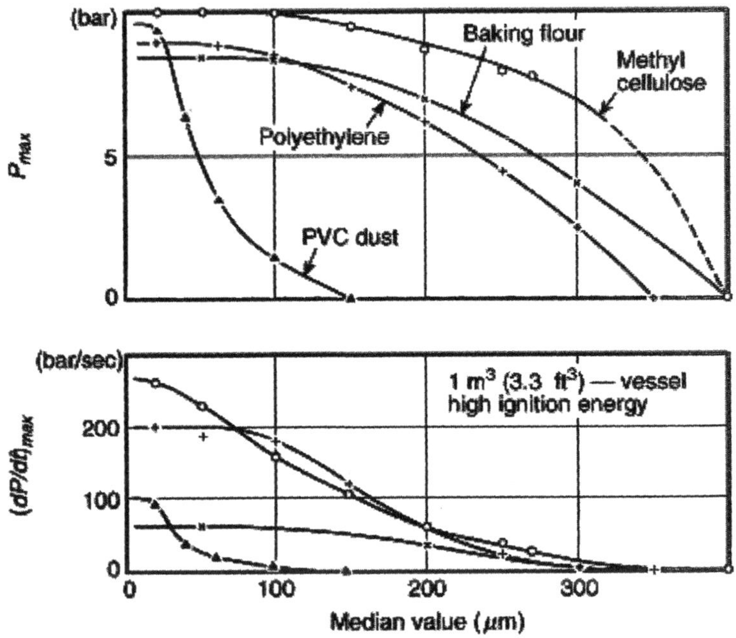

Figure 4-70 Effect of particle diameter on P_{max} and $(dP/dt)_{max}$ of various dusts, reprinted with permission from NFPA 68-2002, Venting of Deflagrations, Copyright © 2002, National Fire Protection Association. This reprinted material is not the complete and official position of the NFPA on the referenced subject, which is represented only by the standard in its entirety.

ticulate handling and process upsets in an operating facility. In lieu of such a relationship, the standardized test method emphasizes the need for each laboratory to verify that it is making measurements using a dust dispersion method and ignition delay time that maximizes the turbulence level at the time of ignition. The standards also emphasize using a strong ignition source consisting of one or two 5 kJ pyrotechnic igniters, as long as there is verification that the strong ignition source does not overdrive the pressure development for low K_{St} dusts. The latter can be checked with some tests in a larger (1 m³) test vessel.

Figure 4-70 shows the effect of particle median diameter on the measured values of P_{max} and $(dP/dt)_{max}$ for four different materials tested in a 1-m³ vessel. In all cases the measured values approach an asymptotic maximum as the particle diameter decreases. However, the diameter at which the asymptote is approached is different for the different materials. With polyethylene, methyl cellulose, and baking flour, the P_{max} asymptote occurs at diameters less than about 100 μm. With the much less reactive PVC, the P_{max} asymptote seems to occur at diameters less than about 20 μm. The $(dP/dt)_{max}$

asymptote for polyethylene and baking flour occurs at about 100 μm, but for methyl cellulose and PVC it seems to occur at diameters less than 20 μm.

The variation of $(dP/dt)_{max}$ with diameter shown in Figure 4-70 has an interesting implication regarding the difference between the particle sizes used in the U.S. and European standards for combustible dust cloud testing. The U.S. standards specify using a sample that has passed through a mesh with a 75-μm opening. The European standards specify using a sample through a 63-μm mesh opening. Should the results be equivalent? According to Figure 4-70, the difference between a 75 μm diameter and a 63 μm diameter is negligible for some dusts but not for others. An empirical curve fit through the data for PVC indicates that the difference in $(dP/dt)_{max}$ between those two diameters can be about 40%. Thus, there may be some significant differences in K_{St} values obtained for some materials using the two different standards, even though many materials should produce comparable results.

When eccentric shaped particles are considered, the diameter does not suffice to represent size effects. Flake particulates are best described in terms of specific surface area, and the data in Figure 4-71 for aluminum flakes show that the K_{St} value seems to be a linearly increasing function of specific surface area. In the case of short fibers (flock), data reported by Bartknecht (1989) and shown here in Figure 4-72 indicates that K_{St} correlates with the product of the flock denier (effectively a surrogate for fiber diameter) and length. In the case of nylon flock with a cut length of about 1 mm, this and other data indicate that flock diameters ≤15 μm have sufficiently high P_{max} and K_{St} to produce potentially destructive deflagrations.

Prior to the advent of the 20-liter spherical test vessel for dust explosion testing, most of the tests were conducted in the 1.2-liter Hartmann (bomb) cylinder. Although there were extensive databases obtained using the Hartmann cylinder, the results cannot be scaled to substantially larger vessels. Eckhoff's comparisons (1984/85) showed that the K_{St} values measured in a 1 m^3 vessel using the procedure described above were often two to three

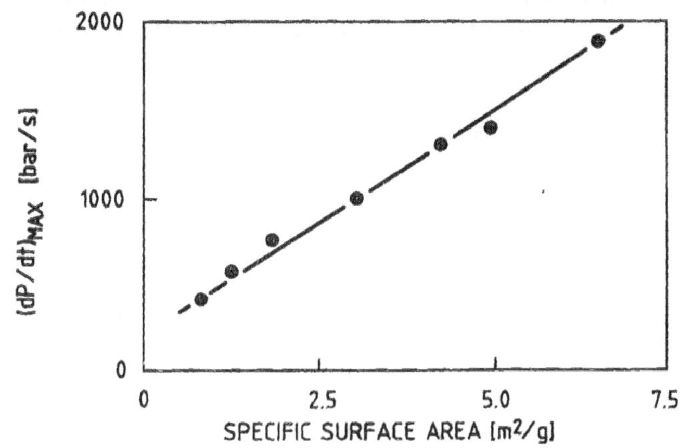

Figure 4-71 Effect of aluminum flake specific surface area on $(dP/dt)_{max}$. (From Eckhoff, R., *Dust Explosions in the Process Industries*, 3rd ed. Amsterdam: Elsevier, 2003.)

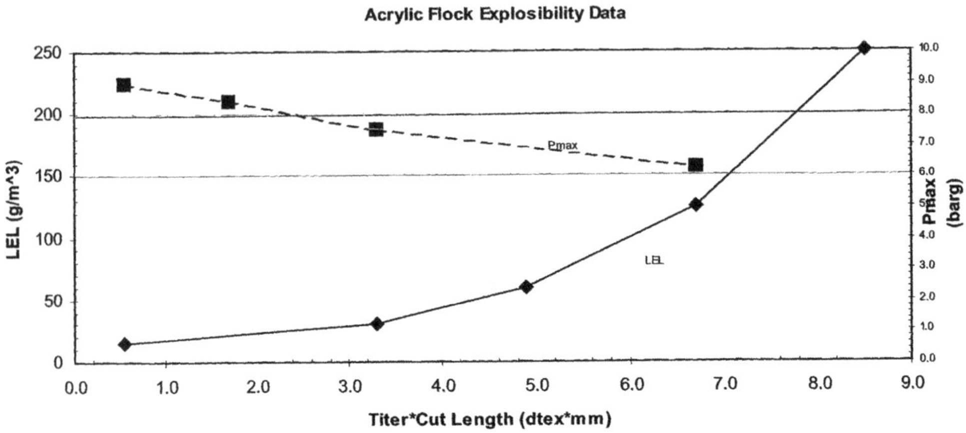

Figure 4-72 Variation of P_{max} and LEL with flock dtex-length.

times as large as those measured for the same materials in the Hartmann cylinder, and were about ten times as large in the case of the less reactive dusts. This is in contrast to the use of the 20-liter sphere, which produces almost the same K_{St} values as those in the 1-m^3 vessel (Siwek, 1996). Limitations of the Hartmann cylinder and its inappropriateness for use in determining K_{St} values for explosion protection design are clearly stated in the ASTM E 789 Standard. Although most testing laboratories no longer use the Hartmann cylinder for these purposes, there are still some laboratories that provide P_{max} and $(dP/dt)_{max}$ data per ASTM E 789 instead of the more appropriate ASTM E 1226.

Addition of small concentrations of flammable vapor or gas to combustible dust clouds can produce unexpectedly high values of P_{max} and K_{St} compared to the corresponding values for the dust alone. This effect is illustrated in Figure 4-73 in terms of the addition of methane concentrations to PVC dust clouds of varying concentrations. For example, the addition of 3% methane (below the 5% LFL) can cause the K_{St} of weakly reactive PVC dust to triple, and can cause surprisingly high values of P_{max} at concentrations well below the MEC for PVC alone. Protection measures for these hybrid dust–vapor explosions requires accurate dust explosibility data in the presence of the added flammable vapor or gas.

Limiting Oxygen Concentration Tests

The Limiting Oxygen Concentration (LOC) of a fuel–oxidant–inert gas mixture is defined as (ASTM E 2079) the oxygen concentration at the limit of flammability for the worst-case (most flammable) fuel concentration. It is a crucial parameter for inerting applications (NFPA 69), and its value depends on the particular inert gas used. Tests to determine the LOC involve first

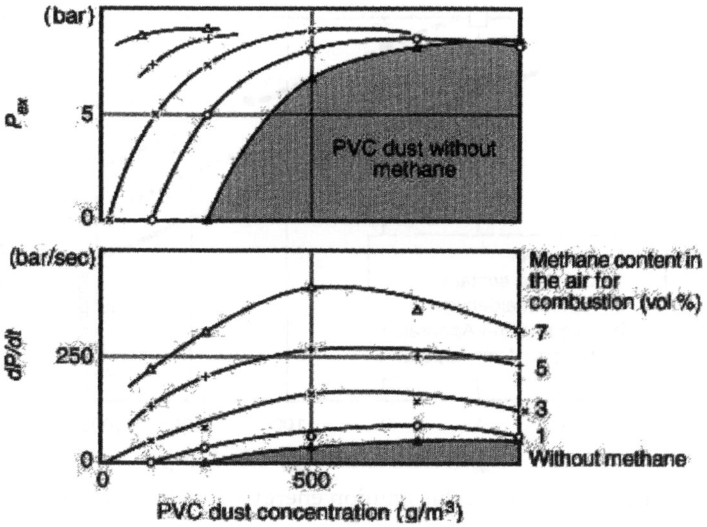

Figure 4-73 Effect of added methane on explosibility of PVC dust, reprinted with permission from NFPA 68-2002, Venting of Deflagrations, Copyright © 2002, National Fire Protection Association. This reprinted material is not the complete and official position of the NFPA on the referenced subject, which is represented only by the standard in its entirety.

making an air–inert gas mixture or oxygen–inert gas mixture in a pressure vessel, then injecting a preweighed dust charge, and trying to ignite the dust cloud after an optimum time delay to allow the dust cloud to develop at the ignition site. According to Siwek (1996), the LOC is determined as the oxygen concentration below which a mixture can no longer be ignited in three successive tests.

The plot in Figure 4-74 indicates that the LOC decreases logarithmically with increasing ignition energy. Data are shown for tests in the 20-liter sphere and in the 1-m^3 sphere. Siwek (1996) states that the LOC should be the value obtained using a 10 kJ ignition energy in the 1-m^3 vessel. However, most LOC dust tests today are conducted in 20-liter spherical vessels using a smaller ignition energy. According to the correlations in Figure 4-74, an ignition energy of 250 J in the 20-liter sphere can be correlated to results obtained with a 10 kJ igniter in the 1-m^3 test vessel. Siwek prescribes use of this correlation to obtain LOC values for extrapolation to larger vessels.

CEN (1998) has developed standardized test procedures for determining combustible dust cloud LOC values. The tests begin with varying the oxygen concentration while using a dust cloud concentration of 250 g/m^3. After determining the limiting oxygen concentration at this dust concentration, the dust concentration is increased to the optimum concentration, and the oxygen concentration is decreased in 1% increments until the LOC for the dust material has been confirmed.

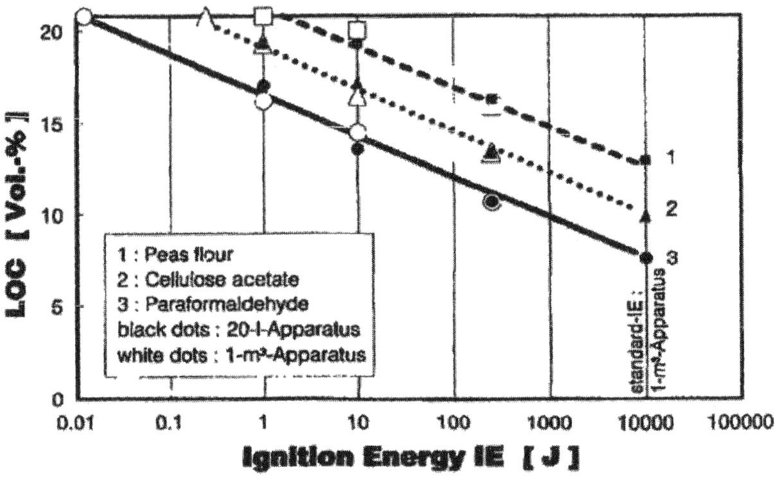

Figure 4-74 LOC versus ignition energy (from Siwek, 1994).

Siwek has correlated his LOC values against the combination of MIE and BAM furnace measured MIT values as shown in Figure 4-75. Based on his correlation, he suggests that rather than measure LOC values, they can be calculated from the following empirical equation.

$$\mathrm{LOC} = 1.62 \log\left[1+\left(\frac{\mathrm{MIE}}{2.73}\right)\right] + 12.9$$

The comparison shown in Figure 4-75 appears to indicate a ±2 volume % maximum deviation between data and the equation. Readers should also recall the scatter of 67% to 150% in the Kuhner round-robin MIE test data. A

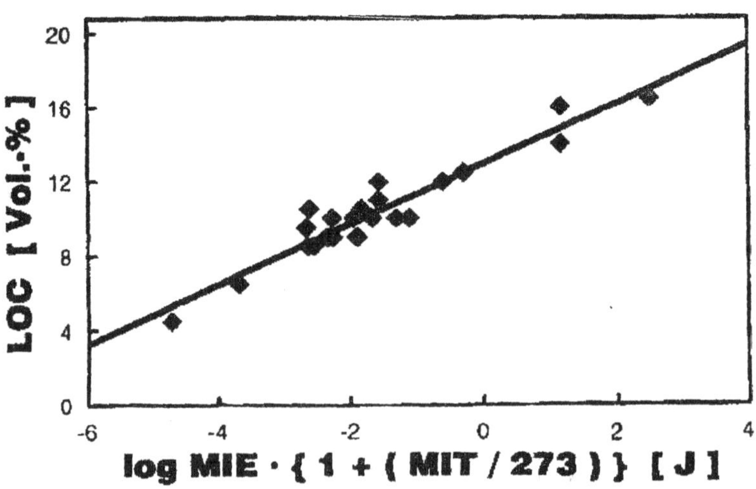

Figure 4-75 Siwek correlation of LOC versus MIE and MIT (from Siwek, 1994).

150% overestimate of the MIE would correspond to a 1 volume % overestimate of the LOC using the empirical correlation with a MIT of 400°C.

Summary of Dust Cloud Explosibility Tests

Table 4-15 provides a simplified summary of the applications, advantages, and limitations of the various dust explosibility tests discussed in this section. Appendix A provides tabulations of commercial laboratories that conduct these tests.

Figure 4-76 is a flow chart showing one suggested sequence for conducting these tests depending on the dust/powder processing and handling conditions. End points in the chart suggest that the test results lead to site-specific dust explosion protection determinations and guidelines. Other flow charts for dust explosion assessments are shown in Section 3.8. Dust explosion protection methods are described in Chapter 6.

4.3.8 Fire Exposure Tests

Concerns with regard to fire exposure are: (1) the particulate material may become unstable and undergo some type of violent self-reaction, (2) materials may spread the fire to the point that it cannot be controlled using conventional fire suppression methods and amounts of suppression agent; (3) the exposure fire may breach the container or packaging and allow the release of toxic or radioactive material; and (4) the material may react violently with the fire suppression agent.

The first concern (material instability) has been addressed in Section 4.3.2, which describes various laboratory tests for thermal instability. Materials that have been shown to be prone to thermal instabilities have special packaging requirements intended to reduce the fire heat flux passing through the container. DOT and U.N. hazardous materials transport regulations describe those packaging requirements.

The second concern (fire spread and suppression difficulty) is addressed, for transportation fire exposures, in the laboratory dust layer combustion test described in Section 4.3.5. This test includes a provision to determine the propensity for the fire to spread beyond the region of prewetted particulate. However, the test is primitive in that there is no measurement of either the critical heat flux for ignition or the fire heat release rate, which is the most important fire hazard parameter in contemporary fire testing and analysis. The most commonly used fire testing apparatus for measuring critical/ignition heat fluxes and heat release rates per unit surface area is the cone calorimeter (Babrauskas, 2002, 2003).

In the case of warehouse storage fire exposures, combustible particulate materials are generally considered to be free-flowing materials that tend to smother the exposure fire when released from their container (NFPA 13),

TABLE 4-15
Dust Cloud Explosibility Test Methods

Test	Standard	Applications	Advantages	Limitations
Minimum Explosible Concentration (MEC)	ASTM E 1515 CEN prEN 14034-3	Prevention via dust concentration control, e.g. in pneumatic conveying	The measured MEC value can be compared to in-situ measurements of suspended dust concentrations in conveyors and other process equipment, and can be used to verify concentration is below MEC.	Measured MEC value is based on a pressure rise of one atmosphere above the pressure due to the igniter; explosions with smaller pressure rises can occur at concentrations below the reported MEC value.
Minimum Cloud Auto Ignition Temp: (MAIT)	ASTM E 1491 IEC 1241-2-1	Safe operating temperatures in heated process equipment.	MAIT is valuable data for both particulate material manufacturing and for post-production processing by other companies and facilities.	BAM (horizontal) oven yields lower MAIT values than Godbert-Greenwald (vertical) furnace. MAIT value depends on the residence time of the dust cloud in the heated equipment, and with the area of a heated surface.
Minimum Ignition Energy (MIE)	ASTM E 2019 IEC 1241-3	Electrostatic ignition hazard evaluations; FIBC material classes	MIE value determines precautions needed in silo/bag filling and other particulate handling operations.	Measured MIE value depends on dust cloud turbulence level as well as amount of inductance in spark generation circuit.
Maximum Explosion Pressure (P_{max}) and K_{St}	ASTM E1226 ISO 6184	Deflagration containment, deflagration venting or suppression	K_{St} is often considered the most important parameter to characterize dust material combustibility.	Measured K_{St} value depends on both ignition energy and dust cloud turbulence level as determined by time delay between dust injection and ignition.
Limiting Oxygen Concentration (LOC)	ASTM E 2079 CEN prEN 14034-4	Inerting per NFPA 69 and NFPA 654	Provides valuable explosion prevention data.	Measured LOC values vary with ignition energy and with particular inert gas used in test.

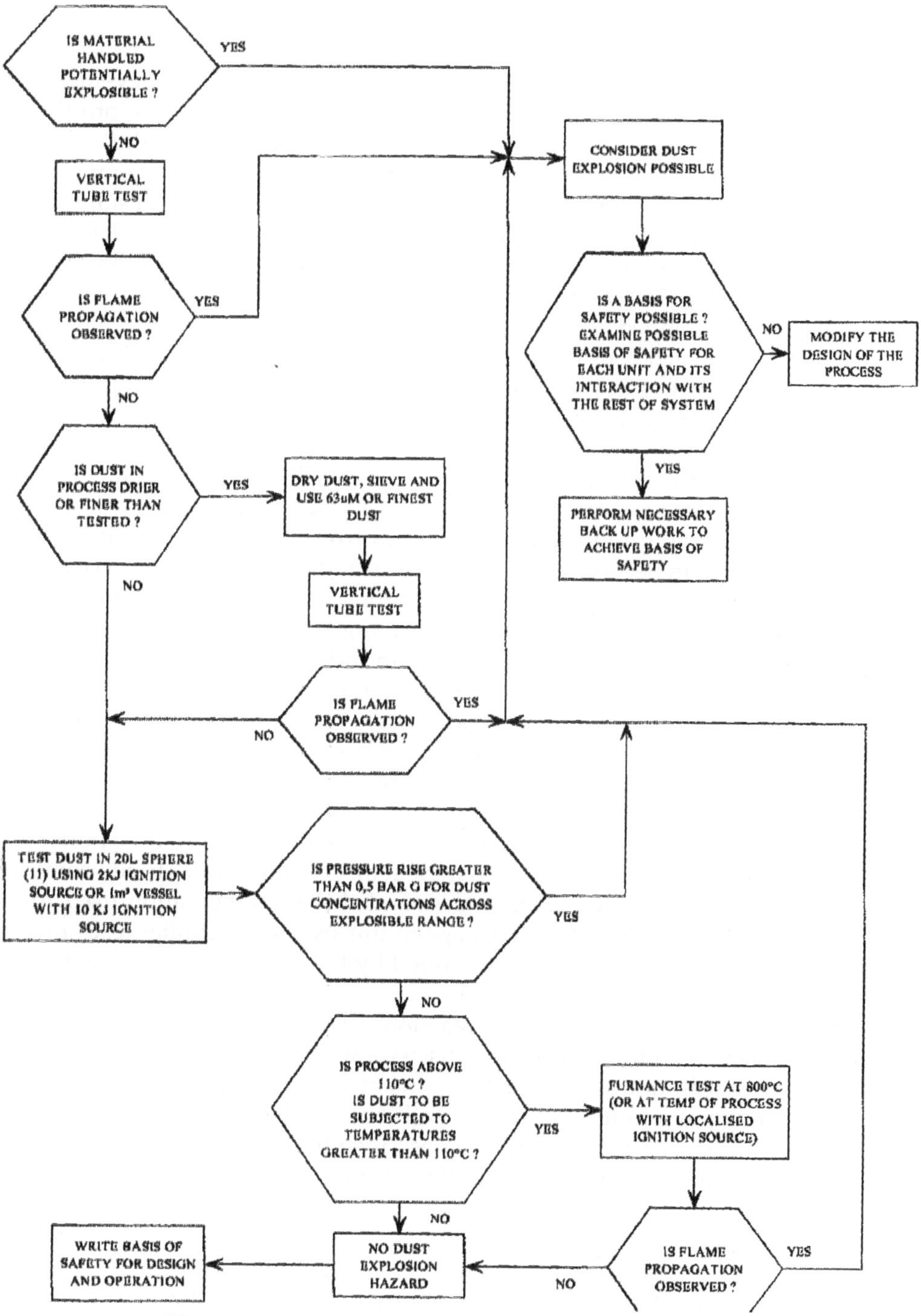

Figure 4-76 EC RACE Dust explosion testing flow chart.

and are therefore classified as Class IV commodities. However, some products and commodities may raise special fire control/suppression issues beyond whether or not they are free flowing. Fire tests have been developed by Factory Mutual (Zalosh, 2003, p. 144) to classify commodities into the NFPA 13 sprinklered warehouse storage categories: Classes I, II, III, IV, and Group A, B, and C plastics. These tests involve eight pallet loads of commodity tested for fire development in the presence of water application simulating ceiling sprinkler discharges. Fire researchers are also trying to develop laboratory-scale tests to correlate with the commodity classification tests (Tewarson, 1995), but additional research is needed before a reliable correlation is developed.

Container breach considerations—the third concern for special hazard materials—are addressed in different ways depending on the application and associated standard. For example, in the case of storage applications, the new International Fire Code (2000) requires stringent isolation of materials that are categorized as either toxic or "highly toxic" based on the inherent toxicity of the material as measured by its LC_{50} value for a 4-hour exposure. The issue of deriving four-hour LC_{50} values from laboratory toxicity test data is discussed in Section 4.3.9. The NFPA standards do not have a corresponding generalization for storage restrictions, but do require prominent display of the NFPA 704 symbol designating the material's instability, flammability, and toxicity categories plus an indication of water reactivity hazards where applicable. There are also separate NFPA standards for certain categories of chemicals such as oxidizers.

The DOT and Nuclear Regulatory Commission require fire exposure tests for packages/containers of certain radioactive materials. The test requires the package/container to be fully engulfed in a hydrocarbon fuel–air fire with an average flame temperature of at least 800°C for a period of 30 mintes (10 CFR Part 71.73). Hovingh et al. (1999) have described some facilities used for conducting such tests and the instrumentation used to characterize the response of the package to this exposure. Similar tests and associated package responses are described by Pinton et al. (1999) and Burgess (1993).

Concern about possible violent reactions of the material with fire suppression agents is addressed in the discussion of Section 4.3.3 of general reactivity hazards, and water reactivity hazards in particular. Many metal particles are known to react violently with both water and some other fire suppression agents. Nelson (2002) has described the development of special suppression agents for metal powder fire suppression.

4.3.9 Particulate Toxicity Testing

The main purpose of toxicity testing is to provide a data base that can be used to assess and manage the risk associated with exposure to a chemical agent.

Toxicity testing results are used to manage risk in a wide variety of ways including (Casarett and Doull, 1991):

- registering and permitting the manufacture and sale of products,
- determining acceptable exposure levels for the public and the environment,
- determining appropriate packaging, labeling, shipping and storage requirements,
- communicating the hazards of the material,
- determining short-term and long-term exposure limits for workers,
- designing health monitoring programs for workers,
- designing manufacturing processes,
- selecting appropriate personal protective equipment for workers,
- development of appropriate antidotes and treatment regimes for poisoning, and
- development of analytic techniques to detect residues of chemicals in tissues and other biologic materials.

The toxicity testing required for a specific particulate solid depends on applicable regulations, company policies, intended use of the material, potential for exposure, known toxic effects of the material, and suspected toxic effects based on known toxic effects of structurally similar compounds. Toxicity testing can be very expensive, especially for long-term exposure testing in multiple animal species; but careful design of the appropriate studies can minimize the cost. Toxicologists should be consulted to determine the toxicity testing required, to design and manage the testing program, and to interpret the test results.

In the United States there are several federal agencies involved in the administration of dozens of statutes involving human exposure to chemicals. These agencies include the Food and Drug Administration (FDA), the Environmental Protection Agency (EPA), the Occupational Safety and Health Administration (OSHA), and the Consumer Product Safety Commission (CPSC). Depending on the particular statute involved, each regulatory agency can require or recommend a variety of toxicity tests. As an example, the EPA New Chemicals Program requires the proposed large-volume manufacturer or importer of new chemicals to submit a Pre-Manufacture Notice for EPA review (http://www.epa.gov/opptintr/newchems/). The type of test data required in a Pre-Manufacture Notice depends on whether the new chemical falls into one of 45 chemical families defined in the EPA New Chemical Program. Some chemical families require human health toxicity test data, while others require environmental toxicity data, and /or safety test data.

Determination of acute oral toxicity is usually the initial step in the evaluation of a material's toxic characteristics. In an acute toxicity study, the toxic effects are determined for a progression of doses administered as a single short-term exposure involving one or more animal species (usually rats). The EPA (2002b) describes a recommended method for determining the median lethal dose (LD_{50}) using animal testing. The acute toxicity tests give a quantitative estimate of acute toxicity for comparison to other substances, identify target organs and other clinical manifestations of acute toxicity, establish the reversibility of the toxic response, and give dose-ranging guidance for additional studies. Acute dermal or acute inhalation studies may be performed if there is reasonable likelihood of substantial exposure via those routes. All small particulates are capable of producing adverse respiratory effects when people or animals are exposed to them at sufficiently high concentrations or doses. One of the primary issues in particulate inhalation testing is separating respiratory system overload effects from the inherent toxicity of the particulate material.

Subacute toxicity tests are performed to obtain information on the toxicity of the chemical after repeated administration of short-term doses, and to help establish the doses for subchronic studies.

Subchronic tests can last for different periods of time, but 90 days is the most common test duration. The principal goals of the subchronic study are to establish a no-observable-effect level and to further identify and characterize the specific organ(s) affected by the test compound after repeated administration by the route of intended exposure (usually oral). The subchronic toxicity studies not only characterize the dose–response relationship of a test substance following repeated administration, but also provide data for determining appropriate doses for chronic exposure studies. If significant exposure to the chemical is likely to be by dermal contact or inhalation, subchronic dermal or subchronic inhalation experiments might also be required. Subchronic inhalation studies are particularly difficult to conduct for particulate solids. Special methodologies must be used to administer accurate doses over long periods of time, and to prevent the material from accumulating on the animal's fur and subsequently being ingested during grooming activities.

Chronic toxicity tests are used to determine the effect of long-term exposure to a substance. The period of exposure is longer than 90 days (in rodents, usually 6 months to 2 years), and the length of exposure is somewhat dependent on the intended period of exposure in humans.

Other toxicity tests that may be required or recommended include:

- Dermal and ocular irritation
- Skin sensitization
- Delayed hypersensitivity reaction
- Carcinogenicity
- Developmental and reproductive toxicity
- Mutagenicity
- Immunotoxicology
- Toxicokinetics

Epidemiological studies may also be required or recommended to determine exposure effects in human populations.

4.3.10 UN Testing Scheme for Classification of Materials as Explosives

The U.N. *Recommendations on the Transport of Dangerous Goods: Manual of Tests and Criteria* Third Edition (1999) describes the testing and associated criteria for designating materials as explosives, and for determining the appropriate category of explosive material. The determinations require a lengthy test program consisting of various series of tests as illustrated in Figures 4-77 and 4-78. The three possible results of these tests are:

1. The material is not designated as an explosive (UN Class 1) and should be tested to determine the appropriate nonexplosive class.
2. The (packaged) material is considered too unstable for transport.
3. The material is a Class 1 and falls into one of the six divisions for explosives within Class 1.

Figure 4-77 is used to determine provisionally if a new material (including new mixtures of previously tested materials) or newly packaged material should be designated as an explosive for transportation purposes. If the material has not been previously tested for this purpose, and if it is not manufactured for use as an explosive, the flow chart begins on the left column with Test Series 1 and Test Series 2. Based on the results of either Test Series 1 or 2, it is possible that the new material may be considered to be sufficiently insensitive or inherently nonexplosive. The specific tests conducted in Test Series 1 are listed in Table 4-16, and those in Test Series 2 are shown in Table 4-17.

In the case of packaged potentially explosive materials, the flow chart in Figure 4-77 begins with Test Series 3, and depending on the test results, may require additional testing to classify the package. Figure 4-78 is used to determine into which of the six categories of Explosive the packaged material should be classified, and/or if the material can be packaged in such a way that it sufficiently reduces the hazard so it should not be classified as a Class 1 material.

4.4 SCALING CONSIDERATIONS IN APPLYING LABORATORY TEST DATA

Some of the tests described in Section 4.3 have been developed from fundamental considerations of scaling up laboratory test data to larger scale industrial facilities. Two examples are the oven tests for spontaneous heating and the near-adiabatic (low φ value) thermal instability and chemical reactivity tests. The theoretical basis for scaling spontaneous heating test data is

TABLE 4-16
Explosives Classifications Test Series 1

Test	Objective	Apparatus	Result Criteria
1a UN Gap Test	Determine the ability of bulk material to propagate a detonation.	Expendable steel tube 4.8 cm diameter, 40 cm long, with 160 g explosive charge at one end and a 3mm thick witness plate at the other end.	Detonation propagation is determined by tube fragmentation or by a hole in witness plate.
1b Koenen Test	Determine the effect of heating bulk material under confinement with a small vent.	Propane flame exposed expendable steel tube 7 cm long, 2.5 cm diameter, with a vented end cap and one of several orifice plates with different size orifices.	Tube fragmentation indicating an explosion has occurred with an orifice of 1.0 mm diameter or larger.
1c (i) Time/Pressure Test	Determine if ignition in a closed tube causes a deflagration with an unacceptably large pressure.	Steel pressure vessel 8.9 cm long and 6 cm in diameter equipped with an electric fuse-head and a pressure transducer.	Pressure rise of at least 2070 kPa (300 psig).
1c (ii) Internal Ignition Test	Determine whether igniting bulk material under confinement causes deflagration-to-detonation transition.	3-inch Schedule 80 steel pipe 46 cm long and 7.4 cm in diameter with a 15-amp igniter.	Pipe or end cap fragmentation.

described in Section 4.3.4. The near-adiabatic instability/reactivity tests may not have a formal scaling procedure, but the rationale is based on the premise that scaling to larger vessels inherently implies going to a more adiabatic system. In other cases, such as the 20-liter sphere dust explosibility testing, there is no inherent theoretical justification, but there are empirical scaling test data to provide some measure of confidence as far as size effects are concerned (turbulence effects do not have a demonstrated scaling basis).

Many other tests described in Section 4.3 are entirely empirical without benefit of any systematic scaling investigation. Readers should use caution in applying the results of these tests directly to large-scale industrial facilities. Since there is no generally accepted scaling basis for many of these tests, it may be necessary to seek expert assistance, or to use due diligence in researching the current thinking as expressed in recent reports, technical papers, and technical presentations. The Center for Chemical Process Safety and the American Institute of Chemical Engineers Loss Prevention Symposia are good resources for obtaining current information of this nature.

TABLE 4-17
Explosives Classifications Test Series 2

Test	Objective	Apparatus	Result Criteria
2a Detonative Shock Sensitivity Test	Determine the detonative sensitivity of confined bulk material	Similar to 1a, i.e., Expendable steel tube 4.8 cm diameter, 40 cm long, with 160 g explosive charge separated by a 5 cm long PMMA gap spacer. There is a 3mm thick witness plate at the other end.	Complete fragmentation of tube or hole in witness plate.
2b Koenen Test	Determine the sensitivity of confined bulk material to intense heat.	Same as 1b, i.e., Propane flame exposed expendable steel tube 7 cm long, 2.5 cm diameter, with a vented end cap and one of several orifice plates with different size orifices.	Tube fragmentation indicating an explosion has occurred with an orifice of 2.0 mm diameter or larger.
2c(i) Time/pressure Test	Determine if ignition in a closed tube causes a deflagration with an unacceptably large rate-of-pressure-rise.	Steel pressure vessel 8.9 cm long and 6 cm in diameter equipped with an electric fuse-head and a pressure transducer.	Time for pressure to increase from 670 kPa (100 psig) to 2070 kPa (300 psig) is less than 30 ms.
2 (d) Internal Ignition Test	Determine whether igniting bulk material under confinement causes deflagration-to-detonation transition.	3-inch Schedule 80 steel pipe 46 cm long and 7.4 cm in diameter with a 10-g black powder ignition source.	Pipe or end cap fragmentation.

4.5 LARGER-SCALE TESTING AND THEORETICAL MODELING

When a particulate material's hazard cannot be assessed solely with existing laboratory-scale or small-scale testing, there is a need to develop a new approach, usually entailing large-scale testing or theoretical modeling. Intermediate-scale or large-scale testing is most appropriate when an empirical hazard assessment is warranted, often for a particular application or facility. Theoretical modeling is most appropriate when the hazard phenomenology is well understood and either a generalization is sought or there is a need to analyze an application that is not amenable to testing.

For example, consider the dust explosion hazard associated with the return air duct from a dust collector. Although the return air duct is usually dust free, a dust explosion in the collector can cause burning dust to be

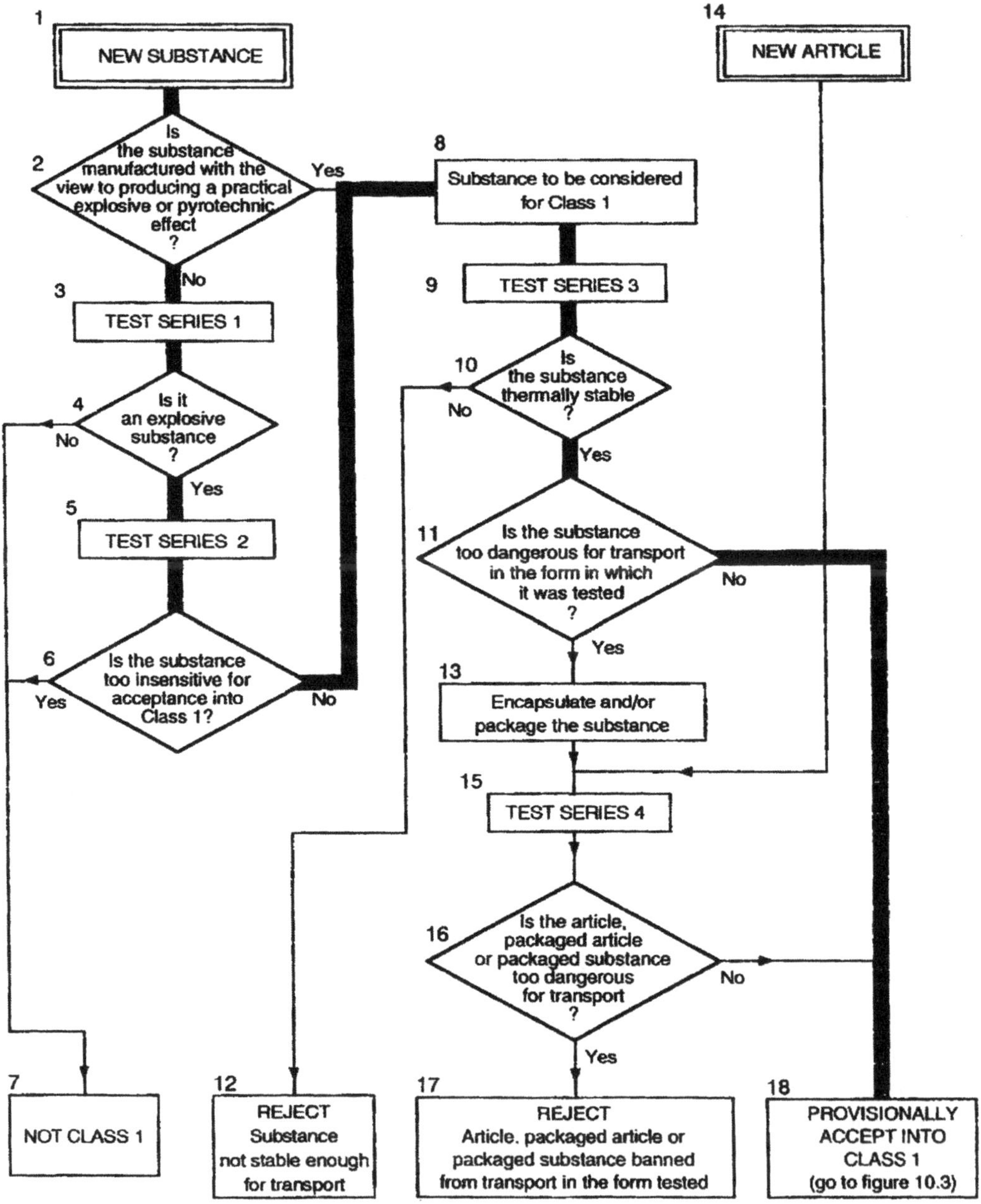

Figure 4-77 UN testing flow chart for potential explosives (UN, 1999).

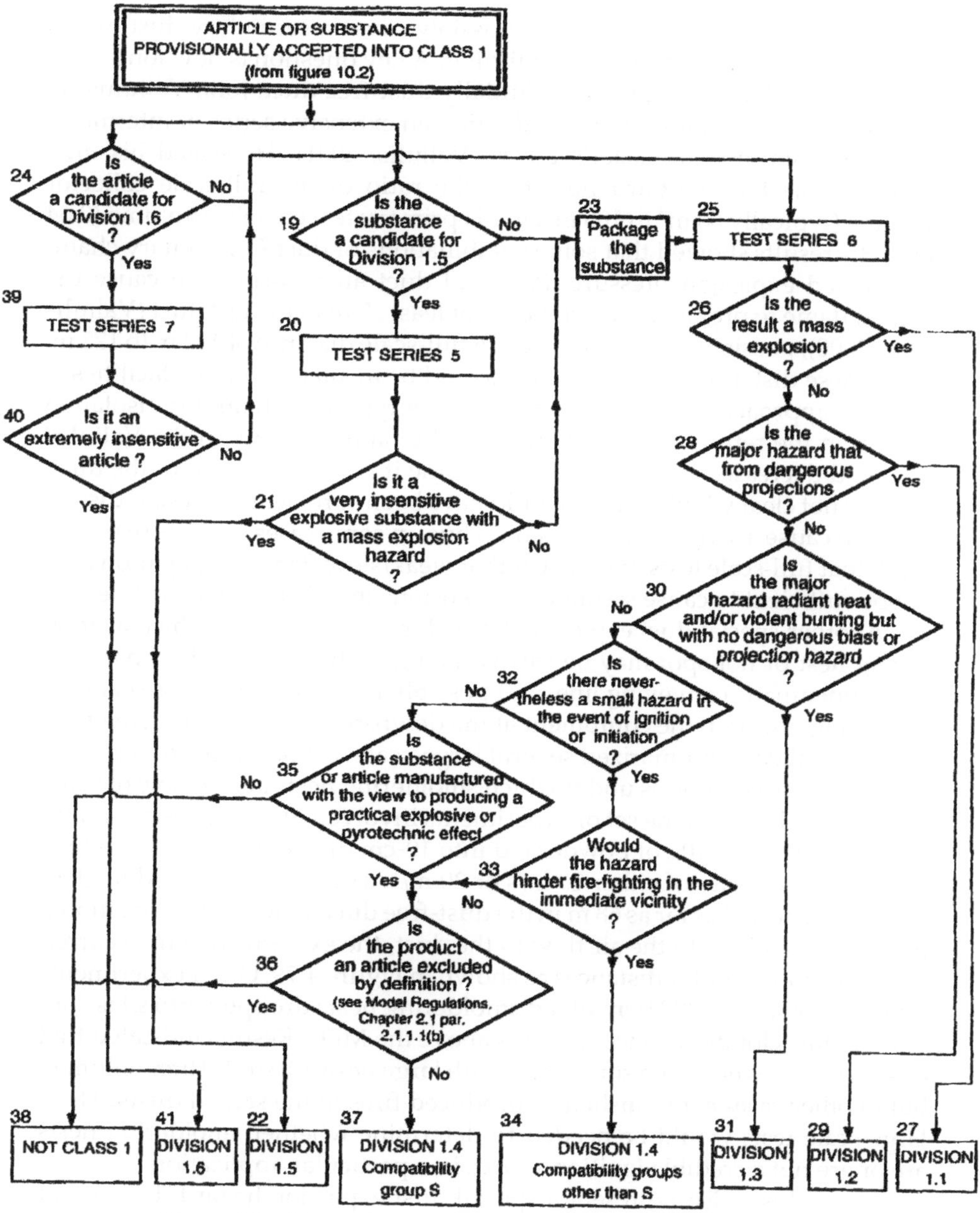

Figure 4-78 UN explosives classification flow chart (UN, 1999)

vented into the return air duct and possibly cause a secondary dust explosion in the particulate processing building. A key question is how long can a vented explosion propagate in an initially dust-free duct. John Valiulus set out to answer that question by conducting an extensive series of intermediate-scale tests for his master's thesis (Valiulis, 2001). He scaled his duct diameters as the one-third power of the ratio of the full-scale collector volume (typically 60 m^3 to 180 m^3) to his primary test chamber volume (0.64 m^3). His results showed that some combinations of dust K_{St}, primary chamber vented explosion pressure, P_{red}, and duct air velocity, can cause the vented flame to propagate a distance of at least 27 m in the test duct (Valiulis et al., 2000), possibly corresponding to a full-scale distance of 122 m to 177 m. Since these distances are larger than would be available in most facilities, a prudent approach would be to install some type of deflagration isolation system in the return air duct. However, the flame velocity in the Valiulis' tests had decreased significantly by the time it reached the end of the duct, and it is not clear whether there would have been sufficient flame and disturbance to cause a secondary explosion. Thus, in some cases, there is reason to think that full-scale tests may reveal that weak/slow flame propagation may not be sufficient to cause secondary explosions in at least some facilities.

Another special dust explosion hazard is the scenario in which a smoldering nest or hot spot may be carried along with the un-ignited particles into a pneumatic conveying line, and possibly ignite a dust explosion in the conveying line or in the equipment at the downstream end of the conveying line. Lunn (2002) summarized several large-scale test programs designed to determine the conditions under which smoldering nests are capable of causing dust explosions. One set of experiments involved glowing nests containing approximately 10g of powder fed into 10-cm diameter ducting in which there an air velocity of either 10 m/s or 20 m/s. The nests continued to glow while transported as far as 68 m in the dust-free duct. However, as additional powder was added to the air flow in the duct, the extinguishment occurred at significantly shorter distances as shown in Figure 4-79. Other experiments reviewed by Lunn (2002) involved smoldering nests dropped into silos containing dust clouds at concentrations above the MEC. Results indicated that dust explosions occurred sometimes with large nests (0.5 to 1.0 liter volume), but in other cases extinguished or produced fires in the settled dusts. Thus, large-scale tests would be needed to determine whether smoldering nests are or are not a credible ignition source in a particular application.

Metal dusts also represent a special dust explosion hazard (because of their exceptionally high flame temperatures and K_{St} values) warranting larger scale testing to determine appropriate protection via deflagration venting or suppression. Going and Snoeys (2002) have described tests in the Fike 1-m^3 chamber to confirm that rapid detection and high suppressant concentrations can successfully suppress aluminum dust explosions. Deflagration venting tests in the 1-m^3 chamber and in a 2.6-m^3 chamber indicate that

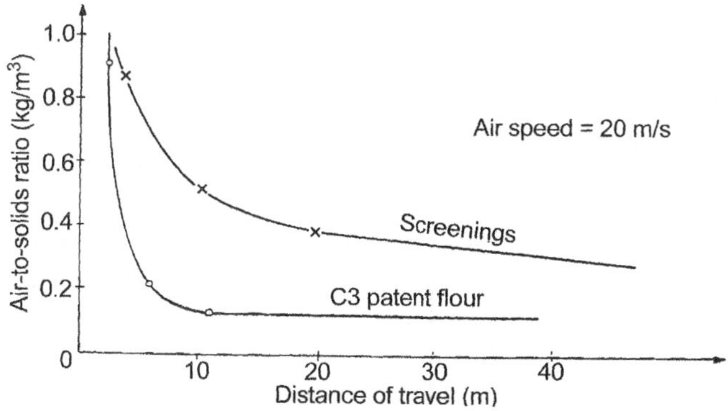

Figure 4-79 Distance-to-extinguishment of smoldering nests transported with dust clouds of varying concentrations in a 10-cm duct (from Lunn, 2002).

the general deflagration venting guidelines may need to be revised to require disproportionately larger vent areas for aluminum dust explosion applications. Similarly, the spatial extent of the vented flame and associated hazard zone is significantly larger than would be predicted based on existing guidelines.

One example of a hazard that is amenable to theoretical modeling is the spontaneous ignition of thermally unstable particulate stored in a large silo. Zoghlami et al. (1997) developed a theoretical model intended to determine whether or not ignition would be expected for various size particles in silos of varying size, geometry, and convective cooling at the silo walls and roof. Their results showed that for a material with a heat of reaction of 300 kJ/mol and an activation energy of 58.24 kJ/mol stored in a 200 m^3 silo, spontaneous ignition would be expected for particle diameters less than or equal to about 7 mm. The time-to-ignition was about 40 days, and did not vary much with particle size, but could be delayed significantly or prevented by increased natural convection at the silo walls, possibly by increasing the silo surface area for a given volume. Readers are cautioned that this type of theoretical calculation entails the use of possibly proprietary computer software that needs to be validated against realistic test data. After successful critical review and validation, the calculations have the potential to reduce conservatisms inherent in simpler analytical formulations and data scaling procedures, such as those described in Section 4.3.4.

Although pyrophoricity is usually considered a hazard requiring empirical (laboratory-scale) testing for evaluation, Glassman et al. (1992) have described a simple theoretical model to determine the critical particle size below which a particular metal becomes pyrophoric. Their hypothesis is that small metal particles will be pyrophoric if/when the initial oxide coating that forms upon exposure to air generates sufficient heat to vaporize the remaining metal and the initial oxide coating. Based on this hypothesis, Glassman et al. developed a steady-state heat balance in which the heat of oxidation generated by the initial oxide layer is just balanced by the heat required to bring the metal and oxide coat to their respective boiling points. This criterion explains that metals such as lithium, sodium, and potassium are pyrophoric because they have relatively high heats of oxidation and relatively low boiling points. Other metals such as the first four listed in Table 4.6-1 can be pyrophoric at particle sizes that are on the order of 0.01 μm. There is good agreement with experimentally measured critical particle diameters for those metals. On the other hand, there is poor agreement for the last two metals listed in Table 4-18. Glassman et al. explain the poor agreement for those metals in that their oxide layers merely crack rather than vaporize to allow continued combustion of the nascent metal. These two exceptions demonstrate the limitations of relying entirely on theoretical analyses without some experimental data. It can be argued that the main value of theoretical analysis is in extrapolation of the experimental data to situations in which additional testing would be difficult or inconvenient. Examination of pyrophoricity effects in nanoparticles is one such application.

At the present time, computer modeling is not a viable substitute for larger scale testing to provide a definitive hazard assessment in the absence of at least some test data. However, advances in computer modeling, particularly in the use of Computational Fluid Dynamics (CFD) models with a particulate phase and a gas phase, offer hope that CFD codes may ultimately be able to do hazard assessments for particulates in the manner they are currently being used for gaseous combustion and reactivity hazard assessment.

TABLE 4-18
Critical Metal Particle Size for Pyrophoricity (data from Glassman et al., 1992)

Metal	Oxide	Oxide Coating Thickness (Å)	Calculated Critical Diameter (mm)	Measured Critical Diameter (mm)
Al	Al_2O_3	25	0.0273	0.03
Cu	CuO	45	0.0091	0.01–.03
Fe	FeO	35	0.0101	0.01–.03
Pb	PbO	31	0.0157	0.01–.03
U	UO_2	25	0.0133	10
Zr	ZrO_2	50	0.0324	3.0

REFERENCES

Abbot, J., 1990, *Prevention of Fires and Explosions in Dryers—A User Guide,* 2nd ed., Rugby, U.K.: The Institution of Chemical Engineers.

ACC/ANSI Z400.1, 1998, 20, "Hazardous Industrial Chemicals—Material Safety Data Sheets—Preparation," Washington, DC: American Chemistry Council; West Conshohocken, PA: American Society for Testing and Materials.

ACGIH, 2003, "2003 TLVs® and BEIs®," Cincinnati, OH: American Conference of Government and Industrial Hygienists.

Anthony, E., and Greaney, D., 1979,"The safety of hot self-heating materials," *Combustion Science and Technology*, 21: 79–85.

ASTM, 2002, "ASTM CHETAH," Version 7.3, West Conshohocken, PA: American Society for Testing and Materials.

ASTM D 1929-96, 2001, "Standard Test Method for Determining Ignition Temperature of Plastics," West Conshohocken, PA: American Society for Testing and Materials.

ASTM D 3523-92, 1997, "Standard Test Method for Spontaneous Heating Values of Liquids and Solids," (Differential Mackey Test), West Conshohocken, PA: American Society for Testing and Materials.

ASTM D 4470-97, 1997, "Standard Test Method for Static Electrification," West Conshohocken, PA: American Society for Testing and Materials.

ASTM E 1491-97, 1997 "Standard Test Method for Minimum Autoignition Temperature of Dust Clouds," American Society for Testing and Materials.

ASTM E 537-98, 1998, "Standard Test Method for Assessing the Thermal Stability of Chemicals By Methods of Thermal Analysis," West Conshohocken, PA: American Society for Testing and Materials.

ASTM E 680-79, 1999, "Standard Test Method for Drop Weight Impact Sensitivity of Solid-Phase Hazardous Materials," American Society for Testing and Materials.

ASTM E 2019, 1999, "Standard Test Method for Minimum Ignition Energy of a Dust Cloud in Air," West Conshohocken, PA: American Society for Testing and Materials.

ASTM E 473-00, 2000, "Standard Terminology Relating to Thermal Analysis," West Conshohocken, PA: American Society for Testing and Materials.

ASTM E 487-99, 2000, "Standard Test Method for Constant-Temperature Stability of Chemical Materials," West Conshohocken, PA: American Society for Testing and Materials.

ASTM E 659-78, 2000, "Standard Test Method for Autoignition Temperature of Liquid Chemicals," West Conshohocken, PA: American Society for Testing and Materials.

ASTM E 1226, 2000, "Standard Test Method for Pressure and Maximum Rate of Pressure Rise for Combustible Dusts," West Conshohocken, PA: American Society for Testing and Materials.

ASTM E 1515, 2000, "Standard Test Method for Minimum Explosible Concentrations of Combustible Dusts," West Conshohocken, PA: American Society for Testing and Materials.

ASTM D 5451, 2001, "Standard Practice for Sampling Using a Trier Sampler," West Conshohocken, PA: American Society for Testing and Materials.

ASTM D 5633, 2001, "Standard Practice for Sampling with a Scoop," AWest Conshohocken, PA: American Society for Testing and Materials.

ASTM D 5680a, 2001, "Standard Practice for Sampling Unconsolidated Solids in Drums or Similar Containers," West Conshohocken, PA: American Society for Testing and Materials.

ASTM D 6668-01, 2001, "Standard Test Method for Discrimination between Flammability Ratings of $F = 0$ and $F = 1$," West Conshohocken, PA: American Society for Testing and Materials.

ASTM E 476-87, 2001, "Standard Test Method for Thermal Instability of Condensed Phase Systems (Confinement Test)," West Conshohocken, PA: American Society for Testing and Materials.

ASTM E 789-95, 2001, "Standard Test Method for Dust Explosions in a 1.2-Litre Closed Cylindrical Vessel," West Conshohocken, PA: American Society for Testing and Materials.

ASTM E 1231-01, 2001, "Standard Practice for Calculation of Hazard Potential Figures-of-Merit for Thermally Unstable Materials," West Conshohocken, PA: American Society for Testing and Materials.

ASTM E 2021-01, 2001, "Standard Test for Hot Surface Ignition Temperature of Dust Layers," West Conshohocken, PA: American Society for Testing and Materials.

ASTM E 2079-01, 2001,"Limiting Oxygen Concentration in Gases and Vapors," West Conshohocken, PA: American Society for Testing and Materials.

ASTM E 968-02, 2002, "Standard Practice for Heat Flow Calibration of Differential Scanning Calorimeter," West Conshohocken, PA: American Society for Testing and Materials.

Babrauskas, V., 2002 ,"Heat Release Rates," Chapter 3-1 in *SFPE Handbook of Fire Protection Engineering,* 3rd ed, Quincy, MA: National Fire Protection Association.

Babrauskas, V., 2003, *Ignition Handbook,* available from www.doctorfire.com and Society of Fire Protection EngineersWeb sites.

Bailey, M., Hooker, P., Caine, P., and Gibson, N. 2002, "Incendivity of Electrostatic Discharges in Dust Clouds: The Minimum Ignition Energy Problem," http://www.safetynet.de/Publications/articles/MIE.PDF

Bartknecht, W., 1989, *Dust Explosions: Course, Prevention, Protection.* New York: Springer-Verlag.

Barton, J. (editor), 2002, *Dust Explosion Prevention and Protection.* Rugby, UK: Institution of Chemical Engineers.

Beever, P., 1995, "Self-heating and spontaneous combustion," in *SFPE Handbook of Fire Protection Engineering,* 2nd ed., Section 2, Chap 12, Quincy, MA: National Fire Protection Association.

Bennett, D., Lunn, G. A., Torrent, J. V., Querol, E., and Fritze, J-P., 2002, "A Test for Electrical Ignitions of Flammable Dust Clouds." Presented at Symposium on Process Safety and Industrial Explosion Protection, Nuremberg, Germany, October.

BIA, 1987 "Brenn-und Explosions-Kenngrö¶en von Stäuben," *BIA-Handbuch* 7.1tg, V1/87.

Bingham, E., Cohrssen, B., and Powell, C. (eds), 2001, *Patty's Toxicology,* 5th ed., Hoboken, NJ: John Wiley & Sons, Inc.

Bowes, P.C., 1984, *Self-Heating Evaluating and Controlling the Hazards*. London: Her Majesty's Stationery Office

Britton, L., 1999, *Avoiding Static Ignition Hazards in Chemical Operations*, New York: AIChE/Center for Chemical Process Safety.

Burgess, M., 1993, "Putting packages to the test," *Atom*, 430: 33–36.

Carson, P., and Mumford, C., 2002, *Hazardous Chemicals Handbook*, 2nd ed., Burlington, MA: Butterworth-Heinemann.

Casarett, L. J. and Doull, J., 1991, *Toxicology, The Basic Science of Poisons*, 4th ed., edited by Mary O. Amdur, John Doull, and Curtis D. Klaassen, New York: Pergamon Press.

CCPS, 1995a, *Guidelines for Safe Storage and Handling of Reactive Materials*, New York: AIChE/Center for Chemical Process Safety.

CCPS, 1995b, *Guidelines for Chemical Reactivity Evaluation and Application to Process Design*, New York: AIChE/Center for Chemical Process Safety.

CCPS, 1998, *Guidelines for Pressure Relief and Effluent Handling Systems*, New York: AIChE/Center for Chemical Process Safety.

CEN, 1998, "Method for the Determination of the Limiting Oxygen Concentration for Dust Clouds," CEN/TC 305/WG/1 SG1,prEN 14034-4, European Commission.

Ciba, 2000,"Safety Tests for Powders and Dusts in Conjunction with Specific Standard Operations," Ciba Specialty Chemicals, Safety Testing Laboratory, October.

Eckhoff, R., 1984/85, "Use of $(dP/dt)_{max}$ from closed bomb tests for predicting violence of accidental dust explosions in industrial plants," *Fire Safety Journal*, 8: 159–168.

Eckhoff, R., 1997, *Dust Explosions in the Process Industries*, 2nd ed. Burlington, MA: Butterworth-Heinemann.

Eckhoff, R., 2003, *Dust Explosions in the Process Industries*, 3rd ed. Amsterdam: Elsevier.

EEC, Seventh Amendment to EEC Directive 67/548/EEC, "Directive on Classification, Packaging, and Labelling of Dangerous Substances," European Union Commission, Brussels, Belgium (June 5, 1992).

EN 14034-1, 2002, "Determination of the Explosion Characteristics of Dust Clouds – Part 1: Determination of the Maximum Explosion Pressure," European Committee for Standardization, Brussels.

EPA, 1996, "Health Effects Test Guidelines," http://epa.gov/docs/OPPTS_Harmonized/870_H_Hlt24123597_Hlt24123598e_3__32__3__32__3__3__32__3__32__3_alth_Effects_Test_Guidelines/Master/870mast.pdf.

EU Directive 92/69 Annex, *Official Journal* L383, vol 35, p. 89.

Factory Mutual Loss Prevention Data Sheet 7-76, 1997, "Prevention and Mitigation of Combustible Dust Explosions and Fires," Johnston, RI:FM Global.

Fauske, H., 2000a, "Properly size vents for nonreactive and reactive chemicals," *Chemical Engineering Progress*, 96: 17–29.

Fauske, H., 2000b, "Easily determine self-accelerating decomposition temperature (TSADT)," *FAI Process Safety Newsletter*, spring. http://www.fauske.com/PSNews/FAInews_Spring_2000.pdf.

Field, P., 1982, *Dust Explosions*, Amsterdam: Elsevier.

FMRC, 1990, "Advances in Commodity Classification, A Progress Report," *FMRC Update*, Vol. 4, No. 1.

Frurip, D. J., Hofelich, T. C., Leggett, D. J. Kurland, J. K., and Niemeier, J. K., 1997, "A Review of Chemical Compatibility Issues," *Proceedings of the 1997 AIChE Loss Prevention Symposium*. New York: American Institute of Chemical Engineers.

Gad, S., 1995, *Regulatory Toxicology*, 2nd ed., London: Taylor & Francis.

Gibson, J. and Weber, J., 1969, *Handbook of Selected Properties of Air- and Water-Reactive Materials* (RDTR), U.S. Naval Ammunition Depot, available from NTIS. An Index for finding specific materials in the *Handbook* is available at www.doctorfire.com/pyrophoric.html.

Gibson, N. and Harper, D.J. (1981) "Frictional and Localized Heat Initiation of Powders Capable of Fast Decomposition—An Exploratory Study," U.K. Inst. of Chem. Engrs. Symp. Series No. 68, Runaway Reactions, Unstable Products and Combustible Powders, pp. 3/R:1–3/R:14.

Gibson, N., Harper, D. J., and Rogers R. L., 1985, "Evaluation of the fire and explosion risk in drying powders," *Plant/Operations Progress*, 4: 181–188.

Gibson, N., Harper, D.J., and Rogers, R.L., 1987, "Hazards from Exothermic Decomposition of Powders," in *Proceedings Runaway Chemical Reaction Hazards Symposium*, IBC, London, England.

Glassman, I., Papas, P., and Brezinsky, K., 1992., "A new definition and theory of metal pyrophoricity," *Combustion Science and Technology*, 83: 161–165.

Glor, M., and Schwenzfeuer, K., 1997, "Occurrence of cone discharges in production silos," *Journal of Electrostatics*, 40–41: 511–516.

Going, J., and Snoeys, J., 2002, "Explosion protection with metal dust fuels," *Process Safety Progress*, 21: 305–312.

Gray, B., 2002,"Spontaneous Combustion and Self-Heating," Chapter 2-10 in *SFPE Handbook of Fire Protection Engineering*, 3rd ed., Quincy, MA: National Fire Protection Association.

Greenfield, W., 2003, Personal communication with ISP Test Summary.

Grewer,Th., 1987, "Dewar Test Methods for Exothermic Reactions," in *Proceedings Runaway Chemical Reaction Hazards Symposium*, IBC, London, England.

Grewer, Th., 1994, *Thermal Hazards of Chemical Reactions*, Industrial Safety Series, 4. Amsterdam: Elsevier.

Hatch, T., and Gross, P., 1964, *Pulmonary Deposition and Retention of Inhaled Aerosols*, New York: Academic Press.

Herrmann, D., 2003, Private communication, January 2003.

Hertzberg, M., 1987, "A Critique of the Dust Explosibility Index: An Alternative for Estimating Explosion Probabilities," U.S. Bureau of Mines RI 9095.

Hofelich, T., 1997, "A quantitative approach to determination of NFPA reactivity hazard rating parameters," *Process Safety Progress*, 16: 121.

Hovingh, J., Thomas, G., Carlson, R., and Shaeffer, M., 1999, "Fire and Furnace Testing of Transportation Packages for Radioactive Materials: Facilities and Measurements," *ASME Pressure Vessels Piping Division Conference on the Transportation, Storage, and Disposal of Radioactive Materials*, New York: ASME.

IEC, 1984, "Electrical apparatus for use in the presence of ignitable dust. Part 2 Test Methods. Sheet 2-2: Method for determining the minimum ignition temperature of dusts—Dust cloud in a furnace at constant temperature." Geneva: International Electrotechnical Commission. http://www.iec.ch/

IEC, 1990, "Electrical apparatus for use in the presence of ignitable dust. Part 2 Test Methods. Sheet 2-5. Method for determining the minimum explosible concentration of dust/air mixtures." Geneva: International Electrotechnical Commission. http://www.iec.ch/

IEC 1241-2-2, 1993, "Method for Determining the Electrical Resistivity of Dust in Layers," Geneva: International Electrotechnical Commission. http://www.iec.ch/

IEC 1241-2-1, 1994, "Method for Determining the Minimum Ignition Temperature of Dust," Geneva: International Electrotechnical Commission. http://www.iec.ch/

IEC 1241-2-3, 1994,"Method for determining the minimum ignition energy of dust/air mixtures," Geneva: International Electrotechnical Commission. http://www.iec.ch/

IEC 61241-10, 2001 (Draft). "Electrical Apparatus for use in the Presence of Combustible Dust—Part 10: Classification of areas where combustible dusts are or may be present,"Geneva: International Electrotechnical Commission. http://www.iec.ch/

ISO/DIS 6184/1, 1985, "Explosion protection systems—Part 1: Determination of explosion indices of combustible dusts in air," Geneva: International Organization for Standardization.

ISO/DIS 6184/2, 1985, "Explosion protection systems—Part 2: Determination of explosion indices of combustible dusts in air," Geneva: International Organization for Standardization.

ISO/DIS 6184/3, 1985, "Explosion protection systems—Part 3: Determination of explosion indices of fuel/air mixtures other than dust/air and gas/air mixtures," Geneva: International Organization for Standardization.

International Fire Code, 2000, available from http://www.icbo.org/

Johnson, R., Rudy, S., and Unwin, S., 2003, *Essential Practices for Managing Chemical Reactivity Hazards,* New York: AIChE/Center for Chemical Process Safety.

Jones, J. C., 1998a, "Towards an alternative criterion for shipping safety of activated carbons," *Journal of Loss Prevention in the Process Industries,* 11: 407–411.

Jones, J. C., 1998b, "Towards an alternative criterion for shipping safety of activated carbons. Part 2. Further data analysis and extension to other substrates," *Journal of Loss Prevention in the Process Industries,* 12: 183–185.

Jones, J.C., 1999, "Recent developments and improvements in test methods for propensity towards spontaneous heating," *Fire and Materials,* 23: 239–243.

Kerfoot, E., Steinmetz, D., and Ozimek, T., 1995, "Industrial Toxicology," in (S. Grossel and D. Crowl, eds.), *Handbook of Highly Toxic Materials Handling and Mnagement,* New York: Marcel Dekker.

King, E., 1990, "Health Risks of Fine Powders," in (M. Rhodes, ed.), *Principles of Powder Technology,* Hoboken, NJ: John Wiley & Sons, Inc.

Kühner, 1991, "Safety Test Equipment: Screening Tests and Procedures," Birsfelden, Switzerland: Adolf Kühner AG, June.

Kotoyori, T., 1999, "The self-accelerating decomposition temperature (SADT) of solids of the quasi-autocatalytic decomposition type," *Journal of Hazardous Materials*, 64: 1–19.

Kotoyori, T., 2003, "Critical Teperatures for the Thermal Explosion of Chemicals," Draft manuscript.

Leggett, D. J., 2002, "Chemical Reaction Hazard Identification and Evaluation: Taking the First Steps." Proceedings, AIChE Spring National Meeting; 36th Annual Loss Prevention Symposium, New Orleans, LA, 2002..

Lewis, R. J., and Irving, N., 2000, *Sax's Dangerous Properties of Industrial Materials*, 10th ed. Hoboken, NJ: John Wiley & Sons, Inc.

Lunn, G., Torrent, J., and Hesener, U., 2001, "Testing Methods for electrical Apparatus Installed in a Dusty Environment with a Potential Risk of Explosions," Synthesis Report for EC Contract SMT-CT98-2273, available at http://www.safetynet.de.

Lunn, G., 2002, "Frictional Ignition of Powders: A Review," http://www.safetynet.de

Melhem, G., 2003, "Reactivity Screening Made Easy," ioMosaic White Paper, www.iomosiac.com.

Miron, Y. and Lazarra, C., 1988, "Hot-surface ignition temperatures of dust layers," *Fire and Materials*, 12: 115–126.

Monakhov, V., 1995, *Methods for Studying the Flammability of Substances*, New Delhi: Amerind Publishing.

Murphy, M., Singh, S., and Shanley, E., 2003, "Computationally evaluating self-reactivity hazards," *Chemical Engineering Progress*, 99(2): 54–61.

Nelson, 2003, "Special Agent Suppression Systems," in *Fire Protection Handbook*, 19th ed., Quincy, MA: National Fire Protection Association.

NFPA 69, 1997, "Standard on Explosion Prevention Systems," Quincy, MA: National Fire Protection Association.

NFPA 232, 1997, "Code for the Storage of Organic Peroxide Formulations," Quincy, MA: National Fire Protection Association.

NFPA 499, 1997, "Classification of Combustible Dusts and of Hazardous (Classified) Locations for Electrical Installations in Chemical Process Areas," Quincy, MA: National Fire Protection Association.

NFPA 13, 2000, "Automatic Sprinkler Systems," Quincy, MA: National Fire Protection Association.

NFPA 230, 2000, "Code for the Storage of Liquid and Solid Oxidizers," Quincy, MA: National Fire Protection Association.

NFPA 704, 2001, "Standard System for the Identification of the Hazards of Materials for Emergency Response," Quincy, MA: National Fire Protection Association.

NFPA 68, 2002, "Guide to Deflagration Venting," Quincy, MA: National Fire Protection Association.

NFPA 484, 2002, "Standard for Combustible Metals, Metal Powders, and Metal Dusts," Quincy, MA: National Fire Protection Association.

NIST, 2001, "Particle Size Characterization," NIST Recommended Practice Guide, Special Publication 960-1. Gaithersburg, MD: National Institute of Standards and Technology.

NOAA 2002, Chemical Reactivity Worksheet, Version 1.5. U.S. National Oceanic and Atmospheric Administration. [http://response.restoration.noaa.gov/chemaids/react.html].

OECD, 1981, *Guidelines for Testing of Chemicals*, No. 113, "Screening Test for Thermal Stability and Stability in Air," Paris: Organization for Economic Cooperation and Development.

OECD, 2002, *Guidelines for the Testing of Chemicals*, PDF Edition (ISSN 1607-310X), Vol. 1, No. 4, May

Pohanish, R. P. and Green, S. A., 2003, *Wiley Guide to Chemical Incompatibilities*, 2nd ed., Hoboken, NJ: John Wiley & Sons, Inc.

Querol Aragón, E., Garcia Torrent, J., and Cámara Rascón, A., 2002, "Spontaneous Combustion Testing," http://www.safetynet.de

Pinton, E., Duret, B., and Berthoud, G., 1999, "Behavior of a UF6 container during a fire—I: Analysis and phenomenological interpretation of Tenerife experimental results," *Nuclear Technology*, 127: 332–351.

Reza, A., Kemal, A., and Markey, P., 2002, "Runaway reactions in aluminum, aluminum chloride, HCl, and steam: An investigation of the 1998 CONDEA Vista explosion in Maryland," *Process Safety Progress*, 21: 261–267.

Sax, I. and Lewis, R., 1987, *Hazardous Chemicals Desk Reference*, New York: van Nostrand Reinhold.

Schwab, R., 2003, "Dusts," in *Fire Protection Handbook*, 19th ed., Chapter 15, Section 8, Quincy, MA: National Fire Protection Association.

Sharma, T., Varshney, B., and Kumar, S., 1993, "Studies of the burning behaviour of metal powder fires and their extinguishment: Part 2—Magnesium powder heaps on insulated and conducting material beds," *Fire Safety Journal*, 21: 153–176.

Sittig, M., 1991, *Handbook of Toxic and Hazardous Chemicals and Carcinogens*, Norwich, NY: William Andrews Publishing, Noyes Publications.

Siwek, R., 1996, "Determination of technical safety indices and factors influencing hazard evaluation of dusts," *Journal of Loss Prevention in the Process Industries*, 9: 21–31.

Siwek, R., and Cesana, C., 1993, "Assessment of the Fire & Explosion Hazard of Combustible Products for Unit Operations," *International Conference and Exhibition on Safety, Health, and Loss Prevention in the Oil, Chemical, and Process Industries*, Burlington, MA: Butterworth-Heinemann.

Siwek, R., and Cesana, Ch., 1995, "Ignition behavior of dusts: Meaning and interpretation," *Process Safety Progress*, 14 (also in 1994 AIChE Loss Prevention symposium).

Stevens, J., and Breckenridge, C., 2001, "Agricultural Chemicals, Regulation, Risk Assessment, and Risk Management" in (S. Gad, ed.), *Regulatory Toxicology*, 2nd ed. London: Taylor & Francis.

Stevick, K., 2003. Private communication, May 2003.

Syngenta Process Hazards Section, 2001, "Thermal Stability Screening Test—Diffusion." Basel, Switzerland: Syngenta International AG.

Syngenta Process Hazards Section, 2001, "Thermal Stability Screening Test—Aerated." Basel, Switzerland: Syngenta International AG.

Tapscott, R., 1997, "Combustible Metal Extinguishing Agents and Application Techniques," *Fire Protection Handbook* , 18th ed., Chapter 6-26, Quincy, MA: National Fire Protection Association.

Tewarson, A., 1995, "Generation of Heat and Chemical Compounds in Fires," *SFPE Handbook of Fire Protection Engineering,* 2nd ed., Section 3/Chapter 4, Quincy, MA: National Fire Protection Association.

Townsend, D., 1981, "Accelerating Rate Calorimetry," *I. Chem. Eng. Symp. Series,* 68.

U.N., 1999a, *Recommendations on the Transport of Dangerous Goods, Model Regulations,* 11th ed. New York: United Nations.

U.N., 1999b, *Recommendations on the Transport of Dangerous Goods, Manual of Tests and Criteria,* 3rd rev. ed. New York: United Nations.

Urben, P.G. (ed.), 1999, *Bretherick's Handbook of Reactive Chemical Hazards,* 6th ed. Oxford, UK: Butterworth-Heinemann. Also available on CD-ROM as *Bretherick's Reactive Chemical Hazards Database, Version 3.0.* Oxford, UK: Butterworth-Heinemann; and on-line by subscription [http://www.chemweb.com].

Valiulis, J., 2001, "Dust Explosion Effects Propagation in Dust Collection System Clean Air Return," Master's thesis, Worcester Polytechnic Institute.

Valiulis, J., Zalosh, R., and Tamanini, F., 1999, "Experiments on the propagation of vented dust explosions to connected equipment," *Process Safety Progress,* 19.

Varshney, B., Kumar, S., and Sharma, T., 1990, "Studies of the burning behaviour of metal powder fires and their extinguishment: Part 1—Mg, Al, Al-Mg alloy powder fires on sand bed," *Fire Safety Journal,* 16: 93–117.

VDI 2263, 1990, "Dust Fires and Explosions, Part 1, hazards, assessment, protective measures; test methods for the determination of the safety characteristic of dusts." Verein Deutscher Ingenieure. VDI—Handbuch Reinhaltung der Luft, Band 6

Zalosh, R. 2003, *Industrial Fire Protection Engineering,* Hoboken, NJ: John Wiley & Sons, Inc.

Zoghlami, H., Laurent, A., Corriou, J., and Adrian, J., 1997, "Safe design of a thermally unstable solid storage container: Natural convection effects on thermal ignition," *Process Safety Progress,* 16: 231–235.

Chapter 5

EQUIPMENT HAZARDS AND PREVENTIVE/PROTECTIVE MEASURES

5.1 INTRODUCTION

Many different types of equipment are used for the storage, handling, and processing of particulate solids (powders, bulk solids, granules, pellets, flakes, etc.). A great number of particulate solids that are in these items of equipment are potentially hazardous as they may be combustible, may self-decompose, may react with air or water, and may adversely affect the health of operators and nonplant people who come into contact with them.

This chapter is concerned primarily with discussing potential hazards associated with various types of equipment, systems, and containers used in particulate solids processes and operations. It also presents information on methods of accident prevention and protection of these items of equipment from the occurrence of fires, explosions, and unwanted health-hazardous emissions. Case histories are presented of incidents to illustrate the types of hazards that may be present which can lead to serious consequences.

Many young, and even more experienced, chemical engineers (process design, production, project, and maintenance) and process safety/loss prevention specialists who will be using this book may have a limited knowledge of the construction and operating principles of the various types of equipment, systems, and containers used in particulate solids processes and operations. Therefore, detailed descriptions and discussions of these are presented in Appendix B, which, it is felt will lead to a better understanding of how equipment construction and operation can contribute to safety problems. A knowledge of this can often help an engineer to select a more appropriate type of equipment which will avoid an accident or, if it should occur, minimize its consequences.

5.2 SAFETY ASPECTS OF BATCH VERSUS CONTINUOUS OPERATION

Whenever feasible, consideration should be given to using continuous-type equipment rather than batch-type equipment for handling and processing of particulate solids as continuous-type equipment is often inherently safer. This is because continuous-type equipment contain smaller quantities of hazardous particulate solids than batch-type, and thus, if a fire, explosion, or unwanted emission occurs, the consequences will be, most likely, less severe. Also, batch-type equipment would normally have more frequent startups and shutdowns, more product changeovers, more frequent cleaning requirements, and more opening and closing of the system. Thus, safe operation can be affected by loss of inerting or difficulty in sealing the system and possible cross-contamination.

Continuous types of equipment are available for blenders/mixers, drying equipment, screening and classifying equipment, size enlargement equipment, and size reduction equipment. Their practicalities should always be evaluated for use in a process before batch-type equipment is selected.

5.3 PARTICULATE SOLIDS PROCESSING EQUIPMENT HAZARDS AND PREVENTIVE AND PROTECTIVE MEASURES

This section discusses the potential hazards of various types of process equipment used for the storage, handling, and processing of particulate solids and presents information on measures for preventing fires, explosions, and unwanted emissions, and protecting equipment (minimizing the consequences) should these incidents occur.

5.3.1 Bag Openers (Slitters)

Fires and explosions can occur in bag slitters due to the generation of dust clouds and electrostatic charging of the solids from the action of the belt conveyor and tearing rollers. A highly critical area is the entrance hood, and installation of a suppression system in this hood will not only quench an explosion, it will also prevent the propagation of an explosion from the machine into the work area where a secondary explosion could have devastating effects. A complementary pressure relief device (rupture disk) on the hood will not only reduce pressure build-up to an acceptable level, but also vent combustion gases and suppressant from the suppression system to the atmosphere. Also, since the protective devices installed in the entrance hood cannot fully prevent an explosion from propagating into the main machine casing, this casing should be fitted with a rupture disk sized in accordance with the mechanical strength of the casing. Figure 5-1 is a schematic of a bag

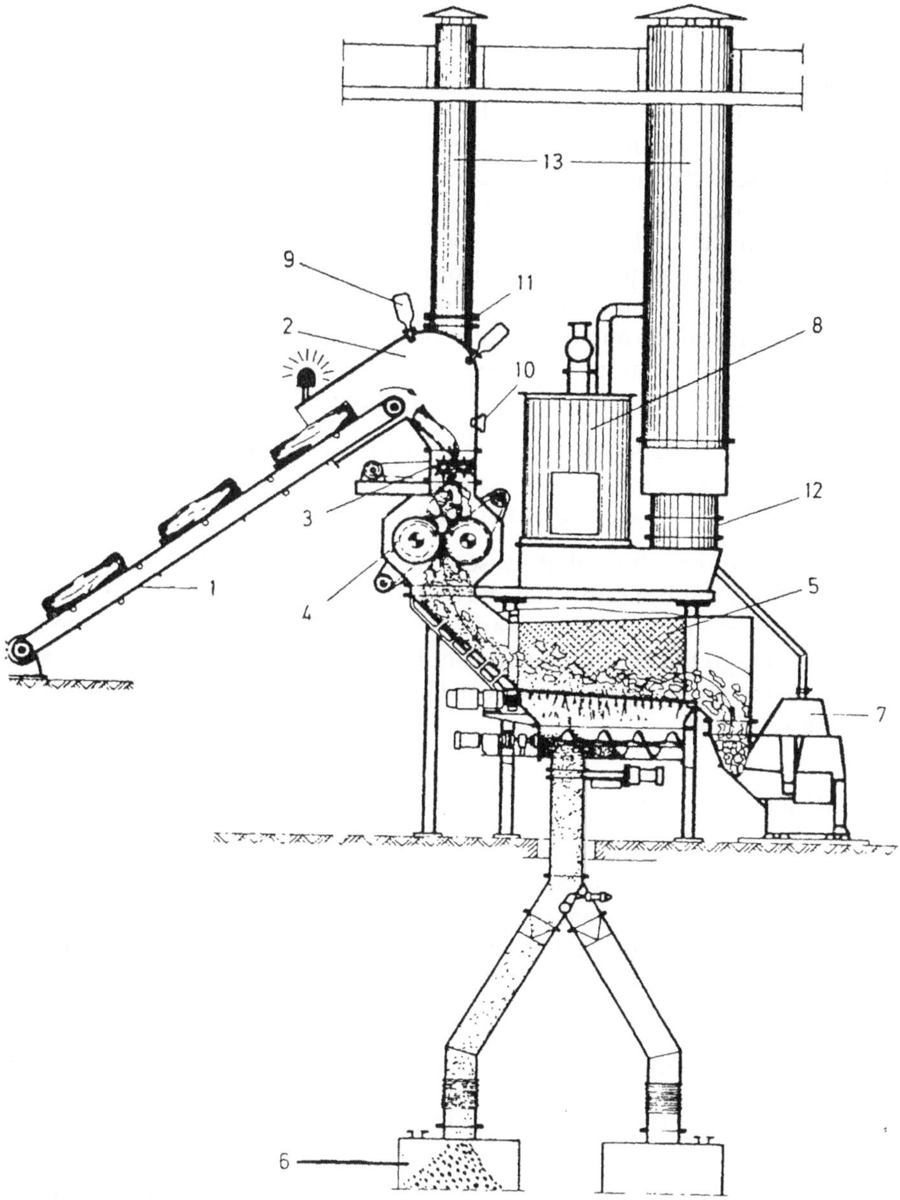

Figuree 5-1 Automatic bag opening device protective features. 1, conveyor belt; 2, intake hood; 3, slitting machine; 4, crushing cylinders; 5, sieve drum; 6, solids receiver; 7, bag compactor; 8, filter; 9, suppressant bottle; 10, pressure sensor; 11, 12, rupture disks; 13, relief vent pipes. (Source: Bartknecht, W. *Explosions: Course, Prevention, Protection.* New York: Springer-Verlag, 1981. Reprinted with permission of Springer-Verlag.)

slitting machine showing the types and location of explosion protection equipment

Bartknecht (1981) presents more details on protection of bag slitters. Proper grounding and bonding of the bag slitter is necessary to minimize the potential for dust cloud ignition by electrostatic discharges, and the motor has to be in accordance with the appropriate National Electrical Code (NEC) area classification (see Section 6.4.1).

The trash (empty bags) compactor should also be properly bonded and grounded and also requires the appropriate motor NEC area classification.

As a preventive measure against secondary explosions from combustible and reactive solids and operator exposure to toxic solids, dust emissions must be minimized and not allowed to accumulate in the work area where the slitter is operating. This can be achieved by:

1. Maintaining a slight negative pressure on the slitting chamber at all times when dust is being generated, and
2. Making the machine casing of dust-tight construction.

For toxic dusts (those that present health hazards, including long-term sensitization), the operator may have to wear personal protective equipment. This decision should be made by a qualified industrial hygiene specialist, although the process design engineer can assist in the final decision by discussing the equipment construction and operation with the industrial hygiene specialist.

5.3.2 *Blenders/Mixers*

Fires and explosions can occur in blenders and mixers for particulate solids for a number of reasons, such as, heat generation due to frictional heating of solids or rubbing of internal parts against each other, electrostatic charging of the solids, dust formation inside of the equipment, etc. A case history is presented below describing an incident in a solids mixer.

Case History of an Explosion in a Conical Orbiting Screw Mixer

Whitmore et al. (1993) present the following case history of an explosion in a blender.

An explosion occurred in a 3.7-m^3 Nautamixer (conical orbiting screw mixer) during the blending of azodicarbonamide (AC) with an aqueous solution of salts to produce an AC formulation. During the batch blending cycle, hot water (80°C) was circulated through the blender jacket for several hours, and the vacuum in the blender was released by purging with nitrogen. The explosion caused the mixer vessel to rupture and two large sections of the top were torn out completely and struck the floor above.

There was extensive damage to the building, windows were broken up to 90 meters away by the pressure wave, and missiles were projected up to 120 meters away. The four people in the plant at the time of the explosion were shaken up, but uninjured, while there were a few cuts to people in nearby buildings due to flying glass. The TNT equivalence of the blast was estimated at 3.3 kg. Subsequent experimental testing indicated that the explosion was caused by a decomposition which reached high rates due to a critical degree of confinement. The initiating source of the decomposition was not primarily identified, but it was assumed the heat was generated by mechanical friction due, for example, to the screw rubbing on the vessel wall. Another possibility is that a small metal item found its way into the vessel and became trapped between the screw and the wall.

Ed. Note: A number of things could have been done to avoid this accident, among them being the following:

1. The material being mixed, AC, should have been tested for the potential for thermal decomposition.
2. An overload trip on the motor should have been installed to shut down the motor if the orbiting screw contacted the mixer wall, which results in an increase in power draw.
3. A deflagration venting system might have prevented the mixer from rupturing.

Discussed below are safety concerns and preventive/protective measures that should be considered to avoid fires and explosions in certain types of blenders/mixers.

1. Tumbling type blenders/mixers can cause electrostatic charging of the powders by the tumbling action, especially in glass-lined units. They can often operate under vacuum, which can suck in air, but can be easily inerted. Tumbling blenders/mixers fabricated of metal should be properly bonded and grounded to minimize the potential for spark ignition.
2. Ribbon, single rotor, and double rotor mixers have the potential to cause friction sparks if the mixing elements come in contact with the trough wall. Tramp metal may also cause friction sparks. Frictional rubbing may also cause localized heating and subsequent fires. They often do not operate with the trough completely filled so that a dust explosion can occur. Inerting, as well as deflagration suppression, is often used to avoid a dust explosion. Explosion venting may be feasible, even if the blender is in a building, if a vent duct is installed to direct the fireball and burning particles outside the building. If instal-

lation of a vent duct is not feasible, a flame-quenching device (see Section 6.6.1) can be installed, which obviates the need to run a vent duct through the building wall. Fugitive emissions can occur at the shaft bearing seals and special mechanical seals or inert gas-purged seals can minimize this problem.

3. Vertical, orbiting screw blenders can cause friction sparks and subsequent dust explosions if the screw comes in contact with the vessel wall (see the case history above). A screw rubbing against the vessel wall can cause gouging and eventually result in severe damage to the vessel or its mechanical failure (creation of a hole in the vessel wall). Rubbing of the screw on the vessel wall will usually result in an increased motor load, and providing an overload trip on the motor to shut it down can prevent severe gouging from occurring. Protective measures such as inerting, deflagration venting, and deflagration suppression can all be used. The bottom bearing for the shaft is a potential spot for wear and subsequent dust emissions and it should be purged to keep erosive solids out of the bearing.
4. In-bin blenders are also susceptible to dust explosions and protective measures such as inerting, deflagration venting, and deflagration suppression can also be applied.
5. Continuous blenders/mixers can have dust explosions due to electrostatic charging of the solids, and rubbing of the mixing elements against the trough, depending on the design. Providing an overload trip on the motor to shut it down can be used to prevent damage to continuous blenders/mixers from rubbing of the mixing elements against the trough. Inerting and deflagration suppression are common protective measures applied. In some cases (e.g., ribbon blenders, Zig-Zag® blender, etc.) it may be feasible to have the blender designed for deflagration pressure containment (see Section 6.6.3).

The following general safety measures should be considered:

- For each type of blender/mixer, consideration should be given to the various types of protective measures that may be applicable. Discussions with the vendor are often invaluable as they may have experience from other similar applications.
- Wherever electrostatic charges can be generated, proper grounding and bonding should be provided for metal units.
- If a blender/mixer is used for a toxic solid (especially if wet with flammable or toxic vapors) great effort must be made to prevent emissions from covers and shaft seals. Covers with heavier flanges, frequent bolting, and thick gaskets should be used to provide tight closures and minimize emissions. Double shaft seals with a gas purge or maintaining a slight vacuum on the unit will reduce emissions. If the solid is toxic, it may be prudent to locate the blender/mixer in a ventilated

enclosure and have the operators equipped with personal protective equipment.

- Provisions should be made to prevent tramp metal from entering blenders/mixers that have rotating internal elements and/or have small clearances between the housing and internal parts. Tramp metal can be excluded by a screen located before the blender/mixer to collect larger solids. Normally, the largest screen opening allowed is 70% of the clearance gap between the agitator and the housing (vessel wall). Other ways to exclude tramp metal are brush sifters, magnets to remove ferrous components, and electric field–actuated diverters.
- Locate agitator bearings outside of the vessel rather than within the body of the blender/mixer to keep hot bearings as far away as possible from the solids being mixed.

The hazards posed by ignition sources in a blender/mixer, even when mixing very easily ignitable particulate solids, can be discounted under the following conditions (Nelson, 2002; Jaeger and Siwek, 1999):

- When filling and emptying a mixer, the same conditions and safety measures apply as for filling and emptying vessels.
- When filling and emptying a mixer, mixer components must only run with tip velocities of ≤1 m/s, safeguarded by technical measures (e.g., motor speed controller and shutdown system).
- During operation in a closed condition, the rotational speed of the mixer parts (internal rotating elements) is not limited if the blades are completely covered by the material or if the filling level is 70 vol. % or more.
- To avoid propagating brush discharges on the inner wall of a mixer, no insulating lining with a high breakdown voltage (more than 4 kV) must be allowed to form; corresponding testing of product build-up should be done during maintenance.
- With orbiting screw mixers, having a bottom agitator support which can heat up in operation, care is advised with solids that can undergo a spontaneous decomposition.
- Circumferential velocities (tip speeds) of up to 10 m/s can be tolerated during filling a mixer that is filled below 70 vol. % if the autoignition temperature (AIT) of the solids to be mixed is above the limits given in Table 5-1.

Discussions of safety hazards of particulate solids blenders and mixers are presented by Eckhoff (2003) and the IChemE (1982).

5.3.3 Drying Equipment

Fires and explosions have occurred in drying operations fairly frequently. The following case history cites one example.

TABLE 5-1
AIT Limiting Values for Products Filled and Emptied in Mixers While Running

MIE, mJ[a]	<1 [c]	1–3	3–10	10–30	30–100	100–300	300–1000	>1000
AIT, °C[b]	Do not process	530	500	465	430	395	360	325

[a] Minimum ignition energy determined in a Kuehner AG MIKE 3 apparatus with 1 to 2 mH inductance in spark discharge circuit. Inductance causes the capacitor storing energy to discharge in a protracted manner.

[b] Autoignition temperature of a dust cloud determined in a BAM furnace (BAM is the German Federal Institute for Testing of Materials in Berlin).

[c] The MIE lies between the highest energy at which ignition fails to occur in 10 successive attempts to ignite the dust–air mixture (1 mJ) and the lowest energy at which ignition occurs within 10 successive attempts (3 mJ).

Case History of a Fire and Explosion in a Batch Rotating Vacuum Dryer

Drogaris (1993) cites an example of a fire and explosion in a batch rotating vacuum dryer used for drying a pharmaceutical powder. An operator had tested dryer samples on a number of occasions without any problems. After the last sampling, he closed the manhole cover, put the dryer under vacuum, and started rotation of the dryer. A few minutes later an explosion and flash fire occurred, which self-extinguished. No one was injured. Investigations revealed that after the last sampling, the dryer manhole cover had not been securely fastened. This allowed the vacuum within the dryer to draw air into the rotating dryer and create a flammable atmosphere. The ignition source was probably an electrostatic discharge (the Teflon coating on the internal lining of the dryer could have built up a charge). No nitrogen inerting had been used. After the incident, the following precautions were instituted to prevent similar accidents from occurring in the future:

- Nitrogen purging is carried out before charging or sampling of the dryer.
- If the absolute pressure rises to about 4 psia, the rotation is stopped, an alarm sounds, and a nitrogen purge starts automatically.

Ed. Note: In addition to the measures taken after the incident, as indicated above (nitrogen purging and absolute pressure monitoring), the SOP should be revised to require the operator to make sure that the manhole is securely fastened before starting up the dryer.

The protection methods that can be used in drying operations depend on the material being dried and the type of dryer. In fact, the type of dryer used depends often on the material to be dried. Based on a knowledge of the chemical structure of a material, plus the results of some simple laboratory tests, substances can be grouped into the following categories (Abbott, 1990):

1. Type 1 materials defined as deflagrating or detonating explosives by United Nations tests (UN, 1999);
2. Type 2 materials, which on heating, exhibit exothermic decomposition with evolution of large volumes of gas, even in the absence of air (e.g., peroxides, blowing agents). If the amount of gas evolved is sufficiently large, there may be a dangerous build-up of pressure with no fire;
3. Type 3 combustible materials, which can undergo exothermic oxidation or decomposition when heated in air. Such materials may present a fire hazard or a dust explosion hazard;
4. Type 4 materials, which do not undergo either exothermic oxidation or exothermic decomposition when heated in air.

As a general rule, it is dangerous to dry Type 1 materials in a general purpose dryer. Special precautions specific to the explosives industry are required and special regulation must be obeyed when handling these materials. It may also be dangerous to dry Type 2 powders by application of heat, depending on the amount of gas evolved and the temperature at which decomposition occurs. As with Type 1 materials, the hazard exists regardless of the environment in which the heating is carried out. Type 3 powders can usually be dried safely by application of heat provided the operating conditions are chosen correctly, an appropriate method of explosion prevention or protection is employed, and rigorous precautions are taken to reduce the risk of ignition. Type 4 materials only present a fire or explosion hazard if they are wetted with a flammable solvent or if the air is heated by direct oil or gas firing.

Some general hazards (applicable to all types of dryers) that occur in drying operations are:

1. In many dryers, the solids contain a flammable liquid that can vaporize and form a hybrid mixture. Hybrid mixtures (combinations of a flammable vapor and a combustible dust) present a greater explosion hazard than that presented by the combustible dust alone. The greater hazard is characterized by (1) the hybrid mixture may explode more severely than a dust–air mixture alone, that is, the maximum pressure and maximum rate of pressure rise may be greater, even if the vapor concentration is below its lower flammable limit, (2) the MIE of hybrid mixtures is usually lower than that of the dust–air mixture alone, and (3) the minimum explosible concentration (MEC) of a dust is reduced by the presence of a flammable vapor even if the latter is below its LFL. Measurable effects are observed for vapor levels as low as 20% of the vapor LFL. Such solids should be dried in inerted dryers or under

vacuum. Another method of explosion prevention, which can be used in spray and fluid bed dryers, is to use ventilation to keep the flammable vapor concentration well below its lower flammability limit (see NFPA 86, 1999).

2. Direct firing of dryers creates a strong ignition source, and for dryers containing flammable vapors direct heating systems are too dangerous to be employed.
3. Other hazards in dryers arise if the unit is not operated according to design conditions. Thus, if the feed rate is too low, the material may become overheated and may ignite. Overheating may also occur if material from a previous batch is exposed to hot air on start-up. Similar hazardous conditions can occur in shutdown or emergency situations. Therefore, the control of dryer operating conditions is particularly important.
4. The hot product from a dryer is another hazard, and it may be necessary to cool it before storage if self-heating is to be avoided. This hazard increases with increasing container size and increasing temperature. One company recommends that the maximum size that can be safely stored is 200 liters (Nelson, 2002).
5. Electrostatic sparks are a common cause of dust and flammable vapor deflagrations. Dryers and drying systems that can generate electrostatic charges must be properly bonded and grounded to drain off these charges and minimize the possibility of deflagrations. Inerting is often used to prevent the occurrence of a deflagration if electrostatic sparks are a possibility.
6. Frictional sparks and heating can result from a number of causes, such as (1) overheated bearings, (2) frictional impact with shovels and scoops used in removing product from a dryer, which may cause ignition of materials with a low minimum ignition energy (MIE), such as peroxides and sulfur, (3) rotating fan blades touching the casing, and (4) tramp metals and stones fed with the solid into the dryer.
7. Electrical equipment such as contact switches, fuses, circuit breakers, etc. can discharge sparks with energy greater than the MIE of the dust or flammable gas, and can cause a deflagration. Electrical equipment must be specified with the appropriate area classification to avoid this problem. They must also be properly installed and maintained or they will not meet code requirements.
8. Material autoignition can be a hazard, and all precautions must be taken to ensure that at no stage of the drying operation, including start-up and shutdown, does the material temperature exceed its autoignition temperature. The exhaust gas temperature is a safe way of controlling the material temperature in a dryer.
9. The MIE of particulate solids can be greatly reduced at elevated temperatures (they become more easily ignitable).

10. Solids having a burning class of 6 (very rapid combustion with flame propagation or rapid decomposition without flame) are very difficult to control once they are ignited, and in the case of such solids, the drying conditions must be specified based on an individual risk assessment. Bartknecht (1989) and Jaeger and Siwek (1999) discuss burning classes and test methods for determining them. Table 5-2 lists burning classes and their characteristics.

Some hazards that are specific to certain types of dryers are as follows:

Spray Dryers

1. Spray dryers and drying systems (including ductwork and associated equipment such as cyclones, dust collectors, etc.) are prone to accumulation of deposits on dryer walls and ductwork. Solids often accumulate on spray devices at the top of these dryers where the highest dryer temperature is often experienced. Frequent cleaning and monitoring may be required to ensure that these deposits do not overheat and autoignite. Tests should be conducted to evaluate the hazards of dust deposit ignitability. The characteristics of materials deposited on walls or other surfaces may change over time when the materials are exposed to high temperatures or other process conditions and may thermally decompose. Thermal stability testing is required to assess this hazard. It should be recognized that some large spray dryers (e.g., milk dryers, which have chambers up to 1200 m^3) may have inlet temperatures very close to the self-heating temperature of the deposited

TABLE 5-2
***Burning Class Definitions for a Powder Layer*[a]**

Test Result		CL	Reference Product [20°C]
No ignition	No spreading of fire	1	Table salt
Brief ignition, rapid extinction	No spreading of fire	2	Tartaric acid
Localized combustion or glowing with practically no spreading	No spreading of fire	3	D + Lactose
Glowing without sparks (smoldering) or slow decomposition without flame	Fire spreads	4	l-Amino-8-naphthol-3,6-disulfonic-acid (H-acid)
Burning like fireworks or slow quiet burning with flames	Fire spreads	5	Sulfur
Very rapid combustion with flame propagation or rapid decomposition without flame	Fire spreads	6	Blackpowder

[a] Source: Bartknecht, W., *Dust Explosions: Course, Prevention, Protection.* New York: Springer-Verlag,, 1989. Reprinted with permission.

milk solids, so that there is always the danger of a fire or an explosion (Tyldesley, 2004).

2. For spray dryers, loss of feed may result in overheating of the solids in the dryer and downstream equipment.
3. The presence of explosive dust–air mixtures must be expected in the lower part of spray dryers, at least. This is true even for spray drying processes for which a calculation of the dust concentration indicates a value below the LFL (MEC) of the dust.

Direct-Heat Dryers

1. As direct-heat (convective) dryers are usually operated at atmospheric conditions (air is always present), the occurrence of explosive dust–air mixtures must be expected as a matter of principle.
2. In pneumatic (flash) dryers dust explosions are most likely to occur in the upper part of the drying tube and in the dust recovery section where the particles are dry. However, the moisture content of the feed to this type of dryer is often comparatively low, particularly when recycled product is blended with the fresh feed, so the dust in the lower part of drying tube may still be combustible. Deposits are likely to occur at the top bend of the drying tube. They are most vulnerable to ignition during start-up and shutdown operations. Deposits can also occur near the dryer feed point due to stickiness of the wet feed. Many pneumatic conveying dryer fires occur in this area.
3. In fluid bed dryers the concentration of dust in the fluidized bed itself is too high to sustain an explosion (above the UFL), but the concentration in a portion of the space above the bed, and perhaps also in the dust recovery equipment, may be within the explosion limits. The most likely places for a dust layer to accumulate are corners and other poorly fluidized areas on the distribution plate. Agglomerates which are too large to fluidize will also collect on the plate. The plate is virtually at the temperature of the inlet air, and any fires starting here will burn rapidly. In continuous fluid bed dryers, dust may also be deposited in the duct connecting the dryer to the dust recovery equipment if the air velocity in the duct is too low. This deposit will only be exposed to the exhaust air temperature during normal operation, but could be exposed to higher temperatures during start-up and shutdown. Tarry deposits can also form at the top of the vessel and in the exhaust air ducting when drying certain materials.
4. Although rotary dryers are not generally used for finely divided material, there is often some generation of fines by attrition and breakage in the rotating drum. These fines are carried out by the exhaust air to dust recovery equipment, which is where the dust concentration may be above the MEC and a dust explosion could occur. If the material

has any tendency to form a deposit adhering to internal surfaces, this can create a fire hazard. The hazard will be greatest at the air inlet end of the drum where the air temperature is highest.

5. In continuous through-circulation dryers the air velocity is usually low enough not to blow dried material off the perforated plate or screen, so there should be no dust explosion risk in this dryer. However, if the dried material is friable and has an appreciable fall into the discharge chute, there could be a dust cloud having a concentration above the MEC in the chute, and a dust explosion could occur.

Indirect-Heat Dryers

1. In agitated vacuum pan, spherical, and filter-dryers, agitation-related events are probably the most frequent causes of problems, such as:
   (a) Agitator drive power is often quite large , and even with no heating present on the jacket, product temperature can rise significantly if moderate speed agitation is maintained for prolonged periods of time. Typically, with heat-unstable materials, the heating system is controlled and alarmed to ensure that the product stays below the AIT or decomposition temperature of the product by a safe margin (50°C is reasonable for some, but not all, products). The agitator should also be interlocked and/or alarmed with product temperature so that its heat work input does not increase the product temperature to where it approaches its AIT or its decomposition temperature.
   (b) Product overheating can be caused by temperature control system failure. The most reliable strategy is to control heat input based on jacket temperature, maintaining the heat transfer fluid at a safe margin below the instability temperature for the product being dried. Filter-dryers sometimes use hot gas recirculation rather than vacuum for drying. Proper temperature measurement location is critical to good control and measurements should be taken on the clean, inlet gas stream, not on the exhaust stream. Where combustible dusts or flammable vapors are present, the gas stream must be inert, and be monitored to ensure that oxygen levels are below the LOC by a safe margin (see NFPA 69). Product temperature measurement is not particularly reliable and should be used for information purposes only rather than for heating system control. Contact measurement probes are flush-mounted with the wall on dryers where axial agitator movement is a feature of the operation. The face of the thermowell can get coated with product, leading to lower than actual temperature readings (invasive thermowells on dryers without axial agitator movement are also prone to this problem). Indirect temperature measurement devices such as infrared

have their own issue such as keeping the "vision" window free of dust (nitrogen sweeping of the window can help) and being set up so that they can distinguish between other hot surfaces, such as the vessel wall or agitator, and the product.

(c) Contact of moving parts can be a significant problem. Because of the small clearance between the agitator and the walls and base, contact is possible, for example, due to deflection under heavy loads such as during startup with a wet product with too high density in the dryer. If a heated agitator is heated while the vessel is still cold, thermal expansion may reduce clearances too much, which in combination with small deflections, can lead to contact and localized generation of heat or sparks which could initiate a dust explosion. With filter-dryers, if the filter medium covering the base is improperly fitted, or the agitator clearance is too small, the medium can contact the agitator and be torn, or quite extensively damaged, causing problems.

(d) Tramp metal can cause severe problems. If foreign objects (nuts, bolts, cleaning spray devices, dust filter hardware, or other tramp metal) fall into the dryer, they can be trapped between the moving agitator and the vessel, possibly resulting in friction heating or sparking, and a subsequent fire or explosion.

(e) Discharge valve leakage can also be a problem. One type of discharge valve is a hinged door and O-ring type that is prone to build-up of hard product on the seal faces after several discharges, which can be a major source of air (oxygen) leakage inward during vacuum operation. Manual cleaning of the valve after each discharge may be necessary to avoid this problem.

(f) In operations when vacuum is not present such as charging and discharging, it may be necessary to do this under an inert gas (usually nitrogen) blanket to avoid ignition of vapors or powder by static charges where the powders have the ability to generate a dust explosion, or are wet with flammable liquids.

2. Screw conveying, paddle, and disk dryers have problems similar to agitated dryers with respect to tramp metal entering the dryer and causing frictional heating. For some solids deposits tend to build up on the shaft and blades, and on the inner surface of the trough. Deposits on the heat transfer surfaces will be approximately at the temperature of the heating medium, and consideration must be given to the possibility of thermal decomposition. This should be assessed by thermal decomposition testing. If, for some reason, the shaft of the dryer stops rotating, the material held up in the dryer will attain the temperature of the heating medium if the latter is not shut off promptly.

Fires and explosions in dryers can be prevented by:

1. Maintaining the concentration of the combustible dust outside of its flammable range (this may not be feasible for all types of dryers).
2. Maintaining oxygen concentration below its limiting oxidant concentration (LOC).
3. Excluding all possible ignition sources (this is not always possible).

Explosion protection can be achieved by:

1. Deflagration venting
2. Suppressing the explosion
3. Designing the equipment so that it is strong enough to contain the explosion.

Some operational and design safety features and recommended practices for several types of dryers are presented below (Field, 1982; Nelson, 2002, Palmer, 1973):

1. Because major portions of some drying systems (spray and fluid bed dryers) operate below the minimum explosible concentration (MEC), partial-volume venting equations may be applicable (see NFPA 68, 2002). However, Tyldesley (2004) cautions against doing this without a thorough analysis of the potential for unexpected conditions. It is true that in steady-state operation, the upper parts of these dryers are probably lean (below the MEC), while the lower parts are likely to be above the MEC. However, dust explosions occur when things are not running normally. For example, deposits often accumulate on the dryer walls, and if these fall for some reason (perhaps as a result of a small explosion near the dryer bottom), there could be enough dust cloud generation to fill the entire dryer chamber. Therefore, the potential for a dust explosion could exist in the entire dryer.
2. A number of dryers (e.g., spray, plate, belt) often can be designed with a closed-loop inert gas system (usually nitrogen).
3. Some dryers (plate, belt, etc.) can be designed in dust-tight and gas-tight construction. This is especially important for dryers handling toxic solids.
4. Some dryers (fluid bed, spray, etc.) can be designed to contain a deflagration pressure.
5. As a general rule, shock-sensitive solids should not be dried in thin film dryers, paddle dryers with inserted metal rods, or flash dryers.
6. Solids that can undergo spontaneous decomposition should not be dried in a drying apparatus holding large quantities of product, that is, paddle dryers, filter-dryers, or fluid bed dryers. If the use of such equipment seems unavoidable, supplementary tests should be done to determine the velocity of the decomposition reaction, and to evaluate the possibility of interrupting decomposition. In addition, the dryer should be equipped with the appropriate protection devices/

systems. Note that inerting or vacuum operation will not stop a spontaneous decomposition.

7. For direct-heat dryers (convective dryers) the following are recommended:
   (a) The temperature within the dryer shall not reach or exceed the ignition temperature of the dry product (as determined for airborne dusts). If the solids are wet with a flammable solvent, then the temperature within the dryer must also be kept below the minimum ignition temperature of the solvent.
   (b) Direct-heat (convective) dryers should be operated with fresh air only (i.e., the air must not be recycled), if the solids to be dried can develop flammable decomposition gases. In this case, a closed loop, inert gas system is a better way to operate the dryer.
   (c) For convective dryers, provide controls to ensure that the fan or blower will continue to operate long enough after turning off the heating system to prevent heat accumulation. In case of a failure or breakdown of the fan, the heating system should be switched off automatically and an alarm should be triggered; it may be necessary to provide additional automatic measures to prevent heat accumulation. Also, in each individual case, it should be established whether special precautions will be required with regard to possible breakdowns in the supply of important utilities (e.g., electricity, water, steam).
8. For indirect-heat dryers (also called contact dryers) the maximum allowable heating medium temperature (based on test data) should be monitored and controlled by instrumentation (redundancy should be provided as indicated by a process risk analysis).
9. Screw conveyor, agitated paddle, and disk dryers handling solids wet with a flammable solvent should be inerted. However, inerting does not provide protection against thermal decomposition with massive gas evolution.
10. Field (1982) suggests that to protect vacuum dryers against possible overpressure resulting from decomposition of the product, the nuts on their doors or covers should be loosened after evacuation and before the dryer is heated up, so that they can act as an explosion vent. This is recommended only if the dryer is inerted to keep any air leakage inward below the LOC. However, if an inert atmosphere cannot be guaranteed at all times, this may allow ingress of air and pose an explosion hazard. It is preferable that the dryer be equipped with a rupture disk explosion vent of suitable size. Vacuum dryers should be heated up after evacuation only. In special cases, e.g., with solids having a strong tendency towards foaming, it may be possible to heat up some dryers, such as paddle dryers, with the doors or covers not fastened, at ambient temperature and up to a heating medium tem-

perature of 80°C, if the product is solvent-free and has a burning class less than 3 (localized combustion with practically no spreading) at 100°C.

11. For solids with a burning class of 4 [glowing without sparks (smoldering) or slow decomposition without flame] or class 5 (burning like fireworks or slow quiet burning with flames), or solids which develop noticeable quantities of flammable decomposition gases at 220°C, it is recommended to apply the following additional measures to vacuum dryers:
    (a) The dryers should be charged only when cold and heated up only after closure or evacuation. Exceptions should be agreed upon with the company safety department.
    (b) At the end of the drying process, as a matter of principle, the vacuum should be broken with nitrogen.
    (c) Vacuum dryers may be aerated and discharged only when the product temperature has fallen below the temperature allowable for atmospheric air-drying or, if that value is not known, below 40°C.
    (d) Vacuum failure must be indicated by an alarm, and the dryer must then be inerted immediately and cooled either automatically or manually.
12. For spray dryers the following are recommended:
    (a) If solids with a burning class of 4 or 5 are to be processed in a spray dryer, it is recommended that the dryer be fitted with a water spray deluge system. This is recommended even for inerted dryers processing solids with a burning class of 4 or 5.
    (b) For spray dryers, tests should be performed for determining the lowest temperature at which an exothermic decomposition in air can be observed; in case of an inerted dryer, a supplementary test in a mixture of 92% nitrogen and 8% oxygen should be carried out (Grewer test). If in this test a decomposition temperature higher than 200°C is observed, the air inlet temperature (for an inerted dryer this is the inlet temperature of the inert gas) should be selected by knowledgeable specialists in accordance with the test results. The air or inert gas outlet temperature should not normally exceed 150°C. Exceptions may be possible with consent of knowledgeable specialists.
    (c) If a spray dryer processing a product with a low MIE (less than 10 mJ) has to be opened (e.g., for inspection), particular care should be taken due to the possible presence of smoldering material . As a minimum precaution, cooling down to less than 40°C is required. The procedure to assure sufficient cooling should be agreed upon with the company safety experts. Explosion safeguards and monitoring of the oxygen concentration (including automatic corrective

measures) may be inactivated only when the dryer has been cooled to room temperature. It is recommended, where feasible, to thoroughly wet all the product still present in the dryer with water prior to opening the dryer or inactivating any safety devices.

13. For fluid bed dryers, tests should be performed to determine the exothermic decomposition temperature in a fresh air stream, and the lowest temperature at which an exothermic reaction can still be observed in a 24-hour hot storage test (or other acceptable test, e.g., aerated solids screening test). For inerted fluid bed dryers, the test should be done in a mixture of 92% nitrogen and 8% oxygen. The maximum permissible air inlet temperature should be specified so that no exothermic reaction will occur in the 24-hour hot storage test. Exceptions would have to be agreed upon after more thorough testing of the product. The maximum air inlet temperature should normally remain below the melting point or melting range of the product. The surface temperature of the air heating elements should be kept as low as possible. Indirect heating of the air by steam, hot water, etc. is recommended. If electrical resistance heating is used, the heating elements should not be installed within the dryer, but in the air supply system at a far distance from the inlet screen. It is recommended that a metallic air filter be installed between the heating elements and the product chamber to prevent entrainment of hot particles (e.g., rust) into the dryer.

In recent years there have been a number of fires and explosions in dryers used for the drying of sewage sludge (Tyldesley, 2004). There are a variety of sludge dryer designs. The ones with large rotating drums have experienced problems and explosions because they create a dust cloud in the drum. Sludge dryers with a heated screw, that are effectively full, probably cannot have an explosion in the dryer section itself, but can produce burning and glowing material which can act as an ignition source in downstream equipment. Sludge dryers tend to self-inert during steady-state operation as the steam evolved displaces air. But a number of explosions that occurred during start-up and shutdown, showed that the material could self-heat to ignition at temperatures close to the operating temperature. A possible solution to this hazard is to provide nitrogen inerting for start-up and shutdown, linked to gas analysis to monitor the oxygen level in the dryer. In a most recent incident in the UK, dry fines were being fed back into the system that was being allowed to cool down. Actually, it started to heat up again, and the fines were fed into a system that was actually on fire for 15 minutes (which was evidently unknown) before the explosion occurred.

Abbott (1990) presents a comprehensive overview of the prevention of fires and explosions in dryers, in which he discusses the hazards, potential ignition sources, safety from explosions, process specification, equipment specification, and operation and maintenance of several types of commonly used dryers (spray, pneumatic conveying, fluid bed, rotary, band, batch atmospheric tray, batch vacuum, and trough dryers). Other good reviews of dryer safety are presented by Field (1982), Gibson et al. (1985), Palmer (1973), Markowski and Mujumdar (1995), and Roques (1996). Bartknecht (1981) presents detailed discussions of the protection of fluid bed dryers and spray drying installations specifically. NFPA 86 (1999) also discusses dryer safety.

5.3.4 Dust Collectors

This section discusses hazards associated with four types of dust collectors (cyclone separators, electrostatic precipitators, fabric filters, and wet scrubbers), as well as preventive and protective measures for them.

5.3.4.1 Cyclone Separators

Cyclone separators are less susceptible to fires and explosions than fabric filters, but they do occur. Often the source of ignition is not at the cyclone itself but dust electrostatically charged in the process upstream of the cyclone. If the dust cloud in the cyclone is above its MIE and its dust concentration is in the flammable range, an explosion can occur if newly entering charged dust is discharged and ignites the cloud. Fires can be caused by ignition of dust deposits on the cyclone walls.

Tyldesley (2004) reports an incident at an aluminum powder production unit containing multiple small cyclones which had caked aluminum dust on their inside walls, which somehow was ignited.

The most common protective measures for cyclones are venting and suppression. Vent sizes should be calculated by the procedures given in NFPA 68 (2002). Deflagration suppression is preferred over venting if the cyclone is handling toxic dust. Also, cyclones may be designed to contain the maximum explosion pressure, which is also preferred over venting if the cyclone is handling a toxic dust. Proper grounding and bonding is critical to avoid build-up of electrostatic charges. Preferably, cyclones should be located outside plant buildings, on the ground or roof, so that vent ducts to route the exhaust gases/vapors from a room, are not required (the vent discharges directly into the atmosphere).

If it is known that the dust being handled can self-decompose or is highly reactive, and has a propensity for caking on walls, then inspection ports should be installed on the cyclone to check on cake formation on a regular schedule, and means provided for removing the deposits (e.g., water or solvent spray nozzles where applicable).

Additional information on protective measures for cyclones is presented by Eckhoff (2003), Field (1982), and Palmer (1973).

5.3.4.2 Electrostatic Precipitators

Electrostatic precipitators may develop fires as a result of either electrical discharges resulting from dust accumulations reducing clearances within the unit to below the voltage breakdown distance, or because of particles ignited outside the unit being collected in the ESP and igniting other dust. These units are often cleaned by rapping mechanically and burning dust dispersed in this way could cause an explosion.

The following protective measures for ESPs handling combustible dusts are recommended by FM Global (FMG 7-73, 2000):

1. Provide automatic sprinkler protection at a minimum of 0.25 gpm/ft^2 for the following:
   (a) Inside ESPs with oil-bath reservoirs.
   (b) Inside ESPs collecting combustible material.
   (c) Over ESP oil seals and associated oil settling tanks, reservoirs and piping and for 20 feet beyond.
2. Provide an interlock to automatically de-energize the ESP on actuation of automatic sprinkler or automatic water spray systems.
3. For ESPs and associated ducts serving two or more pieces of process equipment:
   (a) Automatic open head water spray in the ducts and in the ESP, as follows:
      (i) In the ducts provide a minimum density of 0.2 gpm/ft^2.
      (ii) In the ESP provide a minimum density of 0.25 gpm/ft^2.
   (b) Locate an automatic sprinkler inside each duct penetration at roof level. Locate an automatic sprinkler inside the duct at each floor penetration for multistory buildings. Ensure that the sprinkler temperature rating is 50°F higher than gas temperatures in the duct.
   (c) Provide for drainage of sprinkler water. Slope ducts so that water flow is toward the ESP. Provide a separator tank for ESPs used to collect hydrocarbon mists so that the discharge of the water spray system does not result in a release of hydrocarbons which could expose plant buildings or property. This is especially important if the tank is inside the building.
   (d) Interlock the fan on the exhaust system to shut down on actuation of the water spray detection system.
   (e) Provide access ports in the duct so that detectors can be checked and water spray nozzles serviced.
4. Use special protection systems such as water spray, carbon dioxide or steam where frequent losses have occurred.
5. Arrange ESPs processing hybrid mixture (solids with flammable vapors) to be purged with inert gas before the introduction of the hybrid mixture on start-up and before the introduction of air on shutdown. Provide combustibles/oxygen analyzers to monitor oxygen

concentration in inlet gas. Interlock the analyzer to sound an alarm at 1% concentration of oxygen and shut down power to the ESP at 2% oxygen concentration.

6. Provide flow meters for wet ESPs designed for continuous water washing of collecting surfaces during normal operation. Interlock flow meters to sound an alarm and de-energize the units in the event that water flow drops below a preset minimum value. Automatic sprinkler protection is not normally necessary.
7. Do not use ESPs where dry combustible dust concentrations in air may exceed the MEC due to the possibility of ignition by arcing in the ESP.

5.3.4.3 Fabric Filters

Over the years, fabric filters have experienced many fires and explosions, as well as loss of containment (dust emissions) due to broken bags.

The following two case histories illustrate dust explosion incidents in baghouses:

Case History of an Explosion in a Flash Dryer Baghouse

An explosion occurred in a dust collector used to collect a pharmaceutical product from a hammer mill/flash drying operation. The impact hammer mill had been operating for approximately 10 minutes when the operator heard unusual grinding sounds coming from inside the mill. He immediately shut down the mill just as an explosion occurred within the dust collector, located inside the building on the second floor. The pressure wave caused the explosion vent (a hinged panel) of the dust collector to open, and the explosion products and unburned powder were directed outside the building via a vent duct. However, a screen had been securely fastened at the end of the duct to prevent birds from entering, and as the vent panel swung upward and outward, it struck the screen and opened no further. It is estimated that the screen prevented the explosion vent panel from opening to no more than 50 percent of the vent area.

With the vent partially obstructed, the access door to the dust collector failed under pressure and released a dust cloud into the building, which ignited. The flame front went through the vent duct and followed the dust cloud through the access door, resulting in a fireball at both locations. Also, on the first floor, a fireball was seen exiting the vicinity of the rotary valve outlet at the bottom of the dust collector, which feeds a sifter. There was no secondary explosion on the first or second floors. However, windows were blown out at both floors. The ensuing fire in the dust collector engulfed the wool filter bags (which were burned up) and the remaining powder in the collector hopper, but the fire was quickly extinguished by the automatic sprinkler system inside the dust collector.

A subsequent investigation of the incident revealed that a carbon steel bolt from the inside of the feeder (which feeds wet powder to the hammer mill/flash dryer) had become loose and fell into the hammer mill. The bolt became trapped inside the 3600 rpm mill, where it became heated to above the autoignition temperature of the powder. The hot metal ignited some of the powder in the mill which was pneumatically conveyed into the dust collector. In the collector, a dust cloud created by the blow ring (pulse jet), was ignited by the hot powder conveyed in from the hammer mill. An inspection of the feeder revealed that six 3/8-inch carbon steel bolts were missing.

Ed. Note: The following measures should be taken to avoid this incident from occurring again:

1. The vent duct opening should not be obstructed in any manner.
2. Preferably use a diaphragm-type vent closure with a vent duct rather than a hinged panel type vent closure. The effect of the vent duct on the baghouse P_{red} or vent area should be checked by the method in NFPA 68 (2002).
3. Any interior nuts and bolts in the feeder should be tack-welded to the housing to prevent them from becoming loose due to vibration.

Case History of an Explosion in a Packaging Room Baghouse

An incident of an explosion in a packaging room baghouse is presented by Pickup (2001). A baghouse dust collector, located outside of the building, drawing dust from a packaging room and container packer experienced a dust explosion shortly after a metal collection drum was emptied and returned to its prior position. The drum was transported under the baghouse on a wheeled trolley. The drum cover was connected via a flexible connection to a slide gate valve at the bottom of the baghouse. Normally, the negative pressure in the baghouse would have pulled the cover onto the drum, forming a tight seal. However, in this case, the operator noticed that the drum was not correctly positioned under the baghouse and the cover had not engaged. As the operator reached in to position the drum, the baghouse automatically pulsed. When the operator moved the drum, a dust explosion occurred directly above him in the baghouse, causing the explosion vent to open and send a large fireball out directly above his head. The operator ran into the building to warn his co-workers. However, as he entered, the other operators were evacuating the building due to the quantity of dust that was thrown into the air inside the packaging room. The operators fortunately managed to exit the building to a safe location before the dust cloud found a source of ignition and exploded, approximately 2 minutes after the explosion in the baghouse. Ninety seconds after the second explosion, a third violent explosion occurred that caused significant structural damage to the building. The pressure that occurred during the baghouse explosion was high and was transmitted throughout the "nuisance" dust collection system ducting, and led to the suspension of significant quantities of dust inside the packaging room. The accident investigation concluded that the drum located on the wheeled trolley became charged and was isolated so that an electrostatic spark discharge occurred and ignited the dust that became suspended by the air pulse. The trolley had a high electrical resistance from drum to earth and the drum lid was connected to the dust collector bottom by an elastomeric flexible connector internally supported by a spiral wire. Subsequently, the trolley was removed so that the drum sits on the ground and a flexible metal connector joins the drum to the dust collector.

Ed. Note: A number of safety measures would have prevented this accident, as follows:

1. The slide gate valve should not have been opened until the drum was properly connected to the baghouse.
2. The metal drum should have been bonded and grounded.
3. The baghouse should have been isolated from the rest of the dust collection system to prevent flame propagation back into the system and packaging room. Either a rapid-acting valve or a flame front diverter could have been used.

The main hazard from the presence of dust collected in dust collectors (baghouses and cartridge filters) is an explosion from electrostatic spark discharges. These may be capacitance, brush, and propagating brush discharges. Capacitance and propagating brush discharges are incendive for dusts, whereas brush discharges are incendive for flammable vapors. If hybrid mixtures are present (this can occur under normal operating conditions), then brush discharges are capable of being an ignition source. Dust explosions occur quite frequently in baghouses because the likelihood of the presence of an easily ignitable fine dust atmosphere is high and there is high turbulence, which can causes electrostatic charge accumulation on the dust particles. Another ignition source is the entrance of hot, glowing, particles into the baghouse from upstream equipment.

Isolated conductors that become charged, usually through extended periods of field inductance and a high resistance path to ground, are a common hazard. Typical sources of isolated conductors are wires in cartridge filters, clamps and filter bag supports (cages). Cleaning mechanisms (e.g., air pulsing, shaking) may jostle the charged conductors and make possible closer movement to a grounded part of the collector and potential energy discharge by a capacitance spark.

Brush discharges can arise from the charging effect that occurs when powder is captured on the filter medium surface. However, these are not incendive to dusts having MIEs >3 mJ. In fact, recent studies by Larsen et al. (2001) indicate that brush discharges are incapable of igniting even microfine sulfur (MIE of ~1 mJ), except in oxygen-enriched atmospheres.

Propagating brush discharges can occur during pneumatic conveying into a dust collector when a deflection plate coated with plastic is present to deflect the incoming gas–solids stream. The plastic coating behaves as a dielectric in a capacitor if it possesses a breakdown voltage of greater than 4 kV.

Fabric filters can be protected from fires and explosions by venting, suppression or containment. In the U.S., vents should be designed in accordance with procedures given in NFPA 68 (2002). In Europe, VDI 3673 (2002) is often used for the design of deflagration vents for baghouses. One problem with the design of such vents is that the vent area may become blocked by bags that are displaced by the dust explosion, and full venting is not achieved. Systems for the suppression of fires and explosions should be designed in accordance with procedures given in Chapter 7 of NFPA 69 (2002). If a fabric filter is not too large it can be purchased as a cylindrical unit which can be designed for deflagration containment.

Where frequent accidents have occurred in fabric filters, FM Global recommends the following practices to reduce frequency and minimize damage and downtime (FMG 7-73, 2000):

1. Provide automatic water protection at a minimum density of 0.2 gpm/ft^2 in the bag section, in the clean air plenum, and in hoppers shielded from protection in the above areas. In the bag area provide a maximum 50 ft^2 head spacing. In the hopper area provide one head per hopper and a maximum 100 ft^2 head spacing. In the clean air plenum provide one head per 100 ft^2. This protection can be either of the following:
   (a) Automatic sprinkler protection with heads at a 212°F temperature rating.
   (b) A closed-head water spray system, or open-head spray system activated by an infrared or continuous line type detection system.
2. Interlock the rotary valve at the hopper bottom to stop on actuation of the collector fire extinguishing system to prevent transfer of burning dust into another part of the process.
3. Install high speed infrared detectors in the duct between the process and the collector. Interlock the detectors to actuate an extinguishing system in the duct or in the collector. Arrange the detection system to stop the rotary valve to prevent burning material from being trans-

ferred out of the collector, unless the material can be diverted to a safe location.

4. Subdivide collectors into smaller fire areas (100 bags or less). Install partitions of 30 minutes fire resistance in collectors without sprinklers. Partitions may be constructed of 18 ga. (1.3 mm) sheet metal in collectors with sprinklers.
5. Install a spark-arrester or settling chamber in the duct between the process and the collector. A cyclone collector, scrubber, or similar device that would deflect or extinguish heavier embers or sparks from the gas stream is acceptable.
6. Ensure that manual extinguishing equipment is available to personnel performing maintenance on a collector. For small collectors, portable extinguishers (preferably water-type units) are acceptable. For larger, walk-in type collectors, install 1½-inch hose with a combination water spray straight-stream nozzle near the door outside the collector. Provide access ports for all areas of the collector where necessary for effective manual firefighting.
7. Provide rupture disks, hatches attached with springs, or other reliable devices where it is possible to accumulate enough water from hose streams or sprinklers to result in structural damage to the collector. They should actuate at 1.5 psig or less, and be located as close as possible to the bottom of the hopper.
8. Provide deflagration isolation in the duct between the process equipment upstream of the baghouse to prevent flame propagating from the baghouse back into the process equipment.

Other recommended safety practices are:

1. Preferably locate the fabric filter outdoors so that a vent duct is not required.
2. Properly ground and bond the fabric filter components (housing, tubesheet, cages, clamps, etc.) to dissipate electrostatic charges.
3. Install a broken-bag detector to sound an alarm if a bag break occurs and interlock the broken bag detector to the exhaust fan to shut it down so that dust emissions are minimized. This is especially important when handling toxic dusts.
4. Interlock the exhaust fan to shut down on actuation of the collector fire extinguishing system.
5. Install a high-temperature sensor and alarm to warn of a possible fire in the collector. This can be interlocked with an automated block valve in the water supply piping to the water protection system.
6. For dusts having a MIE of 3 mJ or less or for hybrid mixtures, some companies require that an electrostatically conducting filter medium be used. However, this type of filter medium may actually increase

the ignition hazard if such bags are not properly grounded when installed.

7. All internal walls, on which dust can impact with a high velocity, must have no insulating lining with a high resistance (the breakdown voltage must not exceed 4 kV).

Additional information on hazards of fabric filters and their protection is presented by Bartknecht (1981), Field (1982), and Palmer (1973).

5.3.4.4 Wet Scrubbers

Despite the fact that wet scrubbers operate with large amounts of water, they still can have fires, as shown by the following case history.

Case History of a Fire in a Wet Scrubber

An incident involving a fire in a wet scrubber is described by Ness (2002).

A wet scrubber on a dryer air exhaust sustained damage when a fire occurred in the area of the bottom tray. Full water flow was put on the scrubber, the dampers to the fan were closed, and the dryer train was shut down. A ¾-inch layer of resin had accumulated in an "I-shaped" channel beam that was supporting the bottom tray. A charred layer of resin was found in the deposit. The bottom and middle spray nozzles had become blocked by solids, so there was no water flow from the spray nozzles, allowing the resin to dry out. The resin was known to have thermal stability problems upon aging and on contamination with iron. The scrubber was made of carbon steel and rust was visible, and the deposit clearly built up over a long time.

To prevent a reoccurrence, the following steps were taken:

1. Weekly external inspections were started to check for nozzle blockages.
2. Monthly internal inspections were started to check for rust.
3. A filter was installed on the spray water recycle line to minimize possible blockages.
4. The spray water recirculation pump capacity was increased to create higher flow to reduce the potential for blockages.
5. Flow indicators were installed on each set of spray nozzles.

Wet scrubbers do not usually pose a fire or explosion hazard because the particles are wet. However, if the scrubbing liquid (usually water) flow is stopped, combustible or toxic particles can be emitted. Therefore, low flow or low pressure switches and alarms should be provided and interlocked to prevent operation of the scrubber if the flow of scrubbing liquid is stopped.

FM Global recommends the following practices (FMG 7-73, 2000):

1. Wherever possible, use noncombustible materials for scrubber construction, linings, and packing materials.
2. Provide automatic sprinkler or automatic spray protection inside collectors of combustible construction, or having combustible lining, or packing materials.
3. Maintain the liquid level above the level of sludge (wet solids), or arrange for sludge to be continuously drained and disposed of properly.
4. Remove sludge subject to spontaneous heating from collectors as soon as they are shut down.
5. Arrange for chambers of scrubbers handling dusts, such as magnesium, that produce hydrogen when wet to be vented to the atmosphere. A 1-inch diameter hole in the top of each section of the scrubber will serve to dissipate hydrogen formed during shutdown and usually will not interfere with normal operations.

5.3.5 *Extruders*

Since extruders usually work with wet powder masses, fires and explosions do not often occur. However, fires can result from leakage from lube oil or hydraulic fluid systems associated with extruders. If any fire protection measures are provided, they are usually installed before or after the extruder. However, excessive compaction and overpressure can occur, which can damage the die. Considerable work is expended to force a wetted powder through the die of the extruder and temperature rises of 15°C have been recorded, especially when the die holes are small (i.e., 0.6 mm in diameter), and higher temperature rises are possible. A dryer usually follows an extruder, where more energy is added, and the combination of temperature increases across the extruder and dryer may initiate self-heating degradations in more sensitive powders when the hot extrudates reach a storage vessel downstream of the dryer, or when the product is packaged, and then allowed to remain for extended periods.

Appropriate tests, such as the Bowes-Cameron basket tests (see Section 4.3.4), are necessary to determine the maximum allowable exposure temperatures. If processed material has dangerous properties determined in safety testing, such as the powder burns through the sample without stopping (train fire properties) and the onset temperature of decomposition is sufficiently low, then general rules can be made about "quarantining" the material after completion of the operation for a designated period of time (usually 12 or 24 hours). If there is no self-heating during the quarantine period, the material is safe to ship to customers.

In a small number of cases powder being extruded can react with water, after an induction period, and result in an undesired exothermic decomposition reaction. Such a situation can happen if the process is stopped for an extended period, such as a day, and the extruder is not cleaned out of wetted material. In a worst case, the liberated heat can cause a fire or an explosion. If a fire is detected, the extruder should be cleared of in-process materials, which then have to be put in a safe place and given proper treatment, such as being flooded with water.

Since compaction and overpressure can damage the die, one extruder manufacturer uses a hydraulic clamping system to hold the extrusion die in place. If a high extrusion pressure is generated, the end plate is pushed away from a proximity switch, which is interlocked with the motor and shuts down the extruder. Appropriate electrical area classification for the motor should be considered, particularly when flammable vapors can be present.

5.3.6 *Feeders and Rotary Valves*

Hazards and preventive/protective measures for volumetric feeders (which includes rotary valves) and gravimetric feeders are discussed below.

5.3.6.1 Volumetric Feeders

SCREW FEEDERS: The risk of initiating an explosion in a screw feeder is relatively small, and fires are a more likely occurrence. Excessive local heat caused by metallic contact between the screw and the casing can act as an ignition source. Contact could arise from excessive deflection in long screws, center tubes that are not straight due to welding or manufacturing anomalies, the use of over-tight flight tip clearances and/or alignment problems, and from poor casing tolerances or external damage. End clearance between the screw and the casing end plates should be adequate to allow for differential thermal expansion, tensile stress of weight or end thrust, and for manufacturing and assembly tolerances. Another ignition source may be foreign bodies, or elements of the material being conveyed and becoming trapped between the screw and the casing, and high contact pressure under slip conditions may give rise to excessive friction that generates heat. Care should be taken when handling hard, granular type material, that the working clearance is adequate to accommodate the largest combination of pieces that may occupy this clearance space. It is good practice to install temperature sensors, interlocked with the feeder motor to shut it down, if the heat generation becomes very high, especially for reactive solids.

Bearings and seals can generate heat and they should be closely scrutinized to see that they do not add heat to the material. Intermediate bearings are a particular problem as it is virtually impossible to eliminate the ingress of material into the bearing. Product build-up between the shaft and the bushing is almost inevitable and heat input occurs. Also, center bearings present an obstruction to flow and material has to be pushed past the gap in the flights and through the housing supports. The use of such bearings is not recommended where the material is prone to an explosion. The use of plain end bearings is also not recommended because of the prospect of "hot spots" or seizures developing, and ball or roller bearings should be checked to detect wear and signs of rough running. It is not uncommon in many plants for packed glands to be overzealously tightened. This can cause significant heat input to be carried to the screw shafts and their housings. All bearings should be external, if possible, to the powder carrying components, and provided with temperature sensors that will alarm on high temperature reading.

It is often desirable to install pressure sensors to detect flow blockages that may occur at outlets, transfer points, and any intermediate region along the casing that may have features which prevent the smooth passage of material. Apart from the intensive pressures that may be generated, it is also possible for the screw to be damaged and to rub against the casing or cover in extreme circumstances, as something has to "give" if powder is to move towards a blockage. As the power allowances for screw feeders includes a safety factor (for startup and unusual operating conditions), when a blockage occurs, all surplus power, plus the overload capacity of the drive is concentrated at the final point of contact between the screw and the material. This can cause significant work input and generate high pressures. It may be desirable to provide an explosion vent on the top cover with a vent duct to convey the flame and burning dust to a safe location.

End seals or packing glands are required to contain the product and any fire that may occur, and they may also need to resist the loss of gas under pressure. The inclusion of a lantern ring between gland rings and injection of a purge gas (usually nitrogen) offers a resistance to the egress of particles by gas back-flow into the system where there is a leakage potential. The gas leakage rate can be monitored or checked, if required. A purge gas will also help to keep abrasive particles out of seals and prolong their life.

Proper grounding and bonding is necessary to dissipate any electrostatic charges, and the motor should have the proper electrical area classification.

BELT FEEDERS: Belt feeders are usually enclosed to contain any dust and as the solids only occupy a small fraction of the enclosure, they can be subject to explosions. The belt can generate electrostatic charges, and therefore, should be of anti-static material. Overheating can occur in belt feeders due to a jammed idler roller, or if the belt jams, and the drive rollers continue to run. Therefore, for combustible and reactive solids, the enclosure should have explosion protection, such as venting or suppression, and fire protection such as an automatic water sprinkler system or a deluge water spray system. For toxic dusts the enclosure must be dust-tight, and it may be desirable to have the enclosure connected to an exhaust collection system to contain fugitive emissions.

Details of fire protection systems are given in FM Global Loss Prevention Data Sheet 7-11 (FMG 7-11, 2003).

Proper electrical area classification is required for the motor.

VIBRATORY FEEDERS: Vibratory feeders are not usually prone to explosions as they are usually open (no enclosure), but it may be possible to electrostatically charge the particles as they are projected forward. Therefore, the feeder should be grounded and bonded to dissipate any charges generated. It may be desirable to provide sprinkler or water deluge protection if the solids are reactive. If the solids are toxic, the feeder should be enclosed to minimize fugitive emissions, and it also may be desirable to connect the enclosure to an exhaust collection system.

ROTARY VALVES: Rotary valves by themselves usually are not a hazard; in fact they are often used as a choke to prevent the transmission of a dust explosion (Eckhoff, 2003). Bartknecht (1989) states that rotary valves (rotary air locks) will act as a mechanical flame barrier against dust explosions if the following criteria are met:

1. Two vanes per side are engaged (are near the housing walls),
2. The gap between the rotor and housing is ≤ 0.2 mm, and
3. The vanes (tips) of the rotor are made out of metal.

It is not always possible to maintain the desired small gap between the rotor and the housing as blade tips tend to suffer wear, especially when handling highly abrasive solids. It is possible to specify rotor blades made of special abrasion-resistant metal, which will help to minimize this problem. It also may be desirable to maintain a plug of solids above the valve inlet by means of a low-level sensor and switch in the hopper above the valve which will shut the valve off when the solids level in the hopper gets low. This will maintain the effectiveness of the valve to act as a "material choke."

Eckhoff (2003) presents a nomograph for estimating the maximum permissible clearance between the rotor blades and the housing for prevention of transmission of dust explosions through rotary valves.

Excessive friction in the bearings of a rotary valve can result in the generation of sufficient heat to cause local temperatures above the minimum ignition temperature of the solids being handled. This can result in the solids catching on fire and if the rotary valve is not shut down immediately, the smoldering solids can be transmitted to downstream equipment. If a solid is highly combustible or reactive, it may be desirable to install a temperature sensor in the valve body to detect a fire in the solids, and interlock the valve motor to shut it down if this occurs.

There is a divergence of opinion as to whether a rotary valve has to be vented. If venting is done, and the solid is toxic, the vented air (which may contain toxic dust) should be directed to control equipment (e.g., a fabric filter or a wet scrubber).

Flooding is a problem that is common to all feeders and is associated with the bin or hopper to which the feeder is attached. The basic problem of flooding involves fluidization of the powder. The solid particles lose contact with each other and the internal shear stresses become virtually zero. If flooding is not prevented, excessive solids flow may result in equipment malfunction, plugging, and spillage of solids onto the floor or other adjacent equipment. This could result in a dust explosion or health hazards to operators if the solids are toxic.

Flooding is caused primarily by:

- Funnel type flow pattern in the bin or hopper.
- Exceeding the critical exit velocity from the bin or hopper.
- Insufficient retention time of aerated materials in the bin or hopper.

Decker (1975) recommends the following design guidelines to prevent flooding from occurring:

1. *Silo/Hopper Capacity:* The silo/hopper must have adequate capacity to allow a minimum of 10-15 minutes retention time for deaeration to cushion the impact of the material.
2. *Silo/Hopper Filling:* The type of vessel filling system, filling rate, and the effect of the powder are of major concern. The least amount of aeration must be provided. Impact over the vessel outlet must be prevented.
3. *Venting:* The vessel must be equipped with adequate venting capacity regardless of how it is filled because the solids entering the vessel are continuously displacing air.
4. *Low Level Control:* A minimum level must be maintained in the vessel to cushion the impact of new material and to allow ample deaeration time.
5. *Silo/Hopper Configuration:* The hopper slope angles and wall lining must be selected to provide a mass flow pattern. If a mass flow silo/hopper cannot be provided due to headroom limitations, then a special type of flow-promoting device (e.g., live-bottom or other discharger) may have to be installed to avoid flooding.
6. *Discharge Opening:* The outlet opening of the vessel must be large enough to prevent arching. This dimension can be calculated based on flow properties from shear cell test data, or from experience. Velocity aeration should be prevented and the maximum exit velocity should not exceed 6–7 ft/min. A positive cutoff gate (valve) should also be considered.

7. *Feeder:* The feeder chosen must be matched to the outlet size of the silo/hopper. If provision of adequate retention time is not possible, a combination of a live-bottom and positive feeder seal should be considered.

5.3.6.2 Gravimetric Feeders

The hazards and protective measures for the three types of gravimetric feeders are essentially the same as those for volumetric feeders.

5.3.7 Hoses, Loading Spouts, and Flexible Boots and Socks

Hazards associated with the use of hoses, loading spouts, and flexible boots and socks are discussed in this section, and preventive/protective measures are recommended.

Hoses

Case History of an Explosion Caused by Use of a Polyethylene Hose in Pneumatic Transfer of Acrylic Powder

Luttgens (1985) reports a series of explosions caused by the improper use of insulating polyethylene hose in the pneumatic transfer of acrylic powder from a silo to a railroad hopper car. No injuries were incurred, but two railroad hopper cars were damaged and required repair. The weather was wet and snowy on the day of the accident. The accident investigation revealed that all metallic sections of the transfer line were bonded and grounded properly, but that a section of flexible polyethylene hose had been installed for easier handling of the transfer line. The outside surface of the hose, exposed to the weather, was wet with precipitation. The water served as a conductor, and so the insulating polyethylene material acted like a capacitor, storing sufficient electrostatic energy for a very strong propagating brush discharge that ignited the powder. This hypothesis was verified by subsequent tests performed on another railroad hopper car being filled in the same way, but with nitrogen inerting to prevent an explosion.

Particulate solids flowing through hoses will generate electrostatic charges. Powder transfer through hoses typically generates a greater rate of static charge than liquid flow, abrasion is greater, and the hose or loading spout may frequently contain an ignitable mixture. Conductive hoses should preferably be flexible metal so that bonding does no rely on continuity of a breakable element. Other types of conductive hose should be designed so that end connectors are bonded to the grounding elements in the hose, while breakage of a conductive element cannot create a spark gap. To avoid spark gaps formed in this manner, a semiconductive hose design might be considered. Lightweight semiconductive hoses having no metal bonding elements are commercially available. These have a typical end-to-end resistance of a few thousand ohms per meter, depending on hose size (Britton, 1999).

Nonconductive hoses or hoses with a nonconductive liner can produce propagating brush discharges. When a hose is made with a nonconductive liner or entirely of nonconductive material, both conductive and nonconductive particulate solids passing through the hose will exchange electric charges between the solids and the hose, and even between the solids themselves in some situations. The higher the flow rate of the solids and the more dense the solids phase in the gas phase also present, the greater the charging. The nonconductive liner in a hose acts as a dielectric in a capacitor, and the conductive boundary of the hose (the metal wall, spiral wire, the metal wrapping of insulation, or water wetting) is the plate of the capacitor. The dielectric acts to enable greater charge accumulation on the plate, and through induction, there is a charge separation between the flowing solids and the metal boundary. A dipole is created in the dielectric that tends to counter the field of the capacitor with a smaller field of its own that acts in the opposite direction. When the voltage difference becomes great enough, a weak spot in the dielectric will fail, that is reach its breakdown voltage, and the energy available will concentrate at the breakdown location where a spark leaps the dielectric. The spark can be quite energetic, easily 1 Joule of energy, which is sufficient to ignite the solids in the air present should the degree of turbulence be sufficiently subdued to allow the spark energy to remain concentrated.

Bulking brush discharges (cone discharges) can also be produced by particulate solids flowing through hoses. Charges can develop on particulate solids passing through a hose in the following two circumstances:

1. If the solids are electrically nonconductive, solids flowing through either a conductive or nonconductive hose can become charged.
2. If the solids are conductive and the hose is nonconductive, then a charge can develop in the solids.

When the charged solids fall into a container or vessel and accumulate into a volume that exceeds at least 1 m^3, then enough concentration of charge occurs so that there is the potential for bulking brush discharges to occur. Typically, the phenomenon is greatly enhanced when there is a range of particle sizes present in which there are coarse sized particles and fines. The coarse particles are the charge generators and the fines are the fuel and normally there is potential combustion with the oxygen in the surrounding air. See Section 6.3.1.1. for a more detailed discussion of bulking brush discharges.

Nonconductive hose incorporating an internal bonding spiral is not recommended except for pellets or noncombustible dust service in unclassified areas. This is because if the spiral breaks, internal and possibly external spark gaps may be created. In some cases, hose designs have omitted to bond the end connectors together via the spiral. Sometimes there are two spirals present in the hose, but hoses with more than one internal spiral is not recommended, because determining if one of the spirals has lost its continuity is not possible (NFPA 77, 2000).

Jones and King (1991) recommend the following concerning the use of hoses:

1. Never use nonconductive hose in powder handling operations.
2. Never wrap foil or wire around such hose in an attempt to make it "conductive."
3. Avoid the use of hoses constructed with insulating laminations.

Loading Spouts

Particulate solids flowing through loading spouts likewise generate electrostatic charges depending on their materials of construction. The spouts should be properly grounded and bonded. It may be desirable to have the railcar or hopper truck purged with nitrogen before loading begins and then do the loading under a nitrogen blanket. All electrical components should have the proper NEC electrical area classification.

Flexible Boots and Socks

A nonconductive boot could give rise to either brush discharge or propagating brush discharge. Propagating brush discharge cannot happen with a sock, because of the low breakdown strength of the air gaps in the weave. However, there are conditions where socks can produce brush discharges (e.g., where used with FIBCs). NFPA 77 (2000) recommends the following for boots and socks:

1. For combustible dusts, the end-to-end resistance of boots and socks should be less than 10^8 ohms and preferably less than 10^6 ohms, measured with a megohm meter.
2. Flexible boots and socks should not be depended on for a bond or ground connection between process equipment. Separate bonding and grounding connections should be used (i.e., the bonding and grounding connections should be attached to the upstream and downstream equipment).

5.3.8 *Mechanical Conveyors and Bucket Elevators*

5.3.8.1 *Mechanical Conveyors*

All mechanical conveyors involve the risk of fires and explosions due to ignition from electrostatic charge generation, overheating due to mechanical failure (e.g., frictional rubbing of one part against another or powder against a part of the conveyor), overheating of the powder due to compaction, or thermal decomposition.

Consideration should be given to the need for, and practicality of, providing closed conveyors with some method of explosion protection (e.g., inerting, deflagration venting, suppression, and containment). Some conveyors such as flight or en-masse conveyors, usually do not generate large dust clouds because of their construction, which does not provide much free space; however, if they are handling solids that can self-decompose, they may require explosion venting or be designed for deflagration containment.

BELT CONVEYORS: Belt conveyors have the same hazards as belt feeders and should have the same protective measures (see Section 5.3.6.1). Some additional discussion is presented here.

The primary ignition hazard of belt conveyors is friction from misalignment or at a stalled roller. Belt material is usually combustible (unless noncombustible material has been specified) and therefore, a fire is possible in the belt as well as in the material being conveyed on the belt. This ignition source can be conveyed to the discharge point of the conveyor, where dust and air are in suspension usually, and a dust explosion can occur.

Belt materials can be specified with special fire resistant properties to minimize the potential for a fire to occur (advice from belt manufacturers should be sought about the options available).

Some general safety recommendations are as follows:

1. The transfer point(s) should be designed for minimum dust accumulation, such as sloping all surfaces towards the downstream equipment.
2. Access for removal of solids accumulations should be provided along the entire length of the conveyor, such as by hinged panels along the side of the housing (if the conveyor is enclosed). A program for regularly removing dust from points of accumulation should be instituted. Housekeeping inside the conveyor enclosure should be generally as good as the remainder of the plant.
3. Zero-slippage switches should be provided to automatically stop the conveyor upon detection of belt slipping.
4. Misalignment switches, interlocked to shut down the conveyor, should be considered for long belts where misalignment is judged to be a potential problem.
5. Automatic sprinkler or deluge water spray protection should be provided in accordance with FMG 7-11 (2003) where the belt and/or material being conveyed are combustible.

Conveyor belts with trippers may pose a problem as it is difficult to properly enclose the tripping device to avoid dust release and accumulation on outside surfaces. In such installations, frequent (sometimes daily) dust accumulation removal (housekeeping) may be necessary (Tyldesley, 2004).

SCREW CONVEYORS

Case History of an Explosion in a Screw Conveyor

A case history of an accident with a screw conveyor described by Field (1982) is presented below:

Three employees were killed, and two seriously injured, and a factory building completely destroyed in an explosion involving skimmed milk powder conveyed by a screw conveyor. The milk powder was fed into a screw conveyor from a feed hopper and then carried to a blender. A deformation occurred in the screw conveyor housing, causing parts of the screw flights to grind against the housing. The grinding produced sufficient frictional heat and sparks to ignite the dust–air cloud in the free space of the conveyor. The primary explosion burst the screw conveyor housing, dispersing a significant amount of additional dust into the air from the freshly filled feed hopper. A secondary explosion was then ignited by the flames of the primary explosion.

The same hazards exist as in screw feeders and the same protective measures can be applied to screw conveyors (Section 5.3.6.1).

VIBRATORY CONVEYORS: Vibratory conveyors have the same hazards as vibratory feeders and can be provided with the same protective measures as discussed in Section 5.3.6.1.

CHAIN-TYPE CONVEYORS: Apron conveyors are not normally used for handling of powders or hazardous solids. They also operate at slow speeds so that dust clouds are not formed normally. However, if they are used for conveying combustible or reactive solids, they can be provided with deflagration suppression systems.

Drag flight conveyors can generate electrostatic charges on solids as they are dragged along the surface of the trough. Therefore, proper grounding and bonding is necessary to dissipate the charges. For handling toxic solids operation under a slight negative pressure (connected to a ventilation system and fabric filter) can eliminate most of the leakage from seals or bolted joints. Also, purging with an inert gas (usually nitrogen) can clean deposits of toxic solids from chains and housings. Flight conveyors do not usually generate dust clouds as the free volume is small, even when the flights fold back on the return leg, so that explosion venting or suppression is not normally provided (Tyldesley, 2004).

En-masse conveyors can generate electrostatic charges on the solids as they drag them along the inner surfaces of the housing. Here, too, proper grounding and bonding is necessary to dissipate these charges. They move at fairly low speeds and the free volume is small, so that they do not generate large dust clouds, and dust explosions do not normally occur. However, if the solids being transported can self-decompose, then venting or containment should be considered. Containment can be provided by designing the conveyor casing to be sufficiently strong enough to withstand the full explosion pressure (which can easily be done for tubular conveyors). Nitrogen inerting has also been used for providing explosion protection for en-masse conveyors. For toxic solids, explosion containment or suppression is preferable over venting to avoid exposing nearby personnel to emitted toxic material.

Flooding is a potential problem and hazard with mechanical conveyors as it is with feeders, and the causes and preventive measures for mechanical conveyors are the same as for feeders (see Section 5.3.6.1)

Two useful safety publications dealing with mechanical safety aspects of mechanical conveyors are by ANSI (1996) and Schultz (2000). The ANSI standard has been incorporated into regulations by OSHA

5.3.8.2 Bucket Elevators

Bucket elevators have experienced many dust explosions with devastating effects (e.g., in grain elevators and foodstuffs plants to name two facilities where this has occurred often). A case history below cites one such incident, followed by a discussion of hazards and preventive/protective measures.

Case History of a Bucket Elevator Explosion in a Sugar Refinery

A dust explosion in a bucket elevator in a sugar refinery is reported by Field (1982). A dust explosion caused two injuries and severely damaged the plant. The factory had been shut down for a 9-day period and the explosion occurred within two minutes of restarting the plant. Before the shutdown, all sugar dust had been removed from the pit of the elevator shaft, but during the shutdown sugar had accumulated in the pit via a leaking flap-valve. The bucket elevator ran through all 13 stories of the building, collecting sugar from ground level and transferring it to the appropriate processing equipment. On startup, the bucket elevator was under a load for which it was not designed. The strain caused a tensioning device to fail, and the bucket chain slackened, and the elevator buckets ran out of alignment. The frictional heat produced by the rubbing metal surfaces was sufficient to ignite the sugar dust suspension in the elevator shaft.

Ed. Note: Several measures can be taken to avoid this problem in the future, such as:

- Provide adequate explosion venting in the elevator casing
- Provide explosion suppression in the elevator casing
- Provide sensors to detect metal-to-metal contact inside the elevator and shut it down upon detection
- Provide a strain gauge sensor and an interlock between the chain tensioning device and the elevator motor to prevent it from starting up if the tension is not correct.

Bucket elevators are extremely prone to having fires and explosions. Because of their design, dust clouds are likely to be continuously present during their operation, particularly in the head and boot of the elevator. In addition, they have many moving parts, such as sprockets, chains, belts, and buckets, that can break or come loose; shaft seals or side plates can leak powders. The buckets are regularly subjected to impact and the belt supporting the buckets can slip on the pulleys and generate frictional heat. As a result, a source of ignition and a dust suspension can be present simultaneously, which can cause a fire or explosion. Explosion venting and suppression can be used to protect bucket elevators handling combustible and reactive solids. Explosion suppression is preferable when the bucket elevator is located inside of a building and when the solids being conveyed are toxic.

NFPA 654 (2000) has several specific requirements for the design and operation of bucket elevators.

Modern high-capacity bucket elevators, with separate delivery and return legs, have a reduced risk because of the reduced volume per unit weight of solids conveyed. Palmer (1973) recommends that the use of bucket elevators should be avoided for solids known to be readily ignited by friction (e.g., sulfur). If a bucket elevator is selected for a specific application, it should preferably be located outside of a building and supported by the outside wall.. The intake and delivery points should be isolated from the rest of the solids handling equipment by means of "chokes' or isolation valves to prevent flames from propagating from an explosion in the bucket elevator to equipment on both sides of it. For combustible or reactive solids the elevator casing should be constructed of fire-resistant materials, sufficient to retain a fire, and for toxic solids, it should be dust-tight.

Steps should be taken when designing bucket elevators to minimize the generation of ignition sources. These steps may include the provision of strong anchoring of the buckets to the belt and strong bearings for all shafts, external to the casing, provided with sensors (with alarms) for the detection of overheating. The main drive to the elevator should be external. NFPA 654 (2000) states that no bearings shall be located within the casing. Belt slip within the casing can be detected by belt speed meters, and anti-runback devices should be provided. These measures will reduce the development of friction within the casing.

Holbrow et al. (2001, 2002) present venting guidelines for bucket elevators. This methodology is based on the results from an experimental program carried out by the Health and Safety Executive (HSE) and was a collaborative effort with funding from the HSE and manufacturers and users of bucket elevators through the British Materials Handling Board (BHMB). Two bucket elevators were used in the project—a single leg elevator and a two-leg elevator. Four dusts were used, with K_{St} values up to 211 bar-m/s. Dust clouds were produced both by dust injection and by normal operation. Reduced explosion pressures due to the venting were measured, and guidance criteria were derived from the results. The recommended guidelines are as follows:

SINGLE LEG ELEVATORS: Vent openings should have an area equal to the cross-section of the elevator leg and the minimum requirement is that vents should be fitted in the head and as close as practicable to the boot. This generally means a vent should be located within 6 meters of the boot or within the recommended spacing, whichever is the lesser. The spacing between vents along the elevator is a function of the dust K_{St}, the vent burst pressure (values of 0.05 or 0.1 barg were considered), and the reduced explosion pressure (P_{red}), as listed in a table in the article.

For dusts with K_{St} values up to 160 bar-m/s, a vent spacing of 6 meters will limit the reduced explosion pressure to 300 mbarg, when the static burst pressure is 0.1 barg. For dusts with a K_{St} value of 80 bar-m/s, a vent spacing of 20 meters will limit the reduced explosion pressure to 250 mbarg.

TWIN-LEGGED ELEVATORS: Vent areas and spacing should be the same as for single-leg elevators (as given above). The static burst pressure of the vent closure should not exceed 0.1 barg. The spacing of additional vents depends on the K_{St} of the dust, as follows:

1. With dusts of low K_{St} (<100 bar-m/s) the pressures generated are not significant, and no additional vents are required.
2. Dusts with a K_{St} value of 150 bar-m/s are able to develop significant pressures, although the likelihood of explosion propagation though the elevator is low. Vents additional to those at the head and boot may be required on long elevators if the casing is comparatively weak. Graphs in the article should be used to estimate the reduced explosion pressure for a given K_{St} value and vent spacing.
3. Dusts with K_{St} values above 150 bar-m/s will propagate explosions, and vents additional to those in the head and boot are required on elevators taller than 6 meters. Graphs in the article should be used to estimate the reduced explosion pressure for a given K_{St} value and vent

spacing. The strength of the elevator should then be designed accordingly.
4. No data are available for dusts with K_{St} values greater than 211 bar-m/s.

It is essential that a bucket elevator stop quickly in the event of an explosion. This may be achieved by trip switches on vent panels, but because of the uncertainty of as to which panels may open, a trip on a single panel is not likely to be sufficiently reliable. Either a sensitive pressure switch, or switches, or trips fitted to more than one panel are recommended.

Vents on bucket elevators located inside buildings should not open inside the building unless they are fitted with a device that extinguishes the flame and collects the hot particles, for example, a Q-Rohr™ or FlamQuench II™ device (see Section 6.6.1). Otherwise, the vents should be fitted with vent ducts to convey the fireball and hot particles outside of the building to a safe location where they will not impinge on other equipment or personnel.

Field (1982) and Palmer (1973) present additional discussions of bucket elevator protection.

5.3.9 *Pneumatic Conveyors*

Pneumatic conveying systems have a high rate of risk for fires and explosions (they usually occur in the downstream equipment) for the following reasons:

- Static electricity is generated by contact between particles themselves and between particles and the pipewall.
- Dust concentrations within the explosible range can arise at the delivery point where the solids are separated from the conveying gas (silos, cyclones, baghouses).
- Heated particles which are created during grinding or drying may be carried into the pneumatic conveying system and fanned to a glow by the high gas velocity. These particles can then cause an ignition in the storage or collection system at the end of the pneumatic conveyor. Tramp metal in pneumatic conveying systems may also cause frictional heating or sparks as it passes though the system.
- Segments of conveying piping or tubing can be isolated and sparking is possible between line segments and nearby conductive pieces at different potentials.
- Charged powder can leak from joints to the atmosphere and electrostatic sparking can occur resulting in an explosion.

The following design and operating recommendations and practices for pneumatic conveying systems should be considered. These are based on information from several sources such as Barton (2002), Kirby (2001), NFPA

654 (2000), Palmer (1973), and design guides from several operating companies.

1. If the conveying system is located inside of a building, and air is used for the conveying, vacuum systems are generally preferable over positive pressure systems to minimize dust leakage into the building.
2. For positive pressure systems the conveying piping should be air-tight to prevent the escape of dust from the system into the surrounding area where it might present a fire, explosion, or health hazard. If operating under a negative pressure the system should be air-tight to prevent pulling in air or other contaminants.
3. The conveying system should be strong enough to remain intact and tight under normal operating conditions, including vibration; and in some cases to withstand or contain explosive pressure. Pipe flanges, rather than standard couplings, should be used to provide robustness against explosion pressures that could develop during a deflagration. For most organic dusts and dilute phase conveying at low initial pressure, a piping specification that will safely contain 150 psig is adequate.
4. Pneumatic conveying lines should be made of electrically conductive materials (e.g., metals), and accumulation of electrostatic charges must be minimized by proper grounding and bonding. This includes jumpering across nonconductive insulators such as sight glasses, gaskets, and most types of couplings. Some Victaulic® type couplings are designed to provide continuity across the coupling. Bolted-flange couplings, even with Teflon® coated bolts, provide adequate electrical continuity to protect against electrostatic discharges, but may not protect against low-voltage sources such as stray currents. Where jumpers are not provided across couplings, continuity checks should be done at least yearly and each time the piping is reassembled after being taken apart. Electrically isolated metallic objects within the system may accumulate dangerous static charges. Wire braid within rubber-covered transfer hose may act as a static accumulator. The line segments of the conveying line should be bonded together and the bonded units grounded if it is possible to isolate sections by nonconductive gaskets at flanged joints. Bonding conductors between segments should be visible so that visible checks can be made rather than using resistance measurements for other than periodic preventive maintenance checking.
5. The electrical installation must meet the electrical area classification imposed by the solids being handled as well as the surrounding environment.
6. Materials of construction compatible with the solids being conveyed and the surrounding environment must be used.

7. Screens, magnets, and metal detectors should be installed for the detection or removal of any foreign material that might create hazards in the system.
8. Appropriate special materials, such as nonferrous metals (e.g., aluminum), to minimize mechanical sparking in the event of misalignment or failure of moving parts within the process stream, should be considered.
9. High conveying velocities (3000 to 4000 ft/min) will minimize particle settling and therefore reduce frequency of cleaning. Joints and openings should be suitably located to facilitate cleaning or unplugging.
10. An adequate program of maintenance and inspection must be instituted to assure proper alignment of drives, proper clearances, dust tightness, electrical grounding and bonding, and control of ignition sources. This is now a retroactive requirement of NFPA 654 (2000).
11. System air intakes should be located outside of buildings and in areas where flammable vapors are not present. Also, consideration should be given to providing a flammable gas detection system (alarm and shutdown interlock) in the air inlet piping where flammable vapors might occur.
12. The conveying system should be designed to minimize passage of dust through the fan or blower casing (usually a guard filter is installed ahead of the fan or blower). Particles impacting on the impeller may cause it to experience metal fatigue and subsequently fail. Misalignment of the impeller may cause it to rub against the casing, creating a hot spot. Both of these failures are may create ignition sources.
13. Blower, fan, compressor, or vacuum pump motors should be provided with an overload trip.
14. The need for installation of gas pressure relief valves and check valves should be considered.
15. Consideration should be given to the need for installing detectors at the feed point to sense and extinguish hot metal and/or sparks which will shut down the system if these are detected.
16. At start-up, the conveying gas supply should first be established and then solids fed in at its optimum rate as quickly as possible.
17. At shutdown, the solids supply should be stopped quickly and the conveying gas continued at the optimum rate until all solids have been removed.
18. Pneumatic conveying lines should be designed and laid out so that they do not contain any dead areas where solids can accumulate and subsequently plug the lines.

Pneumatic conveying systems, and connected equipment, can be protected from fires and explosions by the following methods:

1. *Venting*: The design and installation of explosion vents for pneumatic conveying systems should be in accordance with NFPA 68 (2002). If the pneumatic conveyor is located inside of a building, vent ducts should be attached to the vents to direct the fireball and unburned solids to outside of the building to a safe area (where they will not impact on equipment or personnel). Alternatively, conveying ducts can be vented inside of a building if a flame quenching device (such as the Q-Rohr™ or FlamQuench II™) is attached to the vent opening.
2. *Suppression*: Explosion suppression systems for pneumatic conveying systems shall be designed, specified and installed in accordance with NFPA 69 (2002).
3. *Pressure containment*: The piping system can be designed to withstand the maximum explosion pressure that can develop (based on test data). It also may be possible to design the equipment at the downstream end (cyclone, dust collector, and receiving vessel) for explosion containment. If not, then these items of equipment will have to be provided with other explosion protection systems (e.g., venting, suppression).
4. *Deflagration isolation:* The pneumatic conveying system can be isolated from the downstream equipment by the installation of fast-acting valves or flame diverters so that a deflagration occurring in downstream equipment will not propagate backwards into the conveying system.
5. *Spark detection and extinguishing systems:* These systems can prevent the spread of sparks or embers from one part of a pneumatic conveying system to another and minimize the possibility of a fire occurring.
6. *Use of an inert conveying gas:* The potential for a dust explosion to occur in a pneumatic conveying system is minimized if an inert gas (usually nitrogen) is used as the conveying gas rather than air. Because nitrogen is costly compared to air, this is usually done in a closed-loop system so that nitrogen is only lost when makeup is required. Makeup nitrogen is determined by an oxygen analyzer which constantly monitors the limiting oxidant concentration (LOC) in the loop and admits fresh nitrogen when the LOC goes above a pre- selected value. Crouch (2000) presents a detailed discussion of the design of closed-loop, inert gas systems for dilute phase and dense phase pneumatic conveying systems. Sleicher (1998) also discusses the use of closed-loop pneumatic systems for the safe handling of polymer pellets. NFPA 654 (2000) now requires that if an oxygen monitoring system is used, it shall be installed in accordance with ISA S84.01 (1996).

Several inherently safer approaches to pneumatic conveying system safety are:

1. For toxic solids use a vacuum pneumatic conveying system so that fugitive emissions into the surrounding atmosphere will be minimized.
2. For friable solids use a dense phase pneumatic conveying system, which uses lower conveying velocities, so that fines production will be minimized, reducing the potential for an explosion to occur.
3. Use an inert gas rather than air for conveying of dusts if the MIE of the sub-200 mesh fraction is less than 10 mJ, and the sub-200 mesh fraction constitutes more than 5 wt. % of the as-received product. This is one company's practice (Kirby, 2001).

Jaeger (1997) presents another company's practice, which states that there is no need to consider ignition hazards due to electrostatic charges in pneumatic conveying systems if all of the following conditions are present:

- The conveying pipe diameter is less than 1 meter.
- The MIE of the particulate solids is greater than 1 mJ.
- The piping is electrically conductive and is grounded.
- The piping is not lined with an insulating material of more than 2 mm thickness or with a breakdown voltage of greater than 4 kV.

Additional information on safety aspects of pneumatic conveying systems is presented by Palmer (1973), NFPA 650 (1998), and NFPA 654 (2000). Britton (1999) discusses electrostatic hazards of pneumatic conveying systems and protective measures, as does Glor (1988).

5.3.10 Portable Containers

The main safety hazard with portable containers (multiwall paper bags, fiber drums, FIBCs, and RIBCs) is the accumulation of electrostatic charges and their discharge as the containers are filled (conical pile discharge) and emptied. The hazard consists of the possibility of charge accumulation on the product handled (if the material is an insulating one), on the portable containers, on parts of the filling and emptying equipment, and on the operators. In the case of ignition hazards caused by charges accumulating on the solids, filling operations have the highest hazard potential for the following reasons: in the course of the filling process, the solids pass through a dispersion phase (filling by gravity, pneumatic transport, etc.). The particle separation processes in the dispersion/transport phase cause the separation of charges. Subsequently, the solids and thus the charged particles accumulate in a small volume, and the charge cannot drain quickly to ground, even if the receiving container is conductive and properly grounded. Thus, a high density of charge, that is, a strong electric field, can exist in this volume.

Many powders and granules are not electrically conductive—the electrical resistivity is greater than 10^{10} ohm-meters. When these powders or granules are filled into a container with a plastic liner, the charge on the solids is maintained for an extended period of time (multiminutes or hours). This allows charges to be induced on nearby isolated conductors over time. The release of the energy stored on these isolated conductors to a grounded part in the presence of a dust cloud of sufficient density has the potential for causing a dust explosion.

To properly evaluate the hazards of filling and emptying portable containers, data on the following must be known:

- Specific resistivity of the powder (ohm-meters)
- Minimum ignition energy, MIE (mJ)
- Median particle size (mm)
- Volume and shape of the container (m^3 or gallons)
- Volatile (flammable vapor or gas) content

Jaeger and Siwek (1999) present matrixes showing the safety measures that should be taken as a function of the items listed above for container volumes of less than 0.2 m^3 (55 gal), between 0.2 m^3 and 2.0 m^3 (55 and 550 gal), and greater than 2.0 m^3 (550 gal).

The hazard can be more serious if the container has a plastic (nonconductive) liner so that the charge cannot be conducted to ground. Glor (1988) states that brush discharges can be produced on a plastic liner of a bag. Brush discharges cannot ignite a powder by itself, but are able to ignite flammable vapors in the vicinity of the powder. Therefore, if the powder is wet with a flammable solvent, when the container is emptied, these charges can act as an ignition source and cause an explosion. To avoid this, the equipment into which the charged solids are discharged must be properly grounded and bonded, and it may be necessary to inert the vessel in which the solids are being added. Where flammable vapors are present, only properly grounded static-dissipative or conductive plastic bags are acceptable. However, sewn or glued-in nonconductive plastic liners are acceptable in paper bags or fiber drums provided that the bag or drum is grounded, and there is no possibility of the liner falling out of the container (Britton, 1999 and NFPA 77, 2000).

Another hazard is the failure (bursting) of multiwall paper bags and FIBCs from exposure to a fire. When a bag or FIBC fails, there is the potential for release of large amounts of powder, which can form dust clouds, and, if ignited, could cause a severe dust explosion.

Specific discussions of the hazards associated with containers and preventive/protective measures are presented below.

Multiwall Paper Bags

Multiwall paper bags can be ignited and burned up by a fire already present in an area adjacent to them or by such operations as shrink-wrapping. This can result in bag failure (bursting) and emissions of powders and possible dust cloud formation which could lead to an explosion. The following case history illustrates this potential problem.

Case History of a Fire Involving Pallets of Paper Bags

Tyldesely (2004) reports an incident involving 25-kg paper bags containing rubber crumb, stored on pallets. A pallet loaded with 46 bags in a stack 2 meters high was ignited by a flame gun used for shrink-wrapping, but the operator did not immediately notice the fire. As the pallet was being moved using a forklift truck into a storage area, an extremely rapid fire growth then occurred. The fire at this stage had spread to involve the other pallets in the area loaded with bags that totaled 70 metric tons of rubber crumb.

Subsequent tests at the Health and Safety Laboratory (HSL) with piles of bags of rubber crumb showed that under certain circumstances, a fire which starts at the base of a pile of bags made from combustible packaging, can grow up the outside of a stack, causing a series of bags at higher level to fail, and to allow powdered material to flow into the fire. Some of this powder burns before it reaches the floor. This results in a fire, which grows much more rapidly, involving neighboring bags, than would be possible if the burning powder formed a simple heap on the floor. Further HSL tests were undertaken to determine how other powdered products behave in similar circumstances. Not all the tests produced the rapid fire growth that is a consequence of burning of powder that is either falling from a burst bag, or being carried upward on convection currents from the fire. Lower fire growth rates sometimes occurred because the powder softened and would not flow, because it charred and did not flow, or because the inherent cohesive forces within the powder did not permit free flow. As a result of thee tests, it was postulated that under slightly different conditions, any of the products tested could be capable of burning effectively as a dust cloud and cause the rapid fire growth rate seen in the original rubber crumb incident. For a large range of powdered products, it must be assumed that a tall stack of bags, or a release from a tall single container caused by fire at low level, can create a fire capable of growing very suddenly, with a risk to people in the vicinity.

This incident and the subsequent tests show the great need to minimize ignition sources in areas where large quantities of stacked bags are stored, and to be extremely cautious when using flame guns for shrink-wrapping. All easily combustible waste materials must be kept out of the area near the stacks.

Fiber Drums

A major hazard of using fiber drums is the possibility of dust cloud formation and ignition (usually from an electrostatic discharge) when emptying a drum into a process vessel. The following case history illustrates such an incident.

Case History of a Dust Explosion That Occurred during Dumping of Powder from a Drum with a Metal Chime into a Vessel

Pratt (1997) reports the following incident:

An operator was dumping a "dusty" powder from a polyethylene drum into a vessel through a manhole. The drum had a metal chime which was provided to protect the drum from the rigors imposed by its handling. The vessel into which he was dumping the powder was made of metal and was well grounded. As the powder was dumped it slid down the interior of the plastic drum and into the vessel. During dumping, the metal chime on the lip of the drum was ungrounded and free-standing. After dumping, the operator withdrew the drum toward himself and touched the metal chime to the metal vessel. As this metal gap was closed, an incendive spark occurred which ignited the dust cloud.

Ed. Note: To avoid this problem several things could have been done, such as (1) ground the metal chime as well as the metal vessel, and (2) inert the closed vessel before and during the charging of the powder into it, so that the oxygen concentration in the vessel would be below the LOC.

Therefore, when dumping powder from a fiber drum directly into a process vessel, the following actions should be taken:

- Make sure that the metal chimes on the drum are properly grounded.
- Gently dump the drum's contents so that dust cloud formation is minimized.
- Inert the vessel to an oxygen concentration below the LOC of its contents before the drum is emptied into it.
- Use a closed charging system (see Section B19) rather than directly dumping the drum's contents into the vessel.

These precautionary measures are especially important if the process vessel already has a flammable liquid (and associated vapors) in it.

Flexible Intermediate Bulk Container (FIBCs)

Hazards of using FIBCs include electrostatic discharges and failure (busrting) of an FIBC due to a fire, with subsequent release of large quantities of powder into a room, and possible dust cloud formation and explosion.

The following case history describes an incident that occurred during filling of an FIBC.

Case History of a Fire That Occurred during FIBC Filling

The following case history is reported by Philiph (2002):

A company routinely filled FIBCs with a powdered food additive. After a tramp metal detector was installed in the filling nozzle, a series of incidents occurred in which there was an ignition and flash of the powder in the FIBC as it was filling. Although there was no major damage, the incidents charred the material and frightened the workers. Investigation of the incidents revealed that the body of the tramp metal detector was a nonconductive plastic tube that was graphite-coated on the outside surface only. This allowed a static charge to build up on the inside of the plastic tube until it discharged to an adjacent conductive surface and ignited the powder flowing past it (this was most likely a propagating brush discharge). The problem was resolved by replacing the nonconductive plastic tube with a conductive plastic tube and bonding it to the rest of the filling apparatus.

Ed. Note: When installing a metal detector before a FIBC it should be specified to have all components made of conductive material and the detector should be properly bonded and grounded. Also, nonconducting tubing or liners should not be used if they have a breakdown voltage of greater than 4 kV.

The choice of which FIBC should be used and the protective measures that should be taken depends on the MIE of the solids and whether there is a flammable atmosphere inside and outside of the receiving vessel. Table 5-3 lists protective measures that should be taken for FIBCs depending on the solids MIE and operating environment (Jaeger, 1997).

For Type B FIBCs, the following requirements are recommended:

- Any inner polyethylene coating/liner should not be thicker than 20–30 microns.
- No plastic enclosing wrapper on the outside of the FIBC should be used.

For Type C FIBCs, the following requirements are recommended:

- The FIBC must be appropriately labeled to indicate its conductivity and the grounding requirement during charging and discharging.
- The FIBC must have a clearly marked area for the attachment of the grounding clamps.
- The lifting straps must also be made of conductive material and have a leakage resistance of less than 10^8 ohms to the FIBC body. The bag should be permanently grounded during the whole period when the bag is being filled or discharged.

TABLE 5-3
Protective Measures Required for FIBCs (From Jaeger, 1997)

Bulk Material	Environment		
	No Explosible Atmosphere	Explosible Dust Atmosphere	Flammable Gases or Vapors
MIE[a] > 1 J	A	B	C
3 mJ < MIE[a] < 1 J	B	B	C
MIE[a] < 3 mJ	C	C	C

[a]MIE measured without inductance in the electrical circuit.

A = No special requirements

B = Breakdown voltage of the FIBC wall material must not exceed 4 kV in order to prevent propagating brush discharges.

C = The bag material, including the slings, must be electrostatic conductive. The resistance to the ground measured at any bag location (inside and outside) must be less than 100 MΩ (10^8 Ω). The flexible bulk bag must have a grounding tap. The conductivity and the necessity for gorunding must be clearly marked.

Ebadat et al. (2002) present data that indicates that the leakage resistance of 10^8 ohms, as specified for Type C FIBCs, may not be suitable for Type C/D FIBCs. This is because a Type C/D FIBC having a resistance-to-ground of less than 10^8 ohms is likely to give rise to incendive spark discharges when used as a Type D FIBC (electrically isolated from ground). The authors discuss a "surface voltage" criterion for qualifying FIBCs as Type C/D. It is suggested that an FIBC having a resistance-to-ground exceeding 10^8 ohms may nevertheless qualify as Type C/D if, when grounded, the surface potential during filling and emptying does not exceed approximately 4 kV.

Rigid Intermediate Bulk Containers (RIBC)

Conductive or static-dissipative RIBCs should be properly grounded when being emptied to dissipate any electrostatic charges that may have accumulated during filling. If the RIBC is made of plastic this may not be possible, and consideration should be given to inerting the vessel into which the RIBC is being emptied. If a RIBC is mounted on wheels, the wheels should preferably be made of conducting material.

5.3.11 Portable Container Emptying (Unloading) Equipment

5.3.11.1 Drum and Box Dumpers

There are two main problems when using drum and box emptying (unloading) equipment:

1. Electrostatic charging of the solids as they flow out of the drum or box into a reactor or slurry mixing tank.
2. Emissions of fines (dust) into the atmosphere and causing operator exposure.

Electrostatic charging of the solids is increased if the drums or boxes have plastic liners. If the receiving vessel is made of metal this problem can be minimized by proper grounding of all parts of the vessel, and inerting of the vessel. If the vessel is glass-lined or plastic-lined, grounding is not effective, and a closed charging system should be used if flammable vapors are present in the vessel. Usually, volumes of drums and boxes are too small for propagating brush discharges to form, and brush discharges are not incendive to combustible dust clouds (Gravell, 2004).

If the contents of a drum or box are charged to the vessel through an open manhole, emission of dust fines often occur. To minimize the emission of fines a ventilated chute funnel (see Figure B-88) should be used. This funnel is inserted through the manhole and fits snuggly in the opening and is connected to an exhaust system so that any fines emissions are contained and conveyed to a control device such as a baghouse of scrubber. This minimizes operator exposure to health-hazardous solids and the operators should wear specified personal protective equipment, determined from personnel protective equipment assessment, as called for by OSHA 1910.132.

Proper selection of electrical equipment should be in accordance with the appropriate National Electrical Code (NEC) area classification scheme.

An article describing and discussing the use of a drum dumper and a ventilated charging funnel, and their emissions control effectiveness, is presented by Betz (2000). The use of a ventilated charging funnel helps to minimize fugitive emission of fines into the atmosphere. However, when solids are added to a vessel, air also enters the vessel with the solids from the portable container and from induction of air from outside the vessel by the falling solids. If the vessel is already filled with a flammable liquid, or if the solids are wet with a flammable solvent, then there is a greater potential for a fire or explosion to occur (a hybrid mixture now exists). To minimize this hazard potential, the vessel should be purged of air before the liquid and solids are added, and a small flow (a sweep) of inert gas should be applied during the dumping of the solids. The amount of sweep air should be determined by tests. As air can still enter the vessel, it is good safety practice to check the oxygen content of the vessel (to make sure it remains below the LOC of the hybrid mixture) by periodic or continuous oxygen analysis. The inert gas is usually nitrogen, but carbon dioxide, which is heavier than air, can also be used as it is more effective in preventing air ingress.

It should be noted that adding solids through an open manhole into a vessel containing a flammable liquid using a drum or box dumper is inherently unsafe and should be discouraged. A better way to do this is to use a closed charging system (see Section B19) or follow the recommendations of NFPA 77 (2000) which advocates that manual addition of solids through an open port or manhole to a vessel containing a flammable liquid be done only in 25 kg batches. It is also recommended that in this operation the operator be grounded if the MIE of the dust is <30 mJ (Gravell, 2004).

5.3.11.2 Bag Dump Stations

The hazards associated with bag dump stations are essentially the same as with drum and box dumpers. Electrostatic charging of the solids occur as the solids leave the bag, especially if the bag has a plastic liner. Proper grounding of the bag dump station will minimize this hazard. In proprietary units the integral filter and fan control the emissions of fines generated by the bag dumping. If the solids are toxic, consideration should be given to providing additional local exhaust ventilation near the dump station or using an automatic bag opener (slitter). Provide the operator with specified personal protective equipment. Proprietary and "homemade" units can be fitted with bag compactors so that the operator does not have to handle the empty bags.

The fan motor must be specified with the correct NEC area classification.

An evaluation of the dust control effectiveness of bag dump stations is presented by Heitbrink, McKinnery, and Rust (1983).

5.3.11.3 Vacuum Pneumatic Conveyor Unloading System

Electrostatic charging is a concern and proper grounding and bonding of the system components are required to minimize this hazard. Nonconductive, plastic piping should not be used for pneumatic conveying unloading systems. For unloading of open drums or boxes, it is one company's practice to ground the operator if the MIE is less than 25–30 mJ (Nelson, 2002). The grounding of operators should also be considered if flammable vapors are present in the area. In addition, it is recommended to have an area where the portable container is positioned to be unloaded that has an electrostatically conductive floor and the operator must wear conductive shoes, or the operator has a conductive tether attached to him and the ground.

To provide an extra measure of safety, inert gas (usually nitrogen) can be used instead of air as the conveying gas, and the nitrogen is circulated in a closed loop. When ingress of air occurs, nitrogen is vented from the system via a bleed valve which can be controlled by a pressure controller or oxygen analyzer, or on a calculated volumetric flow basis. Some companies use an oxygen analyzer connected to the closed loop system to monitor the oxygen level and automatically admit enough nitrogen to bring the oxygen concentration down below the limiting oxidant concentration (LOC). A rupture disk often is provided to vent the system should a deflagration occur.

5.3.11.4 *Flexible Intermediate Bulk Container (FIBC) Unloading Systems*

Dust cloud and/or flammable vapor fires and explosions may be possible from operations involving unloading of FIBCs if the solids are reactive or combustible, and especially if the solids contain residual flammable liquid, or are being discharged into a vessel containing a flammable liquid. Britton (1993) reports a number of accidents with FIBCs. One of these is presented below to illustrate the possible problems.

Case History of an Incident Involving FIBC Unloading

An antistatic FIBC was used to transfer a vinyl resin to a 6000-gallon mixing tank containing a xylene–MEK mixture. The FIBC was woven polypropylene with a 1 mil internal polypropylene coating. It was equipped with thin conductive wires running lengthwise through the spout and connected to a bare stranded aluminum wire and alligator clip. The FIBC was hoisted above the tank using a fork lift and the resin was dumped through a circular port on a hinged tank cover. The tank was inerted at 15 SCF/min with combustion gas (mainly CO_2) introduced through a flow meter. There was no independent venting of displaced vapor and the tank lid was not gastight. The operator reported that the ground wire was missing from the FIBC but proceeded to dump the FIBC anyway. The hinged portion of the tank lid was open allowing solvent vapor to escape freely into the operating area. Accounts at this point differed as to whether the fire occurred immediately or after the FIBC was about three-quarters empty. In any case, the operator was standing a few feet from the tank and turned away when he observed a flash. The side of his head was singed, the back of his neck was burned, and he received second degree burns on his right arm. The flash was outside the tank and the tank contents did not catch fire. The source of ignition was assumed to be a spark from the ungrounded FIBC during emptying. Since it was known that the vinyl resin had a very high MIE in air, it could be assumed that flammable vapor was a major contributor to the ignition process. Although the operator was not grounded, he was not considered a likely source of a spark owing to his location. Since the operation involved making lacquer for can coatings, antistatic shoes would probably have been ineffective owing to the possibility of a film of lacquer on the floor around the tank.

Ed. Note: The importance of grounding FIBCs was not properly recognized since a false sense of security had been instilled after long experience dumping from 50-pound bags. The powder had only been recently started being supplied in FIBCs rather than bags. Recognition of the increased ignition risk associated with the large transient mass flows from FIBCs would have lead to more thorough training on FIBC grounding requirements that could have averted the accident.

The type of the FIBC used depends on the hazard potential of the solids and has an effect on the safety methods needed during unloading (see Section B11 for a discussion of the various types of FIBCs and their characteristics, and see Table 5-3 for protective measures required). If a FIBC has a liner (they are often required to provide resistance to intrusion of air or moisture, or to meet regulatory requirements), the liner must be analyzed for its potential to create a hazardous situation. Conductive liners inside the FIBC must be grounded; if they are not, they represent an isolated conductor which can become electrostatically charged and subsequently pose an ignition risk.

For free-flowing particulate solids, FIBCs can empty quickly, thus increasing electrostatic charging and subsequent ignition.

FIBC unloading hazards arise primarily from electrostatic charges, as follows:

- From brush discharges when flammable vapors are greater than 20% of the LFL.
- From propagating brush discharges when the FIBC bottom rests on a metal surface.
- From inductive charging of nearby conductive materials as well as people.

Some general preventive/protective guidelines for avoiding accidents with equipment unloading FIBCs is to ensure that:

1. All parts of the installation must be conductive and properly grounded.
2. Use the appropriate type of FIBC as discussed in Section B11.

Where a combustible solid is emptied from a FIBC into a vessel containing a flammable liquid, the vessel will often contain flammable concentrations of vapor. Flammable vapor at only a fraction of its LFL may significantly increase the ignition sensibility of a flammable dust suspension. Although it is common practice to unload a FIBC directly into such a vessel, NFPA 77 (2000) recommends that only batches up to 25 kg be added to a vessel containing a flammable liquid. Transfers should preferably be done via an intermediate grounded metal hopper with a rotary valve or its equivalent (e.g., gate lock valve), keeping both the receiving vessel and the hopper inerted throughout the transfer. These precautions isolate the flammable vapor from the FIBC and the operator. The vessel vent line should be appropriately sized to minimize flammable vapor entry into the hopper.

If the solids are toxic, there should be a containment sleeve from the bag unloading spout to the nozzle of the receiving piece of equipment to prevent emission of toxic dust.

Britton (1993, 1999) presents a good discussion of FIBC hazards and protective measures. Other good sources of information on FIBC unloading system hazards are the papers by Gravell (2001) and Luttgens et al. (2001).

5.3.12 *Portable Container Filling Systems*

5.3.12.1 Small Bag Filling Systems

The filling of small bags, both open top and valve bags, can result in electrostatic charging of the solids, but the potential for an explosion is not usually great if the solids do not contain any flammable solvents. Paper bags are unlikely to produce spark hazards, except in the case where either paper or plastic bags contain a metal layer applied either as a surface film or as an inner-plied foil layer (these are sometimes used as moisture barriers). The conductive metal layer can become charged during powder flow and create a spark ignition hazard unless special grounding measures are taken (Britton, 1999).

One safety (health) problem that does exist is that fugitive dust emissions occur during filling, especially with open top bags. As a minimum, local exhaust ventilation near the bag top should be provided to contain the fugitive emissions. Various types of capture hoods, booth-type hoods, and complete enclosures can be used to provide better dust emissions control (Cooper, 1986). A built-in vacuum system along the bag filling and moving station is often provided to remove spilled solids. If the solids are toxic, the operator also may have to wear personal protective equipment.

5.3.12.2 FIBC Filling (Packing) Systems

FIBC filling systems can also result in electrostatically charged solids entering the FIBC, especially those with plastic liners, but fires and explosions have rarely occurred. It is possible to inert the FIBC with nitrogen before starting filling so the potential for ignition from a static spark is minimized. Electrostatic hazards during filling can occur from the cone of a pile during rapidly filling and when the pile is greater than 1 m^3, the MIE $\leq$10 mJ, the material contains substantial amounts of coarse (particle diameter >1 mm) and fine fractions, and the pile is nonconductive (Glor, 1988).

When filling FIBCs with toxic solids, the operator may require personal protective equipment, and the filling machine should be provided with good ventilation to control and minimize emissions of fines to the surrounding vicinity.

5.3.12.3 Drum Filling (Packing) Systems

Filling of drums can also result in electrostatically charged solids entering the drums, but here too, fires and explosions have not normally occurred. Fugitive dust emissions are quite common due to the relatively large open top. To minimize dust cloud formation and fugitive emissions problems, especially with toxic solids, a bottom-filling probe that minimizes dust cloud formation and an exhaust ventilation system can be used. Designs for the ventilation of drum filling systems are presented in the ACGIH Manual (1998). Here, too, operators may require personal protective equipment.

The metal chimes on the drum ends may have to be properly grounded as they can be isolated conductors. Normally, a dedicated grounding connection to the drum is not required if the conveyor upon which the drum sits is conductive and grounded. However, this should be confirmed by measurement. Grounding of the operator should be considered if the dust MIE is <30 mJ and there is a potential for exposure of the operator to a combustible dust cloud (Gravell, 2004).

5.3.13 Samplers and Sampling Systems

Fires and explosions do not usually occur during sampling of particulate solids from process streams and equipment unless the solids are pyrophoric. Pyrophoric solids should be sampled using an inerted sample container.

However, when a solid is toxic, it is critical that the exposure to the operator be minimized. This can be accomplished by automatic samplers installed in a specially designed enclosure that isolates the operator from the toxic material. Such a system is described in a NIOSH report (SRI, 1980). The sampling device is contained within a ventilated box. A rubber dam on the front of the box is split so that the operator may push a sample container through into the box under the delivery spout of the sampler. The box is vented by a local exhaust system and the vent is discharged to an air cleaning system. This permits sampling with minimal potential for operator exposure and prevents the dumping of large amounts of powder into the workplace during the sampling process. For toxic solids, it may be prudent for the operator to wear personal protective equipment.

If a sampler that is inserted into a stream of flowing solids is comprised of metal and plastic parts, the metal parts should be bonded together and the sampler properly grounded. If this is not done, the sampler can be an isolated conductor and has the potential for an electrostatic discharge. in addition, if there is a sensor on the sampling device, then the wiring should be checked to make sure that it is properly connected. If the wiring becomes disconnected, it could act as an ignition source and cause a dust cloud explosion.

Wang (1986) presents a discussion of the health and safety aspects of solid material sampling.

5.3.14 Screens and Classifiers

The motion imparted to solid particles during screening can result in them becoming electrostatically charged and also results in generation of dust clouds. Static charges can not only act as an ignition source, but also can lead to screen blinding and significantly reduce the efficiency of the screen. Therefore, all metal components of the machine (screens, frame, etc.) must be properly grounded and bonded. This will remove the charges from the machine itself, but a residual charge may persist on the solids. Jaeger (1997) indicates that, in general, when sieving even with very easily ignitable particulate solids, the appearance of ignition sources can be discounted if the following conditions are present:

- All parts of the equipment ate grounded, especially the screens, which in many cases are mounted elastically, and are therefore insulated or are clamped between two insulating gaskets.
- The relative velocities caused by the movement (shaking, rotation, oscillation, etc.) are less than 1 m/s and jamming and heating by foreign objects to the ignition temperature of solids need not be taken into account.
- The receiving vessel connected to the screening equipment is protected against fires and explosions.

Insulating sieve materials may be used as long as the MIE of the solids is greater than 3 mJ (Jaeger, 1997).

Palmer (1973) recommends that for dusts of low MIE, the operator should use antistatic footware and clothing. However, antistatic clothing may not be necessary according to a recent European standard (CENELEC, 2003). This standard states that "In spite of the fact that modern clothing, made from synthetic textiles, can readily become electrostatically charged it is not, in general, an ignition risk providing that the wearer is earthed by means of suitable footwear and flooring. However, clothing should be as close fitting as practical and should not be removed or unfastened in areas where there could be flammable atmospheres (e.g., Zone 0, Zone 1, Zone 20 and Zone 21)." See Section 6.4 for definitions of the various zones used in Europe for electrical area classification. Gravell (2004) recommends that operators grounding should be required when the potential exists for operator contact with a combustible dust cloud having an MIE of <30 mJ.

Stone (1987) recommends several methods for overcoming static problems in vibratory screen separators.

Shaking, vibrating, and oscillating screen separators cannot be hard-piped up to upstream and downstream equipment, and are connected to these by flexible connections (often called "socks"), which are a "weak point" in the system. If an explosion should occur, these "socks" would burst and the fireball and burning solids would be ejected into the workplace surrounding the screen separator, and a secondary explosion might occur. Therefore, consideration should be given to locating these machines in an isolated room at an outside wall of a building, and the wall should have explosion vent panels designed in accordance with NFPA 68 (2002). It is good practice to provide some means of detecting when a flexible connection ("sock") fails. This possibly may be done by installation of position monitors (limit switches) at the top and bottom of the flexible connection to indicate when one or both of the clamps holding a "sock" in place has broken or become loose and the "sock" has become detached. Two other possibilities are as follows:

1. Installation of closed circuit TVs in rooms where it is preferred that operating personnel do not enter.
2. Installation of a sensitive pressure sensor in front of the "sock" to detect a pressure increase due to the ejection of particulate solids.

Centrifugal sifters, because of their manner of operation do not usually shake, vibrate, or oscillate, and can be hard-piped, usually, to upstream and downstream equipment. They should also be properly grounded and bonded.

If an explosion were to occur in a screening device it could propagate to the upstream and downstream equipment, or vice versa, if an explosion were to occur in upstream or downstream equipment, it could propagate to the screening equipment. Therefore, if a hazard analysis indicates that these explosion scenarios were highly likely to occur, consideration should be given to protecting the screening equipment on both sides by isolation devises (e.g., fast-closing valves, rotary valves, deflagration suppression, etc.) as per NFPA 69 (2002).

NFPA 654 (2000) presents some safety criteria for screening equipment. It states, however, that screens and sieves shall not be required to have explosion protection. This is because, as a practical matter, screens are difficult to protect against explosions by deflagration venting or suppression.

For toxic solids, it may be necessary to enclose the screen separator in an enclosure operated under a slight negative pressure (connected to a dust control system) to prevent escape of dust into the workplace which could expose the operator to a health hazard.

Because of the severe mechanical stresses to which these machines are subjected, they should be of strong construction and all components subjected to particularly severe stress, such as bearings, should be outside of the working parts of the unit, and kept dust-free. This is to prevent overheating and ignition of dust which could deposit on them. The drive motor must also be specified in accordance with the appropriate NEC electrical area classification.

5.3.15 *Silos and Hoppers*

Fires and explosions in silos and hoppers have occurred quite frequently over the years, especially in agricultural grain storage facilities. Silos also fail due to errors in design, construction, and utilization.

Eckhoff (2003) describes a number of explosions in silos, two of which are presented here.

Case History of a Grain Terminal Explosion

An explosion occurred at a grain terminal in St. Joseph, MO in April 1980. One person was killed and four injured, and the material damage was estimated at $2 million. The explosion probably started in the dust cloud in one silo of a series of silos that was used for receipt and delivery of grain. The probable ignition source was an electric arc between the electric wires of the lower level indicator in the silo. Repeated filling and discharge of grain had pulled the level indicator from the wall and the electric arc occurred between the bare wires that had subsequently been pulled out of their conduit. There was severe structural damage to almost all of the silos in the head house and moderate damage to most of the head house structure. Most of the head house silo roofs were blown up, destroying the spout floor and the top of the cleaner floor. Rupture of the silos around the edge of the head house caused failures in the outside wall. The casings of all bucket elevators, steel as well as concrete, had opened up in many places. A silo complex comprising 18 cells suffered severe explosion damage to the gangway connecting it to the head house, to the gallery, to the far end of the tunnel, and to a small group of silos centered around an air shaft approximately one-third of the way along the gallery. At the location of the air shaft, the gallery roof and wall had been completely destroyed.

Beyond this point the explosion damage to the gallery was still significant, but not as severe. The exterior concrete silo walls had been extensively shattered, leaving in many places only the reinforcing rods. Concrete fragments from this area of the plant had been thrown about a hundred meters into the adjacent railroad yard.

Case History of an Explosion in a Silo Storing Rape Seed Flour

An accident occurred in a silo in Stavanger, Norway in November 1985 that was storing a solid organic material. The incident was not primarily a dust explosion, but an explosion of combustible gases released from the organic solid during self-heating. The explosion occurred in a fairly modern reinforced concrete silo complex used for storage of various feed stuffs. Pellets of Canadian rape seed flour had been stored in one of the silos for some time when it was discovered that the material in the bottom part of the silo had become packed to a solid mass and could not be discharged through the silo exit nozzle. Some time later, one week before the explosion, flames were observed in the silo. The fire brigade was called and covered the pellets in the silo with foam from above. Various unsuccessful attempts were then made to discharge the pellets mass at the silo bottom. During this phase there was considerable development of smoke, which mixed with the air not only in the silo in question, but also in the silo loft above the other silos. It is probable that the smoke contained combustible gases, e.g., CO, and that the strong explosion that occurred just as the top of the pellets had been covered with foam once more, was mainly a gas explosion. However, any dust deposits in the loft may also have become involved. The entire roof of the building was blown up, and debris was thrown into the surrounding area. Because the explosion occurred in the middle of the night (3 a.m.), and just after the fire brigade had left, nobody was killed or hurt.

The following case history, presented by Carson and Holmes (2001), describes a silo failure due to mechanical causes.

Case History of a Mechanical Failure of a Bolted Fly Ash Silo

At approximately 10:00 p.m. on a cool September evening in 1996 in southwestern USA, a thunderous cracking sound rang out to shatter the calm. The only employee in the vicinity of a new 80-ft. diameter fly ash silo realized that he had just heard the warning sound of imminent danger. In the dark of night, he had only his instincts to lead him at full speed away from the failing structure. The first rays of the next morning's sun revealed the devastated silo and the very spot he'd stood at, not 90 feet away, buried under 20 feet of fly ash.

The purpose of this brand new bolted silo was to store 9000 tons of fly ash from the adjacent power generation station. The silo split apart about two weeks after it was first filled to capacity. Up to this point, no ash had ever been discharged. Curiously, the collapse occurred at night when the silo was being neither filled or emptied.

During the course of the investigation into this failure, several deficiencies were revealed. Calculations showed that the silo was underdesigned and did not identify or account for a phenomenon called thermal ratcheting. The walls of outdoor metal silos expand during the day and contract at night as the temperature drops. If there is no discharge taking place and the material inside the silo is free flowing, it will settle as the silo expands. However, the material cannot be pushed up when the silo walls contract, so it resists the contraction, which in turn causes increased tensile stresses in the wall. The effect is repeated each day that the material sits at rest. The investigation also revealed that some cost-saving measures taken by the silo supplier during the construction of the silo contributed to the failure. The design specified that bolts of a particular classification, size, and strength be used in the construction. Bolts of the specified type have a distinct marking on their head which identifies that the bolts have been tested and meet recognized standards. Fewer than 1% of the bolts that were recovered from the failed silo had the specified marking and none of the marked bolts had been used in the critical vertical seams. Strength tests on the unmarked bolts revealed that some had tensile strengths less than the specified minimum.

Many contributing factors acted together and if any one had not been present, the collapse of the silo might have been avoided. Had the potential for thermal ratcheting been recognized at the design stage and had correct design parameters been selected, the collapse may not have occurred. If proper bolts had been purchased and used, the silo collapse may have been avoided. If the silo had been inspected by an independent silo expert either during the construction or after construction was complete, perhaps the incorrect bolts would have been noticed and corrective action could have been taken. Had the operation of the silo been such that material was discharged more frequently, the condition of accumulated stresses that precipitated the collapse could have been prevented.

When filling a silo or hopper with powders of low conductivity, charges will accumulate in the solids, causing continuous small brush discharges, and under certain circumstances, sometimes discharges will also occur with higher "energy value" on the surface of the solids. Such a discharge (also called "conical pile discharge") presents a potential hazard while handling highly insulating solids. Since the energy of these discharges depends on the geometry of the pile and the particle size of the solids, the following simple rules can be applied (Jaeger, 1997):

1. If the MIE of the solids is greater than 10 mJ, the volume of the solids in the silo or hopper is ≤ 2 m^3, and the particle size is less than 200 microns, there is no hazard due to a conical pile discharge. If the solids volume is less than approximately 2 m^3 there is a slow charge buildup and fast dissipation, and there are no ignition concerns.
2. If the solids volume is greater than approximately 2 m^3, the charge buildup in the pile cannot dissipate quickly, and there is a concern about ignition occurring.

There is a potential for ignition of combustible powders from bulking brush (conical pile) discharges if the MIE of the powder is less than the value calculated from the following equation (Glor and Schwenzfeuer, 1997 and CENELEC, 2003):

$$W = 5.22\,(D^{3.36})(d^{1.462}) \qquad [5\text{-}1]$$

where W is the equivalent energy of cone discharge in mJ, D is the silo diameter in meters, and d is the median of the particle size distribution of the product forming the powder heap in mm.

The exact conditions for a bulking brush (conical) discharge to occur during silo filling are not well understood. However, the following general factors that are known to increase its probability have been identified by Glor (1988):

1. An increase in the resistivity of the powder, greater than 10^{10} ohm-m.
2. An increase in the particle size of the powder, greater than 1 mm.
3. An increase in the charge density of the powder, greater than 1 μC/kg.
4. An increase in filling rate:
   (a) For granules with a diameter greater than 1 to 2 mm, a filling rate greater than 2000 kg/hr.
   (b) For granules with a diameter of about 0.8 mm, a filling rate greater that 20,000 to 30,000 kg/hr.

Inadvertently entrained isolated conductive objects within the silo (e.g., tools dropped into the silo, metal chips contained in the incoming solids, etc.) will be charged and form a capacitor with the silo.

As can be seen from the above two case histories fires and explosions in bins and hoppers can occur from a number of operating conditions and ignition sources. Among these are:

- Dust cloud generation from powders mechanically conveyed or pneumatically conveyed directly into the silo or hopper.
- Electrostatic sparks.
- Glowing particles produced by frictional heating upstream of the silo Smoldering combustion.
- Combustible gas released from the powder by self-heating.
- Flames from welding or cutting during maintenance.
- Electric sparks caused by faulty electrical wiring.

Several things can be done to minimize or prevent the occurrence of the above-mentioned explosion-causing conditions and ignition sources, such as:

1. To minimize dust cloud generation during silo or hopper filling, the solids should first be sent to a cyclone or fabric filter and then into the vessel. It may even be desirable to feed the solids from the cyclone or fabric filter into the vessel via a rotary valve to reduce turbulence and suspension of fines.
2. Provide a magnetic separator upstream of the conveying system to remove any tramp metal prior to conveying and avoid the possibility of frictional heating of the tramp metal.

3. If a silo or hopper requires a plastic lining for corrosion protection, it should be an electrically conducting liner, a static-dissipative liner, or a liner having a breakdown voltage of <4 kV.
4. For combustible and reactive solids use only metal vessels.
5. For metal silos and hoppers properly ground and bond all components of the vessel.
6. Provide instrumentation (level, temperature, pressure) that is in accordance with the required NEC electrical area classification. Install these instruments so that they will not be damaged by the flow of solids along the vessel wall.
7. If it is deemed necessary to see the condition of the inside of the silo, provide illumination installed on the outside the silo.

Even after the above preventive measures have been taken, it is usually necessary to provide protective measures, which can be any of the following: venting, suppression, containment, inerting, and fire protection. These are discussed below.

Venting is probably the most economical and widely used method used for the protection of silos and hoppers containing combustible and reactive particulate solids against explosions.

A general and more detailed discussion of venting is presented in Section 6.6.1. The following paragraphs discuss specific aspects of venting as they apply to silos and hoppers.

Venting should not be normally used for explosion protection of silos and hoppers containing toxic solids as the solids will be ejected into the surrounding atmosphere and could cause a health threat to personnel in the area. If the silo or hopper is located outdoors (the preferable location) then the fireball and the unburned solids can be discharged directly to the atmosphere. If the vessel is located inside of a building, then vent ducts routed to the outside or flame quenching devices inside the building will have to be used (see Section 6.6.1 for more details). For low pressure vessels, such as silos and hoppers, explosion vent design is simplified if the vents are located on the vessel roof. Roof vents in cold climates should be designed for the snow load or shielded to prevent the accumulation of snow. Vents should not be located where personnel could be exposed to the fireball and pressure effects of the vent discharge. For personnel safety or because of space limitations, vent panels are often located on the vessel walls at the top. The panels should be located so that they are accessible for maintenance, and must be above the maximum vessel fill level. The vent discharge should not be severely obstructed by adjacent equipment or structures.

Suppression can be used to protect silos and hoppers containing combustible, reactive, and also toxic solids against explosions (see Barton, 2002 and NFPA 654, 2000). Explosion suppression may not be feasible for some silos and hoppers if the maximum pressure of the suppressed explosion is greater than the design strength of the vessel (some silos and hoppers may have low design pressures). Section 6.6.2 presents more detailed information on explosion suppression systems design and installation.

Pressure containment can also be used to protect silos and hoppers storing combustible, reactive, and toxic particulate solids if the vessel is not too large (larger vessels will require thicker walls, which may make the fabrication cost economically unfeasible). Test data will be required to establish the maximum pressure that can result from an explosion. Pressure containment is an inherently safer approach and may be desirable for toxic solids storage. Section 6.6.3 discusses deflagration pressure containment.

Inerting (also called oxidant concentration reduction) is often used for protection of silos storing combustible, reactive, and toxic solids against fires and explosions. If the atmosphere inside of a silo or hopper is kept below the limiting oxidant concentration (LOC), then a fire or deflagration of suspended solids will not occur. However, it will not necessarily prevent smoldering. It is also often used if the solids stored in the vessel are water-reactive, pyrophoric, or degraded by contact with oxygen. The inerting gas is usually nitrogen, but for some metal powders, argon may be required as they react with nitrogen. If inerting is to be used, it is imperative that a reliable, continuous, supply of gas is available, and that the supply piping be provided with low pressure or low flow sensors and alarms. Section 6.5.2 discusses inerting system design and installation requirements in general for all types of process equipment.

Smoldering fires in silos and hoppers can occur, and may develop into large-scale fires with subsequent serious consequences if they are not detected and extinguished. If large-scale fires develop, they can cause structural failure of a silo or hopper, possibly resulting in emission of large amounts of combustible solids and a dust explosion. Therefore, some protective measures for detecting and extinguishing (suppressing) a smoldering fire should be considered.

Detection of smoldering fires in silos and hoppers can be achieved with methane and carbon monoxide (CO) detectors as recommended in NFPA 850 (2000). The detectors should be alarmed to alert the operators to a potential problem, and to take action. A CO concentration of 1.25% (by volume) should alert plant operators according to NFPA 850. It would also be prudent to install oxygen analyzers per the recommendations of Tuomisaari et al. (1998), because the oxygen monitors provide a means of knowing if and when suppression has been achieved. Locating and monitoring hot spots on the silo/hopper wall (with an infrared camera) is another means of fire detection and confirmation of fire suppression.

It may be desirable in some cases to install water spray nozzles in storage silos and hoppers to extinguish a fire if one should develop (this is not meant to provide explosion protection). Wash systems provided for cleanout between product changes can also provide effective fire protection. However, when considering fire protection for storage silos and hoppers, the hydraulic load that can be placed upon a silo or hopper and its foundation during water spray operation must be taken into account. Most silos and hoppers are severely limited as to the amount of water that can be safely accumulated without causing sufficiently high hydraulic pressures to damage the silo structure; so a way to drain the water must be provided in such cases. Therefore, water discharge into a silo or hopper to extinguish a fire may not always be desirable. It should also be noted that water spray discharge into a silo/hopper can generate clouds of combustible solids, which could possibly result in a dust explosion. It also has been found that low expansion foam has been used successfully to extinguish fires in silos/hoppers both in tests at the Technical Research Centre of Finland (Tuomisaari et al., 1998) and in several incidents (Zalosh, 2003).

The most success in extinguishing fires in silos/hoppers has been achieved with using carbon dioxide, and to a lesser extent with nitrogen, in tests where these gases were applied to barrels of smoldering wood chips and peat (Tuomisaari et al., 1998). Zalosh (2003) describes and discusses their work. Better results (earlier suppression) were achieved with bottom injection of the suppressant gases (compared to top injection) because the gases flowed up through the smoldering material and gradually extinguished the fire.

NFPA 850 (2000) also has some guidance on the amounts of carbon dioxide (CO_2) required to suppress smoldering coal fires in silos. Appendix A of NFPA 850 cites experience at one utility that suggests the necessary amount of CO_2 for fighting a coal silo fire should be 3 m^3 of CO_2 per m^3 of silo volume. This is 3 times the amount recommended by Tuomisari et al. (1998) based on their laboratory tests. The additional CO_2 is presumably needed because of additional leakage and more nonuniformity in large silos.

Care should be taken to ensure that silo/hopper fires be fully suppressed before attempting to remove the particulate solids. A number of silo fire incidents are described by van Wingerden and Alfert (1994) in which explosions occurred because attempts to remove the burning particulates prior to complete suppression actually generated combustible dust clouds within the silo.

Some additional safety considerations for silos and hoppers are as follows:

1. Where an explosion hazard exists, there shall be no intersilo venting, i.e., manifolding of vents from several silos (Section 3.2.3.1 of NFPA 654, 2000).
2. Where an explosion hazard exists, silos/hoppers shall be located outside of buildings, with the exceptions listed in Section 3.2.3.2 of NFPA 654 (2000).

3. Factory Mutual Global recommends that air cannons should not be used to break bridges in silos handling combustible materials if there are particles smaller than about 400 microns within the material (FMG 7-76, 1998).
4. FMG 7-76 (1998) also recommends that when a silo has a dust collector (bin vent filter) on the breather vent, explosion protection should be provided for the dust collector as if the dust collector is a separate and distinct vessel, if either of the following conditions exists:
   (a) The duct connecting the silo to the dust collector has an *L/D* ratio greater than 2.
   (b) The cross section of the connecting duct is smaller than the explosion venting area needed to protect the volume of the dust collector alone.

The type of silo used can result in flow problems and subsequent hazards. For example, funnel flow silos are prone to flow stoppages and irregular flow rates associated with the formation of an arch between the hopper walls. Collapse of an arch can cause the sudden formation of large dust clouds in the hopper and in the surrounding area. Dust clouds of toxic or combustible materials pose an obvious hazard. Problems caused by a funnel flow pattern can be cured by altering the pattern to mass flow, which requires changing the hopper or discharge feeder design. Other methods to fix poor flow include mechanical and chemical flow aid. These flow-corrective techniques are discussed by Purutyan et al. (1998).

Some potential difficulties associated with mass flow hoppers are the higher stresses created at the junction between the straight section and converging section, abrasion of the wall due to higher particulate velocities at the wall, and possible contamination of the particles with wall coatings (Williams, 1990).

As was pointed out previously, silos can also fail from errors in design, construction, and utilization. Carson and Holmes (2001) present an excellent discussion of these causes and also provide suggestions and recommendations on how to avoid or minimize these.

Besides NFPA 68 (2002), additional information on protection of silos and hoppers is presented by Barton (2002), NFPA 654, 2000), and NFPA 850 (2000).

5.3.16 Size Enlargement Equipment

Most size enlargement equipment cause electrostatic charging of the solids and some generate more dust clouds than others due to the turbulence created by the moving internal elements or rotation of the shell. In general, all size enlargement equipment should be properly grounded and bonded.

Some types of size enlargement equipment are more prone to fires and explosions than others. For example, because the fluidizing gas causes a very turbulent condition inside a fluid-bed granulator, electrostatic charging of the solids occurs and there is a continuous dust cloud present. Therefore, these units must be properly grounded and bonded. They are also usually protected against explosions by venting or suppression. Nitrogen, rather than air, can be used as the fluidizing gas to minimize the potential for an explosion. Bartknecht (1981) discusses protective measures for fluid-bed granulators. Detailed information is given on explosion venting and suppression designs for round and square granulators. These units can also be purchased designed and fabricated in accordance with explosion pressure-resistant design and explosion pressure shock-resistant design criteria (Bartknecht, 1989; Eckhoff, 2003).

Equipment that creates high turbulence can create dust clouds and could have the potential for a dust explosion, and should be protected by venting or by an explosion suppression system. A suppression system should be used, rather than venting, if the solids are toxic. Shaft bearings should be located outside of the housing so that they do not become coated with powder which can result in overheating of the powder and a subsequent fire.

5.3.17 Size Reduction Equipment

Fires and explosions have occurred in size reduction equipment quite frequently over the years. Two case histories of accidents presented below are illustrative of such hazards.

Case Histories of a Fire and Explosion in a Grinder for Silicon

A chemical plant that processed silicon-based chemicals experienced a fire and explosion in a grinder. Raw silicon was received in 1- or 2-inch lumps which had to be ground to a 300-mesh powder before being used in the chemical process. The air-conveyed silicon powder discharged from the grinder passed through a cyclone and then through a bag filter. An explosion and subsequent fire occurred in the system. The fire was extinguished within 15 minutes by a water hose stream. The system had explosion relief vents, but no sprinklers. Investigation showed that this incident was caused by hot spot ignition resulting from grinder parts scraping against the inside of the housing. This ignition mechanism was supported by observation of high current draw on the grinder motor before the incident occurred (CCPS, 1998).

Ed. Note: This accident could have been mitigated by measuring current-draw and possibly interlocking current-draw with the motor to shut it off when too high, and/or also with a water deluge system to activate it.

Case History of a Hammer Mill System Explosion

A chemical company had been milling an intermediate stage powder in a hammer mill for 2 years without an incident. Processing involved a man with a scoop feeding the mill which in turn fed a plastic-lined fiber drum. An incident occurred in which the mill and drum exploded with such violence that it caused structural damage to the building, and the operator was injured by the blast. Metal was discovered on the mill screen, giving rise to the possibility of frictional spark ignition. Laboratory tests revealed that the material had a very low MIE and AIT in bulk, and at low temperatures, the powder was found to decompose spontaneously and energetically to liberate gas. Therefore, it appears that this material was not suitable for conventional milling, and that inadequate protective measures were in place. Milling under nitrogen would not have prevented this incident as decomposition could still occur under inert conditions, and the liberation of gas would still have pressurized and ruptured the mill (Anon, 1999).

Ed. Note: This powder should have been tested to see if it could spontaneously decompose before selecting a hammer mill. Using a mill with a gentler motion (slower tip speed) might have been a safer approach. Also, the mill should have had a magnetic separator installed upstream of the mill to remove any tramp metal. In addition, specifying the mill for shock-resistant construction would have prevented the rupture of the mill.

Size reduction equipment must always be regarded as providing ignition sources because of the presence of friction and hot surfaces arising from the energy used in the comminution process. A number of ignition sources can occur during grinding and milling, such as:

1. Friction or impact ignition can be caused by the presence in the feed of tramp metal, stones, etc. These should be removed upstream of the size reduction equipment by the installation of magnetic separators or electric field type detectors for removal of metal or cyclones to remove material of greater density than the feed material. It should be noted that magnetic separators will only work with ferrous or cobalt metals, but not with stainless steel.
2. Hot surfaces may develop in a mill if the rate of feed is far from the optimum, and fire and explosion hazards can occur whether the mill is overloaded or underloaded (Palmer, 1973). Therefore, maintaining the proper rate of feed to a mill can minimize the occurrence of fires or explosions.
3. The powder may have a low melting temperature and the energy added in milling raises the internal energy and temperature to cause melting. Melted material accumulates, continues to heat, eventually degrading, and flammable gaseous degradation products result. Tem-

peratures increase until the AIT of the flammable gases is reached and a fire or explosion occurs.

4. The outlet of the mill becomes blocked, powder accumulates in the mill, and the above degradation sequence occurs with like results.

For some types of size reduction equipment, such as hammer mills, consideration should be given to providing vibration monitoring to detect if the machine has thrown a hammer, main shaft bearing failure, etc., and shutdown interlocks should be provided. Consideration should also be given to providing high amperage or high temperature alarms and shutdown for equipment subject to plugging.

Many types of mills for combustible and reactive solids can be designed and fabricated in accordance with explosion pressure shock-resistant criteria (e.g., ball mills, hammer mills, pin disk mills, fluid energy mills to name several).

Most mills can be protected by venting (not directly on the mill itself always, but certainly on the vessel receiving the milled product). Also suppression is very commonly used to protect size reduction equipment against explosions. ESCIS (1994) recommends that the explosion suppression system be designed for St 2 dusts. Explosions of St 3 dusts and hybrid mixtures of all dust explosion classes can be suppressed only to a limited extent. The mill must be designed for the expected overpressure of generally 0.5 to 1 barg in the event of a suppressed explosion. If this requirement cannot be fulfilled, for example with old equipment, the sizing of the suppression system must be modified accordingly. The response of a suppression system must automatically initiate immediate shutdown of the milling installation. For both venting and suppression, the propagation of an explosion into unprotected parts of the plant (upstream and downstream equipment) must be prevented, usually by special explosion barriers which are activated automatically by an explosion (see Section 6.6.4).

Inerting is also frequently used to protect size reduction equipment against explosions. One of the most common mills used in the chemical and pharmaceutical industries is the hammer mill (e.g., the FitzMill®). This type of mill is not normally enclosed and it can generate a large quantity of fine particles due to high impact of the mill blades. It is suggested that the mill be enclosed in order to control dust emissions and achieve an effective inert gas blanketing. With an enclosed design, the charging hopper for the mill is inerted, and the discharge side of the mill is sealed to a product receiver (hopper or portable container), which is also inerted. The vent from the product receiver can go to a dust collection system or can be recycled to conserve nitrogen consumption. Since the product receiver is not usually rated for vacuum or pressure, a sweep-through purge (see Section 6.5.2) may be the most desirable inerting system in this case. Before starting the milling operation, the oxygen must be purged from the entire milling system (feed hopper, mill, and product receiver) to below the LOC of the solids. Once the system is purged to the desired LOC, a continuous flow of inerting gas is maintained during the entire milling operation.

Fluid energy mills that are used for solids having a severe explosibility can be operated using an inert gas (nitrogen or superheated steam) in a once-through or recycle arrangement. The ground product is separated from the air or inert gas in a dust collection system (often a cyclone and baghouse in series), and the dust collection system should be protected from fires and explosions.

Also, if a particulate solid has a severe explosibility hazard, it may be desirable to locate the mill in a separate room, located on an outer building wall, with explosion vent panels. Entry into this milling room from the main work area should be via a strong door opening inwards, and interlocked with the mill so that operators cannot enter into the room while the mill is running.

For toxic solids, it is imperative that the mill housing be dust-tight (with appropriate bolting and gaskets) to prevent any emissions into the atmosphere. It may be possible to operate the mill itself under a slight negative pressure which will minimize dust emissions. Also, locating the mill in an enclosure connected to an exhaust system will minimize dust emissions into the workplace.

All size reduction equipment should be properly grounded and bonded, and the drive motor should be specified in accordance with the appropriate NEC electric area classification.

Good discussions of size reduction equipment hazards and protective measures are presented by Bartknecht (1981) and Palmer (1973). A very useful report on safety aspects of the milling of combustible solids is published by the Expert Commission for Safety in the Swiss Chemical Industry (ESCIS, 1994). In this report, combustible solids are divided into three safety classes for milling and recommendations for protective measures for each class are presented (including drawings of milling systems and their safety features).

5.3.18 Solids Charging Systems

Charging of solids into a vessel is frequently done without a proper process hazard review and without appreciation of potential hazardous results, which often has lead to fires and explosions. The following two case histories describe such accidents.

Case History of a Fire and Explosion Caused by Open-Manhole Charging of a Reactor during Product Rework

A pharmaceutical powder had been stored in a warehouse in a fiber drum with a plastic liner for a long period of time. The company received an order for this product, and it was decided to rework it to remove any impurities that might have contaminated the product while in storage. An existing batch reactor was selected for this rework, and it was discovered that it did not have a nitrogen line connected to it, so a work order was written by the production supervisor to have this done over the weekend. When the pipefitter arrived to do the connection for the nitrogen line, there was no one there to give him any directions as to where the connection was to be made. So he made a choice, which was unfortunately the wrong place, and tied the nitrogen line into the reactor vent line. When the reactor was started up on Monday morning, the operator did not realize that the nitrogen line was hooked up to the vent line, and started purging the reactor as per the SOPs. The nitrogen, did not go into the reactor and purge it, but went out the vent line. Also, for some unknown reason, the reactor had some residual alcohol vapors in it from a previous run. Two operators brought the drum with the product up on the operating level and tipped the drum up into the manhole and began to empty the drum of the powder when a flash fire and explosion occurred. The two operators on the platform, as well as a third operator, were badly burned, and one of the walls in the building room was knocked back about 6 inches from the explosion.

Ed. Note: This accident might have been avoided if:

1. A management of change review had been conducted of the proposed procedure, with the appropriate drawings and process safety information available, before any work was done.
2. The engineer or production foreman had checked the new piping arrangement before startup to see that the piping changes were properly made.
3. The operator had checked the reactor after the purging was done to determine the oxygen concentration inside the reactor (the reactor was not actually purged because the nitrogen went out the vent line).
4. The system had been modified to include a closed charging system rather than charging through the open manhole.

Case History of a Fire and Explosion Caused by Open-Manhole Charging of a Reactor during Varnish Manufacture

Drogaris (1993) presents the following case history of a reactor charging accident.

During the addition of phthalic anhydride to a varnish kettle which contained a mixture of soya-bean oil, glycerol, and caustic soda at 200°C, an explosion occurred at the charging hatch (manhole). The operator was blown back by the force and broke his arm as he fell to the ground. Two other operators standing nearby were not injured. The charging chute was also propelled upward and damaged the kettle agitator motor. The contents of the vessel were unaffected. The rupture disk in the kettle relief line did not rupture. The steel charging chute was not bonded to the reactor because of the presence of a nonconducting gasket in between them. Therefore, the most likely cause of the explosion was the ignition of phthalic anhydride dust by a static discharge from the unbonded chute.

Since this accident, the company has required that all equipment used to transfer phthalic anhydride powder is bonded and grounded and has amended the operating procedures to ensure that the dangers associated with phthalic anhydride are highlighted.

Ed. Note: The following additional safety measures would have prevented this accident from occurring: (1) the charging should have been done through a closed charging system, and (2) the kettle should have been inerted before and during the charging of the phthalic anhydride.

These two accident case histories point out the great danger from open-manhole charging, and to avoid such incidents, many companies are now using closed charging systems.

If it is decided to still use open-manhole charging, the following procedures are recommended if the powder has a MIE of 10 mJ or less, the powder is wet with solvents, or when charging to a vessel containing a flammable solvent (Pilkington, 2002):

1. Perform a formal hazard/safety review on the charging step.
2. If the vessel previously contained flammable liquids as part of other stages of the process or for cleaning, check to ensure that the atmosphere in the vessel is non-flammable prior to charging the solids. Nonflammable is defined as having a vapor concentration $<25\%$ of the LFL. Solvent vapors layer readily, so it is important to check for vapors at all levels in the vessel.
3. Before solids are introduced into the vessel, perform inert gas purging to achieve the desired oxygen level below the LOC in the vessel (see Section 6.5.2 for discussion of the various inert gas purging techniques).
4. Charging solids introduces air into the vessel and causes the oxygen concentration to increase in the vessel. A continuous inert gas (usually nitrogen) purge is necessary during the entire charging operation to minimize the increase in oxygen concentration. The necessary inert gas flow rate should be quantified based on tests and set at a consistent flow rate for the continuous purge.
5. Where oxygen concentrations are not continuously monitored with in-line oxygen analyzers, and multiple drums or bags are charged, tests should be conducted to establish when to stop charging and

repurge the vessel. This can be determined by charging with different number of drums or bags and measuring the oxygen level in the vessel.

6. Ventilation at the manhole should be considered, not only to minimize the employee exposure during charging, but also to maintain the flammable atmosphere below the LFL around the manhole. Be aware that ventilation systems can affect the inert gas blanket in the vessel by inducing air flow around the manhole. If ventilation is added to an existing process, the inert gas purge rate should be revalidated.
7. When charging solids to an empty vessel, keep vents closed during the charging operation, especially when the vent lines lead to a vent header system (connecting multiple vessel vents), as a mechanical vent system can quickly suck an inert gas blanket out of a vessel and introduce air in its place. Any blowback out of a vessel can be handled with local exhaust ventilation and respirators.
8. When charging solids into a flammable liquid, the vessel vent may need to be open or partially open to help eliminate vapor emissions out of the manhole. However, with this option, it will be necessary to conduct tests to determine the optimum balance for the vent to both lower the vapor emissions and maintain the oxygen level below the LOC.
9. In systems where solids are added to a vessel containing a flammable liquid, consider cooling the solvent to at least 5°C below its flash point as an additional precaution.
10. Avoid preheating of vessels containing solids prior to the introduction of the solvent or other flammable liquids.
11. Assure that all equipment is properly bonded and grounded, and provisions for grounding should also include operators. Operators should not shake plastic bags into the vessel to empty residual powder. Shaking of bags and plastic liners can have a significant impact on the development of static charges and may be hazardous.
12. Plastic drum liners should be of the antistatic type as they can be an ignition source when charging solids in flammable atmospheres.
13. Ensure that operators are educated in the special hazards of solids charging operations.

Holbrow and Tyldesley (2003) discuss and describe a device for installation at the inlet of a charge chute for open-manhole charging that prevents or minimizes explosion propagation (flame emission) from a reactor or mixing vessel (see Figure 5-2). This device was developed and tested by the HSE, but has not been patented by them.

Coal dust, anthraquinone dust, and milk powder were used as test dusts. The following vent areas (charge chute openings) were used:

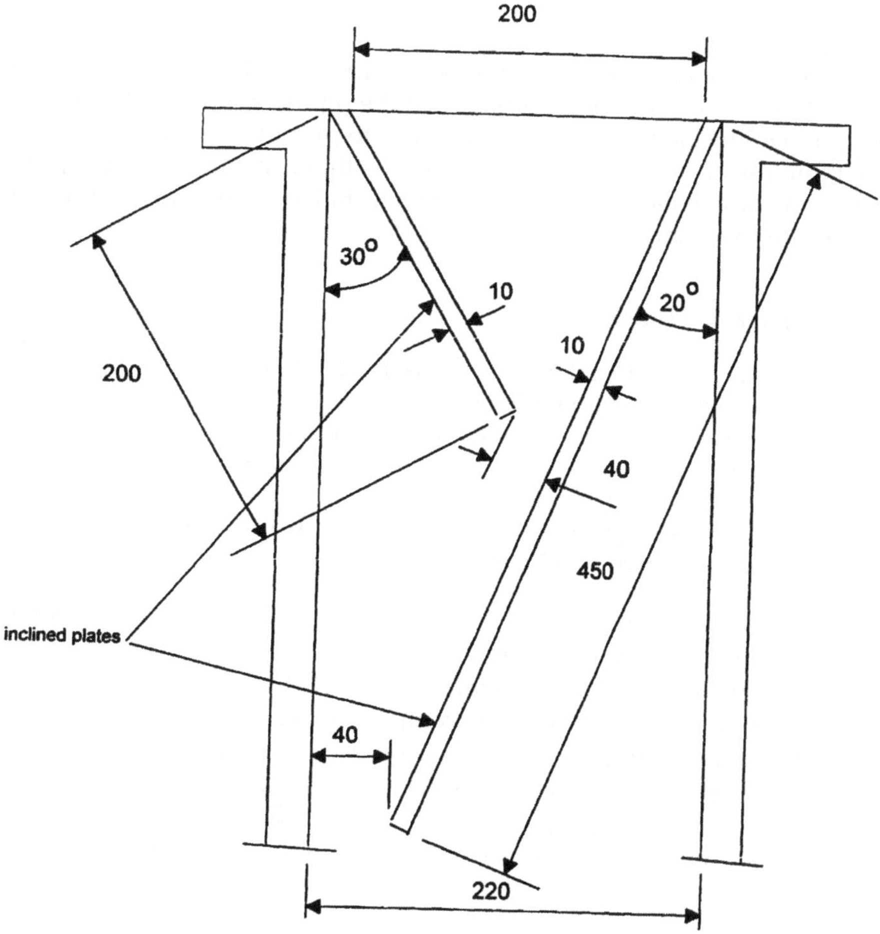

Figure 5-2 Propagation prevention device for charge chutes.

1. For coal dust: 0.13 and 0.26 m^2
2. For anthraquinone dust: 0.26 and 0.6 m^2
3. For milk powder: 0.13 m^2

In the tests with the coal dust and milk powder the device eliminated flame emission from the chute. However, for the anthraquinone tests, there was not complete flame elimination, but the emitted flame lengths were reduced by 75%.

It is possible that electrostatic charges may exist on the solids as they are fed into the hopper above the charging system devices, or they may become charged flowing from the hopper into the charging devices. Therefore, charging systems should be properly grounded and bonded to remove these charges. It is also possible to plug the charging system piping if wet vapors from the reactor or slurry mixing vessels diffuse into the piping, causing the solids to become sticky and "plaster" on the piping walls, and blocking off the piping flow area. It is common practice to provide wash connections on the piping walls to flush out any wet solids that accumulate on the piping walls, and also to have air or nitrogen connected to the piping to dry out the piping after washing.

5.3.19 *Tableting Systems*

Fires and explosions do not usually occur in tableting presses, but dust is created by the "punching" operation and if not controlled could accumulate in the working area around the presses, and be ignited if an ignition source was there. Therefore, tableting presses usually have local exhaust ventilation that removes the dust as it is formed.

Tablet presses for toxic solids (e.g., potent drugs) require special containment features, such as follows (Wood, 2001):

1. Air-pressurized neoprene channel gasket around each access door, for positive sealing while closed.
2. Internal exhaust slots, engineered for efficient pickup at major dust generation points within the compression chamber.
3. Internal compression chamber kept at negative pressurization.
4. Glove ports located and installed in the access doors of the compression chamber, with closures and safety switches for automatic carousel lockout in the event any of the closures were opened.

5.3.20 *Valves for Solids*

Fires and explosions do not usually occur in valves for solids, but emissions can be a problem. It is, therefore, important that the valve seals are properly installed and maintained so that the valves remain dust-tight.

Some valves such as slide valves can become jammed if solids accumulate in the "tracks." To avoid this, connections for air purging can be provided and a continuous small air flow can be applied to "sweep out" any solids accumulations.

Bonding and grounding of valve parts, especially the moving part, is necessary to avoid sparks from ungrounded conductors. Jaeger (1997) recommends that conductive parts of valves and flanges be bonded and grounded in all cases if their size is equal to or larger than a nominal diameter of 100 mm (4 inches).

Britton (1999) reports that due to ball valve electrification sparking can occur from valve handles insulated by the packing; special ball valves have been designed which provide a resistance less than 10 ohms between the ball, spindle, and outer housing. He also recommends that if electrical continuity with the valve body is not found with the handle in all possible positions, the handle should be separately bonded.

Because valves for solids usually encounter more severe erosive conditions than valves for liquids or gases, it is very important that they are on a preventative maintenance program to ensure their mechanical integrity.

For a discussion of rotary valve hazards, see Section 5.3.6

5.3.21 Weighing Systems

Properly designed weighing systems are usually not prone to fires and explosions. Equipment located on load cells must be connected to other equipment by flexible connections. These flexible connections should be of the conducting type (i.e., metal); if they are made of non-conducting material (plastic) they should be provided with a grounding strap, or the equipment on the load cell should be grounded and bonded.

Load cells have to be installed with due regard to mechanical safety. If tension and shear cells are to be used, the installation must take account of the possibility of cells mechanically failing (e.g., shearing of support members), and a means of limiting the fall of the equipment on the cells should be provided. When vessels are mounted outdoors on load cells, there is the potential for lifting due to wind, and the vessel should be restrained with loose tie bolts.

Shock forces applied to a load cell, either during installation or in operation, can cause considerable damage. Therefore, a means should be provided whereby the load may be lowered onto the cell in a controlled way. In situations where shock forces are likely, some means of shock absorption (the use of rubber bridge mounts in conjunction with movement stops) should be installed.

Electronic load cells systems may have to be shielded from stray currents to ensure accuracy. They must also have the proper NEC electrical area classification.

If a weighing system is located in an area where corrosion may be a problem, it should be provided with a corrosion-resistant coating. If a weighing system is located below grade, as for railroad car weighing, the scale pit should have drains so that water does not accumulate and either cause a short-circuit or create a corrosion problem. Low spots can also be the location for accumulation of heavy gases and vapors which may be ignited.

5.4 LOADING AND UNLOADING OF RAILCARS AND HOPPER TRUCKS

5.4.1 *Types of Railcars and Hopper Trucks*

There are usually no major hazards associated with railcars or hoppers trucks themselves. However, there are hazards associated with loading and unloading them, which are discussed below.

5.4.2 *Railcar and Hopper Truck Loading*

Explosions have occurred during railcar or hopper truck loading. A case history of one such incident is presented below.

Case History of a Dust Explosion (Flash Fire) during Loading of a Railcar with Polyvinyl Alcohol (PVA) Powder

Herrmann (2002) reports the following incident that occurred during the loading of a railcar with PVA powder.

A flash fire occurred while loading PVA product into a hopper railcar. Prior to the accident, more than 18,000 hopper railcars had been loaded without incident.

PVA was conveyed to a bag filter located above the railcar using nitrogen conveying. The PVA was separated from the nitrogen, and fed to distribution slides to distribute the PVA to the four railcar compartments. The distribution slides were connected to the railcar by a flexible hose (12-inch diameter and 4 feet long). A small vent hose (4-inch diameter and 12 feet long) was connected to a second opening in each railcar compartment to direct the air displaced by loading the compartment back to the bag filter for dust containment. All four compartments of the railcar fill at the same time. The railcar compartments were empty, but did contain atmospheric air. The railcar was electrically grounded. When the first amount of PVA was introduced into the railcar, a fireball erupted from one compartment of the railcar. The resulting fireball filled the top of the loading building (100 ft × 50 ft). The fireball lasted only a few seconds and quickly subsided. The top edge of the railcar was slightly deformed, indicating about 1 psia pressure developed inside the railcar.

During the investigation of the incident, it was verified that there was no other flammable material in the railcar or unloading system. The flexible hoses were analyzed and it was determined that they could not hold enough electrical charge to initiate a spark. The spiral reinforcing wire of the flexible hose was grounded and could not have been an ignition source. The steel railcar had an internal spray-on liner to protect the PVA from contamination. This liner was tested and found to withstand greater than 10 kV before break-through to the conductive railcar wall. As an insulating material with a large surface areas, the liner can lead to propagating brush discharges having discharge energies of several hundred mJ.

It was postulated that static charges developed during the emptying of the loading bag filter due to the PVA resin flowing through the loading equipment and into the railcar. The PVA became charged when dropping into the railcar allowing a charge to develop on the railcar liner. Once a charge developed on the liner, an equivalent charge was held on the steel wall of the railcar, producing a very large capacitor. The charge found a pinhole or thin spot in the liner and discharged, producing a significant spark. This is the mechanism for "propagating brush discharges." These type discharges are known to have adequate energy for ignition of PVA dust in the presence of air.

To prevent occurrence of such an incident, the railcar loading procedure was revised to include purging the empty railcar compartments with nitrogen and testing to ensure that oxygen was below 6 volume % before introducing PVA into the railcar.

Ed. Note: An additional safety measure would be to specify liners for the railcar having a breakdown voltage of <4 kV, or use antistatic liners.

Ebadat (1999) also presents a case history of this incident.

A number of measures can be taken to prevent explosions and other accidents from occurring during railcar and hopper truck loading, such as:

1. Properly bond and ground the loading spout, pneumatic and mechanical conveying equipment and piping, and the railcar or hopper truck.
2. Do not use coatings on the inside of railcars or hopper trucks unless they are antistatic or have an electrical breakdown voltage of <4 kV.
3. Apply brakes to railcars or install chocks under the wheels of hopper trucks so that they cannot move while loading is being done.
4. Install interlocks so that a hopper truck cannot be started up and driven off unless the loading equipment has been disconnected.
5. If internal components of loading spouts are made of nonconducting materials consideration should be given to having the railcars or hopper trucks purged with an inert gas (usually nitrogen) until the oxygen concentration inside the vehicle is below the LOC of the solids and an inert gas blanket maintained in the vehicle during the whole loading time.
6. If hoses are used for loading of railcars or hopper trucks, use conductive hoses (preferably flexible metal ones), or static-dissipative ones. All hoses should be properly bonded and grounded. It should be recognized that nonconductive hoses incorporating an internal bonding spiral may become a hazard because if the spiral breaks, internal and possibly external spark gaps may be created.

7. If particulate solids contain flammable liquids (hybrid mixtures), inert the railcar or hopper truck before beginning filling.

Some particulate solids may be hygroscopic and prone to "caking," which could result in difficulty in unloading them when they arrive at their final destination. Therefore, before the railcar or hopper truck is loaded, consideration should be given to inerting them, using an inert gas with a low dewpoint, to minimize ingress of any moisture.

5.4.3 Railcar and Hopper Truck Unloading

Explosions have also occurred during railcar and hopper truck unloading. A case history is presented below which illustrates the hazards of unloading a hopper truck.

Case History of an Explosion That Occurred during Unloading Atomized Aluminum Powder from a Hopper Truck

Pratt (1997) presents the following case history of an accident that occurred during unloading a hopper truck.

An explosion occurred when a bulk transport truck (hopper truck) was offloading a consignment of atomized aluminum powder during a one-of-a-kind operation where the operators made up the unloading procedure as they went along.

In normal operations, the truck had always been offloaded into a atmosphere of nitrogen in a closed railcar. The exhaust from the diesel engine of the truck was the pneumatic transport fluid so that the rate of offloading created a concentration of aluminum powder which far exceeded the minimum concentration for a dust explosion. But since the exhaust was oxygen depleted, the atmosphere in the hopper truck had always been inert and there had been no problems.

In this instance, an order was cancelled and the consignment of powder was sent back to the plant for offloading and reclassifying, an operation that had never been previously performed. The plan was to pneumatically move the powder from the truck to the entrance of the plant pneumatic transport system which was some distance away from where the truck could be parked. The 3-inch hoses on the truck could not reach the entrance so an additional hose, of similar construction but of larger diameter, was placed into service. The flexible hoses had metal fittings on each end and were made of rubber with a spiral of heavy wire within the rubber running between the flanges. In this manner, an electrical connection was maintained with the truck, which in this case was properly grounded. However, the flanges on the two different size hoses could not be connected.

The operators devised a connection anyway by inserting the small hose into the larger one and stuffing rags between them to seal the opening. Thus, the last section of added hose was not grounded. The end of the larger, ungrounded hose, was loosely placed into the entrance of the plant pneumatic transport system where the end could move about and bang on the wall of the grounded process equipment.

As if the ungrounded hose were not enough, the operators added another element to the scenario. They recognized that the pneumatic transport system of the truck may not have been powerful enough to adequately move the product through the larger section of added hose. They therefore inserted a ⅜-inch high-pressure plant air hose into the opening between the couplings to "help things along." In so doing, they defeated the inert characteristic of the pneumatic transport fluid (diesel engine exhaust gas).

In normal operation, the pneumatic transport system in the plant was operated in dilute phase mode well below the minimum concentration for ignition of the aluminum dust. The plant system was therefore operated with air as the transport fluid. Air was pulled into the plant system at its entrance where the truck hose had been inserted. The two pneumatic systems were therefore mismatched since the density of the aluminum powder being delivered by the transport medium of the truck was much greater than that of normal operations in the plant, that is, an explosible dust–air mixture was inserted into the plant system.

The offloading operation was started, and within a few minutes an aluminum dust explosion occurred, which propagated throughout the plant. Because of the conditions, an electrostatic scenario for the ignition of the aluminum dust was considered. Calculations by Dr. Pratt for the streaming current, capacitance of the hose, and its resistance to ground showed that an electrostatic discharge was a very credible scenario for the incident.

Ed. Note: to prevent such incidents from reoccurring, avoid using two different size hoses, improperly connected together, which could cause any section of hose to be not properly grounded.

Some general measures that can be taken to prevent explosions or other accidents during railcar or hopper truck unloading are as follows:

1. Properly bond and ground the railcar or hopper truck, the unloading equipment and piping.
2. Apply brakes to railcars or install chocks under the wheels of hopper trucks so that they cannot move while unloading is being done.
3. Install interlocks so that a hopper truck cannot be started up and driven off unless the unloading equipment has been disconnected.

For railcars or hopper trucks that are unloaded using air or inert gas pressurization, care must be taken to ensure that the air or gas pressure is controlled to below the maximum allowable working pressure (MAWP) of the railcar or hopper truck to avoid rupturing it. This can easily be done by providing a safety valve in the pressurizing air or gas line after the pressure reducing valve, set to relieve at below the MAWP of the railcar or hopper truck. This will protect the railcar or hopper truck should the diaphragm in the pressure reducing valve fails open.

In addition, a number of other desirable features and conditions relevant to hazards reduction for unloading railcars and hopper trucks are presented in Table 5-4 (Kraus, 1991).

TABLE 5-4
Desirable Features and Conditions for Unloading Railcars and Hopper Trucks (Source: Kraus, 1991)

Required Operations	Features and Conditions
Car and trailer unloading station	1. Should be in level area, protected from the weather when possible. 2. Spots should be well marked to center the discharge spouts. Use fluorescent paint when spotting is done at night. 3. Street or sidewalk hatch covers should be watertight and lightweight for handling by one man. Covers should be locked when in unprotected areas. 4. Steel blocks should be provided for chocking wheels. 5. Provide warning blinker lights or reflector stanchions to prevent collision by another vehicle. 6. Provide floodlights, where required, for night operations. Aim lights at the top of the car and at car outlet connections.
Connections to cars and trailers	1. Connections should be designed to eliminate the need for men working beneath the vehicle and in cramped spaces. 2. Attachments should be lightweight for handling by one man and should require a minimum of tools for connection to the discharge spout. 3. Hose connections should be snap-on or toggle clamp type with soft-rubber gaskets. 4. Clear distance between discharge hopper of a loaded transport and road or track should be ample for insertion of unloading devices or attachments. 5. Transport discharge spouts should be equipped with easily operated gates above the covers so as to minimize powder spillage when making up connections. 6. Discharge connections should be flexible enough to compensate for the rise of the car on its springs as unloading progresses.
Access to loading hatches	1. Tops of transports should have grab bars at access ladders to assist operator in swinging onto vehicle. 2. Hatch covers on transports should be easily opened without the use of tools and should be fitted with watertight gaskets. 3. Platforms should have nonslip safety treads. 4. Top of transport should clear building canopies or cornices and yet permit passage of operator from one end of transport to the other.

Required Operations	Features and Conditions
Vent filters	1. Vent filters should be lightweight for handling by one man and accessible from top of transport without hoisting. 2. Filter should be easily attached to a loading hatch using the same clamping device used to secure the hatch cover. 3. Filters should be designed so that they may be cleaned after unloading, but before detaching so that dust is shaken down into the transport rather than onto plant grounds.
Flow inducers	1. Transport should preferably empty itself without the use of external flow inducers or poking of material by operator through hatch covers. 2. Vibrators, where required, should be lightweight and easily fastened to one point on the transport most conducive to emptying. There should be no need to shift the vibrator from point to point. 3. Vibrators should cease operation whenever the discharge spout is blocked with material. 4. Vibrators should be operated by air or low voltage electricity. Portable cables and hoses should be attached to retracting reels for safety. 5. Vibrators should be noiseless, if possible, to eliminate neighborhood complaints, especially during night operations. 6. Vibrators should be sized with due regard to the detrimental effects of excessive vibration on the car structure.
Transport internals and clean-out	1. Transport interiors shall preferably be free of structural members and protuberances that may cause hangup of material. 2. Interior of transport should be coated with a releasing agent or plastic which will prevent adhesion of material to slope sheets. 3. All corners, valleys, and knuckle points should have large radius curves to prevent retention of material at these points. Slope sheets should have a minimum angle of 45° to the horizontal, but a slope of at least 60° is preferred for self-unloading cars. 4. Aerating pads, blocks, troughs, or other aerating devices should retain a minimum of residue after transport is emptied. 5. Interior stiffeners, where required, should be designed to shed material and prevent hangup.
Rate of unloading	1. Rate of flow out of transport should be controllable to match conveying rate of in-plant conveying system or capacity of pipeline. 2. Flow withing transport should be free from avalanching.

5.5 INSTRUMENTATION

This section discusses hazards and their preventive/protective measures for various types of instruments used in solids handling operations.

5.5.1 Flow Instruments

Fires and explosions rarely occur in flow measuring devices. However, it is good practice to ground and bond particulate solids flowmeters. Also, all electrical components should be in accordance with the appropriate NEC electrical area classification.

5.5.2 Level Instruments

Several common types of level indicators (in-vessel types) used in silos and hoppers represent electrostatic discharge hazards. This hazard can be caused by the exposed metal structure of probes, slide wires, cables, etc. mounted inside the vessels. Even if the wire is less than ⅛-inch (3 mm) in diameter, there is always the chance that conducting components will become ungrounded due to corrosion, mechanical damage, or negligence on the part of operating personnel. For example, there is the potential for ignition from an electric arc from bare wires (see the first case history in Section 5.3.15) and a deflagration can be caused by an ungrounded capacitance probe (Britton and Kirby, 1989). Also the wrong selection of electrical equipment (e.g., NEMA types) can be the cause of a fire or explosion.

Level instrumentation components inserted inside of equipment can act as sites for charge accumulation and discharge, especially for equipment with nonconducting walls and highly conductive solids.

Jones and King (1991) recommend that it is best to avoid these types of indicators in new designs and instead to specify non-protruding systems such as load cells, ultrasonic sensors, etc.

In cases of existing vessels where retrofitting is impossible, then the continuity of all bonds and the adequacy of all grounds must be assured with a rigorously enforced safety maintenance program.

Fill level indicators of the suspended-weight type should never be operated during filling or emptying of silos and hoppers because the cable can cause cone discharges (also called bulking brush or "Maurer" type discharges). The weight should be made of solid, insulating plastic (never metal coated with plastic) and it should be suspended on a cable no larger than ⅛-inch (3 mm) in diameter (Jones and King, 1991).

5.5.3 Pressure Instruments

In particulate solids systems pressure instruments must be modified to prevent solids from plugging them up. Therefore, for pressure gauges, a conventional Bourdon-tube construction is not suitable, and a pressure gauge with a diaphragm seal should be used. Also, differential pressure systems for pressure drop measurement should also have diaphragm seals, or the pressure taps should be purged with air or nitrogen to prevent plugging the impulse lines.

5.5.4 Temperature Instruments

Thermocouples should be installed in thermowells to protect them against turbulence-induced vibrations which, otherwise, could cause mechanical failure of the thin wires. Thermowells should also be protected from erosion by solids by a baffle or wear-plate arrangement.

REFERENCES

Abbott, J. 1990. *Prevention of Fires and Explosions in Dryers.* 2nd ed., Rugby, UK: Institution of Chemical Engineers.

ACGIH. 1998. *Industrial Ventilation—A Manual of Recommended Practices.* 23rd ed., Lansing, MI: American Conference of Governmental Industrial Hygienists.

Anon. 1999. Dust Fires and Explosions—Learning from Experience. *Powder Handling & Processing,* 11(4): 491.

ANSI. 1996. *Safety Standards for Conveyors and Related Equipment.* ANSI B20.1. New York, NY: American National Standards Institute.

Bartknecht, W. 1981. *Explosions: Course, Prevention, Protection.* Berlin and New York: Springer-Verlag.

Bartknecht, W. 1989. *Dust Explosions: Course, Prevention, Protection.* Berlin and New York: Springer-Verlag.

Barton, J. 2002. *Dust Explosion Prevention and Protection: A Practical Guide.* Rugby, UK: Institution of Chemical Engineers.

Betz, M. R. 2000. "Evaluation of a Drum Lifter in Control of Particulate Emissions." *Proceedings of the 2000 Powder and Bulk Solids Conference,* pp. 61–79.

Britton, L. G. 1993. Static Hazards Using Flexible Intermediate Bulk Containers for Powder Handling. *Process Safety Progress,* 12(4): 240ff.

Britton, L. G. 1999. *Avoiding Static Ignition Hazards in Chemical Operations.* New York: AIChE/Center for Chemical Process Safety.

Britton, L. G. and Kirby, D. C. 1989. Analysis of a Dust Deflagration. *Plant/Operations Progress,* 8(3): 177–180.

Carson, J. W. and Holmes, T. 2001. Silo Failures: Why Do They Happen? *Powder and Bulk Engineering,* November: pp. 31ff.

CCPS. 1998. *Guidelines for Design Solutions for Process Equipment Failures.* New York: AIChE/Center for Chemical Process Safety.

CENELEC. 2003. *Electrostatics-Code of Practice for the Avoidance of Hazards Due to Static Electricity.* Report CLC/TR 50404. European Committee for Electrotechnical Standardization, Brussels, Belgium (June 2003).

Cooper. T. C. 1986. "Dust Controls during Bag Filling." *Proceedings of the 11th Annual Powder and Bulk Solids Conference,* pp. 653–659 (May).

Decker, E. B. 1975. "Special Requirements for Powder Feeders to Prevent Flooding." Paper presented at the Fine Powder and Dust Handling Seminar Workshop, University of Pittsburgh, December 9–11, 1975.

Drogaris, G. 1993. *Major Accident Reporting System: Lessons Learned from Accidents Notified.* Amsterdam: Elsevier Science Publishers.

Ebadat, V. 1999. Investigation: How Dust Exploded during Railcar Loading. *Powder and Bulk Engineering,* 13: 57–64.

Ebadat, V., Espino, P. A., and Mulligan, J. C. 2002. "Testing and Qualification of Type C/D FIBCs." *Proceedings of the AIChE 36th Annual Loss Prevention Symposium,* pp. 151–160

Eckhoff, R. K. 2003. *Dust Explosions in the Process Industries.* 3rd ed., Amsterdam: Elsevier Science Publishers.

ESCIS. 1994. *Milling of Combustible Solids: Safety, Evaluation of the Feed Materials, Protective Measures with Mills.* Booklet No. 5 (English edition). Expert Commission for Safety in the Swiss Chemical Industry, Basle, Switzerland.

Field, P. 1982. *Dust Explosions (Handbook of Powder Technology, Volume 4).* Amsterdam: Elsevier Science Publishers.

FMG 7-11. 2003. *Belt Conveyors.* Loss Prevention Data Sheet 7-11. FM Global, Norwood, MA.

FMG 7-73. 2000. *Dust Collectors and Collection Systems.* Loss Prevention Data Sheet 7-73. FM Global, Norwood, MA.

FMG 7-76. 2001. *Prevention and Mitigation of Combustible Dust Explosions and Fire.* Loss Prevention Data Sheet 7-76. FM Global, Norwood, Ma.

Gibson, N., Harpur, D. J., and Rogers, R. L. 1985. Evaluation of the Fire and Explosion Risk in Drying Powders. *Plant/ Operations Progress,* 4(3): 181–189.

Glor, M. 1988. *Electrostatic Hazards in Powder Handling.* New York: John Wiley & Sons, Inc.

Glor, M. and Schwenzfeuer, K. 1997. Occurrence of Cone Discharges in Production Silos. *J. Electrostatics,* 40–41: 511–516.

Gravell, R. L. 2001. "Safety Issues in the Use of Flexible Intermediate Bulk Containers." Proceedings of the 2001 Powder and Bulk Solids Conference, pp.335–343.

Gravell. R. L. 2004. Peer review comments and personal communication from Robert Gravell, E. I. duPont de Nemours & Co., Inc., Wilmington, DE to S. S. Grossel (February 17, 2004).

Heitbrink, W., McKinney, W., and Rüst, R. 1983. *An Evaluation of Control Technology for Bag Opening, Emptying, and Disposal-The Self-Contained Filter-Bag Dump Station.* NIOSH Report 144-19, November 1983. National Institute of Occupational Safety and Health, Cincinnati, OH.

Herrmann, D. 2002. Personal communication from David R. Herrmann, E. I. duPont de Nemours & Co., Inc., Wilmington, DE (March 14, 2002).

Holbrow, P., Lunn, G. A., and Tyldesley, A. 2001. Explosion Venting of Bucket Elevators. *IChemE Sympos. Ser.* No. 148, pp. 213-225.

Holbrow, P., Lunn, G. A., and Tyldesley, A. 2002. Explosion Venting of Bucket Elevators. *Journal of Loss Prevention in theProcess Industries,* 15(5): 373–383.

Holbrow, P., and Tyldesley, A. 2003. Simple Devices to Prevent Dust explosion Propagation in Charge Chutes and Pipes. *Journal of Loss Prevention in theProcess Industries,* 16: 333–340.

IChemE. 1982. *Guide to Safety in Mixing.* Rugby, UK: Institution of Chemical Engineers.

ISA S84.01. 1996. *Application of Safety Instrumented Systems for the Process Industries.* Research Triangle Park, NC: The Instrumentation, Systems, and Automation Society.

Jaeger, N. 1997. *Static Electricity-Rules for Plant Safety.* Ciba Specialty Chemicals Corp. (available from Larry Floyd, Director CSTL, Ciba Specialty Chemicals Corporation, P. O. Box 113, Ciba Road, McIntosh, AL 36553).

Jaeger, N. and Siwek, R. 1999. Prevent Explosions of Combustible Dusts. *Chemical Engineering Progress,* 95(6): 25–37.

Jones, T. B. and King, J. L. 1991. *Powder Handling and Electrostatics: Understanding and Preventing Hazards.* Chelsea, MI: Lewis Publishers, Inc.

Kirby, D. 2001. Personal communication from David Kirby, Dow Chemical Company, South Charleston, WV to S. S. Grossel (December 2001).

Kraus. M. 1991. *Pneumatic Conveying Systems for Bulk Materials.* 3rd ed., Englewood Cliffs, NJ: Prentice-Hall.

Larsen, O., Hagen, J. H., and van Wingerden, K. 2001. Ignition of Dust Clouds by Brush Discharges in Oxygen Enriched Atmospheres. *Journal of Loss Prevention in theProcess Industries,* 14: 111–122.

Luttgens, G. 1985. Collection of Accidents Caused by Static Electricity. *Journal of Electrostatics,* 16: 247–255.

Luttgens, G., Luttgens, S., and Jordan, E. 2001. Improving Electrostatic Safety of FIBCs. *Bulk Solids Handling,* 21(2): 253–255.

Markowski, A. S. and Mujumdar, A. S. 1995. Safety Aspects of Industrial Dryers. Chapter 4 in *Handbook of Industrial Drying,* Volume 2, edited by A. S. Mujumdar, New York: Marcel Dekker, Inc.

Nelson, M. 2002. Personal communication from Melvin Nelson, Syngenta Crop Protection, Inc., St. Gabriel, LA to S. S. Grossel (January 21, 2002).

Ness. A. 2002. Personal communication from Al Ness, Rohm and Haas Company, Bristol, PA to S. S. Grossel (January 18, 2002).

NFPA 68. 2002. *Guide for Venting of Deflagrations.* Quincy, MA: National Fire Protection Association.

NFPA 69. 2002. *Standard on Explosion Prevention Systems.* Quincy, MA: National Fire Protection Association.

NFPA 77. 2000. *Recommended Practice on Static Electricity. Quincy,* MA: National Fire Protection Association.

NFPA 86. 1999. *Standard for Ovens and Furnaces.* Quincy, MA: National Fire Protection Association.

NFPA 650. 1998. *Standard for Pneumatic Conveying Systems for Handling Combustible Materials.* Quincy, MA: National Fire Protection Association.

NFPA 654. 2000. *Standard for the Prevention of Fire and Dust Explosions from the Manufacturing, Processing, and Handling of Combustible Particulate Solids.* Quincy, MA: National Fire Protection Association.

NFPA 850. 2000. *Recommended Practice for Fire Protection for Electric Generating Plants and High Voltage Direct Current Converter Stations,* Quincy, MA: National Fire Protection Association.

Palmer, K. N. 1973. *Dust Explosions and Fires.* London: Chapman and Hall.

Philiph, J. 2002. Personal communication from Jeffrey Philiph, Monsanto Company, St. Louis, MO. to S. S. Grossel (March 8, 2002).

Pickup, R. D. 2001. Dust Explosion Case Study: Bad Things Can Still Happen to Good Companies. *Process Safety Progress,* 20(3): 169–172.

Pilkington, G. 2002. Personal communication from Gary Pilkington, Abbott Laboratories, North Chicago, IL to S. S. Grossel (June 24, 2002).

Pratt, T. H. 1997. *Electrostatic Ignitions of Fires and Explosions.* New York: AIChE/Center for Chemical Process Safety.

Purutyan, H., Pittenger, B. H., and Carson, J. W. 1998. Solve Solids Handling Problems by Retrofitting. *Chemical Engineering Progress,* 94(4): 27–39.

Roques, M. A. 1996. "Environmental and Safety Issues in Drying." Proceedings of the 10th International Drying Symposium, pp. 61–72.

Schultz, G. A. 2000. *Conveyor Safety.* Des Plaines, IL: American Society of Safety Engineers.

Sleicher, L. M. 1998. "Safe Handling of Polymers in a 'Closed- Loop' Pneumatic Conveying System. Preprints Volume, 2nd Biennial Polymer Producers topical Conference, pp. 153–157, AIChE Spring 1998 National Meeting, March 9–11, 1998, New Orleans, LA.

SRI. 1980. *Control Technology Assessment of the Pesticide Manufacturing and Formulating Industry.* Report prepared by Stanford Research Institute for NIOSH under Contract No. 210-77-0093 (May 1980).

Stone, L. H. 1987. Overcoming Static Problems in Your Vibratory Screen Separators. *Powder and Bulk Engineering,* pp. 22–23 (August).

Tuomisaari, M., Baroudi, D., and Latva, R. 1998. *Extinguishing Smoldering Fires in Silos.* Brandforsk Project 745-961, VTT Publication 339, Technical Research Centre of Finland.

Tyldesley, A. 2004. Peer review comments and personal communication from Alan Tyldesley, HSE, Bootle, UK to S. S. Grossel (February 19, 2004)

UN. 1999. *Recommendations on the Transport of Dangerous Goods: Manual of Tests and Criteria,* Rev. 3, ST/SG/AC.10/11/Rev.3. United Nations. ISBN 92-1-139068-0.

van Wingerden, L. and Alfert, F. 1994. *Detection and Suppression of Smoldering Fires in Industrial Plants.* Research Report CMR-94-F25068. Christian Michelsen Research Institute, Bergen, Norway.

VDI 3673. 2002. *Pressure Venting of Dust Explosions.* English translation. Verein Deutsche Ingenieure-Kommission Reinhaltung der Luft, VDI Verlag GmbH, Dusseldorf, Germany.

Wang, C. C. K. 1986. "Health and Safety Aspects of Solid Material Sampling," Workshop No. 13. 1986 Powder & Bulk Solids Conference, Rosemont, IL (May 1986).

Whitmore, M. W., Gladwell, J. P., and Rutledge, P. V. 1993. Report of an Explosion During the Manufacture of an Azodicarbonamide Formulation. *Journal of Loss Prevention in the Process Industries,* 6: 169–175.

Williams, J. 1990. The Storage and Flow of Powders. Chapter in *Principles of Powder Technology,* edited by J. Rhodes, New York: John Wiley & Sons.

Wood, J. P. (editor). 2001. *Containment in the Pharmaceutical Industry.* New York: Marcel Dekker.

Zalosh, R. G. 2003. *Industrial Fire Protection Egnineering.* Hoboken, NJ: John Wiley & Sons.

Chapter 6

DESIGNING AND INSTALLING SYSTEMS TO PREVENT AND CONTROL COMBUSTION, EXPLOSIONS, UNCONTROLLED REACTIONS, AND RELEASE OF TOXIC PARTICULATE SOLIDS

6.1 INTRODUCTION

This chapter discusses various techniques for designing and installing systems and equipment for the prevention, control, and mitigation of combustion hazards, explosions, uncontrolled reactions, and release of toxic particulate solids. The chapter starts off with a discussion of the causes of fires and explosions, followed by a review of ignition sources (description, control, and removal); electrical equipment hazards and area classification; deflagration prevention methods; deflagration protection methods; siting of equipment and buildings to minimize damage from fires and explosions; blast resistant (damage limiting) construction of buildings; protection of equipment and buildings by water sprinkler/deluge systems; protection of equipment and buildings by foam and other special suppression systems; containment for control of releases of toxic particulate solids; and identification of system-wide design, protection, and prevention requirements.

6.2 CAUSES OF FIRE AND DEFLAGRATION

This section discusses the causes of fires and explosions, including the concept of the "fire triangle" and various types of ignition sources.

6.2.1 The Fire Triangle

Fire (or flame propagation) is a combustion phenomenon, and is defined as the rapid, exothermic oxidation of a fuel. The fuel may be in liquid, vapor/gas, mist, or solid form (dust).

Normally, for flame propagation to occur, three conditions must be met, as follows:

1. The fuel must be within certain concentration limits.
2. An oxidant (usually the oxygen in air) must be above a certain minimum concentration, called the limiting oxidant concentration (LOC). It should be noted that chemicals other than oxygen are oxidants (e.g., chlorine, fluorine, oxides of nitrogen, peroxides, etc.).
3. An ignition source of sufficient temperature, energy, and duration.

All three conditions, as shown on the so-called "fire triangle" (Figure 6-1) must be present for a fire to initiate and propagate. If one or more of the conditions are not met, then a fire cannot happen. It is possible to prevent a fire by changing one or more of the conditions (this will be discussed later on in this chapter). However, it should be recognized that some materials can violently decompose in the absence of an oxidant. For example, some metal powders can react with nitrogen to produce flames. Medard (1989) presents a thorough discussion of solid combustion supporters (oxidants), such as metal oxides and salts of oxygen-containing acids (e.g., nitrates, chlorates, and perchlorates).

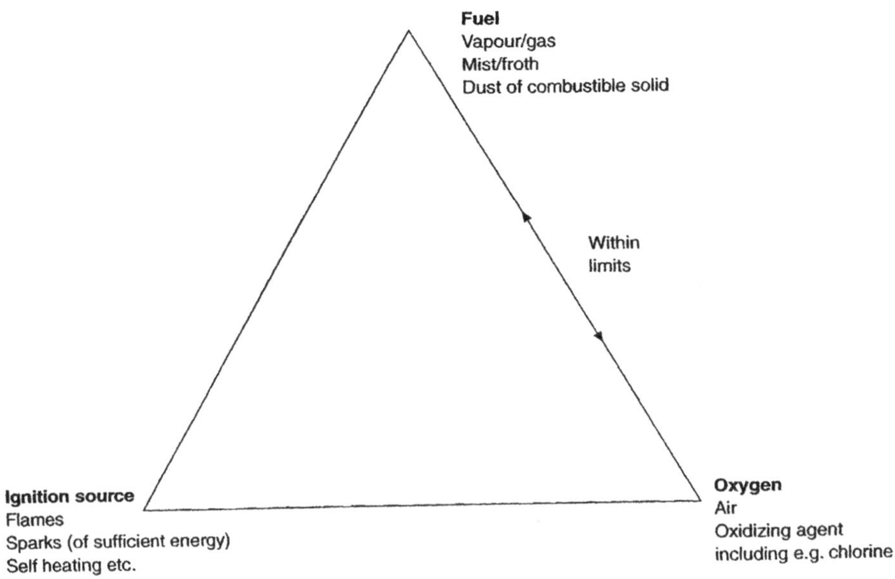

Figure 6-1 The fire triangle.

6.2.2 *Types of Ignition Sources*

A wide variety of ignition sources could be present during particulate solids processing, such as the following:

- Flames (e.g., flares, fired heaters, accidental fires)
- Hot work (e.g., cutting, welding, hot tapping)
- Hot surfaces (e.g., vessels, piping, motors)
- Hot particles (e.g., product from dryers, hot ash)
- Friction and impact (e.g., metal-to-metal contact, jamming of particulate solids, rubbing of conveyor belt on rollers)
- Chemical reactions (e.g., catalysis, reaction with powerful oxidants, reactions of metals with halocarbons, thermite reactions, thermally unstable materials, pyrophoric materials)
- Physical sources (e.g., adsorption heat)
- Hot vented combustion products
- Spontaneous combustion (autoignition or self-heating)
- Static sparks
- Electrical equipment (e.g., arcing across spark gaps)
- Lightning
- Projectiles (e.g., metal fragments from a vessel explosion)

These are briefly discussed below in Section 6.3.

6.3 IGNITION SOURCES: DESCRIPTION, CONTROL, AND REMOVAL

6.3.1 *Electrostatic Hazards and Their Control*

Electrostatic hazards are ubiquitous in the storage, handling, and processing of particulate solids. Among the plant operations where electrostatic charges are generated are mixing, grinding or milling, sieving, pouring, pneumatic conveying, etc.

Electrostatic charges are generated on solids as they are transported due to rubbing of particles against particles or particles against equipment and piping internal surfaces. Individual particles can become charged independently of whether they are of a conductive or insulating nature. Parts of process equipment and piping which come in contact with charged solids, then also become charged. An electrostatic charge by itself does not necessarily represent an ignition hazard. Such a hazard exists only when the charge is so high that an energetic discharge occurs due to the breakdown of a high voltage electric field.

Chapter 5 describes electrostatic hazards related to specific unit operations and types of equipment.

It is important to recognize that the MIE of a dust cloud, and thus its sensitivity to ignition by electrostatic discharge can be significantly altered by a change in particle size, composition, and moisture content of the solids. Therefore, it is important to remeasure the MIE of a powder if a process change has resulted in a change of these properties. Particle sizes can also change in various parts of a process due to handling, which in turn can affect the MIE. The MIE decreases with a decrease in particle size and a decrease in moisture content. Therefore, when evaluating electrostatic discharge hazards, tests should be conducted to establish the particle size and moisture content that will result in the lowest MIE (highest hazard) consistent with product specification requirements. Small amounts of flammable vapors and/or gases mixed with solids (hybrid mixtures) can significantly reduce the MIE.

This section discusses types of electrostatic charges, bonding and grounding, humidity control, ionization and other control methods, and linings and coatings hazards.

6.3.1.1 *Types of Electrostatic Discharges*

Electrostatic discharges are separated into different types by the character of their ionization of air when electrostatic energy is released. The following types of electrostatic discharges can occur in the storage, handling, and processing of particulate solids:

- Corona discharges
- Brush discharges
- Bulking brush (cone) discharges
- Propagating brush (Lichtenberg) discharges
- Spark (capacitor) discharges

These are briefly discussed below.

CORONA DISCHARGES

The corona discharge can be considered as a special case of a brush discharge. If the radius of curvature of a grounded electrode introduced into a high electric field is small (less than about 1 mm), the field will be disturbed only in the immediate vicinity of the electrode tip. A corona discharge may be accompanied by a hissing noise that increases with current and is accompanied by a faint luminosity. Corona discharges do not usually pose an ignition hazard when handling particulate solids.

BRUSH DISCHARGES

Brush discharges can occur when a conductive, grounded, and curved object (electrode) with a radius of curvature typically in the range of 5 to 50 mm, is exposed to a high electric field, for example, emanating from a highly charged nonconductive material. It takes the form of short spark-like discharges from discrete areas of the surface of the non-conductor. The total discharge has a brush-like appearance, hence the name.

For a brush discharge to occur, it is immaterial how the high electrical field is generated. The following are examples of how a brush discharge can occur in solids handling operations:

- approach of a conductive electrode such as a tool or human finger tip to a highly charged insulator surface (e.g., plastic pipe or chutes for the conveyance of powder, plastic bag, plastic drum, filter bag, film web, or conveyor belt).
- emptying of solids out of a plastic bag in the vicinity of metal fittings (e.g., above the manhole of a reactor).
- projection of conductive, grounded internal fittings into a highly charged dust cloud.
- addition of insulating, powdered materials to drums, vessels, or silos; approach of a highly charged dust heap to internal fittings.

Brush discharges can be avoided by eliminating high electric fields through use of conductive materials and grounding them, limiting the surface of nonconductive objects or use of shielding measures. Glor (1988) states that no ignition of a dust cloud clearly caused by a brush discharge has yet been reported, and Britton (1999) states that brush discharges from isolated nonconductors such as plastic sheet or powder beds are only a hazard in the presence of a flammable gas. Recent experiments by Larsen et al. (2201) have shown that even in an oxygen-enriched atmosphere it was not possible to ignite sulfur dust clouds by brush discharges.

BULKING BRUSH DISCHARGES

A bulking brush discharge is a large discharge resulting when a dispersed, charged powder "bulks" when settling in a container, causing a very large increase in its volumetric charge density. It is the type of discharge observed on the cone of a bulked heap of powder; thus it is often called cone or conical pile discharge. Surface flashes up to several feet long are observed in large containers (silos) being filled with powder having a resistivity above 10^{10} ohm-meter (Ω-m), both during, and occasionally for a short time after, the transfer of powder into the vessel. The discharges originate at the vessel wall and propagate across the bed surface. For axial powder feed the discharges appear between the edge of the powder cone and the surrounding walls, while for off-axis powder feed, the discharges appear on the side opposite the powder cone. An accompanying cracking sound has been heard from the top of the silo over the noise of the powder transfer. Bulking brush discharges have an apparent maximum effective energy of 10-20 mJ (with respect to dust ignition), and are believed to be responsible for dust explosions in grounded silos (Britton, 1999).

Recent research results (Glor and Schwenzfeuer 1997) indicate that the maximum energy released in bulking brush (cone) discharges depends on the silo diameter and the median particle size of the products forming the powder heap. For silos with diameters in the range of 0.5 m to 3.0 m and powders with a median range of 0.1 mm to 3.0 mm, they state that the energy released in cone discharges can be estimated using the following formula:

$$W = 5.22(D)^{3.36}(d)^{1.462} \qquad [6\text{-}1]$$

where W is the upper limit of energy of the cone discharge in mJ, D is the diameter of the grounded conductive silo in m, and d is the median of the particle size distribution of the powder forming the cone in mm.

There can be large differences between the actual energy observed in practice for bulking brush discharges (10–20 mJ) and the value calculated by Equation 6-1. For example, if a silo diameter is 3.0 m and the particle diameter is 3.0 mm, then the value of the maximum bulking brush discharge energy calculated by Equation 6-1 is 1043 mJ.

As can be seen from Equation 6-1, cone discharges formed from coarse powder are of much higher energy than those from fine powder. A most important conclusion from the research by Glor and Schwenzfeuer is the fact that cone discharges do not only occur with highly insulating granules, but also with rather fine highly insulating powders.

Britton (1999) makes a number of points regarding the effective energy of bulking brush discharges with respect to dust ignition. Among the many observations are (see book page citations at end of each entry):

1. A rough estimate of the maximum effective energy of a positive brush discharge is of the order 10 mJ. This value agrees with the "spark" MIEs of gas mixtures that have been ignited by large brush discharges. However, there have been no reports of dust ignition by brush discharges in air, even those dusts having very small "spark" MIEs. It appears that the efficiency of dust cloud ignition by brush discharges is only of the order 10% relative to gas mixtures having the same spark ignition energies, presumably caused by disparities in power density and duration of the different types of discharge. Similarly, a typical bulking brush discharge has an estimated effective energy of 100–200 mJ. Assuming the same 10% attenuation factor, a "typical" value of 10-20 mJ is indicated for the effective energy of bulking brush discharges where dust ignition is involved. However, flammable gas mixtures in silos should be at risk of ignition over most of the flammable range. These predictions are consistent with industrial experience (pp. 19–20).
2. Owing to the variety of methods used to measure dust ignition energies, it might be assumed for practical purposes that typical bulking brush discharges have effective dust ignition energies less than that of

lycopodium clavatum dust, whose spark MIE is approximately 20 mJ with a typical reported range of 10–30 mJ (pp. 32 and 63).

3. Experimental test data have been published which suggest that bulking brush discharges have effective energies of 1000 mJ or more with respect to dust ignition. These tests involved collecting the charge from bulking brush discharges and channeling the energy as sparks inside a "Hartmann" dust ignition tube. As noted above, sparks are far more effective dust ignition sources than "nonspark" discharges of the same overall energy, so the technique is invalid and the results are flawed (p. 34).
4. In addition to various factors (mass loading, velocity, etc.) influencing the appearance of bulking brush discharges, the probability of dust cloud ignition is likely to depend on charging polarity. The ignition location is predicted to be close to the container wall at the bed surface. If the fine particles are charged oppositely from that of the bed, the latter will quickly precipitate them from suspension. There is a tendency for fine particles to charge negatively with respect to coarse particles predominating in the settled bed, hence the ignition probability might be greater under conditions where both the bed and the suspension are predominately charged negatively. It has been shown, for example, that polyethylene normally charges in bipolar mode with the bed predominately positive and the dust suspension negative. However, under high humidity conditions both bed and suspension become predominately negatively charged (pp. 34–35). Although the literature states that a combination of coarse particles (yielding a high frequency of bulking brushes) plus fine particles (with low MIE) is the worst possible combination, if bipolar charging occurs, the fine particles may tend to neutralize the bed via electrostatic attraction (pp. 35 and 194).

It should be pointed out that there is a difference of opinion among various experts about the maximum energy achieved in a bulking brush discharge. The latest draft of the new European Guide on Static Electricity states that bulking brush discharges could have the potential to ignite dust clouds with MIEs as high as 100 mJ or higher (Nelson, 2003).

Glor (1988) states that the following conditions favor the occurrence of conical pile discharges:

- Powder of high charge-to-mass ratio as observed in pneumatic transfer at high velocities.
- High filling rates: for granules with a diameter of several millimeters, a filling rate >2000 to 5000 kg/hr (4410 to 11,000 lb/hr) and for particles with a diameter approximately 0.8 mm, a filling rate >25,000 to 30,000 kg/hr (55,100 to 66,100 lb/hr).

Britton (1999) states that bulking brush discharges cannot always be eliminated in practical equipment above about 1 meter diameter where non-conductive powders are transferred, although flow rate reduction may be effective for the more conductive powders in the range of 10^{10} to 10^{11} ohm-meter. Also, the use of active neutralizers has been proposed for relatively small silos. To minimize the probability of ignition, the first consideration should be whether the MIE of the powder being handled can be increased to greater than 20 mJ, such as by minimizing the sub-200 mesh fines concentration, and selecting additives (where product specifications allow this) having an MIE of at least that of the sub-200 mesh product fines, particularly for containers that might accumulate additives. Where such methods are impractical, inerting is sometimes used, especially if powder properties make deflagration venting and suppression undesirable alternatives. Inerting is often considered where the powder MIE falls below the 3–10 mJ range, according to the MIE test method used.

Maurer (1979), Maurer et al. (1989) and Glor and Schwenzfeuer (1997) discuss the hazards of cone discharges in silos.

PROPAGATING BRUSH (LICHTENBERG) DISCHARGES

A propagating brush discharge (PBD) is a discharge along the surface of a thin dielectric (insulating) layer, very highly charged on both sides with charges of opposite polarity. The dielectric layer may be in the form of a separate sheet (film) or a coating on a metal surface. Often the sheet is backed by a conductor, but the essential point is that the sheet is polarized and in the same state as the dielectric of a charged capacitor. If a conductor approaches the nonconductor surface, the resultant electrostatic field promotes ionization across the large area of the surface. A discharge can then take place in which the charge from an extensive area of the nonconductor flows to the initial discharge point through the ionized gas adjacent to the surface. The result is an intensive and highly energetic spark-like discharge (as high as 100–1000 mJ) which can be very dangerous (Britton, 1999).

Britton (1999) states that the principle criteria under which a PBD can be produced are as follows:

1. The layer thickness must be less than about 8 mm or air breakdown will occur above the charged layer before the critical surface charge density of 2.5×10^{-4} C/m^2 can be attained.
2. The breakdown strength of the charged layer must be adequate to attain the critical charge density.
3. Sufficient charge must be available to supply the required critical surface charge density. Calculations show that 50-pound plastic bags and plastic lined 55-gallon drums are too small to represent a credible risk.

The following are examples of how a propagating brush discharge can occur during storage, handling, and processing of particulate solids:

- Pneumatic dust conveying at a high rate through an insulating pipeline or through a conductive pipeline with an insulating inner coating.
- Continuous impingement of fresh dust particles on an insulating surface or on a metal surface with an insulating coating.
- Rapidly moving belt conveyors or drive belts which are either insulating or coated on one side with conductive material.
- Filling of large nonconductive drums and silos, or metal drums and silos having nonconductive linings, with highly charged, insulating particulate solids.

Propagating brush discharges can be avoided by the use of conductive materials or materials of low dielectric strength. If insulating layers (in the form of a coating on a metal surface or a self-supporting wall) with a breakdown voltage of less than 4 kV are used, then no propagating brush discharges will occur at such layers.

SPARK (CAPACITOR) DISCHARGES

A spark (or capacitor) discharge is a discharge between two conductive objects (people, products, and equipment), one of them charged to a high potential and the second one charged to a much lower potential or at ground potential. It occurs when the electric field in the space between the two conductive objects reaches the breakdown field strength.

In a spark discharge, practically all the energy stored on the higher potential charged object is released. Therefore, spark discharges must always be regarded as potentially incendive to dust–air mixtures.

Some situations in particulate solids handling operations where spark discharges can occur are:

- A conductive length of pneumatic conveying piping isolated by seals.
- An operator or maintenance person with insulating footwear
- A metal drum on an insulating base.

Britton (1999) and Glor (1988) provide more detailed discussions of spark discharges and their hazards in systems handling particulate solids.

Spark discharges can be avoided by grounding of all conductive objects (plant equipment, drums, persons, etc.).

More detailed discussions of the various types of electrostatic discharges are presented by Britton (1999), Glor (1988), and Jones and King (1991). Luttgens and Wilson (1997) discuss the origins of static electricity and the various types of electrostatic discharges and present numerous case histories related to these types of discharges.

6.3.1.2 Bonding and Grounding

The most commonly used method of preventing electrostatic discharge ignition of dust–air clouds is charge dissipation by means of bonding and grounding.

Bonding between two bodies, using bonding straps or wires, provides a conductive path through which electrostatic charges can recombine. Therefore, no charge can accumulate, and thus, no spark can occur since the difference in electrical potential is zero. However, Pratt (1997) points out that objects may not be at ground potential and there could be scenarios where there could be a spark between the supposedly grounded objects and ground. In such cases, redundant bonding and grounding is often installed.

Grounding (also called earthing) equalizes the potential difference between objects and the earth. A conductive object can be grounded by a direct conductive path to the earth or by bonding it to another conductive object that is already connected to the ground. To prevent the accumulation of static electricity in conductive equipment, the total resistance of the ground path to earth should be sufficient to dissipate charges that are otherwise likely to be present. A resistance of 1 megohm (10^6 ohms) or less is generally considered adequate. When the bonding/grounding system is all metal, resistance in continuous ground paths will typically be less than 10 ohms. Greater resistance usually indicates the metal path is not continuous, usually because of loose connections or corrosion. A grounding system that is acceptable for power circuits, or for lightning protection, is more than adequate for a static electricity grounding system.

The minimum size of grounding wires (conductors) is dictated by mechanical strength rather than by current-carrying capacity. Flexible conductors should be used for bonds that are to be connected and disconnected frequently. Conductors may be insulated or uninsulated, and some prefer uninsulated conductors so that defects can be easily spotted by visual inspection. If insulated, the conductor should be checked for continuity at regular intervals, depending on operating experience.

Connections may be made to equipment with pressure-type ground clamps, brazing, welding, battery-type clamps, or magnetic or special clamps that provide metal-to-metal contact. All surfaces to which connections are made must be free of paint, grease, oil, or other contaminants that would interfere with good contact. It is important to ensure that the integrity of the bonding and grounding is maintained. This can be done by using an electrical continuity monitoring system (a number of such systems are commercially available). In addition, all grounding devices should be visually inspected on a regular basis.

Personnel grounding is also often used to avoid shock and spark hazards. The simplest type of commercial personnel grounding equipment is a grounding bracelet with built-in resistor, typically giving a resistance to ground of about 1 megohm (MΩ). Where portable containers on wheels are used, footwear and flooring must be made sufficiently conductive. Also, if gloves are used, they should be conductive or antistatic, so that handled items that are troublesome to individually ground, such as tools, are grounded through the grounded person. Operator grounding should be done when the MIE of a solid is ≤30 mJ. NFPA 77 (2000) states that grounding of personnel is achieved by ensuring that the resistance from the skin to ground is approximately 10^8 ohms or less. NFPA 77 also states that workers should only be grounded through a resistance that limits the current to ground to less than 3 milliamps (mA) for the range of voltages experienced in the area. This method is called "soft grounding" and is used to prevent injury from an electric shock from line voltages or stray currents. Britton (1999) discusses personal grounding more fully, as does BS 5958 (1991), and NFPA 77 (2000).

More detailed discussions of bonding and grounding systems, components, and installation practices are provided in Britton (1999), BS 5958 (1991), CENELEC (2003), NFPA 77 (2000), and Pratt (1997).

6.3.1.3 Humidity Control

Where the properties of particulate solids are not adversely affected, humidification of the atmosphere in rooms or buildings where solids are handled and processed may reduce the potential for electrostatic discharges by increasing the surface conductivity of the solids. At humidities of 65% and higher, the surface of most solids will adsorb enough moisture to ensure a surface conductivity that is sufficient to prevent the accumulation of static charges (NFPA 77, 2000). In some cases, localized humidification produced by directing steam or water fog (mist) onto critical areas may provide satisfactory results without the need for increasing the humidity in the whole room. For some polymeric materials an increase in relative humidity can increase the rate of static generation (BS 5958, 1991). The use of relative humidity to control electrostatic charging should be undertaken only after careful consideration of the consequences. Further discussion on humidification as a method of controlling electrostatic hazards is presented by Britton (1999).

6.3.1.4 Ionization and Other Control Methods

Electrostatic charges can be neutralized by devices that ionize the air and increase its conductivity so that charges drain away to ground. In the use of air ionizers, it is important to consider certain factors that can influence their effectiveness, such as environmental conditions (e.g., type of dust and temperature) and positioning of the device in relation to the solids being processed, equipment parts, and operating personnel. It is important to know that these devices do not prevent the generation of static electric charge, but that they provide ions of opposite polarity to neutralize the generated static electric charge.

There are several types of commercial charge neutralizers, of which the two distinct types are those requiring external power ("active types") and those that don't ("passive types"). Charge neutralizers relying on free ions are limited by ionic mean free paths of a few centimeters in air, so they cannot be used to neutralize charge in large volumes such as silos. This limitation can be offset by transferring the charge to small particles which can carry the charge over large distances.

ACTIVE NEUTRALIZERS

These devices typically use AC corona discharges to produce ions of both polarities, often in combination with a blower to help move the ionized air to the charged surface. These AC-powered charge neutralizers must be approved for use in hazardous (classified) areas. Charge neutralizers that use pulsed or steady-state double-polarity DC use a pulsed or steady field to stress the electrodes to produce ions for use in the neutralizing process. If pulsed or double-polarity DC ionizers are used in hazardous (classified) areas, these should be listed for such use.

Another type of an active neutralizer is the radioactive (nuclear) ionizer, which uses ionizing radiation for neutralization of static electric charges. The most common radioactive ionizers depend on the alpha particle generation from the decay of Polonium-210. Performance of radioactive ionizers deteriorates with the decay of the radioactive material. They must be replaced periodically (at least annually) because of the decay. These neutralizers must be registered and installed in accordance with the Nuclear Regulatory Commission regulations. Radioactive neutralizers are often used in conjunction with passive (inductive) neutralizers to control high charge densities. Although these neutralizers are costlier and regulatory compliance issues are associated with them, they are nonincendive, require no wiring, and can reduce static electricity charges to the lowest levels (NFPA 77, 2000).

PASSIVE NEUTRALIZERS

Several different types of passive neutralizers (also called inductive neutralizers) have been used or proposed, depending on specific applications. All types rely on the creation of ions. These types are:

Needle, String, and Tinsel Bar Types. These are used to remove charge from flat surfaces in close proximity to the neutralizer, such as moving belts. Their design is based on or consists of sharply pointed elements arranged for placement in the static electric field near the charged surface.

At the sharply pointed tips an electric field is created (>3 kV/mm) which is sufficient to produce a localized electrical breakdown of the air (a corona) which will inject ions into the air. Although inexpensive and easy to install, these inductive neutralizers require a minimum potential difference between the charged object and the needle tip to initiate corona and the neutralizing process. In the absence of this minimum charge, neutralization will not occur, and a residual potential of a few thousand volts will be left on the material when sharp inductive points are approximately within about 12 mm of the surface (NFPA 77, 2000).

Silo Inlet Neutralizers. Two types intended to reduce the intensity of bulking brush discharges in silos are described by Britton (1999). The first is a pointed discharge rod mounted axially inside the filling pipe outlet so that it points upstream into the filling pipe. The second is a plastic-weighted wire mounted axially within the filling pipe so that it hangs down into the powder heap. Both types should be securely grounded and the diameter of the discharge rod or wire should be 1–3 mm. Britton (1999) points out that there is little information on the success of such devices in large commercial equipment and problems could include mechanical failure exacerbated by drag from incoming and bulked powders, plus decreased effectiveness caused by accumulated powder and coatings on the metal surfaces.

Discharging Rods. These have been widely used to dissipate charge from open powder containers such as drums and tote bins, and comprise a rigid, grounded rod which is inserted close to the bottom of the container before filling of the container begins. The rods can be cylindrical or square. Britton (1999) recommends a square section rod be used as it gives a smaller effective radius of curvature while allowing a greater rod thickness and rigidity.

For conductive powders in plastic or plastic-lined containers it is especially important to insert the discharging rod before powder flow starts since delayed insertion can produce sparks from the powder surface when in an ungrounded or insulated container.

It is critically important that inductive neutralizers are connected to a secure ground, or sparks from the induction bar can occur.

Britton (1999), BS 5958 (1991), Cross (1987), NFPA 77 (2000), and Noll (1995) present more information on ionization and other control methods.

6.3.1.5 Linings and Coatings Hazards

Plastic linings and coatings on the inside of process equipment are usually nonconductive and can accumulate charges, and can result in propagating brush discharges if the breakdown voltage is greater than 4 kV.

Plastics can be made conductive by the addition of carbon into the formulation, resulting in a black plastic. If this is acceptable from a process viewpoint, this is one way to overcome the problem with nonconductive linings and coatings. Such conductive linings and coatings should be grounded. A plastic liner or coating can be made conductive if the thickness can be made thin enough to have a breakdown voltage <4 kV. This may not be feasible if the liner or coating is subjected to erosion by particulate solids.

6.3.2 Spontaneous Combustion: Evaluation and Control

At a sufficiently high temperature, deposited combustible solids may undergo spontaneous heating. This can lead to ignition without the application of any external source of energy such as a spark or flame. Spontaneous combustion is caused by slow exothermic oxidation by atmospheric oxygen. The term "spontaneous heating" is more generic and could include competitive heating processes such as exothermic decomposition.

If the heat generation rate via reaction exceeds the heat loss rate across the boundaries of the dust deposit and a steady state condition cannot be reached, the temperature will increase until ignition occurs. At smaller temperatures, either there is no temperature increase or a moderate temperature increase resulting in steady state. The spontaneous ignition temperature, SIT, is defined as the minimum temperature at which spontaneous heating results in ignition. The most appropriate test method depends on the application. For example, the measured SIT of a dust layer (specified thickness on a standard hot plate) might be directly applied to electrical area classification. Alternatively, thermokinetic models developed either from large-scale test data or adiabatic calorimetry, might be needed to estimate the SIT for bulk material. The SIT for bulk material is a complex heat balance problem with many variables, although (as discussed later), simplified methods are often employed to extrapolate from laboratory data.

It is important to appreciate that there is no single value for the SIT, since this is a solution to a heat balance problem depending on many factors. The SIT for large piles of bulk material often corresponds to a heating time of hours or even days. Where single value SITs are listed, they should be used only in accordance with specific practices referencing a specified standard test method (generally dust layers or cubes). There has been considerable confusion between the terms "spontaneous ignition" and "autoignition." Some data compilations even use the terms interchangeably. Britton (2003a) has recommended that the term "autoignition" be used to describe spontaneous ignition of dust clouds rather than deposited material. Dust cloud tests are commonly performed in the BAM furnace or modified Godbert-Greenwald furnace, which give comparable results. The associated "minimum autoignition temperature," or MAIT, is measured during the short time the dust is dispersed inside the furnace as a cloud. The SIT is usually smaller than the MAIT. If the SIT applies to a large pile of bulk material, the SIT might be close to ambient temperature despite a MAIT of hundreds of degrees Celsius.

Examples of where self-heating of particulate solids can occur are layers of dust accumulating on process equipment, piping, or building structural elements (beams, braces, etc.), storage of coal in piles, or hot material stored in drums or cartons or in bulk (in silos or hoppers). These can undergo spontaneous combustion (self-heating) if left undisturbed for a period of time.

Appropriate tests must be conducted to ascertain the maximum allowable temperature at which particulate solids can be stored to avoid spontaneous combustion (self-heating). These are discussed in Chapter 4.

The SIT of a layer or pile of particulate solids depends on the balance between the rate of heat generation within the layer or pile and the rate at which the heat is lost to the surroundings. A theoretical model originally proposed by Frank-Kamenetskii (1969) highlighted the importance of a dimensionless group of terms, δ, known as the Frank-Kamenetskii parameter. This parameter is fixed by the relevant physical and chemical properties of the solids together with the size of the layer or pile and a reference temperature. All these factors are important; more heat will be generated at elevated temperatures and by highly exothermic reactions, and less heat will be lost from thick layers and large piles with poor thermal conductivity. Solids that are safe in one set of circumstances are not necessarily safe in another.

For a given system, it is generally possible to determine a critical value of δ. This value may be obtained from the literature, calculated by known methods, or derived from first principles by solving the equation for the heat balance. If the value of δ, as evaluated for a given system, is greater than δ_c (a limiting value of δ), then the system will self-ignite (the heat generated at all times exceeds that which is lost). The temperature will rise, slowly at first, and then rapidly until ignition occurs. If the calculated value of δ is less than δ_c, then only moderate self-heating can occur. The theory predicts that the maximum temperature rise that can safely be sustained in a body is low (of the order of a few tens of degrees Celsius in practice). Above this temperature rise, runaway self-heating to ignition will occur. The distinction between ignition and nonignition is, therefore, in principle, sharp. This arises as a consequence of the assumption in the theory that the heat-generating reaction is highly sensitive to temperature. In general, this is true and the distinction between subcritical and supercritical states is also sharp in practice. Refer to Section 4.3.4 for further discussion of this.

Good discussions of the Frank-Kamenetskii theory are presented by Gray (2002) and Lees (1996a).

The two general cases of self-heating usually encountered in industry are powder layer ignition and bulk powder ignition (CCPS, 1993). The essential difference between the two types of self-heating is that the spontaneous ignition temperature (SIT) of powders stored in bulk can be less than normal ambient temperature, and special scaling methods must usually be used to estimate the SIT. Powder layer SIT can be determined experimentally without the need to extrapolate the results.

Ignition of powder layers can occur from heating on one side such as by a light fixture or a hot motor casing. Hot plate tests as described by Beever and Thorne (1982) and by Nagy and Verakis (1983) may be used to determine the SIT and ignition delay time. It is important to recognize that the SIT can change very rapidly with changes in layer thickness, so the experiment should closely simulate the worst-case plant situation. Where layers are heated on both sides, such as inside hot equipment (e.g., dryers), isothermal testing as done for bulk solids should be considered.

Ignition of bulk powders may occur during processing, storage, or transportation, where the initial temperature is equal to or less than that of the surroundings. "Ignition temperatures" based on small-scale hot plate or furnace tests such as ASTM D 1929 are meaningless for such cases. Isothermal test methods described by Beever and Thorne (1982) may be used to determine the SITs of powder contained in mesh baskets of different sizes.

Provided the geometry is held constant, a simple scaling model can be used to extrapolate the data to larger sizes and other geometries. For example, this method has been applied by Bowes and Cameron (1971) to the investigation of spontaneous ignition of activated carbon in ships' holds. The method has also been successfully applied to SITs of agricultural products in large silos, to blocks of polyurethane foam, and to plastic powders in production facilities (CCPS, 1993). The estimation of ignition delay time is usually less accurate.

A special and more difficult case is that of "hot-spot" ignition, where powder is added hot to a cooler container or a mass of powder is heated only locally. Griffiths and Kordylewski (1992) show how to predict the ignition temperatures for "hot-stacked" powders. In many cases, expert consultation is required to address this type of problem.

Another problem causing "hot spots" of material is process blockage, where some moving equipment (e.g., a rotary valve) continues to rotate, but there is no flow of solids. In this case, a small amount of material may be heated until it starts to smolder or burn. It is, therefore, important to identify process blockages promptly, and to stop machinery until the blockage is cleared.

Other useful discussions of self-heating and "hot-spot" ignition are presented by Anthony and Greaney (1979), Eckhoff (2003), Leuschke (1980, 1981), and Thomas (1973).

There are several ways to prevent spontaneous combustion of particulate solids:

- Cool the hot material before sending it to storage.
- Store the solids in several small silos rather than in one large one.
- Highly reactive solids should not be allowed to form dust layers around the plant, and it may be safer to store the solids in airtight containers.
- Purge the silo of air before starting filling and maintain a nitrogen blanket on it during storage to prevent ingress of air.

Field (1982) presents a number of other measures to prevent spontaneous combustion (self-heating), such as:

- Dust should not be allowed to accumulate on hot plant equipment and piping.
- Hot particulate solids should be allowed to cool sufficiently before storage in containers (boxes or drums) or silos. Silos should be pro-

vided with temperature instrumentation, and possibly, smoke detectors.
- For solids being stored in large quantities in silos for long periods of time, recirculation to allow cooling may be necessary.
- In some countries with very hot climates, consideration may need to be given to providing methods of deflecting the sun's rays away from large storage silos or bins containing solids that self-heat.

Bowes (1984) discusses methods for controlling self-heating hazards in more detail.

6.3.3 *Pyrophoric and Water-Reactive Solids*

Some particulate solids can react rapidly with oxygen when exposed to air or water, resulting in fires. These can be categorized as:

1. Those that ignite spontaneously on contact with air; these are called pyrophoric.
2. Those that ignite spontaneously on contact with water; these are called water-reactive.

In both cases, the reaction is exothermic and the temperature is raised to the autoignition point reasonably quickly and hence differs from ignition by self-heating. Some factors that affect spontaneous ignition (reaction) (Bond 1991) are:

- Small particle size resulting in a large surface area for reaction with the air.
- The activity of the solids for a reaction with oxygen.
- Moisture frequently assists in the spontaneous reaction process
- Stress in metal
- Impurities

Pyrophoric Particulate Solids

The term *pyrophoric* is applied to solid substances which, when exposed to air at ordinary temperatures, catch fire spontaneously or oxidize rapidly enough to be raised immediately to incandescence without requiring the presence of a normal source of ignition. Pyrophoric solids thus have zero minimum ignition energy. A number of particulate solids are pyrophoric. Among these are iron sulfides (FeS and Fe_2S_3), unusually reactive metals such as zirconium and uranium, and finely divided solids such Raney nickel catalyst and powdered iron. For pyrophoric behavior to occur, the solids must be of small particle size, often about 1 micron in diameter. On dispersing, a cloud of pyrophoric dust is likely to ignite simultaneously at a number of points, and flame propagation is usually rapid.

Pyrophoric materials may accumulate unexpectedly in a system, as for example, iron sulfides and finely divided metals. Iron sulfides (FeS and Fe_2S_3) may form in anaerobic atmospheres in the presence of hydrogen sulfide. They can produce a hot-spot ignition source upon sudden exposure to air. Finely divided metals such as iron can also ignite upon sudden exposure to air as a result of their large reactive surface area. Ignition of reduced oxides such as ferrous oxide may be possible.

Table 6-1 lists a number of pyrophoric materials (CCPS, 1995). Additional information on pyrophoric materials is presented by Bond (1991), Carson and Mumford (1996), DOE (1994), Kayser and Boyars (1975), and Urben (1995). Johnson et al. (2003) discuss screening methods for pyrophoric materials.

Extreme precautions must be taken when storing and handling a pyrophoric material, due to the need for a highly reliable means of keeping the material isolated from the atmosphere. Typical control measures to reduce the risk from handling pyrophorics include (Carson and Mumford, 2002):

- Handling and storing the minimum quantities necessary at any time.
- Segregation of the material from other chemicals, particularly "fuels" (solvents, paper, cloth, etc.).

TABLE 6-1
Some Pyrophoric Materials

Category	Examples
Finely divided metals (without oxide film)	Aluminum, calcium, cobalt, iron, magnesium, manganese, palladium, platinum, titanium, tin, zinc, zirconium
Many hydrogenation catalysts containing adsorbed hydrogen (before and after use)	Raney nickel catalyst with adsorbed hydrogen
Alkali metals	Potassium, sodium
Metal hydrides	Germane, lithium aluminum hydride, potassium hydride, silane, sodium hydride
Partially or fully alkylated metal hydrides	Butyllithium, diethylaluminum hydride, triethylbismuth, trimethylaluminum
Arylmetals	Phenylsodium
Alkylmetal derivatives	Diethylethoxyaluminium, dimethylbismuth chloride
Analogous derivatives of nonmetals	Diborane, dimethylphosphine, phosphine, triethylarsine
Carbonylmetals	Pentacarbonyliron, octacarbonyldicobalt
Grignard reagents (RMgX)	Ethylmagnesium chloride, methylmagnesium bromide
Miscellaneous	Phosphorus (white); titanium dichloride

- Storage in tightly closed containers or vessels under an inert atmosphere or immersed in a nonreacting liquid (when process specifications allow this).
- Carrying out all transfers and other operations under an inert atmosphere or liquid.
- Immediate destruction and removal of spilled pyrophorics.
- Careful selection and provision of appropriate fire extinguishers in advance.
- Provision and use of appropriate eye/face protection, suits, and gloves.

Special precautions must be taken to safely dispose of pyrophoric materials such as spent hydrogenation catalysts (e.g., Raney nickel). The catalyst should be thoroughly wetted with water before the filter is opened up. Since a further ignition source is not required for combustion of pyrophoric materials, it is not necessary to use electrically classified equipment around such materials (NFPA 70, 2002).

Water-Reactive Particulate Solids

Water-reactive materials can be considered in two categories. Some materials react rapidly and violently with water and have an NFPA reactivity rating of 2 or 3 (see NFPA 704, 2001) based on water reactivity alone. These substances are of particular concern with respect to fire-fighting protection measures. NFPA reactivity hazard category 2 includes chemicals that may react violently with water or form potentially explosive mixtures with water. They have an instantaneous power density (product of heat of reaction and reaction rate) at 250°C of ≥10, but <100 W/ml. NFPA reactivity hazard category 3 includes chemicals that react explosively with water without requiring heat or confinement. They have an instantaneous power density of ≥100, but <1000 W/ml. The NFPA symbol **W** is always shown in the bottom special-hazards quadrant of the NFPA hazard signal for these water-reactive substances.

Other materials react relatively slowly but can generate heat and/or gases that can result in elevated pressure if contained. A water-reactive material of this type would generally be in NFPA reactivity hazard category 1 (having an instantaneous power density of ≥0.01, but <10 W/ml) unless the material also posed a significant instability hazard.

Table 6-2 lists some water-reactive materials (CCPS, 1995). As shown in this table, many metals and metallic compounds are water-reactive.

Carson and Mumford (2002) list a number of measures that can be taken to avoid problems with water-reactive materials:

- Store and use in such a way that accidental ingress of water, or contact with them, is avoided (roofs of storage areas should be regularly maintained to minimize leaks).

TABLE 6-5
Some Water-Reactive Materials

Category	Examples
Alkali and alkaline-earth metals	Calcium, potassium, sodium, lithium
Anhydrous metal halides	Aluminum tribromide, germanium tetrachloride, titanium tetrachloride
Anhydrous metal oxides	Calcium oxide
Grignard reagents	Ethylmagnesium chloride, methylmagnesium bromide
Metal alkyls	Aluminum alkyls, lithium alkyls
Metal amides	Lead amide, potassium amide, silver amide, sodium amide
Metal hydrides	Calcium hydride, lithium aluminum hydride, sodium borohydride, sodium hydride
Nonmetal halides	Boron trifluoride, phosphorus trichloride, silicon tetrachloride
Nonmetal halide oxides	Phosphoryl chloride, sulfuryl chloride, chlorosulfuric acid (inorganic acid halides)
Nonmetal oxides	Phosphorus pentoxide, sulfur trioxide
Low-molecular-weight organic acid halides and anhydrides	Acetic anhydride, acetyl chloride
Other	Calcium carbide

- Provide covered storage, off the ground, away from sprinkler systems, safety showers, overhead water lines, or condensate lines.
- Keep away from water sinks or faucets.
- Store under a chemically inert medium (storage vessels should be checked regularly to ensure that an adequate pressure of inerting gas is maintained at all times).
- Segregate from other flammable materials (e.g., solvents and combustibles).
- Operating personnel should use appropriate eye/face protection, clothing, and gloves when handling these solids.

Since these materials are water-reactive, water cannot be used to fight fires in equipment processing these materials or containers storing them. Other fire-extinguishing agents such as carbon dioxide or Halon substitutes should be used.

Carson and Mumford (1996) and CCPS (1995) present considerable information on water-reactive materials. Johnson et al. (2003) discuss screening methods for water-reactive materials.

6.3.4 Flames and Hot Gases

Flames

Almost any type of flame can ignite a combustible dust. In general, naked flames tend to be more hazardous than most other ignition sources, since they are normally comparatively large, are present for relatively long periods of time, and have temperatures in the range of 1000–2000°C, which is well above the minimum ignition temperature of most dusts. The flames can be present in such process equipment as direct-fired heating equipment (e.g., fired heaters and certain types of dryers), flares, thermal oxidizers, and steam boilers. Even cigarettes and other smoking materials can ignite dust clouds in some cases.

Case History of a Dust Explosion Caused by a Candle Flame

Bond (1991) reports the following interesting case history of a dust explosion caused by a flame from a candle:

On December 14, 1785, a boy was operating a flour sieving machine when the flour bridged in the silo feeding the machine. He dug deeply into the silo to get the flour to flow when the bridge broke and a cloud of flour came out into the room. It was evening, and the candle on the wall ignited the cloud of flour causing an explosion which blew out the front of the shop in Turin, Italy.

In general, process equipment with open flames should not be located in plant or building areas containing particulate solids processes. Direct-fired heating should be avoided wherever possible. Where direct heating is used to provide hot air for a drying process, a separate combustion chamber should be provided to keep flames away from the explosible dust. Direct heating should never be used when a flammable vapor may be present, as in the case of a hybrid mixture. Hot particles from the heater can be a source of ignition also. To avoid this danger the following precautions are recommended (Barton, 2002):

1. Combustion air and dilution air should be drawn from a reasonably dust-free zone and they should be filtered.
2. Burners should be cleaned regularly and they should be operated at the correct air–fuel ratio. Erratic burning and flame blow-off should be investigated and corrected immediately.
3. Powder must be prevented from entering the combustion chamber. This should be borne in mind when considering recirculation of air from the dryer.
4. Precautions should be taken to prevent large particles from entering the dryer. A 3-mm mesh filter, located after the heater, is recommended if large particles can otherwise enter the dryer with the hot gases.

Smoking should be prohibited in areas where there is a potential for a dust explosion. It may be necessary to set up "no smoking" areas, where all smoking materials must be given up before entering these areas. This is especially important where contract workers and staff personnel unfamiliar with the plant are involved, and additional supervision may be necessary.

Hot Vented Combustion Products

When hot vented combustion products (e.g., fireballs) are discharged from rupture disks, safety relief valves, and deflagration vents they pose an ignition source. If these hot gases or fireballs impinge on thin-walled equipment (e.g., adjacent silos) or on containers of particulate solids (e.g., multiwall paper bags or FIBCs) the solids may be ignited. Also, if the solids are unstable, they may undergo a decomposition if impinged upon by hot gases or fireballs.

When installing rupture disks, safety relief valves, and deflagration vents on process equipment they should be located so that the discharge stream will not impinge on equipment containing combustible solids.

6.3.5 Hot Work

Hot work includes such operations as cutting, welding, and hot tapping. These operations produce localized heating of the equipment and piping being worked on as well as sparks, which have been known to cause dust fires and explosions.

These ignition sources are handled administratively by the use of work permits and training to ensure compliance, and are now required by NFPA 654 (2000). Work should only be permitted to start after procedures have been implemented to remove all hazardous conditions internal and external to the equipment. Also, it must be ensured by gas analysis, gas freeing (removal) or inspection that no flammable gases/vapors are present that can be ignited. The surrounding area should also be cleared of any dust layers or accumulations as well as any flammable materials in the vicinity. Wooden floors and structures in the work area should be covered with fire-proofed material or adequately wetted with water before work begins. Flammable materials in adjacent rooms should also be removed if there is a possibility of sparks passing through cracks or openings.

When performing welding, cutting, or other hot work, the recommended practices given in NFPA 51B (2003) should be followed.

6.3.6 Hot Surfaces

Hot surface ignition of dust layers is a function of surface temperature and geometry, contact time, airflow, contamination, and the chemistry and history of the dust layer. Commonly encountered hot surface ignition sources in a chemical process industry plant are hot external surfaces of piping and process equipment (including motors), internal deposits on the walls of process equipment (e.g., dryers), and lighting fixtures and lamps.

The most effective way to minimize hot surface ignition is to prevent the build-up of dust layers on external hot surfaces, by good housekeeping. For internal hot surfaces, such as deposits in dryers, regularly scheduled inspections and cleanings can minimize this hazard. Barton (2002) discusses the various factors influencing hot surface ignition, and methods of evaluating them and what preventive measures can be taken. Field (1982) discusses hot surface ignition sources and their control for several types of dryers.

NFPA 654 (2000) states that the temperature of surfaces external to process equipment, such as compressors; steam, water, and process piping; ducts; and process equipment, within an area containing a combustible dust, shall be maintained below the lower of either 80% of the dust ignition temperature (in °C) or 165°C (329°F). An exception is made that permits maintaining temperatures within 80% of the minimum ignition temperature (in °C) of the dust layers if it is determined to be safe by recognized test methods acceptable to the authority having jurisdiction. Table 500.8 of NFPA 70 (2002) lists recommended maximum surface temperatures of electrical equipment for various dusts.

Lunn et al. (2002) present a thorough discussion of hot surface ignitions of combustible dust accumulations, providing details of the results of a European project in which the ignition behavior due to hot surfaces, produced both electrically and mechanically, has been studied.

Test methods for determining dust layer ignition are discussed in Chapter 4.

6.3.7 *Hot Particles*

Hot particles such as dryer discharge products or hot ash from waste incinerators can cause fires and dust explosions. They often are conveyed from one item of process equipment to another item in the form of incandescent or smoldering particulate solids, and thus cause fires and explosions in the downstream equipment.

Case History of a Dust Explosion Caused by a Hot Lump

Carson and Mumford (1996) describe the following incident of a dust explosion caused by a hot lump of dried powder.

In 1993 an explosion occurred at a milk powder factory damaging vibrofluidizers situated downstream of a spray dryer and fines return cyclones. A fire started some 3 to 5 minutes later in the vicinity of the exhaust fan from these cyclones. The proposed ignition source was an incandescent lump of milk powder which may have fallen off the spray nozzles support piping and into the fluid bed where it continued to heat up and char. It may have broken open on passage through a rotary valve and glowed or ignited within the fluid bed. Subsequent recommendations were to include the exhaust fans from all baghouses into the emergency shutdown system, relocation of butterfly valves to improve containment, consideration of monitoring the temperature probe in the vibrofluidizer exhaust to obtain early warning of any smoldering fire and of incandescent particles in the dryer and/or vibrofluidizers, using an infrared detector.

Ignition by hot particles can be minimized or avoided by their detection and extinguishment. One recent detection and protection system (called a spark detection system by some manufacturers) involves the use of infrared radiation detection sensors which are interlocked to an extinguishing system which precisely injects a small amount of extinguishing agent (e.g., water, steam, carbon dioxide, nitrogen, or other suppressant) into the conveying piping (see Section 6.6.5). Adequate measures must be taken to ensure that the detector cannot become covered with dust which could interfere with its operation. It should be recognized that spark detection systems are not designed to protect processes against the propagation of deflagrations and should therefore not be used without giving consideration to equipment isolation (discussed in Section 6.6.4).

Gibson and Schofield (1977) report that incandescent particles in spray dryers, having temperatures in the range of 600°C to 800°C, will not ignite dust suspensions if their diameter is less than 3–5 mm. Under these circumstances a fine screen of appropriate mesh size (less than 5 mm apertures) installed in the air inlet of the dryer and keeping it free of contamination by appropriate cleaning should eliminate the problem.

Eckhoff (2003) describes research on ignition of dust clouds caused by smoldering particles.

Gummer and Lunn (2003) describe some tests in which clouds of dust with a range of minimum autoignition temperatures (MAITs) were dispersed around dust agglomerations smoldering and flaming at various temperatures. Smoldering nests of dust proved to be poor ignition sources for most dust clouds, failing to ignite dusts even when there is a large difference between the nest temperature and the MIT of the dust cloud. Smoldering nests with temperatures above approximately 700–800°C were, however, able to ignite sulfur clouds. Flaming tests, on the other hand, were able to ignite clouds of dusts up to the maximum MIT used, 600–675°C.

6.3.8 *Friction and Impact*

Friction and impact (mechanical ignition sources) are among the major sources of ignition and have been implicated in 25% of the dust explosion incidents recorded in the United Kingdom (Barton, 2002). Such things as tramp metal, hot bearings, moving vanes and belts are among the most common sources of ignition. There are several different ways in which a dust cloud can be ignited by friction and impact. The heated contact area and the hot particles from the contact materials can ignite dust clouds. The presence of powder in the contact regions can also result in an ignition, particularly in rubbing friction situations, and for this condition, the frictional heating does not have to produce the temperature levels required to ignite the dust cloud, but only the lower temperature levels required to initiate exothermic decomposition of the powder. The progression of the reaction to a red heat condition can result in burning powders.

With hybrid mixture situations, potential ignition mechanisms are the heated contact area produced by rubbing or impact, and any hot particles of the contacting materials that are projected into the gas/air mixture which then can ignite the hybrid mixture.

In the majority of plants handling particulate solids, the consequences of a frictional ignition will be a fire or dust explosion. Certain chemicals (e.g., azides, nitro-compounds, perchlorates, nitrates) can react violently in bulk when processed in equipment where they may be subjected to friction or impact. Eckhoff (2003) discusses test methods available to characterize powders in terms of their reaction in bulk and layers to localized heat sources produced by friction and mechanical impact. There are studies being conducted at the present time in Europe to develop methods for determining the conditions, in terms of the minimum ignition energy (MIE) and minimum velocity between contact surfaces, required for frictional ignition of dust clouds (Barton, 2002).

Case Histories of Explosions and Fires Caused by Frictional Heating

Bond (1991) reports several incidents (fires and explosions). caused by frictional heating. Among them are the following two:

Sulfur passed through a pair of grinding rolls and fell into the base of a bucket elevator where it was raised to the top of a silo. The chain supporting the buckets had broken. During the repairs, the buckets were being pulled by hand when the chain broke and some buckets fell down the elevator casing. Almost immediately there was a dust explosion in the elevator casing, which was almost certainly caused by a friction spark produced by the buckets striking the casing.

A laboratory ventilation system for an animal containment area had metal ducting lined with polypropylene. The exhaust air from the animal section was treated with ozone in the ducting to remove the odors. A hole was being cut in the ducting using a hacksaw when a red glow was seen in the ducting. This quickly became a fire throughout the ducting. The investigation showed that the sawing action could generate temperature of 250–300°C by friction. The duct contained deposits of ammonium nitrate formed by the action of ozone with the ammonia in the atmosphere from the animal section. Additionally, there were organic dusts deposited in the ducting. The ammonium nitrate, being a strong oxidizing agent, reacted with the dust and was ignited by the friction caused by the saw cutting.

For process equipment with inner rotating elements, impact hazards can be minimized by operating at low tip speeds. Tip speed of a rotating element is calculated as follows:

$$v = \omega 2\pi r \qquad [6\text{-}2]$$

where v is the tip speed in m/s, ω is the rotational speed in revolutions per second, and r is the element radius in m.

Bartknecht (1989) presents the following criteria with regard to the ignition capabilities of sparks from rotating steel parts in dust–air mixtures:

$v < 1$ m/s There is no danger of ignition

$v < 1$–10 m/s Every case must be judged separately, considering the particulate solids and material-specific characteristics

$v > 10$ m/s There is danger of ignition in every case

To ascertain safe operating conditions with respect to mixing, refer to Table 5-1 which presents the relationship between minimum ignition energy (MIE) and minimum ignition temperature (MIT) for velocities (tip speeds) less than 10 m/s.

A number of things can be done to minimize the occurrence of friction or impact sparks:

- Use slow rotating equipment (in accordance with the above criteria) so that sparking will be minimized.
- Prevent tramp metal or other foreign bodies from entering process equipment, especially those with rotating internal elements, by installing fine screens or magnetic metal separators upstream of the equipment.
- For equipment with a long rotating shaft such as a screw conveyor or screw feeder, provide intermediate hangers (with purging) so that the shaft is strongly supported and will not deflect.
- Install shaft bearings and mechanical seals on the outside of the equipment so that solids will not collect in them and produce frictional heat-

ing. Air or inert gas purging of the seals will help to keep solids from entering them.

Grinding and conveying equipment can be prevented from overheating by properly controlling the feed rate of material into them. An additional safety measure is to provide an interlock to shut down the equipment when the design load is exceeded. All equipment with rotating inner elements should be regularly inspected to see that all the components are securely attached. Particular attention should be given after maintenance work has been done to ensure that all detachable panels have been reinstalled properly.

Lunn (2000) presents a comprehensive review of frictional ignition of powders, including a discussion of several items of process equipment prone to mechanical ignition problems. Eckhoff (2003) discusses frictional ignition of dust clouds.

6.3.9 Chemical Reactions

The decomposition of a chemical or the reaction of two chemicals may be sufficiently exothermic to raise the temperature to the ignition point. A number of possible routes to ignition exist via localized chemical reactions. Some of these (CCPS, 1993) are briefly discussed below:

- Catalysis
- Reaction with powerful oxidants
- Reactions of metals with halocarbons
- Thermite reactions
- Thermally unstable compounds

Catalysis

In reactive chemical systems catalysis can be a problem and can occur due to materials of construction, migration of catalyst from elsewhere in the system, or catalyzed reaction with either a contaminant or a secondary reactant present at abnormal levels. Decomposition of hot, reactive solids can be catalyzed by a wide range of high surface area materials such as activated carbon, powdered rust, rusty surfaces, etc.

Reactions with Powerful Oxidants

Powerful oxidants, such as peroxides, can react with certain particulate solids to cause their decomposition and result in a fire or explosion. The following case history relates one such incident.

Case History of a Fire Caused by Reaction with a Powerful Oxidant

Ness (2002) reports the following incident caused by a powerful oxidant.

A reaction between t-butyl hydroperoxide (TBHP) and another packaged catalyst caused ignition of the mixture and wooden pallets immediately above it, resulting in over $250,000 damage and a production shutdown of one month.

Pre-weighed buckets of TBHP were "staged" on wooden pallets on the second floor of a process building prior to use. A spill of TBHP onto a cable tray below the grating of the floor in the building resulted in a violent reaction between TBHP and incompatible particulate solids that were also handled in pre-weighed buckets and had accumulated over time due to previous small spills. The mixture ignited and caused wooden pallets located immediately above it to catch fire. The resulting fire of the pallets containing TBHP and other pre-weighed packaged materials caused extensive damage to the plant electrical power wiring, and control and instrumentation wiring. The fire was extinguished by the plant sprinkler system; however, one reactor train was out of service for a week, and two reactor trains were out of service for a month, because the pallets were near an electrical junction box and cable trays.

The following corrective actions were taken as a result of this fire:

1. The amount of TBHP and other package materials weighed up and placed in the process building at any one time was limited to immediate needs only, and no full pallets were allowed in the process building.
2. TBHP and other packaged materials were stored on dedicated, color-coded pallets.
3. Incompatible materials were not staged on the same pallets.
4. Areas were marked for storage of pallets or reactive materials that were away from critical equipment, steam lines, and potential contamination sources. Spill containment was provided for these areas.
5. The buckets for pre-weighed TBHP and other packaged materials were replaced with leak-tight containers.
6. TBHP containers were labeled to clearly differentiate them from other packaged catalysts.

This incident pointed out the importance of reviewing the procedures for handling packages of reactive chemicals as a part of the process hazard analysis. Items to consider are:

- What can happen if the packaged materials come in contact with each other outside of the reactor or in an uncontrolled manner.
- The location of package storage relative to sensitive areas or critical equipment items.

- How to safely handle and dispose of empty packages.

Space for small packaged charges needs to be considered in the design stage to avoid having to "cram" such staging areas into cramped areas or near critical equipment items.

Reactions of Metals with Halocarbons

Halocarbons such as refrigerants, solvents, and lubricants can react with some metals and cause explosions. For example, there have been numerous explosions with aluminum. However, reactions with other metals such as barium, lithium, magnesium, beryllium, and titanium may also occur with halocarbons, and they should be investigated for this possibility. Metal-halocarbon reactions may be hazardous in themselves or provide sources of ignition (CCPS, 1993).

Thermite Reactions

Aluminum, magnesium, titanium, and other light metals and their alloys are capable of undergoing a highly exothermic reaction on frictional contact with certain metal oxides. Other oxygen-containing salts such as nitrates can also react exothermically (Bond, 1991). The glancing impact of stainless steel, mild steel, brass, copper-beryllium, bronze, aluminum, copper, and zinc onto aluminum smears on rusty mild steel can initiate a thermite reaction and cause the ignition of flammable gas and solvent atmospheres and dust clouds formed from certain powders (Gibson et al., 1967). The most common of these thermite reactions is between aluminum and oxides such as iron rust or red lead. In this reaction, aluminum is oxidized and ferric oxide is reduced, releasing a great amount of heat. Analogous reactions can occur in other systems of metals and metal oxides. It is important to recognize that thermite reactions can be hazardous ignition sources even if the metal oxide is present only superficially.

Case History of an Explosion in a Tanker Caused by a Thermite Reaction

Bond (1991) describes the following incident.

An explosion in one of the tanks of the SS *Esso Durham* occurred on January 19, 1961. It was attributed to the impact of a brass object, weighing 36 pounds, onto a magnesium anode on the wall of the tank after a fall of 16 feet. From simulations of possible alternatives, it was considered likely that a grazing impact on the rough, dirty magnesium surface would have smeared the brass, which would then have reimpacted the anode a fraction of a second later, causing a thermite reaction and explosion (ignition of the gas still in the tank). The oxide necessary for the thermite reaction was probably provided by dirt in the area (ignitions were not obtained when impacting a clean magnesium surface). The possibility of ignition from impact onto hardened steel after a 40-foot fall was eliminated by tests.

Thermally Unstable Materials

Many particulate solids decompose and sufficient heat is generated to provide a source of ignition for themselves or other chemicals which may be present. Many unstable solids start to decompose as a result of a small heat input, e.g., friction. One well-known unstable substance is copper acetylide. Silver, gold, and mercury acetylides can also be formed and are unstable. Bond (1991) presents a list of many unstable solid substances.

Case History of a Dust Explosion Caused by Thermal Decomposition

Ness (2002) describes the following incident caused by thermal decomposition of a powder.

A dust explosion, caused by product decomposition triggered by iron contamination, occurred in a baghouse and spread throughout an entire dryer system.

Powder deposits in a baghouse or its inlet duct underwent thermal decomposition due to iron contamination. The thermal decomposition led to a fire and dust explosion in the baghouse.

Explosion panels successfully relieved the explosion, but a pressure wave from the baghouse traveled into upstream equipment. Explosion vents in the other equipment items successfully relieved the overpressure.

The iron contamination probably occurred in the inlet duct, where powder had settled out and was in contact with bands of rust formed where carbon steel support ribs had been welded to the outside of the duct. The powder being dried was known to be thermally unstable, and stabilizing agents were normally added to it. The effect of contamination, especially rust, on the powder's stability was not understood.

After this incident, the following corrective actions were taken:

1. All equipment was checked for contamination and surface rust was removed from all surfaces where found.
2. Insulation was replaced or installed as necessary to prevent condensation inside steel equipment.
3. Inspection procedures were revised to include checks for wet equipment if the dryer train went down for more than 2 days.

Ignition by chemical reactions can be minimized or avoided by knowing what these hazards are and taking appropriate measures to prevent them from occurring. For example, in using acetylene, all equipment and piping (including lubricants and greases) that may contain copper, silver, gold, and mercury are prohibited from being used since acetylides of these metals can form which are thermally unstable.

Extensive data on chemically incompatible substances are listed and discussed by Medard (1989) and Urban (1995).

6.3.10. Physical Sources

Physical adsorption can cause ignition as often the heat of adsorption is quite large. It can occur on adsorbates such as activated carbon, silica gel, and molecular sieves. As indicated in Section 3.1.1, various adsorbates can significantly affect the ignition times for the thermal decomposition of lead azide. If preloading is not carefully done, large exotherms can result. There is also the possibility of exothermic chemical adsorption and catalyzed polymerization of surface reaction, for example with chemisorbed oxygen.

Ignition by heat of adsorption can be minimized by proper preloading (wetting) of the adsorbent and providing temperature monitoring and interlocks to divert the feed stream from the adsorber when a high temperature is detected.

6.3.11 Electrical Equipment

Dust explosions and fires can result from electric sparks (arcing) produced from electrical equipment such as motors, switchgear, circuit breakers, broken wiring, power tools, etc.

Where ignition does occur, it is usually the result of a break in a current-carrying cable and arcing in the ionized air. Another cause of ignition is the overloading of a conductor followed by overheating and arcing due to excessive voltage.

Hot surfaces of electrical equipment can also act as ignition sources for dust layers deposited on them, as they can potentially cause smoldering and burning (fires).

See Section 6.4.1 for a more detailed discussion of electrical equipment hazards.

6.3.12 Lightning

Particulate solids are not usually ignited by lightning if a silo is properly grounded. However, lightning can possibly ignite flammable vapors from a hybrid mixture if it strikes near where wet solids are stored. Ignition can occur by the following four mechanisms (Bond, 1991):

1. Due to the fast current pulses of up to 200,000 amperes, the associated fast changes in magnetic fields are capable of inducing voltages and currents in circuits and structures. Spark-over to grounded points can then occur with sufficient energy to ignite a flammable mixture.
2. A strike on a metal plate could create local heating such that the vapor on the other side of the plate reaches a temperature above the autoignition temperature.
3. By direct entry into the vapor space of a vessel.
4. By sparking of the lightning conducted through piping at a flange joint.

Outdoors process equipment containing particulate solids can be protected from lightning by installation of lightning protection systems as discussed in CCPS (1993), NFPA 780 (2000), and API RP 2002 (1998).

6.3.13 Projectiles

Projectiles (metal fragments from an explosion) that impact and penetrate thin-walled metal vessels, such as silos, can result in the ignition of the solids or vapors from a hybrid mixture stored in the silo. The projectiles give rise to frictional heating and ignition.

Vessels containing particulate solids can be protected against projectiles emanating from adjacent equipment by either designing the adjacent equipment to contain a deflagration (it will not rupture) or by installing a barricade (protective wall or enclosure) around the vessel containing the particulate solids.

More detailed discussions of ignition sources are presented by Bond (1991), Carson and Mumford (2002), CCPS (1993), Palmer (1973), and Lees (1996b). Barton (2002) presents lists of ignition prevention measures for size reduction processes, pneumatic conveying systems, screw conveyors, drag link or en-masse conveyors, belt conveyors, bucket elevators, dryers, storage silos, and dust collectors (filters)

6.4 ELECTRICAL EQUIPMENT HAZARDS AND AREA CLASSIFICATIONS

6.4.1 Electrical Equipment Hazards

As mentioned previously in Section 6.3.11, electrical equipment such as motors, circuit breakers, transformers, and switchgear can produce sparks and ignite dust clouds and hybrid dust/air mixtures in the vicinity. Several other potential hazards relevant to electrical equipment and enclosures are as follows:

1. Ingress of dust into enclosures, and subsequent ignition to cause smoldering or burning (fires). Dust that enters an enclosure will not accumulate as an explosive cloud of particles suspended in the air inside the enclosure, but will settle out as dust layers on internal surfaces and can become heated.
2. Deposition of dust layers on external surfaces of motors and other electrical apparatus that are hot, which can result in smoldering and burning (fires).
3. Electrically conductive dusts can cause short-circuiting when deposited on exposed electrical components and circuits.
4. Abrasive and/or corrosive dusts can damage delicate components of electrical equipment.
5. Electric shock.

A short discussion of types of enclosures and purged and pressurized enclosures is presented below.

Types of Enclosures

To use electrical equipment in a hazardous atmosphere, specific types of enclosures are used to house the equipment. Class II, Division 1 locations require the use of dusttight, ignitionproof enclosures, NEMA Type 9, that exclude all dust from the interior of the enclosure, and at the same time are designed such that enclosed heat-generating devices will not cause external surfaces to reach temperatures capable of igniting or discoloring dust on the enclosure or igniting dust–air mixtures in the surrounding atmosphere.

Purged and Pressurized Enclosures

An option that can be used to allow the use of conventional electrical arcing equipment in hazardous areas is to create an enclosure that is less hazardous (or nonhazardous) by means of dry air or nitrogen purging and pressurization systems. For Class II locations, types X, Y, and Z pressurized enclosures are applicable to both Division 1 and Division 2 locations. NFPA 496 (2003) discusses the various types and operating principles of purged and pressurized enclosures for electrical equipment.

It should be recognized that when pressurized enclosures are shut down (pressurization is stopped), dust ingress may occur.

This dust will not always be removed when the enclosure is repressurized. Dust deposits inside enclosures may cause a fire risk or may be insignificant. However, they are most unlikely to cause an explosion as there is no way to make a cloud.

A number of measures can be taken to avoid or minimize hazards with electrical equipment, such as the following:

1. Electrical equipment, components, wiring, etc. should be specified and installed in accordance with Article 502 of the NEC (NFPA 70, 2002).
2. If available, Approved (e.g., by UL or FM in the United States or a recognized test laboratory in other countries) electrical equipment, components, wiring, etc. should be used. When approved equipment is not available, equipment listed, labeled, or approved by another recognized test laboratory is acceptable.
3. To minimize or remove the potential for an ignition by sparks from electrical equipment, such equipment should be located, wherever feasible, outside of areas in which particulate solids are handled or processed. If this is not feasible, consideration should be given to using intrinsically safe electrical equipment for Class II, Division 1 hazardous locations, or to using nonincendive electrical equipment for Class II, Division 2 hazardous locations.
4. Use dust-tight. ignitionproof electrical equipment in Class II, Division 1 hazardous locations to prevent ingress of dust into enclosures.
5. Use NEMA Type 9 (NEMA STD 250, 2003) enclosures for use in hazardous locations classified as Class II, Division 1, Groups E, F, or G.
6. Do not allow deposits of dust on hot surfaces of electrical equipment (e.g., motors and lighting fixtures) to accumulate. They should be cleaned up as soon as possible.
7. Hazards from electrical equipment and enclosures can be avoided or minimized by correctly classifying area locations and selecting the equipment accordingly (see Section 6.4.2 below).

In Europe, IEC 61241 (published in multiple parts) also provides guidance on the hazards and use of electrical equipment in the presence of combustible dusts. They are discussed by Eckhoff (2003, 2004). However, the requirements in these standards are not always in agreement with that in the NEC.

Several other sources of information on electrical equipment hazards are the books by Buschart (1991), Fordham-Cooper and Jones (1998), Greenwald (2002), and McMillan (1998). Also, FMG 5-1 (2002) contains much useful information about electrical equipment hazards.

6.4.2 Electrical Area Classification

The classification of areas inside a CPI facility that handles combustible dusts is the basis for the correct selection of electrical equipment, components, wiring, and enclosures. Area classifications in the United States are defined in the National Electric Code (NEC), NFPA 70 (2002), Article 502, Class II Locations. In Europe, area classification is defined in accordance with the rules of the International Electrotechnical Commission (IEC) standard IEC 61241-10 (2002).

Locations are classified according to the properties of the material being handled and its surrounding atmosphere. Factors that affect the area classification for particulate solids may include availability of the combustible or explosive solid, combustibility or explosibility of the solid, operating temperature and pressure, autoignition temperature, resistivity of the dust, explosive pressures, dust layer ignition temperature, open or sealed conduit, and ventilation. Hazardous locations are classified by Class, Group, and Division. Class I is for flammable gases and vapors, Class II is for combustible dusts, and Class III is for ignitable fibers or flyings. Class I flammable vapors and gases are divided into Groups A, B, C, and D. In Europe, equivalent groups are IIA, IIB, and IIC, and classification by these groupings is permitted by the NEC.

Class II combustible dusts are divided into Groups E, F, and G, which are defined as follows (NFPA 499, 2004):

- *Group E:* Atmospheres containing combustible metal dusts, including aluminum, magnesium, and their commercial alloys, or other combustible dusts whose particle size, abrasiveness, and conductivity present similar hazards in the use of electrical equipment.
- *Group F:* Atmospheres containing combustible carbonaceous dusts that have more than 8 percent total entrapped volatiles (see ASTM D 3175 for coal and coke dusts) or that have been sensitized by other materials so that they present an explosive hazard. Coal, carbon black, charcoal, and coke dusts are examples of carbonaceous dusts.
- *Group G:* Atmospheres containing other combustible dusts, including flour, grain, wood flour, plastic, and chemicals.

Division designations are characterized in relationship to the probability of the material being within the flammable or explosive regions. For particulate solids (Class II), they are as follows (NFPA 499, 2004):

DIVISION 1 LOCATIONS

A location is considered Division 1 if:

1. If a dust cloud is likely to be present under normal conditions.
2. If a dust layer greater than ⅛inch thick is present under normal conditions.

"Normal" does not necessarily mean the situation that prevails when everything is working properly. For instance, if a bucket elevator requires frequent maintenance and repair, its repair should be viewed as normal. If quantities of ignitable dust are released as a result of the maintenance, the area is Division 1. However, if the bucket elevator is replaced and now repairs are not usually required between turnarounds, the need for repairs is considered abnormal. The classification of the area, therefore, is related to equipment maintenance, both procedures and frequencies. Similarly, if the problem is the buildup of dust layers without the presence of visible dust suspensions, good and frequent cleaning procedures or the lack thereof will influence the classification of the area.

DIVISION 2 LOCATIONS

A location is considered Division 2 if the area is likely to have ignitable dust suspensions or hazardous dust accumulations only under abnormal conditions. The term "abnormal" is used here in a limited sense and does not include a major catastrophe.

As an example of electrical equipment classification for equipment handling particulate solids, consider a bucket elevator conveying an organic dust with a bulk density of 25 lb/ft^3, with hazardous accumulation defined as >$\frac{3}{32}$ inch (see Section 2.2.3.1 of NFPA 654 for adjustments to hazardous accumulation as a function of bulk density). Inside the bucket elevator enclosure the classification would be Division 1 because dust is in suspension and hazardous dust accumulations would be a normal occurrence. Electrical classification outside of the elevator enclosure is a function of enclosure design, dust leakage, and housekeeping. If the enclosure is tight, and leakage of dust from the elevator is not allowed to accumulate in the surrounding area on buildings or equipment to more than about 0.01-inch (this is about the point where the surface under the dust layer is barely discernible if the dust is white), then there is no Division 2 area outside of the enclosure, that is, the area is unclassified, and ordinary electrical equipment is acceptable. A case of a rare breach of the elevator enclosure, resulting in a discharge of dust followed by a prompt clean-up, is not cause for assigning an electrical classification to the surrounding area. If the enclosure has infrequent leaks (a few times a year), resulting in less than hazardous accumulations in the area (up to $\frac{3}{32}$ inch), and if cleaned up promptly (during the shift), the area may be unclassified. If the enclosure leaks frequently or continuously, but good housekeeping keeps accumulations to a depth of no more than $\frac{3}{32}$ inch, and rarely do accumulations exist greater than $\frac{3}{32}$ inch, then the area around the elevator, up to a separation distance of no less than 30 feet (NFPA 654), should be classified as Division 2, and beyond the 30 feet the area is unclassified. If, at the other extreme, the elevator housing leaks profusely and accumulations on buildings and equipment frequently exceeds hazardous levels (>$\frac{3}{32}$ inch), then the hazardous area outside the bucket elevator should be classified Division 1. Adjacent areas beyond the Division 1 area should be classified as Division 2, as described above.

Electrical equipment in chemical process plants does not fail often, but failures occur occasionally. Furthermore, the electrical installation requirement of the NEC for Division 2 areas is such that an ignition-capable spark or hot surface will occur only in the event of an abnormal operation or failure of electrical equipment. Otherwise, sparks and hot surfaces are not present or are contained in enclosures. On a realistic basis, the possibility of process equipment and electrical equipment failing simultaneously is remote.

Walls are more important in separating Division 1 areas from Division 2 and unclassified areas in Class II areas than in Class I areas. Only unpierced solid walls make satisfactory barriers in Class I areas. Closed doors, lightweight partitions, or even partial partitions could make satisfactory walls between Class II, Division 1 areas and unclassified areas. Area classification does not extend beyond the wall, provided it is effective in preventing the passage of dust in suspension or layer form.

Where Group E dusts are present in hazardous quantities, there are only Division 1 locations. The NEC does not recognize any Division 2 areas for such dusts.

In Europe, areas are designated by Zone numbers 20, 21, and 22. These are defined as follows:

- *Zone 20:* Area in which a hazardous flammable atmosphere formed by a dust cloud in air is present continuously or for a long period or frequently, and where dust layers of unknown or excessive thickness may be formed. Note: Areas where piles of dust are present but where dust clouds are not present continuously, or for a long period, or frequently are not included in this zone.
- *Zone 21:* Area in which a hazardous flammable atmosphere formed by a dust cloud in air is likely to occur during normal operation, and in which layers of combustible dust will usually be present.
- *Zone 22:* Area where a hazardous flammable atmosphere formed by a dust cloud is unlikely to occur in normal operation, but if it does occur it will exist only for a short period, or in which accumulations of layers of combustible dust are present.

At the present time, these have not yet been incorporated in the NEC. McMillan (1998) discusses these zone numbers. Eckhoff (2003, Chapter 8 of his book) discusses the European standards for electrical classification of areas containing combustible dusts.

UNCLASSIFED AREAS

Certain areas can be considered as unclassified areas if experience has shown that the release of ignitable dust suspensions from some operations and apparatus is so infrequent that area classification is not necessary. For example, where combustible dusts are processed, stored, or handled it is not necessary to classify the following areas:

- Areas where materials are stored in sealed containers (e.g., bags, drums, or fiber packs on pallets or racks).

- Areas where materials are handled with well-maintained closed piping systems.
- Areas where pelletized materials with minimum dust are handled or used.
- Areas where closed tanks are used for storage and handling.
- Areas where dust removal systems prevent (1) visual dust clouds, and (2) layer accumulations that make surface colors indiscernible.
- Areas where excellent housekeeping prevents (1) visual dust clouds, and (2) layer accumulations insufficient to make surface colors indiscernible.

Dust removal systems that are provided to allow a nonclassified area should have adequate safeguards, with high availability, and warnings against failure.

Buschart (1991), Korver (1995), McMillan(1998), NFPA 70 (2002), and NFPA 499 (2004) present more detailed information on electrical equipment hazards and area classification for combustible dust locations.

6.5 DEFLAGRATION PREVENTION METHODS

This section discusses three methods for preventing dust deflagrations, that is, prevention or mitigation of dust cloud formation, oxidant concentration reduction (inerting), and combustible concentration reduction (air dilution).

6.5.1 Prevention or Minimization of Dust Cloud Formation

Although it may not be possible to eliminate dust cloud formation and dispersion completely, careful design and layout of the plant and equipment can often reduce the potential for dust clouds forming and can minimize their volumes. For example, when pneumatically conveying solids into a silo, it is better to bring the solids into a cyclone or baghouse located above the silo and slowly feed the solids into the silo by a rotary valve than to allow them to fall freely from the entry nozzle. This minimizes dust cloud formation, whereas, filling by free fall results in a large portion of the space above the deposited solids becoming filled with a dust cloud, which could be of an explosible concentration.

Another case where dust cloud formation occurs is at the transfer points in a belt conveyor system. The dust is generated almost exclusively at two points: at the tail pulley where solids are received from preceding equipment, and at the head pulley where solids are discharged. Dust cloud formation can be minimized by proper design of the transfer points, taking into consideration such factors as chute design, skirt boards, troughing and transition, and centering the load on the belt (Swinderman et al., 1991). The transfer points should be enclosed and the enclosure connected to a ventilation system which removes any fines from the air.

For pneumatic conveying systems, attrition and subsequent dust cloud formation can be minimized by selecting a dense-phase (low velocity) conveying systems rather than a dilute-phase (high velocity) system. Dense-phase conveying is gentler and results in fewer fines being produced.

A number of other methods for minimizing dust cloud formation are:

- Reduce heights of solid fall (minimizes dispersion).
- Shield the solids from external air flows.
- Reduce surface area to volume ratio of flowing solids streams.
- Minimize fines production during comminution by using slowly rotating size-reduction equipment.
- Delay size reduction to just before the process requires it.
- Make larger particles by granulation or agglomeration.
- If feasible, spray water or a wetting agent onto stockpiles and flowing streams to increase cohesion at the surface of the solids.
- Immediately clean up dust emissions and spills and do not allow accumulation of dust layers on process equipment, building floors and structural members (beams, cross-bracing, ledges, etc.).

6.5.2 *Oxidant Concentration Reduction (Inerting)*

Oxidant concentration reduction, commonly called inerting, is a very effective method of deflagration prevention. It involves the addition of inert gases to the atmosphere in process equipment or inert solids to a dust itself to prevent the formation of explosible dust clouds.

6.5.2.1 Use of Inert Gases

Oxidant concentration reduction using inert gases can be accomplished by purging and blanketing. Purging involves displacing the oxidant (oxygen in air) from the process vessel before the solids are introduced. Blanketing involves maintaining an inert gas atmosphere above the solids so as to prevent ingress of air into the process vessel.

PURGING

Before filling of process vessels with combustible particulate solids, especially hybrid mixtures, or solvents, the vessels should be purged. Purging can be done in any of the four following ways: (1) pressure purging, (2) vacuum purging, (3) combined pressure-vacuum purging, and (4) sweep-through purging.

Pressure Purging. This method is used for pressure vessels (designed for 15 psig or higher). A commonly used procedure practiced by a number of companies is to pressurize the vessel to 15 psig with inert gas two or three times until the desired final oxygen concentration is reached.

Equation 6-3 below (Pilkington, 1999) shows the relationship between oxygen concentrations, pressures, and number of pressurization purges:

$$C_N = C_i\,(1/P^N) \qquad [6\text{-}3]$$

where N is the number of purges, C_N is the final oxygen concentration in the vessel after N purges (volume %), C_i is the initial oxygen concentration in the vessel (usually 20.7 volume %), P is the purge pressure (atmospheres absolute). This equation ignores any amount of oxygen contained in the purge gas. Table 6-3 shows final oxygen concentrations after N purges as calculated by Equation 6-3.

If the purge gas (nitrogen) is not 100% pure and contains some oxygen (as is often the case), the equation now becomes (Pilkington, 1999):

$$C_N = C_p + (C_i - C_p)[1/P^N] \qquad [6\text{-}4]$$

where C_p is the oxygen concentration of the purge gas, and all the other symbols are the same as in Equation 6-3.

If the equation is rearranged, the number of pressurizations required to reach a desired oxygen level can be calculated by Equation 6-5 below (Pilkington, 1999):

$$N = \log[(C_i - C_P)/(C_N - C_P)]/\log P \qquad [6\text{-}5]$$

where the symbols have the meanings in Equations 6-3 and 6-4.

The above equations are used when the equipment is simple, such as a single reactor or mixing vessel. Where a system is large and contains piping branches, the air in the closed end of the system will be compressed by the inert gas, but is unlikely to mix well. Thus, when the pressure is released, the gas will simply expand, and the oxygen content in the branches will remain similar to that before it was compressed. Therefore, it will be necessary to take account of this branching when calculating the final oxygen content.

TABLE 6-3
Relationship between Number of Purges, Pressure, and Oxygen Concentration

Oxygen Contration after *N* Purge Cycles	
N Cycles	Concentration, %
1	10.39
2	5.14
3	2.55
4	1.26
5	0.62

When the system is very complex, it may be necessary to release the inert gas pressure from each branch in turn to ensure adequate displacement of the original gas/air. If this requires a large number of purges, then vacuum purging may be better (if the equipment can withstand vacuum).

To ensure that the pressure purging has achieved the desired final oxygen concentration in the system, measurements should be taken by a suitable oxygen analyzer the first time the system is purged. Once oxygen measurements have been taken and the oxygen level is found to be acceptable, then it is usually satisfactory for normal operation to infer that the oxygen concentrations are the same as during the initial test, providing that the exact same purging conditions are used as in the test. For process equipment with high speed or close-clearance moving parts (e.g., centrifuges, mills, conveyors, etc.) continuous oxygen monitoring is recommended.

Where a purging system is operated under pressure, any leaks will result in inert gas being emitted into the workplace. Therefore, adequate precautions should be taken to ensure that nearby personnel cannot be asphyxiated by escape of inert gas.

In closed workplaces, adequate ventilation should be provided. Where purged equipment is located in the open air, asphyxiation will only present a risk under conditions of gross leakage.

Vacuum Purging. This method is used for vessels that can withstand full or close-to-full vacuum. This is a more efficient purging method than pressure purging as it uses less inerting gas (usually nitrogen) than pressure purging. In this method, the vessel is connected to a vacuum device, and the pressure is reduced to a low vacuum (below atmospheric pressure), and then the vacuum is broken with an inert gas.

Equation 6-6 below (Pilkington, 1999) can be used to calculate the final oxygen concentration in a process vessel after N number of vacuum purges (this equation includes the oxygen in the purge gas):

$$C_N = C_P + (C_i - C_P)P^N \qquad [6\text{-}6]$$

A calculation will show that after only one vacuum purge to 0.0263 atmospheres absolute (20 mm Hg abs.), the final concentration of oxygen will be 0.55 volume % (based on no oxygen in the purge gas).

The number of vacuum purges to reach a desired level of oxygen can be calculated by Equation 6-7 below (Pilkington, 1999):

$$N = \log\,[(C_N - C_p)/(C_i - C_p)]/\log P \qquad [6\text{-}7]$$

If the process equipment is operated under vacuum during normal operation, then any leaks will allow air to enter the equipment and this will gradually raise the oxygen concentration and make the inerting less effective. The ingress of air can be detected by two methods: the inferential method and oxygen monitoring. The inferential method relies on the vacuum source being isolated and the rate of pressure-rise being measured. Thus it is possible to estimate the maximum oxygen concentration that would occur with time in the system at a given vacuum level. It may be desirable to continuously monitor the oxygen level, which would provide adequate warning that the oxygen level in the system is rising and the operation may become hazardous. For systems operating under vacuum, a safety factor should be applied to the above equations to ensure that any air(oxygen) ingress due to leaks does not compromise safety.

Combined Pressure–Vacuum Purging. If a vessel is designed for vacuum and pressure service and is connected to a vacuum source, then it can be purged by combined pressure–vacuum purging. The procedure can be to pressure purge first and then vacuum purge, or vice versa. When combining vacuum and pressure purging, less nitrogen is used compared to pressure purging, especially if the initial cycle is a vacuum cycle. Crowl and Louvar (2002) present the following equation to calculate the oxygen concentration at the end of the *N*th pressure cycle when using impure nitrogen (contains oxygen):

$$(y_N - y_{oxy}) = (P_L/P_H)^N(y_o - y_{oxy}) \qquad [6\text{-}8]$$

where y_N is the mole fraction of oxygen in the vessel at the end of N purge cycles, y_{oxy} is the mole fraction of oxygen contained in the nitrogen, y_o is the initial mole fraction of oxygen in the vessel, P_L is the low pressure (psia, mm Hg, atmospheres, or bars), and P_H is the high pressure (in the same units as P_L).

Sweep-Through Purging. This method (also called flow-through purging) is used for vessels designed for low pressure close to atmospheric. This is accomplished by adding the inert gas continuously for a period of time and exhausting it to a safe location. The volumetric quantity of inert gas required to reduce the oxygen concentration in a vessel from C_1 to C_2 is Q_vt and can be determined using Equation 6-9, given below (Crowl and Louvar, 2002):

$$Q_vt = V \ln[(C_1 - C_o)/(C_2 - C_o)] \qquad [6\text{-}9]$$

where Q_v is the volumetric flow rate of purge gas (ft^3/min or m^3/min), t is the time required to reach the desired oxygen concentration in the vessel (min), V is the vessel volume (ft^3 or m^3), C_1 is the initial oxygen concentration in the vessel, C_2 is the final oxygen concentration desired in the vessel, and C_o is the oxygen concentration in the purge gas (nitrogen).

Figure 6-2 can be used to estimate the theoretical quantity of sweep-through inerting gas needed to achieve a certain oxygen level. Both Equation 6-9 and Figure 6-2 assume perfect mixing.

Since sweep-though purging does not usually achieve uniform flow and mixing through a vessel, especially a large vessel, it may be necessary to use appreciable more inerting gas than the amount calculated by Equation 6-9 or

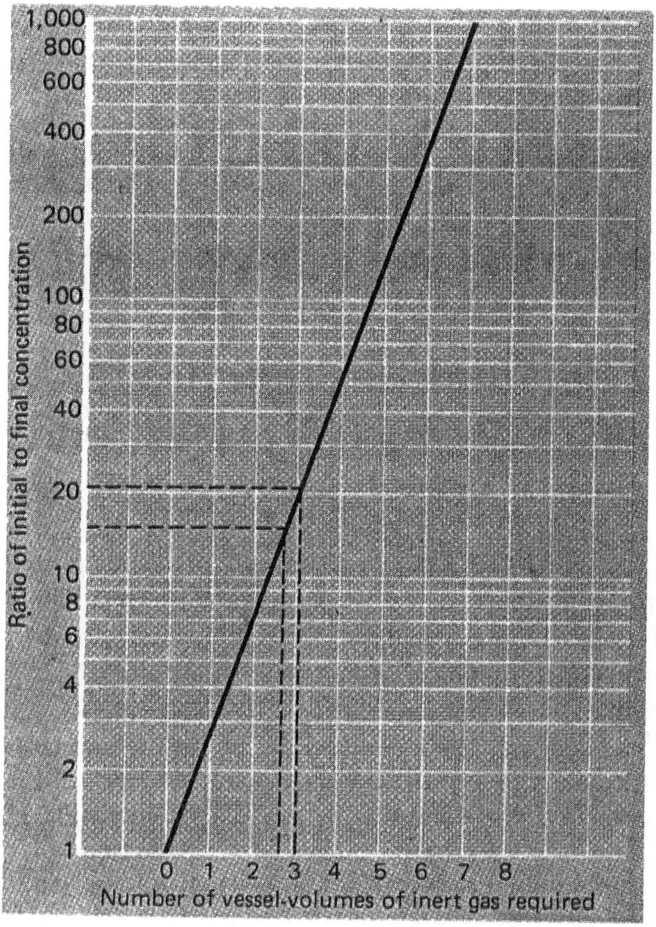

Figure 6-2 Quantity of sweep-through inerting gas required. (*Source*: Blakey, P., and Orlando, G., Using Inert Gases for Purging, Blanketing, and Transfer, *Chem Eng.*, pp. 97–102, May 28, 1984. Reprinted with permission of *Chemical Engineering*.)

obtained from Figure 6-2. Few data exist on defining the degree of mixing, and NFPA 69 (2002) recommends that a correction factor no smaller than 4 be used (i.e., multiply the value obtained from Equation 6-9 or Figure 6-2 by at least 4). Because of this, it is recommended that oxygen concentrations be measured to establish the necessary inert gas quantity during the first time sweep-though inerting is used.

A comparison of the results using Equation 6-9 and Figure 6-2 is presented here, assuming a silo of 300 cubic feet capacity and a desired final oxygen concentration of 3.0 volume percent.

(a) Using Equation 6-9:

$$Q_v t = (300)\ \ln(20.7/3.0) = 300(1.932) = 580\ \text{ft}^3$$

(b) Using Figure 6-2: The ordinate is (20.7/3.00) = 6.9 and the abscissa is 1.95 vessel volumes of inert gas required. The purge volume required is then (300)(1.95) = 585 ft.

The inert gas, if lighter than air, should be introduced in a way that promotes good mixing. For example, for tall vessels (ones with a high L/D) try to introduce the gas at the bottom of the vessel, if possible, and allow it to flow through the entire vessel volume and exit at a nozzle on the top head. If the gas cannot be introduced through the bottom head, then it should be introduced through a nozzle on the top head via a dip pipe that goes down to the bottom of the vessel. The inert gas entry nozzle should be located as far away as possible from the exit vent nozzle. For large volume vessels, multiple inlets and outlets should be considered. If the inert gas is heavier than air (e.g., argon), it should be introduced at the top of the vessel so that the flow is from top to bottom.

Although there are quite a few articles, books, and guidelines available (see the references at the end of this section) that present information on the application and installation of inerting systems, improper inerting still occurs and results in fires and explosions. Morrison et al. (2002) discuss these problems and present six case histories (including two dealing with particulate solids) of fires and explosions that resulted from improper inerting. They also present design considerations for an inerted system to assist engineers in developing a more effective inerting strategy.

BLANKETING

Blanketing, also called padding, is used to prevent the ingress of air into process equipment storing or processing particulate solids. In this method, a pressurized head space is maintained above the solids level, usually 2 to 10 inches of water gauge.

The blanketing can be done in two ways: trickle or balanced-pressure blanketing (Blakey and Orlando, 1984). In trickle blanketing a continuous flow of inert gas is used. This is often done when the stored solids could block vent nozzles and lines. However, while it is simple, this method may not provide a total inert blanket during all operations. A secondary high-rate purge system can be added to begin operating when the vessel is being discharged. Trickle blanketing may use considerably more inert gas than does the balanced-pressure method. Balanced-pressure blanketing maintains a small positive pressure in the vessel headspace, usually by means of a split-range pressure control system. When the vessel is being filled, the pressure sensor/controller opens a control valve to allow the vapors and inert gas in the headspace to be vented, and when the vessel is being discharged, the control valve is opened to allow inert gas to enter the vessel.

Blakey and Orlando (1984) present procedures for calculating blanketing-gas requirements.

TYPES OF INERT GASES

A number of gases may be used for inerting, including nitrogen, carbon dioxide, argon, helium, steam, and flue gases. The choice of the appropriate inert gas for a specific application depends on a number of factors, including:

- Cost
- Availability
- Reliability of supply
- Compatibility with the particulate solids
- Volume effectiveness in reducing explosibility

Table 6-4 shows the relative merits of inert gases.

Nitrogen is the most widely used inert gas as it is relatively inexpensive, it is widely available, and is compatible with most particulate solids. However, it should not be used with certain metal powders and dusts such as magnesium, titanium, uranium, and zirconium as nitrogen reacts with these metals and fires and explosions can occur. Carbon dioxide also reacts with some metal powders and dusts such as aluminum, magnesium, thorium, titanium, uranium, and zirconium. Argon and helium should be used instead of nitrogen or carbon dioxide with these metal powders and dusts. Steam can be used to inert systems containing solvents and vapors/gases, but is not usually used in systems containing particulate solids as it contaminates them or causes other problems. Special health hazards associated with inert gases are discussed in Chapter 8.

LIMITING OXIDANT CONCENTRATION (LOC) AND INERTING

Inerting is provided to maintain the oxygen in a process vessel below the limiting oxidant concentration (LOC). Values of the LOC for a number of particulate solids are given in Tables C.1(b) and C.1(c) of NFPA 69 (2002). The LOC depends on the inerting gas used (e.g., the LOC for a particulate solid is different for nitrogen than for carbon dioxide). Methods for determining LOC are given in Chapter 4.

A safety margin must be maintained between the LOC and the normal working concentration in the system. The safety margin must take into account the fluctuations occurring in the system, the sensitivity and reliability of monitoring and control equipment, and the probability and consequences of an explosion.

NFPA 69 (2002) recommends that where the oxygen concentration is continuously monitored, a safety margin of at least 2 volume % below the LOC should be maintained unless the LOC is less than 5 %, in which case operating at no more than 60% of the LOC is recommended. In equipment where the oxygen concentration is not continuously monitored, the oxygen concentration should be designed to operate at no more than 60% of the LOC, or 40% of the LOC if the LOC is below 5%. If the oxygen concentration

TABLE 6-4
Relative Merits of Inert Gases

Gas	Advantages	Disadvantages
Carbon dioxide	Readily available in compressed form, from proprietary inert gas generators, and in some cases as a waste gas from on-site processes. Effective—higher oxygen levels (percent by volume) are permissible compared with nitrogen. Moderate cost.	Some metal dusts react violently with carbon dioxide (for example, aluminium). Flow of carbon dioxide can generate considerable electrostatic charge.
Nitrogen	Readily available in compressed or cryogenic form, and in some cases as a waste gas from on-site processes. Moderate cost.	Less effective in volume/volume terms thancarbon dioxide. Some metal dusts react with nitrogen (for example, magnesium) at high temperature.
Flue gases	Often readily available as a waste gas from on-site processes or from inert gas generators. Often available at low cost.	Requires additional equipment to cool the gas, remove contaminants, monitor or remove flammable vapors and remove incandescent material. May react with dusts. Storage of flue gas may not be practical, so that adequate quantities may not always be available, for example, during a furnace shutdown.
Argon or helium	Unlikely to contaminate products or react with them.	Expensive.
Steam	May be generated by the process	May not be available during start-up and shutdown. Incompatible with many products. Will condense if temperature falls, leading to loss of inert atmosphere.

is not continuously monitored, the oxygen concentration should be checked on a regularly scheduled basis (some companies check at least once a month). For equipment operating under vacuum conditions, where the oxygen concentration is not continuously monitored, some companies check at least once a week. Section 3.1 of NFPA 654 (2000) states that where oxygen monitoring is used, it shall be installed in accordance with ISA S84.01, *Application of Safety Instrumented Systems for the Process Industries,* unless a documented risk evaluation is performed.

SUPPLY, DISTRIBUTION, AND MONITORING OF INERT GASES

Inert gas must be available from a reliable source that is capable of continuously supplying the amount of gas required to maintain the necessary concentration of oxygen in the process equipment. Inert gas distribution piping should be designed in accordance with recognized engineering standards and practices.

Since most supply headers operate above 15 psig, piping should be designed in accordance with ANSI/ASME B31.3 (Process Piping).

It is good practice to provide inert gas piping systems with filters, screens, or other means of preventing foreign material from entering critical parts of the inerting system, such as pressure regulators, valves, and instrumentation. The inert gas should have a low dew point so as to prevent moisture condensation in the piping system components and the process equipment being inerted.

It is good safety practice to monitor the flow rate and pressure of the inert gas by means of alarmed low pressure and low flow rates instrumentation. In some cases, where inerting is critical, it may be desirable to interlock the low pressure or low flow rate switch with the process/equipment to shut it down on loss of inerting gas. It may also be desirable in such critical cases to provide an inert gas cylinder back-up system connected to the main supply header to supply inert gas from the cylinders on low pressure or low flow indication.

6.5.2.2 Use of Inert Solids

A combustible dust may be inerted by mixing it with a noncombustible dust (diluent dust) such as calcium sulfate, limestone, sodium bicarbonate, common salt, various silicates or stone dust, as long as the noncombustible dust is prevented from separating from the combustible dust during further handling or processing. Noncombustible dusts may act as a heat sink or otherwise interfere with flame propagation. Usually, at least 60% diluent dust is required, and it is necessary for the diluent dust to be mixed with the combustible dust. The quantity of diluent dust required to produce a non-explosible mixture can be determined by tests conducted in accordance with ASTM Standard E1515-00 (2000) or international equivalent test using a 20-liter sphere and a minimum 5 kJ ignition source. Except in coal mines where it is extensively applied, diluent dust inerting is not used very frequently because of the large quantities needed and the consequent contamination issue involved in many cases (e.g., food processing and pharmaceutical manufacturing). Although complete inerting with a diluent dust is seldom used as a deflagration prevention technique, many particulate solid products contain inert additives as part of their formulation. Such additives reduce the risk of handling the product by lowering the K_{St}, increasing the

MIE, etc. Adding such inert ingredients to a process as soon as possible is a way to reduce the risk during further processing.

Dastidar and Amyotte (2002) report on experiments conducted to determine the minimum inerting concentration (MIC) of a solid inertant to prevent a dust explosion. This is the concentration, in g/m^3, of an inertant required to prevent a dust explosion regardless of the fuel (combustible dust) concentration. The combustible dusts investigated were aluminum, anthraquinone, cornstarch, Pittsburgh pulverized coal, and polyethylene, and the inertants used were monoammonium phosphate, sodium bicarbonate, and limestone. The experimental results show that an MIC can be determined in a 20-liter test chamber; however, there is a strong dependence on ignition strength used to initiate the explosion. In the tests, not all combustible dust and inertant mixtures showed a definite MIC, although they did show a strong dependence between inerting level and suspended fuel concentration. As the fuel concentration increased, the amount of inertant required to prevent an explosion decreased. Even though a definitive MIC was not found for all of the dusts, an effective MIC can be estimated from the data.

Tyldesley (2004) reports that one company in the cement industry routinely added rock powder to a coal grinding mill if it had to be stopped without emptying the system of coal. This reduced the risk of coal starting to smolder or causing an explosion on restart of the grinding mill.

Table C.2 of NFPA 69 (2002) lists some data for the amount of diluent dust required for inerting of dust clouds of various combustible solids.

Eckhoff (2003) presents a review of the literature from 1990-2002 on research and development on inerting by adding noncombustible solids to combustible solids (dusts).

Frank (2004) presents an overview of inerting which covers various aspects, including the following: inert gas production, chemical hazards associated with inert gases (compatibility issues), physical hazards associated with inert gases, health hazards associated with inert gases, determination of LOC, purging and blanketing of vessels and equipment, inerting while charging combustible solids into vessels, general design considerations for inerting systems, and design of inert gas jets for purging of vessels with large *L/D* ratios (e.g., silos).

Other detailed discussions of oxidant concentration reduction (inerting) are presented by Bartknecht (1981), Bartknecht (1989), Barton (2002), Eckhoff (2003), ESCIS (1994), and NFPA 69 (2002).

6.5.3 Combustible Concentration Reduction (Air Dilution)

Combustible concentration reduction (air dilution) is a method for reducing the concentration of a combustible particulate solid below its lower flammable limit (LFL), now more commonly called the minimum explosive concentration (MEC). Combustible concentration reduction is more feasible for equipment containing gases and vapors than for systems containing combustible solids because the dust concentration inside process equipment very often varies in unpredictable and uncontrollable ways. Often there is a tendency for the dust to fall out of suspension and settle on internal surfaces, later to be thrown into suspension, forming an ignitable concentration. The success of the combustible concentration reduction method depends to a large degree on using reliable instrumentation for monitoring the dust concentration inside process equipment. Eckhoff (1997) discusses several types of instrumentation that have been used for this purpose.

NFPA 69 (2002) discusses this method and presents the following guidelines:

1. The concentration of the combustible solid should be maintained at or below 25% of the LFL where combustible concentrations are not continuously monitored. However, if automatic instrumentation with safety interlocks is provided, the combustible concentration shall be permitted to be maintained at or below 60% of the LFL.
2. If ventilation is used, the outlets from the protected equipment shall be located so that hazardous concentrations of the exhausted air cannot enter or be drawn into fresh air intakes of environmental air handling systems.
3. Air intakes shall be located so that a combustible material cannot enter the process equipment even in the event of spills or leaks in the vicinity of the intakes.
4. Filters, dryers, or precipitators in the air intakes shall be located so that they are accessible for cleaning and maintenance.
5. Instrumentation shall be provided to monitor the control of the concentration of combustible solids.

When conveying hybrid mixtures, Section 3.5 of NFPA 654 (2000) requires that the percentage of the lower flammable limit (LFL) of flammable vapors and the percentage of the minimum explosible concentration (MEC) of combustible dusts, when combined, shall not exceed 25% within the airstream. Exceptions are permissible if the system is designed and operated in accordance with Sections 3.1.1 (1), (2), (3), and (4) of NFPA 654.

Britton (2003b) discusses some problems with this requirement, as follows:

1. There is an inconsistency with respect to the units of the LFL of the vapor/gas and the MEC of the dust. The LFL is in mol/mol or volume/volume (volume %) units and the MEC is in mass/volume (g/m^3) units.
2. This requirement may not be practical. As an example, consider a pneumatic conveying system where the dust has a MEC of 50 g/m^3,

and the vapor is a mixture of various monomers and solvents with a poorly defined LFL, but say that it is estimated to be 2.8 volume %. Assume that the vapor mixture LFL can be estimated in g/m^3, one is left with a very dilute dust stream indeed, if there is any flammable vapor present, despite the fact that quite high percentages of the vapor LFL are needed for there to be any significant effect on the MIE to result in dust combustion.

Britton suggests the following improvements to the requirements in NFPA 654:

1. Define a hybrid mixture as one containing some minimum percentage of the vapor LFL (e.g., containing at least 10% of the LFL).
2. Use the same units for both the LFL and MEC, g/m^3.
3. Consider some particle size criterion so that coarse granules plus a few thousand ppm of gas/vapor do not trigger an excessive response (deflagration).
4. The percentage of vapor LFL having a significant effect on the MIE is around 50% for high MIE dusts (>100 mJ), and is close to the LFL for very high MIE dusts (>1000 mJ). This should allow some relaxation to the requirements for high MIE dusts such as coal dust or PVC (and co-polymers).
5. The term "explosible dust" is better than "combustible dust" in this context and is consistent with ASTM and other NFPA standards. For example, 400-micron PVC is a combustible dust, but not an explosible one; it burns but does not explode as a dust cloud.

Bartknecht (1981, 1989) and Eckhoff (2003) discuss hybrid mixtures. Bartknecht (1981, p. 50) presents an equation for estimating the MEC of a hybrid mixture. The equation was based on experiments with PVC and methane and propane gases. He states that the validity of this equation for all applications has yet to be confirmed experimentally.

6.6 DEFLAGRATION PROTECTION METHODS

This section discusses the most commonly used methods of deflagration protection, that is, venting, suppression, pressure containment, isolation systems, systems for spark detection and extinguishing, as well as prevention of secondary explosions.

6.6.1 Deflagration Venting

Deflagration venting is the most widely used explosion protection technique because of its simplicity, effectiveness, and relative low cost compared with other explosion protection systems. It is used for protecting process equipment, pipes and ducts, and buildings. Overpressure relief from a deflagration is provided by installing an opening (called a vent) in the wall of an enclosure that is designed to allow escape of burning solids and products of combustion (gases) quickly enough to prevent a greater rise in pressure than the enclosure can withstand.

In the United States, deflagration vents are designed in accordance with the procedures given in NFPA 68 (2002). The present edition uses a different approach than the previous editions. In the previous editions, nomographs, and then equations were used, which were based on the German venting standard VDI 3673, *Pressure Venting of Dust Explosions* (the latest edition, in English as well as in German, was published in November 2002). In the present edition, equations are used which are based on the work of Tamanini and Valiulis (1996).

In England and Europe, vent design is based to a large extent on the German standard VDI 3673 (2002) and a European standard is being developed, also based primarily on VDI 3673 (Barton, 2002).

Vent Sizing Procedures

The sizing of deflagration vents depends on a number of factors, as follows:

- P_{red}. This is called the reduced pressure and is the maximum pressure developed in a vented enclosure during a vented deflagration. NFPA 68 (2002) allows P_{red} to be selected for up to two-thirds of the ultimate strength for equipment provided deformation of the equipment can be tolerated, or it can be selected for up to two-thirds of the yield strength for equipment where deformation cannot be tolerated. It has units of bar (14.5 psi)
- K_{St}. This is the deflagration index of a dust cloud. It is measured by standard tests using either a 1-m^3 explosion test vessel or a 20-liter sphere. The K_{St} value is a volume-independent parameter that characterizes the explosibility of a dust. It has the units bar-m/s, and is calculated from Equation 6-10:

$$K_{St} = (dP/dt)_{max}(V)^{1/3} \qquad [6\text{-}10]$$

  where $(dP/dt)_{max}$ is the maximum rate of pressure rise measured in the standard tests and V is the test vessel volume in m^3.
- P_{stat}. This is called the static activation pressure and is the the pressure at which the vent closure (cover) opens. It has units of bar.
- P_{max}. This is the maximum pressure developed in a contained deflagration as determined in a standard test. It has units of bar.
- The characteristics of the enclosure (equipment, piping, or building), the volume, V, and the L/D ratio. V is the volume (in m^3) of the enclosure and, generally, it is the open volume that is used in the calcula-

tions, e.g., in a dust collector the volume of the filter bags would not be included. *L/D* is the height-to-diameter ratio of the enclosure. The *L/D* ratio has a significant effect on vent area calculations (to be discussed below).

- The characteristics of the vent cover. The type and mass per unit area (inertia) can have an influence on the necessary size of the vent, and an efficiency factor, determined either by explosion testing or calculation, may need to be included in the calculations.

Chapter 7 of NFPA 68 discusses venting of deflagrations of dust and hybrid mixtures. It covers many aspects including vent sizing; effects of partial volumes; effects of initially elevated pressure; effects of vent ducts (discharge piping); venting of dust collectors using bags, filters, or cartridges; flame clouds (fireballs) from dust deflagrations; external pressure effects; hybrid mixtures; and deflagration venting of enclosures interconnected with pipelines. The equations given for dust explosion vent sizing are applicable to buildings (low-strength enclosures) as well as to equipment specified as high-strength enclosures [capable of withstanding P_{red} of more than 1.5 psig (0.1 barg)].

For sizing of vents two equations are given, one for *L/D* values of less than 2, and another equation for *L/D* values greater than 2 and less than 6. The second equation calculates an incremental vent area which has to be added to the vent area calculated by the first equation. As an alternative to calculating the vent area required using the two equations, figures are given which can be used to calculate the vent area (the present printing has figures which are not correct, and they should not be used until new ones are issued). Refer to Chapter 7 of NFPA 68 for details of the procedure for doing the calculations.

This vent sizing method is primarily applicable only to individual vessels and not to vessels interconnected by pipelines (linked vessels). Deflagration isolation methods should be used in conjunction with venting of individual vessels to prevent the flame from the equipment in which the deflagration can occur propagating to connected upstream and downstream equipment (see Section 6.6.4). NFPA 68 (2002) does provide some guidance on how to size vents for linked vessels. Barton (2002) provides more detailed information on this.

Use of Vent Ducts and Their Effect

Process vessels that are located inside of buildings should be placed next to an exterior wall and the vented stream (unburned solids and gases of combustion) should be directed to a safe location in the outdoors via a vent duct (discharge pipe). The use of a vent duct results in an increase in P_{red}, and this increase can be calculated by an equation given in Section 7.5 of NFPA 68. The vent area can be increased to offset this increase in P_{red}. A vent duct should have a cross section at least as great as that of the vent itself. Vent ducts should be as short and as straight as possible. Any bends can cause dramatic and unpredictable increases in the pressure that develops during venting. Barton (2002) discusses three methods for estimating the effect of vent ducts.

As an alternative to using a vent duct when a process vessel is located inside of a building, a flame-arresting/particle retention device can be used. They are also called flameless venting devices. As a deflagration is vented through this device, any burned and unburned dust is retained within the device, combustion gases are cooled, and no flame emerges from the device. In addition, near-field blast effects (overpressure) are greatly reduced outside of the equipment. If such a device is used, the deflagration venting area should be increased to compensate for the reduction of venting efficiency due to the presence of the device. These devices are discussed in Section 9.7 of NFPA 68,and a schematic drawing of one type is shown.

Two types are available in the United States, the Q-Rohr™ (manufactured by Rembe GmbH in Germany and available through Cv Technology, Inc. of West Palm Beach, Florida) and the FlamQuench II™ (manufactured by and available from Fike Corporation of Blue Springs, Missouri). The Q-Rohr™ has been approved by FM Global and the FlamQuench II™ is presently being evaluated by FM Global for approval.

Stevenson (1998) discusses the Q-Rohr™ and Chatrathi and Going (2002) discuss the FlameQuench II™.

Types of Vent Closures

The main requirements for a vent closure device are:

- It must begin to open at the lowest practical predictable pressure, and this pressure should not be greatly affected by changes in ambient temperature.
- It should not open inadvertently, or leak substantial amounts of dust.
- It must be sufficiently strong to withstand the normal operating pressure and normal fluctuations in pressure.
- It needs to open quickly, so its inertia must not hinder its ability to do so (i.e., it must have a low inertia).
- On activation it must be prevented from becoming a dangerous projectile.

Building or room vent closures can be any of the following:

- Hinged doors, window, and panel closures
- Shear and pull-through fasteners
- Friction-held closures
- Weak roof or weak wall construction
- Large-area panels

These are discussed in good detail in Chapter 9 of NFPA 68.

Equipment vent closures can be either hinged devices, spring-loaded doors, or rupture diaphragm devices. Rupture diaphragm devices are used more frequently these days because they are simple in design, burst quickly and reliably, and are relatively cheaper than the other types. These are discussed in Section 9.6 of NFPA 68 and by Barton (2002) and Schofield (1984).

Fireballs and External Pressure Effects

A vented deflagration generates secondary effects, due to fire and blast (overpressure), outside of the equipment being vented.

These can have serious effects on personnel present in the area of the discharge and on adjacent equipment. Estimation of the dimension of the fireball and the maximum external pressure are discussed below.

FIREBALL

Equations are presented by NFPA 68 (2002) and Barton (2002) for the estimation of the fireball dimensions. In both equations the maximum flame length is a function of the cube root of the enclosure volume.

In NFPA 68, the following equation is presented for estimating the length of the fireball if the vented material exits from the vent horizontally:

$$D = 10(V^{1/3}) \qquad [6\text{-}11]$$

where D is the maximum flame distance from the vent (in m or ft) and V is the enclosure volume (in m^3 or ft^3). It is stated that the height of the fireball can be the same dimension, with half the height located below the center of the vent and half the height located above. In some deflagrations, buoyancy effects can allow the fireball to rise to elevations well above the distances estimated by Equation 6-11.

Barton (2002) presents two equations for the maximum flame length, depending on whether the flame discharges horizontally or vertically. For horizontal discharging vents the equation is the same as given by NFPA 68 (Equation 6-11). However, for vertically discharging vents, the following equation is given:

$$X_{fl,max} = 8(V^{1/3}) \qquad [6\text{-}12]$$

where $X_{fl,max}$ is the maximum flame length (same as D in Equation 6-3) and V is the enclosure volume. Barton also states that, in practice, no flame length greater than 30 m has been measured, even for large volumes, and so 30 m should be taken as the upper limit for any estimations of the maximum flame length. An equation is also given by Barton which allows a reasonable estimate of the maximum width attained by the flame in the absence of any obstruction. This equation is:

$$W_{fl,max} \approx 1.3(10V)^{1/3} \qquad [6\text{-}13]$$

The equations in Bartons's book for maximum flame length and width have the following restrictions:

- Vessel volumes: $0.1 \leq V \leq 1000\ m^3$
- Relief vent static bursting pressures: $0.1 \leq P_{stat} \leq 0.2$ bar
- Reduced maximum explosion pressures: $0.1 \leq P_{red,max} \leq 2$ bar
- Maximum material explosion pressure: $5 \leq P_{max} \leq 10$ bar
- K_{St} value of material: $10 \leq K_{St} \leq 200$ bar-m/s

Holbrow et al. (2000) describe a project in which the effects of thermal radiation from vented dust explosions were studied. The aim was to establish the areas around a fireball in which people would be at risk from thermal radiation. Six dusts (coal, cornflour, toner, polyethylene, anthraquinone, and aluminum powder) were tested in two large (18.25 m^3 and 20 m^3) vented vessels and external fireballs were generated under a range of conditions. The fireball geometry and the heat flux from the fireball were studied. A range of material samples (plastic sheets, fabrics, and disposable overalls) were exposed to the fireball, and the safe areas around the fireballs were established for each of the six dusts. Reported fireball characteristics included fireball projected area, total energy emitted, average surface emissive power, and peak temperature. Also, values of incident radiation are reported as function of the distance from the fireball center. Generally, the larger vent areas resulted in the larger fireballs and high heat pulse values. However, the fireball was usually too brief to ignite fabric samples unless they were very close to the fireball. The work has shown that in most cases the safe area was relatively close to the surface of the largest fireball.

EXTERNAL PRESSURE EFFECTS

When a dust deflagration is vented from an enclosure, pressure effects are created in the atmosphere external and adjacent to the vented equipment. Such pressure effects are due to the effects of both the vented products and the further deflagration of excess (unburned) flammable dust. There are usually two pressure peaks, one from the venting process and the other from the deflagration of the dust–air mixture external to the enclosure. Both NFPA 68 (2002) and Barton (2002) present the same equation for calculating the maximum external pressure, which is:

$$P_{max,a} = 0.2 P_{red} (A^{0.1})(V^{0.18}) \qquad [6\text{-}14]$$

where $P_{max,a}$ is the maximum external (atmospheric) pressure in bar, P_{red} is the reduced pressure in bar, A is the vent area in m^2, and V is the enclosure volume in m^3. This is for venting from a cubical vessel. NFPA 68 states that the maximum value of the pressure exists at a distance of about one-fifth of the maximum length of the fireball as calculated by Equation 6-11 (for horizontal venting). Barton recommends that for dusts with high K_{St} values (e.g., aluminum with a $K_{St} \geq 300$ bar-m/s), a safety factor of 2 should be applied to the value calculated by Equation 6-14.

Both NFPA 68 (2002) and Barton (2002) give similar equations to calculate the maximum external pressure if the flame length is larger than that calculated by Equations 6-11 and 6-12.

Bernard et al. (1998) discuss a guideline for determining access restriction zones around vented solids handling equipment based on flame length, fireball volume, and external pressure effects.

The guideline provides a method for ensuring that the discharge from a vented explosion will not cause injury to personnel.

Reaction (Recoil) Forces

When a deflagration vent opens reaction (recoil) forces occur, and the supporting structure for the enclosure should be strong enough to withstand these reaction forces. Both NFPA 68 (2002) and Barton (2002) present the same equation for calculating the reaction forces. This equation applies only to enclosures without vent ducts and is as follows:

$$F_r = \alpha(A_v)(P_{red}) \qquad [6\text{-}15]$$

where F_r is the maximum reaction force resulting from deflagration venting in kN or lbf (SI or English units), α is 120 or 1.2, A_v is the vent area in m^2 or in^2, and P_{red} is the maximum pressure developed during venting in bar or psi.

However, NFPA 68 (2002) and Barton (2002) give different equations for calculating the duration of the pressure pulse after vent opening, and these books should be consulted for these equations. Equations for the impulse and equivalent static force are also presented in both books.

Examples of venting of several different types of process equipment are presented by Bartknecht (1981) and Barton (2002)

6.6.2 Deflagration Suppression

Deflagration suppression is also a very widely used protective measure in the chemical processing industries (CPI). It is often used where it is not possible to vent the contents of process equipment and vessels to a safe place, and is particularly important where deflagration venting could result in the emission of toxic dusts or other substances harmful to people, other process equipment, or the environment.

A deflagration is not an instantaneous event but takes a finite time to build up destructive pressures in a vessel or process equipment. Typically, it takes 30–100 milliseconds (ms) before destructive pressures are reached. This is the basis for the design and operation of a deflagration suppression system. Deflagration suppression requires that the incipient explosion be detected very soon after ignition, and that a sufficient amount of suppressant is discharged into the growing fireball in the enclosure at a fast enough rate to extinguish all flame before a destructive overpressure develops. Figure 6-3 shows the steps in a deflagration suppression.

Deflagration suppression can be applied to suppress (quench) deflagrations in vessels with volumes as low as 0.25 m^3 and up to 1000 m^3. Vessels with volumes larger than 1000 m^3 can also be protected by suppression provided that it is possible to locate suppressant containers within the volume boundaries in appropriate locations.

Deflagration suppression is sometimes used in combination with venting to protect process equipment or vessels where it is not possible to provide suf-

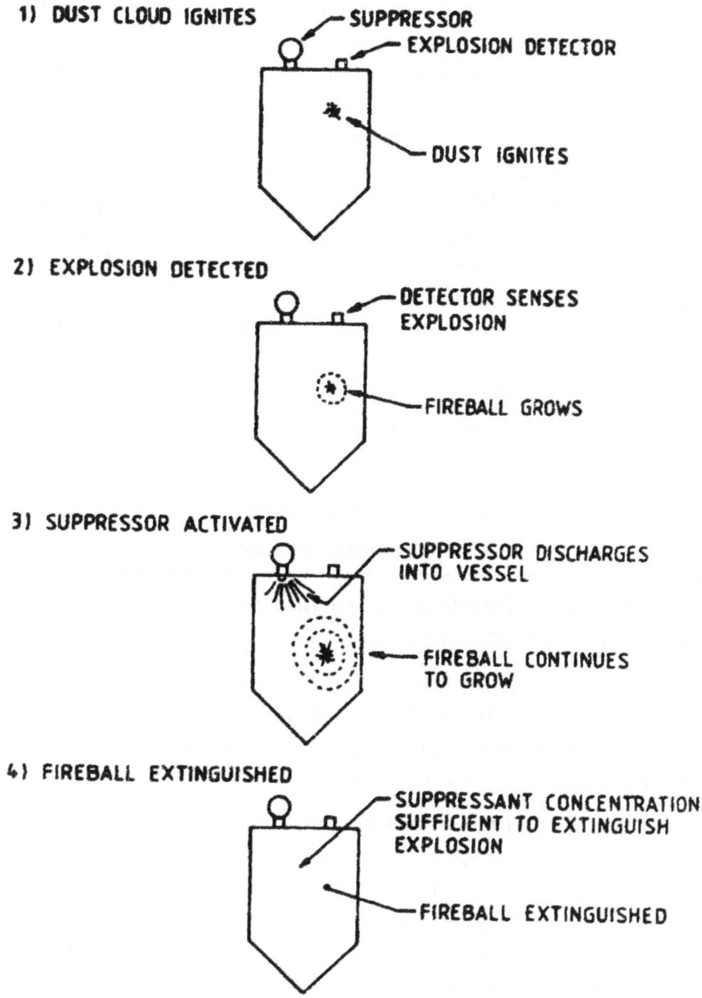

Figure 6-3 Steps in a deflagration suppression.

ficient vent area for venting alone to protect the equipment, where P_{red} with suppression alone is too high for the vessel strength, or where it is necessary to minimize the size of the fireball emitted from the vent. The rate of pressure rise can be reduced by dispersing an suppressant into an enclosure shortly after ignition, thereby allowing smaller vent areas. Also, the size of the fireball can be reduced by means of suppressant injected into the flame front in the vicinity of the vent. Siwek (1992) discussed the results obtained from tests conducted on combined suppression and explosion venting. He conducted tests with Class St 1 and St 2 dusts in a 25 m^3 vessel equipped with two vents (each with a vent area of 0.28 m^2) and four high-rate-discharge (HRD) suppression extinguishers (varying from 20 liters to 45 liters in capacity).

The number of HRDs and vent area were varied during the tests. The explosion suppression was triggered at a very low activation pressure P_a of the suppression system, whereas the vent closures selected did not respond until a static activation pressure P_{stat} value of about twice P_a was reached. The tests indicated that a maximum reduced pressure in the range of 0.1 to 0.2 bar could be achieved with combined suppression and venting, compared to values of 0.5 to 1.0 bar when using only either suppression or venting. The tests also showed that it was possible to minimize the escape of flames from the vent opening or even stop it completely. However, it is cautioned that specialized knowledge is required to design combination vent/suppression protection systems, and their design should be based on experimental data.

Accidents can sometimes occur when using suppression systems, as shown by the following case history.

Case History of an Accident Caused by a Deflagration Suppression System

Ness (2002) presents the following case history of an accident caused by a deflagration suppression system.

A near miss occurred when the door of a solids grinder was blown off into an operating area by the activation of a suppression system after the grinding system was filled with deluge water from an explosion in an item of equipment upstream of the grinder.

A dust explosion occurred in a drying train upstream of a solids grinder. The explosion was successfully vented, however, the deluge system activated by the explosion filled the grinder system as well as points in the dryer train. The grinding system was designed with a suppression system to control potential explosions in it. When the pressure in the grinding system built up enough, the suppression system was activated, injecting Halon gas into the grinding system. With no means to relieve pressure, this caused a "water hammer" which broke the grinder doors open, hurling them 20 feet across the room. No one was in the area at the time, so no injuries occurred.

The corrective actions that were taken included:

- An automatic dump valve was installed in the grinder conveying line to drain deluge water from the system one minute after the deluge system is activated.
- A protection barrier was installed in front of the grinder door to protect people from injury.
- A procedure was written describing when the suppression system needs to be deactivated.
- The Halon suppression system was changed to a dry chemical suppression system.

Fundamentals of Deflagration Suppression

Discharging a liquid or dry powder suppressant into a growing fireball results in a number of complex results, which include:

- Quenching—heat abstraction from the combustion zone by energy transfer.
- Wetting—unburned dust particles are rendered nonexplosible by absorption of liquid suppressant.
- Inerting—concentration of suppressant in suspension in the unburned explosible mixture renders the mixture non explosible.
- Free radical scavenging—active species in the suppressant compete with chain-propagating reactions in the combustion wave.

For dust explosions, quenching is usually the most important mechanism. Explosion propagation is dependent on a heat transfer mechanism between the suspended dust particles, while the combustion of each particle is controlled by combustion chemistry. The energy transfer between the suppressant and the combustion zone is limited by residence time of the suppressant droplets or particles in the combustion zone, by the droplet or particle size distribution, and by the concentration of suppressant in the combustion zone. The heat capacity of the suppressant material, including any latent heat contributions, are the primary influences on suppression efficiency.

Elements of a Suppression System

A deflagration suppression system consists of three subsystems: (1) a detection system, (2) a suppressant injection system (type, number, and location of suppressors), and (3) a control and supervision system. Figure 6-4 is a schematic drawing of a typical suppression system.

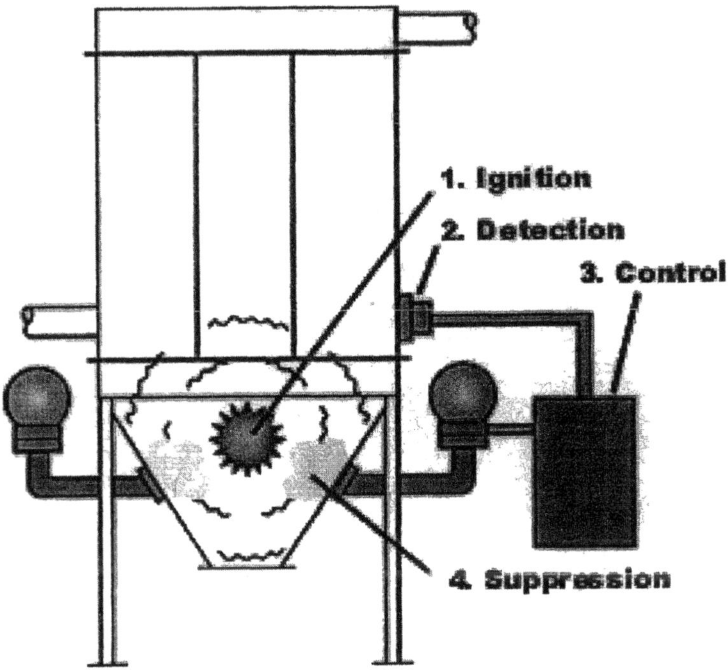

Figure 6-4 Schematic diagram of a deflagration suppression system.

For a given explosion hazard in a vessel or process equipment the reduced explosion pressure for a suppressed explosion depends on:

- The type of detector
- The threshold level of pressure detection at which the explosion is recognized
- The suppression efficiency of the suppressant
- The number of suppressors installed
- The mass of the suppressant
- The discharge rate of suppressant
- The throw and dispersion of the suppressant

The three components of a suppression system are discussed below.

TYPES OF DETECTORS

Incipient dust deflagrations are detected using one of two types of pressure sensors: static pressure or rate-of-rise dynamic pressure detectors. Pressure detectors are designed to alarm and send a signal to the control system when a pre-set threshold pressure or a preset rate of pressure rise is exceeded in the vessel or equipment. The set detection point(s) must be above the process-

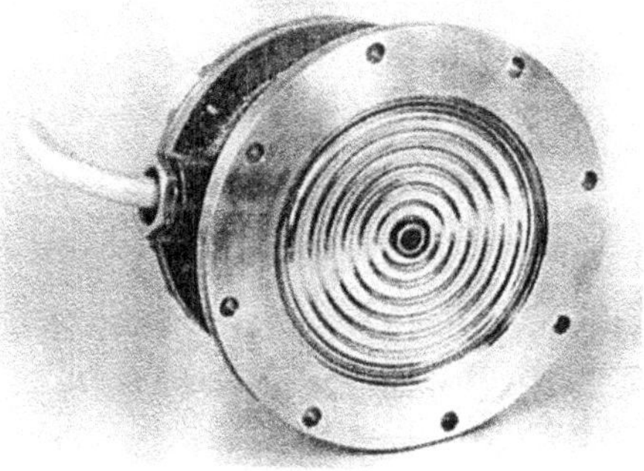

Figure 6-5 Static pressure membrane type detector.

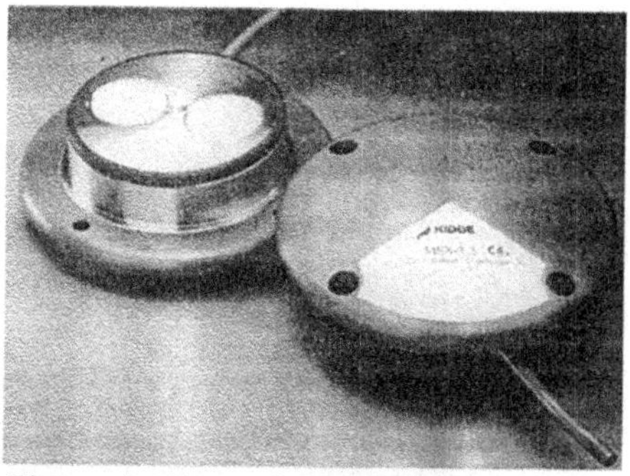

Figure 6-6 Rate-of-rise dynamic pressure detector.

induced pressure fluctuations and be independent of process temperature. Figure 6-5 shows a static pressure type detector and Figure 6-6 shows a rate-of-rise dynamic pressure type detector. The sensing membrane is a large diameter pressure-sensitive diaphragm, used because it will not become occluded by solids nor be effected by build-up of solids on the diaphragm surface.

Static pressure type detectors have performed well in industrial practice. They have, in the past, been subject to spurious (false) trips triggered by shock and other causes. To reduce the frequency of such nuisance alarms, some suppliers deploy two independent detectors, oriented 90° from each other, configured in "AND" logic (both detectors must agree) at each detection station. Other suppliers incorporate two or more sensing diaphragms within each detector device.

The rate-of-rise dynamic type detector is gaining increasing importance because in addition to the normal measurement of pressure, the rate of pressure increase with time can also be used as a tripping criterion for a suppression system. This allows selective utilization of the pressure phenomena occurring in the protected vessel, By these means, the possibility of spurious actuations—for example, through overpressure in a baghouse due to pneumatic bag cleaning—can be completely excluded (Siwek, 1994).

SUPPRESSANT INJECTION SYSTEMS

Suppressant storage and injection systems come in various designs, depending on the manufacturer. They are usually called suppressors or suppressant containers or bottles.

Hemispherical suppressors, as shown in Figure 6-7, are operated by a chemical detonator. The firing of the detonator causes "petalling" of the specially weakened dome, allowing the discharge of the suppressant. The initial velocity of the suppressant (usually water) is about 200 m/s and the suppressor is typically completely discharged in 10–20 ms. These suppressors con-

Figure 6-7. Hemispheric suppressor.

tain relatively small quantities of suppressant, typically up to 5 liters, and are used for liquid suppressants only. They are fixed on the inside of equipment and because of their limited discharge distance (<2.5 m) are mainly used to protect small volumes and ducting. They are not suited for high temperature process operation, that is, >60°C.

High rate discharge (HRD) suppressors (see Figure 6-8) are the most frequently used explosion suppressor type. The HRD suppressor consists of a canister containing a charge of suppressant (liquid, vaporizing liquid, or dry powder) pressurized with nitrogen to a high pressure, usually in the range of 20–120 bar. HRD suppressors have a large diameter valve closure designed to open in a few milliseconds and to provide an unimpeded discharge path

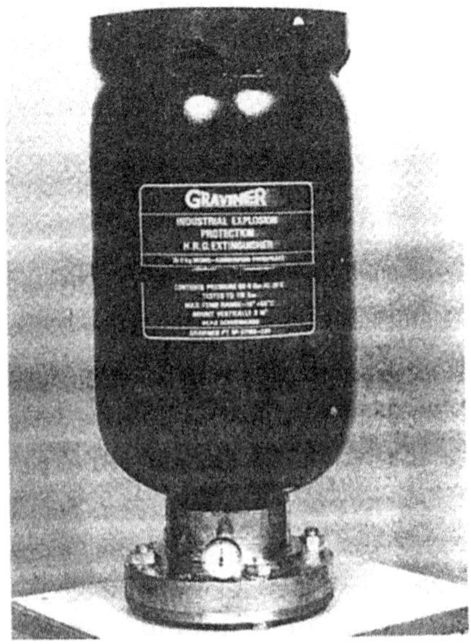

Figure 6-8 High rate discharge (HRD) suppressors (Source: Graviner).

for the suppressant to be expelled from the canister through a discharge nozzle into the equipment being protected. Proprietary HRD suppressors are available in sizes ranging from 3 to 60 liters, and outlet diameters ranging from ¾-inch to 5 inches. A large HRD suppressor may contain 40 kg of suppressant. The discharge valve may be activated by an explosive device (detonator or cartridge), a nonexplosive pyrotechnic actuator or piston actuator, or an electromechanical device such as a solenoid or motor. Some HRD suppressors incorporate a 90° elbow which provides a faster discharge when the suppressor is to be mounted on a vertical surface of the equipment. Other designs use two smaller outlets as an alternative to a single larger outlet to maintain a sufficient mass discharge rate.

HRD suppressors are installed on the outside of protected equipment. Suppressant is delivered through a short outlet spur, sometimes via an elbow, to the inside of the equipment, where it is discharged into the equipment via a nozzle. The suppressor is sometimes sealed from the equipment by means of blow-off cups, rupture foils, or glass disks which are broken when the suppressor is activated; the nozzle then expands telescopically into the equipment. This arrangement prevents ingress of process material into the HRD suppressor fittings and provides a smooth surface that does not impede process flow. The design of the nozzle is dependent on the suppressant type, for example, a simple nozzle can be used for water, but deployment of dry chemicals can be very dependent on the nozzle geometry.

The performance requirements of a HRD suppressor are: (1) high suppressant discharge rate, (2) high suppressant discharge velocity to give effective throw, and (3) good angular dispersion of the suppressant. The mass discharge rate and discharge velocity depend to a great extent on the diameter of the suppressor outlet and on the propelling agent pressure. They also depend on the restrictions to flow such as bends between the outlet and the nozzle, and the angular dispersion of the suppressant depends on the suppressant velocity and on the detailed design of the nozzle. Typically, an HRD suppressor discharges its entire contents in about 100 ms, with the suppressant first entering the equipment within 10 ms of explosion detection, and is capable of throwing the suppressant charge distances of 6–8 m.

CONTROL AND MONITORING SYSTEM

A present-day control and monitoring system employs state-of-the-art electronics and is often called an alarm center. It is the watchdog or operational center of the explosion protection installation. It is constructed in accordance with international regulations and meets all requirements needed to ensure safe working practice. The alarm center records and monitors the state of the pressure detectors, as well as those of other sensors (e.g., temperature). Depending on the evaluation, the alarm center selectively controls the HRD suppressors for explosion suppression as well as other incorporated protection systems, for example, activates connecting piping extinguishing barriers, activates the appropriate rapid-action isolation valves, shuts down equipment, actuates a water deluge system, etc.

The detector zones of the sensors and the control circuits of the suppression equipment are constantly monitored by the internal electronics for open circuits, ground faults, and short circuits. In case of spurious signals, a visual and audible warning is given. Further, all important functions can be recorded and documented by a printer built into some designs especially for this purpose.

SUPPRESSANT MATERIALS

The three most commonly used suppressants are (1) dry powders, (2) water, and (3) fluorinated hydrocarbons (Halon replacements).

Dry chemical powder suppressants are usually variants of the proprietary dry chemical fire extinguishants that have a low median particle size and a high degree of fluidity. Among the powders that are widely used are the alkali metal bicarbonates, ammonium phosphates, and substituted ureas. Where the explosible dust is a pharmaceutical or food product, food-grade compatible suppressants are available. Dry powder suppressants are usually pressurized to quite high pressure (35 bar to 120 bar). This is necessary because the interstitial stored energy between powder particles fluidizes the agent at discharge, thus overcoming any compaction or settling that may have occurred in the HRD suppressor container. Dry powder suppressants settle out relatively quickly after they have been discharged, so that the equipment must be shut down to protect against reignition after the initial activation.

Water has been shown to be a satisfactory suppressant for many dust explosions. A larger amount of water is required than for fluorinated hydrocarbons, because of its lower density and its inability to completely wet the surface of some dusts. Superheated water has the advantage that a proportion of the suppressant flashes to steam during the suppressant discharge, thus adding an inerting effect to the suppression effect. Water-based suppressors mounted on outdoors equipment are subject to freezing in cold climates and some suppression system suppliers offer salt-based antifreeze to provide a measure of protection.

Vaporizing liquid fluorocarbons have replaced Halons as Halons are ozone-depleting substances and are not now admissible under the terms of the Montreal Protocol. The two most widely used fluorocarbons are FE36 (made by duPont) and FM200 (made by Great Lakes Chemical Corp.). They are not as efficient as Halons, on a weight basis, so larger amounts are therefore needed to suppress a deflagration. Fluorocarbons injected late into the hot combustion products of an explosion, rather than the flame front, may decompose resulting in an increased pressure in the protected equipment. In such a case, the explosion is not suppressed. They are not, however, presently being used for suppression of dust explosions.

TABLE 6-5
Relative Properties of Common Suppressants

Consideration	Powder	Fluorocarbons	Water (pure)
Toxicity	No	Slight	No
Contamination	Yes	Negligible	Wets product
Decomposition	Slight	Yes	No
Explosion dust classes	St 1, St 2, some St 3	Some St 1 only	St 1 and St 2

Table 6-5 lists the relative properties of water, dry powder, and fluorocarbon suppressants.

The suitability of a particular suppressant for a given application should be determined by explosion suppression tests.

The correlation between the activation pressure of the suppression system and the reduced maximum pressure has to be established. Also, the suppressant has to be insensitive to the temperature and vibration inherent to the protected system.

It should be recognized that deflagration suppression systems have a number of shortcomings, such as:

- Spurious activation (false trips)
- Servicing problems (testing, the need for bypassing for maintenance, and to bring the system back on-line before starting operations again)
- Clean-up after activation
- Good for only one event and then they have to be refilled

Another shortcoming is that suppression systems can produce pressures of 2–3 psig (0.13–0.2 barg), which may exceed the design strength of some low-strength equipment, such as baghouses, and this must be considered in suppression systems design and application. To prevent permanent equipment damage, the pressure experienced during a suppressed deflagration (P_{red}) should not exceed two-thirds of the equipment yield strength (stress).

As mentioned above, the suppression system should be taken off-line when maintenance has to be done on the process equipment, either by disarming it or shutting off the sensor tubing, and then it must be put back in service before starting up the process again.

More detailed discussions of deflagration suppression are presented by Bartknecht (1981), Bartknecht (1989), Barton (2002), Eckhoff (2003), and NFPA 69 (2002). Moore and Siwek (2002) present an update on the European explosion suppression standards. Two widely used European standards for explosion suppression systems are prEN 14373 (CEN, 2002) and ISO 6184/4 (ISO, 1985). An approval standard for explosion suppression systems has been issued by FM Global (FMG, 1999).

Barton (2002) describes examples of suppression system applications for several types of process equipment.

6.6.3 Deflagration Pressure Containment

Deflagration containment is a technique for specifying the design pressure of a process equipment and appurtenances so that they are capable of withstanding the maximum pressures resulting from an internal deflagration. Explosion containment is used, in particular, when a hazard assessment indicates that an emission of a process material or combustion products could present an unacceptable risk. It may be a valid option, as well, for explosion protection even when emissions are not hazardous. For example, containment is often used to protect process equipment operating at sub-atmospheric pressure (e.g., vacuum dryers), and is also often a suitable option for mills and other equipment of small volume that can be built to be strong enough to withstand the maximum explosion pressure produced by a dust explosion.

Containment for process equipment can be specified as either explosion-pressure-resistant or explosion-pressure-shock-resistant.

Explosion-pressure-resistant equipment should not be ruptured or deformed in any way in the event of a dust explosion which occurs at an initial (pre-explosion) pressure of less than 1.5 psig (0.1 barg). FMG 7-76 (2001) states that all vessels with a design pressure of 87 psig (6.0 barg) or more can be considered as having an explosion resistant design (assuming that the initial pressure is atmospheric).

Explosion-pressure-shock-resistant equipment can be expected to deform, but not rupture in the event of a dust explosion which occurs at an initial (pre-explosion) pressure of less than 1.5 psig (0.1 barg). FMG 7-76 (2001) states that all vessels with a design pressure of 43 psig (3 barg) or more (designed in accordance with ASME Boiler and Pressure Vessel Code, Section VIII), or a yield strength of 87 psig (6 barg) can be considered as having an explosion shock-resistant design (assuming that the initial pressure is atmospheric).

NFPA 69 (2002) presents equations for calculating the design pressure for containment for explosion-pressure-resistant and explosion-pressure-shock-resistant equipment, and also discusses other aspects of this protective technique.

NFPA 69 (2002) allows application of deflagration pressure containment for interconnected (linked) vessels if certain exceptions (criteria) are met, as follows:

- Where interconnected piping is provided with deflagration isolation.
- Where venting is provided for interconnected piping.
- Where interconnected vessels are designed to contain the increased pressure due to the effects of pre-pressurization.

- Where the use of deflagration isolation or venting of one vessel shall be permitted to be used.

Additional discussions on deflagration containment are presented by Bartknecht (1981), Bartknecht (1989), Barton (2002), and Eckhoff (2003). Pilkington (2000) presents equations for calculating the strength of components of weak (low-strength) process vessels, such as silos or baghouses.

6.6.4 *Deflagration Isolation Systems*

Items of equipment in process plants handling particulate solids are often connected together by piping, ducts, chutes, and conveyors. An explosion initiated in one plant item can propagate along these interconnecting links and start a subsequent explosion in other equipment, both upstream and downstream. Isolation methods can be used to interrupt or mitigate flame propagation, deflagration pressures, pressure piling, and flame-jet ignition between items of equipment. A number of isolation methods can be used, as follows:

- Automatic fast-acting valves
- Flame front diverters
- Flame front extinguishing systems (suppressant barriers)
- Material chokes

Other methods are flame arresters, which are discussed in detail by Grossel (2002), and spark detection and extinguishing systems, which are discussed in Section 6.4.5. Dry type flame arresters with internal arresting elements (e.g., crimped metal ribbon, parallel plate, etc.) are not recommended for use in particulate solids processes as they tend to plug up.

6.6.4.1 Automatic Fast-Acting Valves

A variety of fast-acting valves are available, including slide gate, flap (butterfly), and float valves.

SLIDE GATE AND FLAP (BUTTERFLY) TYPE VALVES

Slide gate and flap type valves are actuated (closed) upon a signal from a detector (sensor) in the pipeline between two items of interconnected process equipment. The detector sends a signal to a control device and is relayed to a compressed gas cylinder which then discharges the compressed gas to a mechanism at the top of the valve, thereby closing the valve. A fast-acting slide gate isolation valve is shown in Figure 6-9. With this type of valve, the pipe area is completely open and can be built without pockets and dead corners, so that dust will not settle out or accumulate. Special dampers have been developed in order to absorb the substantial forces from the closing device and to prevent the slide from springing back after closure. The damp-

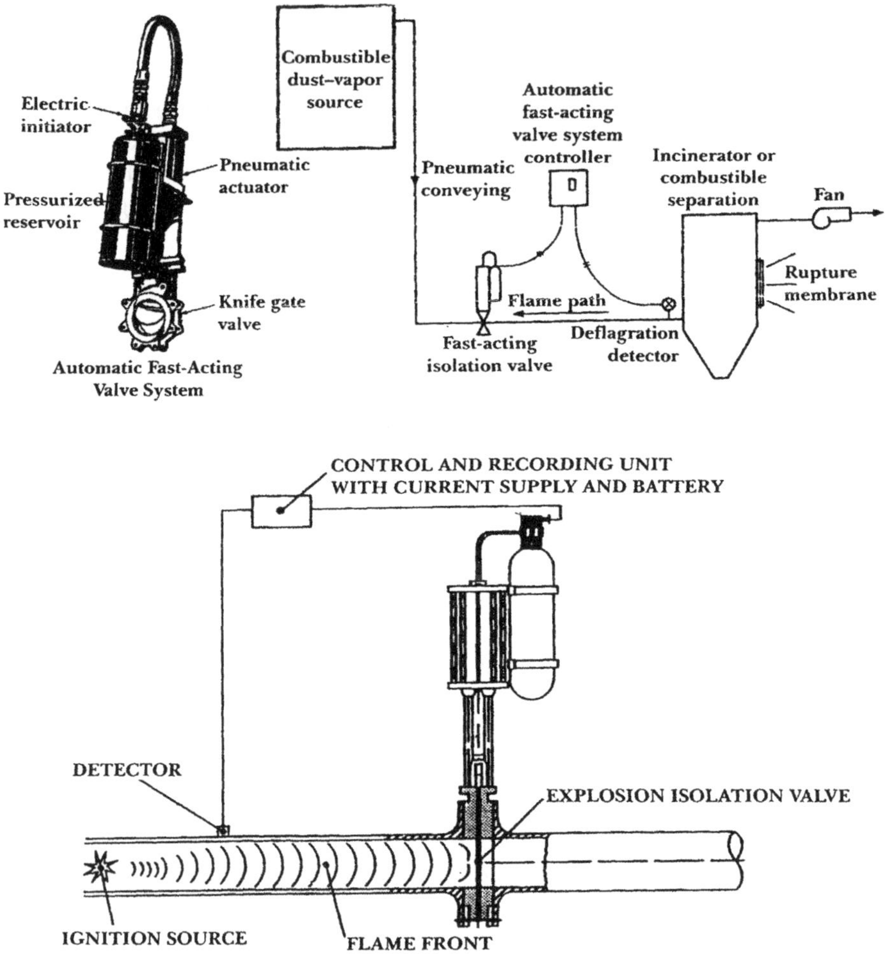

Figure 6-9 Fast-acting slide gate valve isolation system (NFPA 69).

ing elements are exchangeable. As can be seen in Figure 6-9, the valve is connected to a compressed nitrogen cylinder via a high-pressure hose. When an explosion occurs, a signal from a control and indicating device activates a detonator which opens the valve of the compressed nitrogen cylinder. The propellant (nitrogen pressure of 10–40 barg), via a cylinder–piston system, closes the slide gate, which is dampened through the plastic deformation of a braking device. The slide valve can be mounted in vertical, horizontal, or inclined piping.

The measurement of the normal closing time (no explosion) of the valve can be done in place (using standard pressure) with a "testbox" (which can be purchased with the valve) in combination with a pneumatic control valve.

These valves must be tested for flame breakthrough protection and pressure rating in dust explosions. They can meet these requirements for dust explosions, and are effective at shorter distances against dust explosions than for gas explosions. When fast-acting slide gate valves are used to protect against a dust explosion, an optical sensor is used to detect the approaching flame, and to initiate the triggering of the valve closure. If equipment is protected by measures such as containment, suppression, or venting, then the usual explosion pressure sensor, with a corresponding low activation pressure, can also be used to initiate the triggering mechanism for the fast-acting slide valve.

The basis for the design of a fast-acting slide valve is established from experimental tests. The closing time (i.e., the time between the activation of the detonator for the closing mechanism and the complete closure of the valve) depends not only on the propellant pressure, but also on the pipeline diameter and valve, and is generally less than 50 ms. This closing time is the real closing time of the slide valve and does not include the electronic delay time of the sensor used, including the control and indication device. Typical values for electronic delay times are ≤2 ms for an explosion pressure sensor with a control and indicating device, and ≤4 ms for a flame (optical) sensor with a control and indicating device.

There is a definite distance between the protected equipment and the fast-acting slide gate valve required to ensure that the valve will stop the explosion. The minimum distance required depends on the nature of the combustible dust, the closing time of the slide valve (including the electronic delay time of the sensor used and the control and indicating device), the flame velocity, and the maximum (reduced) explosion overpressure in the protected vessel. The calculation of this distance is relatively complex, and should be done only be experts (usually the vendor of the fast-acting slide gate vale should be able to provide assistance on this). Siwek (1996) presents some guidance on the minimum distance required for fast-acting slide gate valves as a function of the protection provided in the vessel (containment, venting with a vent pipe, and venting without a vent pipe or suppression), as shown in Table 6-6. The deflagration sensor (detector) is usually located about 1 meter away from the source of ignition (equipment).

A fast-acting flap (butterfly) valve is shown in Figure 6-10. The principle of operation of this type of fast-acting valve is similar to that of the fast-acting slide gate valve.

Further discussions of fast-acting slide gate and flap (butterfly) valves are presented in the books by Bartknecht (1981, 1989), NFPA 69 (2002), and Siwek (1996).

FLOAT TYPE VALVES

Another type of fast-acting valve is the float type valve (Ventex™ valve). As a certain explosion overpressure is necessary to close such valves, a distinc-

TABLE 6-6
Minimum Distance for Fast-Acting Slide Gate Valves from Protected Vessel

Type of Explosion Protection in the Vessel	Containment	Venting with Vent Pipe	Venting without Pipe or Suppression
P_{max}, $P_{red, max}$	9 bar	3–4 bar	1 bar
DN of the Float Valve (mm)	Minimum distance, *dmin* (m)		
100	ca. 5	ca. 4	ca. 3
400	ca. 9	ca. 6	ca. 5

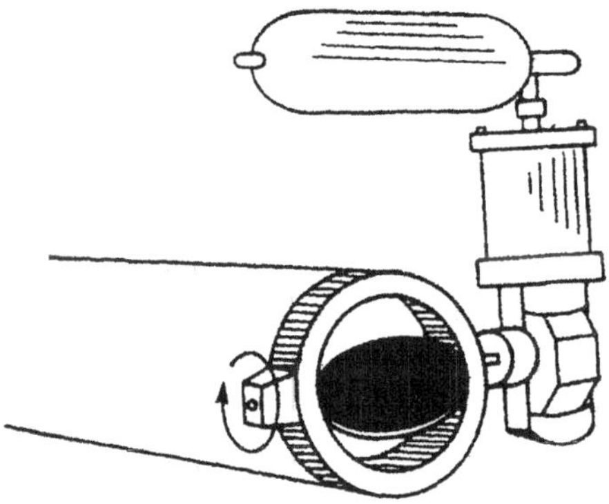

Figure 6-10 Fast-acting butterfly isolation valve.

tion is made between self-actuated and externally-actuated float valves. Both valves can withstand an explosion overpressure within a pipeline of at least 10 barg for pipes of 100–500 mm diamter, and at least 5 barg for pipelines of 700 mm diameter. They have been tested, with satisfactory results, against propane–air and dust–air mixtures. The float valves provide adequate protection against the propagation of dust explosions ($P_{max} \leq 10$ barg, $K_{max} \leq 300$ m-bar/s) and explosions of hybrid mixtures with a maximum concentration of flammable vapor or gas of 50% of the lower explosive limit (Siwek, 1996).

Self-Actuated Float Valves: The interior of the valve contains a valve cone (float) mounted in spherical sockets, and which can be moved axially in both directions; it is held in its middle position by springs. The spring tension is set for a maximum flow velocity of 24 m/s, based on the pipeline cross-sectional area. If an explosion occurs, the valve closes automatically owing to the kinetic energy of the pressure wave preceding the flame front. Here, either the explosion velocity must be >24 m/s or the pressure difference between the front and back of the float must be ≥0.1 bar. The valve float is pressed onto a rubberized valve seat on closing and held in place by a retaining device. In addition, the closed position can be indicated by limit switches. The float valve remains closed until the manual reset device (reset knob) is operated (released from the outside). The self-actuated float valve functions, with the exception of the 700-mm diameter valve, in both directions. A potential problem with this valve is that the rubberized seat may be adversely affected in high temperature environments. Another possible problem is that powder coating accretion on the seal surfaces can prevent a tight seal, and flame breakthrough is then possible.

Since a certain minimum explosion pressure is required to close the float valve, the propagation of an explosion through a pipe will not be stopped if its pressure is lower than the minimum activation pressure of the float valve. To keep this range of uncertainty acceptably small, vessels which are protected by explosion venting or suppression, it must be ensured that the static activation pressure of the venting device is $P_{stat} \geq 0.2$ barg or the activation pressure of the suppression system is $P_a \geq 0.2$ barg. Otherwise, reliable performance of the self-actuated float valve cannot be expected.

A basis for the design of self-actuated float valves is established from experimental tests. A correlation exists between the dynamic activation pressure of the float valve and the momentum of the pressure wave. On the one hand, with increasing dynamic activation pressure, the momentum also decreases, and on the other hand, only a high momentum results in a short closing time. Therefore, self-actuated float valves have more than one closing time.

For the design of self-actuated float valves, minimum and maximum distances from the vessel being protected to the float valve exist. The maximum distance makes sure that no detonation develops in the vicinity of the float valve, and the pressure effect does not exceed the design pressure of the float valve. The minimum distance makes sure that the float valve still closes properly and consequently, no flame breakthrough of the explosion through the float valve can occur. These distances are usually recommended by the valve vendor. Siwek (1996) also presents information on these distances, as shown in Table 6-7. The actual installation distance of the float valve in the pipeline to the protected vessel lies between the minimum and maximum distance, and is dependent upon the explosion overpressure (P_{red} or $P_{red,max}$) in the protected equipment. The installation distance also depends on the nominal diameter of the float valve, and on whether combustible dusts or hybrid mixtures are present. Siwek (1996) discusses this in more detail.

Figure 6-11a shows a schematic drawing of a self-actuated float valve.

Externally Actuated Float Valves. Externally actuated float valves are used when low explosion overpressures are expected, and consequently flame propagation from equipment could occur if a self-actuated flow valve was used. These valves are operated by sensor-controlled gas flow (jets of 120 barg nitrogen from an HRD container) from a hemispherical nozzle, which

TABLE 6-7

Minimum and Maximum Distances from the Protected Vessel for Self-Actuated Float Valves

DN (mm)	Combustible Dust l_{min} (m)	Combustible Dust l_{max} (m)	Hybrid Mixture l_{min} (m)	Hybrid Mixture l_{max} (m)
100	5	12.5	3	5
200–700	5	12.5	5	8

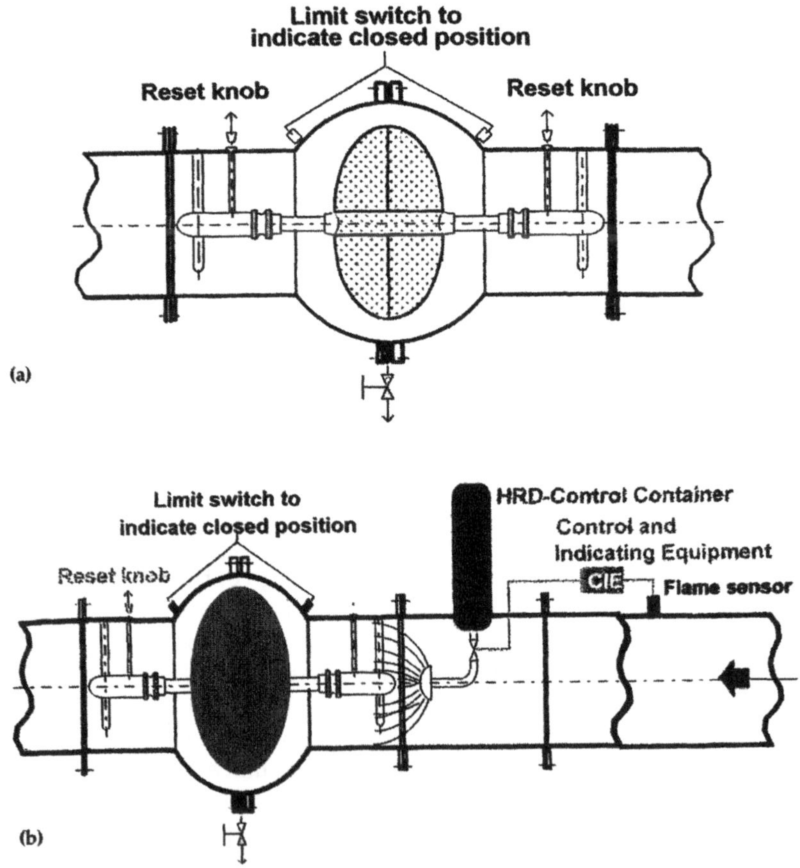

Figure 6-11 Ventex™ (float type) fast-acting isolation valve. (a) Self-actuated float valve. (b) Externally activated float valve.

impinges upon the float (see Figure 6.11b). When an explosion occurs in a vessel, and the burning dust cloud flows through a pipeline, the flame is detected by an optical sensor, which sends a signal to the control and indicating device, which in turn, initiates the flow of the gas from the HRD. If the vessel is protected by means such as containment, suppression, or venting, a pressure sensor in the vessel can be used to initiate the flow of gas from the HRD.

As with the self-actuated float valve, the valve cone (float) is pressed onto a rubberized valve seat on closing and held in place by a retaining device. In addition, the closed position can be indicated by an electrical signal from a limit switch. The float valve remains closed until the manual reset knob is operated from the outside. The externally actuated float valve functions only in one direction.

The design of externally actuated float valves is based on experimental tests. These tests have shown that the closing time depends not only on the number of HRD containers, but also on the nominal diameter of the float valve.

As with the self-actuated float valve, there is a minimum distance required between the protected vessel and the externally actuated float valve to ensure that the valve will stop an explosion from propagating. The minimum distance required depends on the nature of the combustible dust, the closing time of the float valve, the flame velocity, and the maximum reduced explosion overpressure in the protected vessel.

Here too, the calculation of this minimum distance is relatively complex and should be done only be experts (usually the valve vendor should be able to provide this information). Siwek (1996) provides guidance on the minimum distance required between externally actuated float valves and the protected vessel, as shown in Table 6-8. It is assumed that the vessel is protected by a measure such as containment, venting (with and without a vent pipe), or suppression, and that the vessel has an explosion pressure sensor with a low activation pressure (P_a ≤0.1 barg) which is used to initiate the actuation of the float valve.

TABLE 6-8

Minimum Distance for Externally Activated Float Valves from Protected Vessel

Type of Explosion Protection in the Vessel	Containment	Venting with Vent Pipe	Venting without Pipe or Suppression
P_{max}, $P_{red,\,max}$	9 bar	3–4 bar	1 bar
DN of the Float Valve (mm)	Minimum distance, *dmin* (m)		
200	ca. 9	ca. 5	ca. 4
400	ca. 12	ca. 7	ca. 6

Further discussions about float valves are presented in the books by Bartknecht (1981, 1989), NFPA 69 (2002), and Siwek (1996).

EXKOP™ VALVE

Another, more recent, type of fast-acting valve is the Exkop™ valve. The valve trim (internals) is actually a rubber bladder surrounded by an air chamber. In the event of a deflagration, an electrical signal is sent from a sensor, typically mounted on a deflagration relief device, to the Exkop™ valve air tank mounted integral to the valve. The air tank discharges air to the chamber surrounding the rubber bladder and compresses it, which pinches off flow in the pipeline. Because the rubber bladder has relatively little mass, it is both very fast acting and imparts low shock to the piping. The rubber bladder, however, may be adversely affected by high temperatures. One advantage of the Exkop™ valve is that it can be mounted fairly close to the equipment in which the deflagration occurs, usually 15 to 18 feet away from the equipment. It also immediately rearms itself with ordinary plant air and is automatically placed back in service. Figure 6-12 shows details of an Exkop™ valve installation. These valves are often used in connection with the Q-Rohr™ flameless vent device.

6.6.4.2 Flame Front Diverters

Flames can also be prevented from propagating from one piece of equipment to another through interconnecting piping by means of a flame front diverter. The basic principle of operation of this device is that the deflagration is vented at a point where the flow direction is changed by 180°. Due to the inertia of the fast flow caused by the deflagration, the flow will tend tomaintain its direction upward rather than making a 180° turn when the velocity is low (at normal conditions). When the high speed deflagration flame continues upward, it pushes open either a hinged cover or bursts a rupture disk located at the top of the diverter, allowing the flame to be released to the atmosphere. The location of a flame front diverter must be chosen so that the release of the flame does not pose a hazard to people or equipment.

Some flame front diverters have demonstrated the ability to successfully divert deflagration flames by directing them to the atmosphere. However, in some cases, tests have indicated that some diverters have been ineffective in completely diverting a deflagration; but where this has occurred, the deflagration severity has been reduced, that is, pressure piling does not occur or is less severe (NFPA 69, 2002). Therefore, before they are used, it is recommended that they be tested for the desired application. Figure 6-13 shows several flame front diverter designs. Siwek (1996) discusses some design details about flame front diverters.

More information about flame front diverters is presented in NFPA 69 (2002) and books by Bartknecht (1989) and Eckhoff (2003).

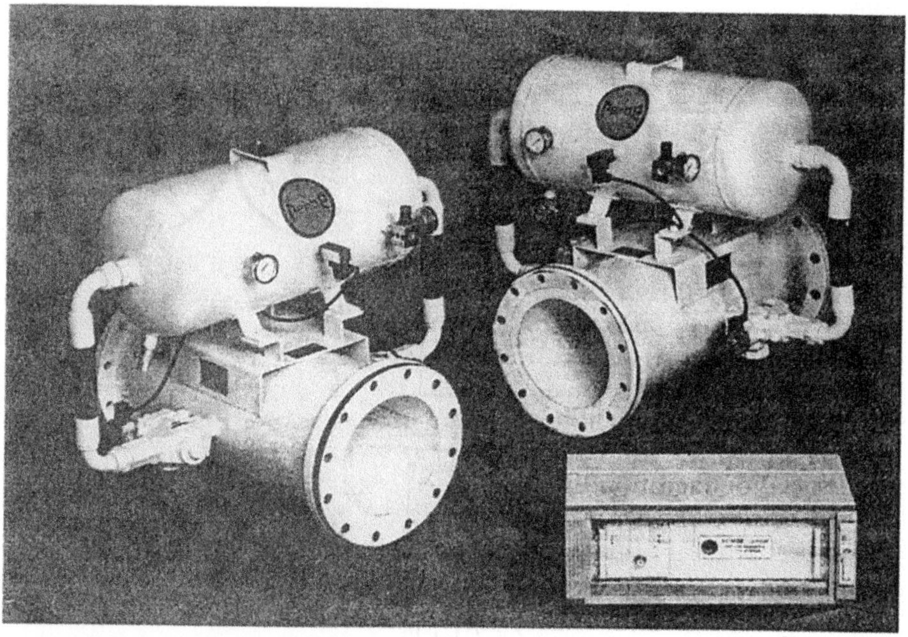

(a) Photo of Exkop fast-acting isolation valve

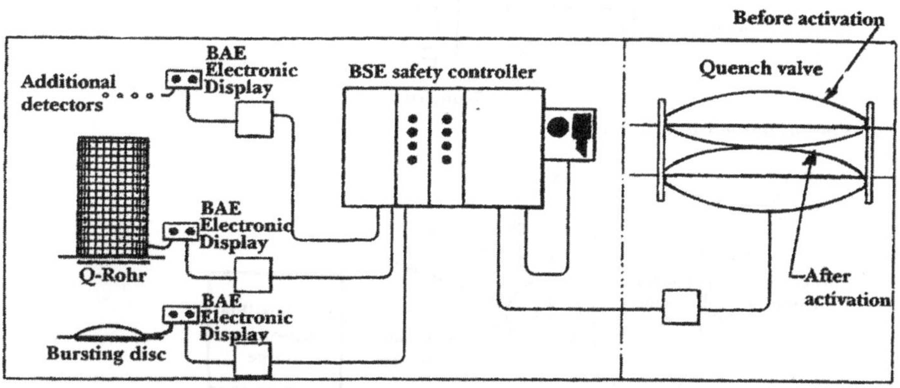

(b) Schematic of Exkop valve before and after activation

Figure 6-12 Exkop™ fast-acting isolation valve (quench valve). (Source: CvTechnology, Inc.)

6.6.4.3 Flame Front Extinguishing Systems (Suppressant Barriers)

This type of isolation device (also called a chemical barrier) is similar to deflagration suppression systems used on process equipment. Suppressant barrier systems are often used where the line diameter is too large for fast-acting valves to be installed (either because of cost or the large structure needed to support the valve).

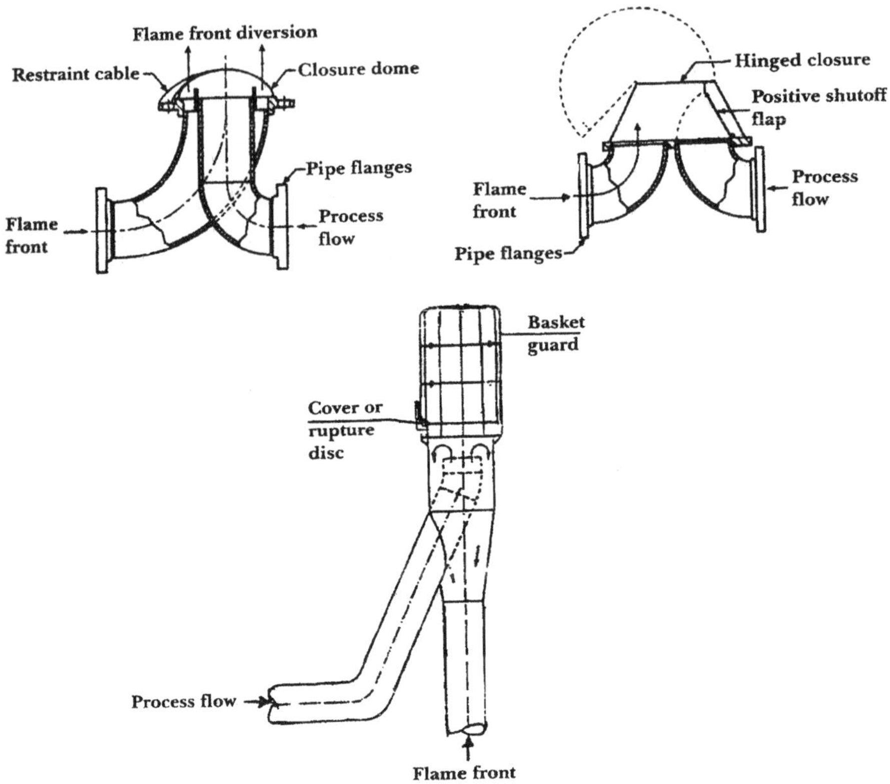

(a) Types of flame front diverters

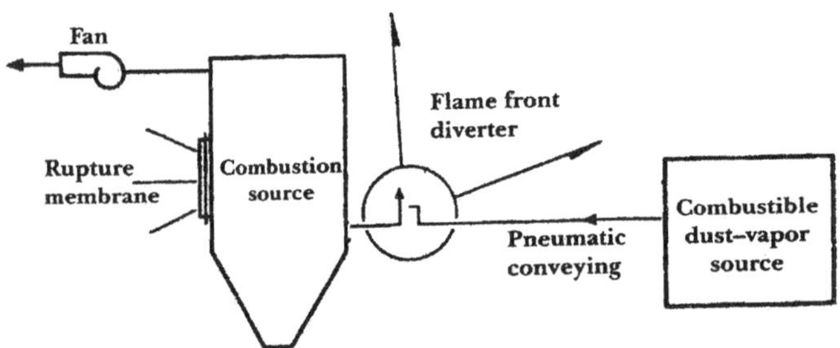

(b) Schematic of flame front diverter installation

Figure 6-13 Flame front diverters (NFPA 69, Bartknecht, 1989).

This barrier system consists of an optical sensor, installed in the pipeline or duct between two items of equipment, which detects an oncoming deflagration flame and emits a signal to a control unit. The amplified signal triggers the detonator-activated valve located in the piping from an HRD suppressant bottle, which injects an extinguishing agent into the pipeline through suitable nozzles. The extinguishing agent, preferably a powder suppressant, is discharged into the pipeline and forms a thick blanket which extinguishes the incipient flame. This type of barrier does not impede product throughput down the pipeline.

Typically, IR-photoconductor sensors, having three photosensors symmetrically installed in the pipe are used. Depending on the pipe diameter, at least two photosensors must be installed. This type of sensor contains a manually or automatically controlled operational test system for checking the photosensor, and a gas (air or nitrogen) flush system for cleaning the optical lens to keep it dust-free. Although optical IR-sensors are used, they have a relatively low sensitivity to daylight, and it is important to prevent daylight from getting into the piping.

Pressure detectors or sensors are not normally used for pipeline barriers since there is no clear distinction between the front of the pressure wave and the flame front, and pressure sensor response times often are too slow for use in this application.

There is a definite distance required between the location of the optical detector and the extinguishing barrier HRD to ensure that the extinguishing agent acts directly on the flame. The amount of extinguishing agent required depends on the nature of the combustible dust, the nominal diameter of the protected pipeline, the flame velocity, and the maximum reduced explosion overpressure in the vessel.

The design of suppressant barrier systems is based on experimental tests. Siwek (1996) describes and discusses tests that were performed to establish explosion velocity (as a function of K_{max}), the suppressant extinguishment sectional density (based on the pipe cross-sectional area), the minimum suppressant charge required, the number of HRD suppressant bottles, the mean flame velocity, and the minimum distance required between the flame sensor and the HRD suppressant bottles.

Suppressant barrier systems have the same shortcomings as cited in Section 6.6.2. In addition, the location of the sensor is critical to the successful isolation of a deflagration flame in a piping system. Bartknecht (1981) states that the flame sensor should be installed at a distance of 1 meter from the ignition source, and the extinguisher nozzles at a distance of 10 meters from the ignition source.

Figure 6-14 is a schematic of a deflagration suppressant barrier system for pipelines.

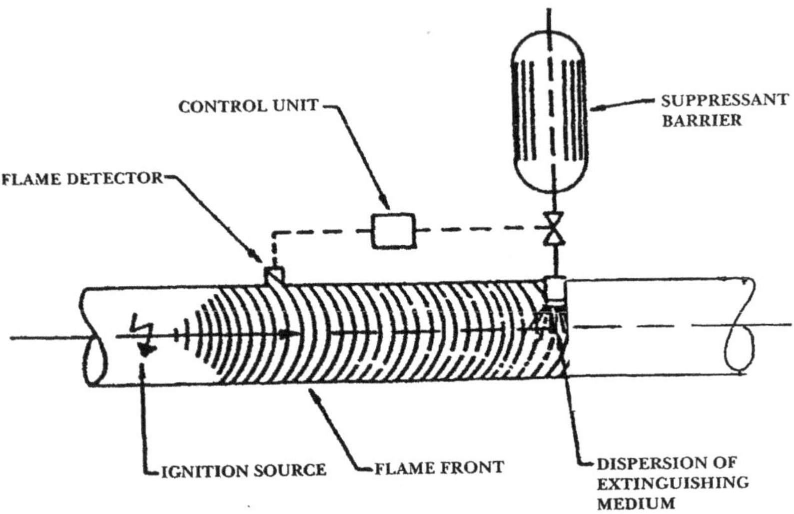

Figure 6-14 Schematic of a deflagration suppression barrier (chemical barrier) for piping

Further information on suppressant barriers can be found in NFPA 69 (2002), Siwek (1996), and the books by Bartknecht (1981, 1989) and Eckhoff (2003).

6.6.4.4 Material Chokes

Flame propagation can also be stopped between process equipment handling bulk solids and powders by judicious selection and design of bulk solids/powders conveying equipment such as rotary valves (rotary airlocks) and screw conveyors. The mass of bulk solids/powders contained in these items of equipment provide a tortuous path through which the gas and flame have to pass, and so can act as a "material choke" when certain design features are implemented.

Testing has shown that rotary valves can be effective in isolating explosion propagation between pieces of equipment if:

- There are three vanes on each side of the valve that are diametrically opposed.
- The valve body has sufficient strength to withstand the explosion pressure developed.
- If close tolerance is maintained between the vanes and the valve body, i.e., the gap between the rotor and housing is ≤0.2 mm (for aluminum powder, the gap between the tip of the rotor blades and the housing should be ≤0.1 mm).
- Two vanes per side are always in contact with the housing.

- The vanes or tips are made out of metal (no plastic vanes) and have a thickness of at least 3 mm.

Rotary valves will generally prevent flame propagation if the above criteria are followed (Bartknecht, 1989; NFPA 69, 2002).

Siwek (1989) presents a nomograph to determine the required minimum gap width between the blade tips and the rotary valve body inside wall that will prevent flame propagation. This nomograph may be used only for organic dusts and wet sulfur.

Wear can increase the tolerances cited above so that the minimum gap necessary to isolate the explosion is no longer present. These devices must be properly maintained to ensure that normal wear and tear do not adversely affect their ability to prevent flame and pressure propagation into downstream equipment. Where it is critical that an explosion not propagate through the rotary valve, the minimum tolerances must be maintained through a good testing and inspection program.

It is also important that the hopper or vessel located upstream of the valve have a minimum inventory of solids to prevent flames passing upward through the solids due to the pressure wave from an explosion in the hopper/vessel. Although there are no experimental data available at this time from which to select this minimum height, Siwek (1996) gives the following rules of thumb:

For solids with bulk density ≥1 kg/m^3: $H = D$

For solids with bulk density <1 kg/m^3: $H = D/BD$

where H is the height of solids in the hopper above the valve, m
D is the valve outlet diameter, m
BD is the bulk density of the solids, kg/m^3

To maintain this minimum height of solids in the hopper upstream of the rotary valve, low level sensors can be provided which are interlocked to shut down the rotary valve before the hopper goes empty, thereby maintaining a level of solids above the rotary valve. This level of solids effectively acts as a seal to prevent flame and pressure transfer into the equipment upstream of the valve. A manual bypass should be provided for the low-level sensor/interlock to allow emptying of the hopper when necessary.

However, the rotary valve can still significantly mitigate explosion propagation even if total isolation is not achieved, by significantly reducing transmission of pressure.

Siwek (1989) presents more information on the use of rotary valves for preventing explosion propagation.

Another type of isolation can be achieved by judicious selection and design of conveying equipment such as screw conveyors. In screw conveyors the removal of one flight turn will ensure that a plug of bulk solids/powder will always remain as a choke. A horizontal screw conveyor needs an adjustable baffle plate to complete the seal of bulk solids/powder with the top of the casing. On an inclined screw conveyor the screw will not normally empty itself below the missing flight even when the supply of feed to the lower end stops.

Siwek (1996) reports that experiments were carried out in which rice meal explosions in a 3.5-m^3 vessel were vented through choked conveyors and through a safety vent at the other end of the vessel. Dust clouds were ejected at the downstream end of the conveyors, but no flame was ejected.

Eckhoff (2003) discusses the use of screw conveyors as material chokes.

One important consideration that should be taken into account when using a rotary valve or screw conveyor as a material choke is that when a deflagration occurs in the equipment upstream of the choke device, it has to be stopped immediately by a suitable detector in order not to pass burning or glowing solids into downstream equipment, where they could then cause a second fire or deflagration. Rotary valves and screw conveyors should be tested for their suitability as flame arresters as well as for their pressure rating with appropriate explosion tests (Bartknecht, 1989).

Additional information on material chokes (rotary valves and screw conveyors) is presented in Bartknecht (1989), Eckhoff (2003), NFPA 69 (2002), and Siwek (1989, 1996).

6.6.4.5 HSE Propagation Prevention Device

Holbrow and Tyldesley (2003) discuss and describe an isolation device that may inhibit flame and pressure propagation between items of equipment. The device was designed for use in a 250-mm-diameter pipe/duct and incorporated one, two, or three cones closely spaced within the duct at a 300-m pitch. The first cone was about 600 mm from the wall of the test vessel. Figure 6-15 shows a typical arrangement with two cones. This device was developed and tested by the HSE, but has not been patented by them.

Tests were conducted with coal dust, anthraquinone, and milk powder, with three vent areas (on the test vessel) of 0.27, 0.66, and 0.96 m^2.

The device was successful in either eliminating or achieving a significant reduction of the flame from the vessel over a range of vent areas. Without the device installed, it was demonstrated that dust explosions in the vented 6.3-m^3 vessel can propagate extensive flames through an open pipe stub attached to the side of the vessel. The pipe stub was approximately the same length as the propagation device (0.3–0.9 m). Flame lengths measured from the pipe stub, without the device installed, were approximately 4 m using milk powder, 1–3 m using coal dust, and 4–5 m using anthraquinone. When the propagating device, with either two or three cones, was installed the flame from the pipe stub was generally eliminated or greatly attenuated.

Flame from the coal and milk powder was completely eliminated using a two-cone device. A three-cone device was successful in the complete elimi-

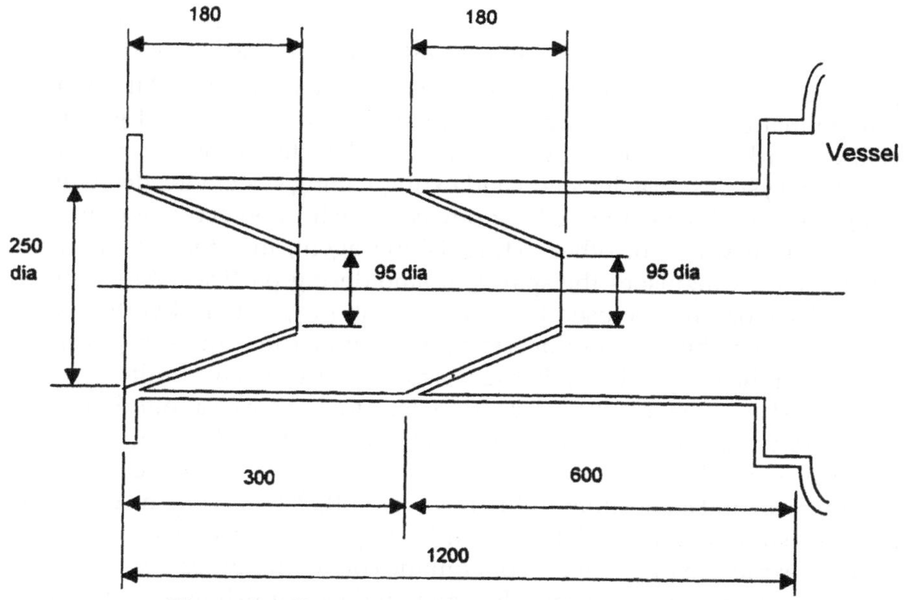

Figure 6-15 Propagation prevention device for pipes.

nation of the anthraquinone flame when a relatively large vent (0.96 m^2)was installed, and 75% reduction in flame length was achieved when a smaller vent configuration was used (0.66 m^2). Even though it was not possible to completely eliminate the anthraquinone flame, the 75% reduction in flame length would reduce the potential of ignition of dust in the pipe. However, the attenuated anthraquinone flame was still capable of igniting the dust in a 5 m long pipe attached to the device and causing propagation of flame down the length of the pipe.

The device, in its present form, is therefore not suitable for reactive dusts similar to anthraquinone (K_{St} = 308 bar-m/s).

However, with further development it may be possible to achieve a more effective device for use with more reactive dusts.

In general, the presence of the device in the pipe acted as an obstruction and resulted in an increase in the reduced explosion pressure within the vessel.

6.6.5 Spark Detection and Extinguishing Systems

Dust explosions can be initiated by sparks or smoldering (glowing) powder particles that are being transported from one item of equipment to another. For example, sparks or smoldering solids that are conveyed to a silo or baghouse containing a combustible solid above its LFL may initiate a deflagration. Milling, grinding, or other highly energetic mechanical action on particulate solids may generate sparks caused by overheating. Alternatively, foreign matter, such as tramp metal, may enter the equipment generating sparks. Sparks and smoldering solids may be detected using infrared optical detection systems. In a detection and extinguishing system the detectors are interlocked with an extinguishing device, which upon activation, injects a fine spray of water, or other extinguishing agent, into the dust-laden air stream and extinguishes the sparks or smoldering particles. Alternatively, spark detectors may be used to activate an abort gate that diverts the fuel, flames, and combustion products to a safe location (no water spray is used to extinguish the hot particles). Because spark detectors have limited fields of view, most systems require at least two detectors to cover a round duct. The detectors are usually situated on the inlet to the protected equipment (e.g., a dust collector). Such systems have been used in the wood processing, food, animal feed, and some other industries (Eckhoff, 2003). Forsyth (1980) presents a discussion of spark detection and extinguishing systems and their use for dust explosion protection in pneumatic conveying processes.

More information on spark detection and extinguishing systems is presented by Cholin (1997) and in NFPA 69 (2002) and Appendix C of NFPA 654 (2000).

6.6.6 *Prevention of Secondary Explosions*

Pressure waves transmitted through the atmosphere during the initial stages of the explosion can cause dust which is deposited on surfaces inside equipment or on building floors and structural members to be thrown into suspension. This disturbance can arise, either because of air movement over the dust deposits, causing the dust to be blown into suspension, or to vibration of equipment or structures so that dust deposits are shaken off and fall into the air. With either mechanism, or both, further dust clouds are formed which can be ignited by the flames already present from the primary explosion. The second stage, in which deposited dust is suspended in air and ignited, is called the secondary explosion. Quite often, the secondary explosion can be more severe than the primary explosion.

The case history below discusses an accident caused by a secondary explosion.

Case History of a Secondary Explosion

The following incident of a secondary dust explosion which occurred in a chewing gum manufacturing plant and involved sodium stearate dust is described in FM Global Loss Prevention Data Sheet 7-76.

A process used to make chewing gum involved applying sodium stearate powder to the surface of the newly finished product. Because the process was not completely enclosed, sodium stearate slowly escaped (migrated) from the system, accumulating in the room. A gum cutting machine failed, causing an explosion within it. The disturbance created by the explosion in the cutting machine dispersed the fugitive sodium stearate dust in the surrounding area, sending it airborne, and the fireball released from the cutting machine ignited the suspended dust. A severe building explosion ensured. Windows were blown out throughout two sides of the building, and several holes were blown through the floor. A cinder block partition wall, located about 10 feet from the production line where the initial equipment explosion occurred, was completely demolished. Mangled metal was all that remained of the machinery in the explosion area.

Although there was no history of a serious fire or explosion at this plant in its nearly 60 years of operation before the incident, the accumulation of combustible dust clearly created a continuing potential for a severe room explosion. Either the process equipment should have been made more dust-tight to prevent dust from escaping into the surroundings, or if dust accumulated very slowly, housekeeping should have been improved to prevent any significant accumulation of fugitive dust.

Secondary explosions can be prevented or mitigated by the following methods:

- Identify and eliminate fugitive sources as soon as they are discovered.
- Establish good housekeeping practices, which means frequent and regular removal of any solids spilled or accidentally discharged.
- Provide isolation devices and systems between interconnected process equipment handling combustible dusts.

NFPA 654 (2000) states that a dust layer is capable of creating a hazardous condition if it exceeds 5% of the building floor area. This would apply for the potential to have a secondary explosion in a building. Annex H of NFPA 68 (2002) discusses the effect of partial volumes of a building (area covered by dust) on the sizing of a building explosion vent. This guide has chosen to apply the layer thickness criteria of $\frac{1}{32}$ inch over 100% of the floor area and other surfaces to be more conservative than NFPA 654.

6.7 SITING OF EQUIPMENT AND BUILDINGS TO MINIMIZE DAMAGE FROM FIRES AND EXPLOSIONS

It is best to install combustible solids processing equipment in outdoor structures if the weather permits this (freezing is not a normal condition). However, in cold or wet (high humidity or high rain frequency) climates, this may not be practical, and the equipment usually has to be installed in buildings. Indoor areas in which combustible solids are processed, handled, collected or produced should be detached, segregated, or separated from other occupancies in order to minimize fire or explosion damage.

With respect to siting of outdoor equipment or buildings containing particulate solids processes and operations there are no published industry or insurance company recommendations similar to the ones for oil refinery and chemical plant layout and spacing (GE GAP, 2001; CCPS, 2003a). Outdoor equipment or buildings should be spaced so as to minimize impacts from fires or explosions. One possible approach to establishing spacing between a building containing a particulate solids process and another process building is to estimate the length of a fireball from a vented dust explosion using Equation 6-11 (from NFPA 68, 2002). According to Barton (2002), no fireball length of more than 30 meters has been measured in practice, so this should be a maximum separation distance with respect to fires. It may be possible to use closer spacing if an effective automatic fire suppression system is installed. This might be feasible in normally unoccupied areas where there is minimal chance of burn injuries. Another consideration is the pressure that the equipment in an outdoor structure or a building would be subjected to from an explosion. The maximum pressure from a vented deflagration can be determined by Equation 6-14. The maximum distance to which this pressure can exist is given in NFPA 68. These values can be used to ascertain what pressure loadings would be on adjacent buildings.

If feasible, equipment for handling and processing particulate solids should be arranged for gravity flow so that conveying of the solids is minimized. Equipment should be arranged so that the material flow follows the process scheme. Figure 6-16 shows the equipment layout for a typical particulate solids process plant. Distances between items of process equipment should be adequate to provide room for operator access and equipment maintenance and removal.

NFPA 654 (2000) states the following with respect to use of separation to protect equipment and buildings:

1. When separation is used to limit the fire or dust explosion hazardous area, the hazardous area shall include areas where dust accumulations exceed $\frac{1}{32}$ inch (0.8 mm) or where dust clouds of a hazardous concentration exist.
2. The required separation distance between the hazardous area identified in item 1 above and surrounding exposures shall be determined by an engineering evaluation that addresses the properties of the materials, the type of operation, the amount of material likely to be present outside the process equipment, the building design, and the

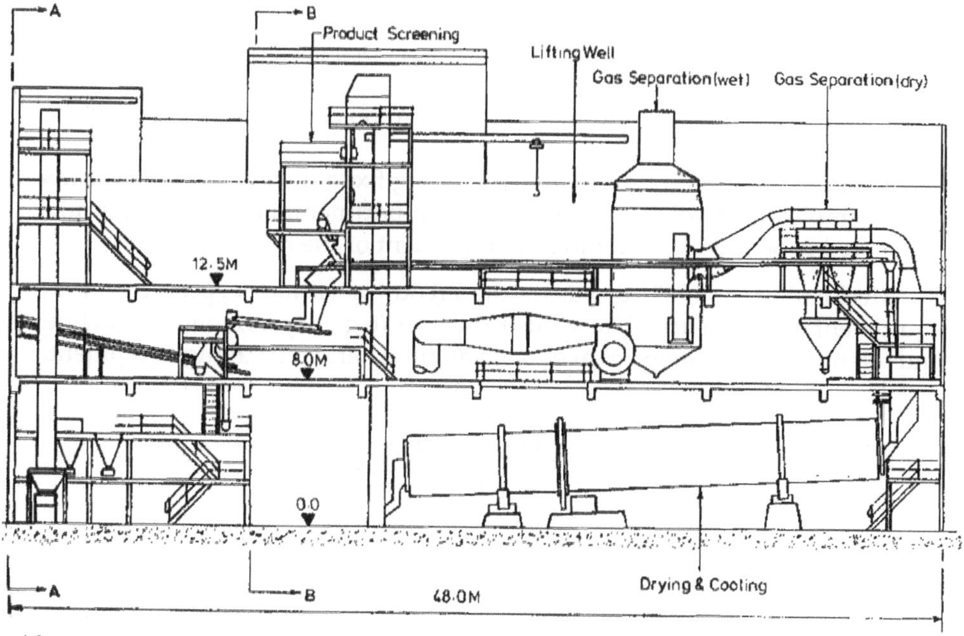

(a)

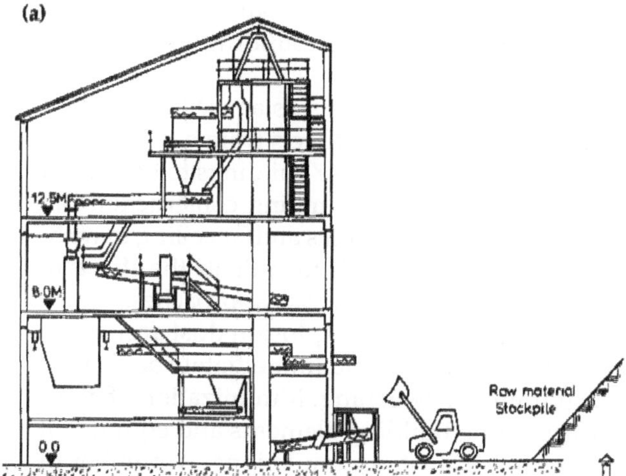

(b)

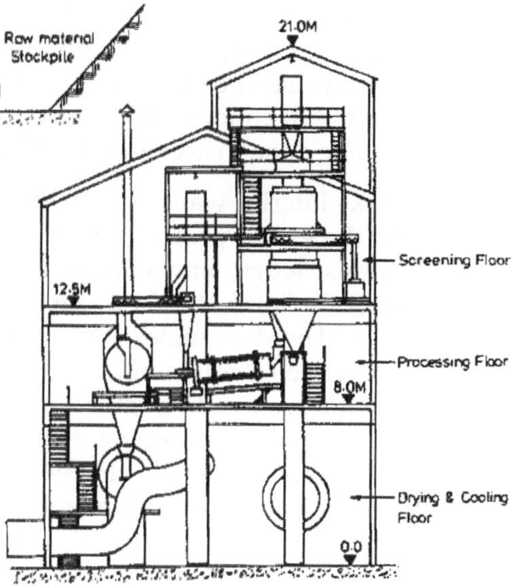

(c)

Figure 6-16 Equipment layout of a typical particulate solids processing plant. (a) Elevation, (b) Section A–A, (c) Section B–B.

nature of the surrounding exposures. In no case shall the distance be less than 30 ft (9 m).
3. When separation is used, housekeeping fixed dust collection systems employed at points of release, and compartmentation shall be permitted to be used in order to limit the extent of the hazardous area.

Mecklenburgh (1985) discusses the layout of various types of process equipment used in particulate solids processes (e.g., dryers, solids reduction and separation equipment, conveyors, and filling and packaging equipment.

6.8 BLAST RESISTANT (DAMAGE-LIMITING) CONSTRUCTION OF BUILDINGS

NFPA 654 (2000) states that all buildings in which combustible solids are stored, handled or processed shall be of Type I or Type II construction, as defined in NFPA 220, *Standard on Types of Building Construction*. Where local, state, or national building codes require, modifications shall be permitted for conformance to these codes. Other information on building construction requirements are also presented in NFPA 654.

For buildings containing particulate solids processes as well as adjacent buildings which could be impacted upon by blast effects from a dust explosion consideration should be given to designing these buildings to be of blast resistant (damage-limiting) construction. Damage-limiting construction can be accomplished by either designing the building walls and/or ceiling to withstand the blast effects or by providing venting panels to mitigate the effects of an internal dust explosion, or a combination of these two approaches.

To minimize the potential for a building dust explosion dust accumulations should be removed or minimized. They should immediately be cleaned up when they are found by vacuuming or sweeping. Minimization of accumulations can also be done by arranging building elements and equipment to reduce the likelihood and amount of dust accumulations. A number of things can be done to accomplish this, such as:

- Make interior walls as smooth as possible with minimal ledges.
- To the extent practical, provide horizontal surfaces such as girders, beams, ledge, and equipment tops with a sloped cover having a smooth finish, to shed dust settling out of the air. Sloped covers should be at an angle of 60° from the horizontal, unless a lesser slope is known to be sufficient.
- Box in overhead structural steel which is out of the reach of normal vacuuming or sweeping operations and which has horizontal ledges (such as I-beams or U-shaped channels in the up or sideways position) with a noncombustible material to eliminate pockets for dust accumulation.

Such measures will not entirely eliminate the room or building dust explosion hazard, but can reduce the frequency and severity of a dust explosion.

In deciding whether or not dust accumulations are excessive enough to create a room or building dust explosion hazard, and warrant immediate cleaning, calculate the proportion of a building volume subject to suspensible dust accumulations. For a 10-ft (3-m)-high building, an explosion hazard exists if more than about 5% of the building area is covered with $\frac{1}{16}$inch (1.6 mm) of suspensible dust. This calculation is based on a dust having a bulk density of 36 lb/ft^3 (580 kg/m^3) (FMG 7-76, 2001).

- For buildings with a different height, or containing a dust with a bulk density significantly different from 36 lbs/ft^3, the thickness of a dust layer required to create a room explosion hazard can be calculated by the following equation (FMG 7-76, 2001):

$$t_{ex} = HA_{tot}/87.5\rho A_{dust} \qquad [6\text{-}16]$$

where t_{ex} is the thickness of dust layer required to create a room explosion hazard (inches), H is the height of the room or building (ft), A_{tot} is the total floor area of room or building (ft^2) [use 20,000 as an upper limit regardless of the actual room or building area (*exception:* if dust is evenly deposited over entire area, can use actual floor area without maximum)], ρ is the bulk density of deposited dust (lb/ft^3), and A_{dust} is the total area (ft^2) of suspensible dust deposits within the room or building volume.

In using this equation, the dust should be considered as "suspensible" if it is located above floor level. Dust accumulated on the floor should be considered as suspensible only if there is some other explosion hazard which is capable of creating a disturbance in the air in the same or adjacent areas (e.g., equipment posing gas, vapor, or dust explosion hazards; adjacent areas with gas, vapor, or dust room-explosion hazards).

Consider overhead beams and ledges in determining the total area of dust deposits. The available surface area for dust deposits on joists or steel beams can be roughly estimated at 5% of floor area. Steel beams might have an equivalent area as high as 10% of floor area, such as when spans between columns are longer than average, or when a large building elevation difference requires a stronger roof structure due to anticipated snow drift loadings. Dust adhering to walls should also be taken into account since this is easily dislodged. Also consider other projections such as light fixtures which can provide surfaces for dust accumulation.

A building explosion hazard can be considered eliminated if the following conditions exist (FMG 7-76, 2001):

1. The fugitive dust escape and accumulation rate is very low, for example, suspensible dust accumulation rate is less than $\frac{1}{16}$inch (1.6 mm)

per a 3- month period [⅛ inch (3.2 mm) for wood and other low density dusts], and

2. The cleaning frequency is high enough to permit one scheduled cleaning to be missed without allowing dust accumulations to reach a level where they create a room explosion hazard.

In designing buildings to withstand the blast effects of a dust explosion inside a building or the blast effects on the outside walls of a building from an external gas or dust explosion, evaluation should include the building response to overpressure, fragments, and induced ground shock. The design of blast-resistant buildings is done by civil/structural engineers and is beyond the scope of this discussion. The principles of blast-resistant design are summarized in Appendix B of *Guidelines for Evaluating Process Plant Buildings for External Explosions and Fires* (CCPS, 1996). The ASCE (1997) also has published a comprehensive book on the design of blast-resistant buildings.

Dust explosions can be mitigated by providing vent panels in building walls and/or roofs. Vent panels should be sized and designed in accordance with the criteria and procedures given in NFPA 68 (2002). Often, physical barriers (internal building walls) are erected to segregate areas where particulate solids are handled and/or processed from other areas where such operations are not conducted. Such physical barriers (walls) shall have all penetrations of floors, walls, ceilings, or partitions sealed dust-tight, and where such structural assemblies have a fire endurance rating, the seal shall retain that rating. Where venting is provided for the room, the physical barrier erected to segregate dust deflagration hazards from other occupancies shall be designed to preclude failure of these barriers before deflagration pressure can be safely vented to the outside.

Some design considerations and practices relevant to room and/or building venting are as follows:

- Make sure that the explosion vent relief (opening) pressure (P_{stat}) is as low as the wind resistance design will permit. In a low wind area, P_{stat} can be as low as 20 psf (0.01 bar), whereas in high wind areas 30 psf (0.015 bar) is more typical.
- Evaluate the use of explosion vents in the roof to provide explosion relief where more than one or two snowfalls occur per year. Locating explosion vents in the walls is the preferred alternative. However, where a thorough engineering study shows that explosion-venting walls cannot provide the needed venting area, roof vents may be accepted if snow and ice are not allowed to build up on the vent. Suitable measures should be taken to prevent or melt away accumulation of snow or ice.
- If explosion venting devices swing out of the way during an explosion, as opposed to rupturing, make sure that they cannot reclose. Use

mechanical devices or gravity to prevent them from reclosing. Tethering cables attached to the tops of the panels may be used, although the cables should preferably be attached to the bottoms.

- If vent closures of the shear and pull-through fastener design or friction-held closures are used, and projectile motion is a concern, tethers to limit the vent closure movement should be provided. Attach no more than two corners to tethering cables, making sure that the tethered corners are adjacent. Tether the panel or closure so that it will not swing back into the vent opening after the explosion.
- Do not attach sprinkler system piping to any wall, ceiling, or roof which is expected to be displaced by the pressure of a room or building explosion, and do not allow such structures to support sprinkler system piping.
- Some companies use prefabricated buildings of light wall construction (e.g., aluminum or masonite panels), rather than buildings constructed of brick or cinder block, as they offer excellent explosion venting design. Alternatively, the building can be constructed so that roof and intermediate floors are supported by a strong frame structure, with the walls being made of light-weight panels that function as vent covers should an explosion occur (Eckhoff, 2003).
- If a building is constructed of reinforced concrete, it can be made sufficiently strong to enable the windows to serve as vents. It is important, however, to ensure that flying pieces of glass do not present a hazard to humans. To avoid this hazard it may be necessary to replace the glass panes by other transparent plastic panes that do not shatter (such as Lexan™ polycarbonate sheets, or equivalent materials).

Explosion vent panel inertia (mass) has an effect on the vent relief pressure and response [see Annex F of NFPA 68 (2002) for a discussion of this effect]. Specifications for explosion vent panel should require a vendor test to demonstrate the response pressure of the relief device.

6.9 PROTECTION OF EQUIPMENT AND BUILDINGS BY WATER SPRINKLER/DELUGE SYSTEMS

Consideration should be given to providing fire protection in facilities storing, handling, and processing particulate solids for extinguishing particulate solids (dust) fires. Fire protection should also be provided in these facilities for equipment containing flammable liquids to prevent a fire in or around them from spreading to adjacent equipment containing combustible solids.

Palmer (1973) presents the following recommendations for fire protection of equipment containing combustible solids or for accumulations of burning solids outside of equipment:

- Water is suitable and effective for extinguishing dust fires unless it reacts with the dust or if there is electrical equipment close by.
- The water should not be applied as a high pressure jet on burning dust deposits, which could raise a dust cloud, but as a low pressure spray which simply dampens the dust deposit. This will cause the dust to become more cohesive, thereby reducing the likelihood of dispersion of the dust.
- The penetration of water into the dust layer can be assisted by the addition of about 2% of a wetting agent such as a detergent. The use of the wetting agent also helps to make the water spread over the surface of the dust layer more rapidly (however, no additional information has been found to corroborate this recommendation).

Fire protection using water can be provided by automatic sprinkler, deluge, or spray systems. These are briefly discussed below.

Automatic Sprinkler Systems

Automatic sprinkler systems are the most commonly used type of fire suppression system. Many U.S. codes and standards recommend or require their installation in facilities processing, handling, or storing large quantities of combustible materials. Although they are used less extensively in other parts of the world, they are often the primary fire protection system in warehouses and in buildings processing solid particulates of various sizes and compositions, providing the materials are not water reactive or otherwise water incompatible.

Most sprinkler systems are wet pipe systems with closed sprinkler heads. Fire pumps are often needed in large industrial facilities to supply the required water flow rates and pressures corresponding to the specified water discharge density (water flow rate per unit floor area covered) and design area of operation (tantamount to the number of flowing sprinklers). NFPA 13 (2002) is the North American standard that dictates the required discharge densities (or flow rates) and design area (or number of operating sprinklers) for different categories of hazard. CEN prEN 12845 (2001) is the draft European standard for sprinkler system design.

The requirements for sprinkler system design depend on the occupancy hazard as given in NFPA 13 (2002). Occupancies are classified according to the hazards associated with the material being stored or processed. These occupancies are listed as Light Hazard Occupancies, Ordinary Hazard Occupancies (Group 1 and Group 2), Extra Hazard Occupancies, and Special Hazard Occupancies. For example, NFPA 13 Appendix A lists "chemical plants – ordinary" as an example of an Ordinary Hazard Group 2 occupancy category, whereas rubber reclaiming, compounding, drying, milling, vulcanizing, and plywood and particle board manufacturing facilities are considered Extra Hazard Group 1 occupancies. An Ordinary Hazard Group 2 occupancy can have a design discharge density as low as 0.15 gpm/ft^2 (6.1 mm/min) for a design area of 4000 ft^2 (372 m^2), while an Extra Hazard Group 1 occupancy cannot have a design discharge density lower than 0.24 gpm/ft^2 for the same 4000-ft^2 design area. If flammable or combustible liquids are present, the occupancy is classified as Extra Hazard Group 2 and the required design area and discharge density are significantly larger. There are also special classifications and associated requirements for facilities doing spray application using combustible materials, handling solid oxidizers, or organic peroxide formulations, and cleanrooms, among others. Since these occupancy classifications have major implications with regard to the sprinkler design and cost, and since classification decisions involve judgments about the relative combustibility and protection challenge of different materials being processed, the classifications and system design should be considered by or with a fire protection engineer.

In the case of storage facilities, the storage commodities (product plus packaging) are classified on the basis of their challenge to the sprinkler system using a different classification system than the occupancy classifications. The NFPA 13 plastic commodity classification decision tree includes considerations of the percentage of the commodity that is foamed (expanded) plastic, the percentage of material with a heat of combustion comparable to cellulosic materials, and whether the commodity is in cartoons or exposed. There is also a special provision for free-flowing plastic materials defined as "those plastics that fall out of their containers during a fire, fill flue spaces, and create a smothering effect on the fire." Since these generic classifications are sometimes difficult to interpret and apply for specific commodities, classification testing methods have been developed, as discussed in Chapter 4.

The fire heat release rates of various commodities and the effectiveness of sprinkler systems designed to protect them are discussed in Chapter 5 of *Industrial Fire Protection Engineering* (Zalosh, 2003).

Besides being installed on a ceiling to provide building protection, automatic sprinkler systems are also installed within certain equipment and structures handling combustible particulates. For example, NFPA 654 (2000) states that sprinkler protection should be provided to protect dust collectors, silos, and bucket elevators. It goes on to recommend that a documented risk evaluation shall be used to determine the need for sprinkler protection inside an enclosure that uses flammable or combustible liquids for processing combustible particulate solids. NFPA 654 also states that "dust accumulations on overhead surfaces shall be minimized to prevent an excessive number of sprinkler heads opening in the event of a fire."

There are several different types of sprinkler systems in addition to the traditional and most commonly used wet pipe system. These include dry pipe systems, deluge systems, water mist systems, ultra-high-speed suppression systems, and pre-action systems. Pre-action systems use closed sprinkler heads, but admit water into the sprinkler piping upon operation of a separate detection system (interlocked pre-action system), or upon actuation of the sprinkler heads (noninterlocked pre-action system). Deluge systems use open sprinkler heads and are described briefly below. More detailed descriptions of the various types of sprinkler systems are given in Section 10 (Water-Based Suppression) of the *Fire Protection Handbook*, 19th edition (Cote and Linville, 2003).

NFPA 13 (2002) discusses in detail the design and installation of automatic sprinkler systems, and should be consulted when considering the use of these.

Water Deluge Systems

A deluge sprinkler system is a sprinkler system designed in accordance with NFPA 13 with open sprinkler heads. Deluge systems utilize multiple open nozzles connected to a water supply and spaced throughout a compartment or around equipment to be protected. Water is held out of the piping system by a special valve (a deluge valve), which can be automatically opened on a signal from a control panel. The system requires a separate fire detection system which provides the signal to the control panel to open the valve and allows water to flow from all nozzles simultaneously. This is similar to a water spray system, but does not use directional water spray nozzles to achieve a specific water discharge and distribution. Deluge protection can be applied over an entire hazard area. Deluge systems are used for protection against rapidly spreading, high hazard fires. Area deluge protection may not be able to wet the underside of some equipment. In this case, water spray nozzles located under these items of equipment spraying onto these dry areas can complement the deluge protection.

For water deluge protection systems, recommended discharge densities can be taken the same as for water spray systems (Table 6-9), unless other information is available.

GE GAP (1994) recommends the following application flow rates for combustible and flammable liquid fires:

- 0.25 gpm/ft^2 over the protected area for fires involving combustible liquids.
- 0.35 gpm/ft^2 over the protected area for fires involving flammable gases or liquids. Where deluge foam-water sprinkler systems are provided to protect against flammable liquid hazards, the required density can be reduced to 0.25 gpm/ft^2.

NFPA 13 (2002) also discusses the design and installation of water deluge systems.

When protecting process equipment by internal water sprinkler/deluge systems (e.g., baghouses), means must be provided to drain the water from

TABLE 6-9
Water Spray Application Rates for Exposed Surface Area[a]

Item	Application Rate (GPM/ft²)
Exposure Protection, General	0.10–0.25
Exposure Protection for Specific Applications	
Air-Fin Coolers	0.25
Compressors	
General	0.25
Compressors in Building	0.30
Cooling Towers	0.15–0.50
Fired Heater Supports	0.25
LPG Loading Racks	0.25
Motors	0.25
Pipe Racks	0.25
Pressurized Storage Tanks	
Radiant Exposure	0.0–0.10
Nonpressure	0.10–0.25
Impingement (Pressure)	0.50 minimum
Process Buildings & Structures	0.15–0.30
Pumps	0.50
Atmospheric Storage Tanks	0.10
Pressure Vessels, Exchangers and Towers	0.25
Transformers	0.25
Turbines	0.25
Control of Burning	0.30–0.50
Extinguishment (Note 1)	
Combustible Solid	0.15–0.30
Combustible Liquid	0.35–0.50
Flammable Liquids	(Note 2)

[a]Source: Adapted from API Pub. 2030 with permission of the American Petroleum Institute.

Notes:

1. Rates should be established by review of relevant test data for the specific materials.
2. May not be desirable or possible. See API Pub. 2030.

the equipment quickly in order to avoid collapsing the equipment from the additional water load, for which the equipment is not usually designed. If the water protection system is external, and water can enter the process equipment somehow, drainage means should also be provided.

Where process equipment is located outdoors in cold climates where freezing can occur, wet pipe type fire protection piping should be traced and insulated, or consideration should be given to providing a dry pipe type fire protection system.

Water Spray Systems

The term "water spray" refers to the use of water in a form having a predetermined pattern, particle size, velocity, and density discharged from specially designed nozzles or devices.

Water spray fixed systems are usually provided for special fire protection systems, since the protection can be specifically designed to provide for effective fire control, extinguishment, prevention, or exposure protection. They may be independent of, or supplementary to, other forms of protection. Water spray fixed systems are most commonly used to protect processing equipment and structures, flammable liquid and gas vessels, piping, and electrical equipment such as transformers, oil switches, and motors. They have also been shown to be effective for controlling fires on many combustible solids.

Fixed water spray systems can be an automatic or manually actuated fixed pipe system connected to a water supply and equipped with water spray nozzles designed to provide a specific water flow rate and particle size discharge and distribution directed over the protected surface or area. Water spray systems discharge water through open spray nozzles of various designs. The piping system is connected to the water supply through a manually or automatically actuated valve that initiates the water flow. An automatic valve is actuated by a detection system installed in the same area as the water spray nozzles. Where an explosion hazard exists, the actuating valve should be barricaded and piping arranged to limit explosion damage.

Process equipment, storage vessels, and structures may be protected with water spray systems. They are extremely efficient at cooling equipment and structural members such as columns and platforms exposed to fire as they absorb heat and reduce temperatures. Adequate cooling will prevent the weakening of metals, collapse of structures, and rupturing of vessels.

- They can also be used for extinguishment of combustible solids fires.
- Design of fixed water spray protection systems will depend on the type of equipment being protected and types of nozzles used.

Water discharge densities depend on the design objectives for the application, the type and nature of the equipment or structure to be protected, and the characteristics of the probable fuel involved. The actual water discharge density used should be selected based on available reference data, judgment, experience, and (in some cases) testing. Recommended discharge densities are given in FMG 4-1N (1998), NFPA 15 (2001) and API Publication 2030 (1998). Table 6-9 (API Publication 2030) lists recommended discharge densities for many types of process equipment exposed to fires. Discharge densities are also given for extinguishing combustible solids fires. Insurers may be able to provide other information on water spray system recommended discharge densities for process equipment handling solids.

More detailed discussions of fixed water spray system design and installation can be found in FMG 4-1N (1998), NFPA 15 (2001), and API Publication 2030 (1998).

There is limited information on water discharge densities for application to equipment storing, handling, or processing combustible solids. However, some recommendations are given for three types of equipment as follows:

- Electrostatic precipitators: see Section 5.3.4.2
- Fabric filters: see Section 5.3.4.3
- Belt conveyors: 0.25 gpm/ft^2 (per FMG 7-11, 2003)

Additional information on water based fire protection systems is presented by Cote and Linville (2003) and CCPS (2003b).

Fire Detectors

Fire detectors are used to actuate automatic water spray and water deluge protection systems, as well as warn operators who can then take action for manually actuated water protection systems. Certain fire detectors may be used for detection and alarm only, or may be utilized to actuate a fire suppression system and/or shut down process equipment. Fire detectors can be thermal detectors, smoke detectors, or optical detectors.

THERMAL DETECTORS

Thermal detectors are available as either fixed temperature, rate-compensated fixed temperature, or rate-of-rise design. Spot-type are the most commonly used. Line type detectors are useful in certain specific applications, such as conveyors and cable trays. The spot-type thermal detector is considered to be one of the most reliable and trouble-free type of thermal fire detectors.

Fixed temperature detectors are designed to operate at a specific temperature, while rate-of-rise detectors respond to temperatures which rise faster than the design rate rather than at a specific temperature. Rate-compensated fixed temperature detectors are a combination of both types and reduce a fixed temperature detector's time lag in rapidly growing fires. Rate-of-rise detectors may react faster than fixed temperature detectors to a rapidly growing fire, but may never operate during a slowly developing fire.

Recommended maximum spacing for thermal detectors inside buildings should be in accordance with NFPA 72 (1999) and UL/FM Approval requirements. Thermal detectors installed outdoors require special consideration, such as closer spacing, and should be addressed on a case-by-case basis. In addition, thermal detectors located outdoors may need special shielding devices to prevent snow and ice accumulation which might delay response.

Additional information on thermal detectors is presented the NFPA Fire Protection Handbook (Cote and Linville, 1997) and NFPA 72 (1999).

SMOKE DETECTION

A number of detectors are available for sensing smoke (airborne particles of combustion), including ionization, photoelectric, and incipient stage/early detection devices. In general, flaming fires are more quickly detected by ionization detectors, and smoldering or slow-developing fires are more quickly detected by photoelectric detectors. Ionization detectors are widely used and are suitable for many applications. Selection of the proper smoke detector depends on the nature of the hazard.

In areas where smoke detection is used to actuate fire suppression systems, cross-zoning or other voting type detection schemes are often used for actuation. Subfloor and above-ceiling areas containing electrical and instrumentation cabling or other combustibles may also need smoke detection. Critical and high-value control cabinets or panels which are not ventilated may require internal smoke detection since an incipient stage fire could escalate beyond control inside a cabinet prior to room detection.

Numerous conditions in chemical plants can have a negative impact on smoke detector performance. These conditions can result in false alarms. Some detectors are sensitive to humidity and dirty atmospheres. Ionization detectors are also sensitive to chlorine, hydrogen chloride, hydrogen fluoride, phosgene, trichloroethylene, and other chemicals.

Smoke detectors are commonly used in electrical equipment rooms, control rooms, and other areas containing electrical equipment. Often, the detectors are located at the ceiling. Electrical equipment failures often generate "cold smoke" which will not rise to the ceiling where the detectors are located, and additional detectors may be required at the bottom of the electrical cabinets or rooms to sense this condition.

More information on smoke detectors is found in the NFPA Fire Protection Handbook (Cote and Linville, 2003) and NFPA 72 (1999).

OPTICAL FLAME DETECTORS

Optical flame detectors are devices which respond to radiant energy. Infrared (IR) radiation and ultraviolet (UV) radiation detectors are the most commonly used types. They have proven to provide prompt detection, but not always been proven reliable as single source detection devices as they will warn if they detect any source of radiation within their sensitivity range. False alarm rates can be high if this kind of detector is applied improperly. However, there are several techniques which reduce false alarms and failures. Two commonly used techniques are dual IR and combination UV/IR detection devices. These dual or combination sensors reduce false alarms from sources such as welding arcs, flickering sunlight, hot bodies, or flashing lamps.

The sensitivity of optical sensors depends upon the cleanliness of the lens. Sensors installed where dust or other materials may accumulate on the lens should be of the self-cleaning type or be located where routine maintenance can be readily performed.

Optical flame detectors are best employed in areas where very fast response is required for actuation of protective or emergency shutdown systems. They find wide acceptance in enclosures where inerting or suppression systems require immediate actuation. The actual detector placement depends on the area or equipment being monitored (physical dimensions, complexity, production importance, etc.), detector cone of vision, and sensitivity level of the sensor. Sensor installation should closely adhere to the manufacturer's recommendations. Optical flame detectors may also be provided with swivel bases to allow for optimum sensor positioning.

Optical flame detectors used as the primary detection device are considered a suitable alternative for standard thermal detection devices. To enhance reliability, additional control safeguards are often utilized, such as cross-zone or voting logic.

Additional information on optical flame detectors is presented in the *NFPA Fire Protection Handbook* (Cote and Linville, 2003) and NFPA 72 (1999).

6.10 PROTECTION OF EQUIPMENT AND BUILDINGS BY FOAM AND OTHER SPECIAL EXTINGUISHING SYSTEMS

Although fire water application is most often used for protecting equipment and buildings from fire, it may not always be the best agent for extinguishment of particulate solids and flammable liquids fires, and it may require special precautions when used on energized electrical equipment. Water cannot be used on water-reactive particulate solids. Other fire extinguishing agents such as foams, dry chemicals, carbon dioxide, and other special extinguishing agents may have application in many particulate solids process facilities, especially those that also contain nearby equipment storing flammable liquids that may leak or have spills that can result in fires. This section discusses foams and other special extinguishing agents and systems.

6.10.1 Foams

In general, foams are usually not used for extinguishing fires in equipment storing, handling, or processing particulate solids. Fire-fighting foams are of limited use on particulate solids (dust) fires as the foams cannot penetrate into the dust deposits, and at best, can form a blanket over the surface. Although they may extinguish flaming and glowing actually on the surface, they are of limited value in controlling fires within the solids. Also, as foams are water-based systems, they cannot be used with water-reactive solids.

However, if there are items of equipment containing flammable liquids in the areas where particulate solids are processed, and they could leak and catch fire, then the use of foam may be feasible to control a flammable liquid fire, depending on the source of the fire (flammable liquid).

Deluge foam–water sprinkler and foam–water spray systems are often preferred to "regular" water deluge or spray systems where a flammable or combustible liquid pool fire is expected. This is particularly true in areas with poor drainage. The foam coverage will greatly reduce the exposing fire, thereby allowing a reduction in the water spray or deluge flow application rates (densities). The extinguishment mechanism is mainly through exclusion of oxygen by smothering, assisted to a degree by cooling water released from the foam blanket.

Modern foams are produced mechanically by proportionately mixing foam concentrate with water and then aerating the solution for expansion. Foams are classified by their expansion ratio. Low-expansion foams have less than a 20:1 expansion ratio (normal expansion ratio ranges between 8 and 12). Medium-expansion foams can have up to 200:1 expansion, and high-expansion foams have expansion ratios between 200:1 and 1000:1 (in practice, the normal expansion ratio is between 750 and 1000).

Several types of foams are available and used in the chemical industry for fire extinguishing and suppression of vapors of spilled volatile chemicals by blanketing. Discussions about them are presented by Hiltz (1993) and Scheffey (1997).

Before the use of a foam is contemplated, foam manufacturers should be consulted as to the feasibility of using a foam for suppressing fires in combustible solids.

NFPA 11 (1998), NFPA 11A (1999), and NFPA 16 (1999) should be consulted for specific requirements for the design, installation, operation, testing, and maintenance of foam systems.

6.10.2 Dry Chemical Systems

Dry chemical (powder) systems are considered the most effective agents for extinguishment of combustible metal fires and other materials that are incompatible with water. When these dry powders are used to extinguish fires of combustible metals, which are of severe explosibility, particular care must be taken to avoid stirring up the burning dust while applying the extinguishing agent. This is usually done manually with long-handled shovels in a gentle manner to avoid disturbance of the burning metal powders. The requirement of manual application means that only small fires can be handled (Palmer, 1973). Dry powders can also be used for extinguishing dust layer fires on electrical equipment.

Tapscott (1997) provides a discussion of various types of dry powders that are used for extinguishing combustible metal fires.

Where flammable liquids are stored near particulate solids, it may be desirous to use dry chemicals (powders) to put out a flammable liquid fire to protect the adjacent solids processing equipment. However, dry chemical systems are usually far less effective in preventing reignition of flammable liquid pool fires than is foam. Therefore, extreme care and judgment must be given to its application. Twin or combined agent systems utilizing foam and a dry chemical can effectively be employed.

Dry chemical fire extinguishers are listed by Underwriters Laboratories (UL) and other testing organizations for use on various types of fires and are rated based on their relative effectiveness. Acceptable extinguisher location (determined by the travel distance between the extinguisher and the hazard and the minimum area capable of being protected by each extinguisher) is a function of the extinguisher rating and the hazard(s). Based on these criteria, portable, wheeled, or stationary dry chemical extinguishers are strategically positioned throughout the protected facility.

There are a number of different types of dry chemical agents that can be used in extinguishing a fire. Among the most effective and commonly used are potassium bicarbonate (Purple K), sodium bicarbonate (regular dry chemical), monoammonium phosphate (ABC dry chemical), potassium chloride, and urea-potassium bicarbonate. Various additives are mixed with these base materials to improve their storage, flow, and water repellency characteristics. The most commonly used additives are metallic stearates, tricalcium phosphate, or silicones, which coat the particles of dry chemical to make them free-flowing and resistant to the caking effects of moisture and vibration. Regular dry powder and Purple K are effective on flammable liquid and energized electrical fires (Class BC), while the ABC or multipurpose type is also effective on ordinary combustible fires (Class ABC). Although dry chemical agents are very effective on electrical fires, their residue after discharge usually requires extensive cleanup.

Fixed dry chemical systems are normally used to protect small hazards and their use in chemical plants depends on the hazard severity and application. Large systems have proven applicable in high risk areas such as flammable liquid storage, process areas, compressor, and pump rooms. Local application dry chemical systems have also been used for vent stack fire extinguishment.

Additional information on the design, installation, operation, testing, and maintenance of dry chemical systems is available in a number of publications. NFPA 10 (2002) has information on portable fire extinguishers, and NFPA 17 (2002) is relevant to dry chemicals extinguishing systems. Also, the *NFPA Fire Protection Handbook* (Cote and Linville, 2003) contains a chapter on this subject.

6.10.3 Carbon Dioxide Systems

Carbon dioxide systems often are used to extinguish fires by displacing air, thus lowering the oxygen concentration and thereby smothering fires. Carbon dioxide may be able to be used to extinguish smoldering fires in silos (see Section 5.3.15 for a discussion of the work by Tuomisaari et al., 1998).

Carbon dioxide can be used in both manual and fixed applications and in local and total flooding systems. Manual (portable) carbon dioxide extinguishers are more often used on electrical fires since the agent is nonconductive and leaves no residue.

Portable carbon dioxide extinguishers are recommended in power generation facilities, control rooms, and switchgear rooms, but due to a limited discharge rate, these extinguishers should only be used indoors. Like other portable extinguishers, carbon dioxide extinguishers are classified and rated for use and effectiveness and are located accordingly.

Use of a carbon dioxide fixed system for a total flooding application typically requires an atmosphere of about 34% carbon dioxide by volume. Atmospheres containing carbon dioxide at concentrations above about 5 volume percent pose significant hazards to exposed personnel. The severity of the hazard increases quickly with increasing carbon dioxide concentration.

See http://www.epa.gov/Ozone/snap/fire/co2/co2report.pdf or http://www.epa.gov/Ozone/snap/fire/co2/appendixa.pdf for a more complete discussion of the hazards associated with carbon dioxide use. Due to the life safety hazards, carbon dioxide flooding systems should be provided with appropriate safeguards to protect personnel. Another hazard is mentioned by Cross (1987), who points out that an electrostatic charge is generated when a gas containing solid particles issues from an orifice. Thus, a carbon dioxide jet being directed from a fire extinguisher into a vessel containing a flammable atmosphere may contain small particles of charged solid carbon dioxide and may have a sufficient charge to ignite the mixture.

NFPA 12 (2000) and the *NFPA Fire Protection Handbook* (Cote and Linville, 2003) contain more information about carbon dioxide systems.

6.10.4 Halon Replacement (Clean) Agents

Various types of Halon were used for many years as effective fire extinguishing agents, but have been found to be detrimental to the ozone layer and are no longer being used as fire extinguishing agents. The various types of Halon are being replaced by what is called "clean agents." Clean fire suppression agents are defined as fire extinguishants that vaporize readily, leave no residue, are nontoxic, and nonconductive.

Halon replacement (clean) agents may be effective for extinguishing certain particulate solids fires, but this should be confirmed with the supplier. NFPA 2001 (2000) states that clean agents shall not be used on fires involving the following materials unless thy have been tested to the satisfaction of the authority having jurisdiction:

1. Certain chemicals or mixtures of chemicals, such as cellulose nitrate and gunpowder, that are capable of rapid oxidation in the absence of air.
2. Reactive metals such as lithium, sodium, potassium, titanium, magnesium, zirconium, uranium, and plutonium.
3. Metal hydrides.
4. Chemicals capable of undergoing autodecomposition, such as certain organic peroxides and hydrazine.

Clean agent systems are useful in extinguishing fires in specific hazards or equipment and in occupancies where an electrically nonconductive medium is essential or desirable, or where cleanup of other media presents a problem. Some typical hazards where total flooding clean agent fire extinguishing systems are used are as follows (NFPA, 2001, 2004):

- Electrical and electronic hazards
- Subfloors and other concealed spaces
- Flammable and combustible liquids and gases
- Other high-value assets
- Telecommunications facilities

Clean agent Halon replacements fall into two broad categories: (1) halocarbon compounds, and (2) inert gases and mixtures. This section will only discuss halocarbon compounds.

These new clean agents are primarily replacements for Halon 1211, 1301, and 2402. Halon 1211 and 2402 are streaming agents which have most often been used in manually applied fire extinguishing equipment and local application-type fixed systems. Halon 1301 was most often used in total flooding-type fixed systems. Halocarbon replacement chemicals consist of primarily carbon, hydrogen, and fluorine (chlorine and bromine are no longer included). Halocarbon clean agents extinguish fires by a combination of chemical and physical mechanisms, depending upon the compound. At the present time the two most commonly used Halon replacement agents are HFC-227ea (sold as FM-200 or FE-227) and HFC-236fa (sold as FE-36) (Barton, 2002). These two clean agents, and others that have been developed and are being developed, are not as effective as Halon 1301, that is, they require much more agent mass (as high as 1.7 to 1.9 times as great) to achieve the same extinguishing effectiveness as Halon 1301.

NFPA 2001 (2004) and the *NFPA Fire Protection Handbook* (Cote and Linville, 2003) contain additional information on Halon replacement agents.

6.11 CONTAINMENT FOR CONTROL OF RELEASES OF TOXIC PARTICULATE SOLIDS

Many particulate solids, such as insecticides and pharmaceutical active ingredients (also known as potent compounds), may be harmful to health in small doses. Therefore, they must be contained (prevented from being emitted from a process or equipment) to avoid harm to operating personnel or contamination of the surrounding environment.

Manufacturing or products from toxic particulate solids involves the transport of solids and liquids into and out of process equipment. These activities have the potential for releasing dust and vapors into the workplace environment. Therefore, there is a need to incorporate suitable control measures into design, procedures, and operational practices. Prime requirements are to avoid operator and work environment exposure to these toxic solids, and prevent cross contamination, bad housekeeping, and product loss. Several principles are used to meet these prime requirements, as follows:

- Eliminate the problem (is there a safer alternative?)
- Enclose the process (containment)
- Reduce stray dust levels at source (removal, dilution, etc.)
- Provide operator personal protection

Some worker health and safety regulations state that operator protection can be used only as the secondary line of defense, so the primary solution must therefore be containment. The amount of containment that must be provided is usually based on a performance-based occupational exposure limit (PB-OEL) as shown in Table 6-10 (Heidel, 2001). For high potency powders, Category 3A, 3B, and 4 are used in the pharmaceutical industry.

TABLE 6-10
Containment Category Characteristics

Category	Potency	Design	OEL Range
1	Low (> 100 mg/day)	Conventional open equipment; incidental contact with compound	>100 μg/m^3
2	Moderate (10–100 mg/day)	Gasketed, flanged equipment; laminar flow/directional laminar flow; enclosed transfers	20–100 μg/m^3
3A	High (0.01–10 mg/day)	Transfers using high-containment valves (e.g., split butterfly valves); containment for every disconnect	20–5.0 μg/m^3
3B		Barrier technology (negative pressure)	<5.0 μg/m^3
4	Extreme (<0.01 mg/day)	Remote operation; fully automated; no human contact	<0.01 μg/m^3

The IChemE (Hirst et al., 2002) has defined five containment strategies, as follows:

Strategy Number	*Required Controls*
1	Controlled general ventilation
2	Local exhaust ventilation and airflow booths
3	Barrier isolation systems whereby open powder transfer within the isolator chamber itself is permitted
4	Barrier isolation systems with contained powder transfer systems or packages within them
5	Fully automated or robot type operation. No direct operator involvement

These containment strategies are discussed below.

Strategy 1—Controlled General Ventilation

When particulate solids are not health-hazardous, it may be feasible to use controlled general ventilation (also called dilution ventilation) in a building or room. This method controls the air environment by removing and replacing contaminated air before concentrations reach unacceptable levels. In general, dilution ventilation is inappropriate for solids of high toxicity where the concentrations fluctuate, or where an operator works close to a point of release of contaminant.

The use of dilution ventilation has four limiting factors:

1. The quantity of contaminant (dust) generated must not be too great or the air volume necessary for dilution will be impractical.
2. Workers must be far enough away from contaminant evolution, or the evolution of contaminant must be in sufficiently low concentrations so that the workers will not have an exposure in excess of the established threshold limit value (TLV).
3. The toxicity of the contaminant must be low.
4. The evolution of contaminants must be reasonably uniform.

The location of the exhaust fan and air inlet require careful consideration in order to provide a suitable air-flow pattern in the room to ensure operator exposure is minimized. A number of rules of practice to achieve these criteria are as follows (HSE, 1975):

1. Locate the exhaust fan near to the source of the contaminant.
2. Ensure that the fresh air movement is from the worker to the contaminant leak source, and not vice versa,
3. Ensure that the air inlet supply is not contaminated with exhaust air.
4. Provide back-up air supply for the air inlet where necessary.

Additional discussions of general (dilution) ventilation are presented in the books by ACGIH (2001) and Burton (1998)

Strategy 2—Local Exhaust Ventilation and Airflow Booths

Local exhaust ventilation (LEV) removes contaminants (dusts) from the source (equipment) directly into a ventilation system, that is, the arrangement of exhaust equipment is such as to envelop the source of contaminants before they can escape into the environment. It does not rely on the dilution or displacement effects of general ventilation. It should be designed to remove the contaminant from around the source rather than from around the operator.

An LEV system may be considered to consist of the following elements:

1. Contaminant source
2. Capture hood (booth)
3. Ducting
4. Contaminant collection/removal equipment
5. Exhaust fan

CONTAMINANT SOURCE

The contaminant source is the process equipment from which dust is emitted. The main problem with designing an effective LEV system is the dispersion of the dust form the source. Once it has escaped from the immediate area of the source, it is very difficult to bring it under control. It is necessary, therefore, to obtain information on the behavior of the dust leaving the source, such as its direction and velocity. The air currents which influence the behavior of the dust should also be studied.

Air currents which are not essential to the process, but are incidental or unconnected, should be eliminated. Important sources of air motion are:

1. Thermal air currents, especially from hot processes or heat- generating operations.
2. Motion of machinery, e.g., belt conveyor, etc.
3. Material motion, as in dumping or container filling.
4. Movements of the operator.
5. Room air currents (which are usually taken at 50 fpm minimum and may be much higher).
6. Spot cooling and heating equipment.

CAPTURE HOOD (BOOTH)

There are two basic types of capture hood or booth systems in a LEV system: (1) enclosures and receptor hoods, and (2) captor hoods. Enclosures may be either total or partial enclosures.

With a receptor (capture) hood the emission source is either inside or at the mouth of the hood. The hood receives dust-laden air brought into it by air currents. These are either hot air currents rising from the process or cold air currents induced by the process. Total enclosure is the most effective arrangement, but is not always practical. Where a total enclosure is used, it needs to be airtight so that it contains the dust handled, and strong enough to remain so. When such an enclosure is used, the space and the equipment inside may be heavily contaminated with dust. There is thus a need for measures to limit exposure when the enclosure has to be opened. These may include measures to remove the dust and safety measures such as entry procedures and/or interlocks to control access. If a partial enclosure is used, it should be no larger than is necessary for the purpose. The velocity of the air entering it should be high enough to overcome any tendency for the air inside to escape. Turbulence may be created within the enclosure by working of equipment inside it, and outside the enclosure by air currents.

The other main type of hood used in LEV is the captor hood. Whereas a receptor hood receives a flow of dust-laden air carried into it, a captor hood draws the air flow in. The two types of hood may sometimes be similar in shape, but the principle of operation is different. A receptor hood operates on the "push" principle while the captor hood operates on the "pull" principle. For a captor hood the air velocity required depends critically on the distance between the hood face and the contaminant source. For a normal shape such as a circle, square, or rectangle (although not for a slot), the air velocity at a point only one hood face diameter from the center of the hood face itself is typically only one-tenth of the face velocity. Thus, a broad rule-of-thumb for a practical system design is that the contaminant source should not be more than one hood face diameter from the hood face. The performance of a captor hood may be enhanced by the use of flanges around the face inlet. The flanges eliminate air flow from ineffective zones where no contaminant exists. The use of flanges can reduce air requirements by as much as 25% (ACGIH, 2001). For most applications, the flange width should be equal to the hood diameter or side to be effective. The volumetric flow of air required for a captor hood can be very much greater than for a receptor hood.

Discussions of LEV systems and their design are presented in the books by ACGIH (2001), Burton (1998), and Burton (2000).

DUCTING

When handling combustible dusts, ducting should be constructed of metal or other noncombustible material, and be of adequate strength and rigidity to meet service (temperature and pressure) conditions and installation requirements. Where exterior insulation is used on duct systems, it should be noncombustible.

Combustible insulation on indoor duct systems is acceptable by some insurance companies if adequate ceiling sprinkler protection is provided.

Ducts of plastic material and plastic-lined metal ducts should be used only for noncombustible dusts. If they pass through an area where combustible materials are stored or processed and a fire could occur, these ducts should be protected by a sprinkler system.

Recommendations for the specification, installation, and fire protection of ducts are discussed in FMG 7-78 (2000).

Ducts conveying particulate solids should be designed for a gas (air or inert gas) velocity high enough to prevent particles from settling out and depositing inside the duct. Accumulation of combustible solids inside of ducts can lead to fires and explosions, and possible system shutdown if the accumulation becomes too great. The conveying duct also may not be sufficiently strong enough (they are often made of thin wall sheet metal) if a heavy solids accumulation occurs, and may collapse, emitting particles into the adjacent areas. The following recommendations for conveying duct design velocities are given in Industrial Ventilation (ACGIH, 2001):

• Dry dusts and powders: (e.g., fine rubber dust, cotton dust, Bakelite molding powder dust)	2500–3500 feet/minute
• Average industrial dusts: (e.g., sawdust, granite dust, limestone dust)	3500–4000 feet/minute
• Heavy dusts: (e.g., metal turnings, brass turnings, lead dust)	4000–4500 feet/minute
• Very heavy or moist dusts: (e.g., moist cement dust, quicklime dust)	4500 ft/minute+

Strategy 3 and 4—Barrier Isolation Systems

The differences between Containment Strategy 3 and 4 have been brought about by the realization that the weak links in any barrier isolation system are the inlet and outlet transfer doors/valves/ports. These could be double transfer ports, split butterfly valves or similar. When these ports are contaminated with the toxic solids it is inevitable that some emission of material will occur due to either poor maintenance or ingress of the powdered material itself into the finely machined faces of these transfer ports. As such, this may bring about a risk of solids emission. Containment Strategy 4 eliminates this risk by requiring that all powder transfers within the isolator body are fully contained. Here, the isolator provides an outer layer of containment and an extra level of safety for the operators. A short discussion of barrier/isolator systems is presented below.

Containment of a potent compound should reflect the systems approach, with the system having at least two of the following four levels of protection in place. The four levels of protection for a containment system are (1) the internal environment of the containment device, (2) the device itself, (3) the external environment in which the containment device is placed, and (4) the structure housing the containment system. An example using barrier/isolation technology would be the pressurization scheme inside the barrier/isolator, the physical structure of the barrier/isolator, the pressurization scheme in the room containing the barrier/isolator, and the wall of the room.

Barrier/isolator systems are made up of four basic parts: the physical structure, the internal environment, the transfer and interaction technologies, and the monitoring systems. Within each of these parts there are a series of choices that will greatly impact the success of the containment.

The physical structure of a barrier/isolator is generally classified by the appearance of the structure and falls into two categories: hard shell and soft shell. The next level of differentiation is the materials of construction, with the choices for the hard shell being plastic, plexiglass, glass, and/or stainless steel, and the choices for the soft shell being a soft plastic or otherwise flexible film. Important issues concerning the construction of the shell are mainly the following three: durability or integrity of the system, ability (or need) to have cleanable surfaces, visibility of internal operations.

The durability or integrity of the barrier/isolator should be consistent with the environment in which the unit will be used, the process and tools involved, and the frequency of use.

The cleanability issue involves the quality of finish, welds, joints, and corners. The finish quality is a balance between a surface smooth enough to allow the removal of residual materials (using, for example, a clean-in-place type of spray-ball action), and the issue of reflectivity of light, which can cause operator eye strain. A reasonable compromise is a No. 4 finish. The issue of welds and joints is important. Joints around penetrations such as viewing areas, air inlets, and exhausts, and other required components should be smooth and free of ledges. Materials used for gaskets need to be checked for durability and compatibility with the powders handled. Effects of exposure to cleaning agents should also be checked.

Visibility is also an important consideration. Pharmaceutical operations require a number of interactive steps, including visible checks of activities.

The internal environment of a barrier/isolator is usually negative pressure, and is a form of secondary containment to the system and should be designed to allow for protection from a breach of the barrier. The most common breach is a glove port or bag ring not properly sealed or torn during operation. The internal airflow of a barrier/isolator designed for containment should allow clear viewing of the operation and removal of any heat generated by the process equipment contained in the system. The airflow system's size is determined many times by the flow requirement of the emergency or breach volume requirements. A number of systems have been developed using two-stage fans, having speeds for normal operation and emergency situations. Lighting and utilities also require planning. Lighting should be located exterior to the barrier/isolator structure with see-through panels allowing the light to enter. Utilities can be provided interior to the barrier/isolator through several means, including fixed panels inside or flexible connection ports. Attention should be paid to potential electrical hazards that could be present from solvents or dust.

Transfer of materials between a barrier/isolator and the outside, without exposure of the potent material, is one of the biggest challenges of any system. The transfer can be accomplished by the following technologies: double posting port, bag rings, airlock, and airflow.

The double posting port is the mainstay of high-integrity transfer technology. It is also called a rapid transport port (RTP). The double posting port design and several variations are manufactured by a number of vendors. The idea of this component is a double door, capable of separating to form two seals for separate enclosures, allowing both to maintain integrity.

These ports enable hazardous materials to be transferred from a container into an isolator and vice versa without loss of containment. Both the container carrying the hazardous material and the wall of the isolator must have compatible port designs: the alpha assembly, built into the wall of the isolator, and the matching beta assembly, which closes off the container. Both assemblies comprise a frame holding a circular door, which cannot be opened unless a matching assembly is docked in place, in which case the two doors open together as one. It works in the following manner:

1. The container is offered up to the locked port door in the isolator wall,
2. The alpha and beta assemblies are docked together, and rotated between 45° and 90° (depending on the design),
3. This locks the port and the container together and interlocks the removable disks that form the closure pieces in each half,
4. Only when fully interlocked can the disks be lifted or hinged out of the way, usually by the operator working through glove ports in the isolator.

Developed for the nuclear industry, the design has proven transferable to the pharmaceutical industry. The advantage of the double-door posting port for transfer of materials, is a high level of containment integrity and a positive sealing of two enclosures for the transport of materials between operations. Disadvantages are the durability of the rotating seals, the rotation required for docking of most systems, the round configuration of the door by most designs, and a small amount of contamination sometimes left on the sealing after use. When using an RTP to charge a vessel, each receiving vessel needs its own small isolator glove box to remove the port doors to permit the flow of powder from the transit container into the receiving vessel.

The split butterfly valve technology (developed by Glatt Systemtechnik in Germany, Serck in the United Kingdom, and Buck in Germany) permits the docking of a precharged container onto a process vessel without the need for a second isolator glove box required for an RTP. The valve provides a way to transfer particulate solids between vessels with a high level of containment, allowing solids to flow out of the bottom opening of one container fitted with one-half of the valve into the top opening of another container equipped with the other half of the valve.

The split butterfly valve, as its name implies, is effectively a single butterfly valve that is split into two parts, generally along the long dimension of the disk to make two flappers. One matching part (called the passive valve) closes off and seals the bottom of the upper container holding the powder, while a matching half (called the active valve) seals the inlet connection of the lower receiving vessel. Split butterfly valves must always be closed when the two parts are separated. Only when both valve halves are docked together, using a centering and locking pin that is vertically positioned, to form a complete connection do interlocks release to permit the combined disk to rotate 90° to open and permit transfer of the powder from the upper to the lower vessel. After transfer is completed, the valve (combined disk halves) are rotated back to the closed position and inflatable seals about the vertical pin, on the passive side of the valve, and about the circumference of the active part of the valve, are inflated to seal. The two containers are then separated by separating the valve into the original two halves that are connected, respectively, to the bottom of the upper vessel and the top of the bottom vessel. Figure 6-17 is a schematic of a split butterfly valve showing its principle of operation.

Bag rings involve the use of plastic sleeves or bags to introduce materials into or out of a barrier/isolator. The method is more procedure dependent in sealing technique, but has proven effective in handling potent compounds. Bag rings can come in a variety of sizes and shapes to meet requirements and offer flexible means of transfer.

The technique to remove a contaminated object from an isolator works as follows. The object to be removed is passed into a bag through a spigot with two grooves positioned around the outer circumference, located one above the other. A rubber retaining ring holds the bag in place by engaging with one of the grooves, normally the one nearer the end of the spigot, to form a seal sufficient to satisfy Containment Strategy 3. When the object is ready to be removed, the bag is tied around it with two seals close together and cut between the ties. An improvement upon this technique is to use heat to seal the bag in three places and to cut across the middle seal. The sealed bag containing the object can then be removed. A new bag is placed over the remnant of the old one around the spigot and secured by another ring, engaging with the groove nearer the isolator. Working through the new bag, the operator disengages the old one from the spigot, allowing its loose end and retaining ring to fall into the new bag. Meanwhile, the ring holding the new bag is moved to engage with the groove nearer the end of the spigot and the new bag is then ready to receive further objects for removal.

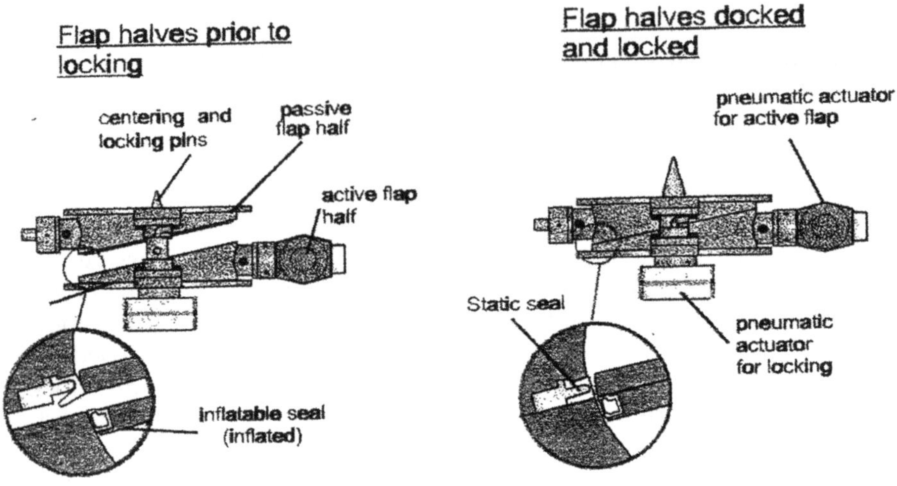

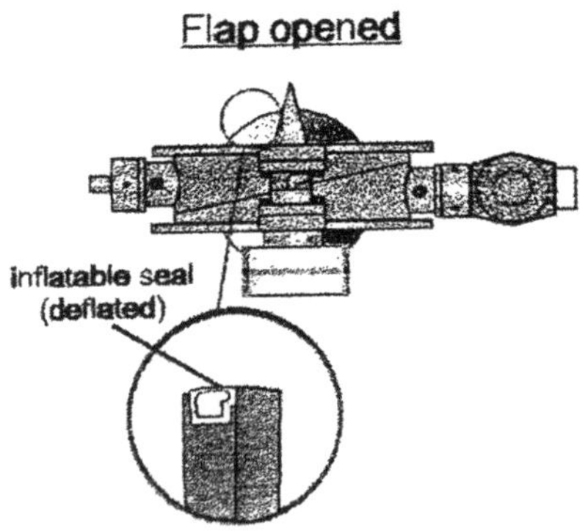

Figure 6-17 Schematic of a split butterfly valve (docking valve) showing its principle of operation. (Courtesy of Glatt Air Techniques, Inc.)

Two disadvantages are longer-term dependability of seals and working with materials that can be torn or cut by sharp objects.

Airlocks and airflow techniques have been used to transfer potent compounds from one container or vessel to another vessel, but they are not as effective a containment device as double posting ports, split butterfly valves, or bag rings.

Allowing people to interact with the process or equipment contained in the barrier/isolator is an important part of pharmaceutical operations. This can be done via flexible membranes, half suits, glove boxes, or robotics.

Monitoring of operations is critical for providing a safe workplace. Developing a plan or strategy for measurement of the levels of potent compounds outside the containment system is necessary. The measurement program should involve both facility and personnel monitoring. Barrier/isolator monitoring can be done by any of the three following methods: pressure detection, air filter leak detection, and gas leak detectors.

Pressure detection is used to determine pressure difference. Devices may be as simple as gauges, with visual readouts or units connected to alarms.

Online feedback is important when dealing with very potent compounds, and this can be provided by air filter leak detection. The pressure drop approach to monitoring filters is not sensitive enough to detect a leak of quantities that are above exposure levels. Online particle counters are recommended after the filtration system to determine integrity of filters for compounds having an exposure limit below 10 micrograms/m^3. These systems can be integrated with facility information systems for record keeping and alarming.

Gas leak detectors, such as oxygen analyzers, can be used to determine proper processing environments and to act as a personnel alarming device. Gases such as helium can be used internally to the barrier/isolator to achieve detection in cases of extremely potent compounds. Integration with information systems for record keeping and safety should be considered.

Some general design practices for a typical facility handling toxic powders are presented below (Dream, 1998):

LOADING DISSOLVERS AND/OR REACTORS

Toxic powders are usually loaded into dissolvers and/or reactors directly from loading compartments (barrier/isolator devices) located above the vessels. Transfer containers holding liner bags filled with powder are transferred from the intermediates store to the appropriate loading compartment. Each loading compartment is equipped with a pivoting transfer container hoist, located in a central position between a pair of dissolvers or bank of reactors. Transfer containers are mounted to the hoist still in the valve-up position. The hoist lifts the transfer container, turns it to the valve-down position, and pivots it to the appropriate alignment location above the chosen transfer tube, before lowering the transfer container onto the transfer position. Following sequential operation of the appropriate valves, and in the case of a liner bag unit, removal of the container base will cause the contents of the transfer container to discharge by gravity. The liner bag may be gently agitated (avoiding rupture) by the operator until it is seen to be empty. Following sequential closure of the appropriate valves and replacement of the of the container base, the hoist is used to return the transfer container to the free-standing valve-up position for transport to the store/cleaning or liner bag disposal station. The facility should be designed for the appropriate cleaning for this type of barrier/isolator.

UNLOADING CENTRIFUGES

Some options for unloading centrifuges are as follows:

- Gravity discharge into rigid intermediate bulk containers (RIBCs) (direct connection with the aid of proper valve design arrangement)
- Gravity discharge into bins (direct connection with the aid of isolator and proper valve design arrangement)
- Gravity discharge into other equipment (e.g., filter-dryer) with direct connection arrangement

Lined bins are not a preferred method since it is often difficult to find liners that are compatible with all products and solvents. Normally, centrifuged products will subsequently be dried. The following recommendations assume that RIBCs will be used. It is proposed that centrifuges be unloaded into stainless steel RIBCs. An RIBC system on its own does not provide a contamination-free "connect/disconnect" method or an assured discharge valve closure unless manually fitted with a valve locking device. Reliable methods of eliminating external contamination must be provided when connecting/disconnecting RIBCs. It is recommended that to enhance cleanability and valve security, all RIBCs should be of the single entry type. RIBCs should be transported and stored in the completely closed/isolated mode.

LOADING DRYERS

When handling toxic particulate solids it is strongly recommended that dryers be loaded directly from centrifuges or from RIBCs via a dust-tight connection device. The method of connection between the dryer and the centrifuge or RIBC depends on the type and model of dryer selected to suit the product. If the dryer is directly fed from a centrifuge, the loading can be done via a connecting pipe or chute so that the product is inherently contained. If the dryer is loaded from a RIBC, the connection must be done by an appropriate barrier/isolator device as described previously.

UNLOADING DRYERS

Most potent pharmaceuticals are produced in small volumes and it is assumed that the dryer is a batch type and is unloaded into transfer containers. The dryer output is monitored by use of a weigh scale under the transfer container. In some cases, this weigh scale will be used under manual control to dispense a predetermined weight. Tray dryers require special consideration as the trays have to be manually removed by an operator and transferred into a transfer container. The options available for unloading this type of dryer are:

- Unload under a laminar flow hood, and transfer trays to another location for discharging
- Unload in an enclosed space (room) that is also used for discharging trays into lined containers and for dryer shell and tray cleaning

The laminar flow hood is not recommended because transferring loaded trays to another area creates spaces to clean without providing a contained space for dryer shell cleaning. Also, transport of exposed trays to the surrounding environment creates a risk of airborne particles. Therefore, an enclosed space working arrangement is recommended. The contained space is seen as a "quarter suit" operation allowing easy operator access for short working periods. Tray emptying into lined transfer containers and tray cleaning would also take place in this contained space, since there are usually a limited number of trays.

POWDER FORMULATION EQUIPMENT

Delumping, granulation, milling, and blending equipment are to be designed with complete containment in mind. Usually, RIBC and proper valve arrangement design is used in this type of operation to contain the release of particulates into the workplace. By design, the powder processing equipment has an inherent dust-generating propensity associated with them. They, therefore, require more than containment design. They require incorporation of dust collection equipment that extracts the airborne dust particles from within the containment environment and surroundings. Also, this equipment requires safety features that minimize and protect against ignition and explosions.

An important concept of containment for control of releases of toxic particulate solids is segregation. Segregation is perceived as a concept with relevance at two levels. The first level relates to the need to provide segregation between processing systems. The second level relates to the segregation between batches. Segregation between process systems is particularly important to the manufacturer because of the high biological activity (health hazard) of many of the products. The current regulatory requirements with regard to operator and environmental safety are very much biased to primary containment and not secondary containment systems (personal protective equipment).

Segregation of batches means that if specific equipment items are used for multiproduct production, cleaning between batches is required. Cleaning can be regarded as a two stage process. The first stage relates to the removal of residual product on completion of the processing step (called decontamination or deactivation). The second stage is the preparation of the equipment for maintenance access on the next use. Decontamination (deactivation) is the operation that is performed to reduce or immobilize the potent compound to the point that the material cannot produce an airborne concentration exceeding the exposure guideline for the compound. This is accomplished by a number of steps and should be validated for individual compounds because each has different characteristics, etc. For example, Chung and Brookes (1997) describe the following cleaning procedure to decontaminate a reactor used to make a potent pharmaceutical. Solvent is first used to dissolve the bulk of the potent compound. The solution is drummed out and sent away for incineration. A "kill solution" is then charged inside the equipment, stirred, and circulated through all transfer lines. The solution is then drained out of the reactor. Then water rinses are applied until all residual "kill solution" has been removed. The final water rinse is then tested for the residual potent compound. A limit was calculated based on the final rinse volume and the equipment surface area, such that the concentration of potent compound in the final rinse corresponds to less than 10 $\mu g/100\ cm^2$. If the limit is exceeded, then the "kill solution" has to be applied again, and repeated until the limit is met.

Validation of cleaning is required for the protection of employees when working with potent compounds. Visual inspection and taking of swab samples of the equipment internal surfaces can be used to check the residual concentration of the potent compound. However, there may be small areas that are hard to reach by the decontamination agent, and thus, these areas have the potential of holding on to residual potent compound even after extensive decontamination and after the rinse solution sample has passed the HPLC test. These areas, such as between flanges and gaskets, can be identified by the fluorescein test. If such areas are found, then caution needs to be taken to further wet down and decontaminate these areas after equipment and piping are taken apart.

For multiproduct pharmaceutical plants producing potent compounds, the cleaning of equipment and transfer lines after each batch is critical to successful operation. The very low level of permissible contamination combined with the wide range of product activities make cross contamination prevention of extreme importance. The ability to clean the plant to very low levels of residuals, with acceptable allowable limits, is therefore critical.

Besides providing barrier/isolator systems for containment of toxic particulate solids, operators often have to use protective clothing ("space suits") when involved in handling potent compounds.

Buildings are also designed with a "clean-dirty" corridor concept. Material flowing in "clean" corridors is contained in either the shipping container in which it has arrived or is contained in a RIBC into which it has been loaded. In either case the exterior of the containers is free of high potency or toxic materials. For example, the design concept presented below is for a typical plant handling and/or producing formulations containing potent compounds. Raw materials go from the weigh staging area to the ingredient weighing/dispensing area. Clean, empty RIBCs also follow this route. The corridor leading to and coming from the dispensing area is considered a "clean" corridor through which unexposed RIBCs, contained raw materials and loaded RIBCs can pass. The exterior of the loaded RIBCs has not been exposed to the high potency compounds because of the material handling techniques employed in the dispensing area. From the dispensing area loaded RIBCs go to the staging area in a clean corridor to await processing in the granulation and drying train. RIBCs entering the granulation area are loaded onto a discharge station and the material is discharged into a mixer/granulator where the granulation solution is added. Empty RIBCs are considered "dirty" at this point because some residue will be left inside the container and will need to be cleaned before refill. This has to be done to avoid cross contamination and preserve batch integrity. Dirty RIBCs are taken from the granulation area down a "dirty" corridor to a "dirty" RIBC elevator to a second floor for cleaning in a RIBC wash station. The corridors through which the RIBC passes and the elevator are considered "dirty" areas. RIBCs enter the wash station from a "dirty" corridor and exit the station into a "clean" area. The RIBCs are then taken through clean corridors to the weigh staging area to repeat the cycle.

Material discharged into the granulator is gravity fed to a fluid bed dryer and then through a mill into a clean RIBC at the bottom of the processing train. Because of the material handling techniques the exterior of the RIBC has not been exposed to the granulated and milled materials, and is considered clean. The RIBC is then taken through clean corridors to a tumbling area and then from there to a mezzanine, via a "clean" RIBC elevator, for discharge into tableting or encapsulating machines. Discharged RIBCs are considered dirty and are removed from the mezzanine using a "dirty" elevator and transported to the RIBC wash station through dirty corridors to be cleaned and repeat the cycle.

Tablets are discharged into RIBCs for transport across the "clean" aisle to the coating rooms. After coating, the active ingredients within the tablets are considered totally contained and are taken along the "clean" aisle out of the manufacturing area for packaging or to storage. Capsules are handled in much the same way as the tablets, except that they are taken from the encapsulating area to a printing area for imprinting and then to packaging or storage.

More detailed information about containment for the control of releases of toxic particulate solids is presented by Dupuis (2001); Fay and Phillips (2002); Fay, Phillips, and Kraus (2002); Hines (1998); Hirst et al. (2002); Liberman et al. (2001); Rahe (1996); and Wood (2001). A recent article by Herreman (2003) reviews containment technology for contained manufacture of pharmaceutical. It contains a useful table that lists exposure levels (Class 1 through Class 4) and guidelines for containment to achieve these levels.

6.12 IDENTIFICATION OF SYSTEM-WIDE DESIGN, PROTECTION, AND PREVENTION REQUIREMENTS

In designing plants for processes involving the storage, handling, and processing of hazardous particulate solids, the design should be based on a system-wide approach. That is, an item of equipment should be designed taking into account its interaction with other items of equipment upstream and downstream of it, as well as its relationship to buildings, structures, and personnel. Some considerations and requirements relating to equipment design, operation, and protection are as follows:

1. When protecting against fires and explosions in one item of equipment, ascertain whether the fire or explosion could be transmitted to upstream and downstream equipment, and if so, provide isolation between the interconnected items of equipment.
2. If an item of equipment is to be protected by deflagration venting, determine if the fireball emitted from the vent can impact on adjacent buildings, structures, and personnel, If this is possible, try to relocate the vent so that the fireball is directed to a safe area.
3. If an item of equipment is located inside a building and explosion protection is to be provided by venting, consider using a flameless venting device (see Section 6.6.1). If a flameless venting device is not used, this will require installing a vent duct onto the vent itself to direct the fireball outside of the building. This may still result in the fireball impacting on adjacent buildings and structures and personnel.
4. Ascertain whether it is feasible to design equipment that must be protected against an explosion by using deflagration containment. This may be the most economical alternate in some cases.
5. When installing explosion relief (blowout) panels on building walls, tether them so that they do not become flying missiles that could cause damage to adjacent equipment and buildings or bodily harm to personnel.
6. When installing interlock systems to shut down equipment upon malfunction or failure, review the system to see if interlock bypasses are needed to be able to start up the equipment after it has been shut down. The provision of interlock bypasses should only be done if deemed absolutely necessary. Administrative procedures must ensure that interlocks are reinstalled after they have been bypassed.

Other considerations and requirements will usually become evident during a process hazard analysis.

REFERENCES

ACGIH. 2001. *Industrial Ventilation: A Manual of Recommended Practice.* 24th edition. American Conference of Governmental Industrial Hygienists, Inc., Lansing, MI

Anthony, E. J. and Greaney, D. 1979. The Safety of Hot Self-Heating Materials. *Combustion Science Technology,* 21: 79–85.

API Publ. 2030. 1998. *Application of Fixed Water Spray Systems for Fire Protection in the Petroleum Industry.* Washington, DC: American Petroleum Institute.

API RP 2003. 1998. *Protection Against Ignitions Arising Out of Static, Lightning and Stray Currents.* Washington, DC: American Petroleum Institute.

ASCE. 1997. *Design of Blast Resistant Buildings in Petrochemical Facilities.* New York: American Society of Civil Engineers.

ASTM Std. E1515–00. 2000. *Standard Test Method for Minimum Explosible Concentration of Combustible Dusts.* Conshohocken, PA: American Society for Testing and Materials.

Bartknecht, W. 1981. *Explosions: Course, Prevention, Protection.* Berlin and New York: Springer Verlag.

Bartknecht, W. 1989. *Dust Explosion: Course, Prevention, Protection.* Berlin and New York: Springer Verlag.

Bartknecht, W. 1993. *Explosionsschutz—Grundlagen und Anwendung.* Berlin: Springer Verlag.

Barton, J. (editor). 2002. *Dust Explosion Prevention and Protection.* Rugby, UK: Institution of Chemical Engineers.

Beever, P. F. and Thorne, P. F. 1982. *Isothermal Test Methods for Assessing Combustible Powders.* Building Research Establishment Current Paper CP 5/82. Borehamwood, Herts., England: Fire Research Station, Building Research Establishment.

Bernard, L., Brodie, F., Ludwig, D., Ness, A., and Weidner, K. 1998. Vent Access Restriction for Solids Handling Systems. *Process Safety Progress,* 17(1): 16–19 (Spring).

Blakey, P. and Orlando, G. 1984. Using Inert Gases for Purging, Blanketing and Transfer. *Chemical Engineering,* May 28: 97–102 .

Bond, J. 1991. *Sources of Ignition-Flammability Characteristics of Chemicals and Products.* Oxford, UK: Butterworth-Heinemann Ltd.

Bowes, P. C. 1984. *Self-Heating: Evaluating and Controlling the Hazards.* London: Her Majesty's Stationery Office.

Bowes, P. C. and Cameron, A. 1971. Self-Heating and Ignition of Chemically Activated Carbon. *Journal of Applied Chemistry and Biotechnology,* 21: 244ff (September).

Britton, L. G. 1999. *Avoiding Static Ignition Hazards in Chemical Operations.* New York: AIChE/Center for Chemical Process Safety.

Britton, L. G. 2003a. Personal communication from Laurence G. Britton, Process Safety Consultant, to S. S. Grossel (May 29).

Britton, L. G. 2003b. Personal communication from Laurence G. Britton, Process Safety Consultant, to S. S. Grossel (May 1).

BS 5958. 1991. *Code of Practice for Control of Undesirable Static Electricity,* Part 1. *General Considerations,* and Part 2. *Recommendations for Particular Industrial Situations.* London: British Standards Institution.

Burton, D. J. 1998. *Hemeon's Plant & Process Ventilation.* 3rd edition. Boca Raton, FL: Lewis Publishers, Inc./CRC Press.

Burton, D. J. 2000. *Industrial Ventilation Workbook.* 4th edition. Salt Lake City, UT: IVE, Inc.

Buschart, R. J. 1991. *Electrical and Instrumentation Safety for Chemical Processes.* New York: Van Nostrand Reinhold.

Carson, P. A. and Mumford, C. J. 1996. *The Safe Handling of Chemicals in Industry, Volume 3.* New York: John Wiley & Sons.

Carson, P. A. and Mumford, C. J. 2002. *Hazardous Chemicals Handbook.* 2nd edition. Oxford, UK and Boston, MA: Butterworth-Heinemann.

CCPS. 1993. *Engineering Design for Process Safety.* Chapter 11– Sources of Ignition. New York: AIChE/Center for Chemical Process Safety.

CCPS. 1995. *Guidelines for Safe Storage and Handling of Reactive Materials.* New York: AIChE/Center for Chemical Process Safety.

CCPS. 1996. *Guidelines for Evaluating Process Plant Buildings for External Explosions.* New York: AIChE/Center for Chemical Process Safety.

CCPS. 2003a. *Guidelines for Facility Siting and Layout.* New York: AIChE/Center for Chemical Process Safety.

CCPS. 2003b. *Guidelines for Fire Protection in Chemical, Petrochemical, and Hydrocarbon Processing Facilities.* New York: AIChE/Center for Chemical Process Safety.

CEN (Comité Européean de Normalisation). 2001. Draft standard prEN 12845, *Automatic Sprinkler Systems, Design and Installation and Maintenance* (English version). European Committee for Standardization, Brussels, Belgium.

CEN (Comité Européean de Normalisation). 2002. Draft standard prEN 14373, *Explosion Suppression Systems* (English version). European Committee for Standardization, Brussels, Belgium.

CENELEC. 2003. *Electrostatics-Code of Practice for the Avoidance of Hazards Due to Static Electricity.* Report CLC/TR 50404 (June 12, 2003). European Committee for Electrotechnical Standardization, Rue de Stassartstraat 35, B-1050 Brussels, Belgium.

Chatrathi, K. and Going, J. 2002. Efficiencies of Flameless Venting Devices. Proceedings of the 36th Annual AIChE Loss Prevention Symposium, pp. 222–246.

Cholin, J. M. 1997. Woodworking Facilities and Processes. Chapter 3–15 in *Fire Protection Handbook,* 18th edition, edited by A. E. Cote and J. L. Linville. Quincy, MA: National Fire Protection Association.

Chung, J. and Brookes, Jr., A. S. 1997. Manufacturing of a Potent Compound in the Pharmaceutical Pilot Plant. Paper presented at the AIChE 1997 annual meeting, November 16–21, Los Angeles, CA.

Cote, A. E. and Linville, J. L. (editors) 2003. *Fire Protection Handbook,* 19th edition, Quincy, MA: National Fire Protection Association.

Cross, J. A. 1987. *Electrostatics: Principles, Problems, and Applications.* Bristol, UK: Adam Hilger.

Crowl, D. A. and Louvar, J. F. 2002. *Chemical Process Safety: Fundamentals with Applications.* Upper Saddle River, NJ: Prentice Hall PTR.

Dastidar, A. and Amyotte, P. 2002. Determination of Minimum Inerting Concentrations for Dusts in a Laboratory-Scale Chamber. *Trans. IChemE,* 80(Part B): 287–297 (November).

DOE. 1994. *DOE Handbook-Primer on Spontaneous Heating and Pyrophoricity.* Report DOE-HDBK-1081–94. Washington, DC: U.S. Department of Energy (December).

Dream, R. F. 1998. Containment Practices in the Pharmaceutical Industries. *Pharm. Engineering*. (March/April): 88, 90–92, 94, 96, 98, 100.

Dupuis, M. R. 2001. Caging the Shark-Bulk Operations Containment. *Chemical Health & Safety*, pp. 9–11 (September/October).

Eckhoff, R. K. 2003. *Dust Explosions in the Process Industries*. 3rd edition. Oxford, UK and Boston, MA: Butterworth-Heinemann.

Eckhoff, R. K. 2004. A Critical View on the Treatment of Combustible Powders/Dusts in the European Union 'ATEX 100a' and 'ATEX 118a' Directives with Special Reference to Design and Use of Electrical Apparatuses. Paper presented at the International ESMG Symposium 2004, held March 16–18, 2004 in Nuremberg, Germany.

ESCIS. 1994. *Inerting: Methods and Measures for the Avoidance of Ignitable Substance-Air Mixtures in Chemical Production Equipment and Plants*. Safety Series Booklet 3 (English translation). Expert Commission for Safety in the Swiss Chemical Industry, Lucerne, Switzerland.

Fay, T. and Phillips, N. 2002. Safe Handling of Potent Compounds. *Chemical Engineering*, pp. 62–68 (April).

Fay, T., Phillips, N., and Kraus, J. 2002. Safe Handling of Potent Compounds. *Chemical Engineering*, pp. 68–72 (May).

Field, P. 1982. *Dust Explosions (Handbook of Powder Technology, Volume 4)*. Elsevier Scientific Publishing Company, Amsterdam, The Netherlands.

FMG 4–1N. 1998. *Water Spray Fixed Systems for Fire Protection. Property* Loss Prevention Data Sheet 4–1N. FM Global, Norwood, MA.

FMG 5–1. 2002. *Electrical Equipment in Hazardous (Classified) Locations*. Property Loss Prevention Data Sheet 5–1. FM Global, Norwood, MA.

FMG 7–11. 2003. *Belt Conveyors*. Property Loss Prevention Data Sheet 7–11. FM Global, Norwood, MA.

FMG 7–76. 2001. *Prevention and Mitigation of Combustible Dust Explosions and Fire*. Property Loss Prevention Data Sheet 7–76. FM Global, Norwood, MA.

FMG 7–78. 2000. *Industrial Exhaust Systems*. Property Loss Prevention Data Sheet 7–78. FM Global, Norwood, MA.

FMG. 1999. Approval Standard for Explosion Suppression Systems, Class Number 5700. FM Global, Norwood, MA (October).

Fordham-Cooper, W. and Dolbey Jones, D. A. 1998. *Electrical Safety Engineering*, 3rd edition. Burlington, MA: Newnes/Elsevier Science (USA).

Forsythe, V. G. 1980. Dust Explosion Protection in Pneumatic Conveying Processes. *Fire Prevention*, No. 135: 25–30 (March).

Frank, W. 2004. Inerting for Explosion Protection. *Proc. AIChE 38th Annual Loss Prevention Symposium*, Paper 3f, pp. 337–383.

Frank-Kamenetskii, D. A. 1969. *Diffusion and Heat Transfer in Chemical Kinetics*. New York: Plenum Press,.

GE GAP. 1994. *Fixed Protection for Oil and Chemical Plants*. GE GAP Guideline 12.2.1.2., GE Global Asset Protection Services, Hartford, CT.

GE GAP. 2001. *Oil and Chemical Plant Layout and Spacing*. GE GAP Guideline 2.5.2. GE Global Asset Protection Services, Hartford, CT.

Gibson, N., Lloyd, F. C., and Perry, G. R. 1967. Fire Hazards in Chemical Plant from Friction Sparks Involving the Thermite Reaction. IChemE Symp. Ser. No. 25: 26–35. Rugby, UK: Institution of Chemical Engineers.

Gibson, N. and Schofield, F. 1977. Fire and Explosion Hazards in Spray Dryers. IChemE Symp. Ser. No. 49: 53ff. Rugby, UK: Institution of Chemical Engineers.

Glor, M. 1988. *Electrostatic Hazards in Powder Handling*. New York: John Wiley & Sons.

Glor, M. and Schwenzfeuer, K. 1997. Occurrence of Cone Discharges in Production Silos. *Journal of Electrostatics*, 40/41: 511–516.

Gray. B. 2002. Spontaneous Combustion and Self-Heating. Chapter 2–10 in *The SFPE Handbook of Fire Protection Engineering*. 3rd edition, Quincy, MA: National Fire Protection Association.

Greenwald, E. K. (editor). 2002. *Electrical Hazards and Accidents: Their Cause and Prevention*. New York: John Wiley & Sons.

Griffiths, J. F. and Kordylewski, W. 1992. The Prediction of Spontaneous Ignition Hazards Resulting from the "Hot Stacking" of Process Materials. *Plant/Operations Progress*, 11(2): 77–84 (April).

Grossel, S. S. 2002. *Deflagration and Detonation Flame Arresters*. New York: AIChE/Center for Chemical Process Safety.

Gummer, J., and Lunn, G. A. 2003. Ignitions of Explosive Dust Clouds by Smouldering and Flaming Agglomerates. *Journal of Loss Prevention in the Process Industry*, 16(1): 27–32 (January).

Heidel, D. S. 2001. Industrial Hygiene Aspects of Pharmaceutical Manufacturing. Chapter 3 in *Containment in the Pharmaceutical Industry*. edited by J. P. Wood, New York: Marcel Dekker, Inc.

Herreman, G. 2003. Current Technology for Contained Manufacturing. *Pharmaceutical Technology*, pp. 132, 134, 136, 138, 140, 142, 144, and 146 (October).

Hiltz, R. 1993. Foam Blanketing: The Use of Foam to Mitigate the Vapor Hazard of Spilled Volatile Chemicals. Chapter 8 in *Prevention and Control of Accidental Releases of Hazardous Gases*. edited by V. M. Fthenakis, New York: Van Nostrand Reinhold.

Hines, J. M. 1998. Conceptual Design of Isolators for Handling and Processing Dry Bulk Pharmaceutical Chemicals. *Pharm. Engineering*, pp. 60. 62, 66, 68, 70, 72–74, 76, 78 (March/April).

Hirst, N., Brocklebank, M., and Ryder, M. 2002. *Containment Systems: A Design Guide*. Rugby, UK: Institution of Chemical Engineers.

Holbrow, P. Hawksworth, S. J., and Tyldesley, A. 2000. Thermal Radiation from Vented Dust Explosions. *Journal of Loss Prevention in the Process Industry*, 16(6): 467–476 (November).

Holbrow, P. and Tyldesley, A. 2003. Simple Devices to Prevent Dust Explosion Propagation in Charge Chutes and Pipes. *Journal of Loss Prevention in the Process Industry*, 16(4): 333–340 (July).

HSE. 1975. *Principles of Local Exhaust Ventilation.* Health and Safety Executive, England, UK.

IEC 61241–10. 2002. *Electrical Apparatus for Use in the Presence of Combustible Dust—Part 10: Classification of Areas Where Dust Are or May Be Present.* International Electrotechnical Commission, 1 rue de Varembe, Geneva, Switzerland.

ISO. 1985. International Standard 6184/4. Explosion Protection Systems—Part 4: Determination of the Efficacy of Explosion Protection Systems. International Standards Organization, Geneva, Switzerland.

Johnson, R. W., Rudy, S. W., and Unwin, S. D. 2003. *Essential Practices for Managing Chemical Reactivity Hazards.* New York: AIChE/Center for Chemical Process Safety.

Jones, T. B. and King, J. L. 1991. *Powder Handling and Electrostatics: Understanding and Preventing Hazards.* Chelsea, MI: Lewis Publishers, Inc.

Kayser, E. G. and Boyars, C. 1975. *Spontaneously Combustible Solids-A Literature Survey.* Report NSWC/WOL/TR-75–159. Naval Surface Weapons Center, White Oak Laboratory, Silver Spring, MD (available from NTIS as PB 244046)

Korver, W. O. E. 1995. *Classifying Explosion-Prone Areas for the Petroleum, Chemical and Related Industries.* Park Ridge, NJ: Noyes Publications.

Larsen, O., Hagen, J. H., and van Wingerden, K. 2001. Ignition of Dust Clouds by Brush Discharges in Oxygen Enriched Atmospheres. *Journal of Loss Prevention in the Process Industries,* 14(2): 111–122 (March).

Lees, F. P. 1996a. *Loss Prevention in the Process Industries.* 2nd edition, Volume 2, Chapter 16, pp. 16/19–16/85. Oxford, UK and Boston, MA: Butterworth-Heinemann.

Lees, F. P. 1996b. *Loss Prevention in the Process Industries.* 2nd edition. Volume 2, Chapter 16, pp. 16/58–16/69. Oxford, UK and Boston, MA: Butterworth-Heinemann.

Leuschke, G. 1980. Induction Times of Dust Deposits Stored at and Above Self-Ignition Temperatures. *Proc. Loss Prevent. Safety Promot.* No. 3, 2: 647–656.

Leuschke, G. 1981. Experimental Investigations on Self-Ignition of Dust Deposits in Hot Environments. *IChemE Sympos. Series* No. 68, Paper 2G.

Liberman, D., Lockwood, C., McConnell-Meachen, McNally, E., Rahe, H., Shepard, K., and Snow, G. 2001. Barrier Isolation Technology: A Safe and Effective Solution for Providing Pharmaceutical Development Facilities. *Pharm. Engineering,* (July/August): 20, 22–30 .

Lunn, G. 2000. Frictional Ignition of Powders: A Review. Paper presented at the Safetynet Electronic Conference and available for downloading from www.safetynet.de.

Lunn, G. A., Torrent, J. C., Hesener, U., and Fritze, J. P. 2002. Hot Surface Ignitions of Dust Accumulations. Paper presented at the International ESMG Symposium 2002, October 8–10, 2002, Nurnberg, Germany

Luttgens, G. and Wilson, N. 1997. *Electrostatic Hazards.* Woburn, MA: Butterworth-Heinemann.

Maurer, B. 1979. Discharges Due to Electrostatic Charging of Particles in Large Storage *Silos. Ger. Chem. Engrg.,* 2: 189–195.

Maurer, B. Glor, M., Luttgens, G., and Post. L. 1989. Test Rig for Reproducible Generation of Discharges from Bulked Polymeric Granules. *Journal of Electrostatics*, 23: 25–34.

McMillan, A. 1998. *Electrical Installations in Hazardous Areas*. Oxford, UK and Boston, MA: Butterworth-Heinemann.

Mecklenburgh, J. C. 1985. *Process Plant Layout*. New York: John Wiley & Sons.

Medard, L. A. 1989. *Accidental Explosions*, Volume 2: *Types of Explosive Substances*. Chapter 17. New York: John Wiley & Sons.

Morrison III, D. R., Carpenter, A. R., and Ogle, R. A. 2002. Common Causes and Corrections for Explosions and Fires in Improperly Inerted Vessels. *Process Safety Progress*, 21(2): 142–150 (June).

Nagy, J. and Verakis, H. C. 1983. *Development and Control of Dust Explosions*. New York: Marcel Dekker.

Nelson, M. 2003. Personal communication from Melvin Nelson, Syngenta, St. Gabriel, LA to S. S. Grossel (May 15).

NEMA STD. 250. 2003. *Enclosures for Electrical Equipment (1000 Volts Maximum)*. National Electrical Manufacturers Association, Washington, DC.

Ness, A. 2002. Personal communication from Albert Ness, Rohm and Haas Company, Bristol, PA to S. S. Grossel (June).

NFPA 10. 2002. *Standard for Portable Fire Extinguishers*. Quincy, MA: National Fire Protection Association.

NFPA 11. 2002, *Standard for Low Expansion Foam*. Quincy, MA: National Fire Protection Association.

NFPA 11A. 1999. *Standard for Medium- and High-Expansion Foam Systems*. Quincy, MA: National Fire Protection Association.

NFPA 12. 2000. *Standard on Carbon Dioxide Extinguishing Systems*. Quincy, MA: National Fire Protection Association.

NFPA 13. 2002. *Standard for the Installation of Sprinkler Systems*. Quincy, MA: National Fire Protection Association.

NFPA 15. 2001. *Water Spray Fixed Systems for Fire Protection*. Quincy, MA: National Fire Protection Association.

NFPA 16. 2003. *Standard for the Installation of Deluge Foam- Water Sprinkler and Foam-Water Spray Systems*. Quincy, MA: National Fire Protection Association.

NFPA 17. 2002. *Standard for Dry Chemical Extinguishing Systems*. Quincy, MA: National Fire Protection Association.

NFPA 51B. 2003. *Fire Prevention During Welding, Cutting, and Other Hot Work*. Quincy, MA: National Fire Protection Association.

NFPA 68. 2002. *Guideline for Venting of Deflagrations*. Quincy, MA: National Fire Protection Association.

NFPA 69. 2002. *Standard on Explosion Protection Systems*. Quincy, MA: National Fire Protection Association.

NFPA 70. 2002. *National Electrical Code*. Quincy, MA: National Fire Protection Association.

NFPA 72. 2002. *National Fire Alarm Code.* Quincy, MA: National Fire Protection Association.

NFPA 77. 2000. *Recommended Practice on Static Electricity.* Quincy, MA: National Fire Protection Association.

NFPA 496. 2003. *Purged and Pressurized Enclosures for Electrical Equipment.* Quincy, MA: National Fire Protection Association.

NFPA 499. 2004. *Classification of Combustible Dusts and of Hazardous (Classified) Locations for Electrical Installations in Chemical Process Areas.* Quincy, MA: National Fire Protection Association.

NFPA 654. 2000. *Standard for the Prevention of Fire and Dust Explosions from the Manufacturing, Processing, and Handling of Combustible Particulate Solids.* Quincy, MA: National Fire Protection Association.

NFPA 704. 2001. *Standard System for the Identification of the Hazards of Materials for Emergency Response.* Quincy, MA: National Fire Protection Association.

NFPA 780. 2000. *Standard for the Installation of Lightning Protection Systems.* Quincy, MA: National Fire Protection Association.

NFPA 2001. 2004. *Standard for Clean Agent Fire Extinguishing Systems.* Quincy, MA: National Fire Protection Association.

Noll, C. G. 1995. Electrostatic Charge Elimination Techniques. Chapter 33 in *Handbook of Electrostatic Processes,* edited by Chang, J.-S., Kelly, A. J., and Crowley, J. M., New York: Marcel Dekker.

Pilkington, G. 1999. *Guidelines for Inert Gas Blanketing.* Abbott Laboratories, North Chicago, IL (January 13).

Pilkington, S. (editor). 2000. *Process Vessels Subject to Explosion Risk-Design Guidelines for the Pressure Rating of Weak Process Vessels Subject to Explosion Risk.* Rugby, UK: Institution of Chemical Engineers.

Pratt, T. H. 1997. *Electrostatic Ignitions of Fires and Explosions.* New York: AIChE/Center for Chemical Process Safety.

Rahe, H. 1996. Consider Barrier/Isolation Technology for Batches. *Chemical Engineering Progress,* 92(12): 72–75 (December).

Scheffey, J. L. 1997. Foam Extinguishing Agents and Systems. Chapter 6–22 in *Fire Protection Handbook,* 18th ed., edited by A. E. Cote and J. L. Linville, Quincy, MA: National Fire Protection Association.

Schofield, C. 1984. *Guide to Dust Explosion Prevention and Protection, Part 1–Venting.* Rugby, UK: Institution of Chemical Engineers.

Siwek, R. 1989. New Knowledge About Rotary Air Locks in Preventing Dust Ignition Breakthrough. *Plant/Operations Progress,* 8(3): 165ff (July).

Siwek, R. 1992. The Combination of Explosion Venting and Explosion Suppression. Explosion Suppression in Small Volumes. Paper presented at EuropEx International Seminar (February).

Siwek, R. 1994. Latest Development in Explosion Protection Technology. Paper presented at the 6th International Colloquium on Dust Explosions (August)

Siwek, R. 1996. A Review of Explosion Isolating Techniques. Proc. 2nd World Seminar on the Explosion Phenomena and on the Application of Explosion Protection Techniques in Practice, held March 4–8, 1996 in Ghent, Belgium, Session 2.

Stevenson III, W. J. 1998. Dust Explosion Mitigation Using the Q-Rohr and Exkop. *Process Safety Progress,* 17(3): 184–189.

Swinderman, R. T., Becker, S. L., Goldbeck, L. J., Stahura, R. P., and Martin, A. D. 1991. *Foundations-Principles of Design and Construction of Belt Conveyor Transfer Points.* Neponset, IL: Martin Engineering Company.

Tamanini, F. and Valiulis, J. 1996. Improved Guidelines for the Sizing of Vents in Dust Explosions. *Journal of Loss Prevention in the Process Industry,* 94(1): 105–118.

Tapscott, R. E. 1997. Combustible Metal Extinguishing Agents and Application Techniques. Chapter 6–26 in *Fire Protection Handbook,* 18th edition. edited by A. E. Cote, Quincy, MA: National Fire Protection Association.

Thomas, P. H. 1973. An Approximate Theory of "Hot Spot" Criticality. Combust. Flame, 21: 99–109.

Tuomisaari, M., Baroudi, D., and Latva, R. 1998. Extinguishing Smouldering Fires in Silos. Brandforsk Project 745–961, VTT Publication 339, Technical Research Centre of Finland.

Tyldesley, A. 2004. Peer review comments and personal communication from Alan Tyldesley, HSE, Bootle, UK to S. S. Grossel (February 19, 2004).

Urben, P. G. (editor). 1995. *Bretherick's Handbook of Reactive Chemical Hazards.* 5th edition. Oxford, UK and Boston, MA: Butterworth-Heinemann.

VDI 3673. 2002. *Pressure Venting of Dust Explosions* (English translation). Verein Deutsche Ingenieure-Kommission Reinhaltung der Luft, VDI Verlag GmbH, Dusseldorf, Germany.

Wood, J. P. (editor). 2001. *Containment in the Pharmaceutical Industry.* New York: Marcel Dekker.

Zalosh, R. G. 2003. *Industrial Fire Protection Engineering.* Hoboken, NJ: John Wiley & Sons.

Chapter 7

PLANT OPERATION AND MAINTENANCE

7.1 INTRODUCTION

This chapter deals with various aspects of plant operation and maintenance that contribute to the safety and mechanical integrity of plants storing, handling, and processing hazardous particulate solids.

7.2 REGULATORY REQUIREMENTS

There are several federal regulations that may bear upon the design and operation of chemical process industries (CPI) plants using or processing hazardous chemicals (including hazardous particulate solids). The following regulations may be relevant and are briefly discussed below:

- OSHA and EPA Clean Air Act (CAA) General Duty Clauses (GDC)
- The OSHA Process Safety Management (PSM) standard

Other relevant regulations and standards

- OSHA and EPA Clean Air Act (CAA) General Duty Clauses (GDC)

Although the OSHA PSM standard does not specifically cover hazardous particulate solids, a facility is subject to the general duty clause of OSHA, and the Clean Air Act (CAA) 112r amendments of 1990.

The OSHA General Duty Clause is Subsection 5 of the OSHA Act of 1970, and states as follows:

"Each employer shall furnish to each of his employees employment and a place of employment which are free from recognized hazards that are causing or likely to cause death or serious physical harm to his employees."

In the Clean Air Act Amendments of 1990, Congress enacted Section 112(r)(1), also known as the General Duty Clause (GDC), which makes the owners and operators of facilities that have regulated and extremely hazardous substances responsible for ensuring that their chemicals are managed safely.

The General Duty Clause applies to any stationary source producing, processing, handling, or storing regulated substances or other extremely hazardous substances. Extremely hazardous substances are any chemical listed in 40 CFR Part 68, or any other chemical which may as a result of short-term exposures because of releases to the air cause death, injury or property damage due to their toxicity, reactivity, flammability, volatility, or corrosivity.

Facilities subject to the General Duty Clause are responsible for, among other things:

- Knowing the hazards posed by the chemicals and assessing the impacts of possible releases;
- Following codes, standards and other business practices to ensure the facility is properly constructed and maintained, and the chemical is managed safely (e.g., NFPA 77, NFPA 654, etc.); and
- Having a contingency planning process, which would involve community responders, if necessary, to aid in an adequate response in the event of an accident.

OSHA Process Safety Management (PSM) Standard

OSHA published the final rule "Process Safety Management of Highly Hazardous Chemicals" on February 24, 1992 (29 CFR 1910.119). The PSM standard has 14 major sections: employee participation, process safety information, process hazard analysis, operating procedures, training, contractors, pre-startup safety review, mechanical integrity, hot work permits, management of change, incident investigations, emergency planning and response, audits, and trade secrets. A brief description of the pertinent sections is given in Louvar and Louvar (1998).

Hazardous particulate solids are not covered in the OSHA PSM standard, but can be covered under the OSHA General Duty Clause (see discussion above).

Other Relevant Regulations and Standards

Listed below are other regulations and standards that may pertain to the safe storage, handling, and processing of particulate solids.

EPA Regulations

- **40 CFR 50.6** National Primary and Secondary Ambient Air Quality Standards for Particulate Matter (Implementation is accomplished through the various State Implementation Plans)
- **40 CFR Parts RCRA** Treatment, Storage, and Disposal
- **264/265** Facilities

NFPA Standards and Guidelines

- **NFPA 45** Standard on Fire Protection for Laboratories Using Chemicals
- **NFPA 61** Prevention of Fires and Dust Explosions in Agricultural and Food Products Facilities
- **NFPA 68** Venting of Deflagrations
- **NFPA 69** Explosion Protection Systems
- **NFPA 77** Recommended Practice on Static Electricity
- **NFPA 86** Standard for Ovens and Furnaces
- **NFPA 120** Coal Preparation Plants
- **NFPA 430** Storage of Liquid and Solid Oxidizers
- **NFPA 434** Storage of Pesticides
- **NFPA 484** Standard for Combustible Metals, Metal Powders, and Metal Dusts
- **NFPA 490** Storage of Ammonium Nitrate
- **NFPA 499** Classification of Combustible Dusts and of Hazardous (Classified) Locations for Electrical Installations in Chemical Process Areas
- **NFPA 654** Prevention of Fire and Dust Explosions from the Manufacturing, Processing, and Handling of Combustible Particulate Solids
- **NFPA 655** Prevention of Sulfur Fires and Explosions
- **NFPA 664** Prevention of Fires and Explosions in Wood Processing and Woodworking Facilities

7.3 MANAGEMENT OF CHANGE

Often, a change in a process may significantly change the physical and safety characteristics of a particulate solid. A management of change (MOC) procedure should then be developed to prevent accidents from occurring. Whenever there is a change in raw materials, process chemistry, process sequence, operating procedures, and process equipment the company should develop and implement documented procedures to manage these changes. Before these changes are instituted a hazard analysis should be done to ascertain if these proposed changes will adversely affect the safety of the operation. In addition, the operators should be trained so that they understand the nature of these changes, and a pre-startup review must be conducted. After all this has been done, then the plant can be allowed to start up.

The following three case histories illustrate problems that can result from process and equipment changes.

Case History of an Increase in MIE due to a Process Change

Ness (2002) reports the following effect on the MIE of plastic particles due to a process change:

A purification procedure for a plastic particle product was modified by reducing the amount of water washing that was used.

A later analysis of the material showed that the minimum ignition energy (MIE) had increased significantly. The residual impurities remaining in the plastic due to the lower quantity of wash water used had led to a change in the electrostatic properties of the plastic particles.

Case History of a Potential Electrostatic Ignition due to a Change in Equipment Components

The following increase in the potential for a dust explosion caused by an electrostatic spark discharge is reported by Philiph (2002a).

During a process hazard analysis (PHA) of a process for herbicide manufacturing, the PHA team was discussing the explosion hazards associated with one of the powdered raw materials. The powder has a low minimum ignition energy and there was some concern about the potential for a static discharge igniting a dust cloud. One of the PHA team members (an operating technician) mentioned that he knew that personnel were being shocked "all the time" when they cleaned up spilled powder with the vacuum system. After a quick investigation, the PHA team discovered that the vacuum system consisted of a metal wand connected to a nonconductive hose, which was connected to a metal pipe. This arrangement allowed a static charge to accumulate on the metal wand. The team learned that the vacuum system originally had a conductive hose, all parts were bonded, and the entire system was grounded. Evidently, someone at some time had replaced the conductive hose with the nonconductive hose. Fortunately, the only consequence in this instance was some discomfort to the technicians who did the cleaning up, but it illustrates the importance of workers understanding the hazards associated with their jobs and how a simple change can create a hazard.

Case History of a Process Change That Changed the Combustibility of a Powder

Kahn (2002) reports the following incident in which the combustibility of a powder was changed due to a change in the process for manufacturing the powder.

An international company has an active ingredient for a formulation that is made in another country other than the United States, which was an ST2 material having a MIE of 30-100 mJ. While the synthesis of the active ingredient is performed in a different country, the end product is formulated in the United States. A bright young chemist in the other country decided that he could improve the yield of the active ingredient by eliminating some impurities. He was quite successful in minimizing the impurities and increasing the yield, which won him special recognition. No one in that country thought that changes in the impurity profile might have implications for changes in the combustibility characteristics of the material. An international management of change announcement was not formally made. The change was mentioned within production circles because now less active ingredient was needed to formulate end products and updates were made to regulatory registrations. The safety group was finally made aware of the change in impurity profiles just weeks before end product formulation was to begin. Testing was done and revealed that the active ingredient was now an ST3 powder with an MIE of 1-3 mJ. The company had to change the shipping container, move the end product formulation to another location which had equipment with different safeguards, and drastically change the level of protection required to safely produce the formulation.

When dealing with changes in particulate solids processes, it is important to consider a number of factors that may influence the safety properties of the solids being processed, possibly for the worse. Listed below are a number of changes that can occur, and should be captured in a MOC procedure:

1. Changes in Particulate Solids Formulation
   a. Change in supplier of formulant, which may result in the material having different constituents, which produce different safety properties.
   b. Active ingredient made by changes in the process or different supplier and the trace components are different, resulting in a change in the safety properties of the active ingredient.

c. Formulation change caused by either a change in an ingredient or a formulant. Usually, the active ingredient has the most dangerous safety properties and the dilution with formulating ingredients reduces the dangerous properties of the product. Reducing an inert ingredient, such as clay, has the effect of making a product more dangerous.
d. Inadvertent change, such as an oil leak into the process.

Typical changes to safety properties that can be caused by the above changes are: MIE, K_{St}, P_{max}, onset temperature for decomposition or other reaction. In the worst case, the results are more easily ignited particulate solids with greater explosion violence or more easily degraded solids as the time-to-reaction is reduced.

2. Change in Particle Size
   a. Use a different mill, modify a mill, or grind for a longer time.
   b. Change the particle size specification
   c. Change the quality control test method for particle size analysis.

All of these changes to particle size will affect the MIE. Smaller particle size lowers the MIE of the processed solids, while increasing the particle size raises the MIE. Lowering the MIE increases the potential for initiation of a dust explosion.

3. Change in Processing Temperature
   a. Make a deliberate process temperature change to raise drying rates or reduce a volatile component.
   b. Make an unintended temperature change due to a faulty instrument calibration or change in instrument location.

Raising or lowering the process temperature will change the MIE. Raising the temperature lowers the MIE, making the potential for ignition greater. Raising the temperature that is sustained by the powder for hours or days may exceed the onset temperature (Bowes-Cameron) that can lead to a decomposition or combustion reactions that likely are exothermic. See Section 4.3.2 for a discussion of tests for determining thermal degradation and instability.

4. Change in Rotational Speed of Equipment in Contact with a Powder
   a. Make a deliberate speed change in equipment such as blenders, agitators, screw conveyors and feeders, or sifters for production rate increases or to make the operation more effective.
   b. Make a replacement of equipment with a different type (not in kind).

Increasing the power input to a process increases the likelihood of formation of hot spots or mechanical sparking which can act as ignition sources.

The failure to conduct a proper management of change (MOC) analysis has resulted in many accidents. Many different types of accidents due to neglecting to do a MOC analysis are presented by Sanders (1999).

7.4 PROCESS HAZARD ANALYSES

A process hazard analysis (PHA) should be conducted for each new process or when a process is changed. A PHA should also be done on a process that has never been reviewed. It is most important (and usually more economical) to do the PHA at the early stages of a design project, because as the project progresses, changes are more difficult to make, and also are more costly.

A PHA can be conducted at several levels of rigor, depending on the complexity of the process and plant. The primary types of PHAs used in the chemical process industries are as follows (CCPS, 1992):

- Safety review
- Checklist analysis
- Relative ranking
- Preliminary hazard analysis
- What-if analysis
- What-if/checklist analysis
- Hazard and operability analysis (HAZOP)
- Failure modes and effects analysis
- Fault tree analysis
- Event tree analysis
- Cause-consequence analysis
- Human reliability analysis

A very thorough discussion of the various types of process hazard analysis techniques, along with worked examples, is presented by CCPS (1992).

When doing a process hazard analysis (PHA) for a particulate solids process, the questions to be asked are somewhat different than for processes involving liquids and gases/vapors. Table 7-1 is an example of a checklist of questions that should be asked in such a PHA.

7.5 HOUSEKEEPING PRACTICES TO PREVENT OR MINIMIZE DUST EMISSIONS AND ACCUMULATION

It is very important to prevent or minimize emissions and accumulation of particulate solids because they can result in secondary explosions where processing equipment is located inside a building or other enclosure.

TABLE 7-1
Checklist of Questions for a Process Hazards Analysis of a Particulate Solids Process

A. Safety and Health Data and Issues

- Are the particulate solids (PS) combustible?
- What are the sizes of particles at various locations in the process (their distribution and/or median size)?
- How do the PS respond to elevated temperatures and what is their response to extended exposure times?
- Are exothermic reactions (combustion or decomposition) accelerated at elevated (beyond ambient) temperatures in bulk powders?
- Are decomposition reactions or onset temperatures known?
- Can the PS undergo auto-catalytic decomposition?
- Are the PS pyrophoric or water-reactive?
- Are the PS shock-sensitive and/or friction-sensitive?
- Are the PS sensitive to air or light?
- What is the minimum ignition energy (MIE) of the PS (for various size fractions)?
- What is the autoignition temperature (AIT) of a dust cloud?
- What are the electrical resistivities of the PS at various locations in the process?
- What is the minimum explosive concentration (MEC) of the PS?
- What are the PS dust explosion class, maximum explosion pressure, and maximum rate of pressure rise?
- Do the PS contain flammable vapors (hybrid mixture)?
- Are the PS corrosive when moist?
- Are toxicological data and/or MSDSs available?
- Is inerting required?

B. Electrical and Electrostatic Issues

- Are electrical classifications suitable?
- Are all solids-handling equipment properly bonded and grounded?
- Are there nonconductive solvents/powders/granules in the process?
- Do items of equipment have nonconductive elements?
- Are there nonconductive surfaces involved in PS conveying systems?
- Is the appropriate type of FIBC being used and is proper bonding and grounded provided when necessary?
- Do cartridge filters and dust collector bags need to be conductive (e.g., in hybrid mixture service)?
- Are all the dust collector internal metal parts properly bonded and grounded?

C. Safety Issues of Equipment with Rotating Parts

- Do moving metal parts contact the PS?
- Can tip speeds of moving parts create a hazard (fire or explosion) by frictional heating or impact?

TABLE 7-1 (*Cont.*)

- Should a fluid energy mill be used instead of a mechanical mill?
- Are protective measures such as explosion suppression, venting, or pressure containment required and provided?
- Is the equipment properly bonded and grounded?

D. Safety Issues with Storage Equipment

- Are volumes of PS greater than 200 liters? Volumes > 200 liters may restrict conduction of heat to the atmosphere from hot solids or exothermic reactions within the solids.
- Are volume of PS greater than 1 m? Volumes > 1 m are minimally capable of having bulking brush (conical pile) electrostatic sparking from filling.
- Are protective systems such as inerting, explosion venting, or suppression required and provided?
- Are silos and hoppers properly bonded and grounded?

E. Safety Issues with Conveying Equipment

- If pneumatic conveying is used, should dense-phase conveying be used rather than dilute-phase conveying to minimize particle attrition and/or electrostatic charging?
- Is it necessary to design the pneumatic conveying system to be dust-tight?
- Should the pneumatic conveying system use an inert gas rather than air as the conveying gas?
- Should the pneumatic conveying system be provided with explosion venting or suppression, or be designed for explosion containment?
- If mechanical conveying is used, what type of conveyor should be selected to minimize particle attrition or conveyor component abrasion?
- Should the mechanical conveyor be provided with a high-high temperature sensor/switch interlocked with the motor?
- Does a pneumatic or mechanical conveyor have nonconductive parts that could cause electrostatic problems?
- Is the mechanical conveyor properly bonded and grounded?
- Should the mechanical conveyor be provided with explosion protection such as inerting, venting, or suppression?

F. Safety Issues with Drying Equipment

- Are the PS capable of deflagrating?
- What is the AIT and MIE of the PS?
- Can the PS undergo an exothermic reaction (decompose)?
- What is the first exotherm temperature (needed to specify the safe operating temperature in the dryer)?
- If a decomposition occurs, what are the gases evolved, and are they toxic?
- Do the PS being dried contain flammable solvents (hybrid mixture)?
- Is an explosion protection system required and provided?
- Is fire protection (sprinkler or deluge system) required internally (to extinguish a smoldering PS fire)?
- Is the dryer properly bonded and grounded?

TABLE 7-1 (*Cont.*)

G. Safety Issues with Dust Collection Equipment

- Is an explosion protection system required and provided?
- Is fire protection (sprinkler or deluge system) required internally (to extinguish a smoldering PS fire)?
- If a baghouse is used, does the bottom hopper have a high- high level sensor and alarm?
- If a baghouse is used, are all the parts of the dust collector properly bonded and grounded?
- If a wet scrubber is used, does the scrubbing water line have a low flow or low pressure sensor and alarm?

H. Equipment Design Issues (General)

- Are the materials of construction properly selected (some trace components in metals may react with PS)?
- Has excessive corrosion been experienced in prior equipment?
- Can the equipment be purchased with explosion-containment design (obviates need for explosion venting or suppression)?
- Is the equipment designed per ASME or other acceptable codes?

I. Instrumentation and Computerization Issues

- Are critical process parameters monitored (with alarms) and interlocked?
- Is redundant instrumentation needed?
- Is periodic calibration needed?
- Is the failure position of valves properly selected and known?
- Is an uninterruptible power supply (UPS) back-up system needed?
- Are events associated with computer failure known?

J. Industrial Hygiene/Occupational Health Issues

- Is appropriate ventilation provided to protect operating and maintenance personnel?
- Are operators provided with appropriate personal protective equipment for health-hazardous PS?
- Are the PS toxic and require containment?

K. Environmental Issues

- Are explosion vents located so that the effluent stream (fireball and pressure wave) is directed to a safe location where they will not impact on plant personnel and adjacent equipment?
- Are toxic and other harmful streams from vents directed to a safe location or to a control device?
- Are liquid waste streams sent to storage and treatment facilities before final disposal?

Many types of solids handling and processing equipment continually leak small quantities of fines when in service. Also, not all piping and ducting systems are dust-tight. With time, dust will deposit on all surfaces in a building or enclosure. If a piece of equipment should experience a dust explosion, or if a shock wave should be transmitted from elsewhere, the deposited dust can be dislodged and form a dust cloud with great explosion potential. An explosion of this dislodged dust is called a secondary explosion, and is usually more severe than a dust explosion inside a piece of equipment, called a primary explosion. Such dust deposits should be prevented by all possible means.

Good housekeeping is essential in buildings and enclosures containing particulate solids handling and processing equipment. First of all, all equipment and piping that leak should be tightened up and consideration should be given to installing a containment enclosure around leak sources (e.g., flanges). Capture and removal equipment which transports leaking dust to a safe location should be provided at dust escape points. It is common practice to provide capture hoods and "elephant trunk' dust collection equipment to pick up emissions from operations such as bagging, bag slitting, tableting, etc., or other points where equipment emits dust into the work area.

NFPA 654 (2000) provides some guidance for the safe conduct of housekeeping operations. The standard cautions against vigorous sweeping or blowing down with steam or compressed air of surfaces, as this can produce dust clouds, and should be permitted only when:

- The area and equipment have been vacuumed prior to blowdown.
- Electrical power and other sources of ignition have been shut down or removed.
- Only low pressure (15 psig maximum) steam or compressed air shall be used.
- There are no hot surfaces in the area capable of igniting a dust cloud or layer.

Dust which does deposit should be cleaned frequently as part of the housekeeping program. If vacuuming is intended as part of the housekeeping program, NFPA 654 requires either the use of a fixed-pipe ("centralized house") system with a remotely located exhauster (vacuum cleaner) and dust collector (properly protected against explosions), or a portable vacuum cleaner listed for use in Class II locations. Grounded, electrically-conducting vacuum hoses should be used. Centralized house vacuum cleaning systems should be installed with hose stations located at strategic points (where dust emissions are known to occur).

Although dust collection systems may operate below the minimum explosive concentration, flashback from an explosion in the dust collector can cause reverse flow and push unburned dust in front of the flame, and potentially result in a fireball coming out of the capture hoods or other collecting points. It is good practice to avoid placing barrels/drums of powder beneath fugitive dust capture hood entrances as there is the possibility of creating a large dust cloud and secondary explosion if a flashback from a dust collector occurs. Similarly, Valiulis et al. (1999) and Valiulis (2001) has shown that flames can propagate far into the clean return air duct; therefore, it is good safety practice that the return duct should not be located near a normally occupied area.

The potential for flashback flame propagation is greater in large diameter ducts. If an organic dust is being handled it is thought that there is a low probability of an explosion flashing back from an explosion-vented dust collector through ducts that are less than 4 inches (100 mm) in diameter. This might not be true if the dust is a metal dust or a hybrid mixture (Kirby, 2002), or if the dust concentration is greater than about 25% of the MEC (FMG 7-76, 2001). Isolation against flashback, such as by fast-acting valves or chemical barrier suppression systems, is often needed, especially if the piping is larger than 4 inches in diameter. For small spillages and accumulations, portable or mobile vacuum cleaners may be suitable, but care should be taken to ensure that the exhaust air does not stir up nearby dust deposits. It is particularly important that elevated horizontal surfaces such as roof trusses be cleaned frequently.

It is also important to clean dust deposits off of equipment that is a source of heat (e.g., dryers) as the insulation provided by the dust may cause overheating of the product inside the equipment.

Good housekeeping can be facilitated at the design stage of a plant by minimizing the area of horizontal surfaces in buildings and by ensuring easy access for cleaning of all parts. For example, horizontal surfaces such as beams should have the tops formed to minimize dust accumulations.

Criteria for dust accumulation thicknesses that could result in dust explosions are discussed in Section 6.8 and Appendix A.2.2.3.1 of NFPA 654 (2000).

In some cases, where there is a potential for personnel exposure to combustible dust clouds having MIEs less than 30 mJ, and where electrostatic charges can be readily built up, conductive floors often are installed and operators wear conductive clothing and static-dissipative shoes in an effort to drain off electrostatic charges and to prevent the operator from becoming an ignition source. ANSI Z41 Type II static-dissipative (SD) footwear is normally used for this purpose. Conductive shoes, defined as those having a resistance of <50,000 ohms to earth, should never be used for this purpose since they pose an electrocution hazard. The specified minimum resistance of one megohm for Type II SD footwear prevents this for contact with low voltage mains (Gravell, 2004). Also, the use of static-dissipative clothing is generally not justified since studies have shown that no advantage is gained by the use of such clothing, provided that they are not removed in a potentially flammable atmosphere (Tewarson, 1974).

Good housekeeping practices should be enforced to prevent powder/dust from accumulating on conductive floors and walking surfaces because they may act as an insulator (resistance) and negate effects to conduct charges away from the operator.

The following case history cites an accident caused by bad housekeeping.

Case History of a Dust Fire Caused by Bad Housekeeping

Philiph (2002b) reports the following case history of a dust fire caused by poor housekeeping (although inadvertently).

A company had a pneumatic conveying system which included a section which ran over a warehouse. In the middle of the night, the conveying piping came apart, dumping combustible powder onto the warehouse roof. The problem went unnoticed until several tons of the material had accumulated on the roof and caused the warehouse roof to collapse. The collapsing roof broke some electrical wiring and the electric arc ignited the material as it fell. Because the collapsing roof also damaged the warehouse sprinkler system, the ensuing fire destroyed the warehouse.

7.6 MECHANICAL INTEGRITY OF EQUIPMENT

The three main aspects of mechanical integrity (MI) are:

1. Inspection and testing of equipment
2. Upgrading and repairs of equipment
3. Written procedures to maintain on-going integrity of equipment

These are briefly discussed below.

7.6.1 Scheduled Inspections and Testing of Equipment

Without scheduled inspections and prompt maintenance, chemical plant equipment and machinery can lapse into a dangerous state from wear, fatigue, or corrosion. Regular inspections and testing are essential to determining the extent of equipment and piping deterioration. Since the inspection results may call for shutting down a plant, the staff responsible for making inspections should, if possible, be independent from the operating department. The person who is responsible for planning inspections in any company may be given the authority to overrule production pressures, espe-

TABLE 7-2
Origins and Types of Equipment Defects

- Raw mateial Defects
  - –stress cracking
  - –gas porosity
  - –slag inclusion
  - –shrinkage porosity
- Defects produced during manufacture
  - –welding defects
  - –machining defects
  - –heat treating defects
  - –residual stress cracking
- Defects produced during assembly
  - –additional welding defects
  - –missing parts
  - –incorrect assembly
  - –additional stress cracking
- Defects produced during service
  - –wear
  - –thermal degradation
  - –creep
  - –fatigue
  - –corrosion

cially when the safety of personnel and integrity of the plant are clearly at stake.

Process equipment should be inspected on a scheduled basis to determine if there are any defects that have occurred that could result in unsafe operation. Table 7-2 (CCPS, 1995b) lists various types of equipment defects and their origins. Inspections can be done by visual means, destructive testing (the equipment is destroyed so that it cannot be used again), and nondestructive testing. Table 7-3 (CCPS, 1995b) lists various nondestructive inspection techniques and their advantages and disadvantages.

Rotary equipment such as rotary valves, sifters, fans, blowers, and screw conveyors may experience more severe wear than stationary equipment, and may have to be inspected more frequently to ascertain if their deterioration could result in frictional heating and sparks.

A good inspection program requires that potential modes of failure for each piece of equipment be identified so that the inspector knows what to look for. For many types of equipment the most significant concern is corrosion and/or erosion. Corrosion may be of several types (see Section 7.7). For example, it may be general or pitting, or other types of corrosion and under

TABLE 7-3
Nondestructive Inspection Techniques

Technique	Advantages	Disadvantages
Radiographic: Used to examine the internal soundness of weldments and metals by bombarding the piece with x-rays or gamma rays.	Sharp picture of defects. Film provides a permanent record	Special personnel protection and training required. Both sides of the piece must be accessible.
Ultrasonic: Uses high-frequency sound waves to locate defects	Very sensitive; can detect very fine surface and subsurface cracks Equipment is portable Only one side need be accessible.	Personnel must be trained to interpret equipment response. Not effective on rough surfaces or welds with backing rings.
Magnetic particle: Used to find surface defects by applying a liquid suspension of fine particles that flow into fine cracks. A strong magnetic field concentrates the particles in the area of the defect, highlighting its size and shape.	Shows fine cracks that are not noticeable in radiographic examination. Shows where and what material must be removed for weld repair.	Cannot be used on nonmagnetic material. Detects surface cracks only. Cannot detect defects parallel to the magnetic field.
Dye-penetrant: Used for surface defects. A liquid dye is applied to a clean, dry surface and allowed to penetrate surface cracks and dry. A developer put over the surface causes the dye to outline the defect clearly	Useful for nonmagnetic materials. Can be used on nozzles and surfaces difficult to inspect radiographically.	Detects surface defects only. Not practical on rough surfaces.

insulation or internal to the equipment. Erosion is usually internal and associated with changes of direction, high velocity, or solids in the process.

The potential failure mechanisms need to be identified on each item of equipment to be included in the MI program. The potential causes of failure are normally known by the specifying engineer, and the information needs to be included in a line list or other document for use by the MI team. This process should not be very time consuming as most equipment can be categorized in fairly large groups. Armed with this information, the MI team can then determine what inspections will need to be performed during operations to evaluate the fitness-for-service (FFS) of the equipment For example, straight beam ultrasonic wall thickness examinations are often performed for general corrosion evaluation. However, if pitting corrosion is considered likely, then through-insulation radiography, or various angle beam ultrasonic inspections may be performed. If fatigue, creep, or embrittlement are the most likely failure mechanisms, then wall thickness checks will provide limited value as these conditions may not be detected by measuring wall thickness. Metallurgical evaluations using replication, portable hardness tests, destructive samples or test coupons are the more appropriate inspection methods. Once the likely failure modes are known, and the preferred method of inspection has been identified, then the preferred baseline data can be specified. Thickness data may be appropriate, or hardness tests on base metal, weld metal, and cladding may be needed, or installation of insulation plugs and taking thickness readings. API RP 579 (2000) can be used to determine fitness-for-service (FFS) for pressurized equipment in particulate solids processing and handling plants as well as in petroleum refineries.

Currently, many CPI companies are using risk-based inspection methods. Risk-based inspection (RBI) is an integrated methodology that factors risk into inspection and maintenance decision making. RBI is both a qualitative and quantitative process for systematically combining both the likelihood (probability) of failure and the consequence of failure to establish a prioritized list of process equipment based on total risk. From the list, the RBI user has the opportunity to design an inspection program that manages (reduces or maintains) the risk of equipment failures. Reynolds (1998) discusses the API methodology for risk-based inspection analysis for the petroleum and petrochemical industry. A detailed discussion of risk-based inspection is also presented in API RP 580 (2002). Mauney and Schmidt (1998) discuss financially structured risk-base methods for inspection and maintenance programs from the ASME Application Handbook on Risk-Based Methods for Equipment Life Management.

Various methods for process equipment inspection and testing are discussed in Chapter 19 of the book by Lees (1996). Also, the American Petroleum Institute (API) has published a number of publications dealing with the inspection of pressure vessels (API RP 572), piping system components (API RP 574), and atmospheric and low pressure storage tanks (API RP 575). NFPA 68 (2002) discusses inspection of vent closures. Types of solids processing equipment which may be considered as pressure vessels are dense-phase pneumatic conveying blowpots and equipment designed for deflagration containment. It is especially important to regularly inspect process equipment, where it is known that erosion or abrasion is possible due to the nature (characteristics) of the particulate solids being handled. If the vessel fails (a hole is formed), a large amount of dust could be emitted, which could then result in a dust explosion.

It is also important to regularly inspect and test process instrumentation and control systems to avoid inadvertent failures and shutdowns. Calibration of such systems is also vital to their effective operation. System designs should address prevention of plugging, erosion, and coating of instrumentation.

7.6.2 *Upgrading and Repairs of Equipment*

When an inspection detects a flaw or condition where potential failure may occur, the equipment should be repaired or upgraded as quickly as possible. The time at which this should be done can be based on prior operating history or can be determined by an FFS evaluation (using API RP 579). Repairs or upgrading are usually done by the maintenance department. Repairs may involve such procedures as replacing welds or cladding equipment where the walls have become thin due to corrosion or erosion. Upgrading may involve complete replacement of an item or equipment with the same material of construction or fabrication with a different material of construction.

7.6.3 *Documentation*

The employer should document each inspection and test that has been performed on process equipment. The documentation should identify the date of the inspection or test, the name of the person who performed the inspection or test, the serial number or other identifier of the equipment on which the inspection or test was performed, a description of the inspection or test performed, and the results of the inspection or test.

7.7 CORROSION, EROSION, AND MATERIALS OF CONSTRUCTION

7.7.1 *Introduction*

Corrosion and erosion can cause equipment failure if the equipment is not monitored and inspected at regularly scheduled times, and/or if the wrong materials of construction are used. It is critical to maintaining long equipment life to select the appropriate materials of construction as this can minimize the effects of corrosion and erosion. Also, using the proper fabrication procedures can minimize corrosion and erosion.

Section 2.3.4 of Chapter 2 discusses some aspects of the corrosivity of particulate solids and how they may cause corrosion of process equipment and piping.

Some aspects of corrosion types, corrosion detection and measurement, corrosion prevention and minimization methods, erosion and its effect on equipment, and selection of materials of construction are briefly discussed below.

7.7.2 Types of Corrosion

Corrosion is a chemical attack on a material (most often a metal, but plastics can also corrode) caused by exposure to an environment. Corrosion may occur at a uniform, predictable, rate, or it may be localized, on the surface, or as a subsurface phenomenon. Corrosion is normally thought of in terms of the internal (i.e., process) environment, but it also can occur on external surfaces of process equipment and piping.

There are many types of corrosion, such as the following:

- General (uniform) corrosion
- Stress corrosion cracking
- Pitting corrosion
- Intergranular corrosion
- Erosion–corrosion
- Galvanic corrosion
- Corrosion fatigue

Several corrosion problems that are related to particulate solids handling are briefly discussed below.

Stress Corrosion Cracking (SCC)

Organic powders which contain chlorine atoms may cause SCC when the powder becomes moist from water. Every alloy is subjected to stress corrosion cracking (SCC) in some environment; however, SCC is commonly associated with austenitic stainless steels. The majority of SCC problems are associated with stainless steel and aqueous chloride salts, but both sulfide and chloride stress cracking are common in the CPI. It occurs when a material has been under tensile stress in an environment containing chloride salts or sulfide compounds for a period of time.

Pitting Corrosion

Pitting corrosion is in itself a corrosion mechanism, but it is also a form of corrosion often associated with other types of corrosion mechanisms. The growth of pits, once initiated, is closely related to another corrosion mechanism, i.e., crevice corrosion. It is characterized by a highly localized loss of metal. In the extreme case, it appears as a deep, tiny hole in an otherwise unaffected surface. Because of its localized and deeply penetrating nature, pitting is one of the most damaging types of corrosion in the CPI. Pits can extend through the material within a short period of time. Pitting is difficult to detect by on-line monitoring. Pitting often occurs or is accelerated when vessels and/or piping are opened for inspection or for other reasons due to exposure to oxygen in air.

Accumulations of particulate solids on equipment surfaces can result in localized corrosion such as pitting. Surface deposits can create local microscopic variations in oxygen concentration, ion concentrations, or pH, resulting in isolated, stagnant regions of relatively concentrated electrolytes. The corrosive microenvironments that form can aggressively attack the underlying equipment surface.

Factors that can contribute to pitting under surface deposits include:

- Trapped moisture
- Wet/dry operating cycles
- Presence of ionic compounds
- Microbial action
- Protective coating defects
- nonuniformities in the material surface

Pitting can be minimized by the following:

- Designing equipment and operating procedures to minimize the formation of moist surface deposits
- Selection appropriate corrosion-resistant materials of construction
- Maintaining the integrity of protective surface coatings

Erosion–Corrosion

Erosion–corrosion occurs when a particulate solid or slurry moves at a high velocity through piping or equipment. The flow of the solids or slurry is not necessarily the cause of this form of attack, but can aggravate the attack by helping to destroy the passive film or corrosion product film formed on a metal by stripping it off by erosion, resulting in more exposure of the metal to corrosion. The mechanism is usually identified by localized corrosion which exhibits a pattern that follows the flow of the solids or a slurry.

Specific types of erosion-corrosion include wire drawing (grooving), impingement, cavitation, and fretting corrosion. Fretting corrosion is a specialized form of erosion-corrosion where two metal surfaces are in contact and experience very slight relative motion causing damage to one or both surfaces.

Again, in the presence of a corrodent, the movement causes mechanical damage of the protective film leading to localized corrosion. The corrosion usually takes the form of pitting-type attack.

More information about types of corrosion are found in books specifically about this subject such as those by Fontana (1986) and Schweitzer (1996). The National Association of Corrosion Engineers (NACE) also has many books on various aspects of corrosion.

7.7.3 Corrosion Detection and Measurement

A corrosion detection (monitoring) program can predict the corrosion rate, identify local corrosion, and estimate remaining service life. This information should be maintained in a computer database as part of the plant maintenance program.

On-line techniques, and other detection and measurement techniques are discussed by Perkins (1996). NACE publishes a number of books and standards dealing with corrosion testing and monitoring.

7.7.4 Corrosion Prevention and Minimization Methods

A number of things can be done to prevent and/or minimize corrosion, such as:

- Select the proper materials of construction
- Use the appropriate equipment fabrication techniques (e.g., welding, heat treatment)
- Apply protective linings and coatings where necessary
- Use cathodic or anodic protection where applicable
- Use an appropriate corrosion allowance

These are briefly discussed below.

Materials of Construction Selection

Selection of the proper materials of construction is the first step in minimizing corrosion. A number of factors have to be considered in making a selection. The selection of the appropriate materials of construction is usually based on corrosion data from pilot plant or full-scale plant experience, or from tables of corrosion data. A more complete discussion of the selection of materials of construction is presented in Section 7.7.6.

Equipment Fabrication Techniques

Using the appropriate fabrication techniques can also minimize corrosion. Many corrosion problems are related to the welding process used during equipment and piping fabrication. Weld cracking causes a large percentage of equipment and piping failures. Poor welding may result in crevices which make cleaning difficult and provides places for deposits to accumulate and for corrosion to start. The welding procedures should be well defined and included in material specifications. The procedures given in the ASME Boiler and Pressure Vessel Code, Section IX (ASME latest edition), and the American Welding Society Handbook (AWS latest edition) should be followed.

Landrum (1989) discusses many design solutions to corrosion problems caused by type of corrosion, fabrication techniques and environmental factors.

Protective Linings and Coatings

The application of protective linings and coatings to the inside and outside of process equipment can help to minimize corrosion. These can include sheet linings (metal and plastic) and liquid-applied coatings.

If plastic linings or coatings are used for equipment in particulate solids processes there is a need to provide protection against propagating brush discharges. This can be accomplished by using linings or coatings that have a breakdown voltage of less than 4 kV, so that there is no possibility of this sort of electrostatic ignition source occurring. It is better if plastic linings or coatings are avoided, if it is not possible to provide ones with a breakdown voltage of less than 4 kV. However, conductive or static-dissipative linings are also acceptable.

Schweitzer (1996) discusses the selection and application of sheet linings and liquid-applied coatings. Also, NACE has a number of books on coatings and linings.

Corrosion Allowance

Although technically not a way to control corrosion, use of a corrosion allowance is a commonly used method to address the problem of general (uniform) corrosion. A corrosion allowance is added to the wall thickness based on the general corrosion rate predicted by previous experience to compensate for the metal expected to be lost over the life of the process equipment or piping. Corrosion allowance cannot be used to compensate for pitting or localized corrosion. Periodic inspection and wall thickness determination must be made and monitored to determine when the corrosion allowance has been used up and the equipment or piping must be derated or replaced.

If a new plant is being designed with no prior experience with respect to corrosion rates, many companies specify a corrosion allowance of 1/16-inch for carbon steel equipment that is expected to be subjected to general (uniform) corrosion.

7.7.5 Erosion and Its Effect on Equipment

Erosion is not technically a form of corrosion, but is a mechanical effect; however, it is a significant factor in materials of construction selection. It can lead to equipment failure if not designed for and monitored, especially in storing, handling, and processing abrasive particulate solids. Erosion is the wearing away of material by mechanical energy. Erosion occurs by impingement of solid particles or liquid drops on a surface. It is seen very frequently in high velocity slurry and pneumatic conveying systems. Erosion can typically be found at equipment inlet and outlet nozzles, on vessel walls opposite inlet nozzles, on internal piping, on piping elbows and tees, and on impingement baffles. A number of things can be done to minimize this problem, such as:

- Selecting a material of construction for process equipment with a higher-than-usual hardness.
- Installation of wear plates and impingement baffles at the inside of equipment connected to a pneumatic conveyor, just after the inlet nozzle.
- Using long sweep elbows in pneumatic conveying systems. These elbows are often made of a hard material such as Ni-Resist, or have an internal ceramic lining. Before selecting internally lined elbows, it should be determined whether this can result in a propagating brush discharge.
- Limiting flow velocities, for example, using low velocity (dense phase) pneumatic conveying systems rather than high velocity (dilute phase) conveying systems.
- Specifying a larger-than-usual corrosion allowance.

Refer to Section 2.2 for a discussion of the effect of particle properties on erosion.

7.7.6 Materials of Construction

The basis for selection of materials of construction is performance under design conditions, that is, how the material will function under the process environment, not only at standard operating conditions, but also under startup, shutdown, and upset conditions. The behavior of a material in a process environment is determined by its physical, chemical, and mechanical properties. These properties determine how the material will be affected by process chemicals as well as how the material will affect the process.

With respect to selecting materials to resist corrosion, the best basis is prior operating plant experience, or corrosion rate data from laboratory, pilot plant, or demonstration plant tests. Also, quite often corrosion rate data are taken from such compilations as those by Graver (1985), Hammer (1975), and Schweitzer (1995). However, it should be pointed out that for many chemicals listed in these compilations the corrosion rate data are for pure chemicals. Therefore, if a process stream of predominantly one chemical has trace components (e.g., acids) the corrosion rate data from these compilations may not be correct for the process stream to actually be handled. In this case, corrosion testing should be done to obtain actual corrosion rate data for the mixture (impure chemical).

Several good books are available as guides for the selection of appropriate materials of construction, such as those by Chawla and Gupta (1993), Dillon (1992), and Hansen and Puyear (1996).

7.8 MAINTENANCE PRACTICES

7.8.1 Introduction

Routine maintenance is essential to ensure that all process equipment is able to perform as designed safely. Before maintenance is begun, plant management must ensure that the process equipment and/or piping is in a safe condition to work upon. Where combustible, reactive, or toxic particulate solids have been present within the equipment or piping, care must be taken to ensure that these materials have been removed. Extra care should be taken where any form of hot working is to be done, because the presence of a film or thin deposit on the walls of equipment or piping can lead to an explosion or fire if sources of heat are applied. See Section 3.1.1 for examples of explosions caused by heat. Also, many organic solids will produce flammable gas degradation products when heated by hot working, and especially in confined spaces, have a possibility of the flammable gases being above their LFL, which could lead to an explosion.

If it is feasible and where arrangements can be made, it is preferable to remove from the plant to a safe area those items of equipment requiring maintenance. Cleaning is still essential before actual maintenance begins. This method should be considered when it is necessary to do hot work on equipment that contained highly combustible or explosible dusts.

For all types of maintenance work, the safe completion of the job depends on proper preparation. The work permit approach to preparing a plant for maintenance allows the hazards involved in a job to be properly determined. Work should never be carried out in a potentially hazardous area without safe-work procedures. These procedures should ensure that persons doing the work and others associated with them are not exposed to danger, that the work does not cause danger to others not directly concerned with it, and that everyone is complying with regulations, technical standards, and codes of safe practice.

The following principles form a basis for a work permit system:

- Complete and secure isolation of equipment
- Complete removal of any residual hazard
- Clear and correct identification of all equipment in the plant
- Training and instruction of maintenance workers
- Monitoring of procedures
- Authorization of changes in procedure, materials, tools, and personnel.

There are two types of routine maintenance: preventive and predictive maintenance. It is important to distinguish between preventive and predictive maintenance. As valuable a tool as preventive maintenance has proven to be, it is inherently subject to several limitations. In an actual operating plant, variable conditions such as temperature and vibration can accelerate or retard process equipment failure. Preventive maintenance schedules do not allow for these variables. Consequently, there is an inherent risk that equipment may be shut down or replaced prematurely, resulting in unnecessary maintenance expense and lost production. Similarly, equipment operated under unusually stressful conditions may fail before scheduled service. When preventive maintenance is not appropriate for a particular piece of equipment, predictive maintenance is often used.

7.8.2 Preventive Maintenance

Preventive maintenance seeks to reduce the frequency and severity of unplanned shutdowns by establishing a fixed schedule of routine inspection and repairs. Usually these schedules are established on the basis of known quantities such as equipment repair history, design life, service intervals recommended by the original equipment manufacturer (OEM) and mean time between failures. The chief advantage of a preventive maintenance program is that it gives maintenance management the flexibility to plan and execute required equipment service with a minimum disruption of essential plant operation.

The importance of preventive maintenance to process safety management cannot be overemphasized. When maintenance is performed before breakdown, it can be planned, trained personnel can be made available, parts and other materials needed can be made available, and many components of emergency shutdown maintenance that result in accidents and injury can be avoided.

Preventive maintenance programs allow first- and second-line supervisors to ensure that the plant and personnel are properly protected when the maintenance is performed. By planning maintenance they can ensure that isolation, lockout/tagout, and other protection is in place before beginning the work. It also allows them to have personnel with the needed training and skills available to perform the work.

Preventive maintenance is the application of systematic attention and analysis to ensure the proper functioning of process equipment and to retard the rate of deterioration of physical facilities. Preventive maintenance activity includes (CCPS, 1995a):

- Operating maintenance: properly operating, caring for, cleaning, and in specified cases, lubricating equipment. This is usually done while equipment is operating.

- Shutdown maintenance: this type of maintenance includes examining, checking, testing, partially dismantling, replacing consumables, lubricating, cleaning, and other work short of overhaul or renovation. Equipment must be shut down to accomplish the inspection or repair.
- Preventive maintenance overhaul: this involves dismantling and examining equipment before a breakdown occurs and replacing or renewing components as they approach a theoretical maximum service limit.

A successful preventive maintenance program requires the following:

- A set of good maintenance records
- A mutual understanding between the operating and maintenance departments
- Maintenance mechanics who perform in a workmanlike manner
- An adequate inspection program
- A good corrective maintenance program

Preventive maintenance planning should begin during the design phase of the plant or when equipment is selected for replacement or a process change. After equipment is installed, inspecting the new equipment to determine if it meets design standards and specifications is the first step in the preventive maintenance program for that equipment.

7.8.3 *Predictive Maintenance*

Predictive maintenance, in contrast to preventive maintenance, draws on real-time condition-based data input to continuously adjust the profile of equipment performance expectations. Besides temperature and vibration analysis, predictive maintenance systems can use other process variables such as flow rate, revolutions per minute, motor current, and oil analysis.

The resulting composite profile is compared to certain program criteria that flag impending service requirements whenever a key parameter is exceeded.

Maintenance supervisors may be responsible for gathering data or for verifying data obtained by the operations department for use in the predictive maintenance program. It is also the responsibility of the maintenance department to maintain and calibrate monitoring equipment so that the information obtained is accurate.

Critical equipment and systems should be included in a predictive maintenance program. Complex electronic vibration and thrust sensing analysis is done on such equipment as large rotating equipment because if a rotor is lost, the process is going to be down for an extended period of time. Monitoring is performed to predict failure so that maintenance can be done before the failure occurs. Rotating equipment generally lends itself more to analysis and predictive maintenance than other types of equipment do. Another important reason to monitor rotating equipment in particulate solids processes is to avoid the generation of frictional hot spots or mechanical sparks due to equipment deterioration.

Process instrumentation and control systems also are often included in a predictive maintenance program. Some electronic instrumentation predictive maintenance is based on statistical failure information which is used to schedule maintenance. One company uses valve analysis programs that impose a spike output signal on a valve and measures the valve response to determine if the valve needs maintenance (Rodgers, 2004).

Predictive maintenance practices for many types of process equipment, including equipment for handling and processing particulate solids, are discussed by Mobley (2002).

Vibration and condition monitoring is discussed in detail by Bloch (1997). A useful book on process plant maintenance has been written by Townsend (1992).

7.8.4 Good Maintenance Practices for Particulate Solids Processes and Equipment

All process equipment manufacturers provide maintenance manuals for their equipment, and these instructions should be followed to ensure that the equipment is properly serviced.

Abbott (1990) presents maintenance recommendations for a number of dryers (spray, pneumatic conveying, fluid bed, rotary, band, batch atmospheric tray, batch vacuum, and trough).

The U.S. Environmental Protection Agency (EPA) has issued several reports that deal with the operation and maintenance of air pollution control equipment (dust collectors). One report (EPA, 1979) discusses maintenance recommendations for baghouses, electrostatic precipitators, scrubbers, and other equipment used with air pollution control equipment (e.g., inlet baffles, fans, exhaust stacks, etc.). Another report (EPA, 1985) deals entirely with electrostatic precipitators. Design considerations for minimizing operation and maintenance problems of particulate control equipment are presented in another report (EPA, 1980).

Holmes and Carson (2002) discuss how to maintain silos and how to spot signs of trouble in order to prevent failures to a silo's function and structural integrity.

Listed below are a number of good practices that should be included in preventive and predictive maintenance programs for particulate solids processes and equipment.

Electrical Systems, Bonding, and Grounding

- For baghouses and filter cartridges check all straps, cables, cages (filter bag supports), and conductive filter elements to ensure that they are bonded and grounded.
- Do continuity checks on all bonded and grounded equipment using megohm meters (also know as "meggers") as necessary.
- If bonding and grounding wires and straps are disconnected for maintenance work, make sure that they are reconnected after the maintenance work is completed, or their purpose will be defeated.
- Check all painted and coated equipment surfaces to make sure that they do not have high resistivities.
- Check for isolated conductors, such as sifter screens, to make sure that they will not produce an electrostatic discharge.
- Check all electrical systems and components (e.g., wiring, conduits, motors, lighting, etc.) to make sure that they are of the correct type for the electrical area classification.

Prevention of Metals from Entering Equipment

- Count all tools before and after doing maintenance work.
- Check to make sure that welding rods have not been left in equipment or piping.
- Check to make sure that all nuts and bolts are installed in flanges and have not fallen into equipment or piping.
- Check to see that welding slag has not fallen into equipment or piping.
- Check that metal pencils or pens have not fallen into equipment or piping.
- Check metal separators (magnets and screens) to see that they are functional and replace them as necessary.

Precautions during Equipment Cleanouts

- Check to make sure that equipment components are not damaged by brushing or washing during cleanouts.
- Properly isolate equipment being cleaned from other connected equipment to avoid cross-contamination or migration of product from one piece of equipment to another.
- When doing equipment cleanouts take measures to avoid damaging relief devices (relief valves, rupture disks, or deflagration vent panels) that are installed on the equipment. This may be done by isolating the devices by line blinds (slip plates) or they may need to be removed. If a line blind is used, the maintenance procedure must ensure that it is removed before restarting operations.

- For equipment provided with deflagration suppression systems, isolate the systems or remove them when equipment cleanouts are done.

Other Equipment Maintenance Considerations

- Check for plugging or coating of instruments such as pressure gauges, thermocouples, thermometers, level instruments, etc.
- When rodding out nozzles and connections that have become plugged, use wood or other nonconductive, nonabrasive materials for the rods.

Worker Safety Considerations

- Make sure that a work permit is issued before any maintenance work begins.
- Make sure that suppression systems installed on equipment are deactivated before maintenance work begins, and then re- activated before the process is started up again.
- Before starting maintenance work, check that equipment and piping is cleared of any particulates that would expose the maintenance worker to harmful materials.
- Make sure that maintenance workers are provided with the appropriate personal protective equipment, if it is deemed necessary, before they enter any equipment.

Inspection Techniques

- Inspect vessel internals for buildup or accumulation of materials before allowing worker entry. This may be done by use of mirrors and port scopes.
- Certain types of equipment may have to be dismantled in order to inspect internal parts. Maintenance workers should use appropriate personal protection equipment, if it is deemed necessary.

7.9 INCIDENT INVESTIGATIONS

The ultimate purpose of investigating incidents is to prevent the reoccurrence of the incident (including near-misses). Most companies have established procedures for reporting and investigating incidents. Table 7-4 shows the steps involved in a typical incident investigation and illustrates the roles played by operations and maintenance personnel involved in the investigation process (CCPS, 1995a). As can be seen from this table, an incident investigation comprises seven parts:

TABLE 7-4
Incident Investigation Process and the Role Played by the Operations and Maintenance Departments

Incident investigation steps	Operations and maintenance role
Incident reporting	Most incidents are reported by operations and maintenance personnel. Along with plant management, the first- and second-line supervisors should try to create an environment in which all incidents are reported.
Initiation of investigation	This involves the formation of a team with a clear definition of the scope and objectives of the investigation. Usually, plant or corporate management is responsible for this activity. Among others, operations and maintenance personnel are also chosen as team members.
Collection of evidence	All pertinent evidence is collected. Because of their process experience, operations and maintenance representatives should be able to collect important evidence.
Witness interviews	Besides other witnesses, many operations and maintenance personnel may be interviewed. The operations and maintenance representatives should try their best to produce an environment in which their fellow employees feel free to volunteer information.
Testing and analysis	Operations and maintenance representatives should study the testing results and other analyses to understand and provide input to the process. Examples of testing and analysis are metallurgical tests and consequence calculations, respectively.
Root cause determination	Operations and maintenance representatives should study the results of the root cause determination to determine if the results agree with their operational experience. If it does not, they should voice their opinion immediately.
Investigation report	The operations and maintenance representatives should make sure that the findings of the investigation are complete and presented in a usable format.

1. Initiation of the incident
2. Incident reporting
3. Collection of the evidence
4. Witness interviews
5. Testing and analysis
6. Root cause determination
7. Issuance of the investigation report

Evidence of near-misses in particulate solids handling facilities include the following:

- Evidence of combustion (i.e., charred particles)
- Activation of suppression systems
- Activation of relief devices
- Smoke and odor
- Discoloration of ductwork or equipment internals

- Reports of sparking
- Personnel shocks
- Pressure/temperature spikes
- Reports of tramp metal
- Major or sudden change in solids flow

A very comprehensive discussion of incident investigations is presented in a CCPS Guidelines book (CCPS, 2003), which also includes a CD-ROM. This Guidelines book covers the following topics: (1) introduction, (2) designing an incident investigation management system, (3) an overview of incident causation theories, (4) an overview of investigation methodologies, (5) reporting and investigating near misses, (6) the impact of human factors, (7) building and leading an incident investigation team, (8) gathering and analyzing evidence, (9) determining root causes-structured approaches, (10) developing effective recommendations, (11) communication issues and preparing the final report, (12) legal issues and considerations, (13) implementing the team's recommendations, (14) continuous improvement for the incident investigation system, and (15) lessons learned. The book also contains a number of pertinent appendixes, including: relevant organizations; professional assistance directory; photography guidelines for maximum results; example case study-fictitious NDF company incident; example case study-more bang for the buck: getting the most from accident investigations; selected OSHA and EPA incident investigation regulations; quick checklist for investigators; additional resources; and contents of the CD-ROM.

The National Fire Protection Association also has published a thorough guide for fire and explosion investigations (NFPA 921, 2004).

REFERENCES

Abbott, J. 1990. *Prevention of Fires and Explosions in Dryers.* 2nd ed., Rugby, England, UK: Institution of Chemical Engineers,

API RP 572. 2001. *Inspection of Pressure Vessels.* 2nd ed., Washington, DC: American Petroleum Institute.

API RP 574. 1998. *Inspection Practices for Piping System Components.* 2nd ed., Washington, DC: American Petroleum Institute.

API RP 575. 1995. *Inspection of Atmospheric & Low Pressure Storage Tanks. American Petroleum Institute, Washington, DC.*

API 579. 2000. *Fitness-For-Service.* Washington, DC: American Petroleum Institute.

API 580. 2002. *Risk-Based Inspection.* Washington, DC: American Petroleum Institute.

Bloch, H. P. 1997. *Practical Machinery Management for Process Plants:* Volume 4—*Major Process Equipment Maintenance and Repair.* Austin, TX: Gulf Professional Publishing.

CCPS. 1992. *Guidelines for Hazard Evaluation Procedures-Second Edition with Worked Examples.* New York: AIChE/Center for Chemical Process Safety.

CCPS. 1995a. *Guidelines for Safe Process Operations and Maintenance.* New York: AIChE/Center for Chemical Process Safety.

CCPS. 1995b. *Guidelines for Process Safety Fundamentals in General Plant Operations.* New York: AIChE/Center for Chemical Process Safety..

CCPS. 2003. *Guidelines for Investigating Chemical Process Incidents.* 2nd edition. New York: AIChE/Center for Chemical Process Safety.

Chawla, S. L. and Gupta, R. K. 1993. *Materials Selection for Corrosion Control.* Materials Park, OH: ASM International.

Dillon, C. P. 1992. *Materials Selection for the Chemical Process Industries.* New York: McGraw-Hill.

EPA. 1979. *Management and Technical Procedures for Operation and Maintenance of Air Pollution Control Equipment.* Report No. EPA-905/2-79-002 (June 1979), Washington, DC: US Environmental Protection Agency.

EPA. 1980. *Design Considerations for Minimizing Operation and Maintenance Problems of Particulate Control Equipment.* Report No. EPA-905/2-80-002 (May 1980). , Washington, DC: US Environmental Protection Agency.

EPA. 1985. *Operating and Maintenance Manual for Electrostatic Precipitators. Report No. EPA/625/1-85/017 (September 1985)*, Washington, DC: US Environmental Protection Agency.

FMG 7-76. 2001. *Prevention and Mitigation of Combustible Dust Explosions and Fire.* Property Loss Prevention Data Sheet 7-76. Norwood, MA: FM Global

Fontana, M. G. 1986. *Corrosion Engineering.* 3rd ed., New York: McGraw-Hill, NY.

Gravell, R. 2004. Peer review comments.

Graver, D. L. (editor). 1985. *Corrosion Data Survey-Metals Section.* 6th ed., Houston, TX: National Association of Corrosion Engineers.

Hammer, N. E. (editor). 1975. *Corrosion Data Survey-Nonmetals Section.* 5th ed., Houston, TX: National Association of Corrosion Engineers.

Hansen, D. A. and Puyear, R. B. 1996. *Materials Selection for Hydrocarbon and Chemical Plants.* New York: Marcel Dekker.

Holmes, T. and Carson, J. W. 2002. Silo Maintenance: It's Up to You. *Powder and Bulk Engineering*, 16(11): 31–39 (November).

Kahn, R. 2002. Personal communication from Russell Kahn, Syngenta, Greensboro, NC to S. S. Grossel (December 10, 2002).

Kirby, D. 2002. Personal communication from David Kirby, Baker Engineering and Risk Consultants, Inc., San Antonio, TX to S. S. Grossel (January 10, 2002).

Landrum, R. J. 1989. *Fundamentals for Designing for Corrosion Control.* Houston, TX: National Association of Corrosion Engineers.

Lees, F. P. 1996. *Loss Prevention in the Process Industries.* 2nd ed., Oxford, England, and Woburn, MA: Butterworth-Heinemann,

Louvar, J. F,. and Louvar, B. D. 1998. *Health and Environmental Risk Analysis: Fundamentals with Applications.* Upper Saddle River, NJ: Prentice Hall PTR.

Mauney, D. A., and Schmidt, M. E. G. 1998. Financially Structured Risk-Based Methods from the ASME Application Handbook. *Proc. CCPS Int. Conf. and Workshop on*

Reliability and Risk Management, pp. 381–398. New York: Center for Chemical Process Safety, American Institute of Chemical Engineers.

Mobley, R. K. 2002. *An Introduction to Predictive Maintenance.* 2nd ed. Woburn, MA: Butterworth-Heinemann.

Ness, A. 2002. Personal communication from Al Ness, Rohm and Haas Company, Bristol, PA to S. S. Grossel (September 12, 2002).

NFPA 68. 2002. *Guide for Venting of Deflagrations.* Quincy, MA: National Fire Protection Association.

NFPA 77. 2000. *Recommended Practice on Static Electricity.* Quincy, MA: National Fire Protection Association.

NFPA 654. 2000. *Standard for the Prevention of Fire and Dust Explosions from the Manufacturing, Processing, and Handling of Combustible Particulate Solids.* Quincy, MA: National Fire Protection Association.

NFPA 921. 2004. *Guide for Fire and Explosion Investigations.* Quincy, MA: National Fire Protection Association.

Perkins, A. 1996. Corrosion Monitoring. Chapter 25 in *Corrosion Engineering Handbook,* edited by P. A. Schweitzer, New York: Marcel Dekker.

Philiph, J. E. 2002a. Personnel communication from Jeffrey E. Philiph, Monsanto Company, St. Louis, MO to S. S. Grossel (October 15, 2002).

Philiph, J. E. 2002b. Personal communication from Jeffrey E. Philiph, Monsanto Company, St. Louis, MO to S. S. Grossel (March 8, 2002).

Reynolds. J. T. 1998. The API Methodology for Risk-Based Inspections (RBI) Analysis for the Petroleum and Petrochemical Industry. *Proc. CCPS Int. Conf. and Workshop on Reliability and Risk Management,* pp. 399–417. New York: Center for Chemical Process Safety, American Institute of Chemical Engineers.Rodgers, S. 2004. Peer review comments.

Sanders, R. E. 1999. *Chemical Process Safety: Learning from Case Histories.* Woburn, MA: Butterworth-Heinemann.

Schweitzer, P. A. 1995. *Corrosion Resistance Tables* (3 volumes). 4th ed., New York: Marcel Dekker.

Schweitzer, P. A. (editor). 1996. *Corrosion Engineering Handbook.* New York: Marcel Dekker.

Tewarson, A. 1974. *A Study to Determine the Electrostatic Hazard of Medical and Surgical Products and Packages. Factory Mutual Research Corporation Report.*

Townsend, A. 1992. *Maintenance of Process Plant—A Guide to Safe Practices.* 2nd ed., Rugby, England, UK: Institution of Chemical Engineers,

Valiulis. J. V., Zalosh, R. G., and Taminini, F. 1999. Experiments on the Propagation of Vented Dust Explosions to Connected Equipment. *Process Safety Progress,* 18(2): 99-106 (Spring).

Valiulis. J. V. 2001. *Dust Explosion Effects—Propagation in Dust Collection System Clan Air Return.* M.S. thesis, Worcester Polytechnic Institute, Worcester, MA.

Wood, J. P. (editor). 2001. *Containment in the Pharmaceutical Industry.* New York: Marcel Dekker.

Chapter 8

OCCUPATIONAL HEALTH AND ENVIRONMENTAL CONSIDERATIONS

8.1. INTRODUCTION

This chapter deals with occupational and environmental considerations for particulate solids processes, and discusses occupational health and environmental concerns, routine operations considerations, and nonroutine operations considerations.

8.2. OCCUPATIONAL HEALTH AND ENVIRONMENTAL CONCERNS

8.2.1 Protecting Employees and the Community

All particulate solids process facilities should be designed and operated in accordance with all applicable regulations to protect the health of plant employees, the surrounding communities, and the environment. If this is not done, and an accident (fire, explosion, or release of a hazardous particulate solid) occurs, then the company operating the facility may be liable to fines, and possibly, plant shutdown. In recent years, some company managers and officers have also been sentenced to jail for allowing harmful incidents to occur.

8.2.2 Regulatory Requirements

A number of US regulations pertain to the manufacture of hazardous particulate solids, such as:

- the Toxic Substances and Control Act (TSCA)
- the National Primary and Secondary Ambient Air Quality Standards
- the Clean Air Act Amendments (CAAA)
- the Clean Air Act (CAA) General Duty Clause
- the OSHA General Duty Clause
- the Food and Drug Administration (FDA) current Good Manufacturing Practice (cGMP).

These are briefly discussed below.

Toxic Substances and Control Act (TSCA)

The Toxic Substances and Control Act (TSCA) empowers the EPA to monitor and regulate the manufacture, processing, distribution, use, and disposal of chemicals and mixtures of chemicals. TSCA requires that all chemicals manufactured or processed for sale be on the TSCA inventory list. At present, this list includes over 67,000 chemicals and mixtures of chemicals. The TSCA regulations also require facilities that handle chemicals to keep records of significant adverse reactions to health or the environment, that is, adverse reactions allegedly caused by the substance or mixture. TSCA defines significant "adverse reactions" as reactions that may indicate a substantial impairment of normal activities, or long-lasting or irreversible damage to health or the environment. An "allegation" is defined by TSCA as a statement made without formal proof that implies that the chemical substance or mixture has caused a significant adverse reaction to health or the environment. Records of allegations of adverse reactions to the health of employees must be kept for 30 years. In addition to having a "recording" obligation described above, TSCA also has a "reporting" requirement, i.e., if a company has information which reasonably supports the conclusion that a chemical or chemical mixture can cause a significant adverse reaction to health or the environment, and this information is not previously known to the EPA, the company may have to actually submit that information to the EPA.

National Primary and Secondary Ambient Air Quality Standards

A National Ambient Air Quality Standard (NAAQS), as presented in 40 CFR 50, is the maximum level that will be permitted for a given pollutant. There are two kinds of such standards: primary and secondary. Primary standards are to be sufficiently stringent to protect the public health; secondary standards must protect the public welfare. The EPA sets these standards after it issues a criteria document and a control technology document on the pollutant in question. Both primary and secondary standards apply to all control regions. Each control region must satisfy the standards set forth and is bound by federal enforcement.

The relevant NAAQSs for particulate solids are:

- 40 CFR 50.6 National Primary and Secondary Ambient Air Quality Standards for PM10
- 40 CFR 50.7 National Primary and Secondary Ambient Air Quality Standards for Particulate Matter

40 CFR 50.6 is the ambient air quality standards for PM10 (particles with an aerodynamic diameter less than or equal to a nominal 10 microns). The maximum level is 150 micrograms per cubic meter over a 24-hour time period, or 50 micrograms per cubic meter annual arithmetic mean.

40 CFR 50.7 has two maximum level requirements:

1. 15.0 micrograms per cubic meter annual arithmetic mean concentration, and 65 micrograms per cubic meter 24-hour average concentration measured in the ambient air as PM2.5 (particles with an aerodynamic diameter less than or equal to a nominal 2.5 microns).
2. 50 micrograms per cubic meter annual arithmetic mean concentration, and 150 micrograms per cubic meter 24-hour average concentration measured in the ambient air as PM10.

Clean Air Act Amendments (CAAA)

The Clean Air Act (CAA) was originally issued in 1970 and required the EPA to establish National Ambient Air Quality Standards (NAAQS). Since the NAAQS objectives were not met in a large number of areas within the U.S., and since EPA had been slow in adding chemicals or materials to the National Emission Standards for Hazardous Air Pollutants (NESHAP) list, a group of amendments to the CAA were added to ensure satisfactory reductions in contaminants and hazardous pollutants within a 5- to 10-year period. These amendments were signed into law on November 15, 1990, and have been termed the Clean Air Act Amendments (CAAA).

The CAAA contain eleven titles (sections), but the ones that are most relevant to emissions from particulate solids process facilities are Title I—Nonattainment and Title III—Hazardous Air Pollutants.

Title I states that before individual companies can be found noncompliant, a nonattainment area must be identified. A nonattainment area is a geographical area that does not meet the national ambient air quality standards for criteria pollutants. Criteria pollutants are sulfur dioxide, carbon monoxide, ozone, nitrogen dioxide, particulate matter with diameters less than 10 microns, and lead. The latter two pollutants can be emitted from a particulate solids process facility. The significant harm level for particulate solids with diameters less than 10 microns is 600 micrograms per cubic meter, and must be monitored over a 24-hour average period. Currently the threshold for a major source classification in a nonattainment area is 100 tons per year of any critical pollutant. EPA may eventually establish lower thresholds and different severity classifications.

Title III is based on a list of 188 chemicals that when released may cause potential hazards to human health and the environment. These chemicals are called hazardous air pollutants (HAPs). A number of HAPs listed in Title III are particulate solids. Facilities that are major sources of HAPs are required to install, use, and maintain maximum available control technology (MACT), which must be described in the permit applications. A major source, according to Title III, is one that emits 10 tons per year of any single HAP or 25 tons per year of any combination of HAPs. The EPA, however, has the authority to set lower thresholds. MACT standards for the chemical industry cover emissions from process vents, storage tanks, wastewater operations, transfer operations, and equipment leaks. For these sources, MACT includes control technology, performance standards, and work practice standards. MACT standards for new sources must be at least as stringent as the most stringent emissions level achieved at similar types of facilities. For existing sources, MACT must provide control of emissions to a level that equals or exceeds that level demonstrated by the best-controlled 12% of similar existing sources. Existing sources must comply with MACT within three years after the standards are issued.

OSHA and EPA Clean Air Act (CAA) General Duty Clauses

As discussed in Section 7.2, particulate solids process facilities may be subject to the requirements of the OSHA and EPA Clean Air Act (CAA) general duty clauses to protect the health and safety of plant personnel and the health of people in communities adjacent to the facility.

FDA Current Good Manufacturing Practices (cGMP)

Pharmaceutical and food plants must be designed, installed, and operated to meet the current requirements of the FDA current Good Manufacturing Practices (cGMP). These requirements, defined in US regulations 21 CFR 210 and 21 CFR 211, are meant to protect the product (maintain its purity and potency), as well as the health of the operators (protect them from health hazards). They are particularly stringent for the manufacture of products in their final product form. Equivalent directives are presently available or are being prepared in the European Union (EU).

Because of the nonprescriptive wording in the US cGMPs, many companies have applied their own interpretations of GMP expectations. In order to help facility designers understand current GMP expectations, the International Society for Pharmaceutical Engineering (ISPE) teamed up with the FDA to create a series of Baseline Facility Guides. These guides will cover different types of pharmaceutical product manufacturing facilities, but are not to be construed as being GMPs. Likewise, they are not intended to be the only workable approach to facility design. In the foreword of each published guide is a letter from the FDA stating this fact. The ISPE Baseline Facility Guide on Oral Solid Dosage Forms (ISPE, 1998) is the most relevant one to particulate solids process facilities.

Willig (2000) presents a good discussion of the FDA cGMPs.

8.2.3 Product Stewardship

Product stewardship promotes health, safety, and environmental protection as integral parts in the design, manufacture, marketing, distribution, use, recycle, and the ultimate disposal of chemical products.

The successful implementation of product stewardship is a shared responsibility of all those involved with a product during its lifecycle, from its initial inception to its ultimate recycle or disposal.

The American Chemistry Council (ACC) Responsible Care® initiative contains a Product Stewardship Code of Management Practices which discusses product stewardship requirements in detail. The product stewardship code provides guidance as well as a means to measure continuous improvement in the practice of product stewardship.

Among the components of a product stewardship program are hazard information and characterization and risk management. The hazard information and characterization component seeks to improve hazard identification and risk characterization at every stage of the product lifecycle by addressing the continually growing wealth of information on chemical products. The risk management component provides information on ways to manage risks at every stage of the product lifecycle. This information should be shared with the product users and the surrounding community (facility neighbors) who could be affected if an accident occurred. These are discussed below.

8.2.3.1 Communication with Product Users

Safety and occupational health hazards are usually transmitted to users by Material Safety Data Sheets (MSDSs). MSDSs for particulate solids products should contain more information than those for liquids and gases because particulate solids products usually have a range of particle sizes, which affect safety properties such as the MIE, the MEC, the P_{max}, and the $(dp/dt)_{max}$. The particle size distribution for a particulate solids product will often change from the original "as received" product during the processing or handling of it due to particle attrition. This can severely change these properties and increase the hazards as smaller particles lower the MIE and the MEC, and increase the P_{max}, and the $(dP/dt)_{max}$. Therefore, values of the MIE and other relevant safety (fire and explosion data, thermal decomposition temperature, etc.) properties for a range of particle sizes should be provided, if available. Information on stability and reactivity should also be provided, as well as handling and storage recommendations, and fire fighting measures.

In addition to safety properties, the MSDSs should contain health hazard properties such as lethal dose-50% (LD_{50}) and lethal concentration-50% (LC_{50}) data, and whether a particulate solid is carcinogenic, mutagenic, or teratogenic. Information on first aid measures and exposure control/personal protection should also be included.

Information on how to handle spills and releases of toxic particulate solids should also be included in an MSDS.

In addition to the MSDS, many companies provide supplemental technical information (e.g., product bulletins, etc.) on safe handling of their products. There are also several industry organizations (e.g., The International Fertiliser Society, NFPA, etc.) that also offer literature on safe handling of solids.

An example of how far some companies go to provide their clients with product stewardship is provided by Kahn (2002). A company produces a herbicide active ingredient (that has an explosion class of ST2 and a MIE of 3-10 mJ) that is sold to fertilizer formulators. They have instituted a product stewardship program that includes a personal visit by a safety specialist to the formulator before they begin their first formulation, a formulator's manual which outlines in simple language the hazards of the active ingredient and precautions needed for safe formulation, and a sign-off sheet in which the formulator acknowledges that they have received and understand the information received in the manual and from the safety specialist during the on-site visit.

8.2.3.2 Community Issues and Communications with Facility Neighbors

A facility manufacturing hazardous particulate solids products should develop mechanisms for outreach with the communities in which they operate (this is required by the ACC Responsible Care initiative). This will familiarize the neighbors about what products are being manufactured at a facility and help to assuage the fears of the neighbors about accidental releases which might affect their safety and health. It is also good public relations to hold meetings with the neighbors (and conduct plant tours, if possible) to show them what safety measures have been taken to protect the plant and them.

The facility should provide information to the Local Emergency Planning Committee (LEPC) or Community Advisory Panel (CAP), and the community fire departments about the hazardous properties of the products produced at the plant, and measures required to protect the neighbors from a release.

8.3. ROUTINE OPERATIONS CONSIDERATIONS

8.3.1 Permitting Issues

In the United States, all facilities that have equipment emitting particulate solids during normal operations will need to obtain an air emission permit from the appropriate local, state, or federal government authority before construction or modification begins (similar to vent permits required for equipment emitting volatile organic compounds). This is required by Title I and Title III of the CAAA.

Vent permits require a listing of the amount of particulate solids leaving in a vent stream. This is quite often difficult to estimate or obtain. For a new facility, an estimate is made of the amount of solids in the vent stream based on pilot plant data, operating experience from an existing similar plant, or data from EPA Report AP-42, Compilation of Air Pollutant Emission Factors (data for some solids are available on the internet at http://www.epa.gov/ttn/chief/ap42/index.html).

8.3.2 Monitoring Emissions from Equipment

It is good practice to monitor process equipment handling or processing hazardous particulate solids where emissions are expected to occur (e.g. vibratory screens, bag dumping, bag filling, etc.). If unplanned emissions occur, monitoring will alert operators to potential hazards such as fires and explosions resulting from the accumulation of combustible or decomposable solids on hot surfaces, and/or exposure of operators to health-hazardous solids.

Monitoring of equipment emission can be done by:

- Visual observation
- Area monitoring techniques (equipment)

Many instruments exist for real-time monitoring of airborne particulates (aerosols). Some of the commonly-used, direct-reading aerosol monitors include the following:

- Light-scattering photometers
- Light-scattering particle counters
- Condensation nucleus counters
- Single particle aerosol relaxation time (SPART) monitors
- Beta attenuation aerosol mass monitors
- Piezoelectric crystal microbalance
- Tapered element oscillating microbalance (TEOM)
- Fibrous aerosol monitors

These are discussed and described by DiNardi (1997).

Area monitoring can also be done using active sampling techniques to collect samples over a period of time for subsequent laboratory analysis. The most common sampling methods for particulates involve the use of a sampling train, which includes a collecting medium, a flow meter, and a vacuum pump. The collecting medium usually consists of filter media that are inserted into a plastic support cartridge. Several types of filter media are available and are generally classified as fiber, porous membrane, or straight-through (capillary) membrane filters. The choice of media depends on the particle characteristics and the analytical technique to be used.

Total dust sampling collects airborne particles of all sizes, but often only particles of a certain size are of interest. Measurement of the amount of particles in the inhalable, thoracic, and respirable fractions can be important when assessing potential health hazards. Size-selective sampling techniques often use some type of preselector at the beginning of the sampling train. The preselector can be filtration-based, sedimentation-based, or impaction-based.

Chapter 2 (Section 2.2.10) also discusses and describes in-situ dust cloud concentration measurement devices.

There is very little information in the open technical literature on fugitive emissions of particulate solids from process equipment. Carson and Mumford (1988) present some data on emissions from vibratory screens, bag dumping, and bagging (bag filling) machines (see Table 8-1). These data are examples of some measurements of emissions of some particulate solids and may not be representative for all types of solids

8.3.3 Employee Exposure Monitoring and Risk Assessment

When handling or processing health-hazardous particulate solids it is common practice to monitor employee exposure. This is necessary to characterize potential exposure of employees which then can be compared to the OSHA Permissible Exposure Limits (PELs) or the ACGIH Threshold Limit Values (TLV®s), or internal company standards. Sargent and Naumann (2001) present a good discussion of occupational exposure limits.

Employee exposure monitoring can be done by using a lapel-mounted filter holder connected to a portable vacuum pump with a flow rate of about 3 liters/minute. Respirable solids can be separated out by use of a small cyclone. In order to ensure uniformity of fractionation, smooth and constant flow rates are essential (Carson and Mumford, 1988).

Exposure risk assessments estimate the dose or the quantity of risk agents, that is, health-hazardous particulate solids, received by workers. They can be used to do the following (Louvar and Louvar, 1998):

- Identify solids that are potentially hazardous to workers and/or the environment. Identify specific populations at risk.
- Identify paths of exposure.
- Quantify the exposure.

TABLE 8-1
Emission Rates[a] from Powder Handling

Type of equipment/handling	Emission rate (mg/s)
Vibratory screens	
Open top	5.5 × top surface area (m^2)
Closed top with open access port	
15-cm-dia. port	0.11
20-cm-dia. port	0.21
30-cm-dia. port	0.44
Closed cover—no ports	Zero
Bag dumping	
Manual slitting and dumping	3
Semiautomatic (enclosed dumping but manual bag entry/removal)	0.2
Fully automatic (includes negative pressure)	Zero
Bagging machines (filling)	
No ventilation	1.5
Local ventilation	0.01
Total enclosure and negative pressure	Zero

[a] The emission rates refer to the total dust level of release. To estimate the respirable level, these numbers should be halved.

Most exposure assessments are complicated because workers may move in and out of an exposed region, or the source of exposure may vary with time. To simplify the process, average exposures are used. The use of PPE will modify a worker's exposure.

Exposure assessments typically use three approaches: analogies, monitoring, or modeling. The use of analogies to predict the characteristics of hazardous solids is similar to the use of chemical structure activity relationships to identify hazards of chemicals. A particulate solid as yet unmonitored and unmodeled, but with a molecular structure analogous to a known solid, will likely have the same toxicological effects. Exposure assessments use the analogy to characterize the dispersion in air and health effects of a new solid based on their physicochemical characteristics. This information is also useful for supporting specific judgments or assumptions used with exposure models.

Actual data from monitoring is the most accurate approach for determining exposure. These data may also be used to calibrate exposure models.

Exposure models are used to simulate the behavior of health-hazardous solids when actual monitoring is not possible or not justified, as when conducting feasibility studies for a future plant. Many such models exist and are discussed by Louvar and Louvar (1998).

Exposure assessment and risk analysis have many uncertainties due to data extrapolation, inaccuracies of sampling and analysis, and inaccuracies of defining the sources. An absolute risk value for a specific particulate solid or accident scenario cannot be determined.

Louvar and Louvar (1998) and Stayner (2001) present more detailed discussions of worker exposure risk assessment.

8.3.4 System Design to Eliminate or Minimize Employee Exposure

Process equipment and closed process systems can be designed to minimize employee exposure by utilizing the following practices:

- Select appropriate equipment to prevent or minimize emissions of health-hazardous particulate solids.
- Install containment systems when necessary.
- Design and install piping systems that are dust-tight.
- Install appropriate and adequate dust pickup ventilation systems.
- Design the process to minimize transfers of solids
- Use inherently safer processes

8.3.5 Health Standards

Several governmental, professional, and industry organizations in the United States publish occupational health standards. These include the Occupational Safety and Health Administration (OSHA), the U.S. Environmental Protection Agency (EPA), the American Conference of Governmental Industrial Hygienists (ACGIH), American Industrial Hygiene Association (AIHA), and the National Institute for Occupational Safety and Health (NIOSH).

International organizations that publish occupational health standards are the Health and Safety Executive (HSE) in the United Kingdom (UK), the World Health Organization (WHO), and the International Labor Organization (ILO).

A brief description of occupational health standards published by some of the above-cited organizations is presented below.

OSHA

A number of occupational health standards are published by OSHA, as follows:

- 29 CFR 1910.94 Ventilation
- 29 CFR 1910.107 Spray Finishing Using Flammable and Combustible Materials

- 29 CFR 1910.120 Hazardous Waste Operations and Emergency Response
- 29 CFR 1910 Subpart 1—Personal Protective Equipment
- 29 CFR 1910.263 Bakery Equipment
- 29 CFR 1910.265 Sawmills
- 29 CFR 1910.272 Grain Handling Facilities
- 29 CFR 1910.307 Hazardous (Classified) Locations
- 29 CFR 1910.1000 Air Contaminants
- 29 CFR 1910.1001 Asbestos
- 29 CFR 1910.1003 13 Carcinogens
- 29 CFR 1910.1018 Inorganic Arsenic
- 29 CFR 1910.1025 Lead
- 29 CFR 1910.1027 Cadmium
- 29 CFR 1910.1029 Coke Oven Emissions
- 29 CFR 1910.1043 Cotton Dust
- 29 CFR 1910.1050 Methylenedianiline

The full standard (regulation) can be found in the Code of Federal Regulations.

OSHA also publishes Permissible Exposure Limits (PEL) which are legally enforceable in the United States under the OSHA Act of 1970.

EPA

A number of EPA health standards are discussed and listed in Chapter 4. (Section 4.3.9).

ACGIH

The ACGIH is a private, nongovernmental professional society, made up primarily of industrial hygienists who work for either a State or Federal government agency. A committee of this group called the Chemical Substances Threshold Limit Value (TLV®) Committee, publishes an annual list of about 600 substances (including particulate solids). While the list has no regulatory authority in the United States, it is used voluntarily by many companies and industrial hygienists since it is the most up-to-date list. It is incorporated into the laws and regulations of many foreign countries.

AIHA

This organization is the largest industrial hygiene society and has a Workplace Environmental Exposure Limit (WEEL) Committee which sets limits for workplace exposure. The WEEL Committee has set limits for about 100 substances and their list does not, in general, overlap the ACGIH TLV® list.

NIOSH

In the course of developing criteria for standards NIOSH sometimes makes recommendations for occupational exposure limits. Despite the fact that NIOSH is an agency of the U.S. Government, these recommendations, called Recommended Exposure Limits (RELs) do not carry the force of law unless they are adopted by OSHA.

HSE

The HSE publishes a series of Guidance Notes, one of which, EH 40-Occupational Exposure Limits, gives a set of limits that have legal status in the United Kingdom. This list includes "control limits" which are known to be achievable and are enforced in all UK industries, and "recommended limits" which present good practice and realistic criteria for the control of exposure.

8.3.6 Employee Precautions When Handling Toxic Particulate Solids

Some particulate solids can be very toxic, and consequently are hazardous to health. Operators and maintenance personnel who come in contact with them have to take precautions to avoid harm to their health. Protection of operators should be provided by engineering control methods, as required by OSHA 1910.134, and as discussed in Section 6.11. However, when engineering controls are not feasible, operators may have to wear personal protective equipment (PPE), such as, for example, when transferring trays of toxic solids from a filter to a batch vacuum tray dryer, or packaging of powders.

It is quite often necessary to manually remove spent filter cake from the leaves of a filter or to dig out the solids "heel" from a batch centrifuge. When these solids are toxic, the operators should wear personal protective equipment to avoid direct contact with the solids. The extent of the protective clothing required will depend on the toxicity of the solids. For example, when the solids can only cause minor skin problems, the operator may need to use only hand and eye/face protective equipment, whereas, in the case of very toxic solids, complete body protection may be needed, e.g., protective clothing such as encapsulating suits. It may be possible to prevent or minimize operator contact with health-hazardous particulate solids when cake removal is needed by using certain types of filters and centrifuges that do not require operator involvement. For example, there are various types of filters available that can be used, such as:

- Pressure filters with horizontal leaves in which the leaves are rotated and the cake is spun off with, or without, the assistance of a sluicing liquid.
- Cartridge filters in which the cake is removed by applying a backflushing liquid.

To minimize the need for operators to manually remove the "heel" from a centrifuge, consideration should be given to using a continuous centrifuge rather than a batch type. Also, an inverting filter cloth centrifuge eliminates the need for "heel" removal.

Maintenance personnel must also be protected against exposure to health-hazardous particulate solids. Before maintenance personnel are allowed to enter or work on equipment containing toxic solids, the equipment should be cleaned to remove as much of the toxic solids as possible. If possible, it should be verified that the equipment is free of toxic solids before maintenance work is allowed to proceed. It may be difficult sometimes to completely remove all toxic solids, and in this case, maintenance personnel may also have to wear personal protective equipment (PPE) when inspecting and servicing such equipment. Preparation of surfaces, prior to hot work, should be done to minimize exposure to hazardous decomposition products.

8.3.7 Selection, Storage, and Maintenance of Personal Protective Equipment (PPE)

Personal protective equipment (PPE) come in a variety of types and materials, depending on the hazard (toxicity of the solid) to which one is likely to be exposed in the work environment. The selection and use of PPE is based on the exposure route of the hazards that are or may likely be encountered, and may be based on OSHA regulations, consensus standards (e.g., ANSI), or recommendations by professional organizations like the American Industrial Hygiene Association.

Personal protective equipment may be provided for the following:

- Head protection
- Eye/face protection
- Skin protection (general protective clothing)
- Hand protection
- Foot protection
- Respiratory protection

Brief discussions of the various types of PPE are presented below.

Head, Eye, and Face Protection

Head protection is important in preventing exposure to toxic particulate solids because uncovered hair can be a collector of dust from the air, leading to a continuous dermal exposure on the scalp. The exposure hazard from dusts falling from overhead building members is also greatly minimized by effective head protection. It is especially important that hard hats be kept clean, particularly the sweat pad on the forehead.

Eye and face protection is also extremely important to prevent toxic particulate solids from coming in contact with them. Toxic solids contacting the eye may be nearly equivalent to direct injection of the material into the blood stream, as systemic absorption by this route may be rapid and complete (SRI, 1980). The local effects of dusts, granules, etc. must also be guarded against. Goggles should be used when handling toxic particulate solids. They provide frontal impact protection and can provide protection against other hazards such as mists and splashes. Goggles come in a variety of styles, lens shadings, and configurations ranging from tight-fitting eyecups to chemical goggles. Ventilation for the goggles is usually provided to prevent fogging. Chemical splash goggles and face shields should be furnished and required in some operations where the potential for liquid splashes exist.

Protective Clothing

Protective clothing is used to prevent potentially harmful chemicals, physical hazards (extreme temperature, radiation), and biological hazards from contacting, and, in some instances, entering the body. General OSHA requirements covering protective equipment and clothing are specified under 29 CFR 1910.132. This standard states that such protective equipment and clothing be provided, used, and maintained in a sanitary and reliable condition wherever it is necessary, by reasons of hazards encountered, to protect any part of the body from injury or impairment through absorption, inhalation, or physical contact from chemical, physical, and biological hazards. Totally encapsulating chemical protective (TECP) suits, as well as gloves, boots, and eye and face protection may be needed for handling some toxic particulate solids, either in dry form or in solution (see Figure 8-1). Protective clothing (beyond the conventional work clothing) requirements vary from process to process, depending on the particulate solids being handled. Some plants require the use of disposable garments (e.g., Tyvek®) as a condition of entry to production, formulation, or packaging areas. Other types of materials commonly used for TECP suits are shown in Table 8-2.

In addition to providing a barrier to toxic exposure, other potential hazards, i.e., static generation, splash protection, flame retardation, burn protection against flash fires, etc., should be considered in the selection of PPE.

Wood (2001) describes the protective equipment used by operators in dispensing a 100% active ingredient from its bulk container (a fiber drum). Two operators were involved and they each wore two sets of protective equipment. The inner layer consisted of a tight-fit respirator and a full Tyvek® coverall suit including connected gloves and booties. The outer layer was another full Tyvek® coverall suit, with overgloves and overbooties, and a Bullard breathing hood with powered breathing air supply. The double layering of PPE extended to double breathing protection. The rationale for double layering was to protect the worker during degowning of the contami-

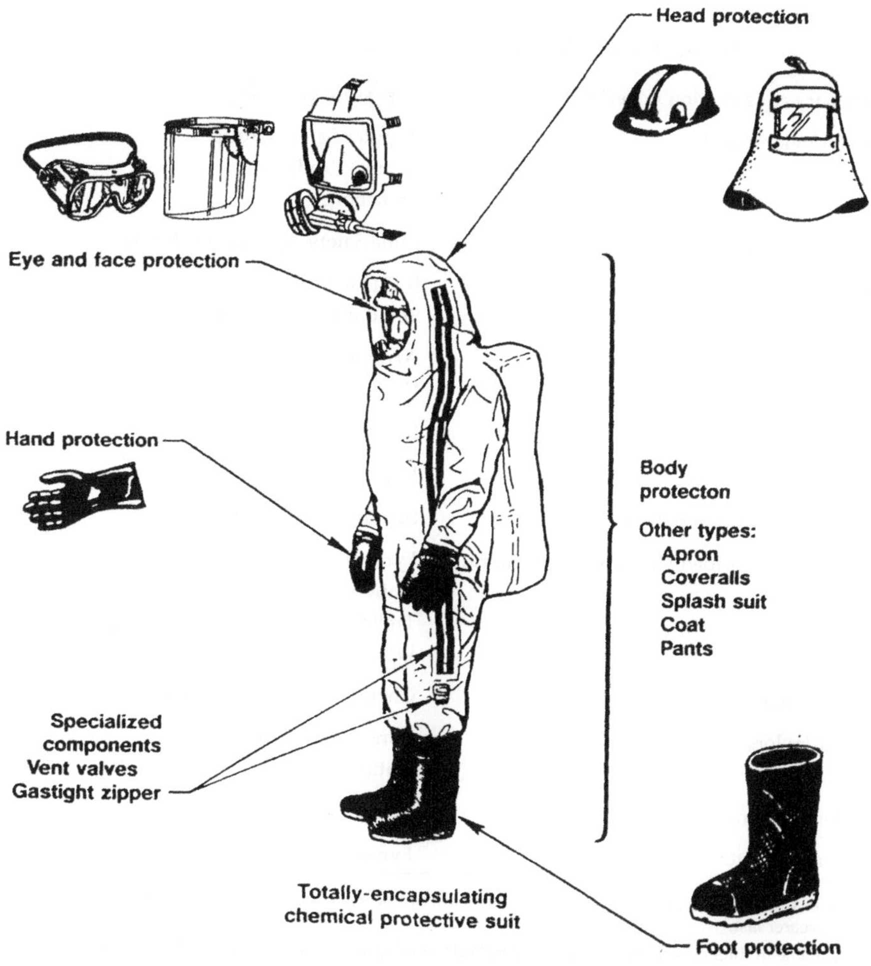

Figure 8-1. Different types of protective clothing.

nated outer layer after dispensing. Gowning and degowning of the outer layer was done in a unidirectional flow room, with room airflow remaining on during this operation. The contaminated outer layer was drummed within the room for later disposal.

EPA also provides recommendations on suitable personal protective equipment for various toxic particulates (see the EPA IRIS program, which is listed on the EPA website cited in Chapter 4).

Hand and Foot Protection

Gloves and boots (or impervious shoes) may be required for handling some toxic particulate solids, especially if they are in solution. Protective gloves

TABLE 8-2
Primary Construction Materials of Commercially Available TECP Suits

Primary construction material[a,b]	Ensemble vendor
Butyl/nylon/butyl	Kappler Safety Group Fyrepel Products Trelleborg
Butyl/polyester/chloroprene	Mine Safety Appliances (MSA)
Chlorinate polyethylene	ILC Dover Standard Safety Equipment
Chlorobutyl/Nomex/chlorobutyl	Arrowhead Products ILC Dober
Chloroprene/nylon	National Draeger
Polyvinyl chloride/nylone	Kappler Safety Group
Polyvinyl chloride/polyester	Standard Safety Equipment
Polyvinyl chloride/nylon/polyvinyl chloride	Fyrepel Products National Draeger Trelleborg Wheeler Protective Apparel
Teflon/ fiberglass/Teflon	Chemical Fabrics Corporation
Viton/nylon/chloroprene	Mine Safety Appliances National Draeger
Viton/butyl/nylon butyl	Trelleborg
Viton/polyester/Viton	Fyrepel Products

[a] The primary construction materials are described with the external surface first and the surface facing the wearer last.

[b] Available in thickness ranges from 8 to 28 mils (1mil is equivalent to 0.001 in. or 0.025 mm).

are available in a wide variety of materials that vary among manufacturers. Before selecting protective gloves, the manufacturer should be consulted as to which material is the best for a specific toxic solid (dry or in solution). The structure of the material has an influence on the permeation across the glove material. Depending on the glove manufacturer, performance data (chemical resistance and physical performance) will vary greatly from material to material, depending on several factors, including material thickness, glove construction, temperature, and the type of material. Where there is a high likelihood of electrostatic discharges occurring, the protective gloves should be conductive or antistatic, so that handled items that are difficult to individually ground, such as tools, are grounded through the grounded person.

TABLE 8-3
Common Construction Materials for Chemical Protective Gloves

Butyl, Isobutylene-isoprene rubber (IIR)
Natural rubber (NR)
Neoprene, chloroprene (CR)
Nitrile, acrylonitrile-butadiene rubber (NBR)
Polyethylene (PE)
Polyvinyl alcohol (PVA)
Polyvinvyl chloride (PVC)
Polyurethane (PU)
Viton[a]
Silver Shield[b]

[a] Du Pont trademark
[b] North trademark

Chemical protective gloves are divided into three categories: unsupported, supported or coated, and impregnated gloves. Unsupported gloves offer several advantages, including a high degree of dexterity and greater selection of polymers, lengths, and thickness versus supported gloves. Supported gloves, or coated gloves, have fabric liners with cured polymer coatings.

For supported gloves, the advantages are increased thermal capacity, abrasion and tear resistance, with the disadvantage of less dexterity.

Gloves are available in a variety of materials. Common construction materials for chemical protective gloves are shown in Table 8-3. Manufacturer test data should be compared when selecting specific gloves, including the structure of the glove materials, the formulation and processing, and the differences between each manufacturer. Specific permeation test data on protective materials vary from one manufacturer to another, and are not always interchangeable. Generally, chemical resistance selection charts provide recommendations for a variety of glove coatings relative to their chemical resistance to exposure to various chemicals. Chemical-resistance selection charts should be obtained from the glove manufacturer, or through a vender, or distributor.

Proper selection of protective gloves should take into consideration the wearer, the workplace conditions, and the protective glove. Employees should be trained in the correct way to put on, wear, and then take off protective gloves to ensure maximum protection.

Gloves should be replaced periodically, depending on frequency of use and permeability to the substance(s) handled. Gloves overly contaminated should be rinsed and then carefully removed after use. There should be adequate facilities for disposal of contaminated protective gloves.

Foot protection against chemical hazards can be of the following two types:

1. Boots constructed of chemical-resistant material to protect the feet from contact with toxic solids and other chemicals.
2. Disposable shoe or boot covers which are made of a variety of materials and slip over a shoe or boot. They protect safety boots from contamination and protect the feet from contact with toxic solids or other chemicals. However, they possibly may electrically isolate the operator from earth ground, and this should be evaluated.

In some cases, electrostatic-dissipative or conductive footwear may also be necessary. These types of footwear are used when it is necessary to ground operators in order to prevent them from becoming electrostatically charged. Their characteristics are:

1. Electrostatic dissipative (ESD) footwear has both an upper and lower resistance requirement of more than 10^5 ohms, but less than 10^8 ohms (BS 5958, 1991 and CENELEC, 2003). NFPA 77 (2000) states that the resistance to earth through ESD footwear and a floor should be between 10^6 ohms and 10^9 ohms. For materials with very low ignition energies, the resistance to earth between footwear and flooring should be less than 10^6 ohms (NFPA 77, 2000). Resistance of ESD footwear can be measured with commercially available footwear conductivity testers. The upper level is low enough to prevent the buildup of electrostatic charge in most situations, and the lower level offers some protection in the event of accidental contact with line voltage. This type of footwear is suitable for general use.
2. Conductive footwear has a very low resistance of typically less than 10^5 ohms (CENELEC, 2003). NFPA 77 (2000) states that this resistance should be less than 10^6 ohms. They are worn, for example, by persons who have to handle sensitive (low MIE) explosives or propellants. This type of footwear should not be worn where a possibility for electrocution by line voltages exists, and is not suitable for general use.

Socks or stockings do not normally adversely affect the properties of either type of the above footwear; however, some types of overshoes can do so (CENELEC, 2003).

Maintenance, Inspection, and Repairs of PPE

The proper storage, inspection, cleaning, and repair of protective clothing is important to the overall protection provided by the clothing to the wearer.

Some protective clothing will have storage limitations, such as a prescribed shelf life as well as requiring protection from ultraviolet light (sunlight), ozone, moisture, temperature extremes, and clothing folding. For example. suits made of natural products are usually susceptible to all of these environmental factors. Many of the encapsulating polymer suits can be damaged if folded rather than allowed to hang upright. The manufacturer or distributor should be consulted for any specific storage limitations its products may have.

Inspection of protective clothing should be performed frequently by the user and co-workers (e.g., on receipt, prior to use, and after each use). As a management policy, it is also advisable to have the worker's supervisor periodically inspect protective clothing items that are reused routinely (e.g., weekly). Inspection criteria will depend on the intended use of the PPE item; however, it would normally include examination for any obvious defects such as tears, holes, imperfections, and degradation.

Protective clothing cleanliness is extremely important, as PPE, if contaminated with health-hazardous solids, may become a source of exposure rather than being a protective measure. Cleaning of protective clothing or reuse must be performed with care. Natural fabrics can be cleaned with normal washing procedures provided that they are not contaminated with toxic materials. Synthetic fibers and materials commonly have cleaning procedure limitations. For example, some flame-resistant treatments will lose their effectiveness if the garments are not properly cleaned. Clothing used for protection against chemicals that are not water soluble cannot usually be decontaminated by simple water and soap washing. For example, tests performed on protective clothing used by pesticide applicators indicate that normal washing procedures are not effective for many pesticides.

Dry cleaning of some protective clothing is not recommended since it is commonly ineffective and can degrade or contaminate the clothing. It is important to consult the manufacturer or distributor of the PPE before attempting cleaning procedures for any protective clothing.

Laundering or otherwise decontaminating personal protective equipment and work garments can be a cost-effective option allowing reuse. Always check with a manufacturer to ensure that your method of decontaminating PPE or laundering work garments will not compromise their effectiveness. Hygienic issues must also be addressed when PPE and work garments will be used by different employees. Decontaminating PPE and work garments will almost always create a waste stream requiring proper disposal. It is important to check with environmental specialists to assure that the waste stream is collected and disposed properly.

If off-site laundering of work garments or decontamination of PPE is chosen as a preferred option, it is imperative to notify the vendor (cleaner) in writing of any possible contaminant that could harm their employees or compromise their waste management system. You may also have to assist them in selecting an appropriate method to protect their employees from the contaminant's hazard.

Most protective clothing is not repairable. Repairs can be made on some limited items such as fully encapsulated polymer suits. However, the manufacturer should be consulted for the proper repair procedures.

Ozimek et al. (1995) present a good discussion of personal protective equipment with many references. CCPS (1995a) also presents several tables of useful information on personal protective equipment. Other good sources of information on personal protective equipment can be found in AIHA (1990), Forsberg and Mansdorf (2003), and Mansdorf (2001).

Respiratory Protection

There are two major classes of respirators: air-purifying respirators (APRs) and supplied-air respirators (SARs). Air-purifying respirators remove contaminants from the air before they reach the respirator wearer's breathing zone. Supplied-air respirators provide clean breathing air from an uncontaminated source.

Different respirators provide different degrees of protection, which are characterized as Protection Factors (PF). PF is the expected level of respiratory protection that would be provided by a properly-functioning respirator or a class of respirators to properly fitted and trained users. For example, if a given respirator has a PF of 100, this means that the concentration of contaminant will be 100 times lower inside the facepiece versus the outside air.

AIR-PURIFYING RESPIRATORS (APRS)

APRs are grouped into three general types: particulate removing, vapor/gas removing, and a combination of the both. Respirators that remove particulates use filter media. Those that remove vapors/gases use chemical cartridges or canisters. Hereinafter, the term "cartridge" implies canister as well. Combinations of filters and chemical cartridges remove both particulates and vapors/gases. All filters and cartridges have an approved label that identifies the contaminants they remove.

Particulate-removing respirators are designed to reduce inhaled nuisance dusts, fumes, mists, toxic dusts, asbestos-containing dusts and fibers, or any combination of these substances. They may or may not have replaceable filters. These respirators may be non-powered or powered air-purifying respirators (PAPR). PAPRs use a blower to pull contaminated air through a filter and then supply the cleaned air to the face.

The National Institute for Occupational Safety and Health (NIOSH) has developed a set of regulations in 42 CFR 84 for testing and certifying nonpowered, air-purifying, particulate-filter respirators (commonly referred to as dust/mist/fume respirators in previous years).

The regulation provides for nine classes of filters (three levels of filter efficiency, each with three categories of resistance to filter efficiency degradation). The three levels of filter efficiency are 95%, 99%, and 99.97% (designated 100). The three categories of resistance to filter efficiency degradation are labeled N, R, and P, which signify:

- N—cannot be used in atmospheres containing oil
- R—removes particulates including oil aerosols, but can only be used for a single shift
- P—removes particulates including oil aerosols, recommended for a maximum of 40 hours or 30 days

The class of filter is clearly marked on the filter, filter package, and/or respirator box. For example, a filter marked N95 would mean an N-series filter that is at least 95% efficient. Chemical cartridges that include particulate filter elements will carry similar markings that pertain only to the particulate filter element.

Vapor/gas removing cartridges are designed with sorbent materials that adsorb certain contaminants from the air to prevent them from entering the breathing zone of the worker.

Protection from a mixture of both particulate and vapor/gas contaminants usually involves a combination of sorbent cartridges specific for a gas or vapor combined with a particulate-removing prefilter, such as a High Efficiency Particulate Air (HEPA) filter.

Air-purifying respirators are available in different facepiece styles. The most common types are half-mask and full-facepiece APRs. Both may be equipped with filters or cartridges. Filter media must be carefully chosen for the specific contaminant for which respiratory protection is required. Using the wrong filter or cartridge for an APR is essentially the same as wearing no respirator at all. Filters may be made of randomly laid nonwoven fiber materials, compressed natural wool, synthetic fiber felt, or fibrous glass that may be loosely packed in a filter container. Pleating increases filter surface area, which can improve filter loading and efficiency and lower breathing resistance.

APRs do not provide any air or oxygen from an outside source. They are strictly prohibited from being used in an oxygen-deficient atmosphere. In atmospheres which are considered to be Immediately Dangerous to Life or Health (IDLH), only supplied-air respirators can be used.

The life of respirator cartridges must be established and monitored, and the equipment must fit well enough to do the job required.

The sorbent material in cartridges can become saturated. At this point, the cartridge can no longer remove the contaminant it is intended to filter. This is called cartridge breakthrough. Breathing rate, contaminant concentration, temperature, and relative humidity are some of the factors that can hasten cartridge breakthrough. Cartridge breakthrough can be avoided by use of end-of-service-life indicators (ESLI) or by implementation of a change-out schedule.

Leakage around the facepiece of negative-pressure respirators permits contaminated air to enter the facepiece. Face seal leakage is addressed by the implementation of fit testing procedures. A good fit is always required to ensure the expected protection from tight-fitting APRs.

SUPPLIED-AIR RESPIRATORS (SARS)

Supplied-air respirators (SARs) supply breathing air to the wearer, independent of the ambient atmosphere. They are used when APRs are ineffective and in atmospheres where airborne concentrations of chemicals are high, the chemicals being used are very toxic, the concentration of the chemical is unknown, an oxygen-deficient atmosphere is present or is IDLH. Supplied-air respirators are also used for fire-fighting and rescue situations.

Self-Contained Breathing Apparatus (SCBA) are respirators that provide the atmosphere (breathing air) independent of the surrounding atmosphere. The wearers carry the Grade D breathing air supply with them. Air is supplied under continuous positive pressure to the facepiece.

Air-line respirators are respirators that provide the atmosphere (breathing air) through a hose which is connected to the wearer's facepiece. Grade D breathing air is provided from compressed-air cylinders or air compressors. The compressor air supply must:

- Be equipped with an alarm to indicate overheating and compressor failure;
- Have the air sampled for carbon monoxide if oil-lubricated, or have a carbon monoxide alarm; and
- Have a flow control valve, regulator, or orifice to govern the rate of airflow to the worker.

Air-line respirators are available in half- and full-facepiece styles. Some facepieces can be configured as APRs, SARs, or a combination of both. SCBAs are equipped with full-facepieces.

There are also different protection factors associated with different types of SARs and different facepieces for each type. Typically, SARs have higher protection factors than APRs, and therefore offer a greater level of protection.

In all cases, supplied-air respirators must provide air which meets or exceeds Grade D breathing air specifications (see Table 8-4). For SCBA cylinders and compressed-air cylinders, certification of Grade D breathing air should be obtained from the vendor when the cylinders are filled. For compressors, the air must be tested. Testing compressor air should be conducted according to the Compressed Gas Association (CGA) Document G-7.1-1989, or by a qualified outside contractor.

TABLE 8-4
Grade D Breathing Air Specifications

Limiting Characteristics	Maximum Concentration for Gaseous Air	Comments
Percent Oxygen Balance Predominantly N_2	atm[a] 19.5 to 23.5	Some states are more restrictive. The oxygen level upper limit in Cafifornia and Minnesota is 23.0 percent.
Water, ppm (v/v)		The water content of compressed air required for any particular quality verification level may vary with the intended use from saturated to very dry. For breathing air used in conjunction with self-contained breathing apparatus in extreme cold where moisture can condense and freeze, causing the breathing apparatus to malfunction, a dew point not to exceed –50°F (63 ppm v/v) or 10 degrees lower than the coldest temperature expected in the area is required. If a specific water limit is required, it should be specified as a limiting concentration in ppm (v/v) or dew point. Dew point is expressed in Degrees Fahrenheit at 1 atmosphere pressure absolute, 101 kPa.
Dew Point, Degrees Fahrenheit		
Oil (condensed) (mg/m^3 at NTP)	5	Not required for synthesized air whose oxygen and nitrogen components are produced by air liquefaction.
Carbon Monoxide (ppm)	10	Not required far synthesized air when nitrogen component was previously analyzed and meets National Formulary specification. Not required for synthesized air when oxygen component was produced by air liquefaction and meets United States Pharmacopoeia specification.
Odor		Specific odor measurement in air is impractical. Air normally may have a slight odor. The presence of a pronounced odor should render the air unsatisfactory for breathing.
Carbon Dioxide (ppm)	1000	Not required far synthesized air when oxygen component was produced by air liquefaction and meets United States Pharmacopoeia specification.

[a] atm (atmospheric) denotes the oxygen content normally present in atmospheric sir; the numerical values denote the oxygen limits for synthesized air.

Only two types of SARs are acceptable for use in IDLH atmospheres:

1. A full-facepiece pressure-demand SCBA certified by NIOSH for a minimum service life of 30 minutes, or
2. A combination full-facepiece pressure-demand supplied-air respirator with auxiliary self-contained air supply.

Air lines can limit mobility for the respirator user. This can be a problem, particularly in confined spaces or any area where movement is restricted or limited by hose length.

SCBAs and air-line systems are typically more expensive than APRs and cost more to properly maintain. SCBAs are heavy and can put a significant physical and psychological strain on the respirator user.

To use SARs properly, especially SCBAs, extensive training, maintenance, and testing are required. Tight-fitting SAR facepieces must be fit-tested in the negative pressure mode.

One should consult with the manufacturer, or review the latest edition of the NIOSH Guide to the Selection and Use of Particulate Respirators Certified Under 42 CFR 84 for the proper selection of respiratory protection equipment for a specific application.

Figure 8-2 shows several types of air-purifying and supplied-air respirators.

Legal codes and regulation for the use of respiratory protection can vary significantly between countries. Safety and industrial hygiene/occupational health specialists should be consulted to determine regulatory requirements for the use of respirators in the country of intended use. Requirements can include a written local respirator program; fit-testing, medical evaluation, and training of respirator users; country-specific listing and labeling of respiratory devices; and workplace evaluation of potential worker exposure.

8.3.8 Normal (Routine) Venting

Vent streams from breather vents (normal venting) are usually directed to the atmosphere (outdoors) by means of vent piping going through the roof or side wall of a building. This is preferable over indoor venting as it avoids dust entrained in the vent stream from depositing on equipment and internal building surfaces (e.g., floors, beams, etc.). As pointed out in Sections 6.8 and 7.5, dust accumulations can be ignited and cause fires and explosions. If the solids are either health- or environmentally hazardous, it is good safety practice to route the vent stream to dust separation or collection equipment prior to final outdoor venting.

8.3.9 Environmental Issues during Maintenance

Quite often maintenance work on equipment storing, handling, or processing particulate solids may result in the release of solids to the surrounding area. This is particularly of concern when the solids are health-hazardous, and could cause harmful health effects to plant personnel. It may also cause health problems for the plant's neighbors if the solids are borne outside of the plant fenceline by winds. Therefore, when maintenance work is planned,

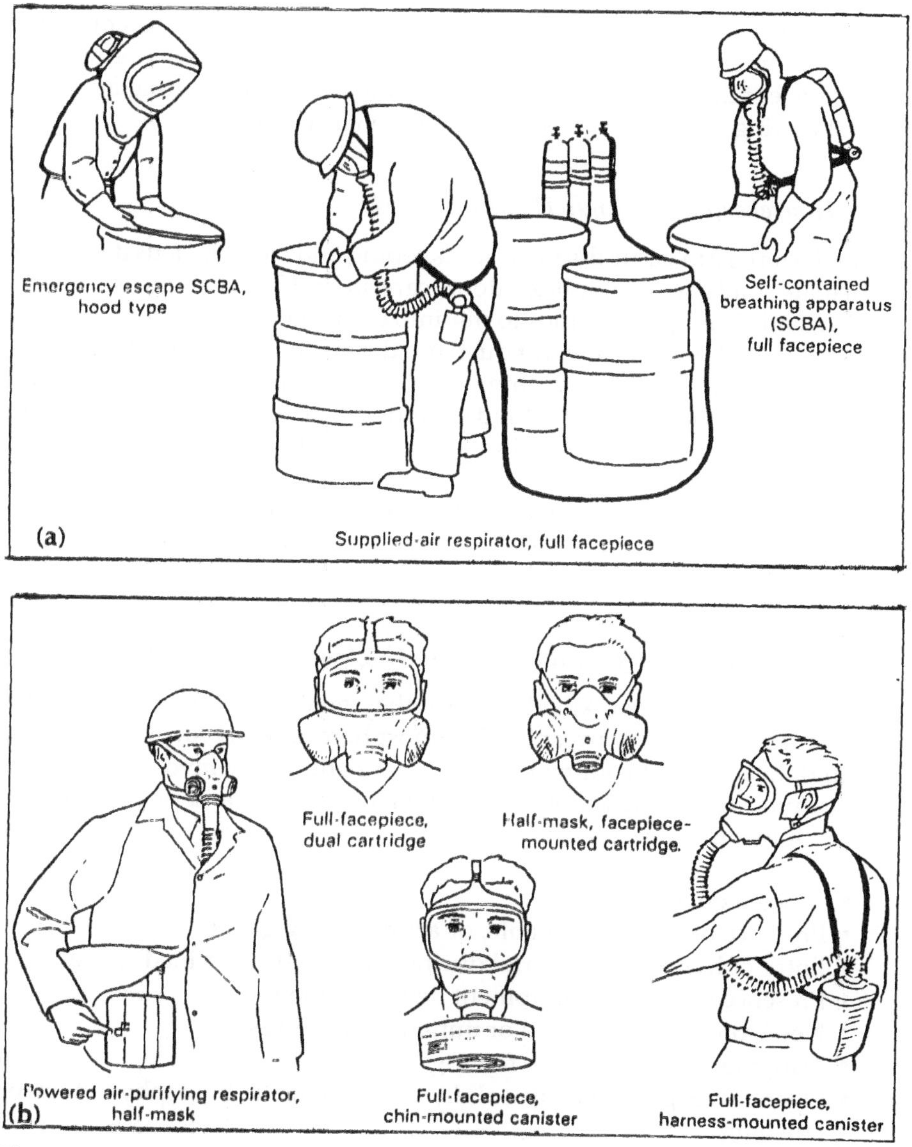

Figure 8-2 Types of respirators. (a) Atmosphere-supplying respirators. (b) Air-purifying respirators.

consideration should be given to the effect of solids emission and steps taken to minimize this from occurring. This may include such measures as:

- Thorough cleaning of the equipment to remove solids.
- Enclosing the equipment by wet tarpaulins to capture any emitted solids.

- Connecting a vacuum source and dust collection equipment to the equipment being worked on (keeping it under negative pressure) to prevent the solids from being emitted into the surrounding area.
- Wearing of appropriate PPE by maintenance personnel.

8.3.10 Housekeeping/Cleanup Health Hazards

Health hazards can occur during housekeeping and cleanup activities. If the particulate solids are health-hazardous, care should be taken during housekeeping and cleanup operations to minimize dust layer disturbances which could result in the suspensions of dust into the air, and thus increase personnel exposure. For example, brushing of floors and building surfaces should be done gently. The cleaning method should be tailored to the hazard. Washing or vacuuming (with HEPA filtration) the floor to remove the solids may be better than brushing.

It may be necessary for personnel that are involved in housekeeping and cleaning to wear personal protective equipment if the solids are health-hazardous.

8.3.11 Hazards of Asphyxiation from Inerting/Safe Vessel Entry

Many persons have been killed because they entered confined spaces that have contained a toxic or irrespirable atmosphere (usually nitrogen used for inerting). Examples of confined spaces whose entry may involve extra hazards are pits, dikes, excavations, control panels, sewers, and process vessels and equipment. Some hazards frequently encountered in confined spaces are:

- Oxygen deficiency
- Presence of toxic, flammable, or asphyxiating gases
- Entry of hazardous material during work in the confined space
- Presence of rotating equipment next to personnel
- Need to work in cramped or unnatural postures
- Tripping hazards or poor footing
- Operator engulfment by solids, causing death by asphyxiation or crushing

Asphyxiation, due either to an oxygen deficiency or to a toxic atmosphere, is the leading cause of death in confined spaces. Many of these fatalities occur among would-be rescuers who try to rescue workers already overcome in confined spaces. Proper training and the use of proper equipment are essential if such incidents are to be avoided. All employees should be trained in the hazards of improper entry and should be aware that the natural impulse to rush to someone's rescue without preparation may only aggravate the situation and may put the rescuer also at risk. The chemical hazards can be overcome by removing dangerous substances and isolating the confined space to prevent entry of hazardous materials.

Ness (2003) describes the following incident in which a maintenance worker was almost asphyxiated during maintenance work on a dust filter (baghouse) of an FIBC filling system.

Case History of Near-Asphyxiation during Maintenance Work

A shift craftsperson was requested to reseat an explosion vent door of an FIBC dust filter (located indoors). The employee approached the vent door via an explosion vent duct that was routed to the outside of the building. When the employee reached the door, he felt light-headed and came back out of the duct. He assumed that he had overexerted himself, and after a pause, re-entered the duct, and the problem reoccurred. He again exited the duct and reported the incident to his supervisor. Oxygen was checked inside the duct and found to be 20 vol. %. But on the inside of the baghouse (other side of the vent door) the oxygen level was found to be only 10 vol. %.

The FIBC filling system was composed of a vacuum blower to transfer powder from a product hopper through a dust filter (baghouse) and then through a rotary valve into an FIBC. The system had been shut down for about two hours with all process flows stopped by their automatic systems (for the rotary valve, blower motor, air and nitrogen solenoid valves). The work had been authorized by a safety permit, which authorized entry to the vent door via the vent duct. The vent duct is a horizontal duct, 1 meter wide, 1.2 meters high, and 3 meters long, open to the atmosphere. It was not considered a confined space entry since no entry inside the baghouse was planned. Entry in this way in the past had been normal practice.

On further investigation it was found that the solenoid valve in the nitrogen supply line to the rotary valve had failed in the open position and was passing nitrogen into the baghouse and then into the vent duct because the vent door was not properly seated. The hazard created by interconnected pieces of equipment was not recognized. Also, the vent duct was not considered a confined space, and appropriate vent duct atmosphere monitoring and nitrogen lockout were not done.

As a result of this incident, the following actions were taken:

1. The plant now treats all duct entries as a vessel entry with full isolation and testing of the baghouse.
2. The plant was surveyed for similar situations. To improve ease of isolation of all plant systems with nitrogen connections they were

reviewed and modifications were made to provide accessible positive disconnection.
3. The incident and action points to be taken will be reviewed with all operating, craft, and supervisory personnel. Shift supervisors and craftspersons were reminded of the requirements for isolation of upstream and downstream connections for confined space entry.
4. The plant reviewed its oxygen testing and confined space entry procedures.
5. The reason why the explosion vent door opens occasionally was analyzed.

OSHA requires permits for confined space entry as stipulated in 29 CFR 1910.146. This standard also describes the requirements for confined space atmosphere testing, personnel duties, personnel training, PPE to be used, and rescue and emergency services.

Oxygen concentrations should be at least 19.5 vol. % (OSHA requirement), unless supplied-air respiratory equipment is used. However, it should not be greater than 23 vol. %. Where possible, positive ventilation should be used and continued for as long as anyone remains in the confined space. Air should be supplied in such a way and at such a location that it must pass the worker(s) before exhausting to the atmosphere.

The air used to ventilate a confined space must come from a contaminant-free source.

When using nitrogen or carbon dioxide as an inerting gas, it should be recognized that if a leak occurs from the equipment being inerted into a closed area, this may expose operators to the potential for asphyxiation.

It should also be recognized that oxygen depletion may occur in a confined space due to reaction of a cleaning solution with solids residue in the equipment. Fatalities have been known to occur when tanks have been entered without appropriate checks for the oxygen level after cleaning with solutions containing caustic soda (sodium hydroxide). The caustic soda reacts with sugar residues and whey solids to form carbon dioxide (Anon, 2002).

Two sources of information on safe entry into confined spaces are ANSI Standard Z117.1 (ANSI Z117.1, 2003) and Hodson (2001).

The U.S. Chemical Safety and Hazard Investigation Board has recently published a safety bulletin on the hazards of nitrogen asphyxiation, which contains statistics, case histories, and good practices for safe handling of nitrogen (CSB, 2003).

8.3.12 Design and Operation of Isolation Rooms

The concept of a clean room (where great effort is taken to keep contaminants out of the work area) is commonly understood. There also exist applications for isolation rooms, which by contrast, are designed to completely contain whatever is manufactured within the confines of the room. Such rooms are sometimes designed for highly potent pesticides. The concept is simple, but the execution demanding. These rooms would have features such as:

- Airlocks at all points of entry and exit
- Maintaining the building or room interior at a slightly lower pressure than atmospheric to have air leaks into the building/room to prevent fugitive losses of potent materials
- Sticky mats on the floor to remove residual material from feet prior to entering the airlock from inside of the building/room
- Specific isolation rooms for processing materials within the building
- Equipment, such as packaging machines, that are isolated by containment enclosures from the rest of building/room, and equipped with self-contained recycle-ventilation containing HEPA filters in the ventilation loop. Some companies have "form and fill" packaging machines that capture the air and fugitive powder displaced in filling the package. Some companies also have two operators present, one redundant, to observe the filling operation to ensure that toxic materials stayed in the package during filling
- Building operators wear PPE, consisting of respirators, oversuits (coveralls), gloves, overshoes (overbooties)
- Change rooms in which workers don and doff protective clothing after they enter the building, but before they enter the isolation rooms
- Complete capture of all waste water produced in the process and, possibly, lavatories, etc.
- Air handling systems that largely recirculate conditioned air and that have very sophisticated filtration and filtration failure alert devices such as light scattering detectors
- Low-maintenance equipment and provisions for decontamination of all equipment prior to removal for maintenance that includes collection of all wastes

8.3.13 Design and Operation of Cleanrooms

Cleanrooms are frequently used in the final finishing steps of the manufacture of pharmaceuticals to prevent contamination of the product. However, since many products handled in cleanrooms are health-hazardous, they also must be designed to safely handle the products to protect the workers.

Certain cleanroom operations involve the use of flammable liquids and hazardous powders that may cause fires and explosions. Fire protection and loss prevention techniques are discussed later on in this section.

ISO 14644-1 (1999) defines a cleanroom as "a room in which the concentration of airborne particulates is controlled and which is constructed and used in a manner to minimize the introduction, generation, and retention of particles inside the room and in which other relevant parameters, e.g., temperature, humidity, and pressure are controlled as necessary."

Cleanroom Classification

Cleanrooms are classified by the cleanliness of their air. Two classification standards are used in the United States: US Federal Standard 209E (FED-STD 209E, 1992) and ISO 14664-1 (1999). Federal Standard 209E classifies cleanrooms by the number of particles equal to or greater than 0.5 microns measured in one cubic foot of air, for example, Class 1 contains 1 particle per cubic. foot. This standard has Classes 1, 10, 100, 1000, 10,000, and 100,000. This standard was administered by the General Services Administration (GSA) in Washington, DC. The GSA assigned the administrative and technical responsibility for updating Federal Standard 209E to the Institute of Environmental Sciences and Technology (IEST). The GSA released a Notice of Cancellation for FED-STD 209E in November 29, 2001, as much of its information is now outdated. The GSA has recommended that ISO 14644-1 supersede FED-STD-209E.

ISO 14644-1 is an improvement over FED-STD-209E in that air cleanliness criteria have been expanded to include such operations as powder milling and filling applications. For example, the ISO Class 9 clean criterion, which allows for up to 35.2 million particles of 0.5 microns and larger per cubic meter (comparable to 1 million particles of 0.5 microns and larger per cubic foot in a controlled space) is appropriate to certain bulk pharmaceutical operations. Conversely, ISO Class 1 and Class 2 air cleanliness criteria allow for the measurement of smaller numbers of 0.1 micron particles required for better quality microelectronic production. ISO 14644-1 has nine classes (see Table 8-5). Table 8-6 shows a comparison of the FED-STD-209E and ISO 14644-1 Classes. Farquharson (1999) discusses contamination control standards presently published, or being developed by the CEN, ISO, France, Germany, Japan, and the United Kingdom.

The required standard of cleanliness of a room is dependent on the task performed in it; that is, the more susceptible the product is to contamination, the more stringent the standard. Table 8-7 gives an indication of the operations carried out in different classifications of cleanrooms.

Most pharmaceutical cleanrooms are designed to an ISO 14644-1 Class 5 (FED-STD-209E Class 100) criterion.

Types of Cleanrooms

Cleanrooms can be divided into four main types, as follows:

TABLE 8-5
Classes of Air Cleanliness According to ISO 14644-1

Classification Number	Maximum Concentration Limits (particles/m³ of air) for Particles Equal to and Larger Than					
	0.1 μm	0.2 μm	0.3 μm	0.5 μm	1 μm	5 μm
ISO 1	10	2				
ISO 2	100	24	10	4		
ISO 3	1,000	237	102	35	8	
ISO 4	10,000	2370	1,020	352	83	
ISO 5	100,000	23,700	10,200	3,520	832	29
ISO 6	1,000,000	237,000	102,000	35,200	8,320	293
ISO 7				352,000	83,200	2,930
ISO 8				3,520,000	832,000	29,300
ISO 9				35,200,000	8,320,000	293,000

TABLE 8-6
Comparison of ISO 14644-1 and U.S. Standard 209E

ISO 14644-1	U.S. Standard 209E
1	—
2	—
3	1
4	10
5	100
6	1,000
7	10,000
8	100,000
9	—

1. Conventional: these cleanrooms are also known as turbulently ventilated or non-unidirectional flow and are distinguished by their method of air supply (see Figure 8-3). As can be seen in the figure, the air is supplied by air supply diffusers in the ceiling.
2. Unidirectional flow: this was previously known as laminar flow. As can be seen from Figure 8-4, clean air is supplied from a bank of high efficiency filters and passes in a unidirectional manner through the room.
3. Mixed flow: as shown in Figure 8-5, this type of cleanroom is conventionally ventilated, but where the product is exposed to contamination, a unidirectional flow cabinet or workstation is used.

TABLE 8-7
Possible Cleanroom Classification Requirements for Various Operations

Class 1	These rooms are only used by integrated circuit manufacturers manufacturing sub-micron geometries.
Class 10	These rooms are used by semiconductor manufacturers producing integrated circuits with lime widths below 2 μm.
Class 100	Used when a bacteria-free or particulate-free environment is required in the manufacture of aseptically produced injectable medicines. Required for implant or transplant surgical operations. Isolation of immunosuppressed patients, e.g. after home marrow transplant operations.
Class 1,000	Manufacture of high quality optical equipment. Assembly and testing of precision gyroscopes. Assembly of miniaturized bearings.
Class 10,000	Assembly of precision hydraulic or pneumatic equipment, servo-control valves. precision timing devices, high grade gearin
Class 100,000	General optical work, assembly of electronic components, hydraulic and pneumatic assembly.

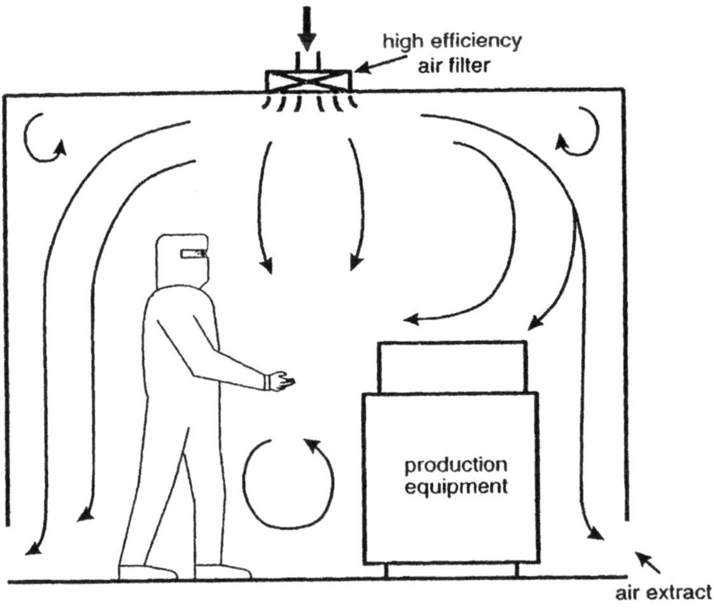

Figure 8-3 Conventional (non-unidirectional) airflow cleanrooms.

4. Isolators or microenvironment: these are used within a cleanroom to give the highest level of protection against contamination (see Figure 8-6). As shown in the figure, the isolator has a unidirectional supply of air, but this may be a conventional turbulent-flow type.

More details about these four types of cleanrooms are presented by Whyte (1999).

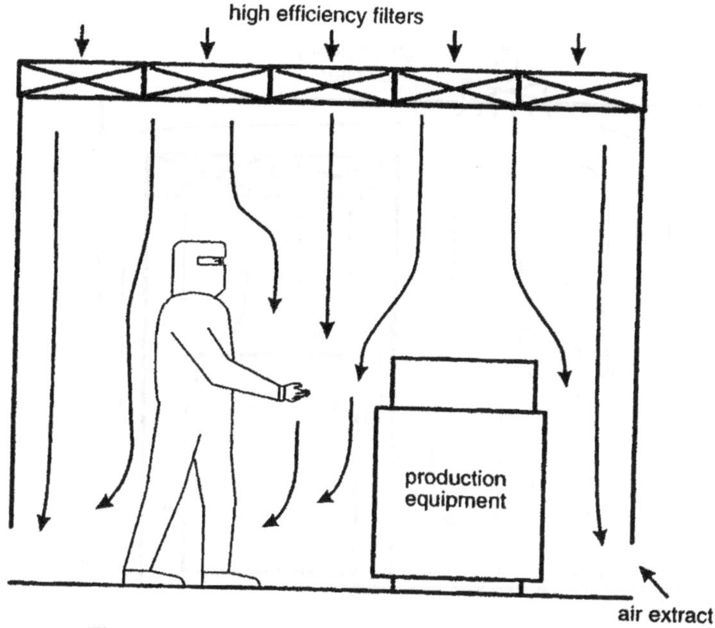

Figure 8-4 Vertical unidirectional flow cleanrooms.

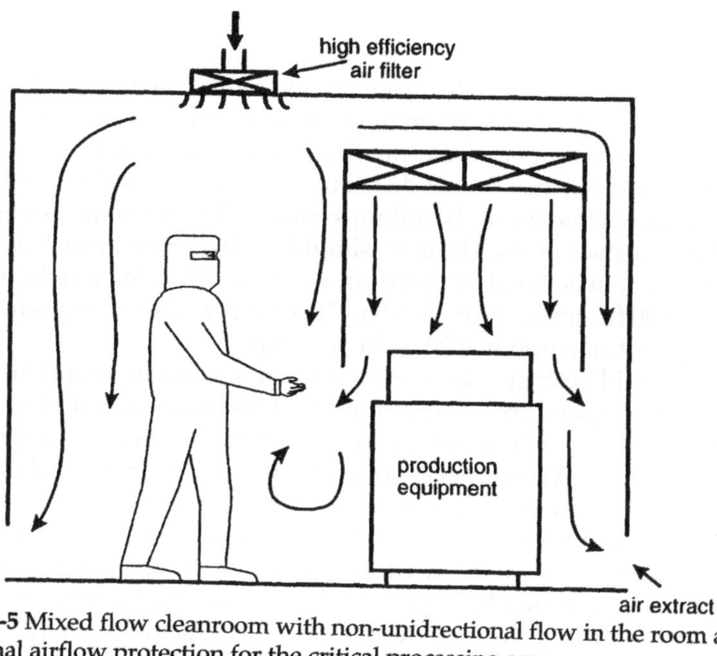

Figure 8-5 Mixed flow cleanroom with non-unidrectional flow in the room and unidirectional airflow protection for the critical processing area.

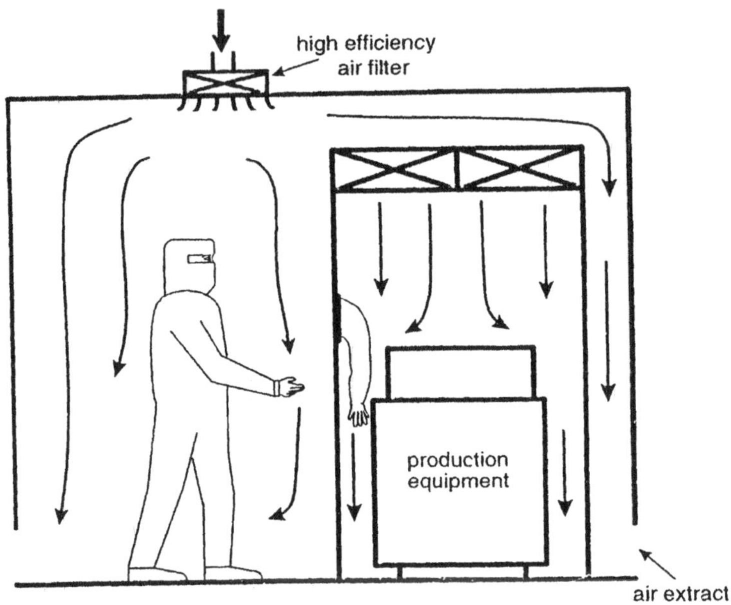

Figure 8-6 Isolator used to protect a process area.

Air Handling Systems

The cleanliness achieved by a cleanroom is dependent on the ability of the air handling system to purge the room of airborne contaminants. This includes not only the effectiveness of the air filters and the number of room air changes per hour, but also the air distribution patterns within the room.

The key to an effective air handling system is low air velocities entering through large surface areas. The exits should be dispersed in such a manner that the entire room is swept by clean air. Air flow velocities should be in the range of 60 to 90 ft/min (0.3 to 0.45 m/s). The number of air changes for a conventional type of cleanroom is 20 to 50 per hour.

The air should be controlled with respect to temperature and humidity. Uniformity of temperature is very important to make sure that necessary operating machinery tolerances are maintained, as well as comfortable working conditions. With respect to operator's comfort, actual body temperature itself will depend a great deal on the clothing worn, and the particular process involved. Dry bulb temperatures may vary from 72°F to 76°F, as far as design conditions are concerned, but once the design temperature has been established, it should be kept to ±2°F of the selected temperature.

Relative humidity is usually kept within the 35% to 55% range. Higher humidity may cause rusting of equipment and lower humidity are not desirable because of the static electricity likely to be generated in such atmospheres, and of course, the operator comfort. Although certain hygroscopic powders may need humidity in the relatively low range of 15% to 35%, cleanroom air may be dehumidified by using commercially available chemical process dehumidifiers.

Air Filters

Cleanroom filters should normally have an efficiency greater than 99.97% in removing particles greater than 0.3 microns from the room air supply.. This is accomplished by using filters known as High Efficiency Particle Air (HEPA) filters. They are usually located at the point of air discharge (entry) into the room. High efficiency prefilters may be installed at their conventional location before spray coil units, but the HEPA filter must be the very last element in the mechanical lineup. This assures that all air to the cleanroom is filtered, and there is minimum possibility of contamination with unfiltered air. HEPA filters are normally used with a maximum pressure drop of 4 to 10 inches (10 to 25 cm) water gauge. This allowable pressure drop must be provided by the fan. Therefore, a fan must be selected that is capable of delivering the desired airflow against the total resistance of the ductwork and the HEPA filter pressure drop. To avoid air volume decrease during operation as the HEPA filter loads up with particles, a fan should be selected which has a steeply rising pressure-volume curve to provide a reasonably constant flow during dust loading.

It is very important that the HEPA filers be installed in such a way as to prevent any possibility of leakage around them. A prefilter is frequently installed upstream of a HEPA filter.

Cleanroom Pressure

Cleanrooms are built to be as "dust tight" as possible. However, to prevent the accidental infiltration of any non-filtered air, it is standard practice to positively pressurize the cleanroom with respect to other "dirtier" areas. A positive pressure of not more than 0.10 inch (0.25 cm) water gauge between the cleanroom and uncontrolled areas will be adequate for most applications. If more than one room is involved, it may be necessary to raise the pressure so that the rooms may be staged from the most closely controlled room to the least controlled room. Where rooms are staged, a 0.05 inch (0.127 cm) water gauge differential pressure is sufficient between rooms. Usually, the operation having the most critical requirements is most highly pressurized, and there is a reduction in pressurization level outward from this most critical area to provide maximum isolation of critical processes.

Since make-up air must be added to cleanroom atmospheres for the physiological needs of the operators, some air will have to be expelled from the system. A minimum of 200 cubic feet of fresh air per hour per operator is usually required. In many areas, local government regulations will mandate the quantity of fresh air that is required.

Pressurization control is achieved by the use of pressure sensing devices which automatically supply additional air to any area of a cleanroom where the pressure may drop because of an open door, or any other cause. Pressure losses can be reduced by the use of air locks between rooms, and between the cleanroom and the rest of the building. When air locks are not used, traffic between rooms may be kept to a minimum by using double-door pass-through boxes for the transfer of material.

Cleanroom Construction

Cleanrooms are built using traditional building construction techniques, as well as specially developed prefabricated systems. The construction should be airtight to prevent any air leaks into or from the room. The materials of construction of the surfaces facing into the cleanroom should have the following characteristics:

- Be free of discontinuities, openings, pits, porosity, crevices, etc. by which contaminating material can be retained or enter. Be resistant to abrasion or other damage in the normally expected usage.
- Resist deterioration leading to particle formation in sizes potentially harmful, either through exposure to the normal environment, or through exposure to decontamination procedures and materials.
- Possess dielectric properties suitable for dissipation of electrostatic charges.
- Provide or permit needed acoustic treatment.
- Provide needed light reflectance qualities, including color.
- Be readily joinable to other materials.
- Be producible in usual shapes and sizes.
- Be repairable or replaceable during the life of the facility.
- Be dimensionally stable.
- Be capable of absorbing building strains or movement.
- Provide needed control of temperature and vapor transmission.

Some of these desired qualities are not attainable in the surface material alone, but can be achieved as a result of proper combinations of surface and substrate, or the proper selection of supporting materials. This is particularly important in matters of acoustic properties, heat transmission, vapor transmission, and dimensional stability. The need to properly support or supplement the surface material requires consideration of the characteristics of all structural materials having a bearing on the performance of the cleanroom.

Dust-reducing features within the basic enclosure include the rounding (filleting) and covering the floor-walls-ceiling joints, proper seals, minimum possible horizontal surfaces, recessed lighting fixtures into the ceiling, and providing services (water, steam, air, etc.) from above. Concealed, but accessible piping should be installed to avoid gathering of any dust on them.

Entryways, doors, and pass-throughs should be of the correct size to permit personnel and required equipment access to the cleanroom. These entryways should be of the double air lock type and should provide air seals sufficient to allow pressurization of the cleanroom.

Anterooms should be provided for cleanroom personnel clothing change area, clothing storage, wash-up facilities, air showers, and other equipment for personnel cleanroom entry requirements. Anterooms may also be provided, as needed, to house cleaning equipment for parts cleaning needs.

High velocity air showers to air brush (clean off) the lint from the personnel clothing should be provided. In critical areas, air showers interlocked with the air locks should be used.

Fire Protection/Loss Prevention

As mentioned at the beginning of this section, certain cleanroom operations involve the use and handling of flammable and/or combustible liquids and powders. These cleanrooms require appropriate fire protection/loss prevention design.

Protection of cleanrooms is discussed in NFPA 318 (2002). Although this standard is specifically concerned with the protection of semiconductor fabrication facilities, many of the recommendations can be applied to the protection of pharmaceutical cleanrooms as well. Another very good source of information on protection of cleanrooms is presented in FMG 1-56 (2000).

A number of good practices from the above two standards for fire protection/loss prevention of cleanrooms are presented below:

- Locate cleanrooms in a fire-resistive or noncombustible building or area.
- Cleanrooms rated Class 100 or cleaner shall have approved, noncombustible components for walls, floors, ceilings, and partitions. Where noncombustible materials cannot be used, use materials that have been certified to achieve a specified level of fire resistance and to have passed certain supplemental tests [e.g., FM Approvals Cleanroom Materials Flammability Test Protocol (Class 4910)]. Most modern fire resistant industrial construction consists of reinforced concrete or insulated steel frame. Commercial insulations that have been certified to achieve a specified level of fire resistance are listed in the Factory Mutual Approval Guide and/or Underwriters Laboratories Fire Resistance Directory, or in certification listings of other testing organizations.
- Separate cleanrooms from adjacent occupancies having no greater than ordinary hazards, by minimum 1-hour fire resistance-rated construction. Locate adjacent occupancies with greater than ordinary hazards in separate buildings. Where there are viewing windows

provided in a cleanroom exterior wall, along an adjacent hallway, construct the hallway of noncombustible materials and use a rated assembly for the hallway wall opposite the windows.

- Provide noncombustible, sealed, interior subdivision walls in cleanrooms to limit the spread of smoke and other contaminants in the event of an accident. In vertical unidirectional flow cleanrooms these subdivision areas should run from the underside of the roof to the cleanroom floor; where a return air space exists below the cleanroom floor, these subdivisions should extend to the return air flow. Where cleanrooms adjoin, use 1-hour rated fire partitions.
- Tightly seal all utility penetrations in solid noncombustible floors and ceilings with approved fire stop materials.
- Seal any penetrations through rated wall systems using approved wall penetration fire stops.
- Avoid any liquid piping and fittings above cleanrooms. When liquid piping above a cleanroom cannot be avoided, locate and protect pipes to prevent liquid damage to the cleanroom in the event of a leak.
- Use noncombustible insulating materials, such as mineral wool or expanded glass, for all needed piping insulation.
- Wet pipe automatic sprinkler protection shall be provided throughout facilities containing cleanrooms and clean zones in accordance with NFPA 13 (2002). The sprinkler system shall be hydraulically designed for a density of 0.20 gpm/ft^2 over a design area of 3000 ft^2.
- Approved quick-response sprinklers shall be utilized for installations within down-flow airstreams in cleanrooms and clean zones. Provide each cleanroom with means to remove smoke during a fire. This will minimize the spread of contaminants within the cleanroom.
- Smoke removal should be provided either by a dedicated smoke control system or by an arrangement of the cleanroom air handling or fume exhaust system. Specific selection and design details are given in FMG 1-56 (2000), Zalosh (2003), NFPA 92A (2000), and NFPA 204 (2002).

IEST and Matthews (2001) discuss global standards for cleanroom design and construction. Other detailed discussions of cleanrooms are presented by Austin (2000), Cole (1998), ISO 14644-2 (2000), ISO 14644-4 (2001), Ramstorp (2000), Whyte (2000), and Whyte (2001).

8.4. NONROUTINE OPERATIONS CONSIDERATIONS

8.4.1 Emergency Venting

Vent streams from emergency vents are normally directed to the outdoors. This is done to avoid causing a secondary explosion inside a building, which could occur if there are dust deposits in the building. Other considerations include protection of personnel and equipment indoors. As discussed in Section 6.6.1, the vent stream should be directed to a safe location where it will not impinge on personnel or other process equipment. Emergency venting can be done inside of a building if the equipment being vented has a flame arresting/particle retention device such as the Q-Rohr™ or FlamQuench II™ attached to the vent opening. Before such a device is used, it should be tested and certified to see if it will effectively quench the flame, control the emission of particles, and cool the gases to a safe temperature (see Section 6.6.1 for a discussion of these devices).

8.4.1.1 Options for Controlling Emergency Releases of Toxic Particulates

The release of toxic or environmentally harmful particulates should be avoided or minimized by employing any of several protective measures, as follows:

1. Provide explosion suppression systems rather than venting (see Section 6.6.2).
2. Design the equipment for explosion pressure containment (see Section 6.6.3).
3. If explosion suppression or pressure containment is not feasible, and venting is necessary, the vent stream should be directed to a catchtank, scrubber, or other containment device to minimize emissions. Special testing and/or analyses may be needed for this type of deflagration vent effluent stream control system design.

8.4.1.2 Control of Products of Combustion and Decomposition

Emission of products of combustion is best avoided by ensuring that combustion does not occur. This often can be accomplished by eliminating sources of ignition, using interlocks to shut down the operation to avoid ignition, inerting, or air dilution (reducing the combustibles concentration). As a final control measure, it may be possible to design the equipment for explosion containment or use explosion suppression so that an explosion does not occur.

The same measures may be taken to control the emission of products of decomposition as can be done to control the emission of products of combustion. However, sometimes a decomposition may evolve hazardous gases which could be combustible or toxic. In this case, appropriate measures should be taken to control these gases from igniting or contacting plant personnel. For example, if a vent stream contains toxic gases in the products of decomposition, the vent stream can be directed to a quench tank that contains a chemical that will neutralize the toxic gases (convert them into nontoxic products).

8.4.2 Measuring the Impact of a Nonroutine Release

When a nonroutine release occurs, an analysis should be done to assess the consequences, such as harmful effects on adjacent equipment, plant personnel, and neighbors in the vicinity of the plant.

To estimate the impact on employees and process equipment from a vented explosion, equations are given in Section 6.6.1 for calculating the fireball length and the maximum pressure external to a vented enclosure. The effects of fire (human injury) can be estimated by equations given in Lees (1996).

If a health-hazardous particulate solid could be released, its impact should be evaluated. Marchello (1976) discusses a method for estimating the dispersion of particulate solids and presents an illustrative sample calculation. Another discussion of toxic particulate solids release scenarios and techniques for the estimation of their dispersion is presented in an EPA report (EPA, 1992).

A number of computer models are available for estimating the dispersion and deposition of particulate solids, such as the TSCREEN model (EPA, 1992), CALPUFF model (Seire et al., 1999) and the Particle Trajectory Model (Vesovic et al., 2001).

8.4.3 Permitting and Reporting Issues for Emergency Vents

Vent permits are not normally required for emergency vents. However, if an emergency venting occurs, the incident must be reported to the appropriate authorities. An investigation of the incident must be done and a report issued.

8.4.4 Emergency Response for Accidents with Powders and Dusts

OSHA and other local regulations require that an employer shall establish and implement an emergency action plan for the entire plant in accordance with the provisions of 29 CFR 1910.38(a). 29 CFR 1910.38(a) lists the designated actions (plan preparation and emergency responses) that an employer must take to ensure employee safety from fire and other emergencies (e.g., explosions; spills; atmospheric releases of toxic, combustible, and corrosive particulates; etc.).

Emergency response may be required for:

- Dust explosions
- Dust fires
- Large releases of particulate solids (loss of containment)

The US Department of Transportation has issued a guidebook for emergency response to hazardous releases, which contains information on what actions to take for many particulate solids releases (DOT, 2000). Another good source of information on emergency response is NFPA 471 (2002). When responding to spills and atmospheric releases of hazardous particulate solids, care must be taken to ensure that the materials used to control the emitted hazardous solids are not incompatible (do not react) with them. A good reference source on information about chemical reactivity and compatibility is the book by Drum (2002).

8.4.4.1 Employee Safety in an Emergency

Employee safety must be ensured during an emergency. This may involve such actions as:

- Shutting down the process in a safe and orderly manner.
- Establishing escape procedures and emergency escape routes and assignments.
- Creating safe havens to which employees can retreat when evacuation is not possible.
- Establishing procedures to be followed by employees who remain to operate critical plant operations before they evacuate.
- Establishing procedures to account for all employees after emergency evacuation has been completed.
- Assigning and training of personnel who are to perform rescue and medical duties.
- Training of employees in the use of personal protection equipment that may need to be used (e.g., when health-hazardous particulate solids are involved).
- For situations involving spills and releases of hazardous particulate solids, emergency response activities should be spelled out in the emergency preparation plan of the facility.

For emergency response activities, three work zone areas should be set up (CCPS, 1995b):

- Hot zone
- Warm zone
- Cold zone

The hot zone, also called the exclusionary zone, is the area where the incident takes place. The size of this area will depend on the size of the release and what mitigation activities will take place. Only properly trained and equipped personnel should work in this area. The warm zone, an area encircling the hot zone, is also known as the decontamination or limited access zone. This is a buffer zone between the hot and cold zones where decontamination will occur. Decontamination occurs in an area of the warm zone referred to as an access corridor. Only trained decontamination personnel and the safety officer can work in this zone. The third work zone is the cold zone (also known as the support zone), which is the area for the command post and staging area. This zone must be secure and only response personnel and necessary advisors should be in this area.

Extinguishing of dust fires is often more complicated than for liquid fires, and the method of extinction often depends on the type of particulate solid. Dust fires may be extinguished by letting the fire burn itself out, by applying extinguishing agents, or by starving the fire of oxygen. Deep-seated fires, such as those that can occur in silos or baghouse hoppers, are often difficult to deal with.

Water is the usual extinguishing agent and is suitable unless it reacts with the dust or if there is electrical equipment involved or in close vicinity to the fire. Water should not be applied as a high pressure jet which could raise a dust cloud, but as a low pressure spray which simply dampens the dust deposit. Flooding with water may not be appropriate to extinguish a fire in equipment if the equipment cannot support the weight of the accumulated water, unless drain connections are available at the bottom of the equipment.

Carbon dioxide may be used in some cases to extinguish smoldering fires (see Sections 5.3.15 and 6.10.3).

Dry extinguishing powders may be used if the dust is one that reacts with water, as some metal dusts do.

Inert gases may be used as extinguishing agents and can be used effectively if the solids are contained in a relatively gas-tight container such as a silo or hopper. Some solids react with nitrogen, so the inert gas should be compatible with the solids. It is necessary, however, not only to cut off the supply of oxygen, but also to provide sufficient cooling to prevent reignition when the air supply is restored. Thus, the inerting may have to be maintained for a long period.

Whichever approach is used, it is essential to avoid disturbing the dust in such a way that would allow a suspension (dust cloud) to form and ignite.

After a fire has been put out, care should be taken when emptying a silo or baghouse hopper to ascertain that the fire is completely out, and that the solids are still not smoldering. This precaution is necessary to prevent a smoldering fire reigniting from contact with air.

If a fire occurs in a baghouse, fans, blowers, or other sources of dust removal should be shut off immediately to cut off the supply of air (oxygen) to the fire.

8.4.4.2 Postemergency Cleanup

Following an incident in which hazardous particulate solids are released into the plant or into the environment, an important factor that should guide recovery and cleanup is time. If too much time elapses, ultimate cleanup cost might increase, if, for example, the hazardous solids may migrate deeply into the ground. Also, prompt attention should be given to cleanup of emergency response equipment that may have become contaminated during the response operation.

Manufacturers may provide emergency response and cleanup procedures in the MSDSs and other company literature, and these procedures should be followed. Federal agencies (DOT and the U.S. Coast Guard) also provide emergency response guidance.

The extent of contamination from particulate solids is typically less significant than other forms (liquids, slurries, mists, or gases/vapors), although the localized levels of contamination may be orders of magnitude higher than liquids or gases/vapors.

Generally, the highest level of contamination will be closest to the point of release. Unlike deposition caused by mist, gases or vapors, most of the contamination from particulate solids will deposit on horizontal surfaces. This form of contamination is easily disturbed by physical contact, rain, wind, and a building ventilation system. This form of contamination, in most cases, is also the easiest to remove.

Before entering a contaminated area to perform cleanup, appropriate work areas must be prepared. These work areas are the same as those used during the emergency response phase, i.e., the hot zone, warm zone, and cold zone. During the decontamination activities a control or check point must be established and entry or exit strictly regulated.

Depending on the size of the release, small-scale or large-scale decontamination activities are required. Small-scale decontamination of facilities and equipment can generally be accomplished using ordinary cleaning and decontamination methods. Such methods include the use of mops, brooms, soap and water, sponges, vacuum cleaners, etc. In some cases, specific decontamination solutions may have to be prepared to deal with contaminants possessing special characteristics. As an example, water is not appropriate to clean up water-reactive solids. However, it is advantageous to use water to clean up solids that dissolve readily in water. Buckets, garden sprayers, scrub brushes, and sponges are commonly used to perform many small-scale decontamination activities. Water used in decontamination and spill cleanup should be disposed of in an appropriate manner.

Large-scale decontamination is a two-phase operation. The first phase will remove the gross levels of decontamination over large areas. This phase may consist of manually removing debris, scraping surfaces with heavy equipment, washing down floors with a fire hose, or using a vacuum cleaner to remove the solids. A round of sampling is normally conducted after this gross decontamination to determine the next step. Typically, the second phase consists of localized small-scale decontamination as described above. In either case, provisions must be made to dispose of debris and hazardous materials in accordance with local regulations.

8.4.5 Determining the Cause of a Protective System Activation

A protective system (e.g., automatic fire suppression system, explosion suppression system, or explosion vent cover) may be activated intentionally to provide protection against a fire or an explosion, or inadvertently due to some malfunction, mechanical fault, or external effect. It is very important to determine whether the activation was caused by an explosion (deflagration), or by an inadvertent effect such as by a "spurious trip." If it was caused by an explosion, this may be indicative of an inherent process hazard, and a thorough investigation should be conducted to determine the root cause of the explosion. An activation cause often can be determined by visual observation of the equipment, internally and externally. For example, an explosion usually leaves tell-tale marks such as coloration of the metal due to the extreme heat, and/or soot deposits.

The following investigation steps should be taken to determine the cause of the activation of an explosion (deflagration) suppression system and an explosion vent cover (Garzia, 2002):

Fire or Explosion (Deflagration) Suppression System

If a fire or an explosion suppression system is activated, an investigation of the causes should be conducted. This is quite often done with the assistance of a field engineer from the suppression systems supplier.

The following should be done by the investigating person for both an activation caused by a failure event and also for an inadvertent activation (Garzia, 2002):

ACTIVATION CAUSED BY A FIRE OR AN EXPLOSION

1. Record all process operating data at the time of the activation, that is, what was occurring at the time of the activation.
2. Record the status of the suppression system control unit (if still energized).
3. Determine which detector(s) activated the system.
4. Check electrical and mechanical conditions.
5. Check interlocks.
6. Record statements from personnel witnesses in the area.
7. Note weather conditions.
8. Take pictures of the accident scene to the greatest extent possible.

ACTIVATION CAUSED BY AN INADVERTENT EFFECT

1. Run pressure tests to duplicate process conditions (if possible).
2. Gather all information to determine the cause, i.e., change in process operating conditions (pressure or air flow), change in process equipment, or any other change.

3. Provide recommendations to correct any problems as soon as possible, before resuming production.

For all investigations, a review should be made to determine that the process equipment and/or the suppression system has not been changed or modified from what is shown on the system drawings.

Explosion (Deflagration) Vent Cover

Explosion vent covers, such as rupture disk diaphragms, occasionally are activated inadvertently (fail prematurely) due to a number of causes, such as:

1. Thermal fatigue (the disk is operated at temperatures above the vendor's recommended maximum).
2. The disk has pinholes in it.
3. The disk has been damaged during installation.
4. The disk is installed incorrectly (e.g., in reverse).
5. Excessive equipment vibration.
6. The disk, which is used in pressure/vacuum service, does not have a vacuum support.
7. Corrosion.
8. Wind buffeting of the vent cover.
9. Installation of incorrectly rated vent cover.

The investigator should consider all of the above possible causes to ascertain the cause of the inadvertent activation. Internal inspection for charring and other evidence of a fire or explosion should also be done. Corrective actions should be taken to avoid this from reoccurring.

8.4.6 Disabling of Protective Systems by an Explosion

An explosion occurring in an item of process equipment possibly can disable protective systems on other adjacent process equipment. For example, the fireball and pressure wave from a vented explosion impacting on an explosion suppression system on another nearby item of equipment can damage and disable it. This is why it is very important that an explosion vent be located so that it discharges into a safe location, that is, one that does not allow contact with other equipment or plant personnel. When any protective system is disabled, the equipment on which it is installed should be shut down, if necessary, and the protective system repaired or replaced, and then reactivated.

An explosion may also disable sprinkler and other safety systems. All protective systems that might have been affected should be checked out and recertified before startup.

REFERENCES

AIHA. 1990. *Chemical Protective Clothing*. Akron, OH: American Industrial Hygiene Association.

Anon. 2002. Unexpected Oxidation of Empty Vessel Causes Fatality. *Loss Prevention Bulletin*, Issue 168, pp. 23 (December 2002). Rugby, UK: Institution of Chemical Engineers.

ANSI Z117.1 2003. *Safety Requirements for Confined Spaces*. Washington, DC: American National Standards Institute.

Austin, P. R. 2000. *Encyclopedia of Cleanrooms, Bio-Cleanrooms, and Aseptic Areas*. 3rd edition. Englewood, CO: Interpharm Press,

BS 5958. 1991. *Code of Practice for Control of Undesirable Static Electricity, Part 2: Recommendations for Particular Industrial Situations, Section 31*. London: British Standards Institute.

Carson, P. A. and Mumford, C. J. 1988. *The Safe Handling of Chemicals in Industry*, Vol. 1, pp. 385. New York: John Wiley & Sons, Inc.

CCPS. 1995a. *Guidelines for Process Safety Fundamentals in Chemical Plant Operations*. New York: AIChE/Center for Chemical Process Safety.

CCPS. 1995b. *Guidelines for Technical Planning for On-Site Emergencies*. New York: AIChE/Center for Chemical Process Safety.

CENELEC. 2003. *Electrostatics-Code of Practice for the Avoidance of Hazards Due to Static Electricity*. Report CLC/TR 50404 (June 12, 2003), European Committee for Electrotechnical Standardization, Rue de Stassartstraat 35, B-1050 Brussels, Belgium.

Cole, G. 1998. *Pharmaceutical Production Facilities: Design and Applications*. New York:Taylor & Francis.

CSB. 2003. Safety Bulletin No. 2003-10-B, *Hazards of Nitrogen Asphyxiation*. June 2003. U.S. Chemical Safety and Hazard Investigation Board, Washington, DC.

DiNardi, S. R. 1997. *The Occupational Environment – Its Evaluation and Control*. Fairfax, VA: AIHA Press.

DOT. 2000. *Emergency Response Guidebook*. U.S. Department of Transportation, Research and Special Programs Administration, Washington, DC.

Drum, D. 2002. *Emergency Responders Guide to Chemical Reactivity and Compatibility*. New York:McGraw-Hill Professional Books.

EPA. 1992. *Workbook of Screening Techniques for Assessing Impacts of Toxic Air Pollutants (Revised)*. Research Triangle Park, NC: U.S. Environmental Protection Agency (December 1992).

Farquharson, G. 1999. Cleanrooms and Associated Controlled Environments. *Pharmaceutical Engineering* (September/ October 1999),1–4

FED-STD 209E, 1992. *Airborne Particulate Cleanliness Classes in Cleanrooms and Clean Zones*. General Services Administration, Washington, DC (available from IEST, Mt. Prospect, IL).

FMG 1-56. 2000. *Cleanrooms*. Property Loss Prevention Data Sheet 1-56. Factory Mutual Global, Norwood, MA.

Forsberg, K. and Mansdorf, S. Z. 2003. *Quick Selection Guide to Chemical Protective Clothing,* 4th ed. New York: John Wiley & Sons.

Garzia, H. 2002. Private communication from Henry Garzia, Kidde- Fenwal, Inc., Ashland, MA. (December 30, 2002).

Hodson, L. L. 2001. Safe Entry into Confined Spaces. Chapter 22 in *Handbook of Chemical Health and Safety,* edited by R. J. Alaimo. Washington, DC: American Chemical Society, and Oxford, UK and New York: Oxford University Press.

IEST and Matthews, R. A. 2001. Global Standards for Cleanroom Design and Construction. *Chemical Engineering* (April), 117–120 .

ISO 14644-1. 1999. *Cleanrooms and Associated Controlled Environments-Part 1: Classification of Air Cleanliness.* Geneva: International Standards Organization.

ISO 14644-2. 2000. *Cleanrooms and Associated Controlled Environments-Part 2: Specifications for Testing and Monitoring to Prove Continued Compliance with ISO 14644-1.* Geneva: International Standards Organization.

ISO 14644-4. 2001. *Cleanrooms and Associated Controlled Environments-Part 4: Design, Construction and Start-up.* Geneva: International Standards Organization.

ISPE. 1998. *Baseline Pharmaceutical Engineering Guide, Volume 2: Oral Solid Dosage Forms.* Tampa. FL: International Society for Pharmaceutical Engineering,

Kahn, R. 2002. Personal communication from Russell Kahn, Syngenta Crop Protection, Greensboro, NC to S. S. Grossel (December 10, 2002).

Lees. F. P. 1996. *Loss Prevention in the Process Industries.* 2nd edition. Section 16.22–Effects of Fire: Injury. Woburn, MA, and Oxford, UK: Butterworth-Heinemann,

Louvar, J. F. and Louvar, D. C. 1998. *Health and Environmental Risk Analysis: Fundamentals with Applications.* Upper Saddle River, NJ: Prentice Hall PTR,

Mansdorf, S. Z. 2001. Personal Protective Equipment. Chapter 46 in *Handbook of Chemical Health and Safety,* edited by Robert J. Alaimo. Washington, DC: American Chemical Society, and Oxford, UK and New York: Oxford University Press.

Marchello, J. M. 1976. *Control of Air Pollution Sources.* New York: Marcel Dekker, Inc.

NFPA 13. 2002. *Installation of Sprinkler Systems.* Quincy, MA: National Fire Protection Association.

NFPA 92A. 2000. *Smoke-Control Systems.* Quincy, MA: National Fire Protection Association.

NFPA 204. 2002. *Smoke and Heat Venting.* Quincy, MA: National Fire Protection Association.

NFPA 318. 2002. *Standard for the Protection of Semiconductor Fabrication Facilities.* Quincy, MA: National Fire Protection Association.

NFPA 471. 2002. *Responding to Hazardous Materials Incidents.* Quincy, MA: National Fire Protection Association.

NIOSH. latest edition. *NIOSH-Certified Equipment List.* Cincinnati, OH: National Institute for Occupational Safety and Health (updated periodically).

Ozimek, T. J., Steinmetz, D. C., and Kerfoot, E. J. 1995. Personal Protective Equipment. Chapter 4 in *Handbook of Highly Toxic Materials Handling and Management,* edited by S. S. Grossel, andD. A. Crowl , New York: Marcel Dekker.

Ramstorp, M. 2000. *Introduction to Contamination Control and Cleanroom Technology.* New York: John Wiley & Sons.

Sargent, E. V. and Naumann, B. D. 2001. Occupational Exposure Limits. Chapter 14 in *Handbook of Chemical Health and Safety,* edited by R. J. Alaimo. Washington, DC: American Chemical Society, and Oxford, UK and New York: Oxford University Press.

Seire, J. S., Strimaitis, D. G., and Yamartino, R. J. 1999. *A User's Guide fro the CALPUFF Dispersion Model (Version 5.0).* Concord, MA: Earth Tech.

SRI. 1980. *Control Technology Assessment for the Pesticide Manufacturing and Formulating Industry.* Report prepared by Stanford Research Institute for NIOSH under contract No. 210-77-0093 (May 1980).

Stayner, L. 2001. Introduction to Risk Assessment. Chapter 11 in *Handbook of Chemical Health and Safety,* edited by R. J. Alaimo. Washington, DC: American Chemical Society.

Vesovic, V., Auziere, A., Calviac, G., and Dauriat, A. 2001. Modelling of the Dispersion and Deposition of Coarse Particulate Matter Under Neutral Atmospheric Conditions. *Atmospheric Environment,* 35(Suppl. 1): 99–105.

Whyte, W. 1999. *Cleanroom Design.* 2nd ed. New York: John Wiley & Sons.

Whyte, W. 2001. *Cleanroom Technology: Fundamentals of Design, Testing and Operation.* New York: John Wiley & Sons.

Willig, S. H. (editor). 2000. *Good Manufacturing Practices for Pharmaceuticals,* 5th ed. New York: Marcel Dekker.

Wood, J. P. (editor). 2001. *Containment in the Pharmaceutical Industry.* New York: Marcel Dekker.

Zalosh, R. G. 2003. *Industrial Fire Protection Engineering.* New York: John Wiley & Sons.

Appendix A

COMMERCIAL TESTING FACILITIES FOR POWDER/DUST HAZARD ASSESSMENTS

Commercial testing facilities listed in this appendix perform the tests discussed in Chapter 4, although no single facility does every test. Compilations of particulates hazards testing services known to the CCPS Committee responsible for this book and to contributing peer reviewers are presented in Tables A-1, A-2, and A-3. Readers should recognize that changes to test facilities, corporate names, etc. render this compilation an incomplete snapshot as of this writing, and updates are required to make them current.

TABLE A-1
U.S. Testing Laboratories for Particulates Hazards

Laboratory and Address	Phone and Fax Number; Email Address	Web Site
Chilworth Technology 11 Deer Park Drive Monmouth Junction, NJ 08852	Tel 732-274-0900 Fax 732-274-1371 Email safety@chilworth.com	http://www.chilworth.com/lab.cfm
Ciba® Safety Testing Laboratory PO Box 113 1 Ciba Road McIntosh, AL 36553	Tel 251-436-2397 Fax 251-436-5008 Dean.Hamel@cibasc.com	http://www.specialchem4polymers.com/sf/Ciba/index.aspx?id=safety www.cibasc.com/cxs
DustTech Creamery Commercial Park U. S. Highway 206 & Rt 15 Box 222 Augusta, NJ 07822	Tel /Fax 973-579-3188 email@hazardsresearch.com	www.hazardsresearch.com
Energetic Materials Research and Testing Center (EMRTC) New Mexico Institute of Mining and Technology 1001 South Road Socorro, NM 87801	—	—
Fike Corporation 704 S. 10th Street Blue Springs, MO 64015	Tel 816-229-3405 Fax 816-228-9277 Email fpssales@fike.com	http://www.fike.com/
Kidde-Fenwal, Inc. Combustion Research Center 90 Brook Street Holliston, MA 01746	Tel 508-429-3190 Fax 508-429-2990 Email crc@kidde-fenwal.com	http://www.Kidde-CRC.com

Laboratory and Address	Phone and Fax Number; Email Address	Web Site
FM Global 1301 Atwood Avenue P.O. Box 7500 Johnston, R.I. 02919	Tel 401-568-6240 Fax 401-568-6241	http://www.fmglobal.com/research_standard_testing/standards/testing.html
Fauske & Associates 16W070 West 83rd Street Burr Ridge, IL 60527	Telephone: 630-323-8750 FAX: 630-986-5481 inforequest@fauske.com	http://www.faichemicaltesting.com/ http://www.bkpublishing.com/fai-testing-svc.htm
Hazards Evaluation Laboratory (HEL) Princess Road Office Park Suite 205, 4 Princess Road Lawrenceville, New Jersey 08648	Tel 609-912-1551 Fax 609-912-1552 Email marketing@hel-inc.com	http://www.helgroup.com/index.php?oURL=home/consultancy
Safety Consulting Engineers, Inc 2131 Hammond Drive Schaumberg, IL 60173	Tel 847-925-8100 Fax 847-925-8120 Email sceinc@sceinc.com	http://www.sceinc.com/
Southwest Research Institute 6220 Culebra Road P.O. Drawer 28510 San Antonio, TX 78228	Tel 210-522-5483 Fax 210-522-3377	http://www.fire.swri.org
Safety Management Services, Inc 1847 West 9000 South, Suite 205 West Jordan, UT 84088	Tel 801-567-0456 Fax 801-567-0457	http://www.sms-ink.com/services_char_thermal.html
TIAX Acorn Park Cambridge, MA 02140	617-498-5061 ralbovsky.p@tiax.biz	http://www.tiax.biz/technologies/pdfs/arc_aptac.pdf

TABLE A-2
Laboratories Outside U.S. Doing Particulate Hazards Testing with Reports in English

Laboratory and Address	Phone/Fax Number and Email Contact	Web Site
Building Research Establishment Garston, Watford England WD 25 9XX	Tel +44(0)1923 664000 enquiries@bre.co.uk	http://www.bre.co.uk/services/Dust_explosion.html
Burgoyne Consultants Ltd. Chantry Drive Iikley West Yorkshire, UK LS29 9HU	Tel +44(0)1 943 817666 Fax +44(0)1 943 817632 Email mktg@buroynes.co.uk	http://www.burgoynes.co.uk
Canadian Explosives Research Laboratory Bldg 12, CANMET Bells Corner Complex c/o 555 Booth St. Ottawa, Ontario K1A 0G1 Canada	Phone: 613-947-7534 Fax: 613-995-1230 plightfo@nrcan.gc.ca	http://www.nrcan.gc.ca./mms/cerl/home_e.htm
Chilworth Vassallo Srl Via Manzoni, 25 23868 Valmadrera – LC, Italy	Tel: 39 0341 200023 Fax 39 0341 201240 Email: info@chilworth.it	http: www.chilworth.it
Chilworth Technology Ltd Beta House, Chilworth Science Park, Southampton, SO16 7NS, UK	Tel: +44 (0)23 8076 0722 Fax: +44 (0)23 8076 7866 Email: marketing@chilworth.co.uk	www.chilworth.co.uk
Chilworth SARL 10 Allee des Charmilles 69360, Serezin, France	Tel/Fax: 33 4 78 02 01 94 Email: jmfrancois@chilworth.co.uk	

Laboratory and Address	Phone/Fax Number and Email Contact	Web Site
Chilworth Safety & Risk Management (Pvt) Ltd. 259/14, Deccan Court SV Road, Bandra (West) [illegible]	Tel: 91 22 2655 7969 Fax: 91 22 2655 7972 Email: info@chilworth.co.in	www.chilworth.co.in
Ciba® Safety Testing Laboratory Macclesfield Chesire, UK SK102NX	Tel: 44 1625888-353	www.cibasc.com/cxs
Ciba® Safety Testing Goregaon, India	Tel: 91 22 28403299	—
Dalhousie University Department of Chemical Engineering 1360 Barrington Street (P.O. Box 1000) Halifax, Nova Scotia Canada B3J 2X4	Phone: 902-494-3976 Fax: 902-420-7639 Email: paul.amyotte@dal.ca	www.dal.ca
GexCon Fantoftvegen 38 P.O.Box 6015 Postterminalen N-5892 Bergen Norway	Tel +47 55 57 43 30 Fax 47 55 57 43 31 Email jrb@gexcon.com	http://www.gexcon.com/index.php?src=dust/laboratory.html
Hazards Evaluation Laboratory (HEL) 50 Moxon Street Barnet, Hertfordshire. England EN5 5TS, UK	Tel +44(0) 20 8441 6778 Fax +44(0) 20 8441 6754 Email marketing@helgroup.com	http://www.helgroup.com/index.php?oURL=home/consultancy

Laboratory and Address	Phone/Fax Number and Email Contact	Web Site
Inburex GmbH August-Thyssen-Str. 1 D-59067 Hamm Germany	Tel +49-2381 271 610 Fax: +49-2381 271 620 Email: Richard.Rogers@inburex.com Info@inburex.com	www.inburex.com
Ineris Parc Technologique ALATA B.P. No. 2 60550 Verneuil-en-Halatte France	Tel + 03 44 55 66 77 ineris@ineris.fr	www.ineris.fr
The Queensland Department of Natural Resources and Mines Mineral House 41 George Street GPO Box 2454 Brisbane, Queensland 4000, Australia	Tel 1-800 803 788 Fax: +61 7 3810 6366 Email: ray.davis@nrm.qld.gov.au	http://www.nrm.qld.gov.au/simtars/mrc. html
Syngenta Huddersfield Manufacturing Centre PO Box A38 Huddersfield Leeds Road West Yorkshire HD2 1FF United Kingdom	Tel 01484 537456 Fax 01484 517067 shella.beattie@syngenta.com 44-1484-486473 michael.bailey@syngenta.com 44-1484-486004	http://www.syngenta.com

TABLE A-3 (A)

Dust Explosivity/Flammability Testing—Test Matrix for Testing Laboratories (Incomplete)

	ASTM E1226 (ISO 6184/2) Max. Expl. Press. (20 Liter)	ASTM E1515 (MEC of Dust Cloud)	ASTM E789 (Dust Expl. in 1.2-Litre Closed Cyl.Vessel)	ASTM E1491 (Min. Autoignition Temp. of Dust Clouds)	A/B Combustibility Screening Test	ASTM E680 BAM Impact Sensitivity	ASTM E2019 (MIE Dust Cloud)	ASTM E2021 (Hot-Surface Ignition Temp. of Dust Layers)	ASTM E4470 Electrostatic Charging	Limit Oxygen Concentration
U.S. Laboratories										
Chilworth Monmouth Jct, NJ	Y	Y	Y	Y	Y	Y	Y	Y	Y	Y
Ciba Safety Test Lab McIntosh, AL	Y	Y	N	Y	Y	Y	Y	N	N	Y
DustTech Augusta, NJ	N	N	Y	Y*	N	Y*	Y	Y*	N	Y
FM Global W. Glocester, RI	Y	Y	N	N	N	N	N	N	N	N
Fauske & Associates Burr Ridge, IL	Y	Y	N	Y	N	N	Y	N	N	N
Kidde-Fenwal – CRC Holliston, MA	Y	Y	N	Y	Y	N	Y	Y	N	Y
Fike Corporation Blue Springs, MO	Y	Y	N	Y	?	N	Y	Y	?	N
Hazards Evaluation Laboratory, NJ	?	?	N	N	Y	N	Y	N	?	Y

	ASTM E1226 (ISO 6184/2) Max. Expl. Press. (20 Liter)	ASTM E1515 (MEC of Dust Cloud)	ASTM E789 (Dust Expl. in 1.2-Litre Closed Cyl.Vessel)	ASTM E1491 (Min. Autoignition Temp. of Dust Clouds)	A/B Combustibility Screening Test	ASTM E680 BAM Impact Sensitivity	ASTM E2019 (MIE Dust Cloud)	ASTM E2021 (Hot-Surface Ignition Temp. of Dust Layers)	ASTM E4470 Electrostatic Charging	Limit Oxygen Concentration
Safety Consulting Engineers Schaumberg, IL	Y	Y	Y	Y	N	Y	Y*	Y	Y	N
Safety Management Services, Inc West Jordan, UT	Y	Y	N	N	N	Y*	Y	N	N	N
Southwest Research Institute San Antonio, TX	N	N	Y	N	N	Y	Y	Y	N	N
International Laboratories										
Building Research Establishment England	Y	Y	N	Y	Y	N	Y	Y	Y	Y
Burgoyne Consultants Ltd. UK	Y	Y	N	N	Y	N	Y	N	?	N
Chilworth Technology UK, etc.	Y	Y	Y	Y	Y	Y	Y	Y	Y	Y
Ciba Safety Testing Labs UK, India	Y	Y	N	Y	Y	Y	Y	N	N	Y
Dalhousie U. CN	Y	Y	N	Y	Y	N	Y	N	N	N

	ASTM E1226 (ISO 6184/2) Max. Expl. Press. (20 Liter)	ASTM E1515 (MEC of Dust Cloud)	ASTM E789 (Dust Expl. in 1.2-Litre Closed Cyl.Vessel)	ASTM E1491 (Min. Autoignition Temp. of Dust Clouds)	A/B Combustibility Screening Test	ASTM E680 BAM Impact Sensitivity	ASTM E2019 (MIE Dust Cloud)	ASTM E2021 (Hot-Surface Ignition Temp. of Dust Layers)	ASTM E4470 Electrostatic Charging	Limit Oxygen Concentration
GexCon Norway	Y	N	N	Y	Y	N	Y	Y	N	N
Hazards Evaluation Laboratory HEL UK	Y	N	N	Y	Y	?	Y	N	?	Y
Inburex GmbH Germany	Y	Y	Y	Y	Y	Y	Y	Y	Y	Y
Ineris France	Y	Y	?	?	Y	?	Y	?	?	Y
The Queensland Dept. of Natural Res. and Mines Australia	Y	Y	N	Y	?	N	Y	Y	?	Y
Syngenta Huddersfield, UK	Y	Y*	Y*	Y	Y	Y	Y	Y	Y*	Y

A Y* indicates that the laboratory conducts a test similar to the ASTM/IEC designated test, but may use a variation of the test apparatus and test procedure.

TABLE A-3(B)

Thermal Stability, Reactivity and Other Miscellaneous Particulate Testing

	ASTM E 487 Isothermal Oven or Grewer Oven	ASTM D 1929 Plastics Ignition Temperature	IEC 1241-2-2 Electrical Resistivity	UN/DOT Division 4.3 Water Reactivity	UN/DOT Division 4.2 Pyrophoric Solids	UN/DOT Division 4.1 Powder Burning Rate	SADT	UN/DOT Division 4.2 Self-Heating Solids	ASTM E 537 DSC, DTA,	Lütolf Oven	Other Stability/Reactivity Test
United States Laboratories											
Chillworth Monmouth Jct, NJ	Y	Y	Y	Y	Y	Y	Y	Y	Y	N	ARC, RC1, Carius Tube, Isothermal Powder layer, Koenen & UN Gap, UN Div 5.1 Oxidizers, Charge Decay
Ciba Safety Testing Lab McIntosh, AL	Y	N	Y	Y	Y	Y	Y	Y	Y	Y	ARC, RC1, Thermal Activity Monitor
DustTech Augusta, NJ	Y	N	Y*	Y	Y*	Y	N	Y	N	N	JANAF thermal stability
EMRTC Socora, NM	?	N	?	Y	Y	Y	Y	Y	Y	?	?
Fauske & Associates Burr Ridge, IL	N	N	N	N	N	N	N	N	N	N	VSSP2, RSST
Kidde-Fenwal - CRCHolliston, MA	N	N	N	Y	Y	Y	N	Y	N	N	N
Hazards Evaluation Laboratory (HEL) NJ	N		N	N	?	N	N	?	Y	N	PHI-TEC II
Safety Consulting Engineers Schaumberg, IL	Y	N	N	Y	Y*	Y	Y	Y	Y	N	UN Division 5.1 Oxidizers Koenen Gap test
Safety	Y*	N	Y*	N	?	N	N	?	Y	N	Fisher-Johns Auto-

	ASTM E 487 Isothermal Oven or Grewer Oven	ASTM D 1929 Plastics Ignition Temperature	IEC 1241-2-2 Electrical Resistivity	UN/DOT Division 4.3 Water Reactivity	UN/DOT Division 4.2 Pyrophoric Solids	UN/DOT Division 4.1 Powder Burning Rate	SADT	UN/DOT Division 4.2 Self-Heating Solids	ASTM E 537 DSC, DTA,	Lütolf Oven	Other Stability/Reactivity Test
Management Services, Inc West Jordan, UT											ignition Test Modified Unconfined Burn (Oxidizers)
Southwest Research Institute San Antonio, TX	N	Y	N	N	N	N	N	N	N	N	-
International Laboratories											
Building Research Establishment UK	Y	N	Y	N	Y	N	N	Y	N	N	N
Burgoyne Consultants Ltd. UK	N	Y	N	N	Y	N	N	?	N	N	?
Canadian Explosives Research Laboratory Canada	Y	N	N	Y	Y	Y	N	Y	Y	N	ARC UN Division 5.1 (oxidizers) UN TDG Division 1 (explosives)
Chilworth Technology UK, etc.	Y	Y	Y	Y	Y	Y	Y	Y	Y	N	ARC, RC1, Carius Tube, Isothermal Powder layer, Koenen & UN Gap, UN Div 5.1 Oxidizers, Charge Decay
Ciba Safety Testing UK, India	Y	N	Y	Y	Y	Y	Y	Y	Y	Y	ARC, RC1, Thermal Activity Monitor
Hazards Evaluation	N	?	N	N	?	?	N	?	Y	N	PHI-TEC II

	ASTM E 487 Isothermal Oven or Grewer Oven	ASTM D 1929 Plastics Ignition Temper ature	IEC 1241-2-2 Electrical Resistivity	UN/DOT Division 4.3 Water Reactivity	UN/DOT Division 4.2 Pyrophoric Solids	UN/DOT Division 4.1 Powder Burning Rate	SADT	UN/DOT Division 4.2 Self-Heating Solids	ASTM E 537 DSC, DTA,	Lütolf Oven	Other Stability/Reactivity Test
Laboratory HEL UK											
Inburex GmbH Germany	Y	Y	Y	Y	Y	Y	Y	Y	Y	Y	Mini-autoclave, Gas evolution, PHITEC, Adiabatic and Pressure Dewar UN/DOT 5.1 Oxidizers
Ineris France	Y	?	?	?	?	?	?	?	Y	?	?
Queensland Dept. of Natural Res. and Mines Australia	Y		N	N	?	N	N	?	N	Y	?
Syngenta UK	Y*	N	Y	Y	Y	Y	Y	Y	Y	Y	Carius Tube, Charge Decay UN/DOT 5.1 Oxidizers

Appendix B

EQUIPMENT OVERVIEW

B1 INTRODUCTION

This appendix contains discussions and descriptions of equipment, systems, and containers used in the storage, handling, and processing of particulate solids (powders, bulk solids, granules, pellets, flakes, etc.).

B2 BAG OPENERS (SLITTERS)

Bag opening can be performed manually using a bag dump station, but automatic bag openers provide greater physical separation from dust generation than manual bag dump stations. However, this equipment is often justified as a labor-saving device rather than as a dust control device, although it does the latter very well. Most of these devices require a worker to manually transfer bags from a pallet to a conveyor which then feeds the bags into an enclosed machine where they are opened (by rotating knives), emptied (spiked rollers separate the bags from the solids), and discarded. The empty bags are often fed into an attached trash compactor, which is also connected to the exhaust ventilation system. The dust generated in these enclosures is usually controlled by local exhaust ventilation. This isolates the worker from the dust generated by the bag-opening process. Figure B-1 shows an automatic bag opener manufactured by one vendor.

Heitbrink and McKinnery (1987) discuss bag-opening machines and provide data on their dust control efficiencies.

B3 BLENDERS/MIXERS

It is useful to classify particulate solids blending/mixing machines into two main groups: active and passive (Kaye, 1997). The active type uses certain moving parts to assist in the randomization of the solid materials or the mixer machine moves about physically in the mixing process. In a passive mixer system the randomization of the materials is achieved by directed

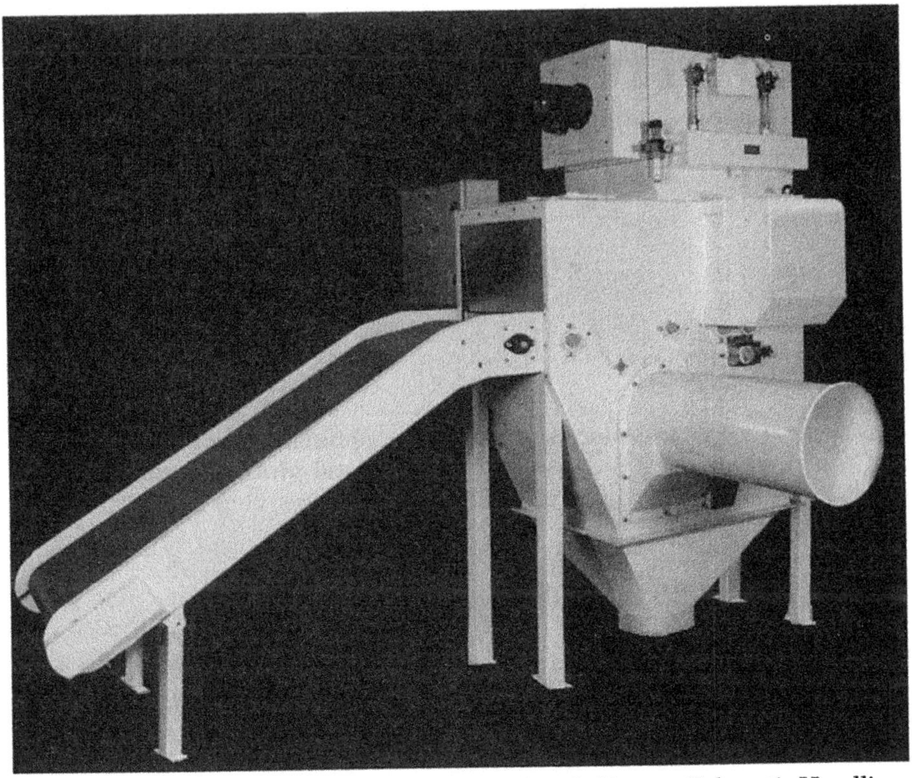

Figure B-1 Automatic bag opening device (bag slitter). (Source: Palamatic Handling USA, Inc.)

flow of the powder streams by baffles, etc. (i.e., no moving parts), as they move through the mixing device. Examples of active blenders/mixers are ribbon blenders, tumbler blenders, and high shear mixing and multi-mechanism blenders. Examples of passive blenders are baffled and gravity in-bed mixing devices. Blenders and mixers are also classified by their mode of operation: batch or continuous.

B3.1 Batch Type Blenders/Mixers

A wide variety of batch-type mixers are available and are briefly discussed below:

Tumbling Type Blenders/Mixers: These are suitable for gentle blending, are easily cleaned, suitable for dense powders and abrasive materials. They are not suitable for breaking up agglomerates, and in fact, may cause agglomerates (balling). Examples are double-cone, twin-shell (V-type), hori-

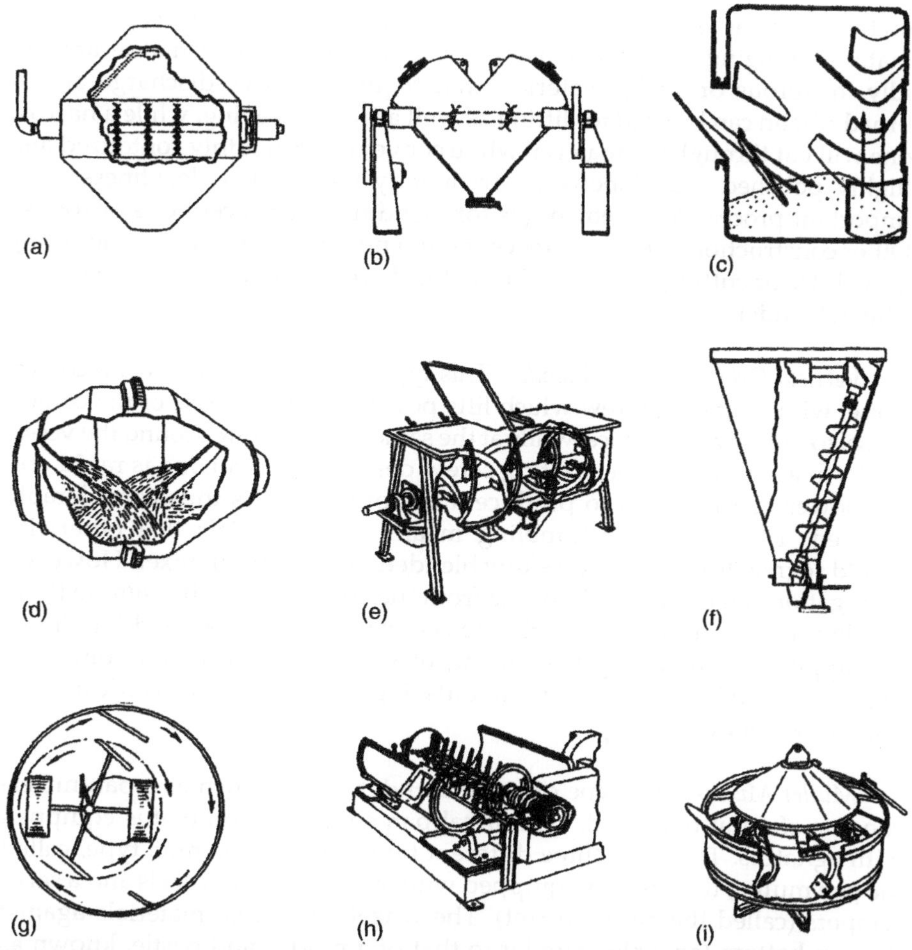

Figure B-2 Various types of batch blenders/mixers. (a) double cone; (b) twin shell (vee); (c) horizontal drum; (d) double cone revolving around long axis (with baffles); (e) ribbon; (f) vertical orbiting screw; (g) batch muller; (h) single rotor; (i) turbine. (Source: *Perry's Chemical Engineers' Handbook,* 7th edition. Reprinted with permission of McGraw-Hill.)

zontal rotating drum, and double-cone revolving around long axis blenders (see Figure B-2a, b, c, and d). These blenders can be provided with blades and baffles to break up agglomerates.

Ribbon Blenders: These are perhaps the most widely used solids blenders. A ribbon rotates within a static trough or open cylinder and the particles are conveyed by the moving ribbon. The ribbon cross section and pitch, clearances between the outer ribbon and the trough wall, and number of spirals on the ribbon are some of the features which can be varied to accommodate materials ranging from low-density, finely divided materials that aerate rapidly to fibrous or sticky materials that require positive discharge aid. A broad ribbon can be used for lifting as well as for conveying, while a narrow one will cut through the material while conveying. Smoothly contoured and highly polished cast ribbons are frequently used when cleanliness is an important process requirement (as for blending pharmaceuticals or foods). Other construction variations are center or end discharge, and the mounting of paddles or cutting blades on the center shaft. Figure B-2e shows a typical ribbon blender.

Vertical Orbiting Screw Blender: This type of blender consists of a conical vessel with a rotating screw which lifts powder from the base of the conical vessel to the powder surface, while at the same time orbiting around the vessel wall. An advantage of the orbiting screw design is that mixing is rapid and the power consumption to produce a uniform blend is thus considerably reduced. The absence of tumbling means that materials that otherwise would segregate can be successfully blended in this type of mixer. However, segregation can occur on discharge from the conical vessel if a simple flow-out device is used. A vertical orbiting screw blender can be used for almost any application requiring the mixing of solids, as well as for some that require the mixing of solids with liquids. Figure B-2f shows a vertical orbiting screw blender.

Muller Mixers: This type of batch blender (also known as a pan mixer) consists of a flat-bottomed, cylindrical pan-shaped vessel that is equipped with agitators. There are three different types. In one common type, called the pan muller, the vessel is equipped with large, rotating wheels and a set of scrapers (called the muller turret). The wheels compress materials against the pan bottom (an action similar to that of a mortar and pestle, known as mulling) while the scrapers move the material from the pan's sides and center to flow under the wheels. In another common design, the vessel is equipped with a set of scrapers and a rotating agitator, which is positioned off-center in the pan. The pan rotates and brings material to the rotating agitator, while scrapers keep moving material into the mixing zone. In a third design the pan rotates clockwise and the muller turret rotates counterclockwise.

A muller mixer is useful for mixing problems requiring certain types of aggregate breakdown, frictional anchorage of particles to one another, and densification of the final mix. It is also often used for mixing solids with liquids to produce pastes. The muller mixer should not be used for materials which are excessively fluid or sticky. The muller mixer is generally used for batch operations (see Figure B-2g), but can be also designed for continuous operation, as shown in Figure B-6e.

Drum-Type Blenders: Drum type blenders with the axis of rotation horizontal to the center of the drum are in common use. Usually they are provided with internal baffles or helical plates to improve the crossflow of solids. Charging or discharging is done through an adjustable chute, similar to that on a concrete mixer. This type of blender is generally not recommended for precise blending nor applications where there is a cleaning problem. See Figure B-2c for a schematic of a drum-type blender.

Single Rotor Blender/Mixers: this type blender consists of a single shaft with paddles inside of a cylindrical shell. These blenders usually operate at high tip speeds (6000–9000 ft/min), although in some cases lower tip speeds are used. They are used for intensive dispersion and disintegration. Figure B-2h shows such a blender.

Turbine Blender/Mixers: These consist of a circular trough with a housing in the center around which revolves a spider or a series of legs with ploughshares or moldboards on each leg. The ploughshares spin around through the circular trough which causes the material to circulate rapidly from the center to the walls, then upwards and down to the center. This type of blender is suitable for free-flowing dry materials or semi-wet materials that do not flow well and is also adaptable for liquid–solid mixing and coating operations. Figure B-2i shows a turbine blender/mixer.

In-Bin Blenders: these blenders are essentially silos with various means of moving the stored solids to achieve blending. There are a number of proprietary blenders available which use several different ways of accomplishing this blending. In one type of in-bin blender pellets or granules from various points in the bin fall by gravity through a perforated pipe or multiple pipes and are then pneumatically conveyed through internal or external pipes to the top of the bin. Another type of in-bin blender achieves mixing by aerating the mass in the bin. Still another type induces blending by introducing pulsed air which creates intense turbulence and causes intermittent fluidization and movement of the material in the bin. It has been found that in-bin blenders using air for mixing do not usually work well with granules or pellets because they are too porous. Figure B-3 shows schematic drawings of the three proprietary in-bin blenders described above. Articles by Johanson (1970), Stein (1990), and Wilms (1992) discuss in-bin blenders.

Other Types of Batch Blenders/Mixers: There are numerous other types of specialized batch blenders/mixers used in the chemical process industries, many of them in the pharmaceutical industry.

Among these are the Littleford-Lodige ploughshare blender (Figure B-4a), the Moritz turboblender (Figure B-4b), and the BULS (Bulk Unit Load

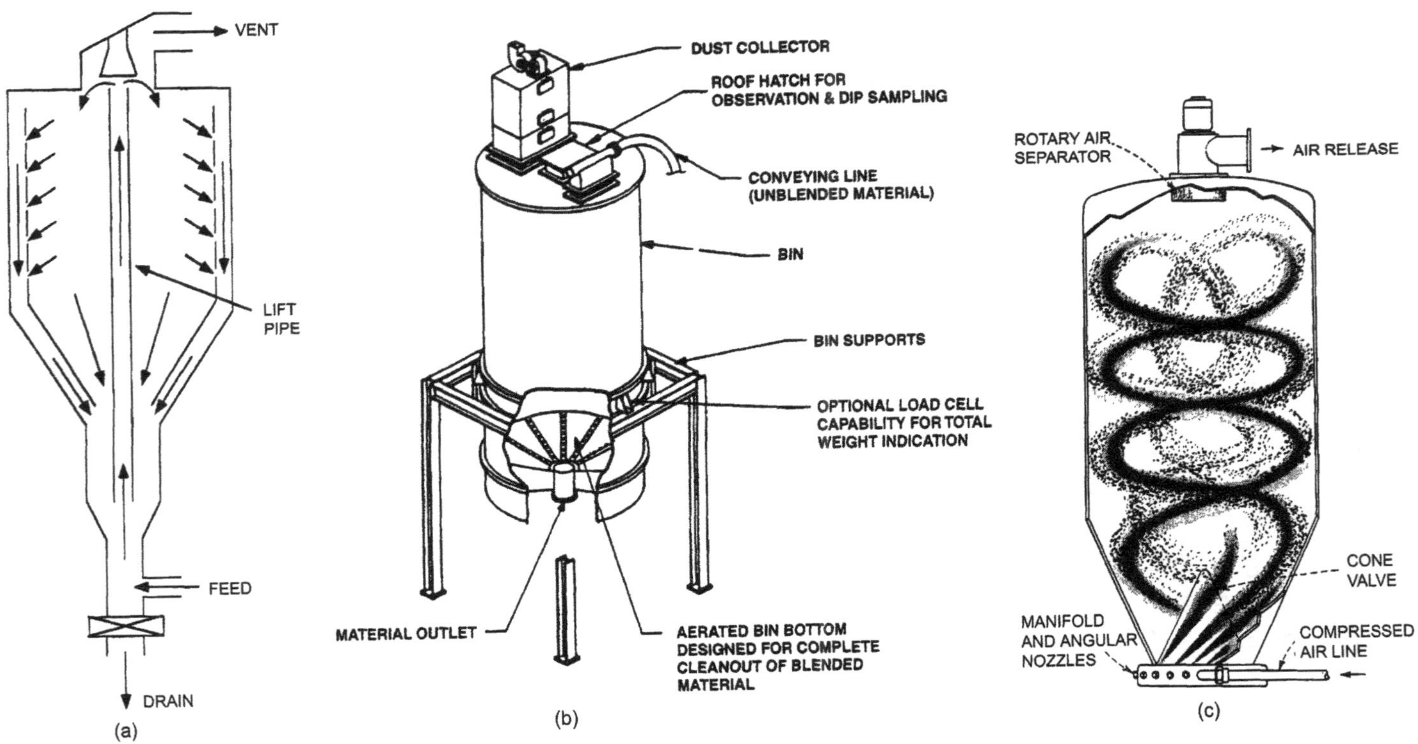

Figure B-3 Several types of in-bin blenders. (a) gravity flow blender; (b) aerated mass blender; (c) pulsed air blender.

(a)

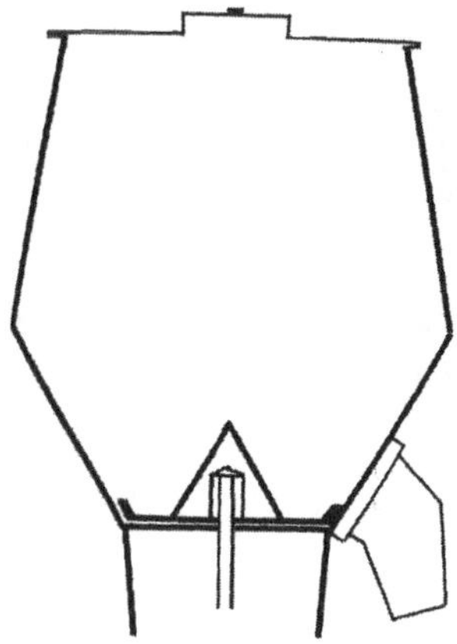

(b)

(c)

Figure B-4 Specialized batch mixers. (a) Littleford-Lodige ploughshare blender; (b) Moritz turboblender; (c) BULS blender.

Systems) blender (Figure B-4c), which can rotate a portable drum or cube in either an end-over-end or an asymmetric motion.

Also widely used are high speed blenders whose rapidly rotating paddles whip up the air and the materials to create a fluidized zone in which intense turbulence intermingles the solids. Two types are the Forberg or Pella mixer, which contains a twin paddle system, and the aerating mixer made by Littleford and Processall (see Figure B-5).

(a)

(b)

Figure B-5 High speed aerating batch blenders. (a) Forberg or Pella mixer; (b) Processall mixer.

B3.2 Continuous Type Blenders/Mixers

Unlike a batch blender/mixer which continues to agitate individual batches of solids for relatively long periods until they are adequately mixed, a continuous blender/mixer agitates and moves material through the equipment, mixing them in one quick pass (as short as one-tenth the time required in a batch blender). To mix materials adequately in one pass, most continuous blenders agitate small amounts of solids very intensely. Because the materials in a continuous blender have such a short residence time and only pass through the blender once, they must be fed into the blender in the exact proportions that the finished mixture should have. This requires proportioning equipment such as screw, belt-gravimetric, or loss-in-weight feeders to accurately meter materials into the blender.

In order to enable the blender to mix materials in one pass, a continuous blender is usually long, narrow, and enclosed. This design ensures that the materials are mixed together in the time it takes the materials to enter the blender's inlet, pass through the blender, and discharge from the outlet. The blender is usually fitted with one or more agitators designed for the materials being mixed. For example, when two agitators are used, the agitator paths can be either overlapping or tangential to each other.

Basic continuous blender types include pug mills, twin-screw blenders, ribbon blenders, pin mixers, static mixers, and free-fall impeller mixers. Specialized designs include the Zig-Zag® blender, the Poly-Kneader®, and the continuous muller. These are briefly discussed below.

Pug Mills: A pug mill consists of a drum- or trough-shaped vessel equipped with a double-shafted or, less often, single-shafted agitator, with paddles mounted on each shaft. The paddles are usually adjustable so that their forward and backward motion can be varied during mixing. A pug mill can handle very large volumes and is often used for very crude solids blending, such as preparing clay for brick forming.

Twin-Rotor Blenders: A twin-rotor (also called a twin-screw) blender is basically a pug mill, but it is manufactured to more exact tolerances and has a very sophisticated twin-screw agitator design. As its name indicates, it consists of two shafts with either screws or paddles attached and encased in a cylindrical vessel. The vessel is designed like a closed figure eight and is machined to fit very closely around the screw agitators. The screw agitators can be overlapping or tangential, and corotating or counterrotating, depending on the application; for example, overlapping screw agitators can eliminate dead spaces in the vessel. The agitators can also be fitted with paddles to intensify the mixing action. A twin-rotor blender is used primarily for mixing solids with liquids to produce pastes and plastic compounds. It can be adapted to provide heating or cooling by attaching a jacket to the shell

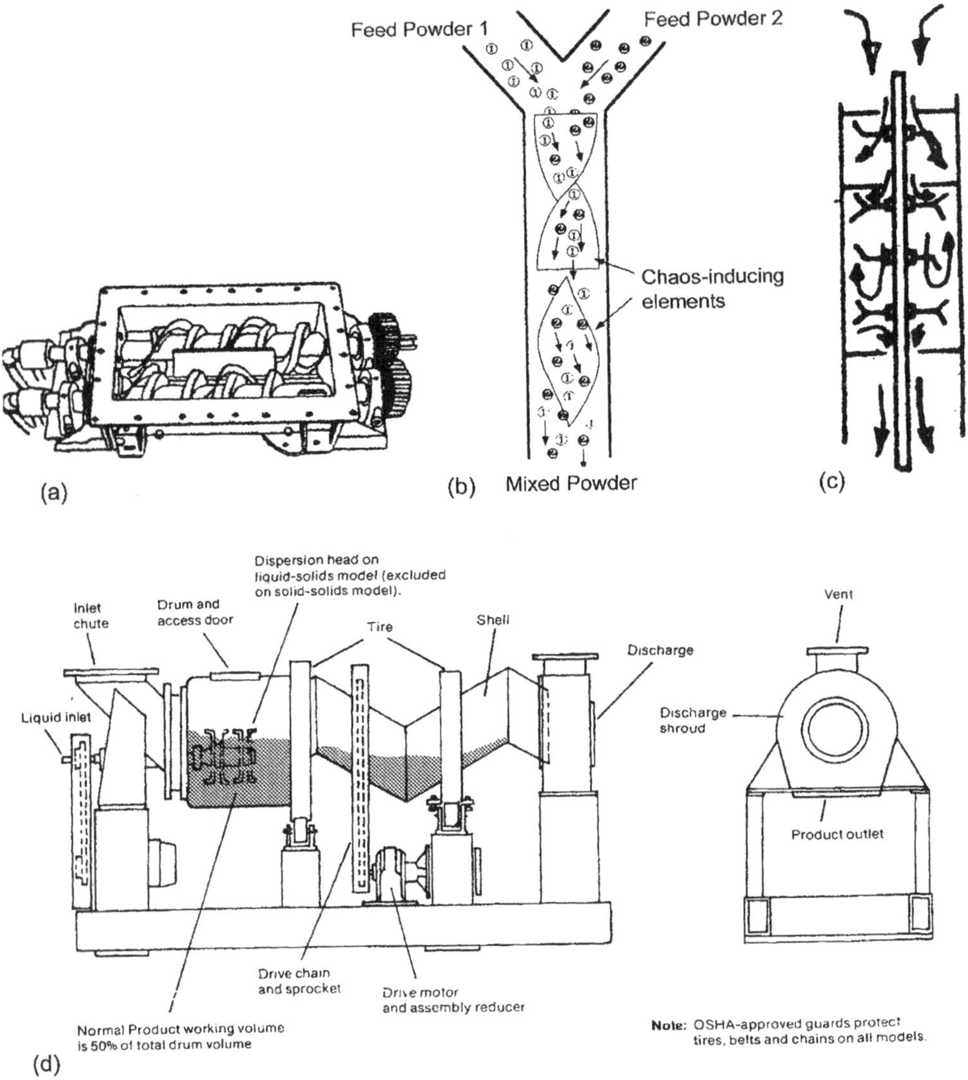

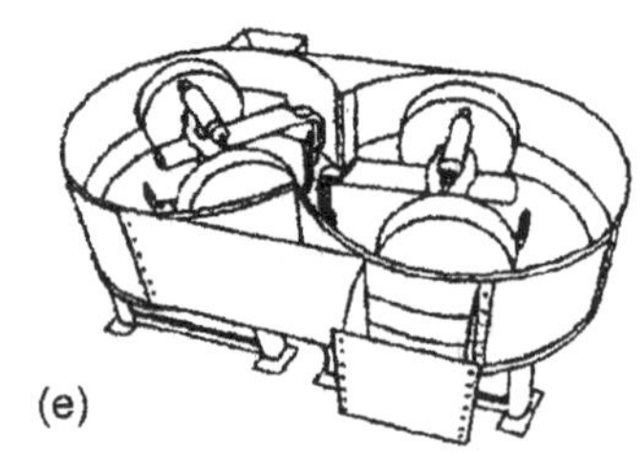

Figure B-6 Several types of continuous blenders/mixers. (a) twin-rotor blender; (b) static mixer; (c) free-fall impeller blender; (d) Zig-Zag blender; (e) continuous muller.

and/or making the shaft hollow to convey heat transfer fluids. See Figure B-6a for a picture of a twin-rotor blender.

Ribbon Blenders: Continuous ribbon blenders are similar to batch blenders in that the materials are moved back and forth by the ribbons to create a folding action, but most of the outer ribbons are designed to move the material from the inlet to the outlet nozzle. A continuous ribbon blender is used primarily for mixing solids with solids, but can also handle mixtures incorporating a small amount of liquid, as long as the finished mixture flows freely.

Pin Mixers: A pin mixer consists of a cylindrical vessel with a single shaft. The shaft, which rotates at high speed, is fitted with radial pins or narrow paddles that intensify the mixing action. A pin mixer is used primarily for mixing small amounts of liquids with solids to produce microsized pellets of material or to dedust incoming solids.

Static Mixers: This type of blender/mixer is also called a "motionless" mixer which consists of a cylindrical chamber in which chaos-inducing elements are located along the axis of the mixer. The solids pass through the mixer by gravity flow and the elements cause the particles to move in different directions. They are suitable for blending such solids as grain and plastic chips. See Figure B-6b for a schematic drawing of a static mixer.

Free-Fall Impeller Blender: This type of blender/mixer consists of a vertical cylinder with blades mounted on a central shaft. As the materials fall through the cylinder by gravity flow the high speed rotation of the blades causes the blending. Typical uses are for the blending of detergents, fat/flour mixtures, and animal feedstuffs. See Figure B-6c for a schematic drawing of this blender.

Zig-Zag® Blender: This blender is based on the design of a tumbling twin-shell V-shaped blender. Essentially, this consists of a number of V-shaped blenders (called legs) joined together. It operates strictly on tumbling action, without internal scrapers, stirring devices, screens, vanes, or paddles. The motion produced is a rolling one, so that particles are not crushed, smeared, or subjected to excessive mechanical work. As the blender revolves, some of the material tumbles forward into the next leg while some tumbles backward, but the overall flow is forward through the unit. The flow rate depends on the speed of rotation and on the angle of tilt at which the axis of rotation is set, which is adjustable. The Zig-Zag® blender can be used for mixing solids with solids or mixing solids with liquids; liquids are added through a dispersion head in the drum that turns the liquid into a mist of controlled-size droplets, which then can be quickly incorporated into the solids. See Figure B-6d for a picture of a Zig-Zag® blender.

Continuous Muller: This blender is a series of two or more batch pan mullers. The pan mullers are arranged so that some material in the first muller flows to the second, some material from the second muller flows to the third, and so on, until the finished mixture discharges from the last pan muller. The continuous muller is used for mixing solids with liquids to form damp solids or pastes. See Figure B-6e for a drawing of a continuous muller.

Poly-Kneader®: This blender consists of a closed cylindrical vessel equipped with stationary pins that project from the housing. The agitator is an interrupted-flight screw (the screw's edges have gaps, or openings, in them). During mixing, the shaft rotates and reciprocates so that the screw moves around the housing's stationary pins, which blends the materials. The Poly-Kneader®, like the twin-screw blender, is used primarily for mixing solids with liquids to produce pastes.

Other Types of Continuous Blenders: Other proprietary continuous blenders are available from various vendors. These include such blenders as the Ko-Kneader, the Falls continuous processor, the Turbulizer, the Kneadermaster, and the Oakes continuous mixer, to name several. Discussions about them are presented by Mead (1964). Others are also available and vendors should be consulted.

B3.3 Blender/Mixer Selection

There are many factors which go into selecting the best blender/ mixer for a specific application, such as the properties of the particulate solids to be blended with either other solids or liquids, process requirements (homogeneity), and the efficiency of the blender. A blender selection decision chart is presented in Figure B-7 which is useful for arriving at a reasonable choice (Harnby, Edwards, and Nienow, 1985). This chart is a guide only and is not a substitute for conducting tests on a selected blender/mixer.

Discussions and descriptions of various types of batch and continuous blenders and mixers for particulate solids are presented by Kaye (1996), Kaye (1997), and Weinekotter and Gericke (2000).

B4 DRYING EQUIPMENT

Industrial dryers may be broadly classified in two ways, as follows: (1) by the method of transferring heat to the wet solids to dry them, and (2) by the handling characteristics and physical properties of the wet material. The first method is based on differences in dryer design and operation, while the second method is most useful as a guide for the selection of a group of dryers for preliminary consideration for a specific drying application. Figure B-8 is a classification chart for drying equipment based on the method of heat

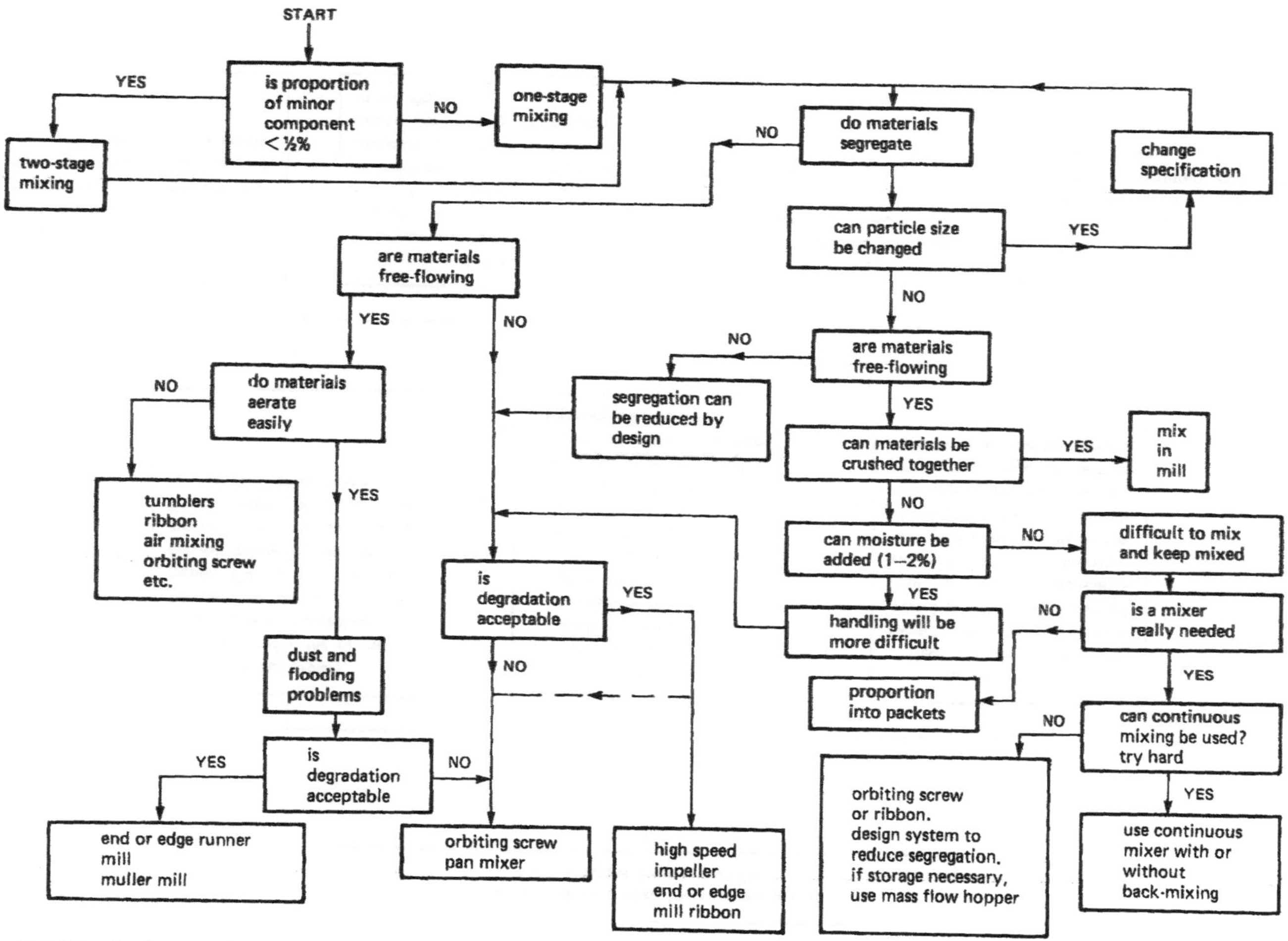

Figure B-7 Blender/mixer decision tree. (Source: Harnby et al., *Mixing in the Process Industries,* 1985.

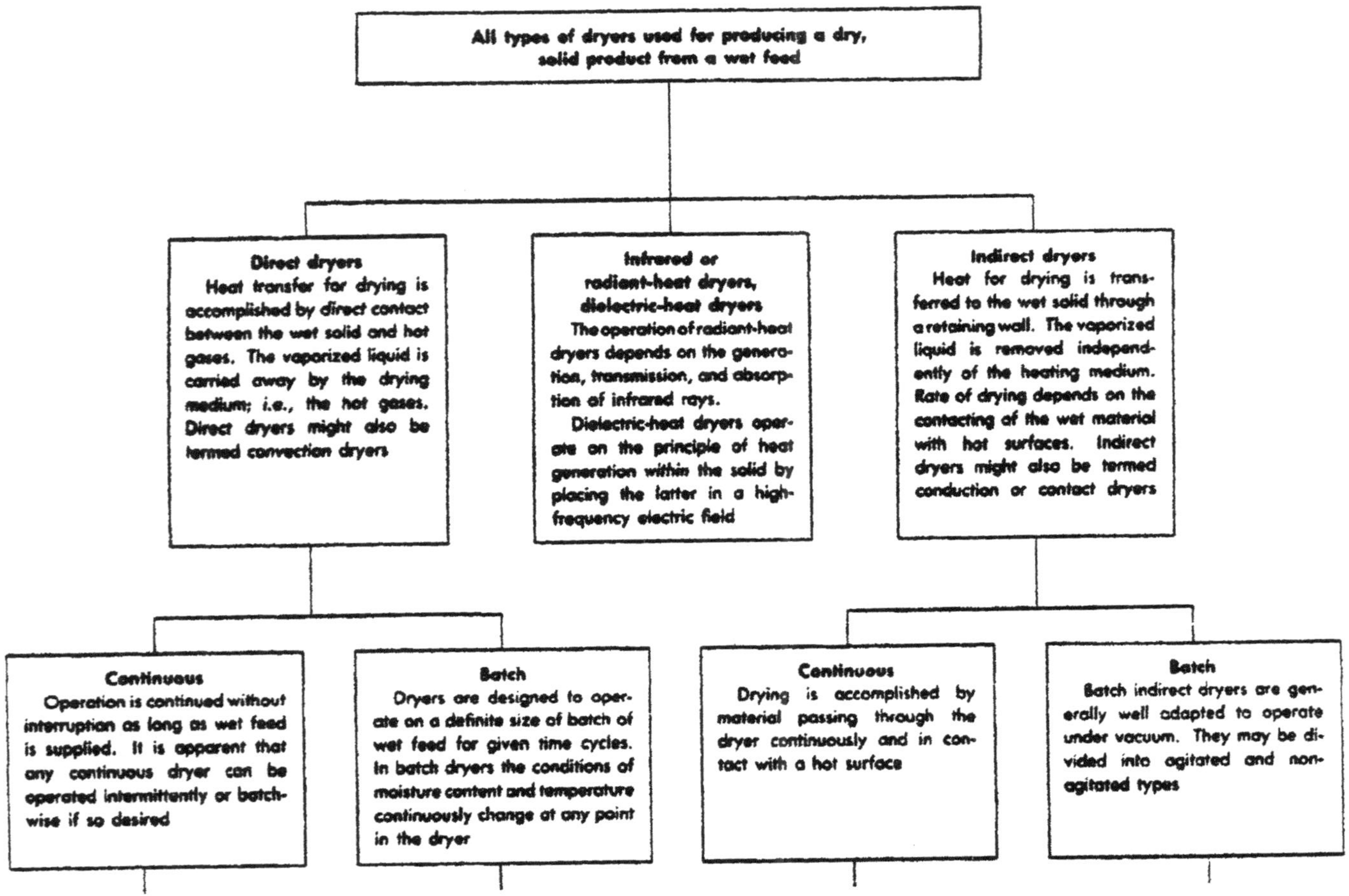
All types of dryers used for producing a dry, solid product from a wet feed
Direct dryers
Heat transfer for drying is accomplished by direct contact between the wet solid and hot gases. The vaporized liquid is carried away by the drying medium; i.e., the hot gases. Direct dryers might also be termed convection dryers
Infrared or radiant-heat dryers, dielectric-heat dryers
The operation of radiant-heat dryers depends on the generation, transmission, and absorption of infrared rays.
Dielectric-heat dryers operate on the principle of heat generation within the solid by placing the latter in a high-frequency electric field
Indirect dryers
Heat for drying is transferred to the wet solid through a retaining wall. The vaporized liquid is removed independently of the heating medium. Rate of drying depends on the contacting of the wet material with hot surfaces. Indirect dryers might also be termed conduction or contact dryers
Continuous
Operation is continued without interruption as long as wet feed is supplied. It is apparent that any continuous dryer can be operated intermittently or batchwise if so desired
Batch
Dryers are designed to operate on a definite size of batch of wet feed for given time cycles. In batch dryers the conditions of moisture content and temperature continuously change at any point in the dryer
Continuous
Drying is accomplished by material passing through the dryer continuously and in contact with a hot surface
Batch
Batch indirect dryers are generally well adapted to operate under vacuum. They may be divided into agitated and nonagitated types

Direct continuous types

1. **Continuous-tray dryers** such as continuous metal belts, vibrating trays utilizing hot gases, vertical turbodryers
2. **Continuous sheeting dryers.** A continuous sheet of material passes through the dryer either as festoons or as a taut sheet stretched on a pin frame
3. **Pneumatic conveying dryers.** In this type, drying is often done in conjunction with grinding. Material conveyed in high-temperature high-velocity gases to a cyclone collector
4. **Rotary dryers.*** Material is conveyed and showered inside a rotating cylinder through which hot gases flow
5. **Spray dryers.** Dryer feed must be capable of atomization by either a centrifugal disk or a nozzle
6. **Through-circulation dryers.** Material is held on a continuous conveying screen, and hot air is blown through it
7. **Tunnel dryers.** Material on trucks is moved through a tunnel in contact with hot gases.
8. **Fluid beds.** Solids are fluidized in a stationary tank. May also have indirect-heat coils

Direct batch types

1. **Batch through-circulation dryers.** Material held on screen bottom trays through which hot air is blown
2. **Tray and compartment dryers.** Material supported on trays which may or may not be on removable trucks. Air blown across material on trays
3. **Fluid beds.** Solids are fluidized in a stationary cart with dust filter mounted above

*Certain rotary dryers may be a combination of indirect and direct types; e.g., hot gases first heat an inner shell and then pass between an inner and outer shell in contact with the wet solid.

1. **Cylinder dryers** for continuous sheets such as paper, cellophane, textile piece goods. Cylinders are generally steam-heated, and rotate
2. **Drum dryers.** These may be heated by steam or hot water
3. **Screw-conveyor dryers.** Although these dryers are continuous, operation under a vacuum is feasible. Solvent recovery with drying is possible
4. **Steam-tube rotary dryers.** Steam or hot water can be used. Operation on slight negative pressure is feasible to permit solvent recovery with drying if desired
5. **Vibrating-tray dryers.** Heating accomplished by steam or hot water
6. **Special types** such as a continuous fabric belt moving in close contact with a steam-heated platen. Material to be dried lies on the belt and receives heat by contact

1. **Agitated-pan dryers.** These may operate atmospherically or under vacuum, and can handle small production of nearly any form of wet solid, *i.e.*, liquids, slurries, pastes, or granular solids
2. **Freeze dryers.** Material is frozen prior to drying. Drying in frozen state is then done under very high vacuum
3. **Vacuum rotary dryers.** Material is agitated in a horizontal, stationary shell. Vacuum may not always be necessary. Agitator may be steam-heated in addition to the shell
4. **Vacuum-tray dryers.** Heating done by contact with steam-heated or hot-water-heated shelves on which the material lies. No agitation involved

Figure B-8 Dryer classification by method of heat transfer and mode of operation. (Source: *Perry's Chemical Engineers' Handbook,* 7th edition. Reprinted with permission of McGraw-Hill.)

transfer to the wet solids and subclassified on the basis of batch or continuous operation (Moyers, 1997). Table B-1 is a guide for preliminary selection of dryers based on the types of materials handled (Ashworth, 1982). Moyers (1997) presents additional guidelines for dryer selection based on the types of materials handled.

Descriptions are given below for a number of industrial dryers.

B4.1 Direct-Heat Dryers

In direct-heat dryers (also called convective drying systems), heat to evaporate the moisture is provided by heated air or gas flowing over the surface of the material and the evaporated moisture is carried away by the drying medium. Air (most common), inert gas (such as nitrogen for drying solids

TABLE B-1
Classification of Commercial Dryers Based on Materials Handling

Dryer Type	Operation	Fluids Liquid Suspension	Pastes Dewatered Cake	Powders	Granules Pellets Extrudates
Forced convection stove (cross airflow)	Batch	(c)	(b)	(b)	(a)
Forced convection stove (through-flow)	Batch	(d)	(d)	(d)	(a)
Agitated pan (subatmospheric)	Batch	(b)	(b)	(b)	(c)
Agitated pan (atmospheric)	Batch	(b)	(b)	(b)	(c)
Double cone tumbler (subatmospheric)	Batch	(d)	(c)	(b)	(c)
Fluidized bed (through-flow)	Continuous	(d)	(d)	(a)	(a)
Conveyor band (through-flow)	Continuous	(d)	(b)	(d)	(a)
Rotary (indirect)	Continuous	(d)	(c)	(a)	(b)
Rotary (direct)	Continuous	(d)	(b)	(b)	(a)
Film drum (atmospherics)	Continuous	(a)	(b)	(d)	(d)
Pneumatic or flash	Continuous	(d)	(b)	(a)	(b)
Spray	Continuous	(a)	(d)	(d)	(d)

Application rating: (a) good; (b) fair; (c) unsatisfactory; (d) not applicable.

wet with organic solvent), direct combustion gases, or superheated steam (or solvent vapor) can be used in such dryers.

Batch Compartment Dryers: these are also often called tray dryers because the wet material is loaded onto trays or shelves. Tray dryers may be of the tray-truck or stationary-tray type. In the former, the trays are loaded on trucks which are pushed into the dryer compartment; in the latter, the trays are loaded directly into stationary racks within the dryer compartment. The compartment enclosure is comprised of insulated panels designed to limit exterior surface temperatures to less than 50°C. An important design requirement is to ensure gas flow uniformity, top-to-bottom of the compartment and back-to-front. Unless the material is dusty, gas is recirculated through an internal heater (usually a steam-heated coil) located before the circulating fan. Only enough purge is exchanged so as to maintain needed internal humidity. For inert gas operation, purge gas is sent through an external condenser where solvent vapors are condensed and the inert gas is returned. Figure B-9 illustrates a two-truck tray dryer.

Turbotray Dryers: the turbotray dryer is a continuous tray dryer composed of a stack of circular trays rotating slowly inside a vertical, insulated, cylindrical housing. Each rotating tray has uniformly spaced radial slots through which material is discharged to the tray below by a stationary plough once per revolution. Material falling through a slot is leveled to a uniform depth on the tray below by a stationary rake. Circulating fans are mounted on the central rotating shaft. Gas reheaters are mounted on the housing walls and gas flows across the trays parallel to the material surface. Figure B-10 is a schematic of a turbotray dryer.

Continuous Web (Sheeting) Dryers: web (or sheeting) dryers are used for drying polymer films, paper, cloth, woven fabrics, printed and coating films, and printed fabrics. Gas impinges on or flows parallel to the moving mate-

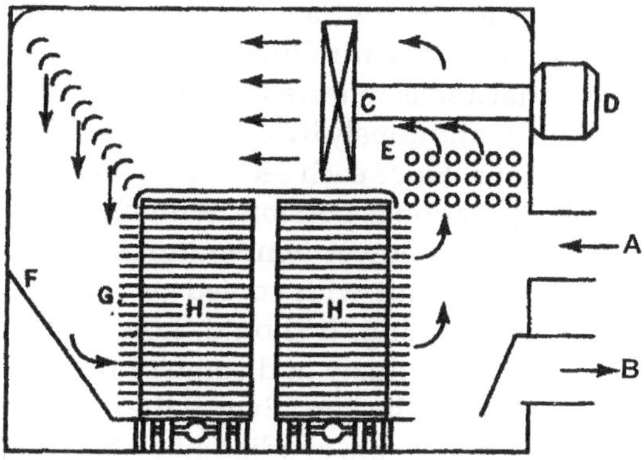

Figure B-9 Two-truck tray dryer. A, air inlet duct; B, air exhaust duct with damper; C, axial flow fan; D, fan motor, 2–15 kW; E, air heaters; F, air distribution plenum; G, distribution slots; H, wheeled trucks and trays. The arrows indicate air and vapor flow pattern.

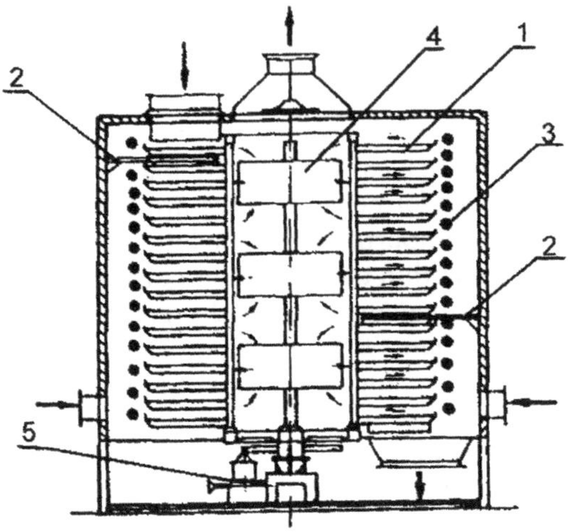

Figure B-10 Turbo tray dryer. 1, shelves; 2, blades; 3, heater; 4, fan rotors; 5, shelf and fan drive. (Source: Mujumdar, *Handbook of Industrial Drying,* 1995. Reprinted with permission of Marcel-Dekker, Inc.)

rial, called a web, that is supported by various methods. On a festoon conveyor the web is draped over sticks or rods that are carried on chains through a heated enclosure. The web is unrestrained and free to shrink or stretch. Gas flow must be comparatively gentle to avoid excessive material movement. On single or multipass roll conveyors the web is conveyed either vertically or horizontally over a series of driven rolls, while web tension is controlled by differential roll speeds. The rolls are slightly crowned to hold axial alignment, but there is no restriction to lateral shrinkage. Because the web is restrained axially, however, high velocity gas impingement slots or nozzles may be used on one of both faces. Radiant heaters are used in these dryers. Figure B-11 is a schematic drawing of a continuous web dryer for airfloat drying of impregnated and coated materials.

Through-Circulation Dryers: in these dryers, permeable materials are conveyed through enclosures (housings) on perforated plate or screen conveyors. The enclosures are comprised of a series of independent compartments, each having its own fans and heating coils. Humid air is removed at the material feed end of the enclosure; fresh dry air is introduced at the dry end. The air circulates through the wet material and is reheated before reentering the bed. A portion of the air is exhausted continuously by one or two exhaust fans, which handle air from several sections. Since each section can be operated independently, extremely flexible operation is possible, with high temperatures usually at the wet end, followed by lower temperatures further along toward the dry end. In some cases, a unit with cooled or specially humidified air is employed for final conditioning. It should, however, be noted that as the material dries it may become dusty and some of the dust may be carried with the recirculation air into the heaters. Even if the quantity of dust is very small, over time it may accumulate on, or adjacent to, the heat-

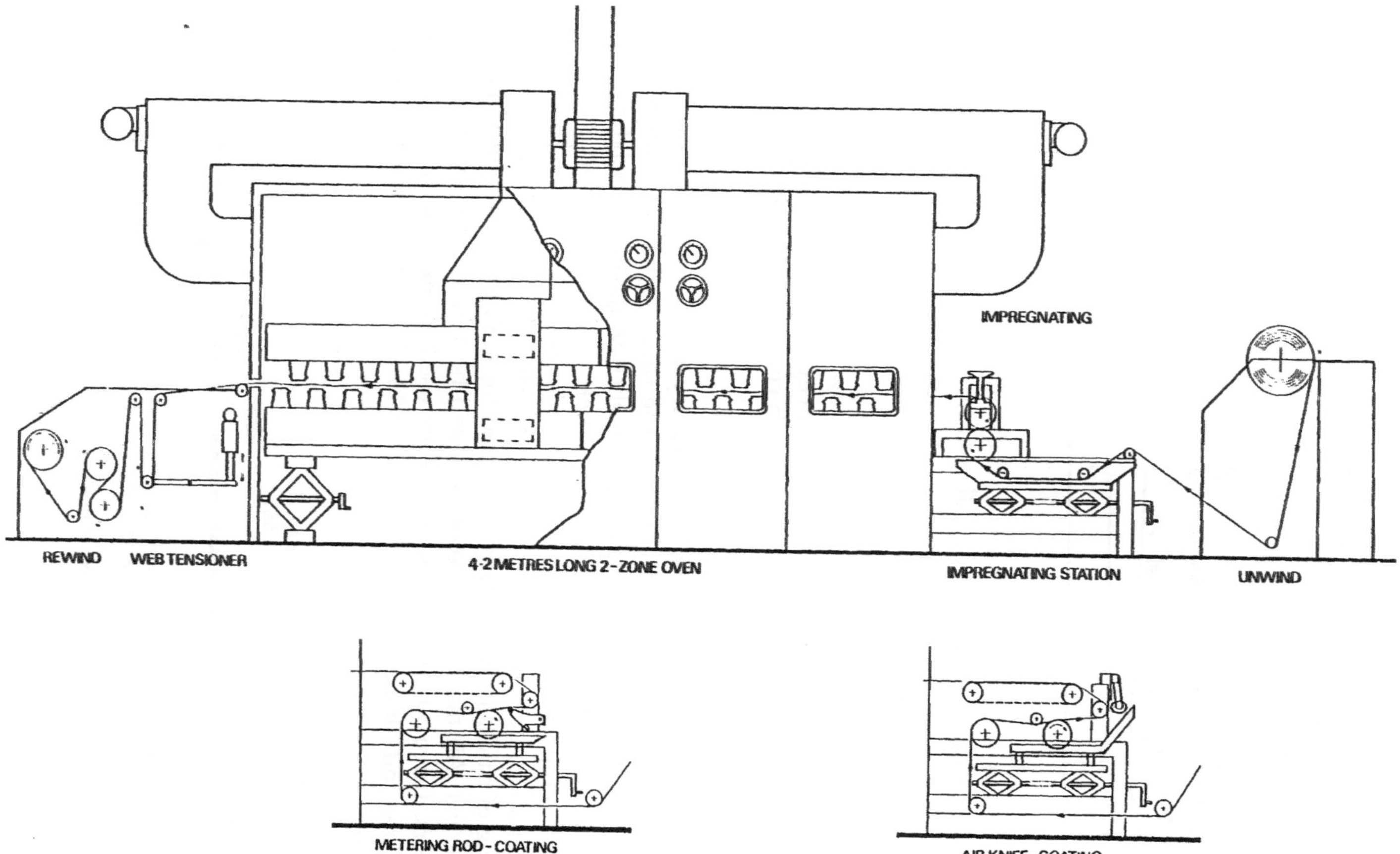

Figure B-11 Continuous web dryer for airfloat drying of impregnated and coated materials.

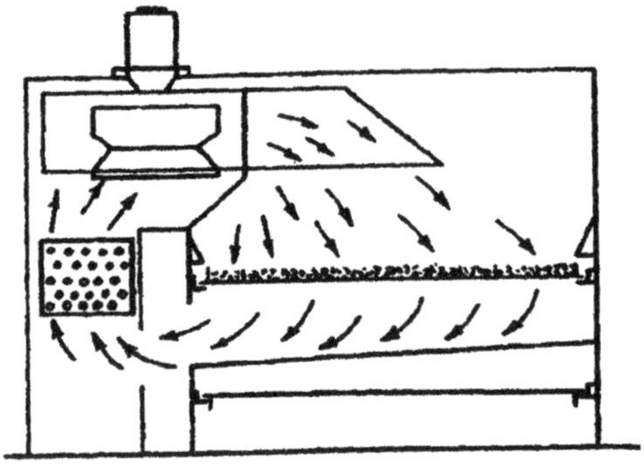

Figure B-12 Through-circulation dryer. (Source: Proctor & Schwartz, Inc.)

ing surfaces, creating a fire hazard. These dryers require that the wet material be in a state of granular or pelletted form so that hot air or gas may be readily blown through it. Figure B-12 illustrates a perforated plate or screen through-circulation dryer.

Another type of through-circulation dryer is the perforated drum dryer, which is specially suited for fiber staple, tow, and nonwoven fabrics. Material is continuously supported and conveyed on a series of perforated screen-covered suction drums installed in compartments similar in form to horizontal conveyor compartments.

Continuous Tunnel Dryers: These are dryers in which batch truck or tray compartments are operated in series. The solids to be dried are placed in trays or on trucks which move progressively through the tunnel (housing) in contact with hot gases. The number of trucks in such a dryer is variable, and can be as high as 100. The operation is semi-continuous as truck loads of wet material are moved at intervals into one end of the tunnel, and the whole string of trucks is periodically advanced through the tunnel until these are removed. Air movement, circulation, and heating methods vary in tunnel dryers. Three different arrangements are used, namely, counterflow, parallel flow, and combined flow (see Figure B-13). These dryers are used for food drying.

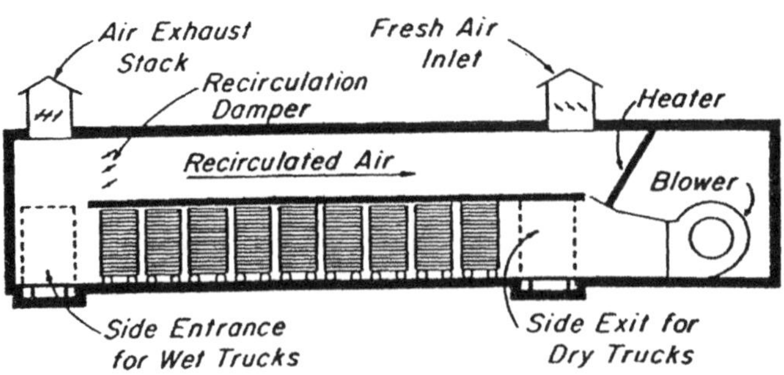

Figure B-13 Continuous tunnel dryer. (Source: Mujumdar, *Handbook of Industrial Drying*, 1995. Reprinted with permission of Marcel-Dekker, Inc.)

Rotary Dryers: A direct-heat rotary dryer is a horizontal cylinder rotated upon bearings, and usually slightly inclined towards the discharge end, through which gas is blown to dry material that is showered inside. They can be both batch and continuous. Batch dryers are usually one or two diameters long. Continuous dryers are at least four to ten diameters long. At each end, a stationary hood is joined to the cylinder by a rotating seal. These hoods contain the inlet and exit gas connections and the feed and product conveyors. One hood also attaches to the inlet gas heater. An array of flights are attached to the inside of the cylinder to provide showering of the material and assist the drying of the wet solids. Knockers are also installed to dislodge wet material that sticks to metal surfaces. Gas flow to these dryers may be either cocurrent or countercurrent. Cocurrent operation is preferred for heat-sensitive materials because gas and product leave at the same temperature. Countercurrent operation allows a product temperature higher than the exit gas temperature and the dryer efficiency may be as high as 70%. Some dryers have enlarged cylinder sections at the exit end to increase material holdup, reduce gas velocity, and minimize dusting. Indirectly heated tubes are installed in some dryers for additional heating capacity. To prevent dust and vapor escape at the cylinder seals, most rotary dryers operate at a negative internal pressure (–0.5 to –1.0 cm of water).

Rotary dryers are used primarily for large particles such as granules, pellets, and extrudates, and are a workhorse in the minerals processing industry. Figure B-14 illustrates the component parts of a countercurrent direct-heat rotary dryer.

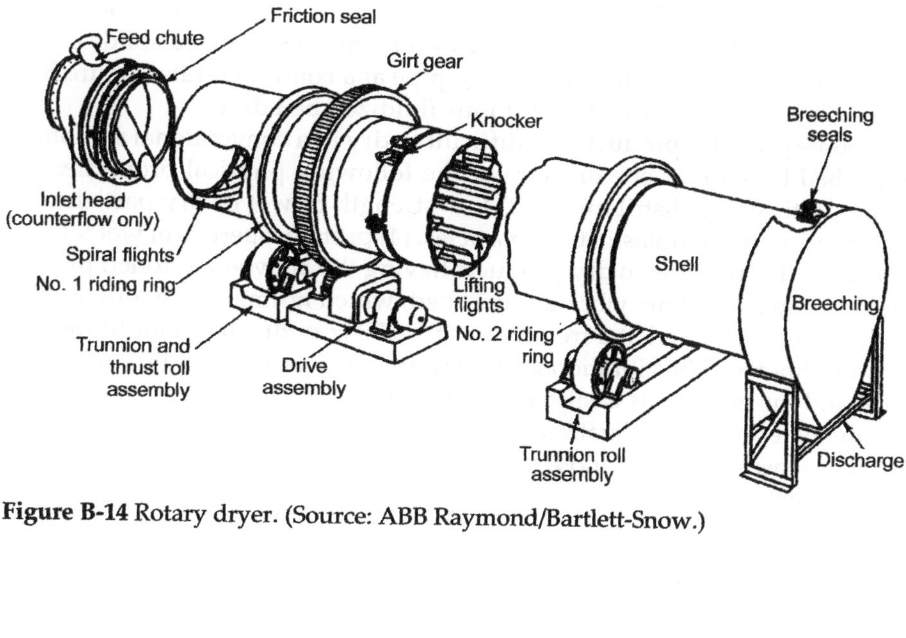

Figure B-14 Rotary dryer. (Source: ABB Raymond/Bartlett-Snow.)

Fluidized and Spouted Bed Dryers: Fluidized bed dryers are designed to introduce a hot gas stream at the base, whether in a container or a cylindrical vessel, through a dispersion plate, generally of the same diameter as the bed enclosure. For drying, the upward gas velocity is less than the terminal velocity of the particles, so that few particles are conveyed out of the bed. At the same time, gas bubbles rise fast enough to lift the particles directly above them. Particle motion is violent and a fluid bed exhibits intensive splashing at its surface. Since each particle is surrounded by gas, very little attrition of particles occurs and abrasion of the container is usually minimized. Because of the violent solids motion, mixing is substantially instantaneous and heat transfer rates are of a very high order, and drying is very fast compared to other types of dryers. Temperature throughout the bed becomes extremely uniform and it is not unusual for the material and the gases leaving the bed of a continuous fluidized bed dryer to be at the same temperature. A substantial freeboard is included above the top of the fluidized bed to allow particle disentrainment and fall-back into the bed. For proper fluidization, it is essential that sufficient pressure drop be taken across the gas distributor so that the gas is distributed uniformly across the entire bed area independently of the bed depth or bed behavior.

Fluidized bed dryers can be batch or continuous types. In batch types, the wet material is introduced into the dryer as a batch of material in a container, and is fluidized by a stream of heated air (or nitrogen) at a selected temperature. When the material is sufficiently dry, it is removed from the container.

Batch fluidized bed dryers are used when the capacity is small and it is preferable to operate batchwise because of quality assurances procedures, as in the pharmaceutical industry. Figure B-15 is a schematic drawing of batch fluidized bed dryer. In a continuous fluidized bed dryer, a continuous feed of wet material is introduced into the dryer, and the fluidized solids can be withdrawn from a overflow discharge pipe at a controlled rate. Figure B-16 is a schematic drawing of a continuous fluidized bed dryer.

Some granular products are difficult to dry in a conventional stationary fluid bed because of one or more of the following physical properties: (1) wide particle size distribution, (2) low strength of wet or dry particles, (3) stickiness or thermoplasticity of particles, (4) pasty properties of wet feed. In this case, it is possible to dry products having the above-mentioned properties by using a shallow vibrated fluid bed, which normally is a long rectangular trough vibrated at a frequency of 5–25 Hz with a half-amplitude of a few millimeters. The vibration vector is applied at an angle (0-45°) to the vertical and the material is easily transported through the dryer because of the combined effect of fluidization and vibration. Vibrated fluid bed dryers are used extensively in the food and pharmaceutical industries. Figure B-17 shows such a unit.

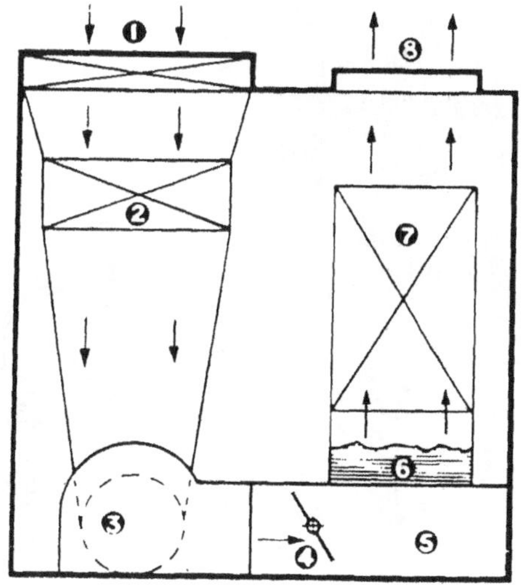

1 Fresh Air Inlet Filter.
2 Steam or Electric Heater.
3 Air Circulating Fan.
4 Air Volume Control Damper.
5 Air Diffuser.
6 Tub containing material.
7 Nylon Filter Bags.
8 Air Exhaust.

Figure B-15 Schematic of a batch fluidized bed dryer.

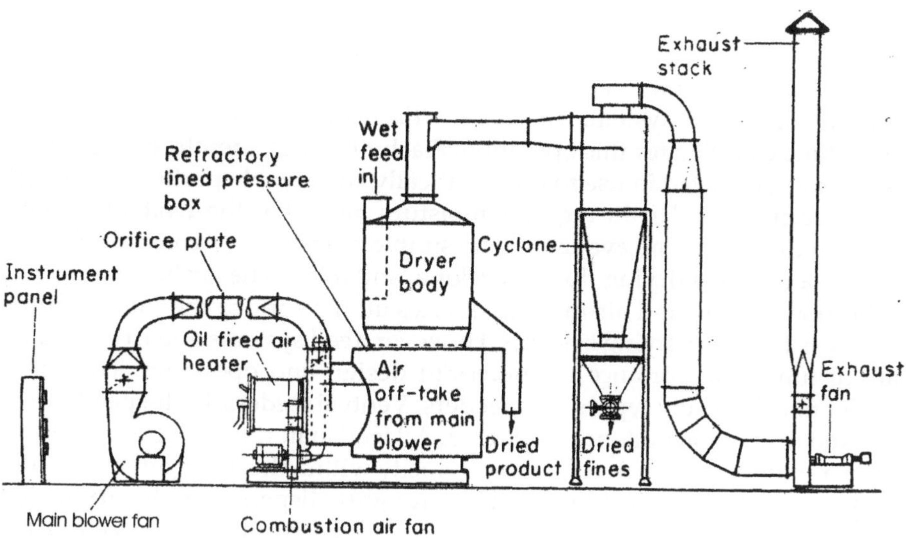

Figure B-16 Schematic of a continuous fluidized bed dryer.

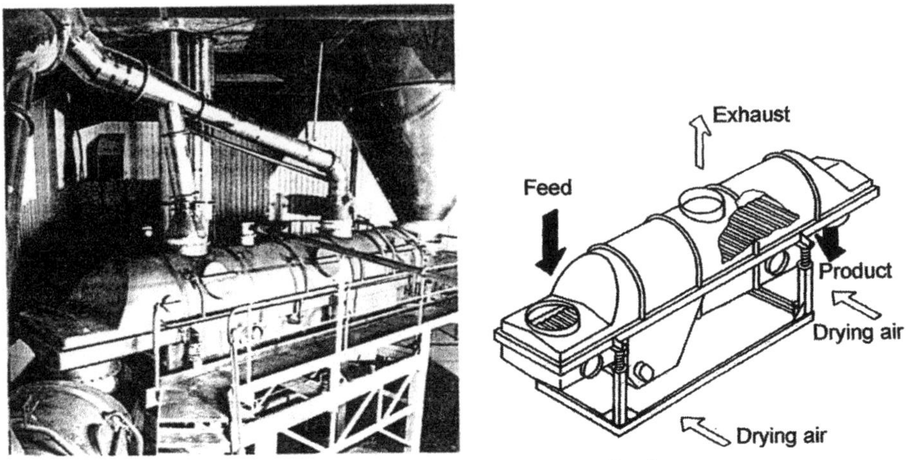

Figure B-17 Vibrated fluidized bed dryer. (Source: Niro, Inc.)

Spouted bed dryers are used for coarse particles that do not fluidize well. A single, high velocity gas jet is introduced under the center of a static particulate bed. The jet entrains and conveys a stream of particles up through the bed into the vessel freeboard where the jet expands, loses velocity, and allows the particles to be disentrained. The particles fall back into the bed and gradually move downwards with the peripheral mass until re-entrained. Particle-gas mixing is less uniform than in a fluidized bed.

Pneumatic Conveying Dryer: This type of dryer, also called a flash dryer, is in concept very simple. Wet feed material is introduced into a duct carrying hot drying air (or nitrogen), which conveys the drying material to the dry product collection system. The name "flash dryer" comes from the very short time duration for this operation. Flash drying is, therefore, a cocurrent operation, restricted in use to pneumatically conveyable particulate materials that can be dried to the required moisture content by the available driving force in the short time available. The simplest flash dryer consists of an air heater, a feeder, a drying duct, a cyclone, and a fan. The air heater, product collection system, and air-moving devices all have much the same requirements as for other drying systems. However, feed systems and drying ducts are of utmost importance to successful design and operation. The solids feeder may be of any type; screw feeders, Venturi feeders, high-speed grinders, and dispersion mills are employed. All feed material must be accelerated from a static condition to conveying velocity within the drying duct. It is also necessary to have the feed material thoroughly dispersed in the drying gas stream as rapid and complete dispersion allows drying to proceed faster, and more effective use is made of the available dryer residence time. Fans may be of the induced-draft or the forced-draft type. However, the former is usually preferred because the system can then be operated under a slight negative pressure. The main conveying and drying duct may be of circular or rectangular cross-section, but in either case should be extremely well finished internally and free from projecting edges which can interfere with gas flow or which may cause build-up of material.

Pneumatic (flash) dryers may be single-stage or multistage. The former is employed for evaporation of small quantities of surface moisture. Multistage dryers are used for difficult drying processes, for example, drying of heat-sensitive products containing large quantities of moisture, and drying materials initially containing internal as well as surface moisture. Figure B-18 is a schematic drawing of a simple single-stage flash dryer. A number of different flash dryer designs are presented in the book by Williams-Gardner (1971).

Spray Dryers: A spray dryer consists of a large cylindrical, usually vertical chamber through which hot gas is blown and into which a solution, slurry, or pumpable paste is sprayed by means of a suitable atomizer. Three atomizers are commonly used: (1) two-fluid pneumatic nozzles for very fine particles, between 10 and 100 μm, at rates less than two tons/hour; (2) single-fluid pressure nozzles for large particles, 125–150 μm, for dust-free products, and (3) centrifugal atomizers for various particle sizes at rates up to 150

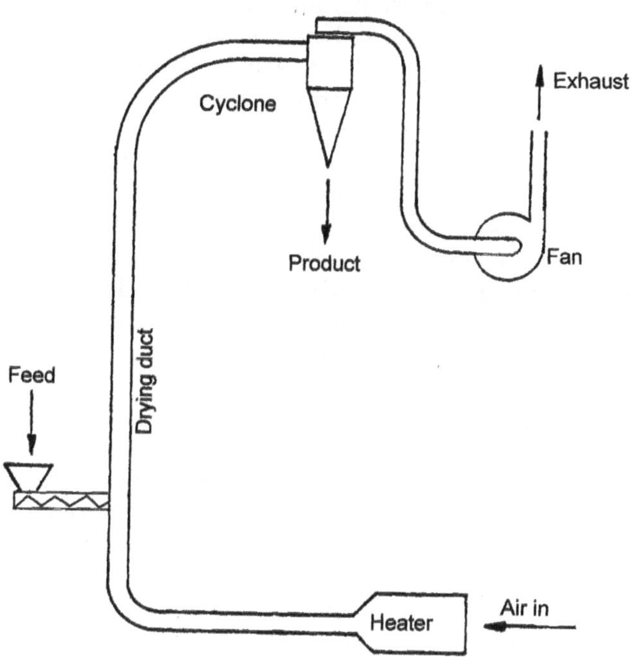

Figure B-18 Schematic of a pneumatic conveying (flash) dryer.

tons/hour. Because all drops must reach a nonsticky state before striking the chamber wall, the largest drop produced determines the size of the drying chamber. Chamber shape is determined by nozzle or disk spray pattern. Nozzle chambers are tall towers, usually having height to diameter ratios of 4–5. Disk chambers are large diameter and short, height being fixed by the fact that the discharge cone slope must be at least 60°, preferably 70°, to discharge dry product accumulation on the sloping wall. The exhaust gas and dry product then go to a product collection system which may be comprised of cyclones, baghouses, scrubbers, and fans.

A spray dryer may be cocurrent, countercurrent, or mixed flow. Cocurrent dryers are used for heat-sensitive materials because relatively high inlet gas temperatures, up to 800°C, may be used while holding the exit gas and product near 100°C.

Three types of spray drying systems are available: (1) open-cycle layout, (2) closed-cycle layout, and (3) semiclosed-cycle layout. The open-cycle layout is by far the most used in industrial spray drying. The open-cycle layout involves intake of drying air from atmosphere and the discharge of the exhaust air to the atmosphere after product collection (see Figure B-19a). The closed-cycle layout is used for drying solids wet with flammable solvents. Closed-cycle dryers ensure complete solvent recovery, and prevention of (a) explosion and fire risks associated with solvent handling, (b) solvent vapor and powder emissions, and (c) oxidation/degradation to the dried product during drying. Closed-cycle systems are gastight installations operating with an inert drying medium (usually nitrogen). These drying systems operate at a slight pressure to prevent possible air leakage inward (see Figure B-19b). Semiclosed cycle dryers are available in two variations: (a) partial recycle and (b) vented "closed" cycle. In partial recycle dryers the exhaust air is partially recycled, utilizing the waste heat in this stream. The amount of recycle depends upon the permissible water vapor content level in the inlet drying air, and the effect of this level on the required outlet drying temperature to achieve the desired dried product moisture content. In the vented "closed" cycle design the majority of the drying gas is recycled. A semiclosed-cycle dryer is shown in Figure B-19c.

B4.2 Indirect-Heat Dryers

In indirect-heat dryers, heat is transferred mostly by conduction, but heat transfer by radiation is significant when conducting surface temperatures exceed 150°C. For jacketed vessels, steam is the most common heating medium because the condensing-side film resistance is insignificant compared to solids material-side resistance. Hot water is circulated for low temperature heating. Heat transfer fluids are often used for high temperature heating. Indirect-heat rotary dryers and calciners operating at temperatures exceeding 200°C usually are furnace-enclosed. The cylinders are heated

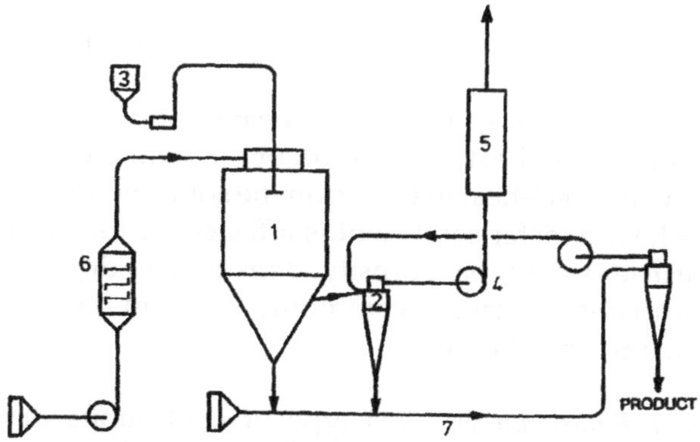

1. Drying chamber with atomizer
2. Dry collector
3. Feed system
4. Exhaust fan
5. Scrubber
6. Heater
7. Pneumatic conveying system

(a)

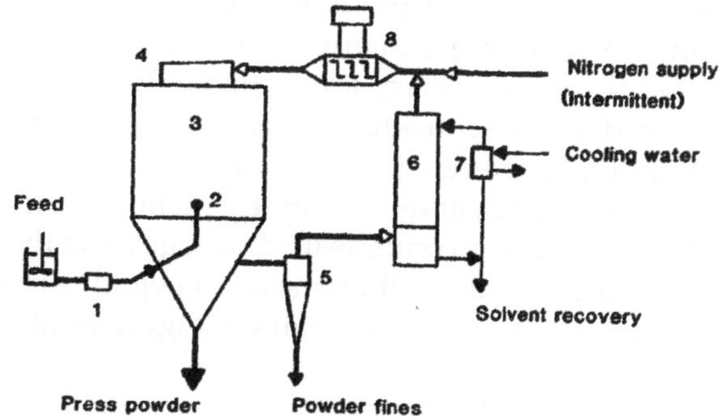

1. Feed pump with agitated tank
2. Nozzle atomizer
3. Drying chamber
4. Gas disperser
5. Cyclone
6. Scrubber condenser
7. Solvent cooler
8. Indirect drying inert gas heater

(b)

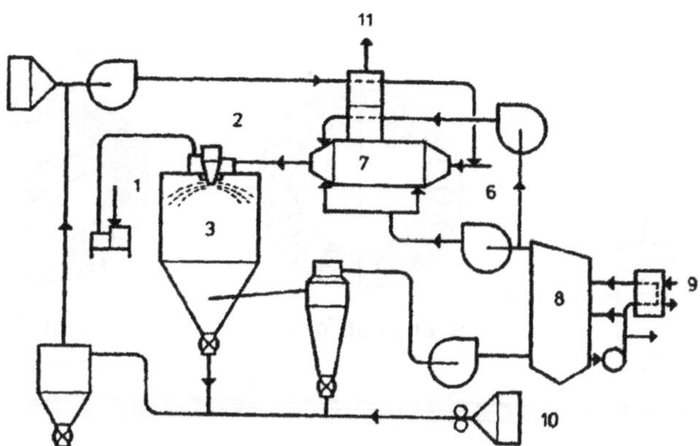

4

1. Feed system
2. Atomizer
3. Drying chamber
4. Product outlet
5. Combustion air
6. Fuel burner
7. Heater (direct fired)
8. Scrubber/condenser
9. Cooling medium
10. Conveying air
11. De-activated air bleed

(c)

Figure B-19 Spray dryer systems. (a) open-cycle layout; (b) closed-cycle layout; (c) semiclosed-cycle layout externally by electric or gas-fired radiant heaters and circulating combustion products. Regardless of heating medium or method, the primary heat transfer resistance in indirect-heat drying is on the solids material-side. The solids material-side heat transfer coefficient is affected by the rapidity of material agitation, particle size, shape, porosity, density, and degree of wetness.

Some indirect-heat dryers commonly used in the chemical process industries are described below.

Steam-Tube Rotary Dryer: This type of continuous dryer is used for drying of granular or powdery solids that cannot be exposed to ordinary atmospheric or combustion gases. It consists of a horizontal rotating cylinder in which are installed one or more circumferential rows of steam-heated tubes. These tubes extend axially the length of the cylinder and are connected to a steam and condensate manifold. Steam is introduced and condensate is removed through a rotary joint attached to the manifold at the product discharge end. Feed is introduced and purge gas usually is removed through a stationary throat piece attached to the rotating cylinder by a sliding seal at the front end. The cylinder is slightly inclined to the horizontal to direct material flow and air and condensate drainage from the tubes. To prevent dust and vapor escape at the cylinder seals, a negative internal pressure (–0.5 to –1.0 cm of water column) is maintained. Figure B-20 shows a steam-tube rotary dryer.

Screw Conveyor Dryers: This type of dryer is limited to materials that will not pack and change desirable physical properties under pressure or coat the walls or screws. Indirect-heat screw conveyor dryers consist of a trough with

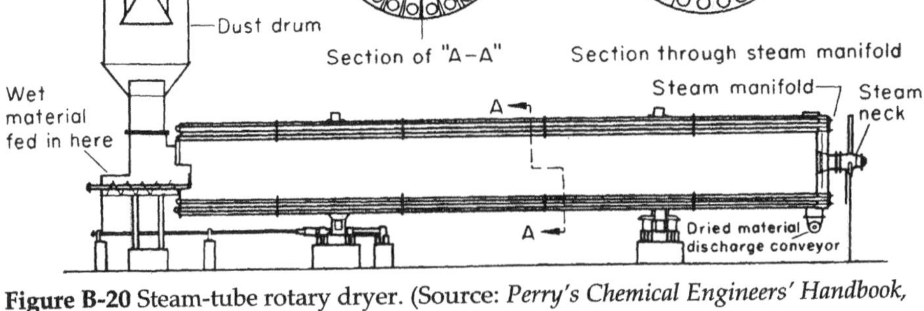

Figure B-20 Steam-tube rotary dryer. (Source: *Perry's Chemical Engineers' Handbook*, 7th edition. Reprinted with permission of McGraw-Hill.)

one or more hollow screws heated with steam, hot water, or heat transfer fluid. The trough is often jacketed to provide additional heat transfer surface, but the trough surface is only a small fraction of the total heating surface. One trough may carry as many as four screws. A popular arrangement used for both batch and continuous drying consists of two screws that convey in opposite directions in a single trough. This internal recycle arrangement is used for drying slurries and solutions. Several passes can be arranged one above the other so that the discharge from one trough feeds the one below. In this way, different zones of drying, temperature, and heating can be arranged. A gentle stream of air or nitrogen can be passed countercurrent though the trough or a slight vacuum can be pulled on the trough to assist in the removal of moisture. In a continuous dryer arrangement, product is discharged through an overflow weir on the opposite side from the feed inlet. Figure B-21 shows a two-screw, two-high screw conveyor dryer.

Agitated Paddle and Disk Dryers: These dryers come in several designs and can be batch or continuous and can operate at atmospheric and vacuum conditions. They consist of a trough in which hollow paddles or disks are mounted on a rotating shaft. The trough is also often jacketed. Steam, hot water, or heat transfer fluid is used as the heating medium. The paddles or disks usually rotate at a low speed (10–40 rpm). The paddles or disks scrape the product from the walls and mix and transport the solids along the dryer length. Disk dryers often contain adjustable scrapers for continuous disk doctoring to maintain a clean heating surface. These units may have one or two parallel shafts and often have stationary, lump-breaker bars that intermesh with the moving paddles or disks.

Figure B-21 Screw conveyor dryer. (Source: Hosokawa Bepex.)

In a continuous unit, the wet feed is fed into the top of the dryer at one end, and as the powder is agitated slowly by the heated rotating paddles or disks, the moisture is conveyed out by a flow of hot air or nitrogen. They can be operated in a closed-cycle mode where organic solvents are the moisture, and the solvents can be recovered in a condenser.

Paddle dryers are used for drying lumpy materials and polymer pellets, and disk dryers are suitable for granular, essentially free-flowing materials. Figure B-22 shows an agitated paddle dryer made by one manufacturer.

Rotating Double-Cone Vacuum Dryer: This type of dryer is used very often in the pharmaceutical industry because it provides a gentle, low-temperature drying operation. The use of vacuum for this dryer, as well as for other vacuum dryers, permits operation at lower temperatures than at atmospheric pressure, which is desirable and often necessary for heat-sensitive materials. It is similar to the rotating double-cone blender in construction, but has an external heating jacket and an internal stationary vapor-outlet tube, usually with a "sock" on its end to prevent dust from entering the vapor-outlet pipe. Vacuum is applied to the dryer through hollow trunnions with suitable packing glands. The vacuum source is nowadays often a liquid ring or mechanical vacuum pump, which have replaced steam jets because of environmental pollution concerns. The solvent vapors, along with the vacuum air, pass through a condenser, located upstream of the vacuum

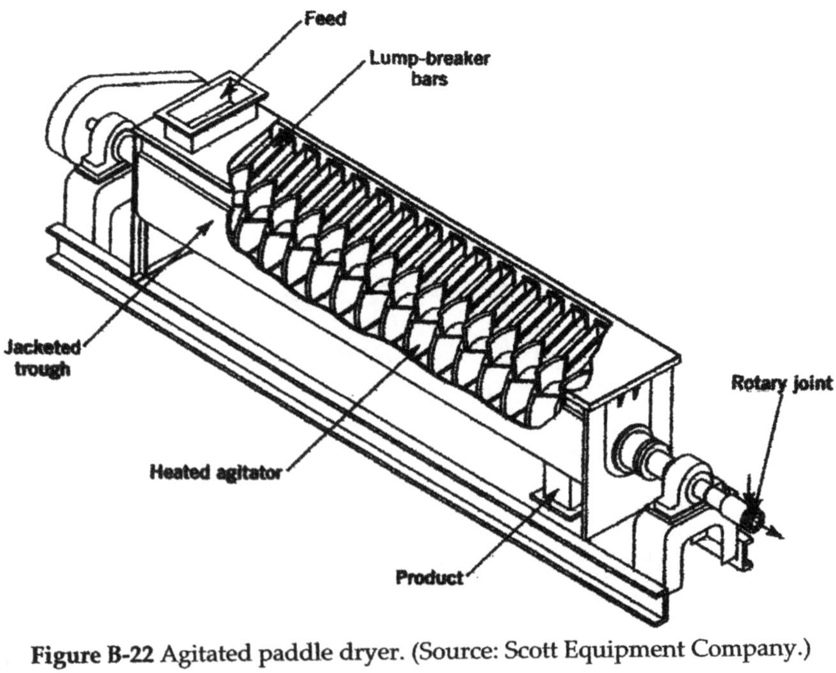

Figure B-22 Agitated paddle dryer. (Source: Scott Equipment Company.)

equipment. This dryer is built with a clean, smooth interior surface, since a sliding action occurs between the product and the walls of the vessel. Smoothness assures complete discharging of the product and eliminates the necessity for extensive cleaning. Materials of construction are usually stainless steel or glass-lined carbon steel. This type of dryer is only suitable for free-flowing solids that do not have a tendency to form lumps or bridges. Figure B-23 is a schematic drawing of such a unit.

Vacuum-Shelf Dryers: Vacuum-shelf dryers are indirect-heated batch dryers consisting of the following components: (a) a vacuum-tight chamber usually constructed of cast iron, steel, stainless steel, or lined/clad steel plate, (b) heated supporting shelves within the chamber, (c) a vacuum source, and (d) usually a condenser. The chamber has one or two doors, depending on its size, which are sealed with resilient gaskets of rubber or similar material. Hollow shelves of flat steel plate, fastened permanently to the inside walls of the chamber, are connected in parallel to inlet and outlet heating medium headers. The heating medium, generally steam (in the range of 100 psig to subatmospheric pressure for low-temperature operations), enters through one header and passes through the hollow shelves to the exit header. Low temperatures can be provided by circulating hot water, and high temperatures can be obtained by circulating hot heat transfer fluid. Some small dryers use electrically heated shelves. Wet product is loaded into drying trays which are then placed on the shelves, vacuum is applied, the shelves are heated, and the product is dried without agitation.

Vacuum-shelf dryers are used extensively for batch drying of pharmaceuticals, temperature-sensitive or easily oxidizable materials, and materials so valuable that labor cost is insignificant. All kinds of moist solids such as pastes, powders, or coarse lumps can be dried. These dryers are particularly useful for drying small batches of material wet with highly flammable or valuable solvents. They are used primarily in pilot plant operations. This

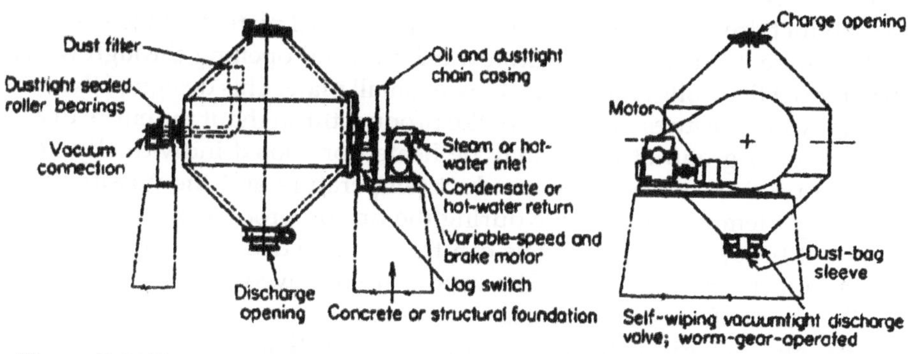

Figure B-23 Rotatiung double-cone vacuum dryer. (Source: Stokes Vacuum, Inc.)

Figure B-24 Vacuum shelf dryer. (Source: Steri Technologies, Inc.)

type of dryer is also used for freeze-drying operations. Figure B-24 shows a vacuum-shelf dryer.

Plate Dryers: The plate dryer is an indirect-heat, continuous dryer available in three modes of operation: atmospheric, gastight, or full vacuum. In these dryers, the product to be dried is metered and continuously fed onto the top plate. A vertical rotating shaft provided with radial arms and self-aligning plows conveys the material in a spiral pattern across stationary plates, which are heated by steam or other heat transfer medium.

The drive unit is located at the bottom of the dryer and supports the central-rotating shaft. Small plates with internal rims and large plates with external rims are arranged in alternating sequence. This arrangement makes the product drop from the outside edge of the small plate down to the large plate, where it is conveyed again toward a central opening through which the product then falls onto the next lower small plate. This design of the conveying system ensures plug flow of the product throughout the entire dryer. Each plate or group of plates may be heated or cooled individually, thus offering precise control of the product temperature and the possibility of adjusting a temperature profile during the drying process. Thermal degradation of heat-sensitive materials can thus be avoided, and cooling of the final product can be achieved. The vapors are removed from the dryer by a small amount of heated purge gas or vacuum.

The atmospheric plate dryer is a dust-tight system. The dryer housing is an octagonal, panel construction, with operating pressure in the range of ±2 inches water column gauge. An exhaust fan draws the purge gas through the housing for removal of the vapors from the drying process. The vapor-laden exhaust air is passed through a dust filter or scrubber (if necessary) and is discharged to the atmosphere. The purge air velocity through the dryer is in the range of 0.10 to 0.15 m/sec, resulting in minimal dust entrainment.

The gastight plate dryer, together with the components of the gas circulation system, forms a closed system. The dryer housing is semicylindrical and is rated for a nominal pressure of ±20 inches water column gauge. The evaporated moisture vapors go to a condenser, where they are recovered, and the purge gas (typically nitrogen) is recirculated back to the dryer via a blower and heat exchanger.

The vacuum plate dryer is provided as part of a closed system also. This dryer has a cylindrical housing and is rated for full vacuum operation (typical pressure range of 0.4–4 psia). The exhaust gas is evacuated by a vacuum pump and is passed through a condenser for solvent recovery. There is no purge gas system required for operation under vacuum. For a vacuum plate dryer, specially designed feed and product discharge locks (rotary valves usually) are required to allow continuous operation of the dryer under near-full vacuum.

The plate dryer is limited in its scope of application only by the consistency of the feed material (the products must not be friable, must be free-flowing, and must not undergo phase changes) and drying temperatures up to 320°C. Applications include pharmaceuticals, specialty chemicals, foods, polymers, pigments, etc. Initial moisture can be as high as 65%. Figure B-25 is a schematic of a plate dryer.

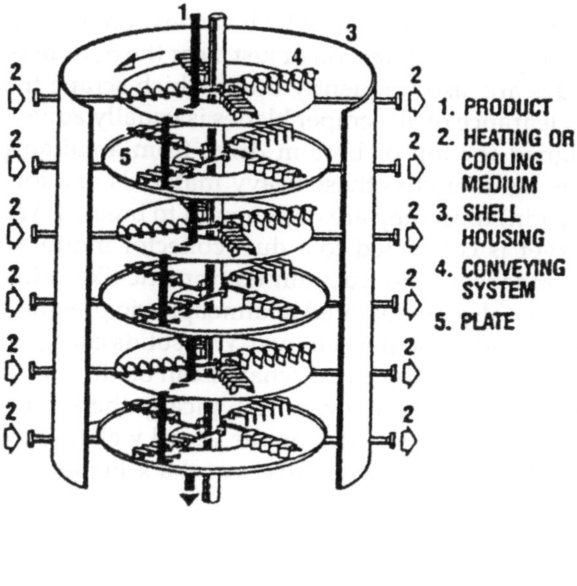

Figure B-25 Schematic of a plate dryer. (Source: Krauss-Maffei Process Technology, Inc.)

Vertical Orbiting Screw Vacuum Dryers: This type of batch dryer is similar to the vertical orbiting screw blender, but a heat transfer medium (steam, hot water, or heat transfer fluid) is circulated through the hollow screw and the vessel walls in a half-pipe coil or dimple jacket. To ensure that no vapor condensation will occur in the top head of the dryer, the head is heated by either a half-pipe coil or heat tracing. This dryer has a bottom-screw drive as opposed to the top-screw drive unit of the blender, thus eliminating any mechanical drive components inside the vessel. It makes maximum use of the product-heated areas—the filling volume of the vessel (up to the knuckle of the dished head) is the usual product loading. The vapors evaporated from the solids are evacuated by a vacuum system through a condenser for recovery of the solvent.

As the top head is free of drive components, this allows space for additional process nozzles, manholes, explosion vents, and a temperature lance for direct, continuous product temperature measurement in the vessel. A dust collector is usually mounted directly on the top head of the dryer, thus allowing any entrapped dust to be pulsed back into the dryer. Because there are no drive components inside of the dryer the risk of batch contamination from leaking gear lubricants are eliminated.

This type of dryer is frequently used in the pharmaceutical industry, as well as for drying polymers, additives, inorganic salts, and many specialty chemicals. Figure B-26 is a schematic of a vertical orbiting screw vacuum dryer.

Agitated Vacuum Pan Dryers: This type of batch dryer is used widely for drying heavy, sticky, and dough-like materials that would overload or break the ribbons or paddles in paddle or disk vacuum dryers. The dryer consists of a vertical, jacketed cylinder (the pan) having a jacketed dished or flat bottom and a mechanically driven vertical shaft which carries radial arms supporting scraper blades (ploughs) which rest on or just clear the bottom of the pan. Vertical scraper blades are also frequently fitted which scrape the vertical sides of the pan. The clearance of all scraper blades is usually adjustable. The shaft may be designed for top or bottom drive from a suitable reduction gear and drive. The agitator stirs these heavy materials at a low speed (about 1–4 revolutions/min) until they are dry enough to break down into particulate form. The dryer is connected to a dust collector mounted above the vapor nozzle, a condenser, and vacuum equipment like other vacuum dryers. These dryers can also be operated at atmospheric pressure, and in this mode, a purge gas is used for vapor removal. Wet solids charging is through a nozzle on the top, and dry product discharge can be through a center-mounted discharge valve in the base, or more typically these days with later designs, through a side-mounted valve on the vessel wall. Figure B-27 shows a picture of an agitated vacuum pan dryer and a schematic of top-drive and bottom-drive units.

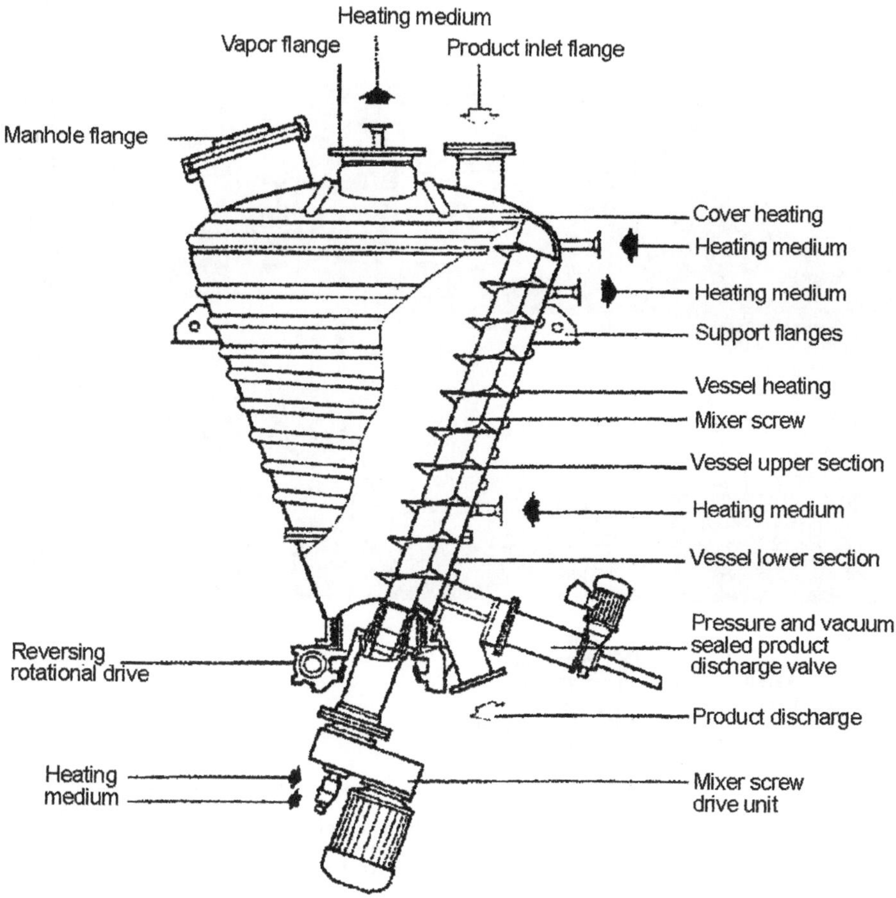

Figure B-26 Schematic of a vertical orbiting screw vacuum dryer. (Source: Krauss-Maffei Process Technology, Inc.)

Agitated Spherical Vacuum Dryer: This type of dryer is also known as a Moritz dryer (other manufacturers now make this unit). It is used primarily in the pharmaceutical industry. It consists of a stationary, spherical, jacketed housing equipped with a high-torque agitating device and rapidly rotating knives (called choppers). The moist material is whipped up by the turbine and possible lumps are reduced into smaller particles by the choppers. Heating is provided by steam, hot water, or heat transfer fluid circulating through the jacket. The solvent vapors pass through a dust collector mounted above the vapor nozzle, go to a condenser, and then to vacuum equipment. Wet solids are charged through a nozzle on the top and dry product is discharged through a bottom nozzle. This type of dryer is used for drying of free-flowing solids, or even pasty products. Figure B-28 is a schematic of an agitated spherical vacuum dryer system.

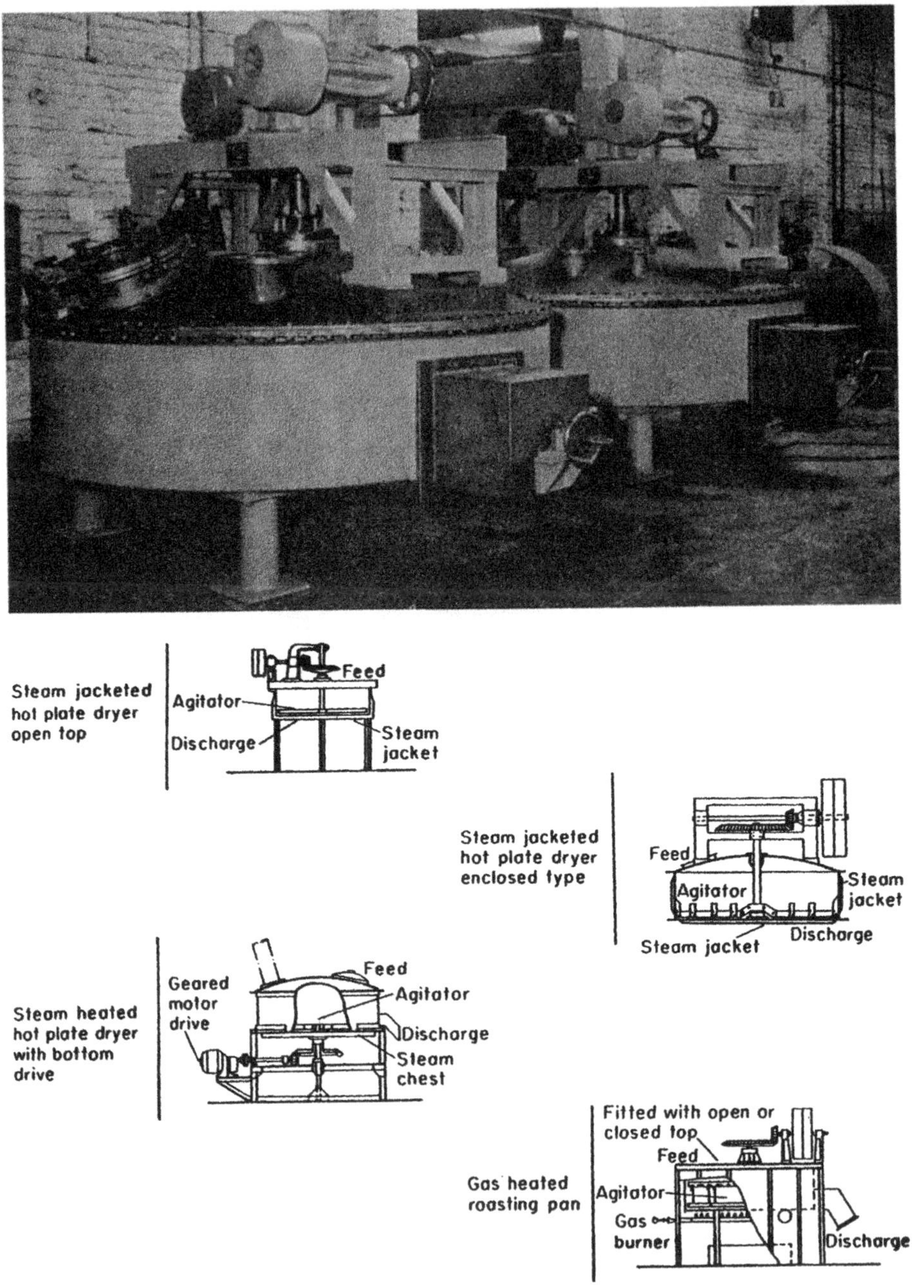

Figure B-27 Agitated vacuum pan dryers. (Source: *Industrial Drying*, by A. Williams-Gardner, CRC Press, Cleveland, Ohio, 1971.)

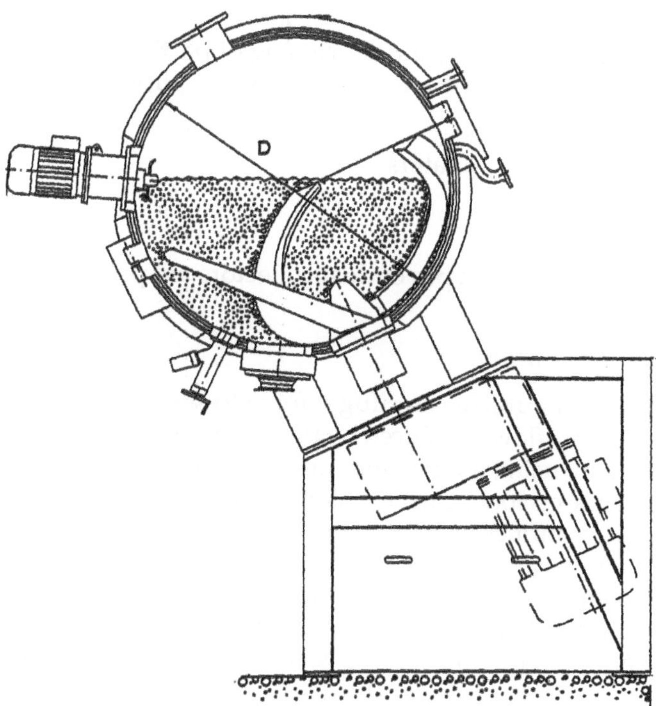

Figure B-28 Agitated spherical vacuum dryer. (Source: Rosenmund Division, De Dietrich Process Systems, Inc.)

Agitated Vacuum Filter-Dryer: This type of dryer combines the functions of filtering and drying in one vessel. This device is essentially a Nutsche-type (single plate) filter equipped with a paddle-type agitator, and heating provided through the heated paddles, vessel jacket, and filter plate. The agitator paddles are streamlined blades with a large surface area, each equipped with mixing/smoothing shovels or paddles. The unit has a dust collector mounted directly on the dryer housing, and the solvent vapors go to a condenser and then to the vacuum equipment. After the filtration cycle is completed the cake is compressed by smoother arms on the paddle, and the cake is washed and reslurried (if needed), and then dried by a heating medium circulated in the heat transfer components. Wet solids are charged through a nozzle on the top head and dry product is removed through a nozzle on the side of the vessel. See Figure B-29 for a schematic of a typical filter-dryer system.

More detailed discussions of the dryers described in this section, as well as many other types of dryers, are presented by Mead (1964), Moyers (1997), Mujumdar (1996), and Williams-Gardner (1971).

B5 DUST COLLECTORS

Dust emitted from handling and processing of particulate solids can be collected and controlled by the following types of equipment:

- Cyclone separators
- Electrostatic precipitators
- Fabric filters
- Wet scrubbers

Discussions of these types of dust collectors are presented below.

B5.1 Cyclone Separators

Cyclone separators utilize centrifugal force to separate particulates from a gas stream. Vortex flow is induced by the design of the gas-inlet duct. This is accomplished by using a tangential inlet or an axial inlet with swirl vanes. After entering a cyclone, carrier gas in which particulates are suspended undergoes two confined vortices—one ascending and one descending. The ascending vortex carries the clean gas out as centrifugal forces drive the heavier particulates toward the cyclone wall. Figure B-30 shows a schematic diagram of a typical cyclone with its vortex-flow patterns.

Although cyclones do not have very high collection efficiencies in relation to particle size, they can be used effectively for relatively coarse particles—generally larger than 10 μm in size. The primary advantages of cyclone separators are simplicity in construction, design, and economy, and

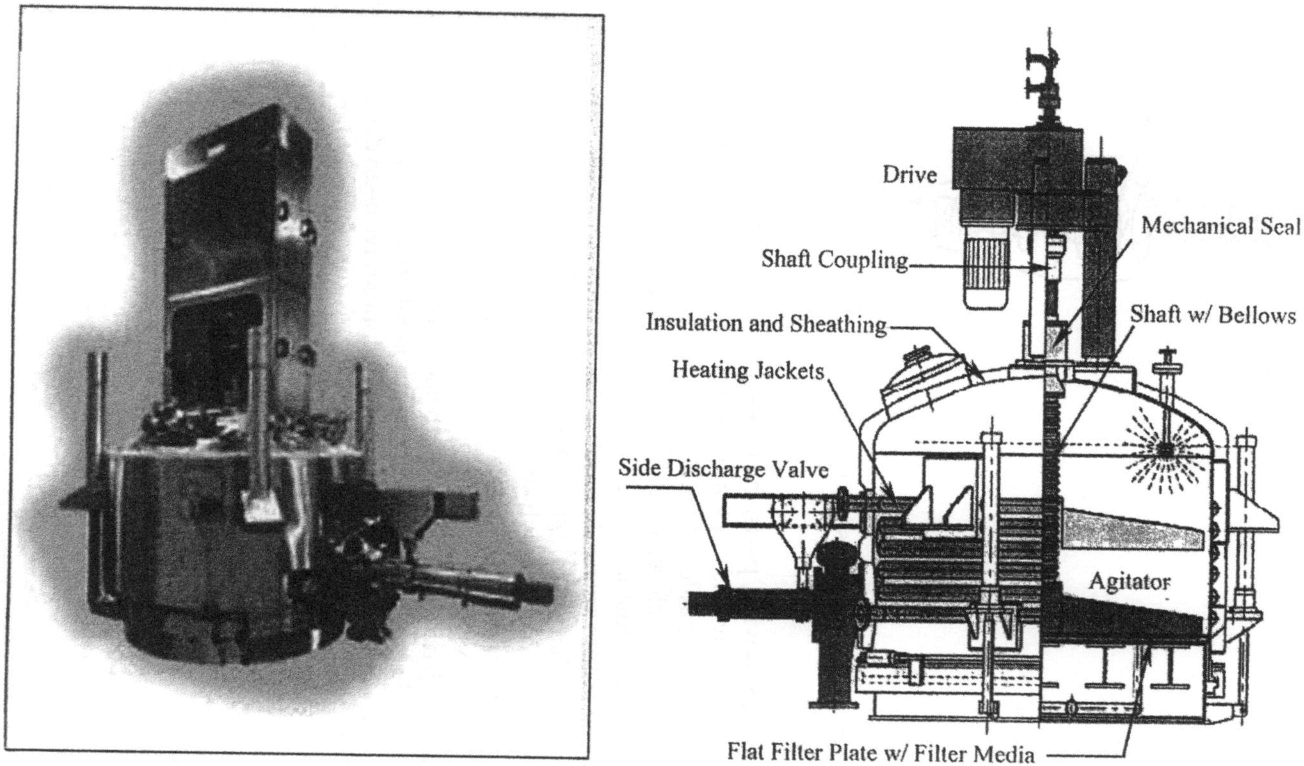

Figure B-29 Agitated vacuum filter-dryer. (Source: Rosenmund Division, De Dietrich Process Systems, Inc.)

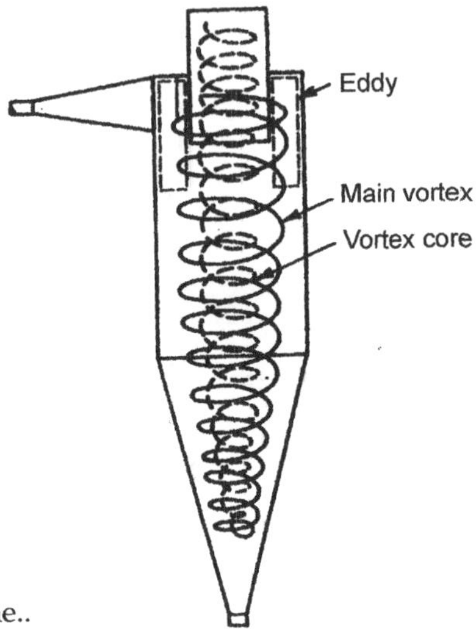

Figure B-30 Vortex and eddy flows in a cyclone..

since there are no moving parts, cyclones are relatively maintenance-free. Pressure drops are low and they are efficient for large particles (above 10 μm), being additionally able to handle fairly large dust loadings. Drawbacks include, as mentioned before, lower efficiency for small particles and sensitivity to variations in dust loading.

Cyclone separators are made in four basic types: (A) tangential gas inlet with axial dust discharge, (B) tangential gas inlet with peripheral dust discharge, (C) axial gas inlet through swirl vanes with axial dust discharge, and (D) axial gas inlet through swirl vanes with peripheral dust discharge. Figure B-31 shows a schematic diagram of these types. Although there are differences in construction of these four basic units, the fundamental operating principles and chief constructional components remain the same.

Over the years, many different types of cyclones have been designed and built. However, the Type A (also called reverse-flow or cone-under-cylinder design) shown in Figure B-32 is the type used most often in industry. In this design, the gas-solids stream enters the cyclone at the cylinder top where the shape of entry causes the gas to spin. Tangential, scroll, and swirl vane entries have been used as shown in Figure B-33; tangential entries are the most common.

To adequately design a cyclone, eight dimensions must be specified, as follows (see Figure B-32):

a = height of inlet duct
b = width of inlet duct
D_e = diameter of gas-outlet tube
S = depth of outlet tube

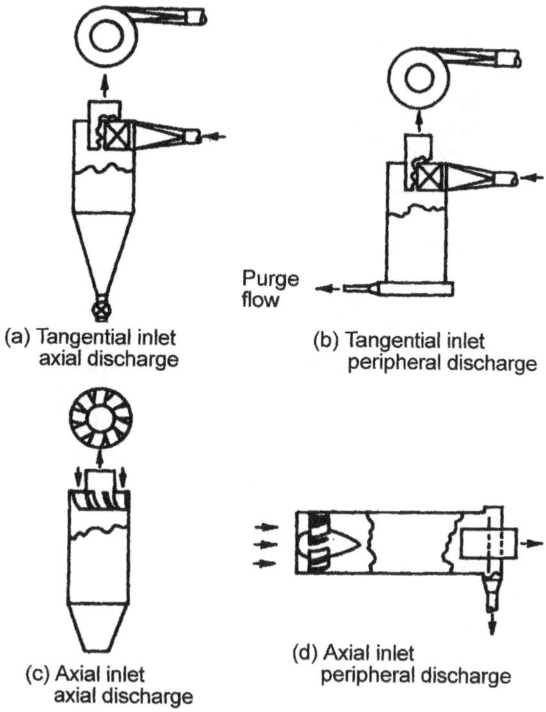

Figure B-31 Types of cyclones in common use.

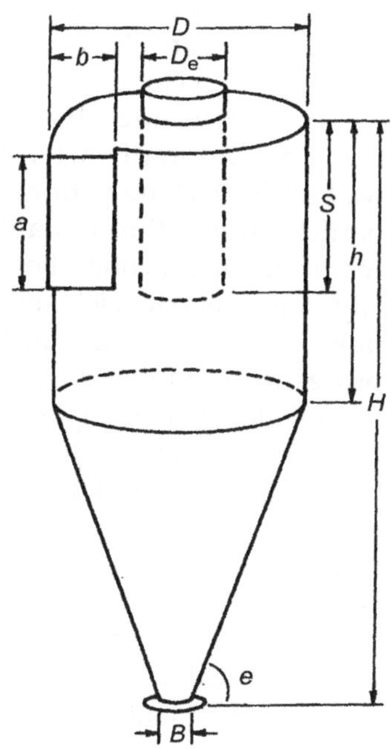

Figure B-32 Reverse flow cyclone with tangential gas inlet.

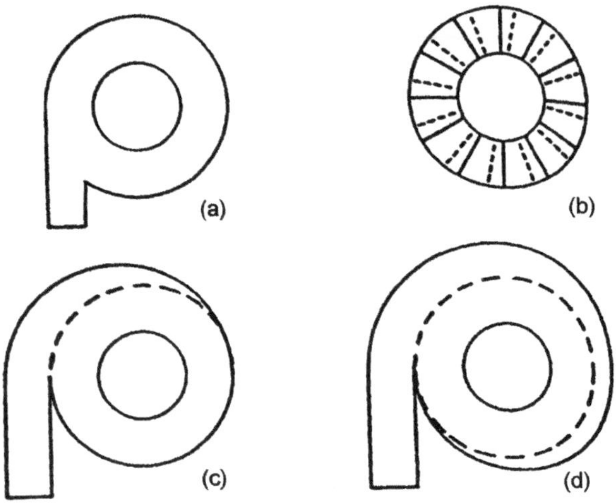

Figure B-33 Types of cyclone entries. (a) Tangential; (b) swirl vane; (c) half scroll; (d) full scroll.

h = height of cylindrical part of cyclone
H = total height of cyclone
B = diameter of solids-outlet tube
D = diameter of cyclone barrel

Table B-2 presents values of these dimensions and other design values as recommended by a number of researchers in cyclone design and operation (Leith and Jones, 1997).

Cyclones are also classified as (1) high-efficiency cyclones and (2) high-throughput cyclones. For high-efficiency cyclones, the inlet gas velocity is higher, thereby imparting a higher centrifugal force. They are generally less than one foot in diameter and have long cones. Heavy particles reach the wall of the cyclone with much smaller angular movement, whereas lighter particles travel through a much greater angle to reach the wall, therefore requiring longer cones. For high-throughput cyclones, diameters are generally larger, efficiencies are moderate to lower, and they can handle larger flowrates. Applications include operations such as grinding, buffing, fiber processing, and wood chip separation. In high-throughput cyclones, particles greater than 50 μm are collected with great efficiency.

Cyclones can be built as either single units or in multiples, and can also be arranged in parallel or in series. When high-efficiency cyclones are used, it is customary to operate a number of them in parallel in order to achieve practical gas volumes for each cyclone. If the number of cyclones in parallel is small, each cyclone should have its own inlet and its own dust collection chamber. However, when the number of cyclones in parallel is large, the only practical arrangement is to use a common inlet plenum chamber, a common dust collection chamber, and a common outlet plenum chamber

TABLE B-2

Standard Dimensions for Reverse-Flow Cyclones

Source	Recommended Duty	D	a/D	b/D	D_e/D	S/D	h/D	H/D	B/D	ΔH	Q/D^2 (m/h)
Stairmand	High-efficiency	1	0.5	0.2	0.5	0.5	1.5	4.0	0.375	6.4	5500
Swift	High-efficiency	1	0.44	0.21	0.4	0.5	1.4	3.9	0.4	9.2	4940
Lapple	General-purpose	1	0.5	0.25	0.5	0.625	2.0	4.0	0.25	8.0	6860
Swift	General-purpose	1	0.5	0.25	0.5	0.6	1.75	3.75	0.4	7.6	6680
Stairmand	High-throughput[a]	1	0.75	0.375	0.75	0.875	1.5	4.0	0.375	7.2	16,500
Swift	High-throughput[a]	1	0.8	0.35	0.75	0.85	1.7	3.7	0.4	7.0	12,500

[a]Scroll type gas entry used.

Source: Fayed, M. E., and Otten, L. *Handbook of Powder Science and Technology* 2nd edition, 1997, Table 15.1. Reprinted with permission of Kluwer Academic/Chapman-Hall.

Figure B-34 Typical multiple parallel cyclone system.

(see Figure B-34). Cyclones in series are used when separation with higher efficiency is required, higher flow rates of carrier gas are encountered, or when it is required to protect a smaller high-efficiency cyclone from larger abrasive particles. Large particles are initially collected in lower-efficiency or large cyclones, followed by a high-efficiency cyclone.

More detailed information on cyclone separator design and operation are presented by Stern (1977), Bhatia and Cheremisinoff (1977), Leith and Jones (1997), Heumann (1997), and Zenz (2001).

B5.2 Electrostatic Precipitators

Electrostatic precipitators (ESPs) are used where very high collection efficiency is required as they are capable of collecting fine dusts at concentrations well below 10 g/m^3.

ESPs remove dust or liquid from a gas stream by utilizing the force resulting from an electrical charge in the presence of an electrical field. The particles are given an electric charge by forcing them to pass through a corona, a region in which gaseous ions flow. The electrical field that forces the charged particles to the walls comes from electrodes maintained at high voltage in the center of the flow lane.

The basic components of an ESP are the discharge electrode, the collecting electrode, the high-voltage power supply, and the electrode cleaning system. Figure B-35 is a schematic diagram of an ESP showing the various components.

There are five types of ESPs: (1) the plate-wire precipitator (the most common type), (2) the flat-plate precipitator, (3) the tubular precipitator, (4) the wet precipitator, and (5) the two-stage precipitator. These are briefly described below.

Plate-Wire Precipitator: Plate-wire ESPs are used in a variety of industrial applications, including coal-fired boilers, cement kilns, solid waste incinerators, electric arc furnaces, just to name a few. In a plate-wire ESP, gas flows

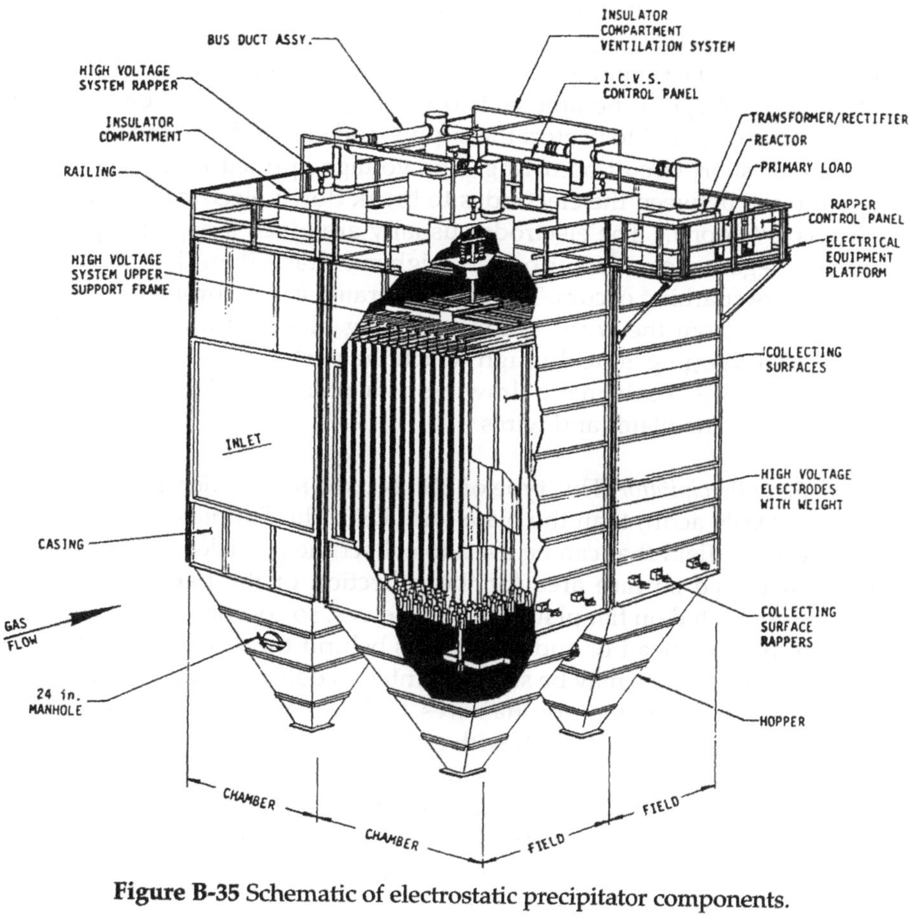

Figure B-35 Schematic of electrostatic precipitator components.

between parallel plates of sheet metal and high-voltage electrodes. These electrodes are long wires weighted and hanging between the plates or are supported there by mast-like structures (rigid frames). Within each flow path, gas flow must pass each wire in sequence as it flow through the unit. The plate-wire ESP allows many flow paths to operate in parallel, and each lane can be quite tall. As a result, this type of precipitator is well-suited for handling large volumes of gas. The need for rapping the plates to dislodge the collected material requires the plates to be divided into sections, often three or four in series with one another, which can be rapped independently. The power supplies are often sectionalized in the same way to obtain higher operating voltages; further electrical sectionalization may be used for increased reliability. Dust also collects on the discharge electrode wires and must be periodically removed similar to the collector plates.

The power supplies for the ESP convert the industrial AC voltage (220 to 480 volts) to pulsating DC voltage in the range of 20,000 to 100,000 volts, as needed. The supply consists of a step-up transformer, high-voltage rectifiers, and sometimes filter capacitors. The unit may supply either half-wave or full-wave rectified DC voltage. There are auxiliary components and controls to allow the voltage to be adjusted to the highest level possible without excessive sparking and to protect the supply and electrodes in the event of a heavy arc or short circuit occurring. The voltage applied to the electrodes causes the gas between the electrodes to break down electrically, an action known as a "corona." The electrodes usually are given a negative polarity because a negative corona supports a higher voltage than does a positive corona before sparking occurs. The ions generated in the corona follow electric field lines from the wires to the collecting plates, and thus each wire established a charging zone through which the particles must pass.

Figure B-36 is a schematic drawing of a plate-wire ESP showing the arrangement of the plates and wires.

Flat-Plate Precipitator: This type of ESP is used for smaller gas flow rates (100,000–200,000 acfm) than the plate-wire ESP. The flat plates increase the average electric field that can be used to collect the particles, and they provide an increased surface area for the collection of the particles. Corona cannot be generated on flat plates by themselves, so corona-generating electrodes are placed ahead of and sometimes behind the flat- plate collecting zones. These electrodes may be sharp-pointed needles attached to the edges of the plates or independent corona wires. Unlike plate-wire or tubular ESPs, this design operates equally well with either negative or positive polarity.

Flat-plate ESPs seem to have a wide application for high-resistivity particles with small (1–2 μm) mass median diameters. For example, fly ash has been successfully collected with this type of ESP, but low gas velocity appears to be critical for avoiding high rapping losses. Figure B-37 shows a schematic drawing of a flat-plate ESP.

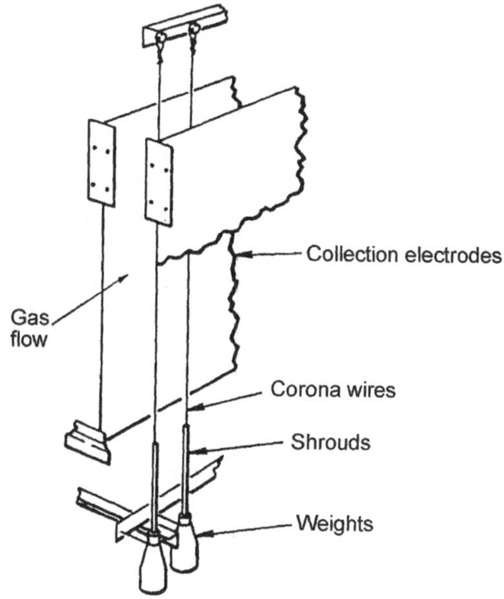

Figure B-36 Schematic of a plate-wire electrostatic precipitator.

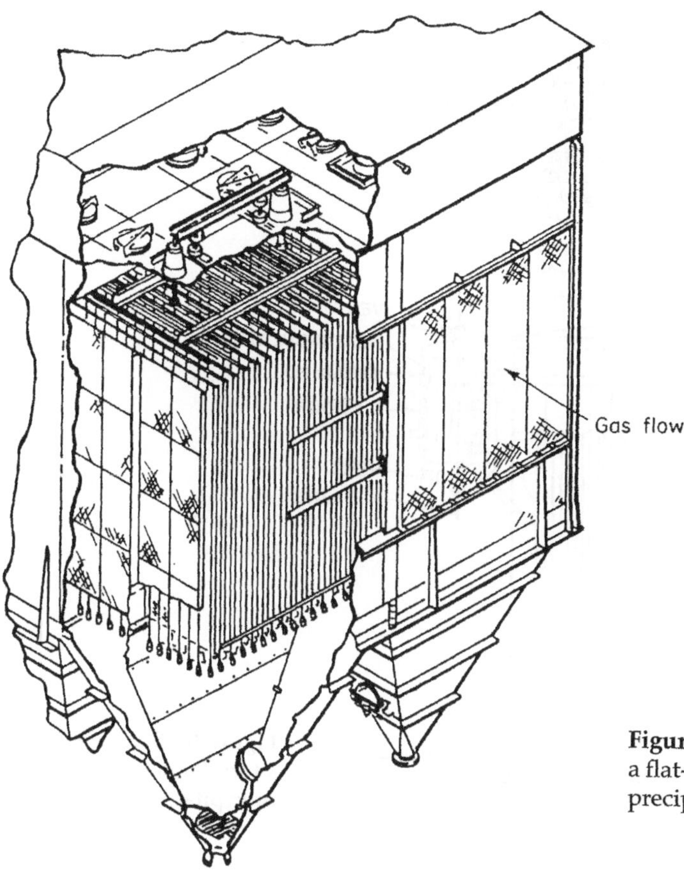

Figure B-37 Schematic of a flat-plate electrostatic precipitator.

Tubular Precipitator: the tubular precipitator is essentially a one-stage unit and is unique in having all the gas pass through the electrode region. The high-voltage electrode operates at one voltage for the entire length of the tube, and the current varies along the length as the particles are removed from the system. No sneakage (particle bypass) paths are around the collecting region, but corona nonuniformities may allow some particles to avoid charging for a considerable fraction of the tube length.

Tubular ESPS are typically used for collecting mists (acid, tar, oil, etc.) and where particulates are either wet or sticky. The unit can be constructed of many tubes operating in parallel for handling large gas flows. The tubes may be formed as a circular, square, or hexagonal honeycomb with gas flowing upward or downward. A tubular ESP can be tightly sealed to prevent leaks of material, especially valuable or hazardous products. They are usually cleaned with water, have re-entrainment losses of a lower magnitude than do the other dry-type ESPs. Figure B-38 shows a tubular ESP designed for collecting tar mist.

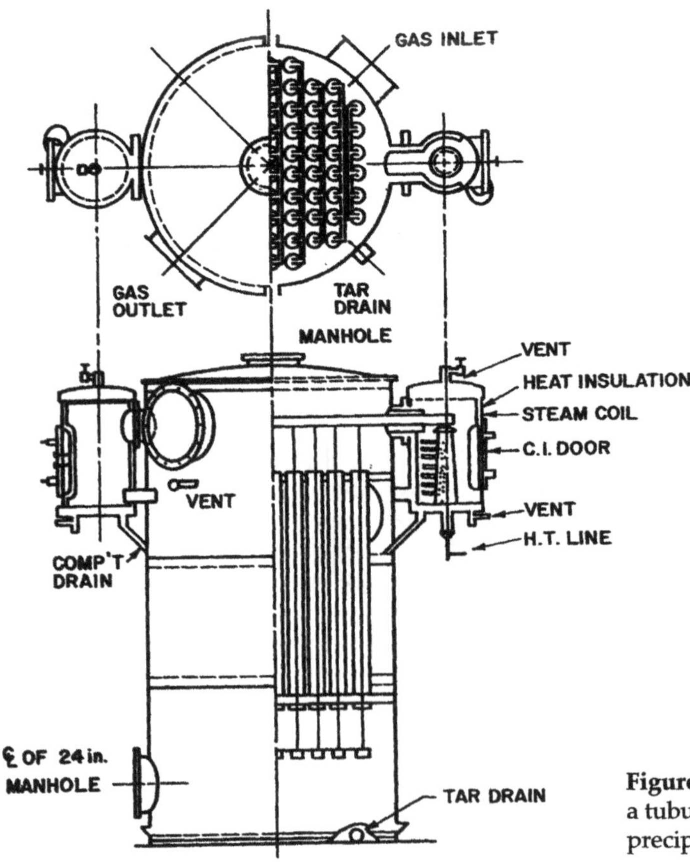

Figure B-38 Schematic of a tubular electrostatic precipitator.

Wet Precipitator: Any of the precipitator types described above can be operated with wet walls instead of dry ones. The water flow may be applied intermittently or continuously to wash the collected particles into a sump for disposal or recovery. The advantage of the wet wall ESP is that it does not have problems with rapping reentrainment or with back corona. The disadvantage is the increased complexity of the wash and the fact that the collected slurry must be handled more carefully than a dry product, adding to the expense of the disposal. Figure B-39a shows a plate-wire wet ESP and Figure B-39b shows a tubular wet ESP.

Two-Stage Precipitator: The previously described ESPs were all parallel in nature, that is, the discharge and collecting electrodes are side-by-side. The two-stage ESP is a series device with the discharge electrode, or ionizer, preceding the collector electrodes. For indoor applications, the unit is operated with positive polarity to limit ozone generation. In this ESP the particles are charged in the first field and collected in a second noncorona stage. Advantages of this configuration include more time for charging, less propensity for back-corona, and economical construction for small sizes. This type of ESP is generally used for gas flow volumes of 50,000 acfm and less, and is applied for submicron materials such as oil mists, smoke, fumes, or sticky particulates because there is little electrical force to hold the collected particles on the plates. Modules consisting of a mechanical prefilter, ionizer, collecting-plate cell, afterfilter, and power pack may be placed in parallel or series-parallel arrangements. Preconditioning of gases is normally part of the system. Cleaning may be by water wash of modules removed from the system, or automatic, in-place detergent spraying of the collector followed by air-blow drying.

Two-stage ESPs are considered to be a separate and distinct type of device as compared with large, high-gas-volume, single-stage ESPs. The smaller devices are usually sold as pre-engineered package units.

Additional information on electrostatic precipitators is presented by Katz (1989), Ogleby and Nichols (1969), and White (1963).

B5.3 Fabric Filters

Fabric filter dust collectors can be either baghouses or cartridge filters. They remove dust (particulate solids) from a gas stream by passing the stream through a porous medium.

Dust particles form a more or less porous cake on the surface of the medium. It is normally this cake that actually does the filtration. Baghouse dust collectors either use woven or felt media and cartridge filter dust collectors use nonwoven filter media.

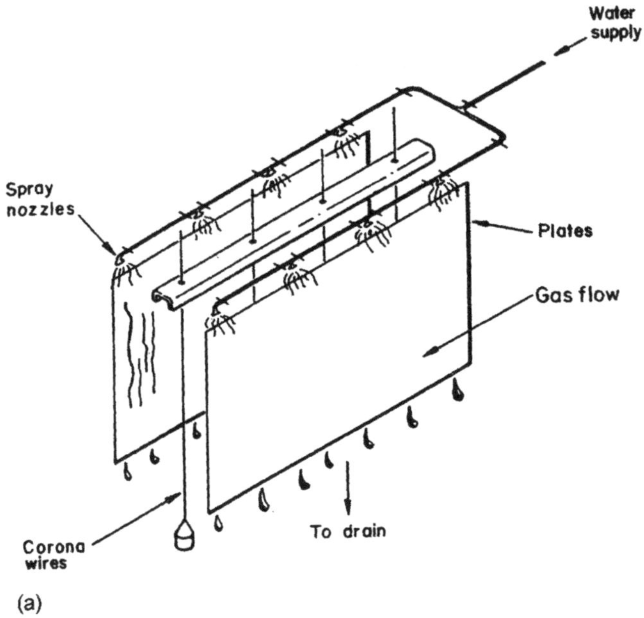

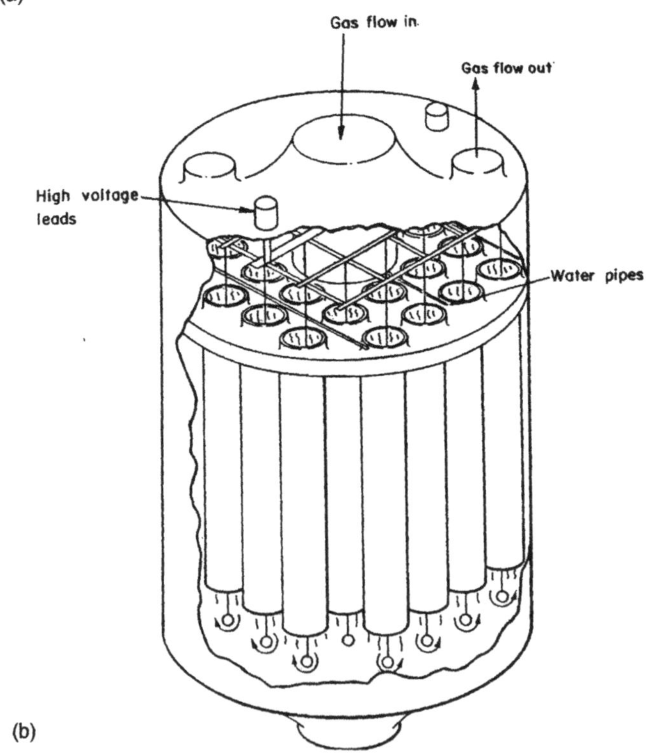

Figure B-39 Schematic of wet electrostatic precipitators. (a) Plate-wire type; (b) tubular type.

Baghouses

Baghouses are composed of a number of compartmented filter bags (sometimes mounted on cages depending on the dust removal method) and a bag-cleaning mechanism; these are enclosed in a shell and provided with a hopper to receive the collected dust. In addition, the gas resistance (pressure drop) resulting from the bag and filter cake is overcome by a fan that may be located either downstream (the more usual arrangement) or upstream of the baghouse.

Baghouses are usually classified by the method of dust removal, of which there are four types: (1) shakers, (2) reverse-flow, (3) pulse-jet, and (4) Hersey-type (Blow-Ring). These are discussed below.

Shaker Baghouses: There are two broad categories of shaker dust collectors: those for intermittent use and those for continuous use. The difference is that continuous-use collectors are compartmented to allow a portion of the whole collector to be taken offstream for cleaning while dirty gas continues to be filtered by the rest of the unit. Intermittent-use collectors are usually single compartment and smaller in size than continuous-use ones. The shaking is done by the use of a motor driving an eccentric which in turn has a rod connected to the bags (see Figure B-40 for a schematic drawing of a shaker-type baghouse). The bags are usually open at the bottom, being fixed in the tubesheet at the bottom and attached to the shaking mechanism at the top. The bag normally contains no rings or cages. The important parameters affecting the efficiency of cleaning are frequency, oscillation, and amplitude. Table B-3 shows the significant parameters for shaker baghouses.

Shaker baghouses usually require the use of woven bags rather than felt bags as the stiffness of felts designed for pulse-jet cleaning prevents the proper flexing of the fabric to dislodge the dust cake. Common bag diameters are 5, 8, and 12 inches. In the U.S. shaker baghouses normally employ gas/cloth ratios (gas or vapor flow rate/cloth area) below 4:1.

Reverse-Flow Baghouses: Reverse-flow cleaning is perhaps the gentlest cleaning method. In this method, the dust is removed from the bags by back-flushing with low-pressure (a few inches water gauge) reversed flow. Cleaning airflow is provided by a separate cleaning fan that is normally much smaller than the main system fan, since only a fraction of the total system is cleaned at any one time. More often, reverse-flow baghouses are comprised of a number of isolatable compartments. The gas/cloth ratios usually employed are less than 4:1. Normally, the dust is collected on the inside of the bag, the bag being open at the bottom and closed at the top. The bag contains rings to keep it from collapsing completely during flow reversal. Cleaning is accomplished both with and without flexing (partial collapsing) of the bag. Figure B-41 shows a schematic drawing of a reverse-flow baghouse.

TABLE B-3
Shaker-Type Baghouse Cleaning Parameters

Frequency	Usually several cycles/second; adjustable
Motion type	Simple harmonic or sinusoidal
Peak acceleration	1–10 g
Amplitude	Fraction of an inch to a few inches
Mode	Off-stream
Duration	10–100 cycles, 30 seconds to a few minutes
Common bag diameters	5, 8, 12 inches

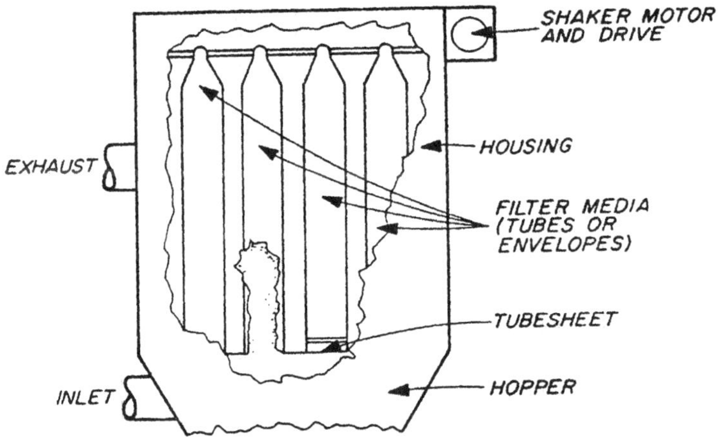

Figure B-40 Schematic of a shaker-type baghouse.

Reverse-flow baghouses often use woven glass bags as it provides the ideal property of high tensile strength (necessary in very long tubes) at very high temperatures. Tube lengths up to 30 feet are not uncommon. Table B-4 shows the significant parameters for reverse-flow baghouses.

Pulse-Jet Baghouses: Pulse-jet cleaning is a method that uses high-pressure (60 to 120 psig) compressed air, with or without a Venturi, to back-flush the bags vigorously. This method creates a shock wave that travels down the bag, knocking the dust away from the bag surface. Normally, this method is employed using felted-filter media and the gas/cloth ratio is generally higher than that of the shaker and reverse-flow cleaning methods. The duration time of cleaning is lower than that of the other two methods; generally the pulse lasts only a fraction of a second.

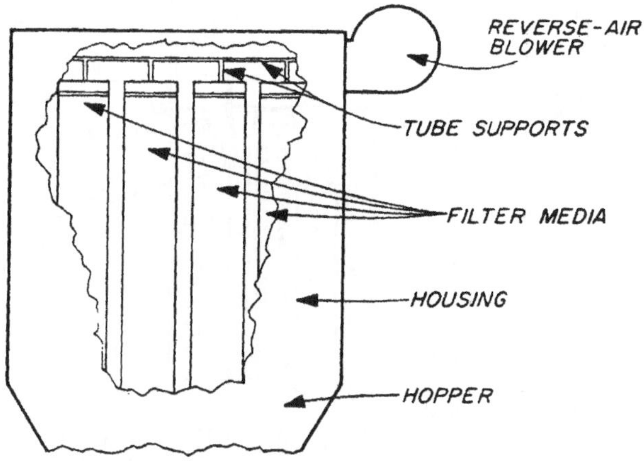

Figure B-41 Schematic of a reverse-flow-type baghouse.

TABLE B-4
Reverse-Flow-Type Baghouse Cleaning Parameters

Frequency	Cleans one component at a time. sequencing one compartment after another; can be continuous or initiated by a maximum-pressure-drop switch
Motion	Gentle collapse of bag (concave inward) upon deflation; slowly repressurize a compartment after completion of a back-flush
Mode	Off-stream
Duration	1–2 minutes, including valve opening and closing and dust settling periods; reverse airflow itself normally 10–30 seconds
Common bag dimensions	8, 12 inches; length: 22, 30 feet
Bag tension	50–75 pounds typical, optimum varies; adjusted after on-stream

The most usual configuration has the bag closed at the bottom and open at the top as shown in Figure B-42. A metal cage is employed within the bag to keep the bag from collapsing. In the normal mode of operation, the dirty gas enters via a nozzle in the hopper section and proceeds to the bags. The dust is collected on the outside of the bags and the cleaned gas exits through the top of the bags and leaves through the plenum. Normally, a row of bags is cleaned simultaneously by introducing compressed air briefly at the top of each bag via a Venturi nozzle. The shock wave thereby created drives the dust off the outside of the bags and down into the hopper. Continuous discharge of the collected dust is often employed using a rotary valve. Cleaning parameters are shown in Table B-5. This system has no internal moving parts and allows for removal of the bags from the clean side of the housing since the bags are usually connected only at the top.

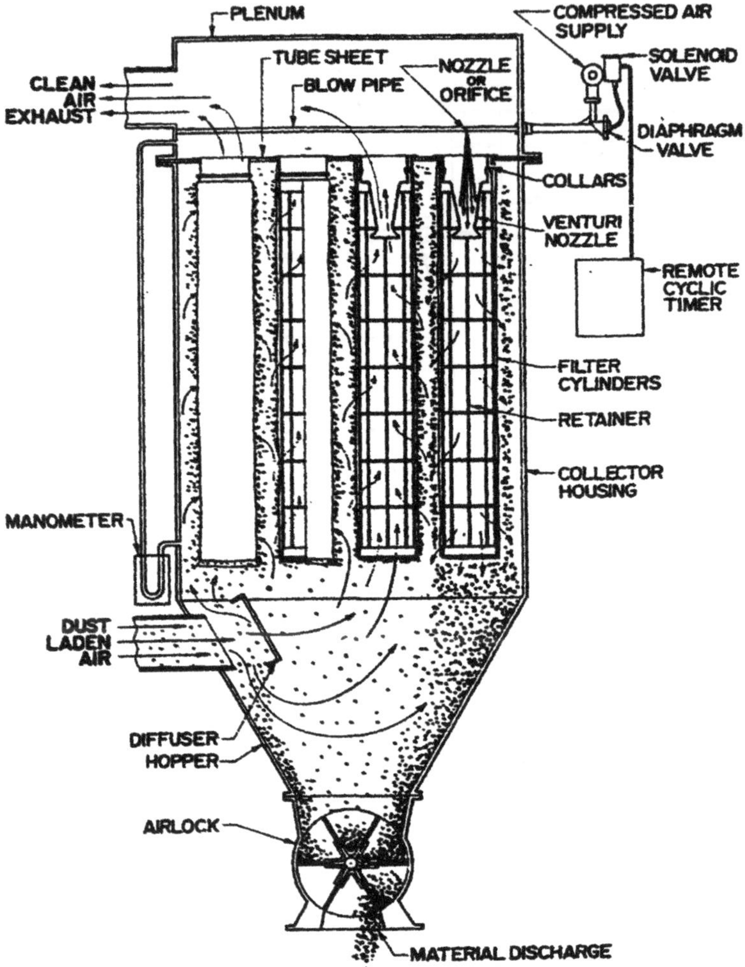

Figure B-42 Schematic of a pulse-jet-type baghouse.

TABLE B-5
Pulse-Jet-Type Baghouse Cleaning Parameters

Frequency	Usually, a row of bags at a time, sequenced one row after another; can sequence such that no adjacent rows clean one after another; initiation of cleaning can be triggered by maximum-pressure-drop switch or may be continuous
Motion	Shock wave passed down bag; bag distends from cage momentarily
Mode	On-stream; in difficult-to-clean applications such as coal-fired boilers, off-stream compartment cleaning being studied
Duration	Compressed-air (100 psi) pulse duration 0.1 second; bag row effectively off-line
Common bag diameters	5–6 inches

Pulse-jet dust collectors are available in sizes ranging from 25 to tens of thousands of square feet of cloth area. Most pulse-jet units use tubes from 4 to 6 inches in diameter and from 2.5 to 14 feet long. Perforated cage-diffusers permit filter bag lengths up to 20 feet on 6-inch-diameter tubes and 40 feet on 12-inch-diameter tubes.

Hersey-Type (Blow-Ring) Baghouse: This type of dust collector is seldom used nowadays and will not be discussed. Kennedy and Cheremisinoff (1977) present information about it.

The filter medium is the key to dust collector performance. All other design and application factors hinge upon the filter medium selected. The selection of filter material and configuration should be based upon experience, testing, new development evaluation, and performance guarantees to determine the best material for a specific application. Table B-6 is a list of commonly used filter fabrics and their characteristics that may be used as a guide for the selection of filter media for baghouses and cartridge filters.

Cartridge Filters

Cartridge filters are comparatively new as they were first introduced in 1974. They are becoming more and more widely used now because they can provide a great deal of filter area in a small element as compared with a filter bag of the same area.

With few exceptions, they use nonwoven filter paper rather than the woven or felt media used in baghouse collectors. The cartridge construction

TABLE B-6
Baghouse Filter Media Characteristics

Fiber	Max. Temp, °C; continuous	Abrasion Resistance	Tensile Strength (durability)	Chem. Resistance (acids)	Chem. Resistance (alkalis)	Supports Combustion
Polyethylene	65	Exc.	Exc.	Exc.	Exc.	Yes
Cotton	70	Good	Fair	Poor	Good	Yes
Polypropylene	90	Exc.	Exc.	Good	Good	Slow
Wool (dry)	95	Av.	Poor	Poor	Poor	No
Nylon (polyamide)	95	Exc.	Exc.	Poor	Good	Yes
Orlon (acrylic)	130	Good	Fair	Good	Av.	Yes
Dacron (polyester)	135	Exc.	Exc.	Good	Good	Yes
Nomex (polyaramid)	220	Good	Good	Fair	Av.	No
Teflon (PTFE)	230	Fair	Fair	Exc.	Exc.	No
Glass fiber	260	Poor	Exc.	Av.	Poor	No

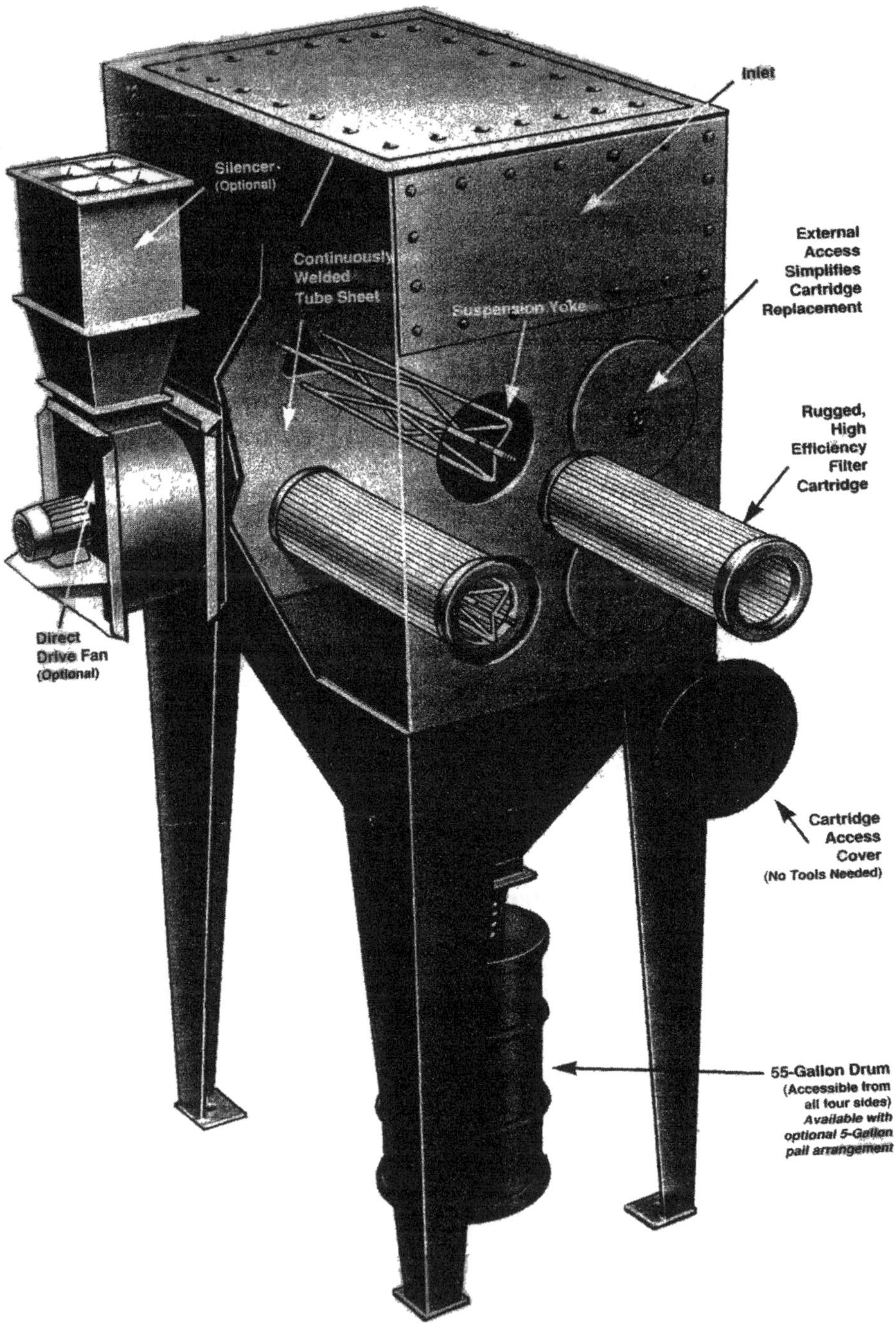

Figure B-43a Horizontal cartridge filter dust collector.

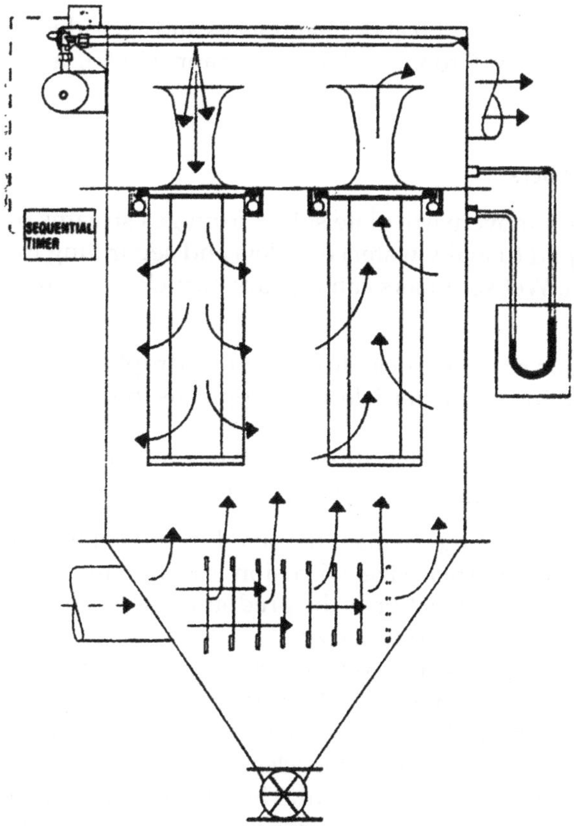

Figure B-43b Vertical cartridge filter dust collector.

is made by pleating or folding the filter media in multiple accordion folds which permits a greater filter area to be used in the same space as a filter bag.

A cartridge filter dust collector consists of a housing with the filter elements mounted either vertically, or more recently, in a horizontal and slightly downward position (see Figure B-43). In the latter arrangement, the airflow is directed in a downward direction, which increases filter life and assists in pushing the dust off the cartridge and into the collection hopper. One of the major advantages of cartridge filter collectors is that usually less headroom and floor space is required than for baghouse collectors. They are also more efficient than baghouses in that they can achieve 99.9% submicron removal versus 99.0% for baghouses. Washable cartridge filters, with open pleat construction, and electrostatically conductive filter cartridges for grounding are among new developments that are now available.

The cartridges are presently available in the following materials: (1) cellulose based, (2) polyester, (3) Nomex, and (4) PTFE membrane.

Additional information about baghouses and cartridge filter dust collectors is presented by Croom (1995), Heumann (1997), and McKenna and Turner (1989).

B5.4 Wet Scrubbers

Wet scrubbers remove particulate solids from gas streams by capturing the particles in liquid (usually water) droplets and separating the droplets from the gas stream. Wet scrubbers can capture particles by any of the following mechanisms:

1. Impaction of the particle directly into a droplet.
2. Interception of the particle by a droplet as the particle comes near the droplet.
3. Diffusion of the particle through the gas surrounding the droplet until the particle is close enough to be captured.

The predominant mechanism involved in the collection of a particle is determined by the particle size. Diffusion is dominant for particles smaller than 0.2 μm. Inertial impaction is the predominant mechanism for collection of particles from 0.5 to 10 μm in size.

There are a great variety of wet scrubbers available in a wide range of designs, sizes, performance capabilities, and both capital and operating costs. Choosing the right scrubber for a particular job requires an understanding of the type and size of particles to be removed and the alternatives which various wet scrubbers present. The various types of wet scrubbers are categorized as follows: plate tower, packed bed, spray chamber, Venturi scrubber, cyclone (centrifugal) spray scrubber, mechanically aided scrubber, and moving bed scrubber. A brief description is given below for each of these.

Plate Tower Scrubbers: These are similar to tray towers used for mass transfer operations (distillation, absorption, extraction). The plates are usually sieve or bubble-cap trays. In some designs, impingement baffles are placed a short distance above each perforation on a sieve plate forming an impingement plate. The impingement baffles are located below the liquid level on the sieve plate and in this way are continuously washed clean of collected particles. Particle collection is also aided by the atomization of the liquid flowing past the holes in the plate, and the subsequent Venturi effect. Particle collection efficiency is generally good for particle larger than 1 μm in diameter.

Packed Bed Scrubbers: There are three basic types of packed bed wet scrubbers: (1) countercurrent-flow scrubbers, (2) cocurrent-flow scrubbers, and (3) cross-flow packed scrubbers. Countercurrent-flow packed scrubbers are similar to those used for mass transfer operations, in that the dust-laden gas enters at the bottom of the tower and scrubbing liquid (usually water) enters at the top. In the cocurrent-flow scrubber, both the dust-laden gas enters at the top and liquid is sprayed downward above the packing and then both streams pass downward together to the bottom outlet. Cross-flow packed scrubbers are horizontal chambers in which one or more beds of packing are located. The dust-laden gas enters at one end and moves horizontally through the packed bed and is washed by scrubbing liquid which flow vertically down through the packing. Two types of design are available: parallel and cross-flow. In the parallel flow design both the dust-laden gas stream and the scrubbing liquid enter at the same front end. In the cross-flow design, the gas enters at one end and the scrubbing liquid is sprayed along the length of the chamber by a header system of spray nozzles. Figure B-44 shows drawings of these two designs.

The packings used in these wet scrubbers are pretty much the same as for mass transfer operations, as well as special types such as Tellerettes. Particulate solids 5 μm and larger are effectively removed by impingement and scrubbing in cross-flow scrubbers.

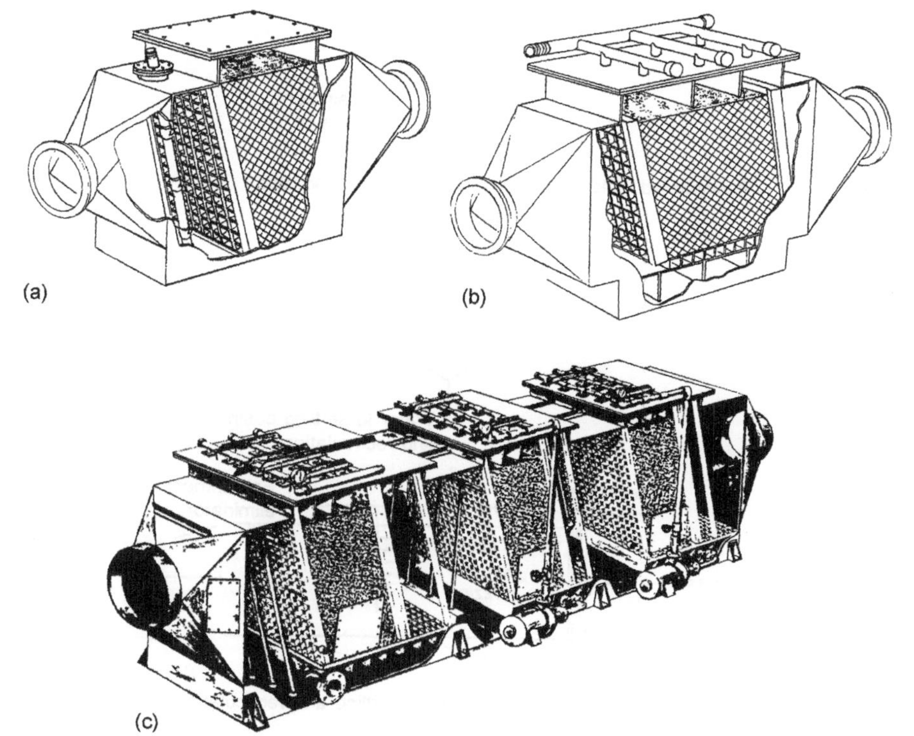

Figure B-44 Cross-flow packed scrubbers. (a) Parallel-flow type; (b) cross-flow type; (c) cross-flow type with the beds in series.

Spray Chamber Scrubbers: These scrubbers are empty chambers, either vertical cylindrical or rectangular horizontal, in which the dust-laden gas is contacted with a liquid spray produced by spray nozzles (see Figure B-45). Droplet size is controlled to optimize particle contact and to provide easy droplet separation from the gas stream. Nozzles providing a cone spray geometry are the most effective. Spray chamber scrubbers are capable of 90% efficiency for particles larger than 8 μm. They also have a relatively low pressure drop compared to other wet scrubbers.

Venturi Scrubbers: Venturi scrubbers are one of the most widely used types of wet scrubber. They consist of a Venturi section for contact of the dust-laden gas and scrubbing liquid, followed by an entrainment separator. They may be used as either high- or low-energy devices, but are most commonly employed as high-energy units. Most large Venturi scrubbers in current use are really not Venturi devices, but are variable orifices of one type or another. Current designs for large units generally use the vertical downflow of gas through the Venturi contactor. There are three design variations of the

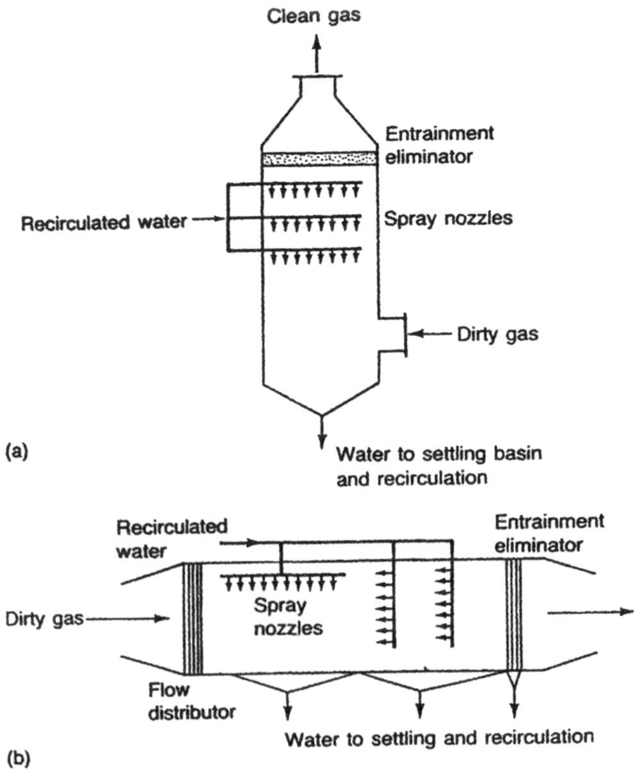

Figure B-45 Schematic of spray chamber scrubbers: (a) Vertical spray chamber (countercurrent flow); (b) horizontal spray chamber (cross-flow).

Venturi contactor: (1) a "wet-approach" or "flooded-wall" entry section, to avoid dust buildup at a wet-dry junction, (2) an adjustable throat for the Venturi (or orifice), to provide for adjustment of the pressure drop, and (3) a "flooded elbow" located below the Venturi and ahead of the entrainment separator to reduce wear by abrasive particles. The Venturi throat is sometimes lined with refractory to resist abrasion by dust particles. Various types of adjustable throats, which may be under manual or automatic control, permit maintaining a constant pressure drop and constant efficiency under conditions of varying gas flow. The Venturi scrubber can achieve collection efficiencies above 95% for particles larger than 0.30 μm. Figure B-46 shows one type of large Venturi scrubber.

Another type of Venturi scrubber, used for smaller flow rates, is the ejector-Venturi scrubber (see Figure B-47). The principle of operation involves a jet effect created by a water spray nozzle. The nozzle is located at the top of the jet scrubber and creates a hollow cone-shaped spray which induces gas flow through the Venturi. The dust-laden gas and liquid enter the throat,

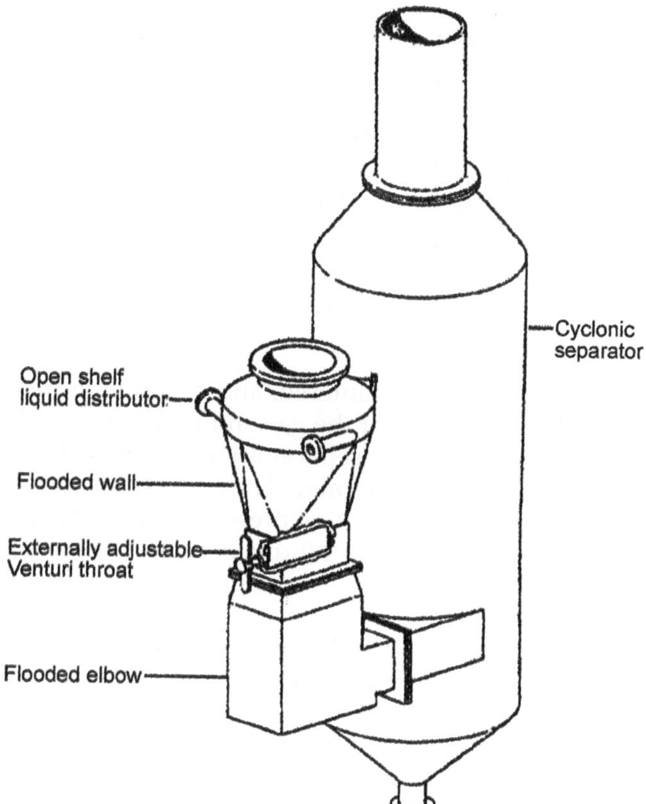

Figure B-46 Schematic of a Venturi scrubber.

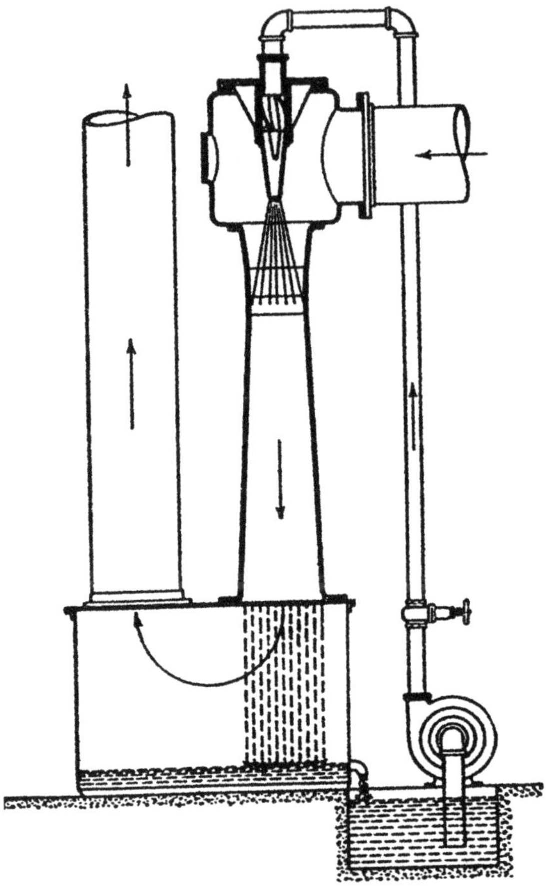

Figure B-47 Schematic of an ejector-Venturi scrubber.

where extreme turbulence is encountered, and continue through a diffuser section where partial separation of the gas and liquid occurs. The partially separated mixture then enters a separator/sump where final separation occurs. The scrubbing liquid is recirculated from the sump by a pump which supplies it to the nozzle at relatively high pressures (20 to 80 psig). Collection efficiencies of 92% are possible for 1.0-μm particles. These scrubbers are available as package units.

Centrifugal (Cyclone) Spray Scrubber: This type of wet scrubber is usually cylindrical in shape, and imparts a spinning motion to the entering dust-laden gas by introducing the gas to the scrubber tangentially or by directing the gas stream against stationary swirl vanes. Liquid spray is introduced into the rotating gas from an axially located manifold in the lower part of the unit equipped with spray nozzles. The atomized fine-spray droplets are caught in the rotating gas stream, and are, by centrifugal force, swept across

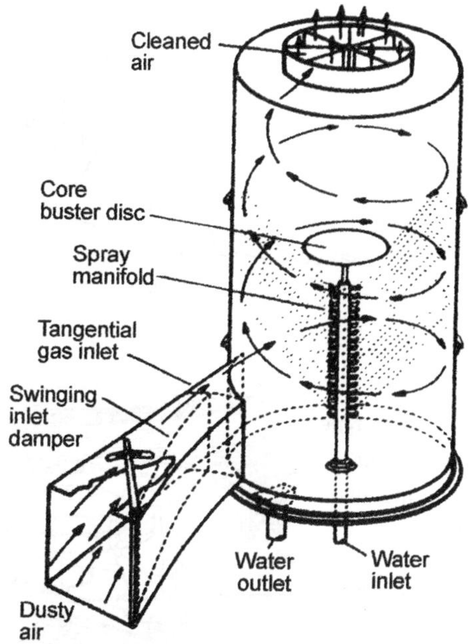

Figure B-48 Schematic of a centrifugal (cyclone) spray scrubber.

to the walls of the cylinder, colliding with and collecting the dust particles en route. The scrubbing liquid and collected particles run down the walls and out the bottom of the vessel; the clean gas leaves at the top.

Centrifugal (cyclone) wet scrubbers are generally used for removal of particles greater than 1 to 2 μm in diameter. Figure B-48 is a schematic of such a scrubber.

Mechanically Aided Scrubbers: These wet scrubbers incorporate a motor-driven device that produces the fine spray and the collection of gas and liquid. The simplest commercial devices of this type are essentially fans upon which water is sprayed. Figure B-49a is a schematic drawing of such a scrubber. Particles are collected by impaction upon the fan blades, and usually liquid is introduced at the hub of the rotating fan blades. The liquid then runs over the blades, washing off the collected particles. A separator/sump has to be provided at the fan outlet to collect and separate the gas and liquid streams. Particle collection efficiency is good for particles larger than 1 to 2 μm in diameter.

Another type of mechanically aided scrubber is one that uses a motor-driven impeller located at the gas–liquid contacting zone (see Figure B-49b). The impeller atomizes liquid into small droplets, which fly from the impeller across the gas stream, collecting particles. This type of mechanically aided scrubber is efficient in removing submicron particles but has a high operat-

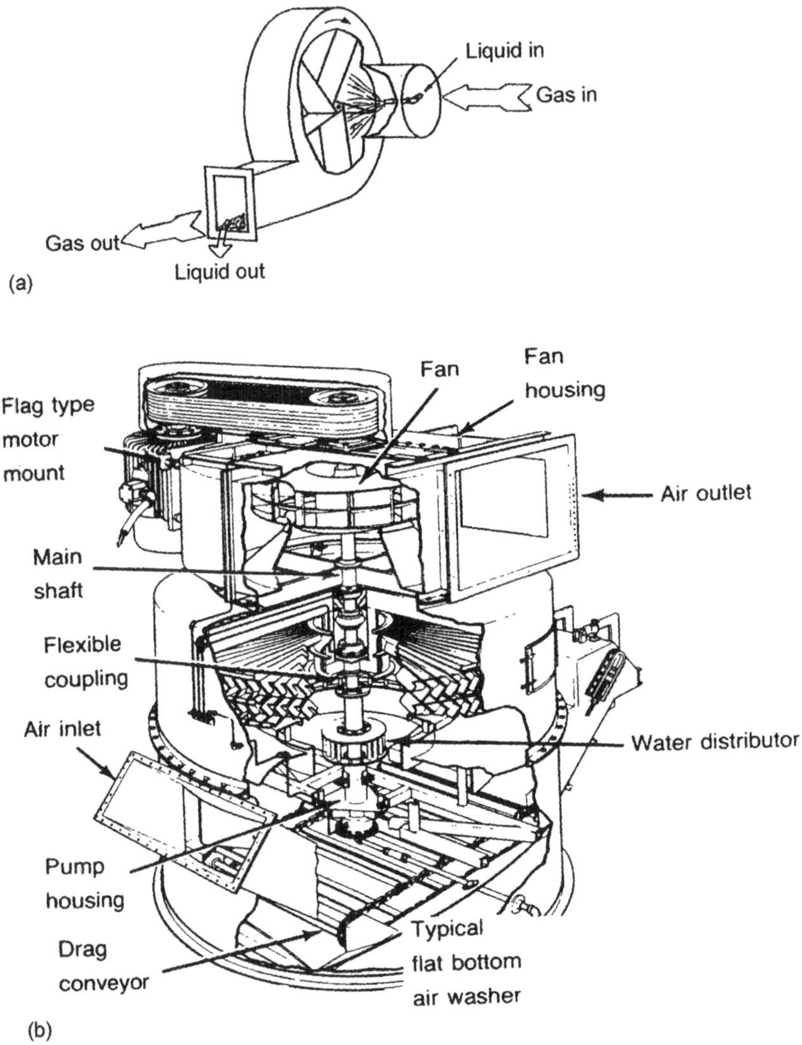

Figure B-49 Schematic of mechanically aided (disintegrator) scrubbers. (a) Fan type; (b) motor-driven impeller type.

ing cost due to the energy needed to drive the impeller at relatively high speeds.

Moving Bed Scrubbers: these wet scrubbers are usually cylindrical in shape with one or more beds of low-density spheres that are able to move between upper and lower retaining grids. The spheres are commonly 1-inch or more in diameter and made from rubber or a plastic such as polypropylene. The plastic spheres may be solid or hollow. Gas and liquid flows are countercurrent and the spheres are fluidized by the upward-flowing gas. Gas velocities are sufficient to move the spheres around when the scrubber is operating, which aids in making the bed turbulent, and keeps the spheres clean. The dust particles are collected primarily by the liquid droplets. Particle collection efficiency can be good for 1-μm particles.

Besides the wet scrubbers described above, other types are also commercially available. Additional information on wet scrubbers is presented by Calvert et al, (1972), Calvert (1984), and Heumann (1997).

B6 EXTRUDERS

An extruder is a device used for the production of noodle-shaped agglomerates (pellets) such as foodstuffs, dog food, catalysts, solid rocket propellants, and vitamin formulations, to name a few. There are three types of extruder commonly used. They are: piston-type, roll-type, and screw type, and are discussed in the following paragraphs.

The piston-type extruder consists of two main parts, the driving mechanism and the press head. The press head contains the mold channel (die). At the rear portion of the press head is a feeding hopper and entrance, through which the solids are fed to the mold channel. The material is pushed forward by the stroke of the piston and compacted by a hammer-like impact against previously-formed pellets in the mold channel.

The roll-type extruder (also known as a "pellet mill") operates by feeding material between a roller and a perforated plate or ring die, a method that forces a moist formulation through the die. There are three basic designs, as follows:

Type 1—A ring rotates around one or more rollers, installed inside the cylindrical die chamber, each of which rotates on its stationary axis (see Figure B-50a). Multiple rollers can be used to distribute or balance the forces and to increase capacity. All rotating elements turn in the same direction. Feed material is introduced onto the inside surface of the ring die and is pressed outward by the rollers. The orientation of the perforated cylinder is horizontal, sometimes with a slight inclination to facilitate feeding.

Type 2—The roller or rollers are mounted on the inside of the ring die and material is fed from a hopper, occasionally with a screw, into the region between the roller and the die (see Figure B-50b). Material is extruded into the center of the ring die and flows out one end. The roller and the die move in opposite directions. The orientation fo the ring die is horizontal or slightly inclined.

Type 3—Rollers are positioned above and roll along the surface of a flat, stationary die plate (see Figure B-50c). This design resembles a muller with a perforated base plate, rather than a solid one.

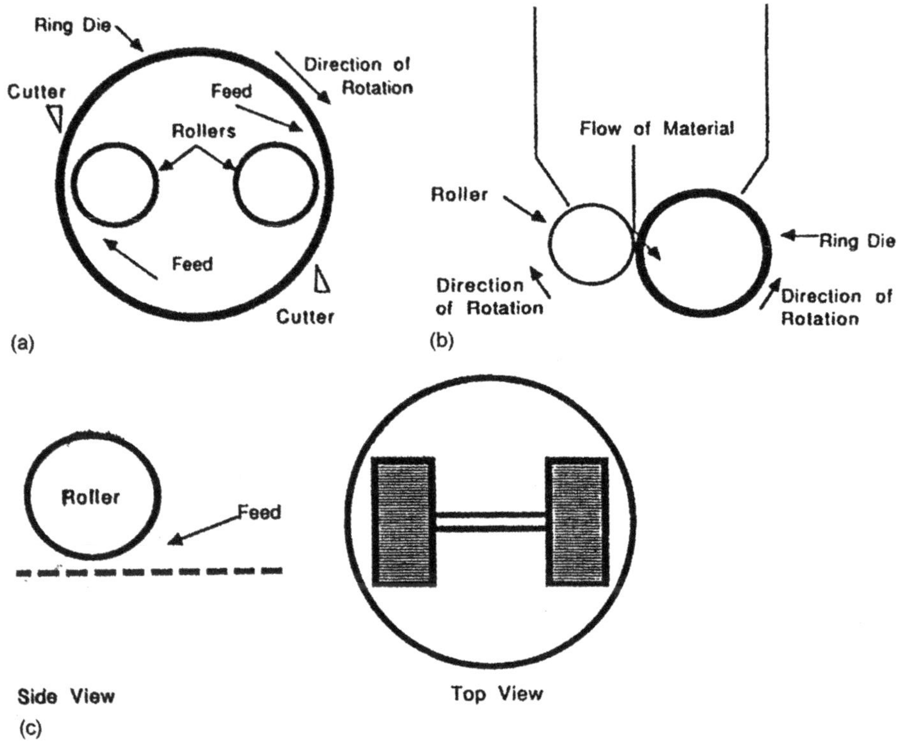

Figure B-50 Roll-type extruders. (a) Pellet mill with internal roller; (b) pellet mill with roller external to die; (c) pellet mill with rollers on flat die plate.

The screw-type extruder, as its name implies, uses a screw to develop the necessary pressure to force material through uniform openings, producing uniform strands or extrudates. The screw extruder has three zones that are defined by the principal mechanical operation being performed: feed zone, transport and compression zone, and extrusion zone (see Figure B-51). The feed zone is the area where material is first introduced into the extruder. It consists of a hopper to channel the flow of material into the chamber where the screw(s) are located. In some units, a conditioning device is located in this zone so that liquid can be introduced into the powder and the material can be kneaded into a moist, homogeneous mass. Material in the transport zone is moved by the auger-like screw(s) from the feed zone into the compression zone. Most extruder manufacturers have both single-screw and twin-screw machine designs. The twin-screw feeder has the advantages of less "bridging" at the feed zone and better transport into the extrusion zone. The single-screw extruder generally producess an extrudate that is slightly more compact (higher density).

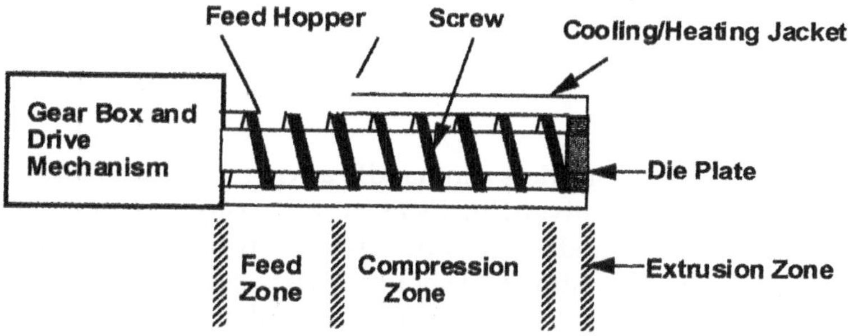

Figure B-51 Schematic of a screw-type extruder.

The die openings in the screen or die plate may be of several basic designs. The shape of the opening varies wit the application. If a more dense product is needed, a thicker die plate or screen is required to withstand the greater extrusion pressure used. Screen or die plate holes can be either cylindrical, tapered inlet and/or outlet, and conical. For thin screens or die plates, the hole is typically straight with a slight neck or taper at the entrance due to the punching method.

The extruders described here are normally unheated or are jacketed and water-cooled. Other types of extruders that are used for making plastic extrudates are heated, but are not discussed in this section.

Additional information about extruders are presented by Capes (1980), Hicks and Freese (1989), and Pietsch (1997).

B7 FEEDERS AND ROTARY VALVES

A feeder is a device that is used whenever there is a need to transfer solids at a controlled rate from a container (bin, silo, or process vessel) to another piece of equipment in a process or to a hopper truck or railroad car. There are several types of feeders available to handle particulate solids, but these can be divided into two categories: volumetric and gravimetric. A volumetric feeder discharges a certain volume of material as a function of time. This type of discharge is adequate for many solids feeding applications. Feed uniformities in the range of ±2 to 5% on a minute-to-minute basis are easily achieved using most volumetric designs. A disadvantage of volumetric feeding is that the feeder does not recognize bulk density changes. A gravimetric feeder relies on weighing the material to achieve the required rate. Feed accuracies of ±0.25% are obtainable using a properly designed gravimetric feeder. However, a gravimetric feeder is usually more expensive than a volumetric feeder.

A number of volumetric and gravimetric feeders are discussed below.

B7.1 Volumetric Feeders

Some types of volumetric feeders are screw, belt, and vibratory feeders and rotary valves.

Screw Feeders: Screw feeders can handle a wide range of materials from lumps to powder, are relatively inexpensive, are easily enclosed to be dust-tight, and easily accommodate slot openings in bin/silo bottoms. However, they can cause particle attrition and should not be used if this is a problem. A screw feeder is composed of a series of flights that are wound around a common shaft. The flights have a particular diameter and pitch (the distance between flights). Some screws have constant pitch flights, while others vary. A standard screw feeder has a pitch-to-diameter ratio of 1. This ratio is satisfactory only for withdrawing uniformly from bin/silo openings where the maximum dimension does not exceed 1 to 1.5 pitches. If the bin/silo opening exceeds this, solids flow will occur at the back end of the screw causing a channel or funnel flow to occur at that point (see Figure B-52). There are several screw configurations that can be used to promote uniform withdrawal from bin/silo slot openings (see Figure B-53).

A comprehensive book on the specification, design, and use of screw feeders has been written by Bates (2000).

Belt Feeders: Belt feeders are used when the bin/silo opening is a long slotted one. They can handle friable, coarse, fibrous, elastic, sticky, or very cohesive solids. Because belts are available in widths up to 98 inches and unrestricted lengths, they can be designed for very large bin/silo outlets. The belt is usually a fabric or elastomeric covered fabric reinforced band riding on a slider bed or rollers. Improperly designed feed hoppers over the belt can cause solids compaction, belt wear, and high horsepower usage. Belt feeders having hoppers designed so that the opening diverges in the direction of

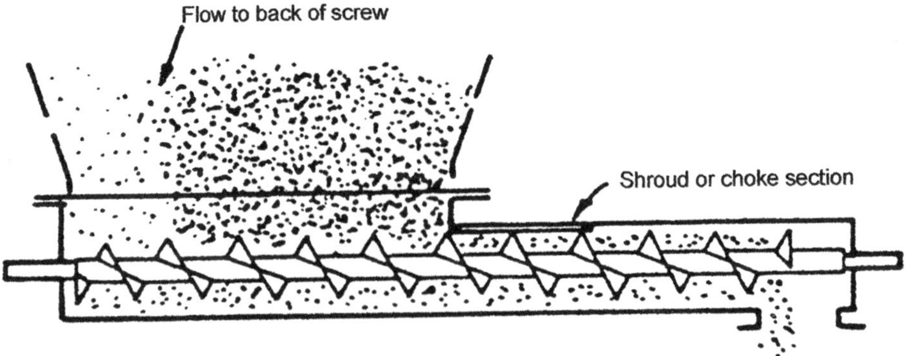

Figure B-52 Screw feeder with uniform pitch.

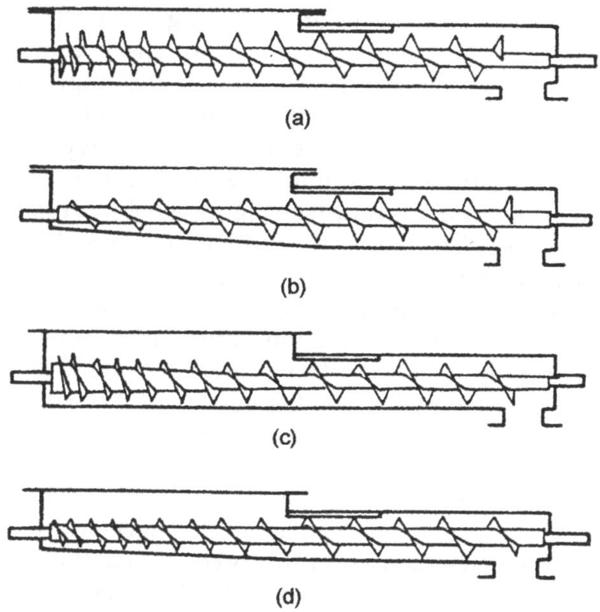

Figure B-53 Various screw feeder geometries producing improved flow patterns under slot openings. (a) Increasing flight pitch; (b) increasing flight diameter; (c) increasing pitch with decreasing shaft diameter; (d) increasing pitch with increasing shaft diameter.

flow have proven successful for handling a variety of granular and powdery materials through long slot openings (see Figure B-54).

See Section B9.1 for a description of belt conveyors constructional features, which are similar for belt feeders.

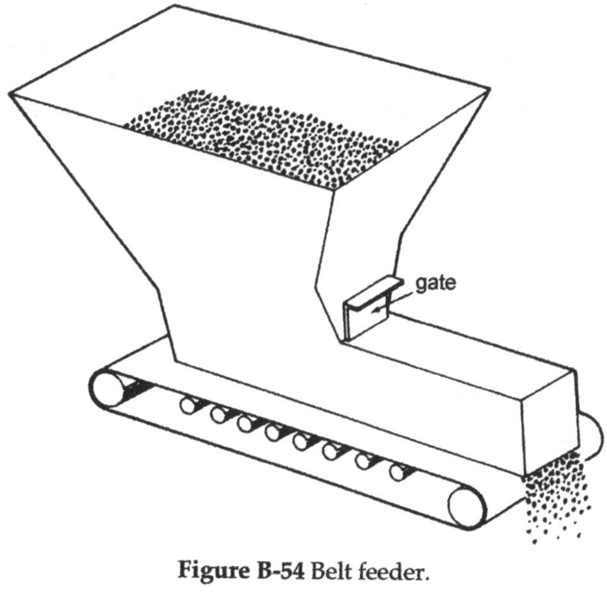

Figure B-54 Belt feeder.

Vibratory Feeders: Vibratory feeders consist of a carrying trough (also called a pan), supporting legs or springs, and a drive system. These feeders provide precise feed control, handle material gently, are self-cleaning, and can handle hot materials. They normally operate at frequencies from 12 to 60 cycles per second and strokes to about 10 mm. There are two general types of vibratory feeders: the direct force (single mass) machine (see Figure B-55a) and the indirect force (tuned two-mass) machine (see Figure B-55b). A rotating counterweight or reciprocating piston cause the motion in a direct force feeder. Essentially a constant rate machine, it is low cost and can handle a variety of particle sizes from lumps to damp fines, but does not provide precise flow control. In an indirect force feeder the vibrating forces from an exciter are amplified by a spring mass system to vibrate the trough. This design is most commonly used since it provides the best control of solids flow, uses the least power, and normally requires less maintenance than the direct force machine.

Two excitation systems are commonly used for indirect force feeders: electromagnetic and electromechanical. In electromagnetic feeders, an alternating or pulsating direct current drives a vibrator that is coupled to the pan through metal or fiberglass leaf springs. In electromechanical feeders, an

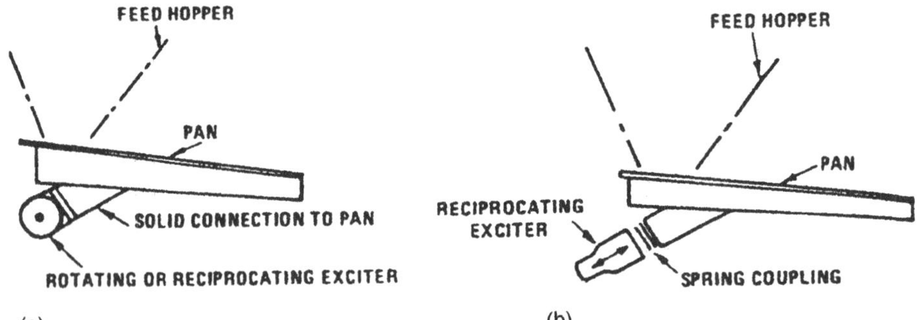

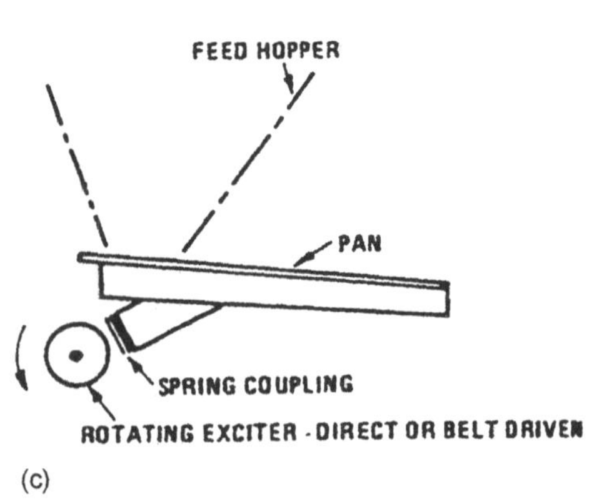

Figure B-55 Types of vibratory feeders. (a) Direct force (single mass) type; (b) indirect force (electromagnetic) type; (c) indirect force (electromechanical) type.

electric motor-driven eccentric weight, coupled through mechanical elastomer, or pneumatic springs, drives the pan (see Figure B-55c).

The performance of a vibrating feeder is more sensitive to particle properties than the other types of feeders. Special consideration should be given to fine powders that tend to aerate. These powders often move at very low rates on vibrating pans and can deaerate on vibrating surfaces and only the top layer of material will move forward.

Rotary Valve Feeders: these feeders (also called rotary vane feeders or star valves) can be used as a volumetric feeder and/or a gas pressure seal (air lock) to pass solids from one pressure environment to another. They are normally used under smaller diameter circular, rectangular, or square outlets. They are particularly useful when discharging materials to a pneumatic conveying system where a seal is required to prevent air flow through the bin/silo outlet. These valves are basically simple in concept, consisting of a bladed rotor in a fixed housing. The discharge rate is set by the speed of the vanes or pockets of the valve.

There are three commonly used types of rotary valve feeders: (1) the "drop-through" valve, (2) the offset valve, and (3) the "blow-through" valve. The "drop-through" (or standard) valve is the most commonly used one and is generally suitable for a wide range of free-flowing products (see Figure B-56a). A variation of the standard type of valve is the offset (or side entry) feeder (see Figure B-56b). Here the inlet and outlet ports are offset from the vertical center line of the valve, and are intended primarily for handling large granular products and plastic pellets. This is designed, usually with a controlled supply of solids, so that the rotor pockets do not become full, and therefore the chance of material becoming nipped between the rotating blades and the housing is virtually eliminated. Another variation to the standard feeder is the "blow-through" valve (see Figure B-56c). Here the conveying air passes through and purges the discharging pockets so that product entrainment into the conveying pipe takes place in the valve. These valves are primarily intended for use with more cohesive materials that might tend to "hang up" in a standard type of rotary valve feeder.

Rotors normally take one of two forms, that is, either "open-end" or "closed-end." In the "open-end" pattern the blades are welded directly to the driving shaft, whereas with the "closed-end" type disks or shrouds are also welded to the shaft and blade ends to form enclosed pockets (see Figure B-57). Although open-end rotors are less expensive they have the following disadvantages: (1)with more abrasive materials, wear of the housing end plates is possible since the product is in constant contact with them, and (2) they are not as rigid as the closed-end types since they have only one edge secured to the drive shaft.

There are basically two rotor pocket configurations in widespread use (see Figure B-58). The most common type (a) which has deep pockets and

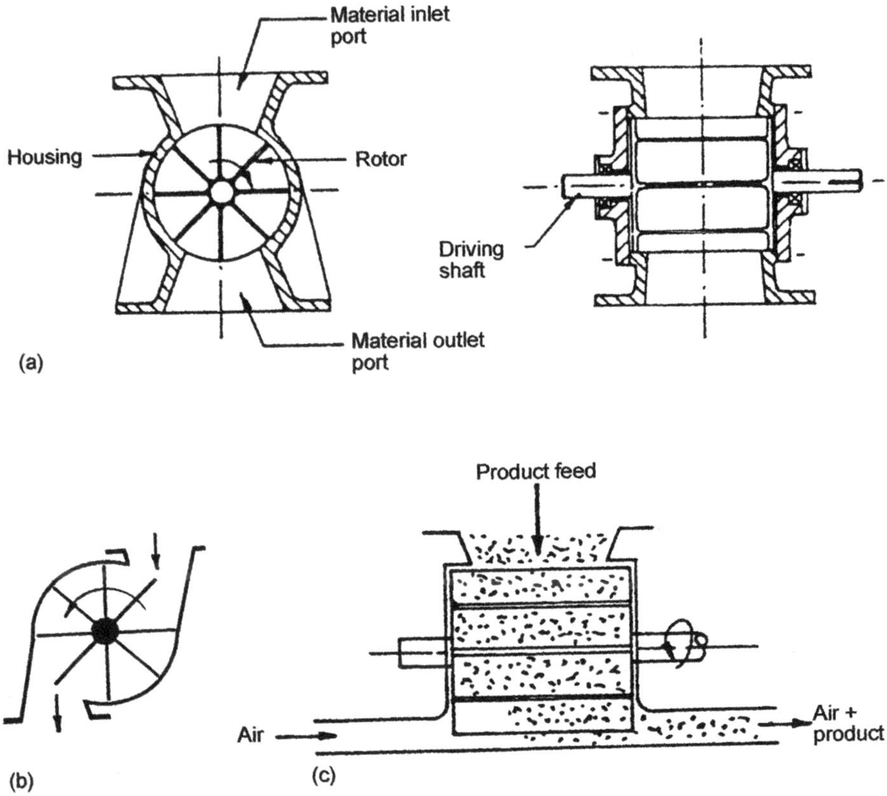

Figure B-56 Rotary valve feeder type. (a) Drop-through type; (b) offset type; (c) blow-through type.

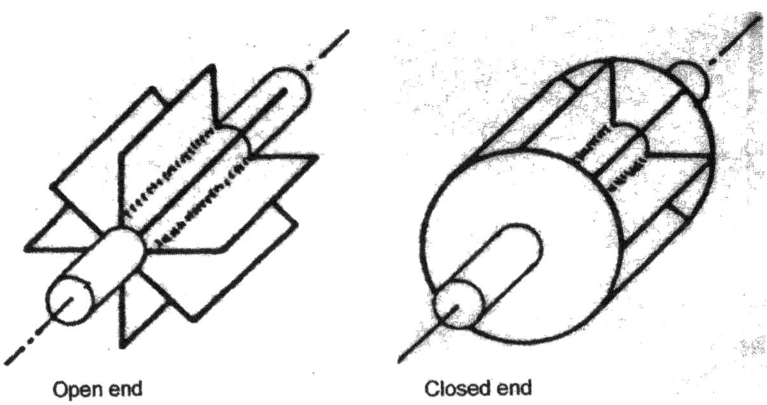

Figure B-57 Rotary valve rotor construction.

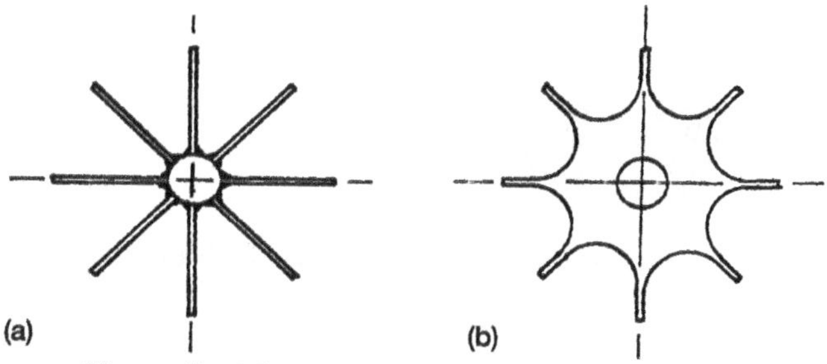

Figure B-58 Rotary valve pocket configurations.

therefore maximum volumetric displacement. It is more suited to handling free-flowing products. Type (b) has shallow, rounded pockets and, therefore, its volumetric displacement is less than type (a), but this configuration has been successfully used with the more cohesive types of solids that could tend to stick in deeper pockets. A Type (b) rotor also provides a more even flow than a Type (a) rotor in situations where this is important. For abrasive materials, rotors with replaceable, adjustable solid or flexible tips are available.

B7.2 Gravimetric Feeders

Examples of gravimetric feeders are weigh belts, loss-in-weight systems, and gain-in-weight systems. Gravimetric feeders rely on weighing the material to achieve the required discharge rate. A gravimetric feeder would be used when accuracies of less than 5% are required, particularly over short time periods, when the material's bulk density varies, or when the weight of material used for a particular process needs to be recorded. There are basically two systems: continuous and batching. A continuous gravimetric system controls the weight per unit time, whereas a batch system simply controls the weight of material discharged, such as 40 kg of material to a mixer.

Weigh Belt Feeder: This type of gravimetric feeder typically is used in continuous feeding applications as opposed to batches (see Figure B-59). A belt feeder can be used as a weigh belt by locating weigh idlers under the belt downstream of the outlet. Load cells weigh the material crossing them and send a signal to a controller where it is integrated with the belt speed and compared to a set point, and the speed of the drive is adjusted accordingly to regulate the discharge rate.

Loss-in-Weight Feeder: The loss-in-weight (LIW) gravimetric feeder is used when feed accuracy is essential. It measures the loss of weight of mate-

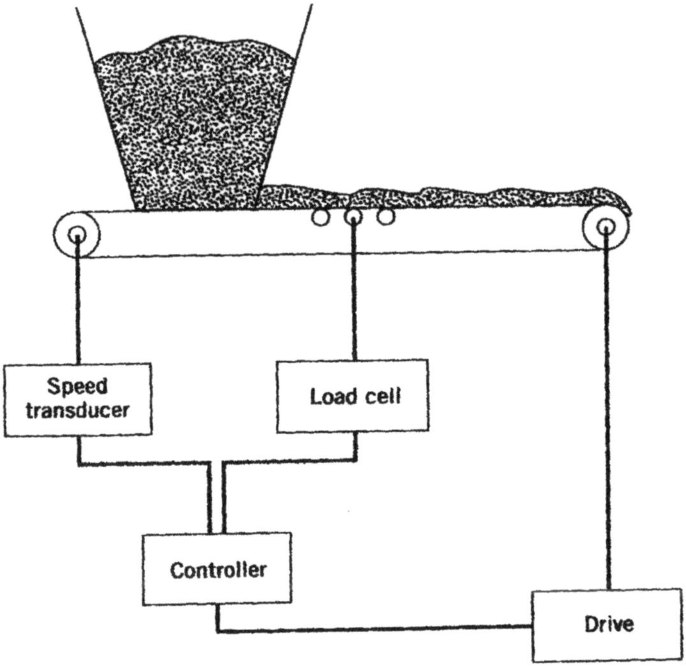

Figure B-59 Schematic of a weigh belt feeder.

rial discharged from the system. As such, it can be used both in continuous and batching systems. Load cells are attached to a bin or hopper that are capable of weighing the bin, feeder, and the contents. These load cells sense changes that take place as material is discharged and send a signal to a controller to speed up or slow down the feeder.

A disadvantage of LIW systems is that they cannot weigh while being filled. A typical LIW system switches to volumetric feeding while the filling process occurs, and when the bin is filled, it switches back to gravimetric. Some systems are available with "no-freeze" designs in which one bin and feeder discharges while another is being filled, and these are switched back and forth. To control the feed rate, feeders can be screws, belts, rotary valves, or vibratory pan feeders.

Gain-in-Weight Feeders: These types of feeders are used only for batching applications. The receiving container rests on a scale or load cells and the system controls the discharge from the filling bin, which can use a volumetric feeder to control the rate. A batch accuracy of ±0.25% at two standard deviations is not unusual.

Additional information on the feeders discussed in this section, as well as other types of feeders are presented by Colijn (1985), Cooksey (1977), Pittman (1985), Vetter (1998), and Woodcock and Mason (1987).

B8 HOSES, LOADING SPOUTS, FLEXIBLE BOOTS AND SOCKS

Hoses, loading spouts, and flexible boots and socks are often used to transfer particulate solids from one item of equipment to another, or from equipment to containers. These items are described and discussed below.

B8.1 Hoses

Hoses are used for conveying solids from mobile (nonstationary) equipment to other mobile equipment or to stationary (fixed-in-place) equipment, for example, railroad hopper cars to discharge adapters (called "unloading sleds" by some vendors) or hopper trucks to silos. Hoses are available in rubber (several types), PVC, and metal (galvanized steel, stainless steel, and aluminum). The type of hose to be used depends on factors such as the nature of the solid material to be conveyed (granules, pellets, powders, abrasive, food grade, etc.), pressure or vacuum operation of the conveying system, and temperature of the solids and conveying gas.

B8.2 Loading Spouts

Loading spouts (also called loading chutes) are used to load railroad hopper cars and hopper trucks from a silo in a dust-free manner. A typical loading spout, as shown in Figure B-60, consists of a retractable spout through which the solids flow, connected to a vacuum system and dust collector. Some loading spouts have a spreader device at the bottom of the spout designed so that the incoming air accelerates the material in a outward pattern with a slightly upward velocity. This obviates the need for a mechanical spreader inside of the railcar or hopper truck. Loading spouts have internal parts, such as inverted cones or vanes, that are made of plastic or metal (often stainless steel). The outer sleeve is usually made of fabric such as rubber-coated nylon or urethane, with inner and outer metal support rings. The inner cones or vanes are often connected with a metal wire, which can be used as a grounding strap. In one design, the spout has internal vanes installed inside to maintain a straight flow of the material. This eliminates swirling of the material which keeps it from mixing with the negative-pressure air stream that draws out the dust. A number of other features of a loading spout are the following:

1. Remote controlled, single and dual direction, spout positioners which allow the operator to properly position the spout.
2. A level sensing device to alert the operator that the filling cycle is nearly completed or the desired product level has been reached.
3. A switch to insure that the negative air system is "on" when the loading spout is extended, and "off" when the spout is retracted in the "full-up" position.

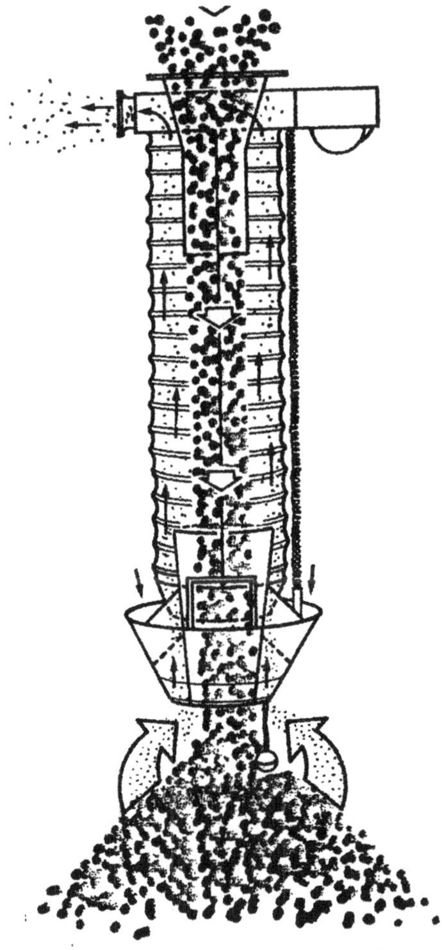

Figure B-60 Schematic of atypical loading spout.

4. Hatch adaptor kits to seal the loading spout into square, round, or slotted railcar hatches.
5. Explosion-proof electrical components.

Variations of the design shown are provided for different loading applications by loading spout manufacturers. These include open truck loading, barge loading, and enclosed ship compartment loading.

B8.3 Flexible Boots and Socks

Flexible boots and socks are commonly used to connect two pieces of equipment during gravity flow operations (e.g., for unloading the cake from a centrifuge to a product receiver). Boots are typically made of plastic or rubber, while socks are typically made of woven fabric. They are usually held in place by clamps or metal bands.

B9 MECHANICAL CONVEYORS AND BUCKET ELEVATORS

B9.1 Mechanical Conveyors

Mechanical conveyors are among the most commonly used particulate solids handling equipment in CPI plants. They are used primarily where conveying distances are short because they have both capital and energy cost advantages over pneumatic conveyors. There are five major types of mechanical conveyors: belt conveyors, screw conveyors, vibratory conveyors, chain-type conveyors, and bucket elevators. The characteristics of the particulate solids are a major factor in the selection of the appropriate mechanical conveyor.

A short discussion and description of various types of mechanical conveyors is presented in the following paragraphs.

Belt Conveyors: belt conveyors are best suited for conveying medium to large amounts of bulk materials (capacities up to 20,000 tons per hour have been successfully conveyed). They are similar to belt feeders in configuration, but operate at much higher speeds (1000–1200 ft/min, compared to under 100 ft/min for belt feeders). They also are designed for much longer distances; for example, single flight belts spanning 7 miles have been built. They may be open, covered, totally enclosed, or even sealed. The support roller system forms a trough to which the belt conforms, thus enabling more material to be carried than on a flat belt. The belt is secured at the discharge end by a driven head pulley, and the feed end by an idler pulley. Belt tension is maintained by a gravity or spring-loaded "take-up." The belt is flattened as it passes over the head (discharge) pulley. Figure B-61 is a schematic

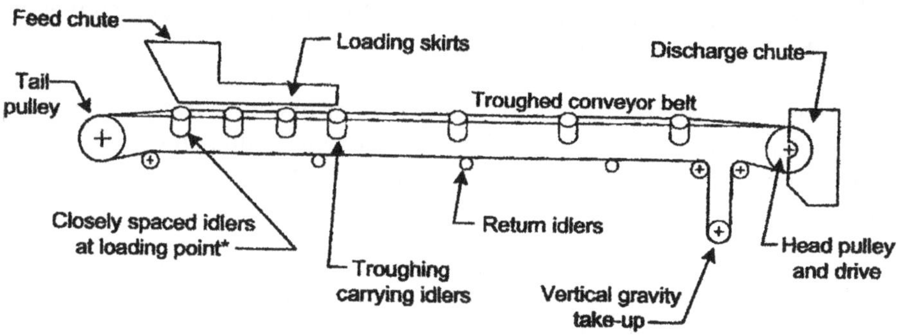

Figure B-61 Typical belt conveyor.

drawing of a typical belt conveyor. The belt can be loaded at a number of points, and unloaded at more than one point by use of a tripper. Belt conveyors can be arranged to follow a number of profiles and/or paths of travel such as horizontal, inclined, declined, with inclusion of concave and convex curves.

A belt is composed of three main components: (1) the carcass, (2) the top cover, and (3) bottom cover. The carcass provides the tensile strength necessary to move the loaded belt, absorbs the impact forces of material being loaded onto the belt, provides lateral stiffness necessary to support the load between idlers, and provides adequate strength for mechanical fasteners to hold on to the carcass. The top cover protects the top of the belt from the material being conveyed by absorbing abrasive wear at loading, and protecting the carcass from any adverse property of the material being conveyed as well as any environmental conditions to which it is exposed. The bottom cover is a lining at the bottom of the belt and protects the carcass from possible wear from idlers and pulleys. It is usually thinner than the top cover.

A number of fabric yarn materials are used for the carcass, such as cotton, rayon, fiberglass, nylon, polyester, steel, and Kevlar.

Belt cleaning devices are critical to avoiding damage to the belt and prolonging belt life. When conveying materials that have a tendency to stick to the surface of the belt, it is essential to employ some sort of cleaning technique at the head end to minimize the buildup of material on snub pulleys and return idlers.

There are several types of nonstandard belt conveyors that are available which are designed to handle a specific problem or can handle applications which standard belt conveyors cannot. Among these are elevating belt conveyors having walled and cleated belts, curved belt conveyors, air-supported, wire mesh belting, enclosed and folded belts, to name some.

Belt conveyors can handle a variety of materials which can range from very fine dusty powders to large lumpy ores, stones, coal, or logs. Materials can also be friable, abrasive, or corrosive.

Materials that might cause sticking or packing if conveyed by other conveyors are often handled successfully on belt conveyors. Also, hot materials such as sand, coke, and iron ore pellets can be conveyed on belt conveyors using the right carcass fabric.

Screw Conveyor: A screw conveyor is similar in configuration to a screw feeder. Screw conveyors must be fed at a controlled uniform rate. This may be from an upstream process, feeder, or other conveyor. A screw feeder section can be added to a screw conveyor section to form a single unit operating with one drive.

The feeder section is buried in the material and removes it from a silo or hopper at a controlled rate. The feeder advances the material to the conveyor section where the bulk solid is conveyed at the appropriate loading depth. Figure B-62 is a schematic drawing of a typical screw conveyor. Depending

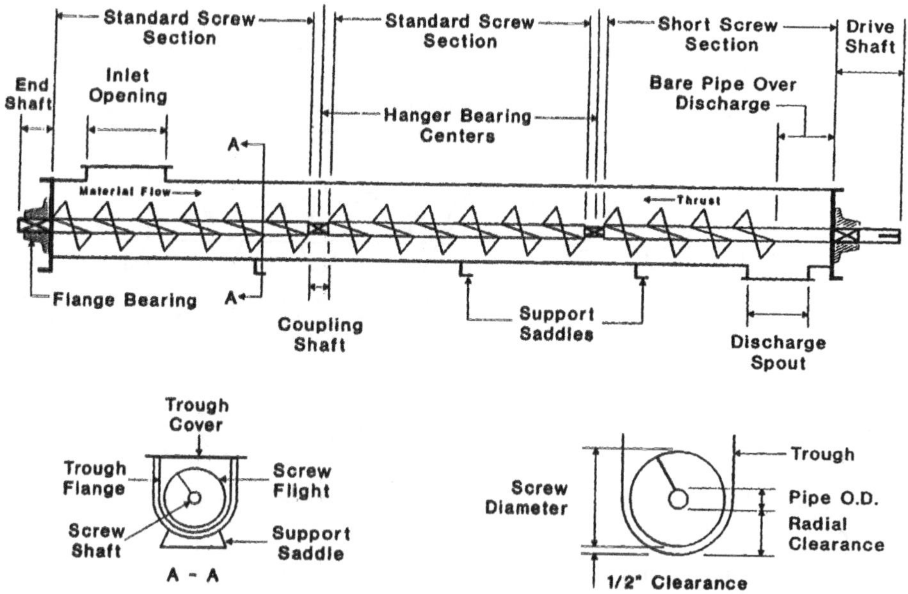

Figure B-62. Typical screw conveyor.

on the material, the trough is operated from 15% to 45% full. Trough configurations are usually U-shaped (with flanges), but other designs are available such as flared, drop-bottom, rectangular, tubular, and jacketed.

Various types of screw flights are available (see Figure B-63), and various pitch and diameter combinations are used (see Figure B-64). Multiple feed and outlet locations can be provided. Lengths are made to customer specification. Very long units can be constructed by bolting together shorter sections of screw and trough and the use of intermediate hanger bearings. Intermediate hanger bearings may not always be acceptable due the potential for product build-up and subsequent frictional heating which could cause an explosion or fire. Purging of these bearings with inert gas may prevent product build-up, and should be discussed with the vendor. In the most common sizes (6- to 12-inch screws) a span of 10 to 15 feet without intermediate hanger bearings is possible.

The advantages and disadvantages of screw conveyors can be summarized as follows:

ADVANTAGES

1. Low investment cost compared to other conveyors of comparable capacity.
2. Compact design, comparatively easy to seal against water or dust passing in or out, and the solids being handled can be blanketed with a dry or inert gas where necessary.

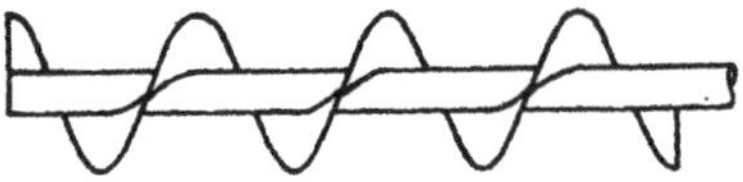

(a) Regular helicoid flighting

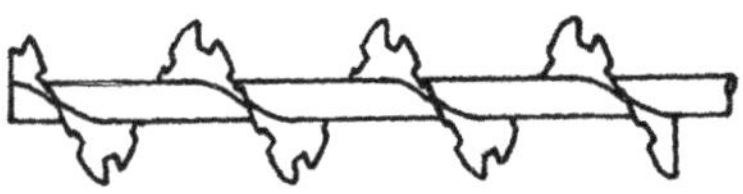

(b) Cut screw flighting. Used for conveying, cooling and moderately mixing materials, simultaneously

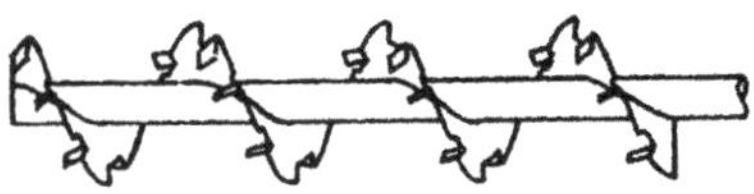

(c) Cut and folded screw flight. Continual lifting and tumbling of the material by the folded flights improves aeration and promotes mixing

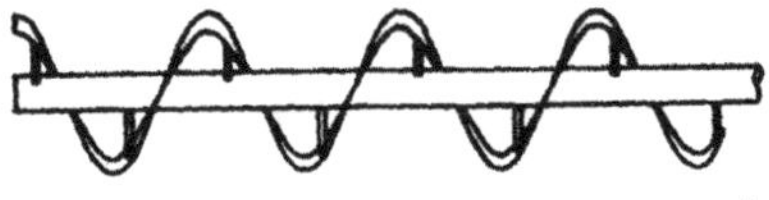

(d) Ribbon flighting. Used for conveying substances that are sticky, gummy or viscous

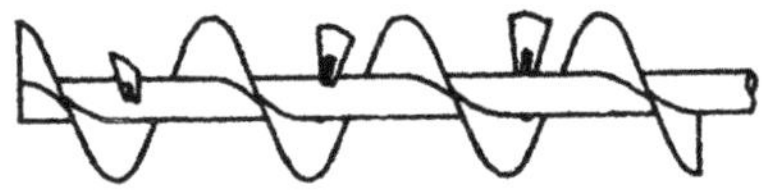

(e) Regular screw flighting with mixing paddles. Used to mix materials where the conveyor length provides time for proper mixing

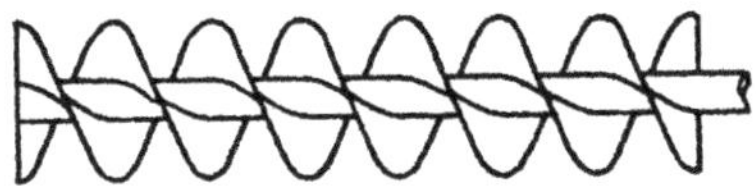

(f) Double flight conveyor screw. Used to promote a smooth and gentle flow of material

Figure B-63 Various types of screw conveyor flights.

3. Fairly simple fabricating with unsophisticated equipment; a high degree of part standardization exists within the industry.
4. Generally low maintenance than with most types of mechanical conveyor; there are fewer moving parts to wear out or get out of order.
5. Ability to handle a wide range of solids.

DISADVANTAGES

1. Lumpy, fibrous, or sticky materials may cause problems.
2. Lengths are limited by the allowable torque capability of the drive and coupling shafts.
3. Power requirements can be high with solids that tend to pack.
4. Conveying efficiency is considerably reduced when screw conveyors are inclined or mounted vertically.
5. Can cause compacting of solids and possible overheating.

Two specialized designs of screw conveyors are available for vertical lifts on a limited number of solids. They are the screw lift (see Figure B-65) and the flexible screw lift (see Figure B-66). The flexible screw lift has the advantage that the "screw" is actually a flexible coil mounted in a flexible plastic housing, making it adaptable to tight and unusual layouts. These two vertical lift screw conveyors should be applied only to free-flowing materials.

Vibratory Conveyors: A vibratory conveyor is composed of the same components as a vibratory feeder: a trough, drive section, and support base (see

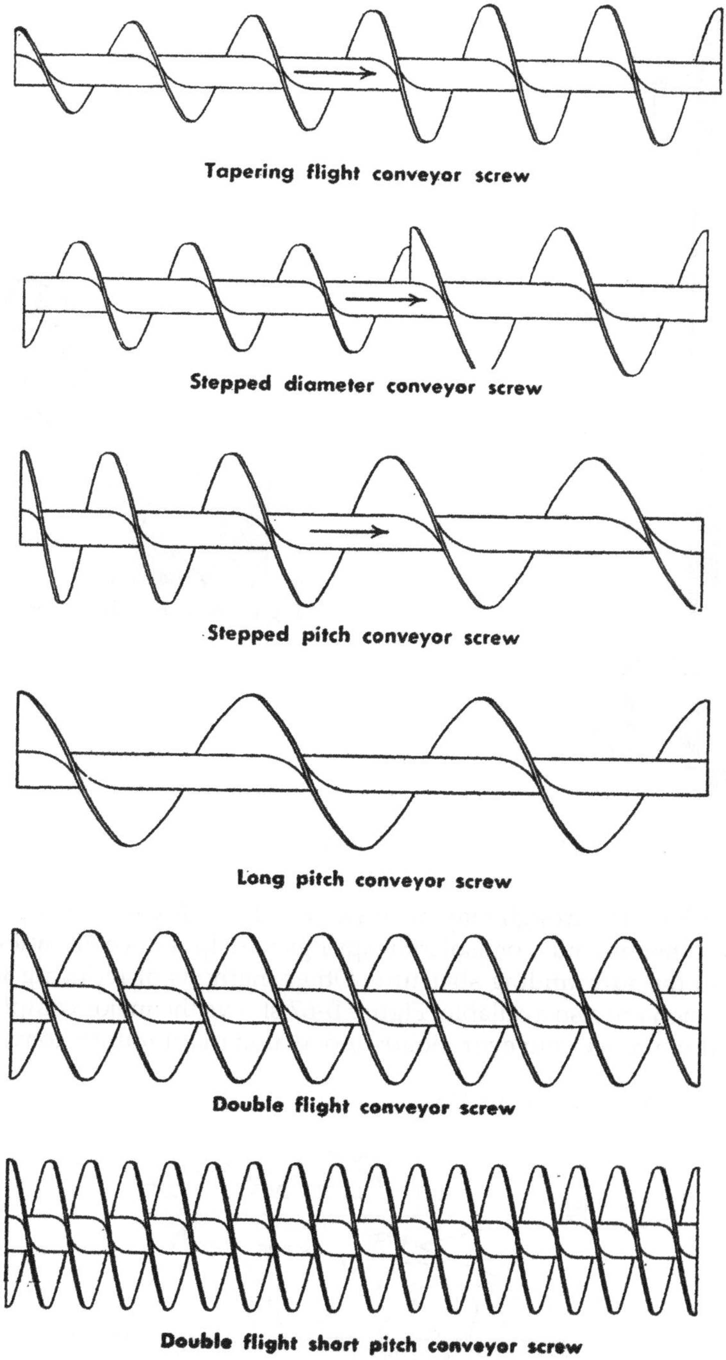

Figure B-64 Various screw conveyor pitch and flight diameter combinations.

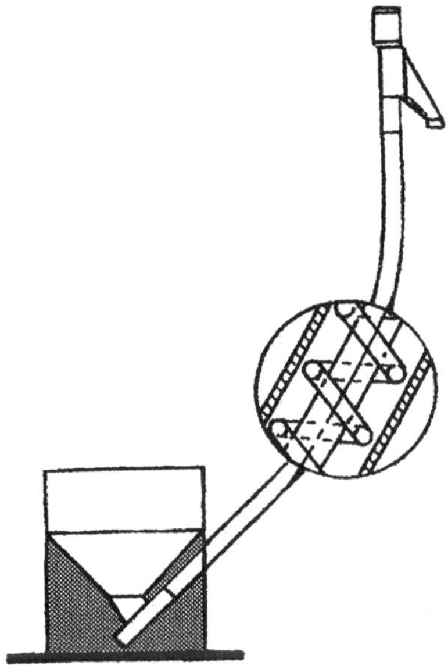

Figure B-66 Flexible screw lift conveyor.

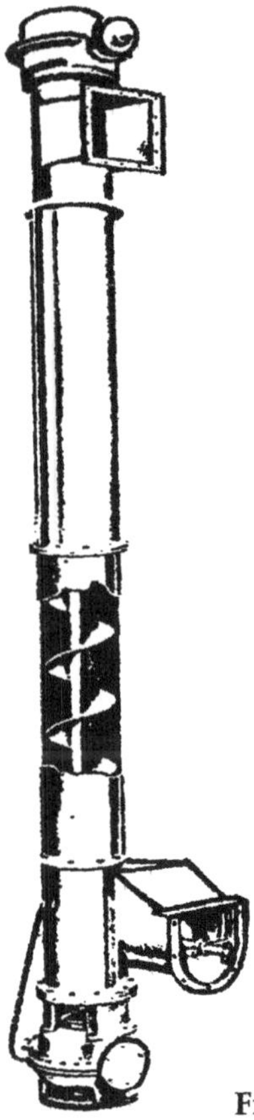

Figure B-65 Screw lift conveyor.

Section 5.3.6). The trough may be open or fully enclosed. The support base frame is often mounted on isolation springs to help reduce vibrations transmitted to the surrounding structure. Other methods of reducing vibration transmission are also available. Figure B-67 shows the major components of a typical vibratory conveyor. Solids move along a vibratory conveyor in a

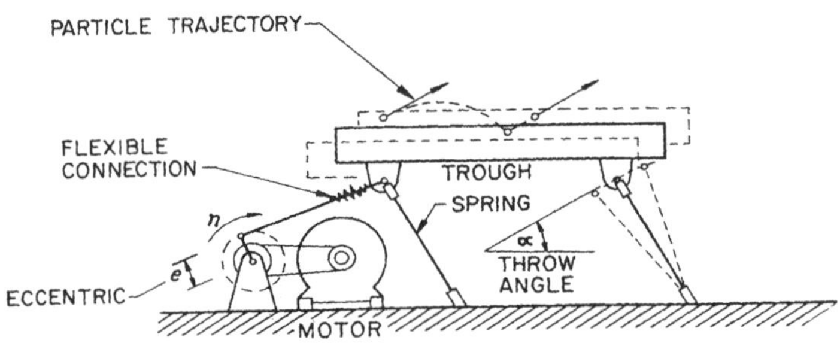

Figure B-67 Schematic of typical vibratory conveyor.

series of "throws and catches." The pan in its forward stroke accelerates the particles off the pan, in a trajectory determined by the suspension system, and the particles return to the pan as it is completing its return stroke. The solids handling capacity of a vibratory conveyor is determined by the relationship: pan width × depth of material × bulk density × linear speed of the material on the pan. Linear speed is a function of stroke and frequency of the driving unit.

The drive systems used for vibratory conveyors are the same as for vibratory feeders. However, there are some differences between the two with respect to acceleration, operating frequency (cycles per minute), and operating stroke lengths, as follows:

- *Acceleration*: Conveyors have a total acceleration of 1 to 4 *g* while feeders have a total acceleration of 3 to 13 *g*.
- *Operating frequency*: Normal operating frequencies for conveyors are between 300 to 700 cycles per minute while feeders generally operate between 900 to 7200 cycles per minute (most common operating frequencies are in the range of 1200 to 3600 cycles per minute).
- *Operating stroke length*: From 4 to 0.25 inches for conveyors and from 0.25 to 0.035 inches for feeders (varies inversely with frequency).

Vibratory conveyors can handle the same types of solids as vibratory feeders, that is, they can handle a very wide range of materials, although, in general, granular materials (say 20 mesh and above) handle better than pulverized, and flat and irregular shapes better than spherical. Also, solids that aerate can be difficult to convey satisfactorily. On the positive side, friable products can be conveyed gently and without excessive degradation.

Chain-Type Conveyors: Chain-type conveyors are conveyors in which endless chains of various types travel the entire length of the conveyor, transmitting the pull from the driving unit, and in some cases, carrying the whole weight of the materials to be transported. Particulate solids may be carried directly on chains, on special attachments fitted to the chains, and by flights pushed or towed by the chain. They are particularly suited for those conveying applications that require complete enclosure for dust containment, small cross-sections, intermediate or multiple filling or discharge capabilities, combinations of horizontal and vertical paths, handling materials at elevated temperatures, and situations that require improved safety. Chain-type conveyors are often used in special situations where the other mechanical conveyors have serious drawbacks. They can handle medium-sized capacities for distances up to 300 feet in length but they are most commonly used for distances under 100 feet.

Among the chain-type conveyors are: apron, drag, plain chain, scraper, flight, and en-masse. Characteristics of various chain-type conveyors are

TABLE B-7
Characteristics of Various Chain-Type Conveyors

Conveyor Type	Material Action	Chain Action	Chain Isolated from Material	Distance	Speed	Attachments
Apron	Carried	Rolling	Yes	Long	Fast	Pans
Drag	Sliding	Sliding	No	Short	Slow	None
Scraper	Sliding	Sliding	No	Short	Slow	Scrapers, sidebars
Flight	Sliding	Sliding or rolling	Yes	Short	Slow	Flights
Plain chain	Carried	Sliding or rolling	No	Long	Slow	Claws, hooks, rollers, or nothing
En-masse	Sliding	Sliding	No/yes	Long	Slow	Skeleton flights

Source: Faye, M. E., and Skocir, T. S. *Mechanical Conveyors—Selection and Operation,* Table 6.1. Reprinted with permission of CRC Press.

presented in Table B-7. Presented in the following paragraphs are brief discussions and descriptions of chain-type conveyors.

Apron Conveyors: An apron conveyor carries material on an endless series of jointed, overlapping apron pans, suspended between two strands of chain. Several different shapes of pans are available. Figure B-68 is a schematic drawing of an apron conveyor; the figure also shows different shapes of pans. The chain (usually with a roller) guides the apron along the track and around bends. The conveying path can be horizontal or a combination of horizontal and inclined. The chain is driven by a head sprocket and guided on the return run by a tail sprocket. Apron pan widths range from 18 to 60 inches, and pan speeds range from 100 fpm (for 18-inch pans) to 60 fpm (for 60-inch pans).

Apron conveyors can handle a wide range of materials such as granular and very lumpy solids (e.g., ore, stone, coal), fine materials (e.g., sand), and large irregular shaped materials (e.g., scrap materials, industrial refuse). They can handle material temperatures to 600°F on most pans, and up to 2000° on some special pans.

Drag Conveyors: A drag conveyor is a type of conveyor having one or more endless chains which drag materials along the bottom of a trough to transport the material to the outlet and returns up and around the tail wheel and over return rolls to the head wheel (see Figure B-69). The trough is usually steel, cast iron, or a combination of steel with hard iron liners, and often has replaceable wear plates or bars. Concrete troughs are also used.

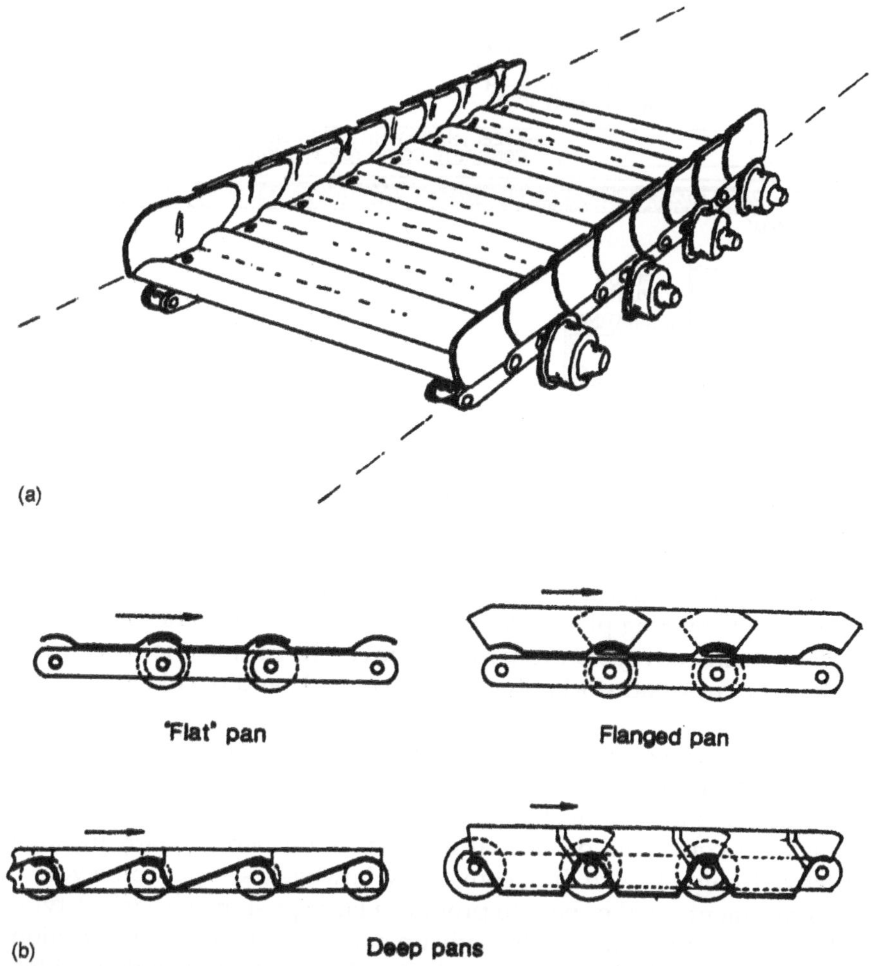

Figure B-68 Typical Apron conveyor. (a) Schematic drawing; (b) some different shapes of apron conveyor pans.

Drag conveyors are used most commonly on cement clinker (hot or cold), ashes, coal, wood chips, sawdust, pulpwood, and for log handling. They are usually limited to speeds of 50 fpm and relatively low capacities of up to 1000 cubic feet per hour for 20-inch-wide chains.

Plain Chain Conveyors: These conveyors move materials by carrying them on one or more strands of conveyor chain. Both rolling- and sliding-type chains are used. These conveyors are typically designed for continuous service and require very little maintenance. Due to the nature of their use, plain chain conveyors are always custom-designed for a specific application.

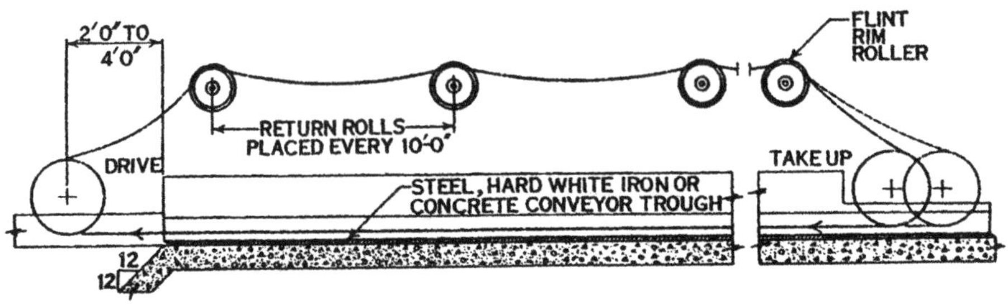

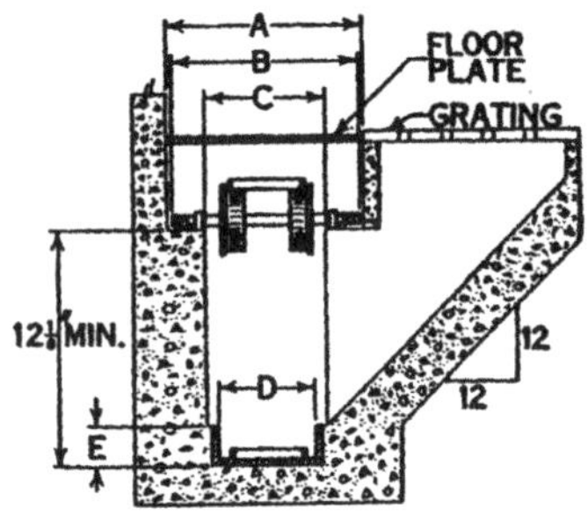

Figure B-69 Schematic of a typical drag conveyor.

Depending on the application, they may consist of only the chain or the chain may have flat covering plates, claws, hoods, and various other types.

Plain chain conveyors are used to convey large unit pieces or objects such as logs, cut wood stock, plywood, sheet rock, and the like.

Scraper Conveyors: Scraper conveyors consist of one or more strands of endless chain, to which are attached spaced transverse scrapers made of either malleable iron or steel plate. Scraper conveyors operate in the same way as drag conveyors do in that material is pushed forward through a restrictive trough. Unlike drag conveyors, however, scraper conveyors use wide attachments on the chain to provide a large pushing surface. A typical single chain scraper conveyor is made up of a drive and discharge section, a take-up section, and sections of troughs (see Figure B-70). The chain rides in

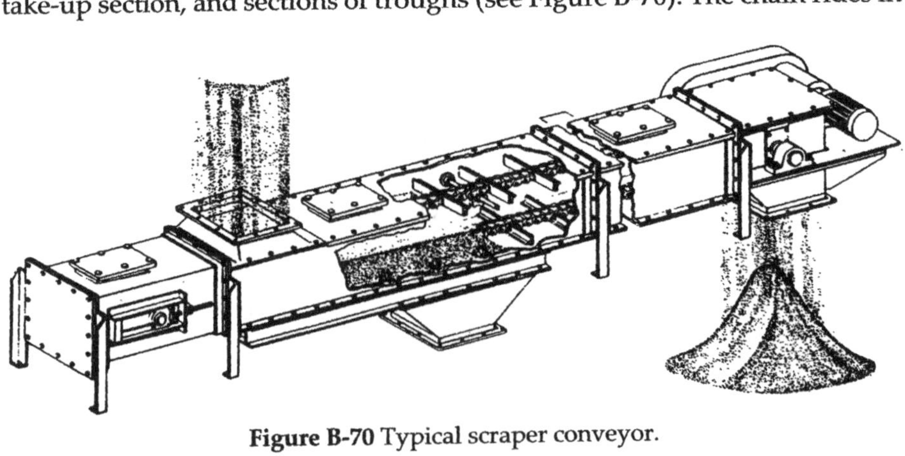

Figure B-70 Typical scraper conveyor.

the trough and the scraper blades continuously scrape any material off the bottom of the trough. They are often used with a flooded inlet acting in a feeder capacity. Scraper conveyors can operate horizontally or on inclines up to 45°, and in some cases, a combination of both. The trough of a scraper conveyor can be constructed of steel or concrete or lined with a variety of abrasion-resistant materials. They can be open, enclosed, dust-tight, and air- or water-cooled. Speeds of up to 100 fpm are possible with capacities as high as 300 tons per hour. Scraper conveyors can have several inlet and outlet ports.

Some materials commonly conveyed with scraper conveyors include bagasse, calcined coke, clay dirt, coal, fly ash, slag, various grains, and moderately abrasive materials.

Flight Conveyors: Flight conveyors generally utilize one or more roller or sliding chains riding on a track above the material in the trough. This feature effectively isolates the chain from the material being conveyed. This is particularly important when using roller chains because it will keep material from fouling the internal chain components. An isolated chain is also important when conveying hot materials as keeping the chain from being directly exposed to the heat will decrease the rate of chain elongation. Troughs for flight conveyors are constructed similarly to those for scraper conveyors. Flight conveyors can also convey horizontally or on an incline, and have several inlet and outlet ports. Speeds and capacities are also similar to scraper conveyors. Figure B-71 shows profiles of various flight conveyors.

Flight conveyors are suitable for handling free-flowing granular or lumpy materials. They are often used to convey very hot materials like cement clinker at temperatures as high as 2000°F and also mildly corrosive materials. However, sticky materials that will build up on trough surfaces of flights should be avoided.

En-Masse Conveyors: An en-masse conveyor consists of an endless chain or cable pulling a series of skeleton or solid plug flights through an enclosed casing. The particulate solids are pulled through the casing by the flights in a continuous stream with the casing cross-section nearly full. They are a highly flexible form of conveyor capable of conveying, elevating, and feeding. They are very compact and are relatively efficient.

En-masse conveyors offer a unique advantage over other types of conveyors as follows:

1. They completely enclose the material, protecting it against contamination, and the conveyor can be vapor tight and gas blanketed.
2. The complete enclosure protects against dust emissions.
3. Many materials are handled with a minimum of particle degradation.
4. The conveyor cross-section is smaller than most conveyors and often is easier to fit into a plant layout.

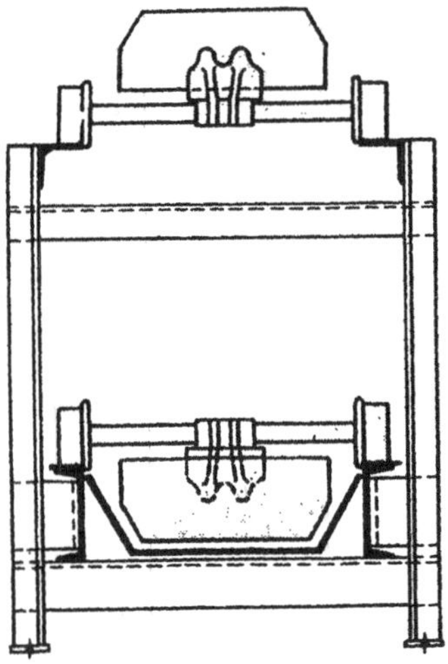

Single strand flight conveyor with roller-suspended flights

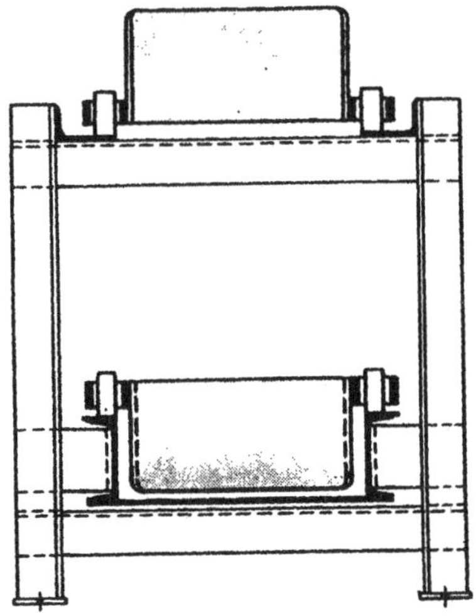

Double strand flight conveyors with roller-suspended flights

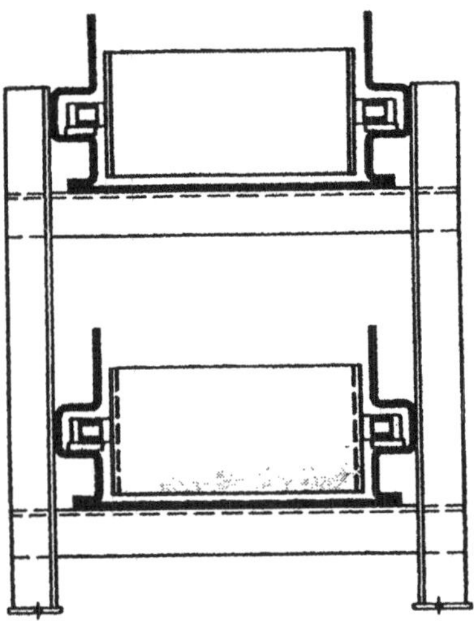

Double strand flight conveyor with sliding chain-suspended flights

Figure B-71 Profiles of various flight conveyors.

5. They can have multiple inlets and discharge openings.
6. They can have an L-shaped or Z-shaped path, eliminating transfer points that are required by a combination of a horizontal conveyor and vertical elevator.
7. They can be self-feeding, in other words, a combined feeder and conveyor.

The casing can be rectangular (see Figure B-72) or tubular (see Figure B-73). The flights can take various shapes (see Figure B-74). The tubular construction type is usually for lighter duty applications than the rectangular type. The tubular type has a lower cost and less sophisticated construction. The tubular type has two general types of construction: an endless series of

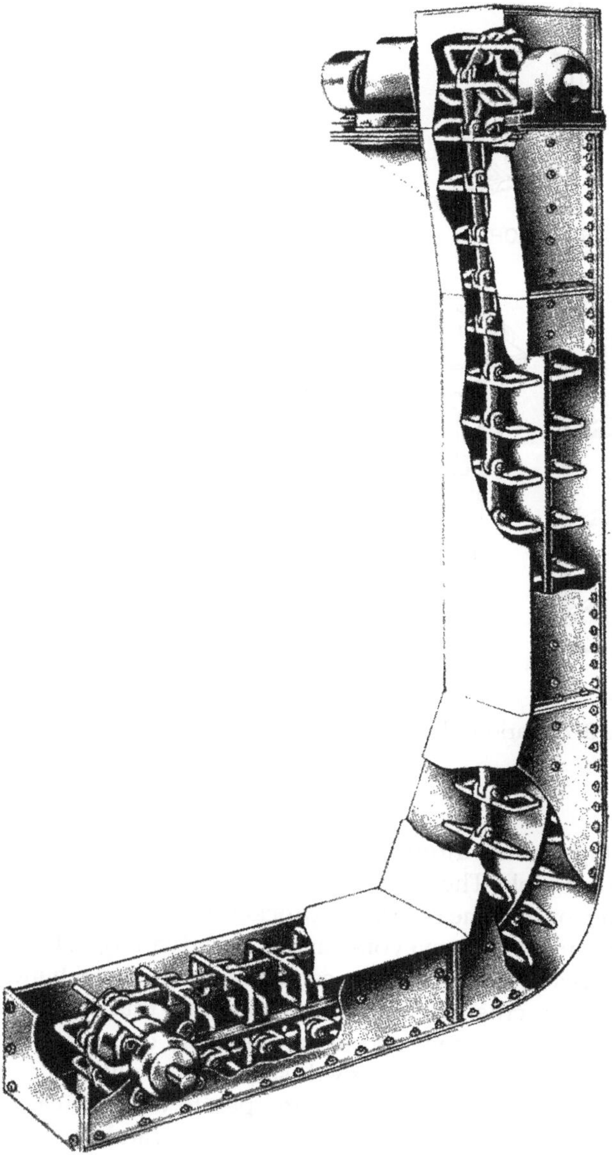

Figure B-72 Typical en-masse conveyor with square/rectangular casing.

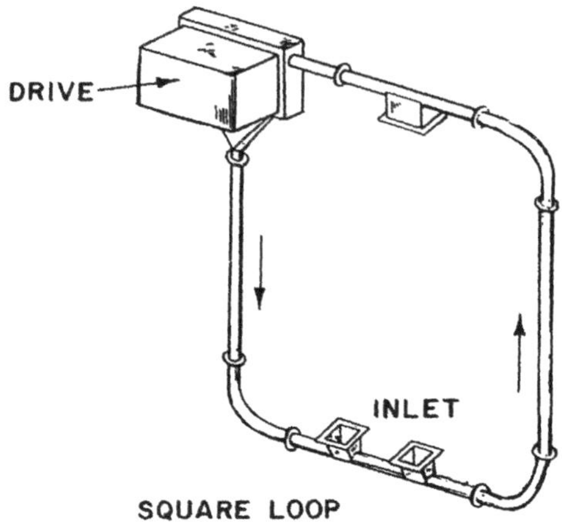

Figure B-73 Typical en-masse conveyor with circular casing.

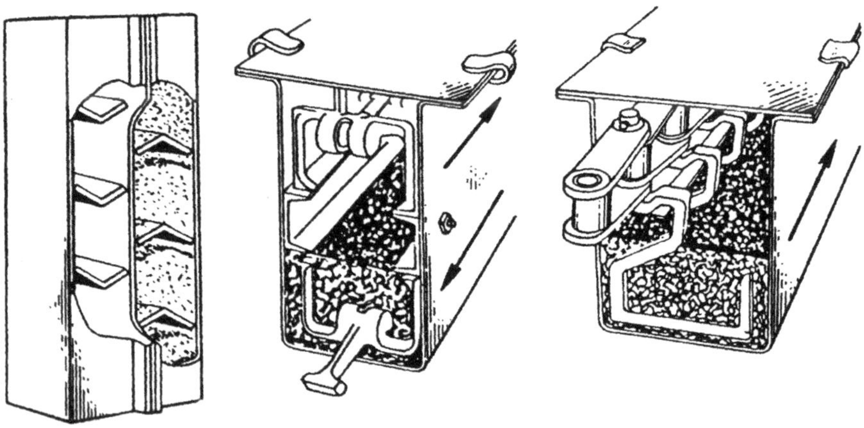

Figure B-74 Various flight configurations for en-masse conveyors.

polymeric plugs pulled through the casing by a continuous chain, or plugs pulled by a steel cable. The plugs and chain are guided around bends by the tubular casing, or in some cases, by sprocket wheels.

Although en-masse conveyors have many advantages, they also are limited by the types of materials they can convey. They can handle only free-flowing, granular, crushed, or ground materials that are noncorrosive or nonabrasive. They are often used to handle dusty or aeratable materials, as the closed casing easily contains any dust generated without requiring modification. Some en-masse conveyors can operate with the casing 100% full, which eliminates the presence of air so that they can handle materials that might present an explosion hazard in other types of mechanical conveyors. They have also been used successfully in handling friable materials as the conveying action is very gentle. Only a small percentage of the material conveyed is actually sliding against the casing walls, and the material moves as a compact immobile mass.

Comprehensive discussions on mechanical conveyors are presented by Colijn (1985), Fayed and Skocir (1997), and Fruchtbaum (1988). A less comprehensive but good review is also presented by Woodcock and Mason (1987).

B9.2 Bucket Elevators

A bucket elevator is a conveyor that is used for vertical lifting of solids, and is the most simple, efficient, and reliable conveyor for this application. In a bucket elevator, a series of buckets attached to an endless chain or belt pass by a loading section where the buckets are filled with material, are elevated to pass around a head pulley where the buckets dump the contained material, then circulate back to be reloaded. The drive is usually situated at the top of the conveyor, which is called the head section. The bottom of the elevator is called the boot section, and it contains the lower sprocket or pulley, a material loading chute, and access panels. The head and boot section are enclosed in a casing structure. Most units are completely enclosed and self-supporting. Bucket elevators are not self-feeding, but must be fed at a controlled rate to avoid overloading or damaging the machinery.

There are four general types of bucket elevators, named by the method by which they discharge materials: centrifugal, continuous, positive, and internal discharge (see Figure B-75). They are briefly discussed and described in the paragraphs that follow.

Centrifugal Discharge Type: This is the most commonly used type of bucket elevator. These elevators consist of buckets spaced at intervals mounted on either a chain or belt. The buckets are spaced to avoid interference during loading and unloading. This type of elevator is best suited for handling loose, free-flowing, fine, or small lump materials (e.g., grain, crushed coal, sand, clay, sugar, and dry chemicals). The buckets scoop the material out of the boot, and any material that resists digging in any way usually poses a problem with this type of conveyor. Large lumps can jam into bucket edges causing severe wear and high chain stress. Sluggish materials produce similar effects. Buckets are operated at speeds high enough to ensure material is discharged by centrifugal force. Most of these elevators operate in the range of 200 to 460 fpm.

Centrifugal force is developed as the buckets pass over the head pulley or sprocket, and the material is discharged under the combined effect of the centrifugal force generated and gravity. The speed of the chain or belt is very important as it affects the operational efficiency of the elevator. At higher speeds, increased centrifugal force will cause material to leave the buckets

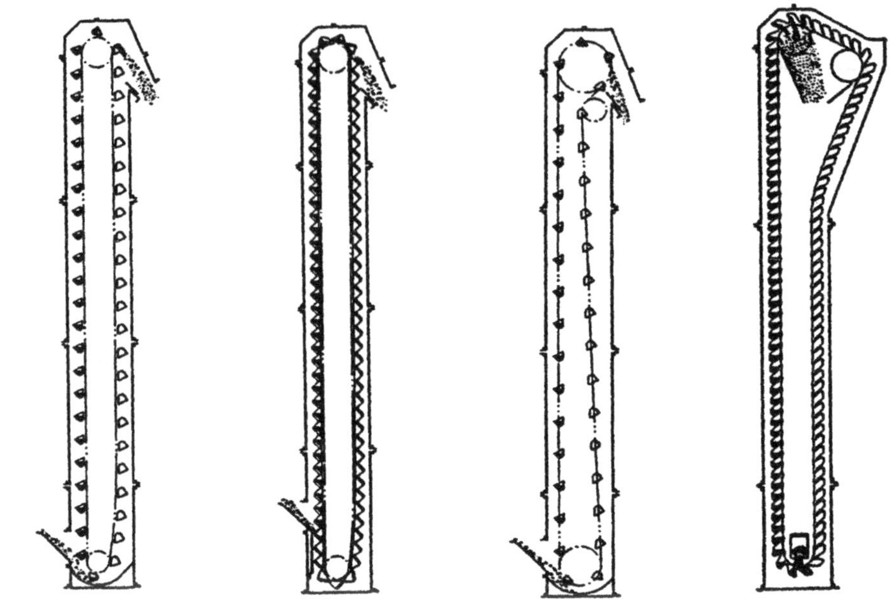

Figure B-75 Various types of bucket elevators.

too soon, resulting in material falling back down the bucket return side ("back leg"), thus reducing capacity. If the speed is too low, there will not be sufficient centrifugal force to hold the material in the buckets, and material will spill out of the buckets at the top of the head sprocket before it reaches the discharge opening.

Continuous Discharge Type: In this design, the buckets are mounted end-to-end on the chain or belt. The material is not scooped from the boot, but fed directly from the inlet chute into the rising bucket. A "loading leg" formed by vertical plates in the boot in close proximity to the rising buckets helps direct the material into the bucket and minimizes side spillage into the bottom of the boot. The material discharges from the elevator by gravity as the bucket passes over the head pulley. The material slides over the back of the bucket ahead of it. The back of the bucket has projecting sides that form a chute to contain the sliding material. Continuous discharge elevators operate at lower speeds than the centrifugal discharge type. Operating speeds range from 100 to 160 fpm, and this lower speed is necessary for proper loading and discharging.

Continuous discharge elevators handle a wide range of material, from light to heavy solids, fine materials as well as lumpy materials, and pulverized fluidlike solids. The slower operating speeds and gentle method of loading and discharging minimize breakage, making this type of elevator ideal for fragile or friable solids. Fine mesh and fluffy materials that tend to fluidize also handle well with slower speeds.

Positive Discharge Type: In this design, spaced buckets mounted between two strands of chain are loaded by a combination of scooping and chute feeding, and discharged by snubbing the chain by additional sprockets after the head pulley, so that the buckets are inverted over the discharge spout. These units operate at speeds about 120 fpm, and the slow operating speed, combined with the spaced buckets, results in relatively low capacities compared with the centrifugal and continuous type elevators.

These elevators are used for handling materials that do not discharge easily in other types of elevators. Materials that are light, fluffy, or sluggish exhibit this type of behavior (e.g., powdered carbon black, feldspar, ground cork, ebonite, and alfalfa meal).

Internal Discharge Type: This design uses a series of overlapping, inwardly facing, continuous buckets side-mounted on two strands of chain, similar to a positive discharge elevator. The inwardly facing buckets must be internally loaded from a chute extending in from either side of the casing. The advantage of inwardly facing buckets is that any material spilled during loading will fall into the open buckets below the loading leg.

A unique feature of this type of elevator is that there are no sprockets or wheels in the boot section. The chains travel around a guide that also provides take-up force.

This design is useful for gentle handling of small particles such as plastic pellets, granular chemicals, agricultural products (seeds, shelled nuts), and even mechanical parts (bolts, rivets, etc.).

Various styles of buckets exist for the different types of bucket elevators. Some bucket styles are designed to handle materials with specific handling properties, and they come in a wide range of sizes since the overall elevator capacity is a function of bucket width.

Colijn (1985), Fayed and Skocir (1997), and Woodcock and Mason (1987) present good discussions of bucket elevators.

B10 PNEUMATIC CONVEYORS

Pneumatic conveying systems are used where particulate solids have to be transported over long distances that cannot be economically done by mechanical conveyors. Pneumatic conveying involves the transportation of a wide variety of dry powdered and granular bulk solids in a gas stream. In most cases the gas is normally air, however, there are special cases (e.g., risk of explosion, fire, health hazards, etc.) where different gases, such as nitrogen, are used. Pneumatic conveying offers the user the following advantages:

1. Dust-free transportation of a variety of particulate solids.
2. Flexibility in routing-can be transported vertically and horizontally by the addition of a bend in the pipeline.
3. Distribution to many different areas in a plant and pick-up from several areas.
4. Low maintenance and low manpower costs.
5. Multiple use—one pipeline can be used for a variety of materials.
6. Ease of automation and control.

However, there are also some disadvantages, which include:

1. High power consumption.
2. Wear and abrasion of equipment in some applications.
3. Incorrect design can result in particle degradation.
4. Limited transfer distance.
5. By virtue of the complex flow phenomena which take place, there is a requirement for high levels of skill to design, operate, and maintain systems.

A pneumatic conveying system is made up of four distinct zones, each one requiring its own specialized hardware to achieve the desired operation. These are: the prime mover; the feeding, mixing, and acceleration zone; the conveying zone; and the gas–solids separation zone. These are briefly discussed in the following list:

- *Prime Mover*: The prime mover is an essential element in a pneumatic conveying system. A wide range of compressors, blowers, fans, and vacuum pumps are used to provide the necessary energy to the conveying gas.
- *Feeding, Mixing, and Acceleration Zone*: This zone is considered one of the most critical areas in any pneumatic conveying system. In this particular zone the solids are introduced into the flowing gas stream. By virtue of the fact that the solids are essentially at rest, a large change in momentum occurs when the solids are mixed with the flowing gas. Associated with this momentum change is the need to provide an acceleration zone. Normally, this zone consists of a horizontal length of pipe designed so that the solids are accelerated to some "steady" flow state. Commonly used feeding and mixing devices are Venturis, rotary valve feeders, screw feeders, and blow tanks.
- *Conveying Zone*: Once the solids have passed through the acceleration zone, they enter into the conveying zone, which consists of piping or tubing. The selection of the piping or tubing depends on a number of

factors including the abrasiveness of the solids, pressure requirements, etc. The piping or tubing system can have a number of bends and diverter valves in order to change flow direction.

- *Gas–Solids Separation Zone*: In this zone the solids are separated from the gas stream in which they have been conveyed. The selection of an appropriate gas–solids separation device is dependent upon a number of factors such as solids loading, particle size and size distribution, and particle properties (e.g., stickiness, corrosivity, abrasivity, etc.). Common gas–solids separation devices are cyclones, baghouses, and cartridge filters.

There are two basic modes of pneumatic conveying: dilute phase and dense phase. For simplicity, each phase is categorized in terms of the mass flow ratio (the ratio of the mass of solids to the mass of conveying gas). Dilute phase systems have a mass flow ratio of up to 15, while dense phase systems have a mass flow ratio greater than 15. Dilute phase (sometimes called lean phase) systems, in general, employ large volumes of gas at high velocities (>20 m/s). The gas stream carries the solids as discrete particles by means of lift and drag forces acting on the individual particles. Dilute phase systems are the most widely used of all pneumatic conveying systems. When the conveying velocity is less than that required to keep the solids in suspension (the saltation velocity in horizontal flow) and the particles begin to settle to the bottom of the pipe, the flow is said to be in a dense phase mode. Dense phase systems use a low conveying velocity (<5 m/s). Flow patterns in the dense phase mode can vary from being very unstable to stable or an intermediate unstable/stable regime. Flow patterns in the dense phase mode can vary from conditions in which the solids completely pack the pipe and move in a continuous dense plug to situations where the solids on the bottom of the pipe move as a series of "dunes" with a dilute phase layer of solids flowing above the dunes.

In addition to classifying pneumatic conveying in terms of their modes, a further classification is used to identify the type of system by whether it operates at low or high pressure. Low pressure systems are positive pressure, negative pressure (vacuum), combined negative/positive, and closed loop systems. High pressure systems use blow tanks to achieve the high pressure needed to convey the solids. These are briefly described in the following paragraphs:

Positive Pressure System: The positive pressure system is the most extensively used configuration in industry. These systems are well suited to multiple discharge applications in which material is picked up from a single point and delivered to several receiving vessels (bins, silos, etc.) (see Figure B-76). The change from one vessel to another is effected by means of multiport diverter valves. Of particular concern in the case of positive pres-

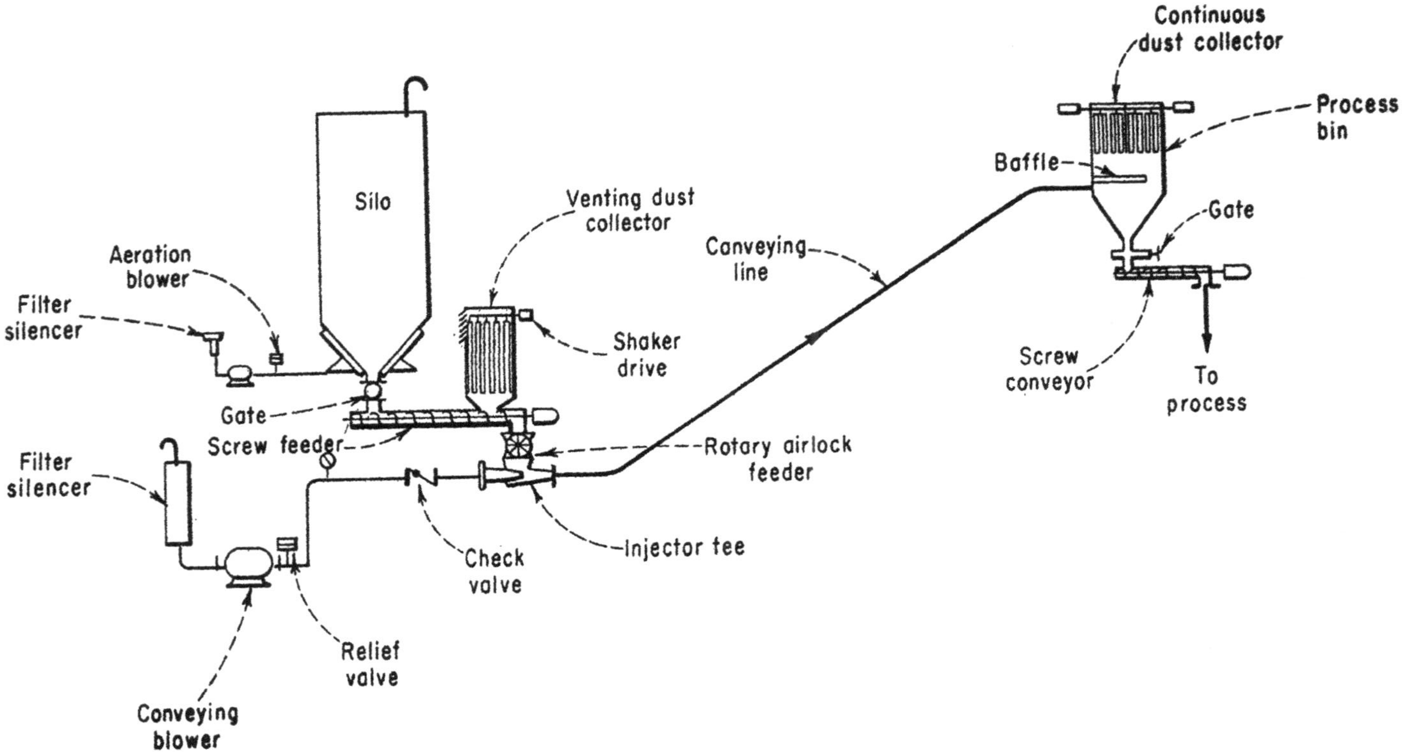

Figure B-76 Schematic of typical positive-pressure pneumatic conveying system.

sure systems are the feeding of solids into the conveying line, the gas-solids separation device (baghouse) and/or vent lines on the receiving vessels (bins/silos) and the elimination of leakage from the system to the surroundings (especially if the solids are toxic materials). Problems with irregular feeding of material into the pipeline, which can cause undesirable pressure surges and even complete blockage of the line, may be the result of poor feeder design, but can also occur as a result of holdup in the supply vessel (bin/silo). Obstruction of air flow by a clogged baghouse or blocked vent line can also be the cause of localized high pressures leading to air leakage into the atmosphere or unreliable solids flow through the system.

Negative Pressure (Vacuum) System: In general, negative pressure (vacuum) systems are used for the transport of material from several feeding points to a common collection point (see Figure B-77). Since the operation involves the employment of exhauster devices, such systems are limited in conveying distance and capacity. Negative pressure systems are extensively used in the conveyance of toxic and hazardous solids. Vacuum pneumatic conveying systems using an inert gas (usually nitrogen) in a closed loop arrangement are often used when the solids are very combustible. These systems permit dust-free feeding and also provide an additional safety feature as they do not allow leakage of solids out of the piping into the atmosphere. It is still necessary to minimize air inleakage since inflowing air could result in unwanted contamination of the conveyed product and possible explosion if the air concentration exceeded the limiting oxygen concentration. It would also tend to reduce the air necessary for conveying at the inlet (feed) end of the pipeline. In this type of system it is especially important that the solid material is adequately separated from the gas as the conveying gas passes through the exhaust fan or blower. Therefore, a high-efficiency separator must be provided.

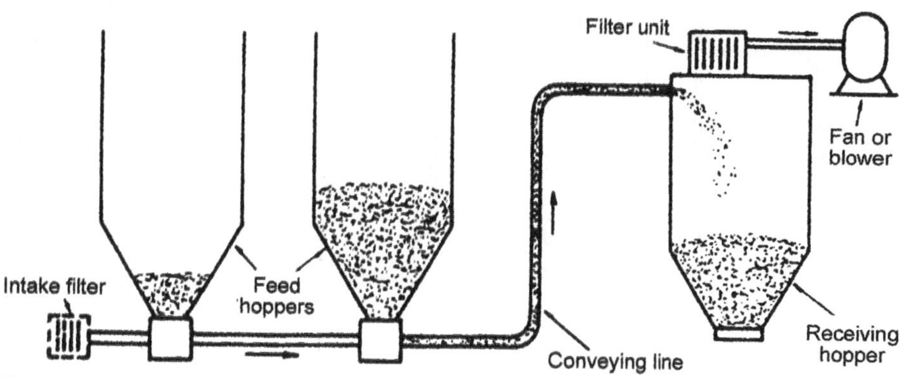

Figure B-77 Schematic of typical negative-pressure (vacuum) pneumatic conveying system.

Combined Negative/Positive Pressure System: This type of system (also called a "pull–push" or "suck–blow" system) is used when there is a need for solids to be collected from a number of different sources and discharged to several delivery points (see Figure B-78). Because of the potential for damaging the gas mover by passing solids-laden gas through it, it is usual to separate the solids from the gas stream and then refeed it to the gas mover to raise the pressure of the gas (hence the need for an intermediate storage vessel).

Closed Loop System: A closed loop conveying system is used where the conveying gas is inert (such as nitrogen) and it is necessary to minimize the wastage of the gas, or where it is essential to avoid pollution of the surrounding atmosphere, as when the solids are toxic. In this system the conveying gas is recirculated in a closed loop (see Figure B-79). A special precaution to be observed with closed loop systems is that high efficiency filtration should be provided, in order to ensure that unacceptable levels of dust are not returned to the suction of the gas mover. Also, when conveying under an inert gas such as nitrogen, it is important to monitor the oxygen level in the loop so that additional nitrogen can be injected when necessary to make up for losses through the feeder (often a rotary valve) and at the solids discharge point. Although not shown on Figure B-79, it may be necessary to install a cooler in the gas recirculation line to remove the heat of compression.

High Pressure System: This type of pneumatic conveyor consists of one or more blow tanks, utilizing high pressure air, usually, to transfer the solids from the blow tank to the receiving vessel. Blow tanks are generally

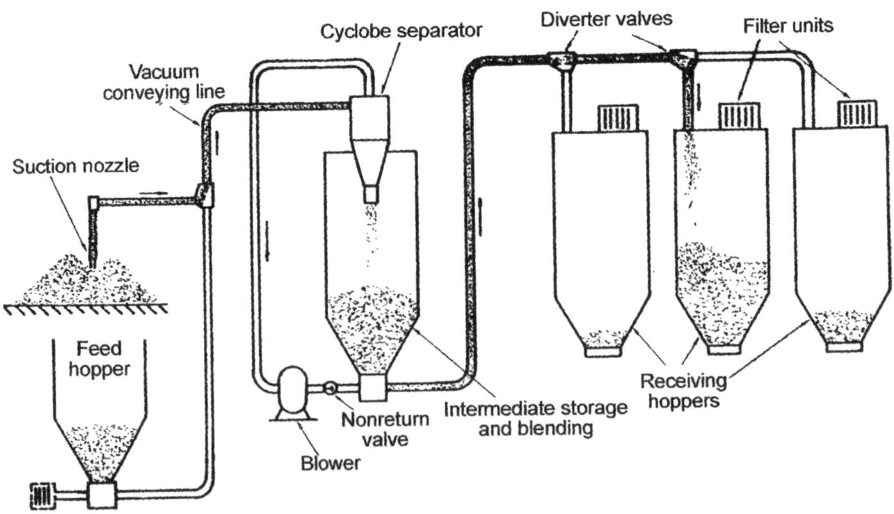

Figure B-78 Schematic of typical combined negative- and positive-pressure pneumatic conveying system.

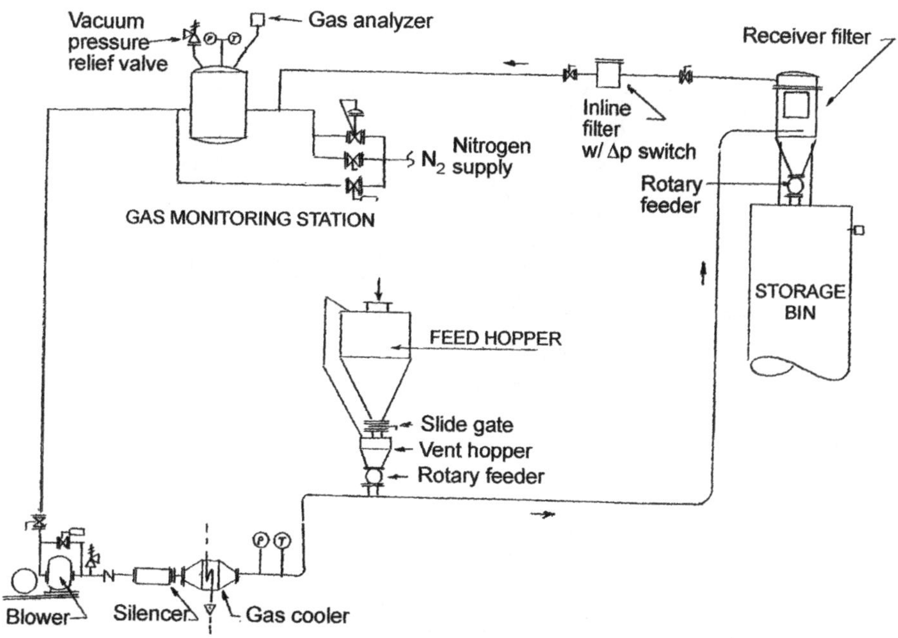

Figure B- 79 Schematic of typical closed-loop pneumatic conveying system.

employed in conveying batches, in the dense phase mode, but it is possible, with two blow tanks in series, to achieve continuous conveying. A particular problem with blow tanks is that they usually require more headroom than other pneumatic conveyors, and this is often the case with blow tanks arranged for continuous conveying. Since there are no moving parts, blow tanks are particularly suitable for handling friable and abrasive solids. It is not even necessary to have a discharge valve on a blow tank. Single blow tanks are available in two designs, as top discharge (without a discharge valve) or bottom discharge (with a discharge valve), as shown in Figure B-80. Numerous configurations of twin blow tank systems have been utilized in recent years. One of the most common is the parallel arrangement in which one blow tank is filled while the other is discharged (see Figure B-81). With an appropriate control system and valves almost continuous conveying can be achieved.

Air-Assisted Gravity Conveyors: Another type of pneumatic conveyor is the air-assisted gravity conveyor (see Figure B-82). In this conveyor, a gas plenum and an enclosed solids flow channel are provided by bisecting a rectangular duct with a porous medium, such as woven cotton, polyester, sintered plastic and metal, and woven steel laminate. Air flows through the lower channel up through the porous medium into the solids above and

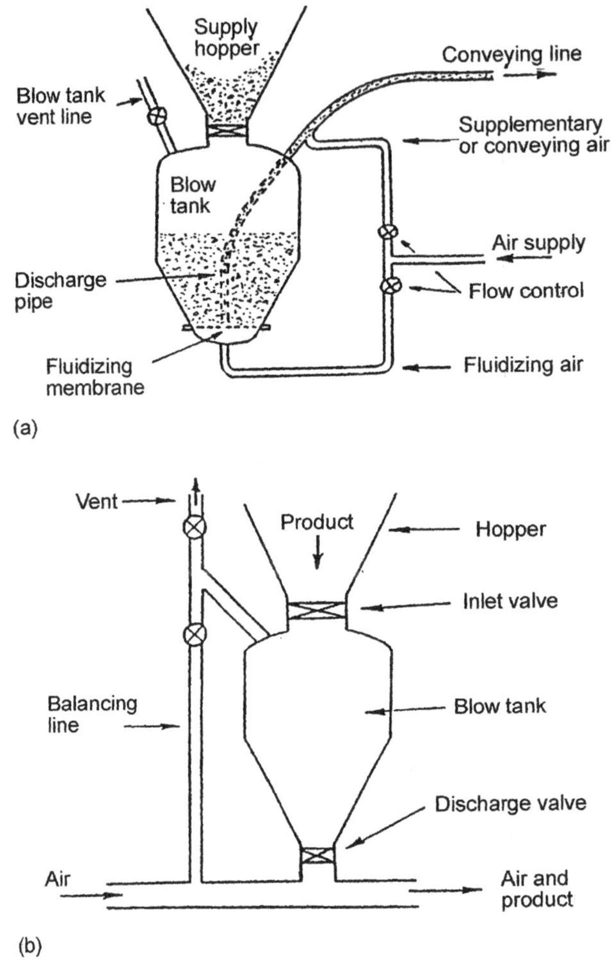

Figure B-80 Schematic of typical single blow tank pneumatic conveying systems. (a) Top discharge blow tank; (b) bottom discharge blow tank

fluidizes the solids. The conveyor is sloped (usually 3° to 8° from the horizontal) which assists in the gravity flow of the solids. Capacities depend on the width of the channel, particle characteristics, and air flow rates. Usual air requirements are 3 to 4 cfm per square foot of channel area at 4 to 30 inches water column pressure. Provided that the continuous downward slope can be maintained, there is generally no limit to the length of conveying channel that can be used [air-assisted conveyors of 330 feet (100 meters) or more in length are not uncommon]. However, with very long conveyors, it is necessary to arrange the air supply so that a uniform pressure exists beneath the

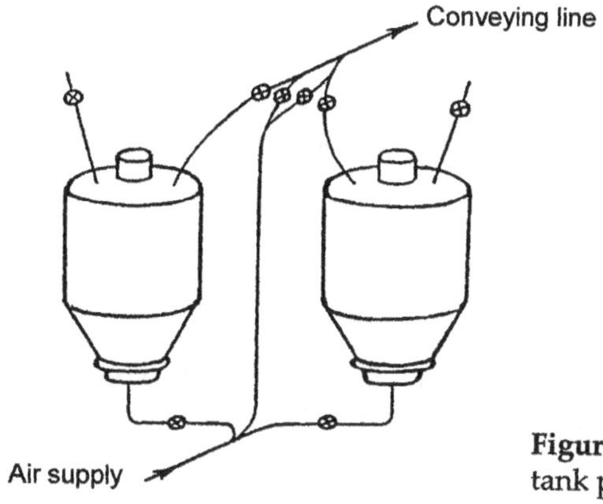

Figure B-81 Schematic of typical twin blow tank pneumatic conveying system.

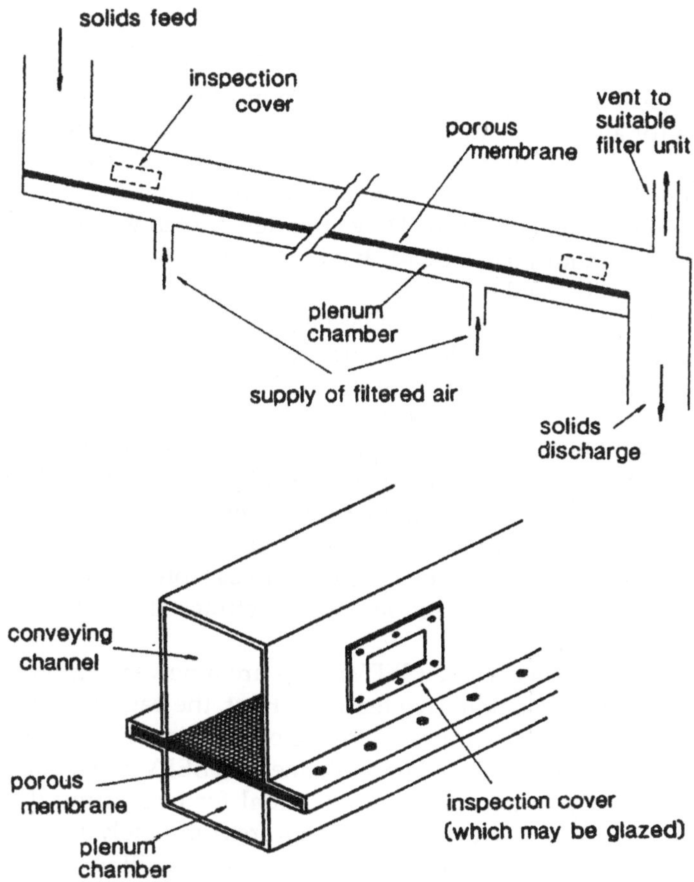

Figure B-82 Air-assisted gravity conveyor.

porous medium, which is often done by providing air inlets at several points along the length.

Good discussions of pneumatic conveyors are presented by Kraus (1991), Marcus et al. (1990), Mills (1990), and Woodcock and Mason (1987).

B11 PORTABLE CONTAINERS

The most commonly used portable containers for particulate solids are multiwall paper bags, fiber drums, flexible intermediate bulk containers (FIBCs), and rigid intermediate bulk containers (RIBCs). These are briefly discussed in what follows.

B11.1 Multiwall Paper Bags

These bags made from plies of kraft paper or from combinations of kraft and special-purpose papers and plastics are the most common type of portable containers for almost any pelleted or powdered solids. There are three primary types of multiwall paper bags: sewn open mouth bags, stepped end pasted valve bags, and pinch-style bags. Each of the three types has evolved in order to meet a specific need. They are often stretch wrapped.

Sewn open mouth bags (SOM) have been used for a long time and are currently normally used to package low cost, commodity products such as seed, animal feed, and fertilizers. SOM bags are open mouth bags which utilize sewn closures for both the top and the bottom. They may be manufactured in either gusseted or flat tube styles. Both types result in a point at the corner of the bag that is unsightly and sticks out when palletized, and are less than desirable for stretch wrapping. SOM bags are filled through the open top, and the top closure is sewn after filling. The bag arrives with the bottom closure sewn by the manufacturer. They can use a variety of barrier plies, and are available in a wide range of sizes. They are the simplest type of bag and least expensive. However, needle holes from the sewing operation preclude good moisture protection and sometimes allows product fines to escape (known as sifting).

Stepped end pasted valve (SEPV) bags are a newer bag type and were developed in order to obtain two features. First, the bags are fast to fill and, second, the bags do not require a closing operation. SEPV bags result in a square, uniform appearance when palletized. In SEPV bags, paper plies are glued individually to minimize the potential for bag sifting (they are not completely sift proof, however). SEPV bags are completely sealed when purchased except for a valve in one of the corners. For filling, the bag valve is placed on the spout of the bag packer (filling machine). Internal sleeves are normally incorporated into the bag valve and are designed to be self-closing.

The extent to which the bag's internal sleeve will actually close and completely eliminate sifting is a function of (1) the relationship between the product volume and the bag volume, (2) the characteristics of the product, and (3) the construction of the internal sleeve. External tuck-in sleeves also can be provided on SEPV bags. They provide a more positive seal, although in most instances, the tucking operation is performed manually at a reduced packaging rate. This style can also use a variety of barrier plies, and the filled bag has a good palletizing shape. The maximum thickness (bottom size) is 8 to 8.5 inches. This style bag is used for virtually every type of product.

Pinch-style (PS) bags are another recent development and represent the highest possible level of product integrity for multiwall paper bags. PS bags are used primarily to package more expensive or sensitive products than the other types of bags. They are gusseted open mouth bags with heat-sealed closures for both the top and the bottom. A properly sealed pinch style bag can be considered to be air- or watertight. Depending on the product and the bag size, PS bags may provide a square, uniform appearance when unitized. The bottom of the bag is heat sealed by the manufacturer prior to shipment. These bags are filled through the open top and then the bags are sealed. Bag sealing is done by reactivating hot melt adhesive which has been preapplied by the manufacturer. This style bag can be packed and closed on automatic equipment, which can be operated and maintained by unskilled labor. These bags also offer a print area on the ends of the bags for product identification while the bags are lying flat. This style is limited in practical thickness (gusset size) to a maximum of 6.5 inches. Recently, PS bags have begun to replace sewn bags for products such as pet food, fertilizer, and flour.

There are a few other multiwall paper bag types manufactured, but they are primarily combinations of the above three types. Some of these other types may be obtained from all bag manufacturers, whereas others are available only from a single, proprietary source.

B11.2 Fiber Drums

Fiber drums are often used for particulate solids and are constructed from solid fiberboard, which is tough and water resistant. The most common type of construction consists of a multiple-ply kraft body with a steel bottom and a reinforcing top hoop (also called a chime) crimped to the drum. For vapor protection, barriers are incorporated among the plies, or liners are used as the first ply in contact with the solids.

There are two primary styles for fiber drums. Fiberpak styles utilize a solid fiberboard slip-on top which slides down over the top of the drum body. In most cases, this top is taped to the body around the seal line, prior to shipment. Lock-ring type fiber drums are commonly larger, up to and including 55-gallon drums. Lock-ring types utilize a separate plastic, fiber, or metal lid, together with a lock-ring that secures the closure. There are a number of options for fiber drum lids, linings, and gaskets which can be provided to meet a particular packaging requirement. Some common barriers and liner materials are polyethylene, aluminum and steel foil, polyesters, and silicones.

Fiber drums are sized in gallons (even when used for solids), and for a particular capacity, the manufacturer uses a specific caliper of fiberboard. Although only a few sizes are common, fiber drums can be made to order in almost any size and diameter–length combinations for volumes of 0.75 to 75 gallons and for weights ranging from 60 to 550 pounds.

Advantages of fiber drums are protection of contents, ease of reclosure, and appreciable reuse–resale value. A serious limitation of drums is the inefficient use of space because of the cylindrical shape, which results in high storage and transportation costs. To overcome this, a fiber drum with a square cross-section has been developed.

B11.3 Flexible Intermediate Bulk Containers (FIBCs)

FIBCs are an important development of the 1970s and are used for providing particulate solids in large quantities in a large variety of volumes, up to 4000 pounds usually. They are usually made from woven polyolefins (e.g., polypropylene) or other materials, and can be equipped with a thermoplastic liner to protect the product against moisture or other contamination.

To choose the right bag for a specific need several design factors must be taken into consideration: bag size, lifting loops, filling inlets, discharge outlets, liners, and single-trip or multitrip styles. A short discussion of some of these factors follows.

Lifting Loops: These are usually located at the top four corners of the bag, and they can also cross the corners. A bag's lifting loop must be strong enough to support the bag when it is filled. Ten-inch loops are typical, and 12-inch loops may be required if forklifts with fork tines are used to pick up and elevate the bag. Pop-up loops stand up to simplify inserting the fork tines into the loop, but are expensive.

Filling Inlets: The bag's filling inlet is tailored to fit the filling machine. A filling spout is the most common type of bag inlet and is used with a filling machine that feeds material through a tube or pipe. While 14-inch spout diameters are common with unlined bags, a lined bag's filling spout diameter can be increased to 20 to 25 inches if the filling machine spout is clamped only to the bag's liner. This larger diameter allows material to fill the shoulders of the bag without bridging in the bag's neck, which is especially important with powders.

Discharge Outlets: The bag's discharge outlet should be big enough to discharge the material easily, which depends primarily on the type of discharger used. The most common type of outlet is an outlet spout. Outlet spout diameters can range from 10 to 25 inches; typical outlet spouts are 14 inches in diameter and 18 inches long. The outlet spout must be long enough, at least 15 inches, to be tied (usually by a drawstring) and, in most case, "goosenecked." If a bag outlet is too long, though, it can get caught in the discharger's feed system. Other bag outlets work with particular materials or dischargers that standard outlet spouts cannot, such as hygienic outlets, superhygienic outlets, total-opening bag bottoms, and flat or plain bag bottoms.

Figure B-83 shows several FIBC bag filling and discharge outlet styles. Davidson (1989) presents a more detailed discussion of FIBCs.

A very important factor in the selection of a FIBC bag is its classification with respect to electrostatic hazards. There are four types, as follows:

Type A: These are general type 100% plastic bags of a woven structure that cannot be grounded. They should only be used for non-combustible products in nonflammable environments.

Type B: These are thin-walled plastic types that cannot be grounded. This type of bag must have a breakdown voltage less than 4 kV across the wall to prevent any possibility of a propagating brush discharge. Most woven polypropylene bags on the market will meet this specification, pro-

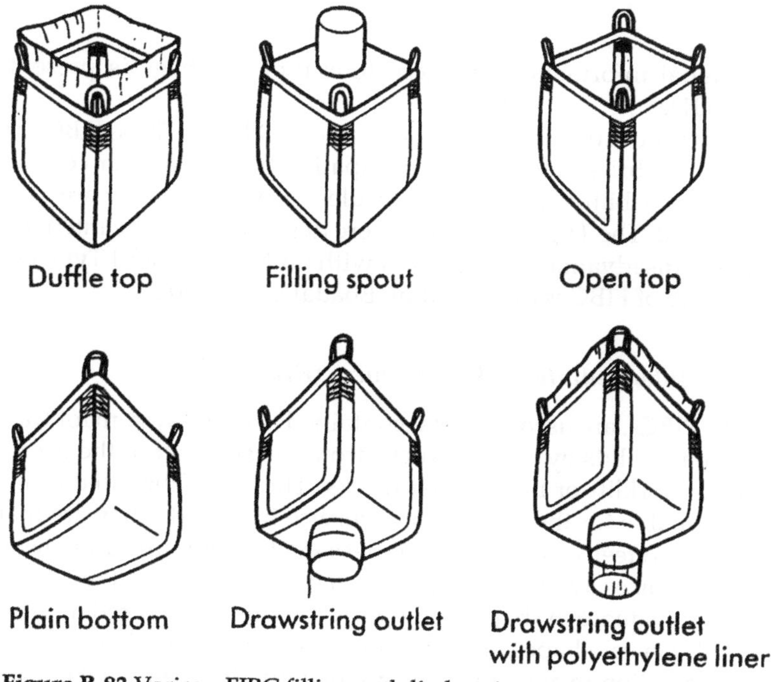

Figure B-83 Various FIBC filling and discharging spouts (connections).

vided that they do not contain an additional internal liner. The Type B bag can be used for combustible powders provided that no flammable vapor concentrations are present inside or outside of the bag.

Type C: These are designed to be electrically conductive. They typically contain a matrix of conductive fibers within the synthetic fiber weave which are terminated in a conductive strip that must be connected to ground. Some bag designs carry the conductive grid into the loops so that the bag will be grounded when hung on a grounded loading or unloading station. The resistance from any point on a Type C bag must be less than 100 megohms. Type C bags are suitable for use with combustible particulate solids and/or where flammable vapor concentrations are possible inside or outside of the bag.

Type D: these bags are of electrostatically dissipative construction. No grounding is required since special static dissipative fibers are used to safely dissipate electrostatic charges via a combination of corona discharge and/or enhanced surface conductivity. There is no specification on the fabric electrical properties for a static dissipative bag; rather, each design must be qualified by special test procedures in which the incendivity of "worst case" discharges to flammable gases of known ignition energies is investigated. Static dissipative bags are suitable for use with combustible solids and/or where flammable vapor concentrations are present, provided that the minimum ignition energy of the vapor is greater than that of the gas used in the qualification tests.

Additional information on these four types of bags is presented by Gravell (2001).

A recent innovation is the so-called Type C/D FIBC. These FIBCs are constructed from fabrics having static dissipative threads and/or coatings that control the probability of ignition regardless of whether they are grounded or ungrounded. The Type C/D FIBCs were designed with the intent of counteracting the disadvantages associated with both Type C and Type D FIBCs. This new type of FIBC is discussed by Ebadat et al. (2002).

B11.4 Rigid Intermediate Bulk Containers (RIBCs)

RIBCs, like FIBCs, are used as a total contained, storage system for transporting particulate solids by road in trucks from the supplier to the user, and in processes by forklift or hand pallet trucks. They are fabricated of metal or plastic as suitable for the product and service intended. They are manufactured in carbon steel, coated steel, stainless steel, aluminum, and low density polyethylene. Sizes available range from 6 ft^3 to 100 ft^3, and they are usually rectangular or square in shape. Some RIBCs are manufactured in cylindrical or inverted conical shapes, and mounted on legs with wheels for moving from one operation to another. They are usually fitted with a combined filling hole and inspection manhole in the top. Discharge is usually by one of two methods: (1) through a large side-hinged flap or door from which the material is discharged to the process equipment, or (2) through a conical bottom. If the type 1 design is used, then the RIBC must be placed onto a special frame to tilt it through 45° for discharging. If the type 2 design is used, the conical bottom is fitted with a simple slide or butterfly valve. Other, more sophisticated, RIBCs have various types of assisted discharge devices and systems.

B12 PORTABLE CONTAINER EMPTYING (UNLOADING) EQUIPMENT

B12.1 Drum and Box Dumpers

Dumpers are equipment that empty the contents of drums and boxes by inverting them. Most dumpers are hydraulically powered, but some are gear-driven, cable-powered, or chain-powered. Today, dumpers are available in capacities up to 6000 pounds, dumping at heights up to 8 feet, and for lower capacities, up to 18 feet. Special designs to exceed these figures are available. Electrical classifications are possible from the everyday NEMA 1, to watertight NEMA 4, explosion-proof NEMA 7 or 9, dust-tight NEMA 12, or oil-tight NEMA 13. Materials of construction and finishes are available to fit almost every situation.

Dumpers have a chute or framework into which the drum fits and holds it in place. A variety of options for the chutes are available. Chute variations can be provided to accommodate practically every shape container. Chutes may be floor loaded or loaded above the floor. They may be flat-bottomed, fork-based, or have roller conveyors in the base. The chutes also may be straight, tapered, or funneled. Funnels (or cones) are also available for attachment to the drums. The funnels (cones) may be open or closed with a variety of valve types at the end: slide, butterfly, or iris valves, to name just a few. The funnels can be made just to fit to the open top of the container to be dumped and to direct the flow or particulate solids, or made to be dust-tight. The outlet of the funnel can also be made to mate to a connection on the receiving vessel. The dumpers can be stationary or portable. In most cases, unless the dumper is quite simple, it may need to be secured by some means prior to dumping, and there are many methods available to accomplish the securing. Additional options which are required in some applications are electrically or air-powered vibrators and magnetic separators to remove "tramp" metal.

Dumper types are available in several designs: the low-level drum dumper, the low-level extended rotation dumper, the high-level drum dumper, the rectangular (box) dumper, and the high-level rectangular (box) dumper.

The low-level drum dumper has as its basis a half-cylinder type chute to accommodate either metal or fiber drums. It has adjustable hold-downs to ease loading or unloading. The dumper simply rotates about a fixed pivot to dump the contents down the chute into a receiver (see Figure B-84). This unit is available in a variety of dumping heights of 36 inches to 60 inches, and capacities up to 1500 pounds with dump angles up to 45°. It is usually powered up and gravity-lowered down.

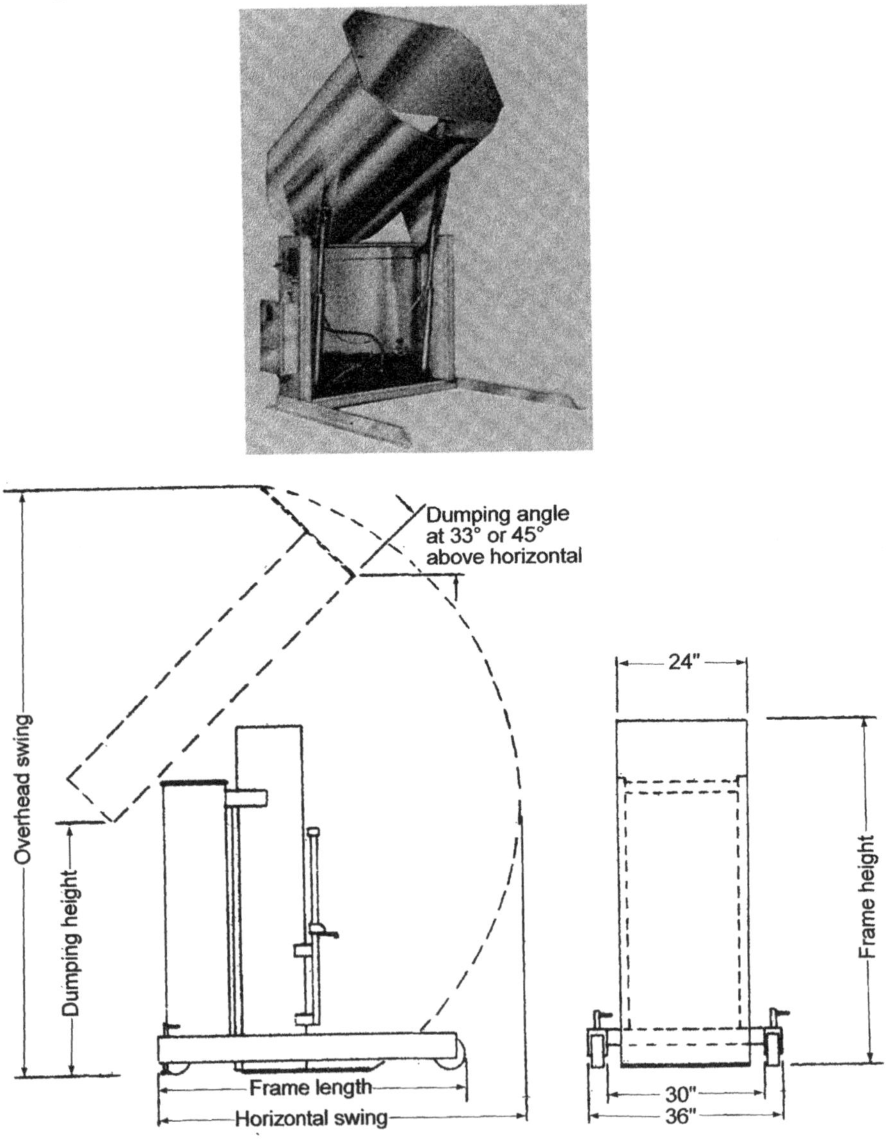

Figure B-84 Low-level drum dumper.

The low-level extended rotation dumper allows the same control as the low level drum dumper but through the added power transmission linkage, it provides positive rotation beyond 135° up to 180°. This unit is powered in both directions.

The high-level drum dumper performs the identical task as does the low-level dumper, with the exception that it is first elevated via a hydraulic cylinder and chain device and caused to rotate at the top through 135° of rotation to give elevated dumping heights. Normal dump angles are 45° or 60°, but specially designed units can provide dump angles of 90°. This unit is powered up only and depends on gravity for down movement. Dump heights of 72 inches to 120 inches are standard up to 1500 pounds capacity. Figure B-85 shows a high-level drum dumper.

The rectangular or box dumpers are built in a variety of frame types. The basic unit simply rotates about a fixed pivot similar to the low level drum dumper, and as its name implies, the chute is box shaped. This unit is available up to 6000 pounds capacity with dumping heights normally up to 60-inches and dump angles up to 60°. It may be powered to go up and down. Figure B-86 shows a low-level rectangular box dumper. As is the case with all dumpers, they may be stand-mounted to allow a higher dump height, but to this, they must be loaded above the floor. A variation of the box dumper is what is called the lift and tilt dumper. In this case, the dumper may be loaded on the floor, then the load is hydraulically raised vertically as much as 36 inches, then rotated to the dump position. For both cases, extended reach and rotation are possible up to 180°. The lift feature of the lift and tilt dumper is a key part of any dumper that has dust-tight requirements.

The high-level rectangular box dumper is a combination of the high level drum dumper and other box dumpers. These units are twin column (cylinder) units that rely on the hydraulic cylinder chain lifting principle, similar to that used on forklifts, with the added feature that the lifting sides are mechanically interconnected to insure even lifting and to give support to the load in case one side fails. These units are powered to lift up and depend on gravity to come down. Capacities up to 4000 pounds and dumping heights up to 15 feet are available. Dumping angles up to 45° are standard. The extended rotation allows rotation up to 180°. As in the case of the high-level drum dumper, the final part of the rotation is used to store mechanical energy to return the unit so gravity can take over on the lowering part of the cycle. Figure B-87 shows a high-level rectangular box dumper.

When dumping particulate solids from a drum or box into a vessel through an open manhole, some companies use a ventilated charging funnel (see Figure B-88) to control and minimize the emission of dust fines into the atmosphere.

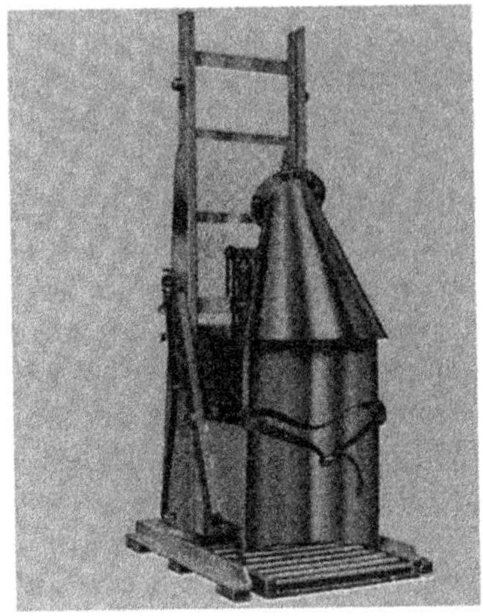

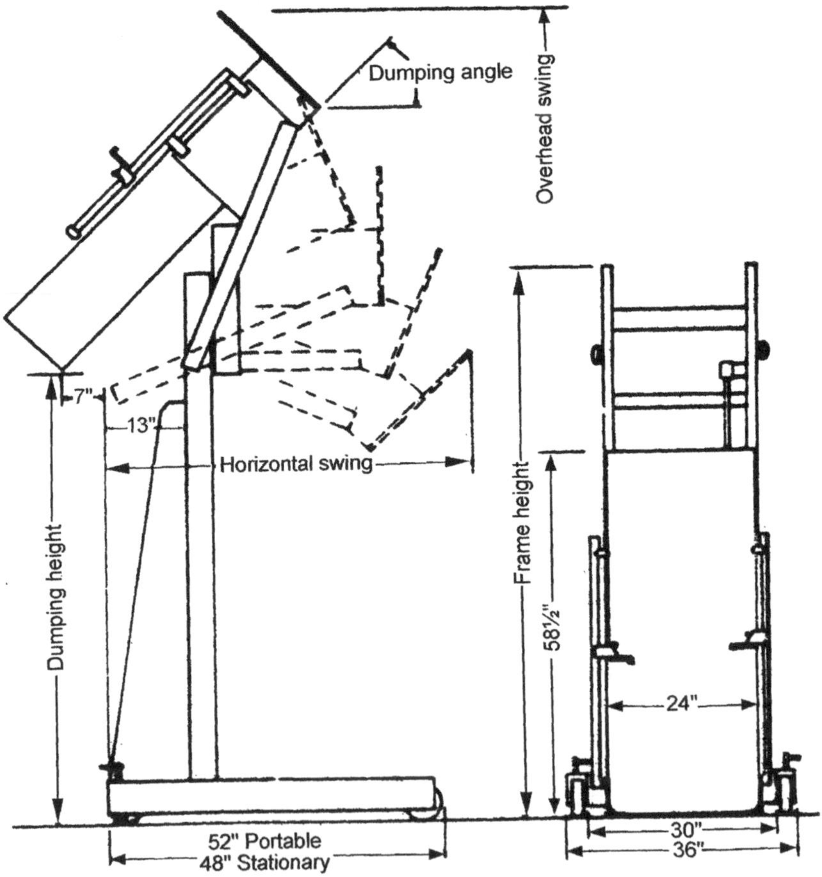

Figure B-85 High-level drum dumper

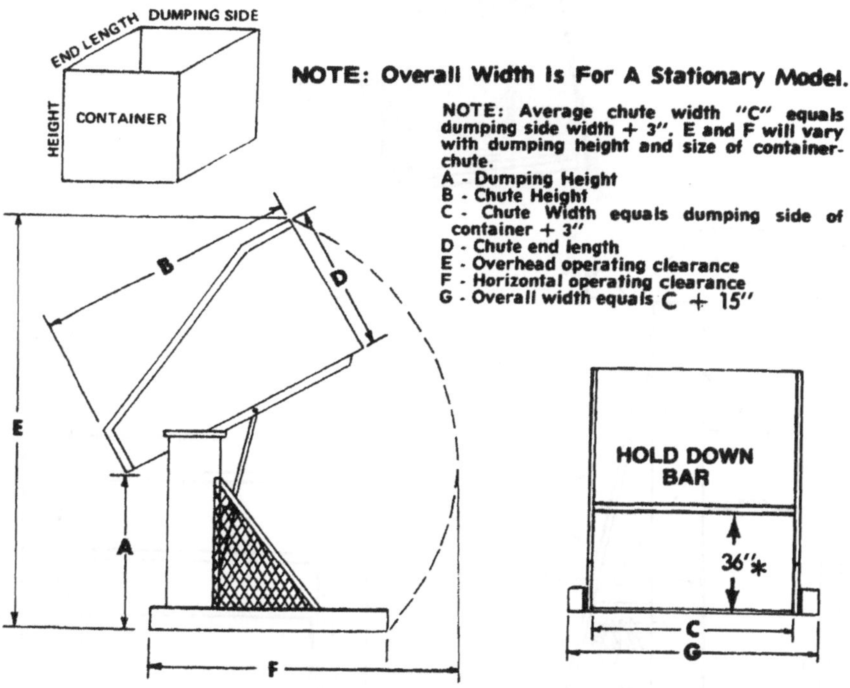

Figure B-86 Low-level rectangular box dumper.

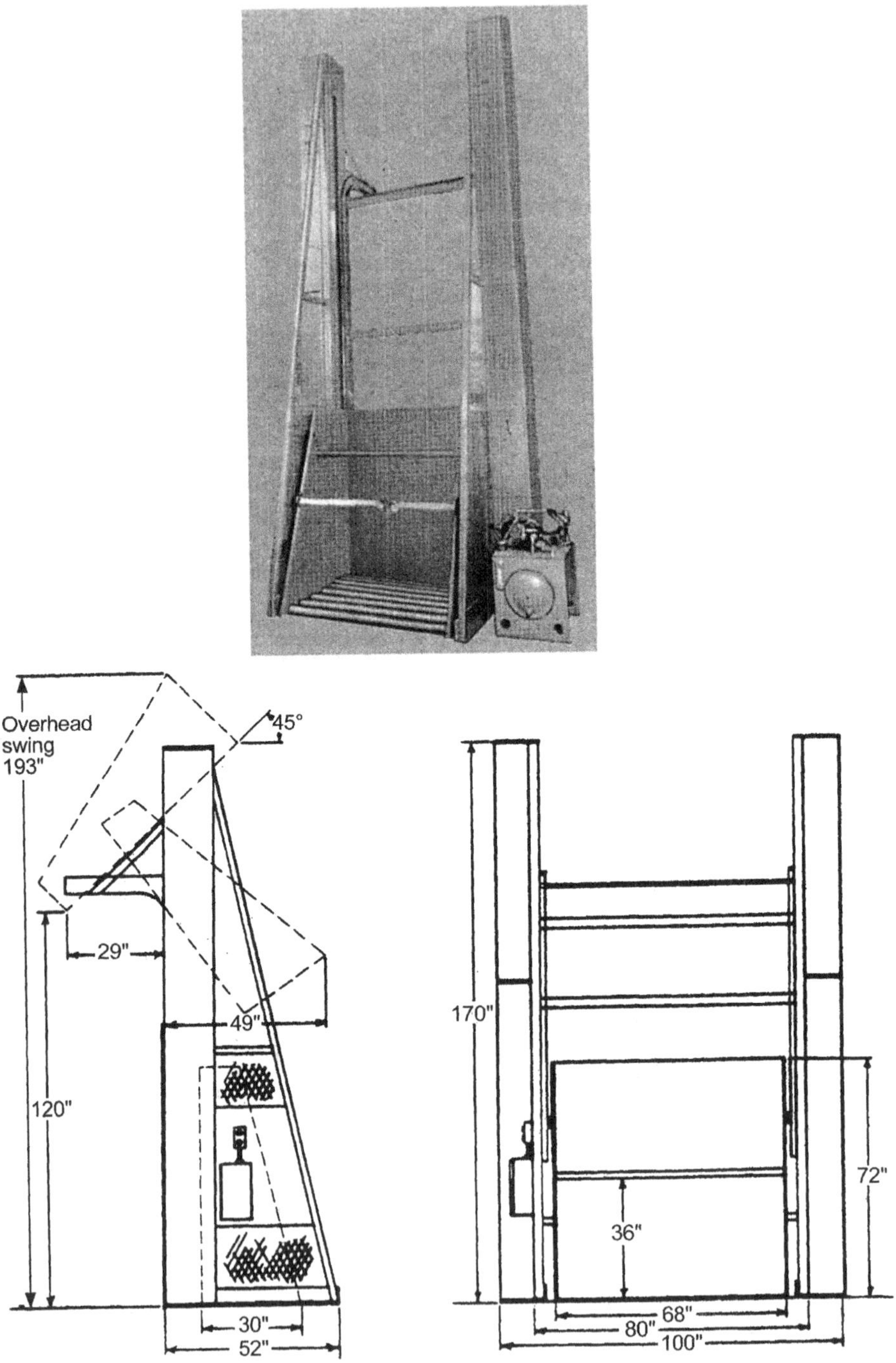

Figure B-87 High-level rectangular box dumper. (Source: Hercules Industries.)

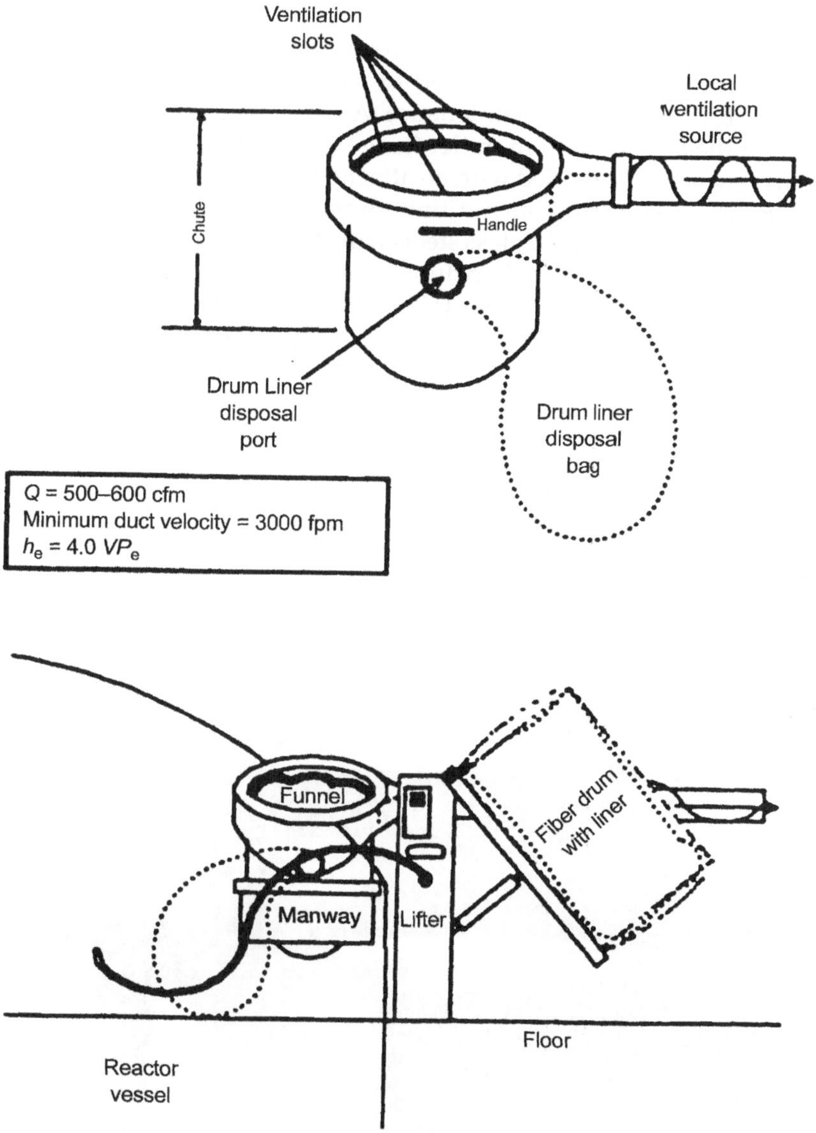

Figure B-88 Ventilated charging funnel.

B12.2 Bag Dump Stations

Bag dump stations are items of equipment used to contain and control the emission of dust when bags are emptied. The bags can be manually or mechanically introduced into the dump station, and are then charged into reactors or process vessels by gravity flow or conveying equipment. Bag dump stations can be purchased from many vendors as self-contained bag dump/filter units. The unit consists of the receiving chamber with a grate at the bottom, a hopper section, a filter, an exhaust fan, and an electrical control unit. The filter is usually composed of cartridge filters, which are easily accessible for removal and maintenance. A valve or conveying equipment is located at the hopper bottom outlet flange (often by the purchaser). Dump stations can also be provided with an attached compactor for empty bag disposal. Figure B-89 shows such a unit. Standard materials of construction include carbon steel and stainless steels with a variety of finishes or coatings. NIOSH has evaluated such a bag dump station and found it to be very effective in controlling dust emissions (Heitbrink, McKinnery, and Rust, 1983).

A simpler "home made" type of bag dump station can be designed by users based on the criteria given in Industrial Ventilation (ACGIH, 1998) (see

Figure B-89 Self-contained bag dump station. (Source: The Young Industries, Inc.)

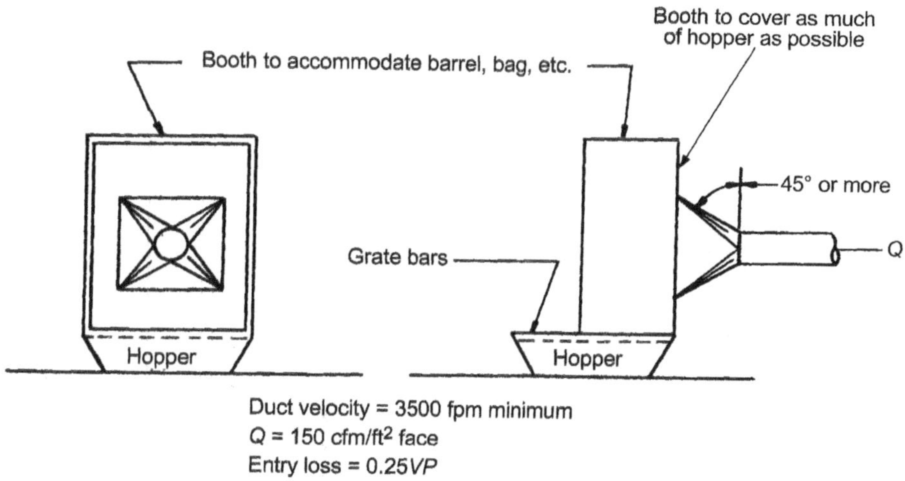

Figure B-90 Self-designed bag dump station.

Figure B-90). In this design the bag dump station does not have an integral filter and exhaust fan, so that the unit must be piped to a dust collector (e.g., baghouse or scrubber) and exhaust fan.

B12.3 Vacuum Pneumatic Conveyor Unloading System

Multiwall bags, FIBCs, boxes, and drums can also be emptied by use of a vacuum pneumatic conveyor system. Such conveyor systems are available from a number of suppliers. In these systems the portable containers are unloaded by inserting a wand into them and removing the solids by vacuum, and transferring them to a receiver/filter located above a reactor or slurry mixing vessel.

The main components of a vacuum pneumatic conveyor unloading systems are:

- A vacuum producing source
- A filter/receiver
- A discharge valve (flapper, rotary. etc.) at the outlet nozzle of the conical bottom of the filter/receiver
- A wand to insert into the portable container
- Conveying hoses, tubing/piping and fittings
- Control panel

Figure B-91 is a schematic of a typical vacuum pneumatic conveyor unloading system for drums. These systems are similar in operating principle to negative-pressure pneumatic conveying systems discussed in Section

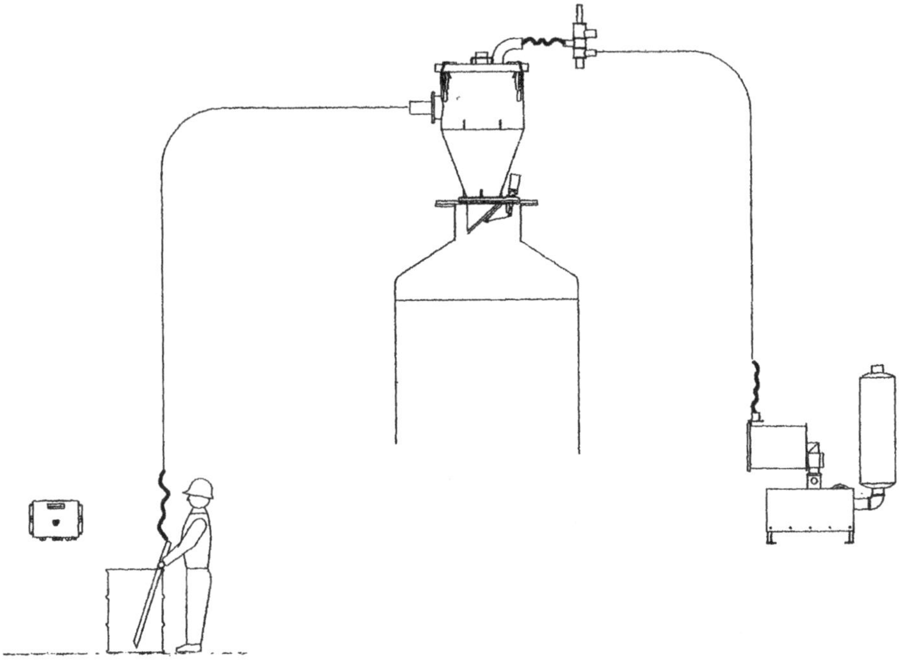

Figure B-91 Typical vacuum conveyor unloading system. (Source: Vac-U-Max.)

B10, but convey solids over comparatively shorter distances and do not have a feeding device.

The vacuum-producing unit can be a vacuum pump (mechanical or liquid ring), a regenerative blower, or an air- or nitrogen-powered Venturi power unit. If an air- or nitrogen-powered Venturi unit is used, air or nitrogen at 60 psig is normally required at a rate of 35 scfm per Venturi, which will result in an induced flow in the conveying line of 110 scfm (Zoppa, 2001).

Materials of construction are normally carbon steel or stainless steel. Pharmaceutical-grade surface finishes are available.

B12.4 Flexible Intermediate Bulk Container (FIBC) Unloading Systems

FIBC unloading (discharging) systems are devices or machines that enable an FIBC (bulk bag) to be easily and cleanly discharged with minimum dust and maximum safety.

There are two basic types available:

- Those that require the filled bag to be placed on a table or platform and have various devices that "massage" the bottom of the bag to maintain the flow of powder from the outlet.
- Those that require the filled bag to be placed into a coned hopper which "squeezes" the bottom of the bag, thereby breaking the bridge of powder over the outlet. The flow of powder is maintained when necessary by vibration.

A number of factors must be taken into consideration when choosing a bulk bag unloading (emptying) system, which are as follows:

1. A means of positioning the bulk bag from a forklift truck or crane onto the frame.
2. A support for the bag from its four loops during unloading.
3. Containment of any dust that might occur when lowering the bag onto or into the unloader and during the discharge. Cone hopper type unloaders also have a rubber diaphragm or membrane stretched across the top of the hopper to act as a dust seal when the bag is lowered into position. The diaphragm can also surround a fixed knife fitted into the hopper so that as the bag is lowered no dust escapes as the knife cuts the bottom of the bag.
4. Easy access for the operator to untie the outlet with a means of dust containment.
5. Dust collection system
6. Ability to shape the bulk bag bottom into a cone shape to promote flow and complete emptying.
7. Facility to allow the emptying bag to "grow" as it empties without the outlet spout or liner becoming entangled in any feed or weighing

system beneath the unloader. The unloader outlet must be designed to allow this to occur.

8. A means to induce flow from the bag (vibrator or other means)
9. Liner tensioner to restrict liner movement out of the bag outlet. Lined bags frequently give the most problems at emptying as the liner tends to slip through the outlet, particularly toward the end of the emptying. To attach the liner to the bag by sewing or adhesive is expensive and to secure the liner to the rigging frame still allows the liner to "grow," sometimes as much as 400 mm below the outlet. To overcome this, a liner retention device is necessary, which will pull the liner out of the emptying bag and eliminate this problem. In the case where the liner needs to be clamped to the feed mechanism below the unloader prior to discharge (as in the case of some food or pharmaceutical quality bags), the liner retention device will ensure that the liner is completely emptied and collapsed by pulling the liner against the clamp. This enables the liner to be tied off prior to release, sealing it against any powder spillage when the empty bag is removed. It also enables the tied-off liner to be removed from the empty bag and rolled up for disposal by stopping dust escaping as the air is exhausted during the liner disposal. This is particularly important with health-hazardous and pharmaceutical powders.
10. An alarm to signal when the bag is empty.
11. An appropriate type of device to convey the powder away. Many conveying devices are available to fit under the unloader, such as: screw feeders, vacuum conveyors, blowers, weighing devices, loss-in-weight feeders, vibratory tray feeders, etc. All these devices are tailored to suit the material being conveyed, the method of feed required, and the process for which the powder is needed.
12. An easy means of removing the empty bag.
13. An appropriately ventilated location for empty bag folding and disposal

FIBC unloading systems are available in a number of designs. The major components of an FIBC unloading system consist of the following:

- Bag support dish and frame
- Bag lifting frame
- Bag tensioning frame
- Bag massaging system
- Liner clamp
- Bag spout pinch bars
- Adaptor chute/hopper
- Powder discharger

Figure B-92 Typical FIBC unloading system. Source: Flexicon Corporation.)

Figure B-92 shows an FIBC (bulk bag) unloading system manufactured by one vendor.

An informative article on bulk bag unloading (emptying) systems is presented by Davidson (1989). Becker (1999) discusses technology developments in bulk bag discharging.

B13 PORTABLE CONTAINER FILLING SYSTEMS

This section discusses and describes small bag filling systems, FIBC filling (packing) systems, and drum filling (packing) systems.

B13.1 Small Bag Filling Systems

The type of bag filler used depends on several factors such as (1) the type of bag, (2) the product (powder, bulk solids, flakes, etc.), (3) the level of accuracy desired, and (4) the required filling rate.

There are two major types of bag fillers (bag packers):

- open mouth bag fillers
- valve bag fillers

Open Mouth Bag Fillers

Open mouth bag fillers monitor the flow of particulate solids into open mouth bags in order to provide individual amounts in the desired quantity. They incorporate a gravity or mechanical product feeder, which is selected based on the specific product to be handled. In the past, most open mouth bag fillers were gross weigh machines. However, presently most chemicals are packaged on net weigh open mouth bag fillers, although some inexpensive commodity chemicals are packaged on gross weigh machines.

Gross weigh open mouth packers support the empty bag and the weight of the product in the bag, such that flow of the product is cut off when the desired weight is reached. The product is fed directly into the bag from a feeder. The height of free fall is usually shorter compared to a net weigh packer, and the product is streamed into the bag. This will result in superior dust control compared to a net weigh packer. Gross weigh packers normally incorporate mechanical weighing systems. Interlocks are usually provided to ensure that this type of packer will not operate except when the bag is in position.

Net weigh open mouth packers have an internal weigh bucket that preweighs the desired amount, using a combined bulk and dribble flow process to achieve accuracy. After the bag is placed into position, the weighed amount is discharged and the process is repeated. A series of interlocks ensures that the bag is in place, open, and empty, before it will drop the charge of material. Net weigh open mouth bag packers normally incorporate electronic load cell weighing systems. Although almost all types of particulate solids are handled by open mouth bag packers, these machines are best suited for free-flowing materials that do not dust significantly. Net weigh fillers are faster than gross weigh fillers, as the internal weigh bucket can immediately be refilled after it has discharged into the bag.

A very critical component of both gross weigh and net weigh open mouth bag fillers is the feeding device as its speed and accuracy determine the bagging production rate. A number of different types of feeders are used, as follows:

- *Gravity feeder*: This type of feeder is used for continuous feeding of free-flowing materials, such as plastic granules, granulated fertilizer, grain, rice, sugar, etc. Two-speed feed is achieved with a two-position gate. Dribble feed is usually adjustable by varying the gate opening.
- *Gravity feeder with agitator*: This feeder is the same as the gravity feeder, but a motor-driven agitator assists flow of materials that may not move as freely. Typical materials are pellets, bran, and animal feed.

- *Belt feeder*: The belt feeder is used for non-free-flowing materials that need to be handled carefully to maintain their structure such as more fragile pellets. This feeder is best for high accuracy and medium speed applications.
- *Vibratory chute feeder*: This feeder is preferably used for difficult materials that need careful handling such as non-free-flowing materials, powdery materials, course ground materials, flaky materials, pelleted materials like resin, coal, coke, and other minerals.
- *Screw feeder*: This type of feeder is often used to continuously feed non-free-flowing, mealy, or powdery materials. Usually, a single screw feeder is used.

Valve Bag Fillers

Valve bag fillers (packers) are usually gross weigh devices which operate with the weight of the empty bag and product suspended from the spout. There are several types of valve bag packers, and some products can be handled successfully with only one type of packer; however, most products can be handled with any or all of the available packers. The four major types of valve bag fillers (packers) are:

- Air packers (pneumatic packers)
- Auger (screw type) packers
- Impeller packers
- Vacuum packers

Air packers (also called pneumatic packers) are the most common type of valve bag packer and are used for a variety of granular materials and powders. Air packers incorporate a chamber into which air is introduced prior to the discharging the product into a bag through its valve. The sequence of operations is as follows: (1) the chamber is opened, (2) material is allowed to flow into the chamber, (3) the chamber is then pressurized forcing the material to flow to the bag. All air packers generally come with two-speed fill rates, bulk and dribble, for accuracy. Pressurizing-type air packers are used for materials that are free-flowing. However, for less free-flowing materials, fluidizing-type air packers are used, which aerate the material and cause it to flow from the packer into the bag.

Auger (or screw type) packers are used for fine powders which are prone to aeration. Auger packers minimize the amount of air entrained into the product through the use of a small, high speed auger inside the spout. Auger packers, when they are set up properly, are the most accurate way to meter material into a bag or other container. Auger packers are generally slower than air or impeller packers, and there must be ample room in the bag or container for the desired amount of material if the highest accuracy is to be achieved.

Impeller packers are primarily used for minerals and materials that behave like cement (fine powders that fluidize nicely and can be transferred easily), and for which it is not necessary to achieve an extremely high level of accuracy. Impeller packers propel the product through the valve into the bag from a rotating, mechanical wheel within the machine. Bag filling times with impeller packers are usually quite short.

Vacuum packers are specialized valve bag packers which are used for packaging extremely low density products. These packers pull a vacuum on the complete bag, while it is enclosed in a chamber, so as to densify the product as it is introduced into the bag.

Centrifugal belt-type and gravity-type packers are two other valve bag packers that are not as frequently used as the ones described above. Raymus (1997) provides information about them.

Figure B-93 shows a filling system for small bags.

Figure B-93 Small bag filling system. (Source: Chronos Richardson.)

A bag packaging system often has other auxiliary equipment associated with it such as a bag placer, bag sealer (closer), take-away conveyor, bag flattener, accumulation conveyor, pacing conveyor, checkweigher, reject station, palletizer, and filled pallet conveyor.

B13.2 FIBC Filling (Packing) Systems

Filling systems (packers) for FIBCs (bulk bags) are somewhat different from fillers used for small bags because of the size and design of bulk bags and their internal liner. The fillers usually consist of a frame on which the bags are hung by the four loops on the bag, a weighing system (a scale, load cell, or loss-in-weight system), a flow assist aid (vibrator), a bag inflator, and a control panel. The bags are often manually mounted, but can also be mounted via a forklift, hoist, or trolley.

Figure B-94 shows a FIBC filling system manufactured by one company. Other companies have different design variations.

A FIBC filling system works in the following manner. Empty bulk bags are hung by their lifting loops from hooks at the top of the filling machine (see Figure B-95). A filling device (e.g., a screw feeder) feeds material into the filling machine's surge hopper, which controls the flow into the bag through a flexible connection. The vibration table (which also supports the bag bottom and adjusts to different bag sizes) raises to support the bag bottom before filling starts. This causes material to flow into the sides of the bag bottom during filling and prevents the bag bottom from forming a rounded,

Figure B-94 Typical FIBC filling system. (Source: Flexicon Corporation.)

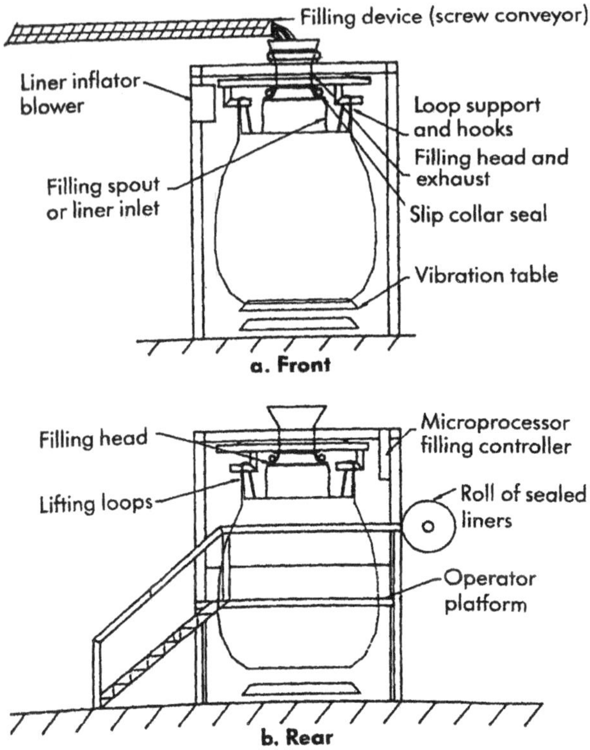

Figure B-95 Schematic of Typical FIBC filling machine for hang filling, densification, and suspended weighing.

unstable shape. When the bag is about one-third full, the table begins to vibrate the bag and densify the material; this operation is repeated for several vibration/stretching cycles, depending on how densely packed the material must be. The vibration table lowers between cycles. When the bag is filled and densification is complete, the table lowers to disengage the bag. The self-releasing hooks then release the bag and a forklift truck or other device is used to move the filled bag through the side of the filling machine's frame. Davidson (1989a) presents a more detailed discussion of filling operations.

B13.3 Drum Filling (Packing) Systems

Filling of drums consists of three operations: setting up, filling and weighing, and closing. The setting-up operation (container positioning) can be performed either manually or by an automated system. In an automated system, the drum positioning is performed by machinery which automatically brings the empty drums to the filling equipment, positioning the drum so that the material may be easily filled directly into the drum, and then releases the filled drum for conveyance to the closing devices. The drum positioning equipment is interlocked with the filling scale to assure that the drum is in the correct position and a discharge of the material into the drum at the right instant. Automated drum positioning equipment releases drums, one at a time, to the filling scale by using indexing devices on the drum indexing conveyor. The filling scale is equipped with an automatic dust cover, when required, to seal against the drum top during the material discharge. Such a system is also used for filling boxes with particulate solids. Figure B-96 is a schematic drawing of a drum or box filling system.

B14 SAMPLERS AND SAMPLING SYSTEMS

Sampling of particulate solids is often done to ascertain product purity or degree of homogenization, and can be done either manually or automatically. Various types of manual mechanical and automatic samplers are used at points in materials handling systems such as vertical or inclined gravity chutes, bulk loading stations, bins and silos, pressure and vacuum pneumatic conveyors, screw conveyors, and belt conveyors. Materials that can be sampled include pellets, granules, powders, and flakes. The material may be uniform, mixed with a tendency to segregate, friable, abrasive, or toxic.

Automated samplers are available in three basic sampling modes and three common operating types. Sampling mode refers to how the sampler collects the sample, as follows: (1) the spot sampler enters the material stream and takes a sample only at the point where it stops; (2) the strip sampler enters the material stream and takes a sample from a narrow portion of the steam all the way across the stream and back; and (3) the cross-cut sampler takes a sample from the stream's entire cross-section. The cross-cut sampler will give the most representative sample of material.

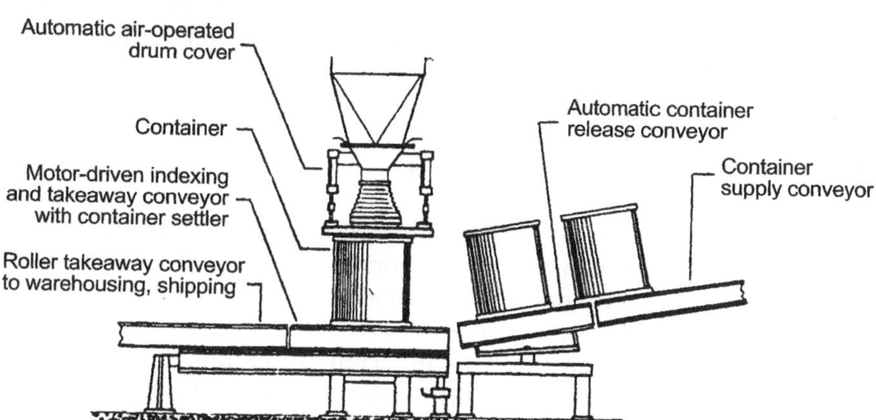

Figure B-96 Typical drum filling system (Source: Chronos Richardson.)

Sampler types are classified by their material-collecting mechanisms. Three commonly used samplers are described below.

The first type is the tube or tube probe sampler and is the simplest and most common type. Basically, it is a tube with a hole near one end (there are variations, such as the cup sampler in which a sampling cup, rather than a tube enters the material stream). The hole-end of the tube moves into the material stream, material flows into the hole, and the tube retracts from the material stream.

The second type is the tube-and-auger sampler, which is the second most common type. It, too, is a tube with a hole in it, but in this sampler an auger rotates inside the tube and helps direct the flowing material into the sampler.

The third type is the pelican diverter or pelican sampler. It is called this because of the vague similarity between the sample cutter (the component of the sampler that passes the material stream and cuts out a sample) and a pelican's beak. The pelican diverter is a three-dimensional box, somewhat narrower at the top, with the bottom edge angled at 45° and with an opening at the top edge and a discharge at the bottom. The box fits inside a rectangular housing that is wide enough to cross the entire material stream, and allows room for the pelican diverter outside the stream on each side. The pelican diverter resides on one side of the stream. Then, when activated, it cuts across the entire material stream and ends up on the other side.

Figure B-97 shows the three types of automatic samplers described above. In all three types, a motor or pneumatic device activates the sampler, causing it to enter and withdraw from the material stream. The material flows into the sampler and then is released either mechanically or by gravity into a sample container that is manually or automatically removed for analysis.

Samplers introduced into pneumatic conveying lines should be only located in vertical runs because of solids stratification in horizontal runs.

Bassett (2001) discusses samplers and sampling systems in more detail.

B15 SCREENS AND CLASSIFIERS

One of the most common methods for separating solids on the basis of size or shape is screening, which is the separation of a mixture of various sizes of particulate solids into two or more portions by means of a screening surface. The screening surface acts as a go/no-go device, and the final portions consist of more uniform size than those of the original mixture. The purposes of screening are: (1) to remove fines from material before grinding equipment, (2) to scalp oversize material or impurities, and (3) to produce commercial- or process-grade product to meet particle size specifications.

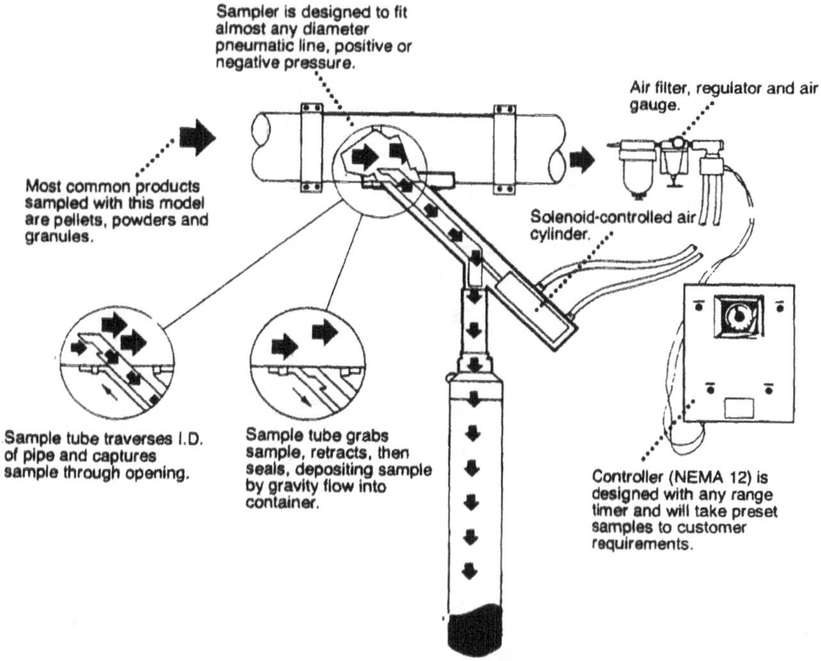

Tube or Tube Probe Sampler

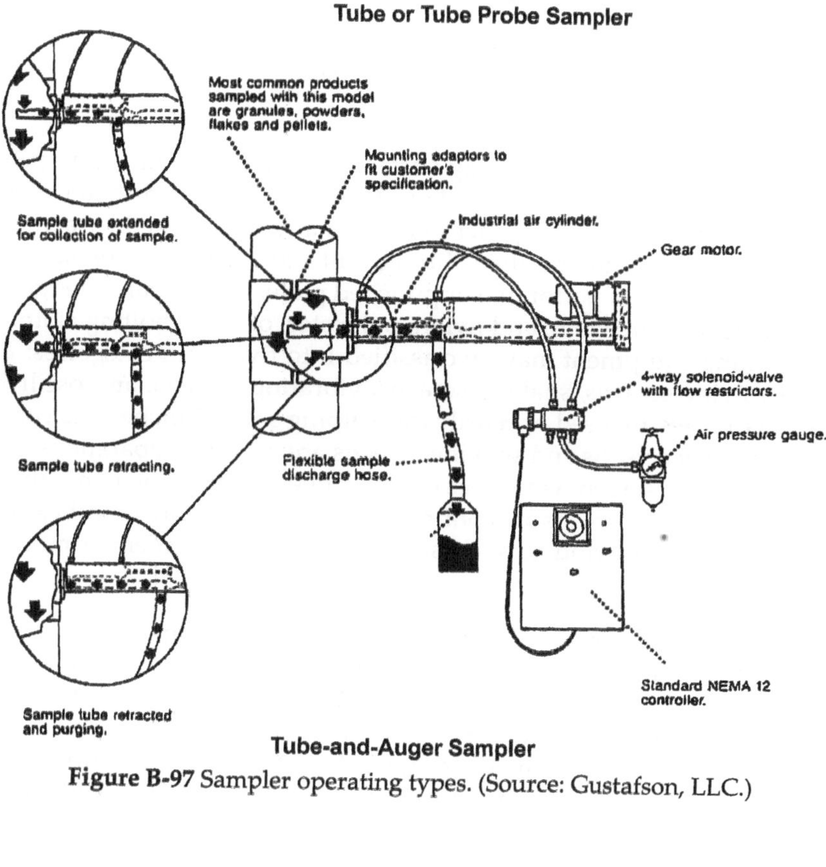

Tube-and-Auger Sampler

Figure B-97 Sampler operating types. (Source: Gustafson, LLC.)

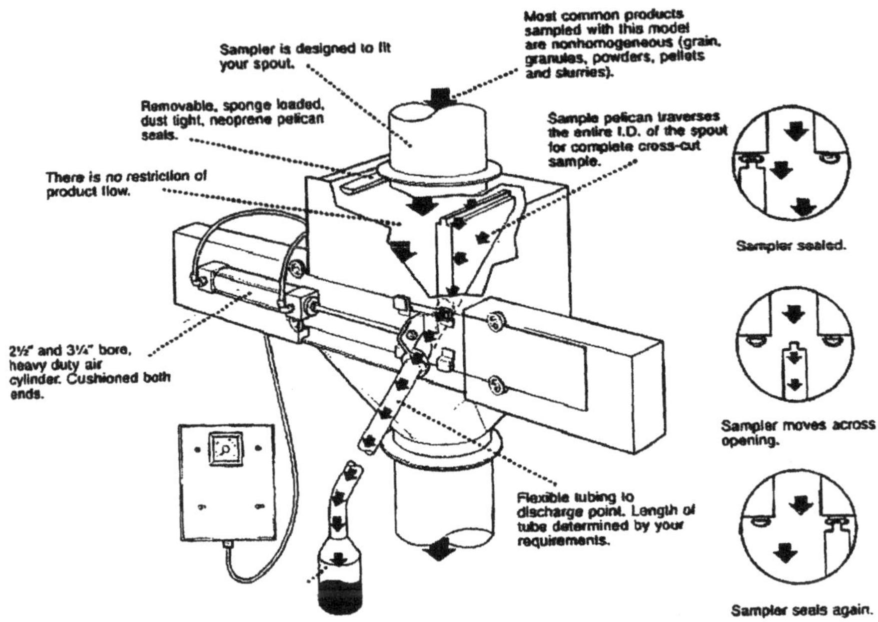

Pelican Diverter Sampler

Fogure B-97 *continued.*

Selection of a proper screening surface is very important, and opening size, wire diameter, and open area should all be carefully considered. Screen cloth area and screen opening sizes are normally determined by a vendor, based on test data and specified throughput requirements. The screening surface may consist of perforated or punched plate, grizzly bars, wedge wire sections, woven wire cloth, and nylon, polyester, or other bolting cloth.

Screening equipment may be classified into five main categories: grizzlies, revolving screens, shaking screens, vibrating screens, and oscillating screens. Grizzlies are used primarily for scalping solids 2 inches and coarser, while revolving screens and shaking screens are used for separations above ½ inch. Vibrating screens cover this coarse range and also down into the fine meshes. Oscillating screens are confined in general to the finer meshes below 4 mesh. A short description of these various types of screening machines follows.

Grizzlies: These consist of a set of parallel bars held apart by spacers at some predetermined opening. Bars are frequently made of manganese steel to minimize wear. A grizzly is widely used before a primary crusher in mineral processing plants. A grizzly can be a stationary set of bars or a vibrating screen. Stationary grizzlies are the simplest type of screening equipment and the least expensive to install and maintain. They require no power and little maintenance. However, it is difficult to change the opening between the bars, and the separation may be not sufficiently complete. Vibrating grizzlies are simply bar grizzlies mounted on eccentrics so that the entire assembly is given a back-and-forth movement or a positive circle throw.

Revolving Screens: These are also called trammel screens or centrifugal sifters. They consist of a cylindrical screen surface of perforated steel plates or wire cloth that is mounted on rollers, is inclined slightly downward from the feed end, and is rotated around its longitudinal axis at speeds of 15 to 20 rpm. Material is fed by gravity or pneumatically into the feed inlet and redirected into the cylinder sifting chamber by means of a feed screw. Rotating, helical paddles within the chamber continuously propel the material against the screen, while the resultant centrifugal force on the particles accelerates them through the screen openings. These rotating paddles, which never make contact with the screen, also serve to break up soft agglomerates. Oversized particles and trash are ejected through the oversize discharge port. These screens are simple and compact, and typically easy to disassemble and maintain. They have no vibration problems, but have relatively high power requirements, they apply high stresses to the screen, and are not typically suited for precise separations of near-sized particles. They have low capacities and low efficiencies compared to vibrating screens.

Shaking Screens: These are also called reciprocating screens. They consist of a rectangular frame which holds wire cloth or perforated plate and is slightly inclined and suspended by loose rods or cables, or supported from a base frame by flexible flat springs. The frame is driven with a reciprocating motion. The material to be screened is fed at the upper end and is advanced by the forward stroke of the screen, while the finer particles pass through the openings. The advantages of shaking screens are low headroom and low power requirement. The disadvantages are high cost of maintenance of the screen and the supporting structure due to vibration and low capacity compared to vibrating screens. These screens are used extensively in the United States and are standard equipment in many chemical and processing plants for handling fine separations, even down to 300 mesh, and are especially good for accurate sizing of large lumps. They are used to handle a variety of chemicals, usually dry, light, or bulky materials.

Vibrating Screens: Vibrating screens are the most widely used screens and are especially good where large capacity and high efficiency are desired. They can be mechanically or electromagnetically vibrated. Vibratory motion is characterized by a short stroke and high frequency. They can be horizontal (good for operations with low headroom) or inclined rectangular machines or round machines. Inclined screens normally have rectangular openings to compensate for the foreshortening of the slope; oblong particles may, therefore, filter through the inclined screen that would normally leave with the oversize on a horizontal screen. High-speed vibrating screens can make size separations for some particles smaller than 200 mesh, which is the usual limit for separation with the lower speed screens, although at much reduced capacities. Several screening surfaces are used if multiple separations are required. In circular vibrating screens, each screen is typically mounted in a round frame; each screen and frame together form a screen deck. Each screen deck also includes a large-opening wire mesh, called a backwire, that is mounted below the screen to hold a set of screen cleaners, such as rubber or plastic balls or cubes that bounce against the screen's bottom surface during screen operation to prevent screen blinding.

Oscillating Screens: These screens are characterized by low speed oscillations [5 to 7 oscillations per second (300 to 400 revolutions per minute)] in a plane essentially parallel to the screen cloth. Screens in this group are usually used for separating particles from ½ inch to 60 mesh. Some light free-flowing materials, however, can be separated at 200 to 300 mesh. Silk cloths are often used for the screening surface.

The screens of these separators are susceptible to premature breakage, which can be caused by fatigue, shock, corrosion, and abrasion. Knowing the causes of premature screen breakage and taking protective measures can lengthen screen life, reduce maintenance costs, improve production cost-effectiveness, and prevent emissions of particulate solids into the surrounding area. Alamzad (2001) discusses these problems and presents precautions (protective measures) to minimize them.

Figure B-98 shows some commonly used screening machines.

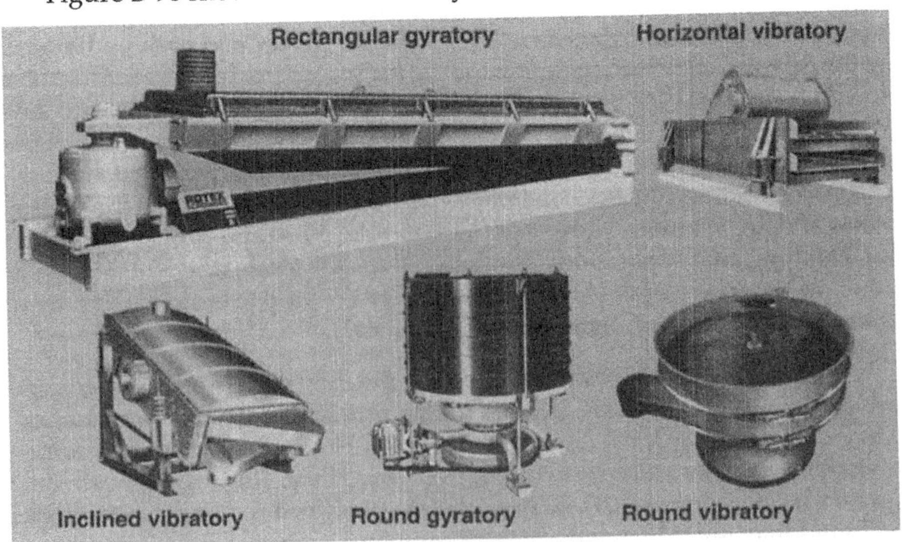

Figure B-98 Several commonly used screening machines. (Source: A. J. DeCenso, Dry Screening of Granular Solids. *Chem. Eng.* April 2000, pp. 76–83. Reprinted with permission of *Chemical Engineering*.)

More detailed discussions of screening machines are presented in Section 19 of *Perry's Chemical Engineers' Handbook* (Green and Maloney, 1997), and articles by DeCenso (2000) and Ricklefs (2000).

B16 SILOS AND HOPPERS

Silos (also often called bins) and hoppers are vessels used for the storage of particulate solids. Normally, silos (or bins) are large vessels used for long term storage either at the beginning of a process to store raw materials or at the end of a process to store products. Hoppers are usually the name given to small storage vessels used to accumulate small quantities of solids for smoothing out feed to a processing step. The term "hopper" is also used as the name of the conical section at the bottom of a silo or bin which connects the straight side to the discharge opening.

Most silos used for in-unit storage of either final or intermediate products in chemical processing units are of modest size (3000 cubic feet or less). They are normally constructed of welded metal and have only one centrally-located discharge opening. Although most silos are circular, square or rectangular silos are often used because of fabrication ease and space economics. Silos of modest size are mounted on skirts or legs, or hung by lugs in structures. Very large silos are usually located on foundations. Welded metal silos can be fabricated of carbon steel, glass-lined or plastic-lined carbon steel, stainless steels, other metals and alloys, and aluminum. The choice of material of construction depends on the corrosivity of the products stored (often due to moisture). For the storage of large quantities of raw materials (e.g, coal, cement, and product from large-scale plastic manufacturing plants), less expensive poured concrete, concrete stave, bolted metal, or spun metal silos are often used. Because of the large diameters, multiple discharge openings are often required. If corrosion resistance is required, coatings are used. Mounting on foundations or skirts is the norm.

Silos are designed from a process (flow) and a mechanical perspective. The process flow perspective relates to the ability to discharge the solids from the silo without arching, hangup, or other problems. Two different flow patterns can develop in a storage silo when it is being emptied: mass flow and funnel flow. In mass flow, all of the material in the silo is moving at all times, although not necessarily at the same velocity; there is normally very little difference in velocity profile across the silo, but the velocity in the center of the hopper section is higher than that near the walls. In funnel flow, a central core of material flows, and is refilled continuously by material sloughing off the top of a stagnant annulus. These normal flow patterns are shown in Figure B-99a and b. There are also two flow pattern deficiencies which can cause flow to cease, both occurring because the discharge opening is too small. Figure B-99c illustrates arching over the discharge opening and

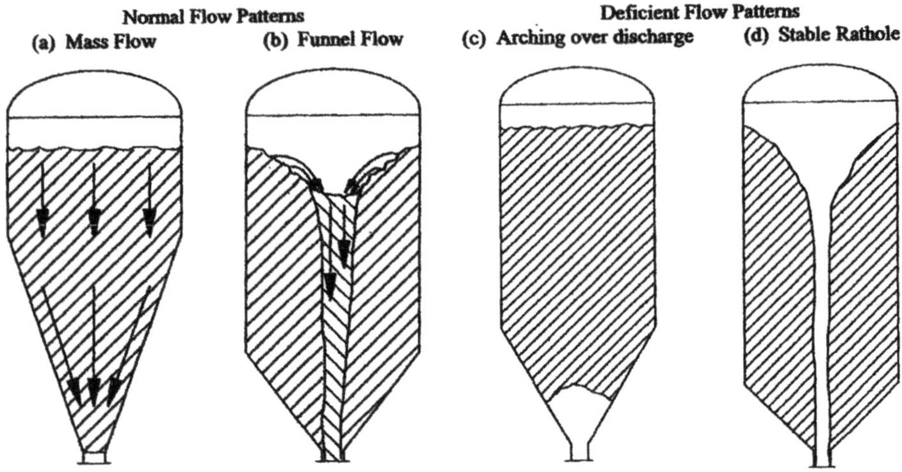

Figure B-99 Flow patterns in mass flow and funnel flow silos.

can occur in either mass or funnel flow silos. Figure B-99d shows a stable rathole in a funnel flow silo where flow occurs only in a central core. Mass flow silos usually require much steeper angles in the hopper section that are required for funnel flow silos, and thus are more costly. Mass flow bins, however, are often required because of four major problems with funnel flow silo operation, which are:

1. Many particulate solids gain strength when they are compressed under load without flowing. If this tendency is severe, such solids can solidify in the nonflowing portions of a funnel flow silo.
2. The flow in the central channel is erratic for some materials, which can cause large quantities of material to fall suddenly. Damage to the silo and auxiliary equipment can occur.
3. Funnel flow causes air to be entrained in the material being discharged. For some materials of small particle size flooding (rapid uncontrolled flow) can occur; entire silos have been known to empty like a tank of liquid.
4. Where the material to be stored has a wide particle size distribution, segregation by size results when the bin is discharged. This is often undesirable in downstream operations.

Determination of the proper discharge outlet diameter to ensure trouble-free flow from a silo is based on the work of Jenike (1970). The data required for the calculations are the flow function, internal angle of friction, sliding angle of friction (on bin wall material) which are obtained from a shear test cell. Details of this calculation method are given by Thomson (1997) and Woodcock and Mason (1987).

From the mechanical perspective, stresses on silo walls must be determined to design the silo to withstand the loads. Stresses on silo walls are caused by combinations of static and dynamic conditions that occur during filling and discharge of a silo. Experimental measurements on models and industrial size silos have shown that the distribution of wall stresses changes significantly when flow begins after the initial filling, and these stresses remain after the outlet is closed. Details of the theory and equations needed to calculate stresses on silo walls are given by Brown and Nielsen (1996), Rotter (2001), and Thomson (1997).

Where mass flow silos cannot be installed and funnel flow silos are used they are often provided with devices to promote gravity flow. A number of these flow-promoting devices have been developed for specific applications. Some commonly used devices are the following:

The "bin activator" (also called a vibrating bin bottom) is widely used and consists of a conical hopper mounted beneath the opening in a silo and suspended from the silo bottom by elastomeric-bushed links. Elastomeric bands connect and seal the inlet to the silo above and to the discharge feed device or chute below. Motor-driven eccentric weights mounted on the vibrating hopper, cause it to gyrate in an elliptical path on a horizontal plane. Vibration frequencies vary from 15 to 50 Hz, but 15 to 30 Hz are most commonly used. The use of the bin activator is often a mixed blessing because vibration applied indiscriminately to a cohesive powder tends to increase its already high strength and makes it even more resistant to flow. In practice, the vibrator must often be cycled on and off during use to prevent the silo from plugging solid either above or below the core. Obtaining the right on/off cycle and the correct vibration strength and frequency for a specific solid is difficult. The best situation for use is where a large amount of material is removed from the silo at a high rate. The bin activator should never be applied to solids which tend to solidify in storage; chunks resulting from breaking up of such a mass may not pass through the gap between the core and the bin wall. Figure B-100 is a schematic drawing of such a device.

Several vibrating louver-type devices are available such as the Silleta® and Superfeeder® discharger-feeders. In their design, a feed tray is suspended from a frame fastened to a silo outlet. A row of fixed position, inclined blades, mounted in a feed tray, divides the flow area into a series of powder feed slots. The feed section reciprocates in response to an electromagnetic or electromechanical vibrator to provide, in theory, an infinite variability in feed rate. The fixed blade dimensions, inclinations, and spacing are determined by tests to ensure that the powder will flow during vibration, and stop when the vibrations stop. These devices combine the function of a bin discharger and a feeder to regulate the flow. They are fabricated to accommodate round or square openings ranging from 0.15 m to 1.5 m in

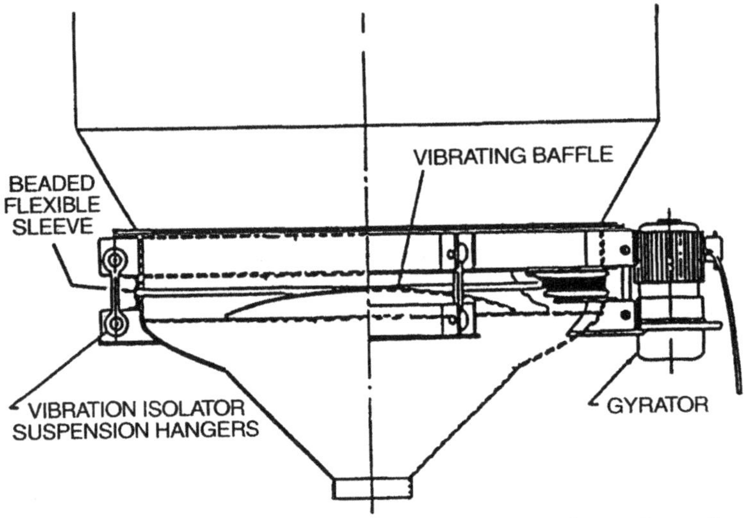

Figure B-100 Schematic of a vibrating bin bottom ("bin activator").

diameter or width. Another type, the Hogan® discharger, is similar to the two devices described above, except that in addition to varying the vibrator stroke, the blade positions can be adjusted to any position between closed (zero flow) to fully open (maximum flow), by manual, electric, or pneumatic actuators, while the unit is operating. Figure B-101 shows schematic drawings of these three types of vibrating louver-type devices. A number of devices and systems are available that fluidize the solids in a hopper and cause aerated flow. They can be effective for cohesive solids which are easily aerated and where solid bridges are absent. These devices include a fluidizing cone (hopper), aeration pads welded to the silo hopper walls, air injection nozzles installed on the inside of silo hopper walls, and other variations. These devices operate at low noise levels, require little maintenance, are relatively low cost, and can handle large volumes of solids with low gas flows. Where material in a silo may be solidly compacted, air blasters (or "air cannons") may be used to break up the solidified mass.

A more complete discussion of these flow-promoting devices and others is presented by Brown and Nielsen (1996) and Thomson (1997)

Silos may have other auxiliaries such as level-monitoring instrumentation, a bin vent filter, ladders, platforms, and a railing around the roof.

For more detailed information about storage in and flow from silos the chapter by Thomson (1997) and the book by Arnold et al. (1979) are recommended. For the mechanical design of silos, the books by Brown and Nielsen (1996), Gaylord and Gaylord (1984), and Rotter (2001) are recommended.

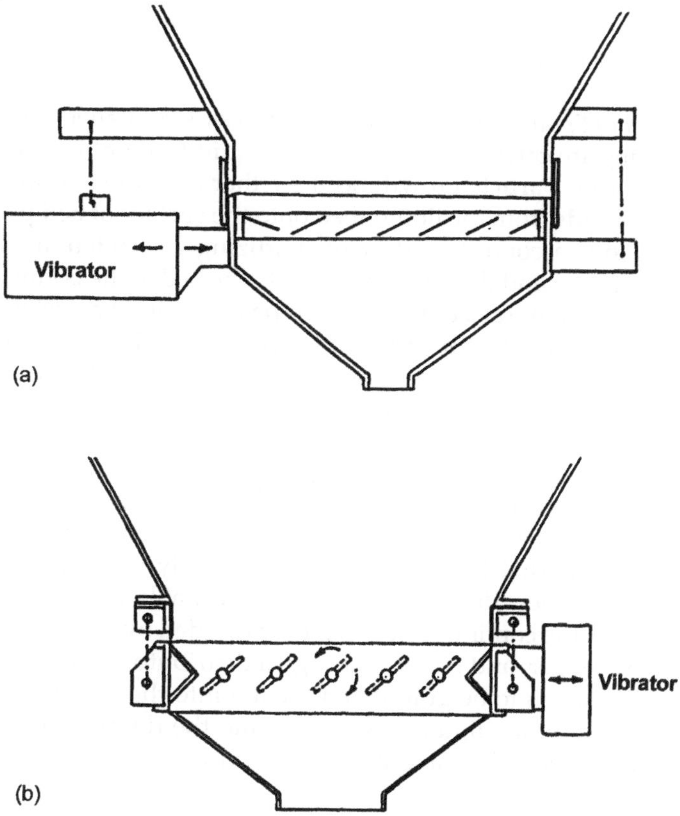

Figure B-101 Schematic of vibrating louver-type silo discharger feeders. (a) Siletta® and Superfeeder® fixed blade designs; (b) Hogan® adjustable blade design.

B17 SIZE ENLARGEMENT EQUIPMENT

Size enlargement of particulate solids, also called agglomeration, is usually achieved by granulation and pressure compaction. Other techniques, which will not be discussed here, are spray drying and prilling, and dispersion techniques such as flocculation and crystallization.

Granulation can be done by low shear agitative granulation, high shear agitative granulation, and high pressure compaction techniques. Low shear agitative granulation is done in equipment such as fluid-bed, pan, and drum granulators. High shear agitative granulation is done in equipment like batch high shear mixers, in-line continuous mixers, and hybrids such as fluid-bed mixers. High pressure compaction techniques are wet and dry extrusion, briquetting, and tableting. Extrusion and tableting are discussed separately in Sections B6 and B20.

Discussions are presented in the following paragraphs of several types of size enlargement equipment:

Fluidized-Bed Spray Granulators: In this process, simultaneous drying and particle forming are carried out by spraying liquid feed onto a bed of essentially dry particles. Particle growth occurs either by particle coalescence or by layering of solids from the liquid feed onto the surface of bed particles. In these granulators, a controlled flow of fluidizing gas (often air) is heated externally to the bed and then introduced into the bed at the bottom through a suitable distribution device. The air then passes though a cyclone or filter to recover entrained solids before being discharged. There are two types: batch (see Figure B-102) and continuous (see Figure B-103). In the batch type, the solids are contained in a product container with a distributor plate at the bottom through which the fluidizing gas passes. The product is sprayed with a fluid (water or organic liquid) by an adjustable spray nozzle. The air is sucked through the container distributor plate by a fan at the top of the unit and then exhausted to the atmosphere. Any entrained solids with the air are collected by an internal filter, which is periodically shaken to return the solids to the container. Batch-type units are available in square or round shape designs. In the continuous type, air is provided by a blower, heated in an external exchanger, and passed through a distributor plate in the bottom of the unit. The solids to be granulated are fed into the unit at a controlled rate by a feeder, and the liquid is sprayed onto the fluidized bed at a controlled flow rate. The granulated product is continuously withdrawn at the

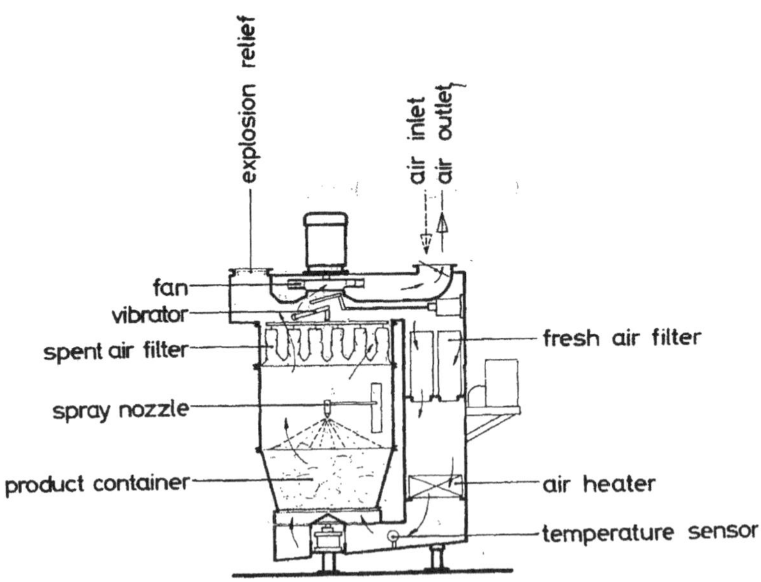

Figure B-102 Schematic of a typical batch fluidized bed spray granulator.

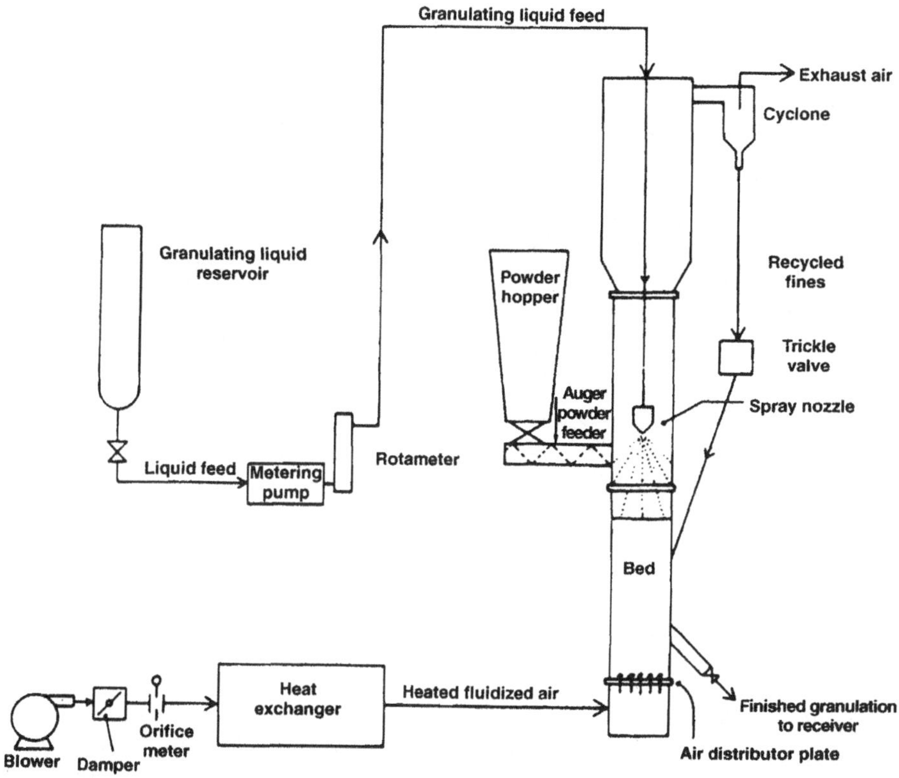

Figure B-103 Schematic of a typical continuous fluidized bed spray granulator.

bottom of the bed, and the exhaust air goes to an external cyclone where the entrained solids are collected and returned to the bed.

Drum and Inclined-Disk Granulators: Although a wide variety of agitation equipment is used in the CPI to produce agglomerates, rotary drums or cylinders and inclined disks or pans are the most important equipment in terms of tonnages produced. Rotating drum granulators (see Figure B-104) consist of an inclined rotary cylinder powered by a fixed- or variable-speed drive. Feed material containing the correct amount of liquid phase agglomerates under the rolling, tumbling action of the rotating drum. The pitch of the drum (up to 10° from the horizontal) assists material transport down the length of the cylinder. A retaining ring is often fitted at the feed end to prevent spillback of the feed. A dam ring may also be used at the discharge end to increase the depth of material and residence time in the drum. Liquid may be introduced, usually by spray nozzles, either before or after the solids

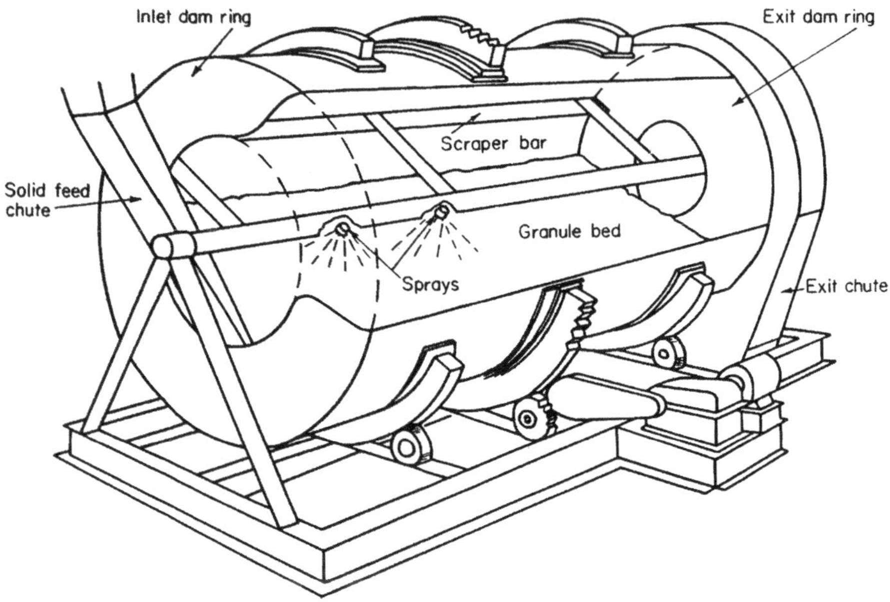

Figure B-104 Schematic of a rotating drum granulator.

enter the cylinder. Various types of internal scrapers are used to limit buildup of material on the inside surface and to provide a uniform layer to promote the correct rolling, tumbling action in the drum. Cylinder length-to-diameter ratios of 2 to 3 and speeds of 10 to 20 rpm are used. Recommended speeds of about 50% of the critical speed of the dry material are used so that adequate cascading then occurs and the range of particle size distribution is narrowed.

An inclined-disk granulator consists of a tilted rotating pan equipped with a rim to contain the agglomerating charge (see Figure B-105). Solids are fed continuously from above or from the front onto the central part of the disk, and product agglomerates discharge over the rim. Water or other binding agents can be sprayed onto the solids by spray nozzles at various locations on the disk surface. Adjustable scraper and plows maintain a uniform protective layer of product over the disk surface and also control the flow pattern of material on the disk. The disk angle can be adjusted from 40 to 70° from the horizontal to obtain the best results, and both constant-speed and variable-speed motors are available as disk drives. Rotating speeds are usually 10 to 30 rpm. Dust covers can be fitted when required. In an inclined-disk granulator the finer particles settle to the bottom; the largest particles remain at the top and then overflow the rim and constitute the product. Because of the size stratification, the product from an inclined-disk granulator is more uniform in size than that from a drum granulator.

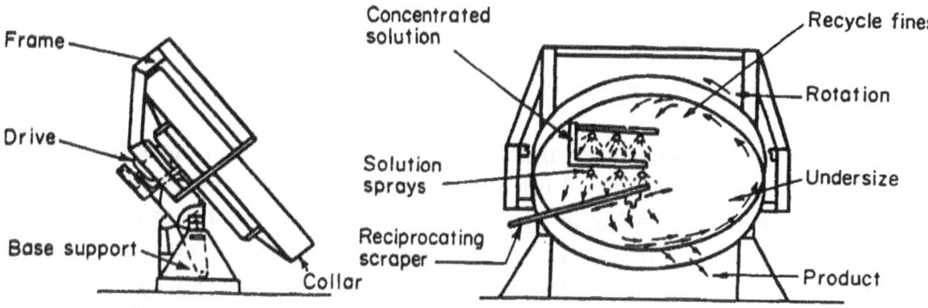

Figure B-105 Schematic of a rotating inclined-disk granulator.

Low Speed Mixers: Pug mills and paddle mixers are used for both batch and continuous applications. These mixers consist of a horizontal trough with a rotating shaft to which mixing blades or paddles of various designs are attached. The vessel may be of a single-trough design although a double-trough arrangement is most popular. Twin shafts rotate in opposite directions throwing the materials forward and to the center to achieve a kneading, mixing action. Typical shaft speeds are less than 100 rpm. They are often used in a premixing step or in a finishing densification step. Construction is robust, with the body of heavy plate (6.4 or 9.5 mm thick) and hardened agitators or tip inserts. Figure B-106 shows a double-shaft mixer–granulator. Operational features include fume hoods, spray systems, and stainless steel construction. Provisions can be made to feed materials at different points along the mixer as well as at the end to ensure that the entire mixing length is used and to add processing versatility.

Figure B-106 Double-shaft mixer–granulator.

Batch planetary mixer–granulators are used extensively in the pharmaceutical industry in typical batches of 100 to 200 kg, using 5 to 20 kW power. They do have long mixing times of typically 20 to 40 minutes.

High Speed (High-Shear) Granulators: High speed mixers include continuous-shaft mixers and batch high-speed mixers. Continuous-shaft mixers have blades or pins rotating at high speed on a central shaft, and both horizontal and vertical shaft designs are available. Two types commonly used are the vertical Schugi Flexomix® and the horizontal pin or peg mixers. These mixers operate at speeds of 200 to 3500 rpm to produce granules from 0.5 to 1.5 mm in size with a residence time of 5 to 30 seconds. These mixers are used for making detergents, agricultural chemicals, clays, ceramics, and carbon black. Batch high-shear mixer–granulators are used extensively in the pharmaceutical industry. There are vertical designs (manufactured by Diosna, Fielder, and Gral) and horizontal designs (made by Lodige). Plow-shaped mixers rotate on a shaft at speeds of 60 to 800 rpm. Separate high-speed cutters or choppers rotate at a much higher speed (500 to 3500 rpm) and act to break up wetted powder and limit the maximum granule size. Granulation times are of the order of 5 to 10 minutes, which includes both wet massing and granulation stages operating at low and high impeller speeds, respectively. Discussions and pictures of these high-speed mixers are presented by Sherrington and Oliver (1981).

Roll Briquetting and Compacting Machines: Both briquetting and compacting machines use roll presses in which particulate solids are compacted by squeezing as they are carried into the gap between two rolls rotating at equal speed. This is probably the most versatile method of size enlargement because most materials can be agglomerated by this technique with the aid of binders, heat, and/or very high pressures. This method generally requires less binder than other methods and, therefore, there is little or no requirement for drying the agglomerates. In briquetting machines, pillow shapes are formed by matching indentations in the rolls. Precise design of these pockets based on practical experience is important to ensure optimum briquette density, minimum incidental feather (fines) production, and dependable pocket release of the finished briquettes. In compaction machines the agglomerated product is in sheet form as produced by smooth or corrugated rolls. The compacted product can remain in sheet form or can be granulated into the desired particle size in conventional size reduction equipment.

Roll presses consist of the frame, the two rolls that do the pressing, and the associated bearings, reduction gear, and fixed or variable-speed drive. Spacers between bearing housings prevent roll contact and allow adjustment of roll spacing. The frame of the press is designed so that all forces are absorbed internally. The rolls are forced together by a hydraulic system, which may incorporate a safety valve to prevent overpressure if foreign material intrudes between the roll faces. The rolls consist of a continuous roll shaft, the roll body, and attached molding equipment. The molding surface may be either solid or divided into segments. Segmented rolls are preferred for hot briquetting, as the thermal expansion of the equipment can be controlled more easily. Segmented rolls can be made more resistant to wear than can one-piece rolls. For fine powders that tend to bridge or stick and are of low bulk density, some form of forced feed, such as a tapered screw feeder must be used to deaerate, precompact, and pressurize the feed into the nip between the rolls. Large machines are available with up to five screw feeders to spread the flow across the rolls, and vacuum hoppers are also used to remove air when densifying low density feeds.

Pellet Mills: Pellet mills differ from briquetting and compacting machines in that the moist particulate solids are compressed and formed into agglomerates by being pushed through holes in dies of various shapes rather than by squeezing as they are carried into the nip between two rollers. The friction of the solids as they are pushed through the die holes supplies the resistance necessary for compaction. The action of the roller and die assembly produces a shearing and mixing action which yields a plastic mix which is pushed through the die. Binders, plasticizing agents, and lubricants may be used to facilitate the process.

Several types of pellet mills are available. The die may be a horizontal perforated plate with rollers acting on its surface to press material through the plate. Rolls may be either side-by-side with material extruded through one or both of the rolls, or one or more small rolls may be fitted inside a large die roll. In yet another design, two intermeshed gears are used and the material is extruded through the die holes located in the gear root. Probably the most popular design of pellet mills is the one which utilizes a ring-type die and two or three rollers mounted in a vertical plane on the inside of the die. Power is applied to the die to rotate it around the roller assembly, which has a fixed axis. Pellet quality and capacity vary with properties of the feed such as moisture, lubricating characteristics, particle size, and abrasiveness, as well as die characteristics (hole size and shape) and speed (usually from 75 to 300 rpm).

Further detailed discussions of size enlargement theory, equipment, and applications are presented by Capes (1980), Messman and Tibbetts (1977), Parikh and Parikh (1997), Pietsch (1997, 2002), Sherrington and Oliver (1981), and Stanley-Wood (1983).

B18 SIZE REDUCTION EQUIPMENT

Size reduction (also called comminution) may be split into two generic categories, crushing and grinding. There is some overlap, but crushing usually means reducing coarse lumps or aggregates, sometimes up to several feet

TABLE B-8
Types of Size-Reduction Equipment

A. Jaw crushers
   1. Blake
   2. Overhead eccentric
B. Gyratory crushers
   1. Primary
   2. Secondary
   3. Cone
C. Heavy-duty impact mills
   1. Rotor breakers
   2. Hammer mills
   3. Cage impactors
D. Roll crushers
   1. Smooth rolls (double)
   2. Toothed rolls (single and double)
   3. Roll press
E. Dry pans and chaser mills
F. Shredders
   1. Toothed shredders
   2. Cage disintegrators
   3. Disk mills
G. Rotary cutters and dicers
H. Media mills
   1. Ball, pebble, rod, and compartment mills
      a. Batch
      b. Continuous
   2. Autogenous tumbling mills
   3. Stirred ball and bead mills
   4. Vibratory mills
I. Medium peripheral-speed mills
   1. Ring-roll and bowl mills
   2. Roll mills, cereal type
   3. Roll mills, paint and rubber types
   4. Buhrstones
J. High peripheral-speed mills
   1. Fine-grinding hammer mills
   2. Pin mills
   3. Colloid mills
   4. Wood-pulp beaters
K. Fluid-energy superfine mills
   1. Centrifugal jet
   2. Opposed jet
   3. Jet with anvil
   4. Fluidized-bed jet

Source: *Perry's Chemical Engineers' Handbook*, 7th edition, 1997. Reprinted with permission of McGraw-Hill.

across, to a size no smaller than ½ inch or ¾ inch (13–19 mm). Grinding (often called milling), in contrast, usually refers to reducing small particles, no larger than ½ inch to ¾ inch to sizes as small as several microns. Crushing is usually done dry; grinding may be done wet or dry.

There is a very large variety of crushing and grinding equipment in use in the CPI (see Table B-8). The selection of the appropriate type of equipment for a specific application depends to a large extent on the feed material size and the hardness of the feed material. Table B-9 is a general guide to the selection of crushing and grinding equipment (referring to the types listed in Table B-8). It should be emphasized that Table B-9 is merely a guide and that exceptions can be found in practice, so that tests should be performed to determine the suitability of the equipment initially selected.

TABLE B-9
Guide for the Selecton of Crushing and Grinding Equipment

Size-Reduction Operation	Hardness of Material	Size[a] Range of Feeds, in.[b] Max.	Min.	Range of Products, in.[b] Max.	Min.	Reduction Ratio[c]	Types of Equipment
Crushing							
Primary	Hard	60	12	20	4	3 to 1	A to B
		20	4	5	1	4 to 1	
Secondary	Hard	5	1	1	0.2	5 to 1	A to E
		1.5	0.25	0.185 (4)	0.033 (20)	7 to 1	
	Soft	60	4	2	0.4	10 to 1	C to G
Grinding							
Pulverizing							
Coarse	Hard	0.185 (4)	0.033 (20)	0.023 (28)	0.003 (200)	10 to 1	D to I
Fine	Hard	0.046 (14)	0.0058 (100)	0.003 (200)	0.00039 (1250)	15 to 1	H to K
Disintegration							
Coarse	Soft	0.5	0.065	0.023	0.003	20 to 1	F,1
Fine	Soft	0.156 (5)	0.0195 (32)	0.003 (200)	0.00039 (1250)	50 to 1	I to K

[a]85% by weight smaller than the size given

[b]Sieve number in parentheses, mesh per inch.

[c]Higher reduction ratios for closed-circuit operations.

Note: to convert inches to millimeters, multiply by 25.4.

Source: *Perry's Chemical Engineers' Handbook*, 7th edition, 1997. Reprinted with permission of McGraw-Hill.

A short description is presented below for a number of crushing and grinding equipment types.

Jaw Crushers: Both single-toggle and double-toggle jaw crushers are widely used. In both, the principle of operation is the same, with feed material being squeezed between a stationary jaw and a reciprocating jaw, driven by an eccentric shaft. The double-toggle machine is generally more rugged and better suited to deal with very hard, tough, or abrasive materials. This is because it has a more uniform crushing motion than the single-toggle machine, as the jaw moves at right angles to the feed flow, minimizing sliding. The end particle-size distribution can be varied by adjusting the width of the outlet gap. Jaw crushers are used in primary size reduction of minerals, and feed openings up to 2.5 × 2.5 meters are available.

Gyratory and Cone Crushers: Both of these designs utilize the principle of an eccentrically driven rotor crushing material against a stationary mantle. The gyratory crusher consists of a cone-shaped pestle oscillating within a larger cone-shaped mortar of bowl. The angles of the cones are such that the width of the passage decreases towards the bottom of the working faces. The pestle consists of a mantle which is free to turn on its spindle, and the spindle is oscillated from an eccentric bearing below. Cone crushers work like gyratory crushers but differ in important ways. Gyratory motion is still applied at the base of the shaft supporting the cone, but the upper bearing is within the body of the machine and below the cone. The size of the end product can be varied by changing the clearance between the rotor and the mantle. This is usually achieved by raising or lowering the rotor and mantle. These machines are used for medium-coarse size reduction, often following a jaw crusher in minerals-processing plants.

Roll Crushers: Most roll crushers have two moving rolls made of steel; some have only one roll. In a double-roll model the rolls rotate in opposite directions to nip and squeeze particles that fall between them. In the single-roll type, feed is crushed between the roll and the breaker plate. This plate has a shock absorber or spring to limit stress and prevent damage in case tramp metal is mixed in with the feed material. The roll surfaces may be smooth, or they may be toothed, corrugated, grooved, or serrated. Smooth rolls make fine-textured products; toothed rolls are used to obtain a more cubic product from slablike materials; serrated or waffled surfaces may be used in later stages of crushing such cubes. Product size depends on the gap between the rolls, the roll diameter, and the angle of the nip. The gap may be adjusted by shims or screws or hydraulically. Manufacturers typically design for nip angles of 11–25°; 20° is fairly average. The minimum gap for a roll crusher is about 1 mm, so this will be the minimum product size. If the crusher is choked, that is, if feed is piled up above the rolls, it will produce smaller particles but also a wider size distribution. Single-roll crushers of the toothed type can handle larger feed sizes than the double-roll type. They work best on material that is not too abrasive or hard, and on wet or sticky materials.

Roll mills or presses are similar to roll crushers, but unlike ordinary roll crushers, which crush individual particles, the roll press is choke fed and acts on a thick stream or ribbon of feed. Particles are crushed mostly against other particles, so wear is very low. The product is in the form of agglomerated slabs. Roll mills are used in a wide variety of applications for medium-coarse down to fine size reduction, such as for flour and cement milling.

Ball Mills and Rod Mills: These are tumbling mills in which the basic grinding action is the tumbling of feed material and grinding media inside. Ball mills are horizontal, cylindrical, short tumbling mills (the length is close to the diameter) in which loose grinding balls are filled up to a certain level (usually 35–45% of the total volume). Size reduction is achieved by rotating the cylinder so that the balls either roll against each other or, if the speed is sufficient, they are lifted and fall. In general, the action is a combination of rolling and lifting. Conventional ball mills are lined with alloy steel and use forged or cast alloy steel balls. When iron contamination is to be avoided, silica linings and flint pebbles may be used; alumina balls are a costlier alternative. Feed size is usually ½ inch or less, and product size ranges from 500 μm to below 5 μm. Ball mills work well on hard, brittle materials. They are not as suitable for grinding elastic, ductile, plate-like, or fibrous materials or for very fine grinding in general.

Rod mills are cylindrical tumbling mills filled with steel rods to about 35–45% by volume. Used for coarse grinding of feed up to ½ to 1 inch in size, their L/D ratio is typically between 1.5:1 and 2.5:1. Rod mills are usually run open circuit, receiving feed from secondary or tertiary crushing and providing a coarsely ground product that can be fed to ball mills. The largest particles wedge the rods apart, so they get ground first while smaller particles are shielded. Rod mills are usually run wet, and they can handle up to about 50% solids in a slurry. The rods are typically 2–4 inches in diameter, and about 6 inches shorter than the internal length of the cylinder to avoid jamming. Cylinder diameters range from 3 to 15 feet, and the throughput depends on the diameter and size reduction ratio desired.

Impact Mills: In this size-reduction group, stress is applied by transferring kinetic energy by either particle–particle contact or machine-particle contact. Impact mills can be broadly separated into mechanical types where high speed beaters impact the material to apply stress, and fluid energy mills, where particles are accelerated by the surrounding medium and impact against each other or a target. Mechanical impact mills include crushers, hammer mills, pin disk mills, turbine mills, and mills with air classifiers.

Fluid energy mills include spiral jet mills, fluidized-bed jet mills, and opposed jet mills. Brief descriptions of mechanical impact and fluid energy mills are given below.

Impact Crushers: Feed material is introduced through a feed opening into a rotor moving at between 25 and 50 m/s. The rotor of these machines is a cylinder to which are affixed tough steel bars (see Figure B-107). The initial impact by the rotor causes some size reduction, and the material is accelerated up to the speed of the rotor and flung against impact plates, where further size reduction occurs. Free impact breaking is the principle of these machines, and it does not rely on pinch crushing or attrition grinding between rotor hammers and breaker plates. It is possible to wear-protect these units quite well so that abrasive materials can be handled. The final end particle size can be varied by the inclusion of an outlet grid to vary the residence time in the machine.

Hammer Mills: One of the most versatile, economical, and widely used impact mills is the hammer mill (see Figure B-108). The rotor shaft may be vertical or horizontal, generally the latter. The shaft carries hammers, sometimes called beaters. The hammers may be T-shaped elements, stirrups, bars, or rings fixed or pivoted to the shaft. The grinding action results from impact and attrition between lumps or particles of the material being ground, the housing, and the grinding elements. A cylindrical screen or grating usually encloses all or part of the rotor. The fineness of product can be regulated by changing rotor speed, feed rate, or clearance between hammers and grinding plates, as well as by changing the number and type of hammers used and the size of the discharge openings. Many variations are produced, with special types available for specialized applications, for example, quick screen change for animal feed, heavy duty for minerals, and light construction for wood chips. The principle employed is similar to that of the impact crusher;

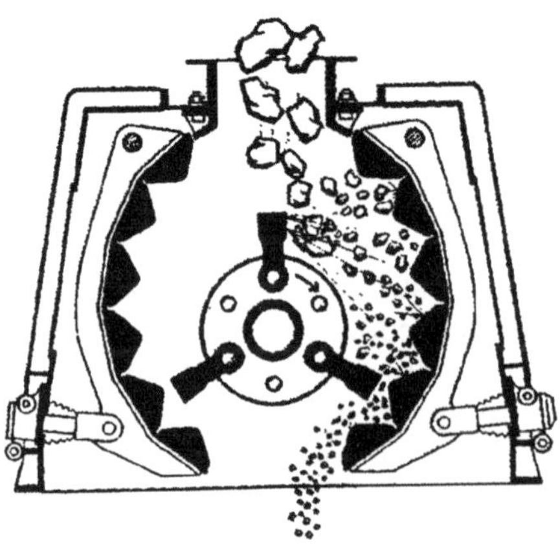

Figure B-107 Impact crusher. (Source: Pennsylvania Crusher Company.)

through combs. This unit also has trailing comb assemblies, and screens may or may not be used. A second type is a dual counter-rotating shaft machine which employs either inward or outward rotation at low speeds. This type of unit also has breaker bar assemblies, with or without extra breaking pins, and often uses a screen or perforated plate to hold back material for specific sizing requirements. A somewhat similar design is the Triskelion Lumpbreaker™ which is like a finger lump breaker except that instead of fingers, triangular blades are used, which gives more of a shearing action than the crushing action that fingers would give. This creates less dust and the blades help avoid the problem that occurs with fingers breaking off. This lump breaker design can be used for wet or dry material. Another type is the Rotocage Lumpbreaker™ in which a rotating cage rotates on a horizontal shaft within a housing. The cage tends to either wear the lumps down or cuts them or breaks them without abrasion if they are lightly fused pellets which must be friable. There is also a Double Rotocage Lumpbreaker™ which is similar to the single Rotocage Lumpbreaker™ except that it can handle wet material that would plug up the latter unit. It normally consists of two counter-rotating cages. Designs may rotate towards the nip between the cages or away from it depending on the application.

All of the above types of lump breakers are available in carbon steel, stainless steel, abrasion-resistant steels, Hastelloy and other alloys. They are also available in sanitary designs for food and pharmaceuticals applications, and also with spray-type cleaning mechanisms.

There are many other types of size reduction equipment not described here; these are discussed and described by Snow (1997), Austin and Trass (1997), Prasher (1987), and Mead (1964). Clement and Purutyan (2002) also describe a number of types of size reduction equipment and discuss factors that play a role in selecting the optimum crusher or mill.

B19 SOLIDS CHARGING SYSTEMS

Hazardous particulate solids often have to be charged into a batch or semi-batch reactor or a slurry mixing tank which already contains a flammable liquid. In the past, it was common practice to have an operator dump a bag or drum of solids into the vessel through the open manhole, which frequently resulted in a fire, explosion, or injury, and even death, to the operator. In more recent times, many companies are charging hazardous solids into vessels containing flammable liquids via closed solids charging systems. A number of different systems are used in the CPI, with some being self-designed and others available as proprietary devices. These are briefly described as follows.

One self-designed charging system consists of a hopper with a rotary valve at the bottom connected to the reactor or slurry mixing vessel by a charging chute (usually a length of pipe). There is a block valve at the end of the chute that sits directly on the vessel nozzle. Figure B-112 is a schematic drawing of this type of charging system. The charging chute often has connections for wash liquid and nitrogen. Another type of self-designed charging system consists of a hopper with an airlock valve device (two valves with an intervening chamber between them) connected to a charging chute which sits on a block valve on the reactor (see Figure B-113). The valves can be butterfly valves or other types. They are alternately opened and closed by a timer, so that as the upper valve is opened, the bottom valve remains closed, and the intervening chamber is filled with solids. Then the timer shuts the upper valve and the bottom valve is

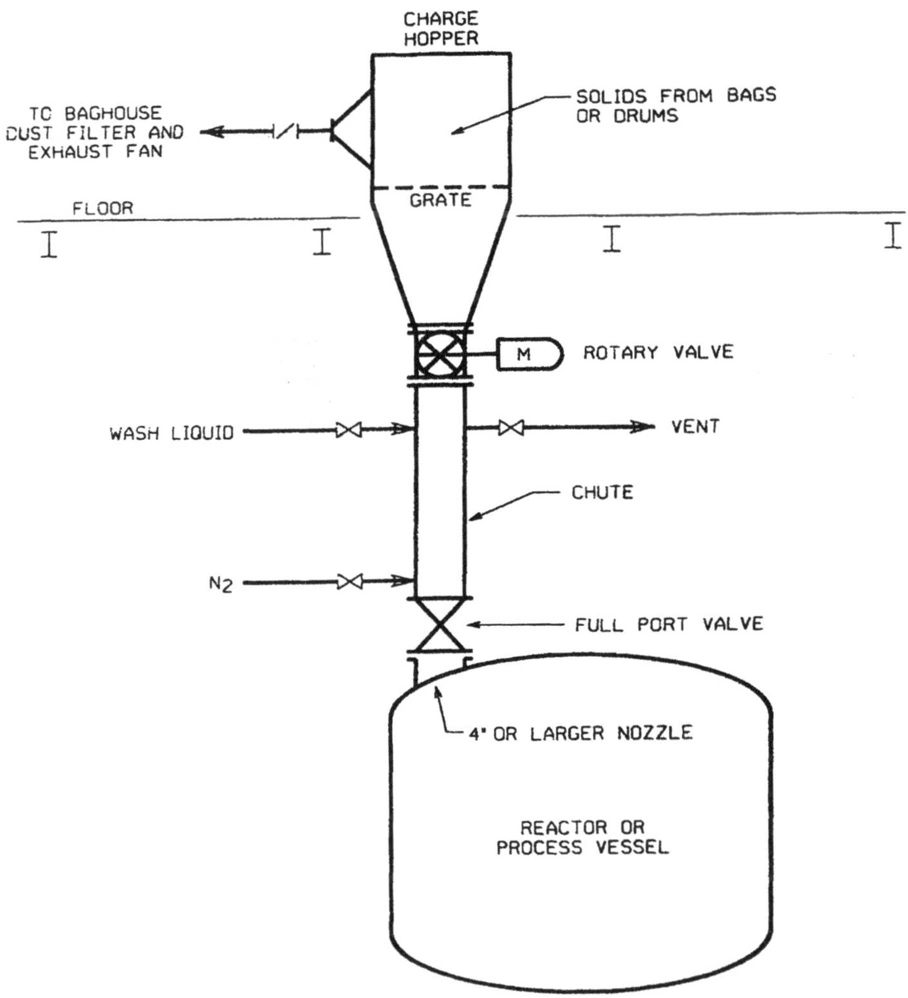

Figure B-112 Solids charging system with hopper and rotary valve.

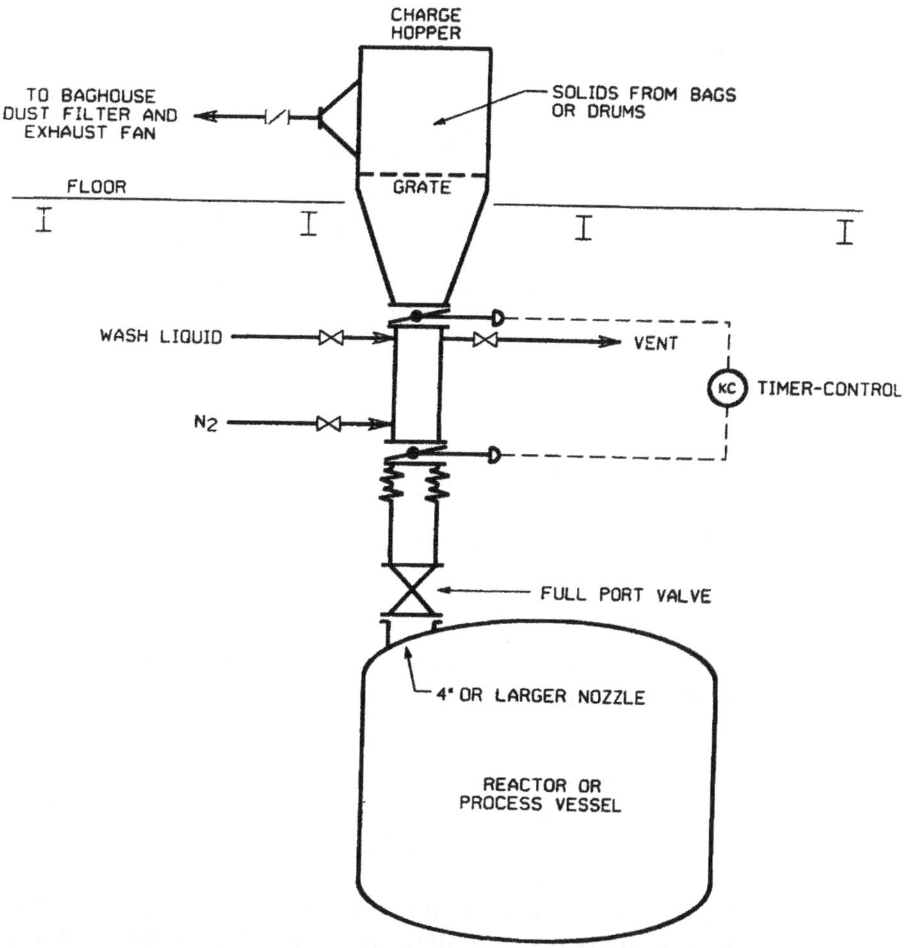

Figure B-113 Solids charging system with hopper and airlock (double dump valve) device.

opened, admitting the solids into the vessel. Airlock valves are available in cylindrical or rectangular configurations.

A different type of solids charging system is a proprietary design called the PTS (powder transfer system). This system consists of a cylindrical chamber connected to a block valve on the vessel feed nozzle (see Figure B-114). The operating principle is as follows:

1. Solids get sucked up into the PTS chamber under vacuum. A special filtration membrane separates the solids from the conveying air or nitrogen.

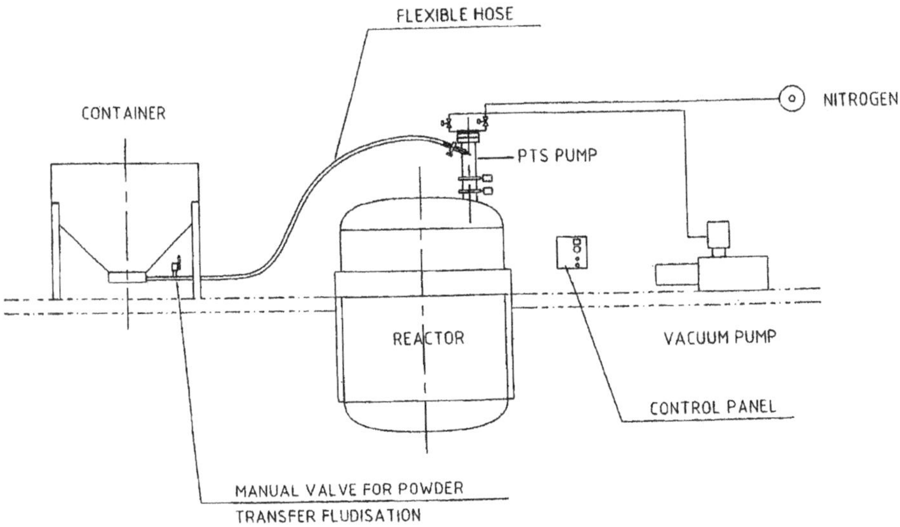

Figure B-114 Powder charging system (PTS) for reactors and mixing vessels

2. When the chamber is full, it is pressurized by nitrogen and the solids are discharged into the receiving vessel; and the vacuum cycle begins again.

To ensure continued optimum performance, the filtration membrane is instantaneously cleaned during the each discharge cycle by the pressurized conveying air or nitrogen. The rate of transfer and discharge of solids can be regulated by the control system.

Solids charging into reactors and mixing tanks also can be done using other proprietary systems (as described in Section B12.3) which unload solids from portable containers by vacuum into the vessels.

Other closed charging systems utilize screw conveyors (both horizontal and inclined) and bucket elevators that connect to a block valve on the vessel fill nozzle. However, they may not adequately contain vapors from the flammable liquids in the vessel and are difficult to inert.

B20 TABLETING SYSTEMS

Tableting is an operation used primarily in the pharmaceutical industry to produce a tablet (dosage form) in a tablet press. The tablet contains an active ingredient (the drug) as well as other components such as binders, fillers, coloring material, and the like, which have been preblended. All tableting presses employ the same basic principles: they compress the granular or powdered mixture of ingredients in a die between two punches, the die and its associated punches being called a "station of tooling." Tablet machines can be divided into two distinct categories on this basis:

1. Those with a single set of tooling-"single-station" (or "single-punch") presses or eccentric presses.
2. Those with several stations of tooling-"multistation" (or "rotary") presses.

The former are used primarily in preparing small amounts of tablets in the laboratory for developing new products or reformulating existing products. The latter, having higher outputs are used in most production operations. A brief description of each is given in the following paragraphs.

Single-Station Presses: All commercial types have essentially the same basic operating cycle, where filling, compression, and ejection of the tablets from the die is done by punch movement utilizing cam actions. Material is fed to the die from a hopper via an oscillating feed shoe, the position of the lower punch at this point determining the tablet weight. The feed shoe then moves away and the upper punch descends into the die to compress the tablet, the extent of this movement controlling the level of compression force. As the upper punch moves upward, the lower punch rises and in so doing ejects the tablet from the die. At this point, the feed shoe moves in, knocks the tablet out of the machine, and the lower punch moves to its bottom position ready for the next press cycle. Sizes of machines in this group vary widely from small ones capable of making tablets up to 12 mm in diameter at rates of 100–150 tablets per minute. Figure B-115 shows a single-station (single-punch) tablet machine.

Figure B-115 Single-station tablet press. (Source: Key International, Inc.)

Multistation (Rotary) Presses: In this type of machine the operating cycle and methods of realizing the filling, compression, and ejection operations are basically different from single-station machines. More specifically, the dies are set in a rotating, circular, metal table (called a turret) and the punches ride in appropriately designed cam tracks or channels in the head and foot areas of the press to achieve the necessary upward and downward stroking action. The central shaft mechanism drives these rotating components in synchrony, producing the designed number of tablets in each cycle. The granulation is fed from the hopper to the dies, passing below the feeder frame at a point when the lower punches are in their lowest position. The frame may contain some devices, such as rotating spindles, to induce or force granulation into the die, as a means to ensure more accurate and uniform fills. Pressure-release devices allow a lift release if an overload at the die occurs. Depending on the number of dies per punch unit, standard rotary presses can produce 5000 tablets per minute, and even more with a double-sided rotary press. The newest high speed presses can achieve 12,000 tablets per minute. Figure B-116 shows a double-rotary tablet press.

Figure B-116 Double-rotary tablet press. (Source: Key International, Inc.)

High speed presses are commonly instrumented and equipped with strain gauges at key points in the overall feed-compress-eject cycle. These measure compression and ejection forces to strictly control tablet weight and ensure a uniform dosage form.

Tablet presses are also provided with exhaust ventilation systems to control dust emissions resulting from the compression step.

Further discussions of tableting machines and tablet production systems are presented by Carstensen (1983), Cole (1990), and Ridgeway-Watt (1988).

B21 VALVES FOR SOLIDS

Rotary valves are described and discussed in Section B7. A variety of other valve designs are available for handling of particulate solids. They can be classified as follows:

- Gate (slide gate, knife gate, and conduit gate)
- Diaphragm
- Pinch
- Butterfly
- Rotating disk
- Standard ball
- Segmented ball
- Gate lock
- Iris diaphragm
- Diverter

These are briefly described as follows:

B21.1 Gate Valves

There are three different types:

1. Slide gate (linear, rotary or radial, and membrane)
2. Knife gate
3. Conduit gate

Slide Gate Valve: The linear slide (or parallel slide) is based on the principle of a gate valve while ensuring that the valve bore is smooth and without pockets in which solids can accumulate. Intermittent guides may be positioned on the side walls, but the bottom of the bore is smooth. When the valve is fully open, the slide is removed completely from the flow path. Seals and wipers on both sides of the slide prevent solids migrating into the body. There are several versions available, with distinct variations in the method of opening and closing the slide. Some operate by withdrawing the slide through a stuffing box and out of the valve. To assist sealing when handling small particles, the slide edge may be fitted a replaceable elastomer insert. They are available in aluminum, carbon steel, and stainless steel in circular and rectangular versions. These valves can be operated manually by a handwheel or by an actuator driven pneumatically or electrically. These valves are used primarily in controlling the flow of powder and granular materials from outlets of silos (bins), hoppers, mixers, screw conveyors, etc. and can be tailor-made to suit the application. Figure B-117a shows a linear slide gate valve. In the rotary (or radial) slide valve the slide rotates about a hinge pin and slides radially out of the pipe bore. The bore is completely clear when open. A stainless steel slide is totally contained within a cast aluminum body and supported by nylon inserts. These valves are designed primarily for hygienic applications. They can be operated manually by a handlever or by a quarter-turn actuator driven pneumatically or electrically. Figure B-117b shows a rotary slide valve. A variation of the slide valve is the membrane slide valve. The valve eliminates the side seals and wipers by enclosing the slide in a rolling membrane. When fully open, the membrane rolls into the body leaving an unobstructed flow path. As the valve closes, the membrane unrolls until it completely seals the flow path. Leakage of fine powder into the valve mechanism is thus eliminated. The use of a membrane, to isolate the working parts of the valve from the solids, has removed all of the problems inherent with maintaining tight seals and dry solids operation.

The forces required to move the slide are greatly reduced. However, elastomers can fail due to excessive bending. This type of valve can be made for circular pipes as well as square and rectangular ducts. Standard materials include Type 304 and Type 316 stainless steels. For hygienic applications, the metal surfaces can be electropolished. These valves can be fitted with a quick-release mechanism to allow removal from the system for cleaning. They have been used successfully in the chemical, food, and pharmaceutical industries for a variety of materials including wet bulk solids. Figure B-117c shows a membrane slide valve.

Knife Gate Valve: The knife gate valve is one of the simplest and cheapest isolation valves available. Basically, it is a slim valve using a thin blade with a beveled (knife) edge on the bottom passing though a rectangular stuffing (packing) box. The blade cuts through the particulate solids and lands on a seat which is flush or protrudes slightly into the valve bore. Standard valves are usually cast or fabricated in carbon steel or stainless steel with purge connections as standard to allow steam to be admitted. The replaceable seat is usually carbon steel or Type 304 stainless steel. To minimize abrasive wear, hard facing can be applied to the seats. Knife gate valves of the outside rising stem type are capable of differential pressures up to 70 psig when handling safe dry solids. Figure B-118 shows a schematic of a knife gate valve.

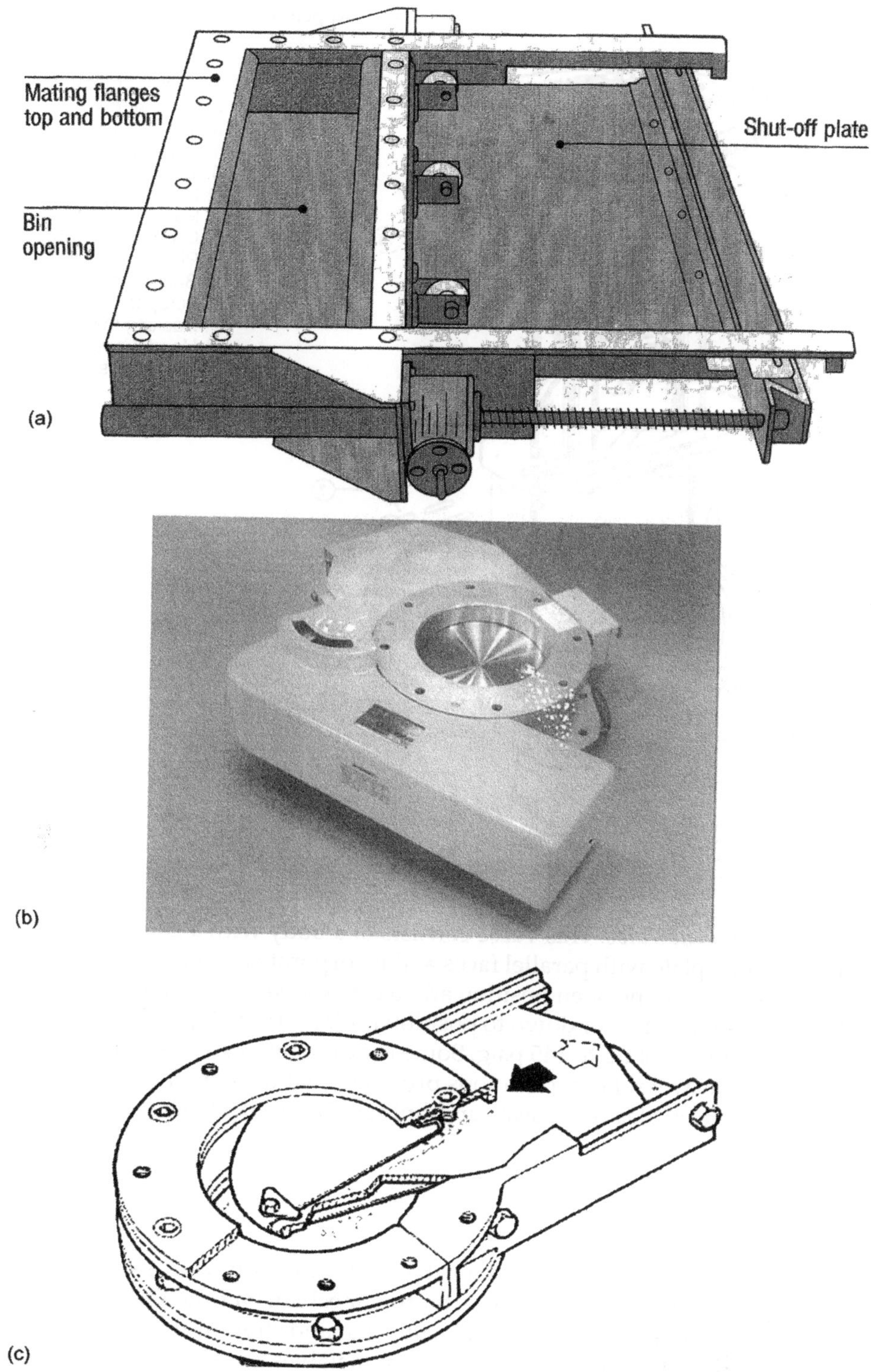

Figure B-117 Slide Gate Valves. (a) Linear slide valve; (b) rotary slide valve; (c) membrane slide valve.

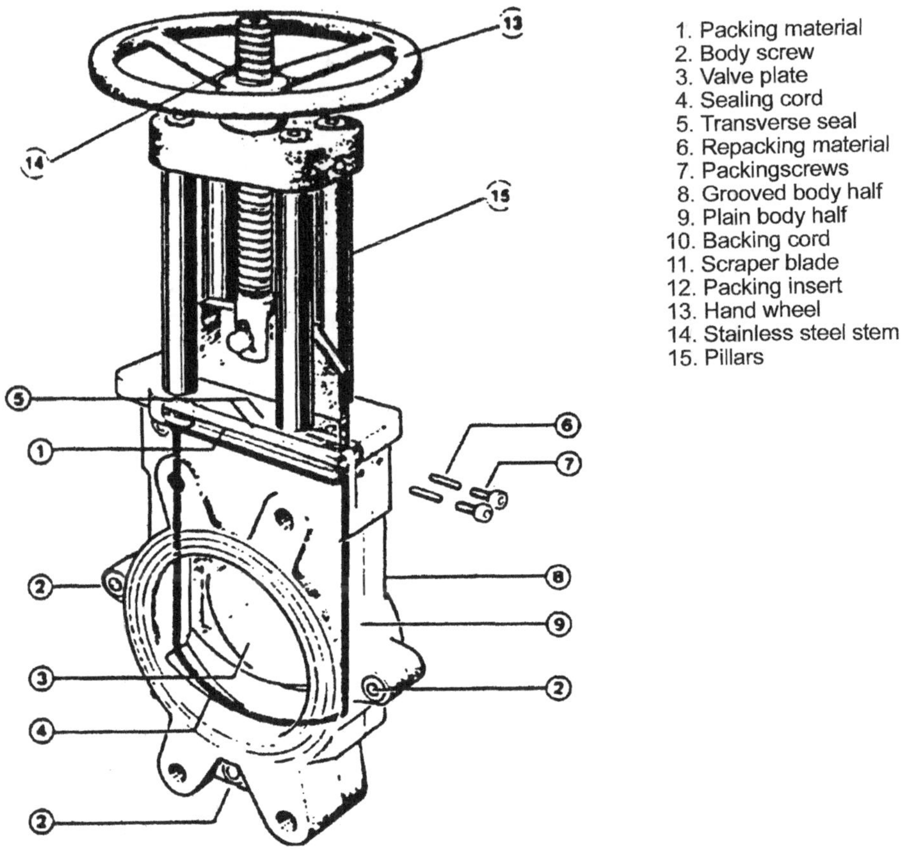

Figure B-118 Knife gate valve.

Conduit Gate Valve: This valve consists of a body with a gate which is in the form of a plate with parallel faces and incorporates a circular aperture as the valve bore for the open position and a blank area for the closed position. The plate is moved up or down to position the bore (opening). This valve has a pressure rating of up to 145 psig. Some conduit gate valves are suitable for dry and some wet solid applications provided that they are fitted with combined reinforced elastomer seats and linings. Figure B-119 shows a conduit gate valve.

B21.2. Diaphragm Valves

The diaphragm valve consists of a valve body assembly with a simple flexible diaphragm which isolates the actuating mechanism from the flow. There are two basic types available, namely, the weir and straight-through designs. Only straight-through (full-bore) diaphragm valves have been used success-

Figure B-119 Conduit gate valve.

fully for handling particulate solids. Standard cast iron valves are used provided that the diaphragm is made from soft rubber, butyl rubber, or neoprene, and the body is lined with an elastomeric material or glass. Other full-bore diaphragm valves made of different materials may be suitable for certain solids and are available in plastics (PVC, PP, PVDF, ABS, and CPVC). Figure B-120 is a schematic of the two types of diaphragm valves.

B21.3 Pinch Valves

The pinch valve consists essentially of an elastomeric tube (also called a sleeve or boot) contained in a steel body with an external clamp to pinch off

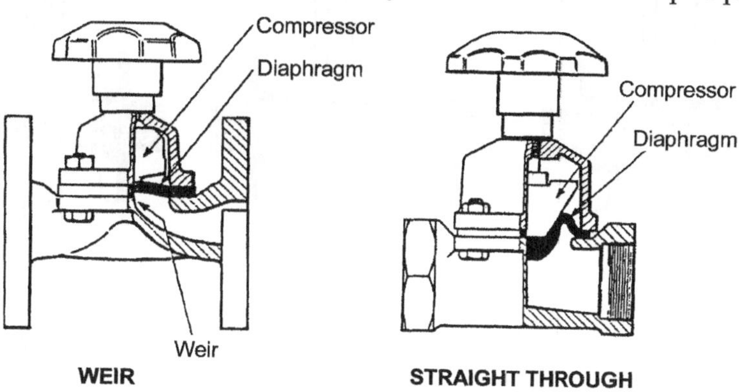

Figure B-120 Diaphragm valves.

the flow. The method of flow control is achieved using a double-acting mechanism to pinch off the valve tube (sleeve). Pinch valves may be operated manually, pneumatically, or electrically. The method of actuation adopted is dependent on the application, the size of the valve, and the pressures against which the valve must close. The pinch valve is suitable for dry and some wet solids handling up to pressures of 175 psig. Powders can be handled without problems, but handling larger solids depends upon the valve size and the tube material and construction. Figure B-121 shows a pinch valve.

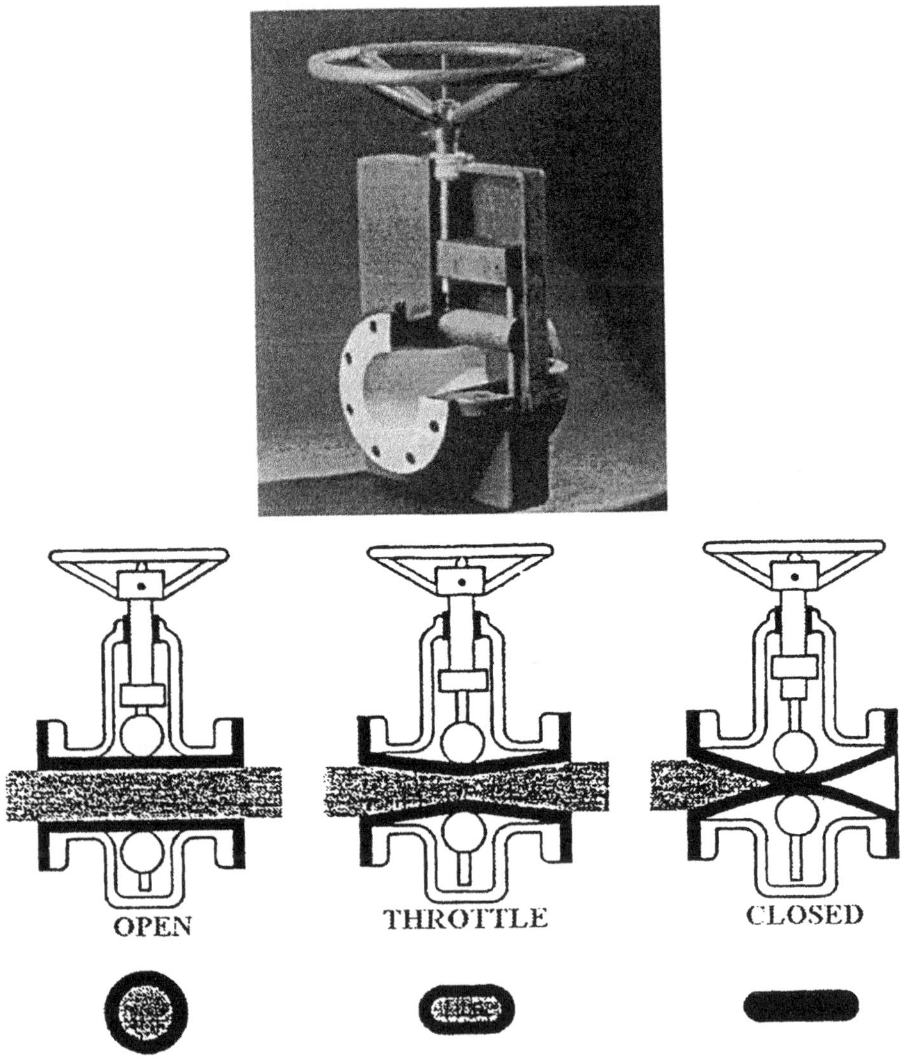

Figure B-121 Pinch valve.

B21.4 Butterfly Valves

The butterfly valve consists of a circular shaped disk which rotates about a diametral axis at right angles within the cylindrical bore of the valve body. Only a 90° turn is needed to move the valve from the fully closed to the fully opened position. In the fully-opened position, the only restriction to flow is that caused by the disk which is edge-on. In the closed position, it is rotated against a seat which may be located in the body of the valve or the periphery of the disk.

Essentially, there are two types of butterfly valves, namely,

- The standard (symmetric disk) in which the disk is totally concentric to the stem. Most of these valves have a single-piece resilient seat or liner that wraps over the end-flange sealing surfaces.
- The high-performance (eccentric disk) valve in which the center of the disk is offset slightly from the centerline of the shaft with a spherical seating surface on the disk and the shaft offset slightly from the centerline of the valve body.

These two types have been used successfully on dry and some wet solids applications provided that the disk is coated and the body is lined with suitable wear-resistant materials. A particular feature available on valves larger than 3 inches is the inflatable membranes on the face of the disk. These membranes can be inflated and/or pulsed to loosen sticky material attached to the disk. Figure B-122 is a schematic drawing of the two types of butterfly valves.

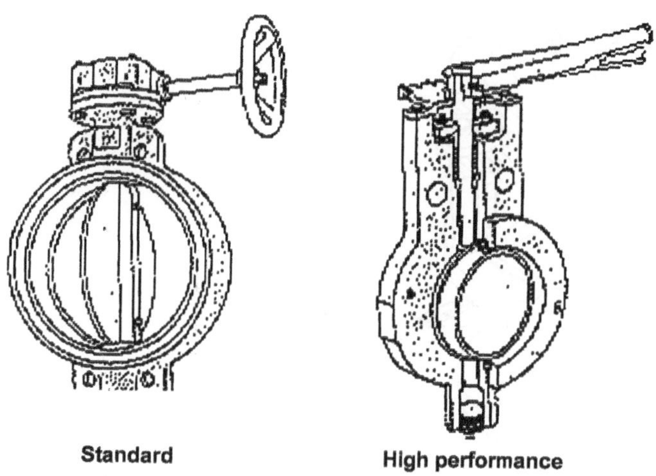

Figure B-122 Butterfly valves.

B21.5 Rotating Disk Valves

The rotating disk valve utilizes a metal–metal seated disk that slides across the face of the piping opening. The disk is both spring and pressure energized to force the metal–metal lapped seats together in a very close tolerance seal with potentially very low leakage rates. The disk and seat materials are usually cast Stellite. Each actuation causes the valve disk to rotate on its seat. This has the dual effect of shearing and wiping away surplus solids while grinding in or lapping the seat surface. Due to the spiral flow through the valve, the particles move over the seat tangentially instead of striking the seat face directly. Because of this flow path, wear arising from direct particle impact erosion of the seat face is minimized. Moreover, seat wear due to the erosive nature of the flow is confined to the inner diameter of the seat, thus preserving the flat lapped sealing surface of the seat. Hence, the seat life is greatly lengthened even in the most severe applications. Another advantage is the innate ability of the valve to heal itself if a large particle should score the seat surface. Rather than the scar becoming a source of further deterioration, the lapping action of the disk will eliminate the scar. The valve is suitable for the most abrasive, high temperature (900°C) and high pressure [ANSI Class 2500 (5800 psig)] applications up to sizes of 24-inch diameter. This valve has been used for handling a wide variety of materials including alumina, ash, coke, diatomaceous earth, limestone, tar pitch, shale, titanium ore, and wood chips. However, it tends to be expensive in comparison with other solids handling valves. Figure B-123 is a schematic diagram of a rotating disk valve.

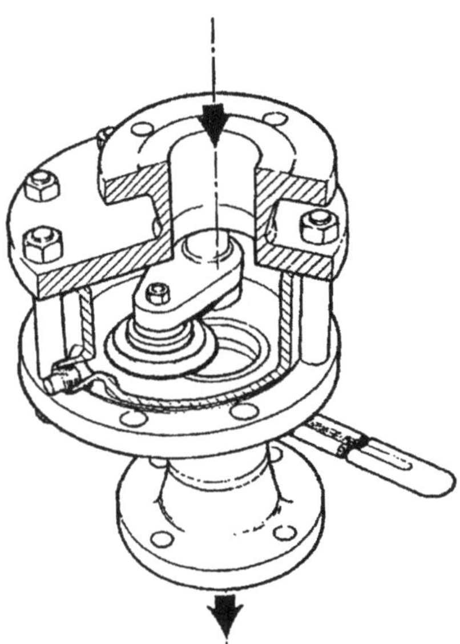

Figure B-123 Rotating disk valve.

B21.6 Standard Ball Valves

Standard ball valves, similar to those used in liquid and gas/vapor service, are often used for particulate solids. However, they are specified as full-bore valves with cavity fillers so that solids cannot enter into the cavity and "jam up" the valve. They can be specified with several options for reducing the wear caused by solids compacting behind the seats, such as:

1. Use of metallic scraper rings on either side of the seat, together with a hardened ball (Stellite or ceramic). These features are expensive, but they can solve the seat/ball abrasion problem quite effectively.
2. Use of inflatable seats where the seat rings deflate for ball movement, then are re-inflated to seal once the ball has come to rest.

B21.7 Segmented Ball Valves

The segmented ball valve (or V-valve), which was developed from the ball valve, is a quarter-turn valve that operates in a similar manner to the eccentric plug valve. This valve has a hollowed-out spherical segment (a partial ball) that is supported by a shaft and bearing at the actuator end of the valve. On the other side is a post and bearing. The segment rotates about an eccentric pivot point to form a single, continuous upstream seal in the closed position, retracting into the body cavity when open. The seal is obtained via a flexible metal or elastomer seat ring pressing against the outside of the spherical segment. Shutoff is assisted by the line pressure which pushes on the seat while the shaft and post hold the spherical segment tightly in place. The valves are supplied with a cast one-piece stainless steel body in flanged or flangeless construction. The spherical segment is frequently hard-faced using, for example, Stellite. The seat ring and pressure-energizing systems are simpler versions of the more complex arrangements used on conventional ball valves. The valves are compact and have good control characteristics. Because of the spherical segment, the valve provides very high rangeability (up to 300 to 1) and a somewhat symmetrical triangular opening that is not as susceptible to clogging as other valves. They are designed to provide good throttling control on dry and some wet solids, and are suitable for high temperature service. Figure B-124 shows a segmented ball valve.

B21.8 Gate Lock Valves

These valves are primarily used to isolate two pieces of equipment such as in solids-charging systems (see Section B19), but are sometimes used to feed positive-pressure conveying systems. They are also known as double-flap valves, double-door discharge gates, or double-dump valves. They consist basically of a housing with two doors or gates which open and close alternately by a rotating cam to permit passage of the material from one vessel

Figure B-124 Segmented ball valve.

into the other. These gates may be motor-driven, cam or air cylinder-operated or may work under gravity, and are controlled by a timer mechanism. They are available as cylindrical, square, or rectangular units. Standard gate lock valves are fabricated in carbon steel or stainless steel (or other corrosion-resistant materials as needed). These valves can be cast or fabricated, depending on the size and material of construction. They are suitable for powders and granular abrasive solids. Figure B-125 shows a cylindrical and a rectangular gate lock valve.

Figure B-125 Gate lock valves. (Sources: Gemco and Wm. W. Meyer & Sons, Inc.)

B21.9 Iris Diaphragm Valves

The iris diaphragm valve consists essentially of a continuous flexible tube (also called a diaphragm) fitted rigidly to the top and bottom of the valve body. By twisting one end of the diaphragm through 180° relative to the other, the valve can be adjusted from fully-closed, to any intermediate bore size, or to fully open. This movement is similar to that of a camera iris. These valves may be operated manually or by a handlever, handwheel through a gear box, or by a half-turn actuator driven pneumatically or electrically. The flexible tube (diaphragm) can be made in various materials to suit the application, for example, elastomers, fabrics, and elastomeric-coated fabrics. Elastomers and fabrics provide a wide range of chemical resistance, abrasion resistance strength, and the ability to seal the valve from he smallest particle sizes. For extremely abrasive solids, the diaphragm can be fitted with a replaceable liner. To reduce the effects of fugitive dust and loss of material, the diaphragm can be extended through the valve to allow attachment to containers when used for filling operations. The metal valve components can be aluminum, bronze, stainless steel, or carbon steel coated with PTFE. These valves can be fitted with a quick-release mechanism to allow the valve to be removed quickly. These valves are well suited for the handling of dry and wet fine powders, granules, lumpy and abrasive solids, and fragile products, and are used in the food processing, pharmaceutical, fertilizer, and pigment industries. Figure B-126 shows the internals of an iris diaphragm valve.

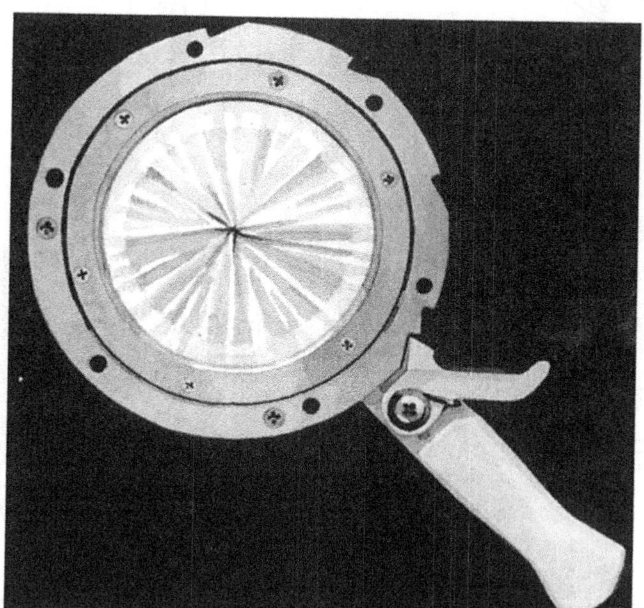

Figure B-126 Iris diaphragm valve.

B21.10 Diverter Valves

Diverter valves are used to change the path of a solids flow, for instance, from a pneumatic conveying system to several different silos, or from one piece of equipment such as a silo or hopper to two parallel items of processing equipment. These are available in various designs such as the flap (blade) valve, the slide valve, the plug valve, the non-jamming gate valve, and the swinging hopper valve type. Diverter valves are available in a variety of sizes, actuation options, and materials of construction (carbon steel, Type 304 and Type 316 stainless steels). Standard diverter angles (30°, 45°, and 60°) and mounting patterns as well as special angles and mounting patterns are available upon request. Figure B-127 shows several types of diverter valves.

B22 WEIGHING SYSTEMS

Scales of various types are available in different forms for weighing operations in industry. Bench scales are small units, generally with a platform, on

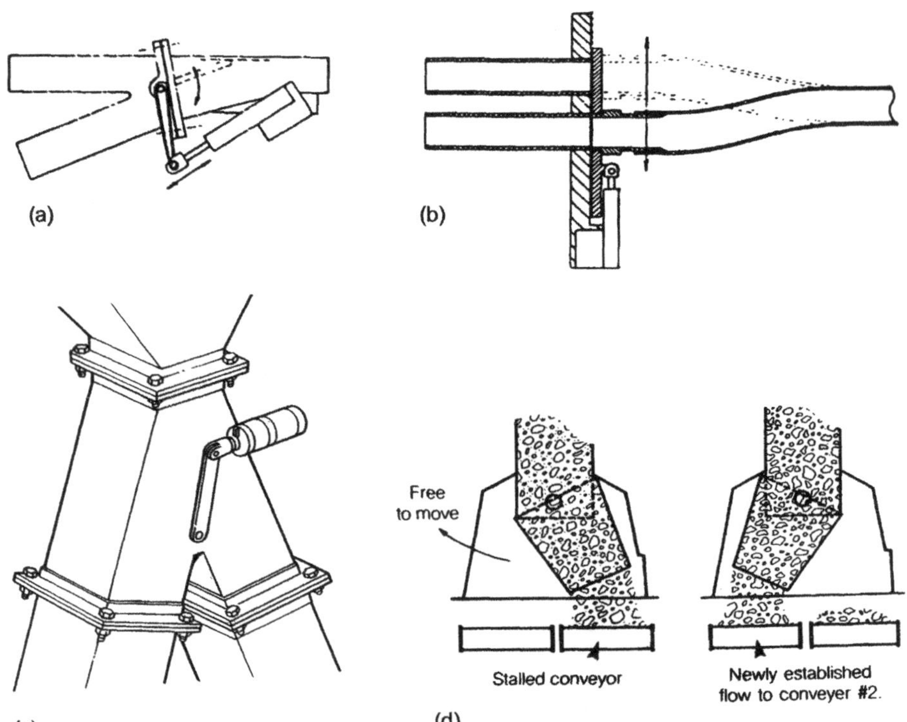

Figure B-127 Diverter valves. (a) Flap type; (b) slide type; (c) blade "pants legs" type; (d) swinging hopper type.

which small quantities can be placed by hand or moved across the scale on a roller conveyor installed on the platform. Portable scales generally have a platform near the floor, and the scale is on wheels so that it can be moved from place to place. Bench scales can also be placed on a wheeled stand for similar use. Floor scales are platform scales installed in the floor. Motor-truck scales are used for weighing hopper trucks and are often installed in a pit. Railroad-track scales are installed in a pit and support the proper length of rail to weigh a railroad hopper car. For static weighing of railroad cars, the rails are generally long enough to support the complete car, although with two-draft weighing, the car can be weighed one end at a time. Some railroad car weighing is done with the cars coupled and in motion. In this case, the rails of the scale may be long enough to weigh the pair of axles at each end of the car, or short, weighing a single axle at a time. Some sort of calculator system is used to obtain total car or train weight.

There are four major types of scales: (1) mechanical scales, (2) hydraulic load-cell scales, (3) pneumatic load cell scales, and (4) nuclear scales. The relative advantages and disadvantages of these four types of scales are presented in the paragraphs that follow, which can serve as a guide for selection of the most suitable type of scale for a specific application.

Mechanical Scales: The advantages of mechanical scales are: high accuracy; relatively inexpensive, particularly in smaller capacities; comparatively simple maintenance because of mechanical construction; no electricity required; they lend themselves to simple control equipment when a simple contact such as a magnetically operated mercury switch is used for the control element; very accurate primary element for use of other controls such as pneumatic or electronic controllers; and they are capable of high overload capacity with relatively little sacrifice of accuracy.

Disadvantages are: relatively complicated and expensive installation for large sizes; damage from such causes as corrosive fumes or liquids; remote indication adds to cost; excessive vibration may cause damage and difficulty of reading; and motion of the platform or weighing container is relatively high, particularly for small-capacity scales.

Hydraulic Load-Cell Scales: The advantages of hydraulic load-cell scales are: they are compact and self-contained units when a single load cell suffices, such as on a crane or hanging scale; inexpensive installation; relatively easily protected from corrosion; effects of vibration can be eliminated by damping; and motion of platform or weighing container is relatively small.

Disadvantages are: additional equipment, either mechanical or electrical, is required if it is necessary to totalize the load on several load cells, such as for a large platform scale; and expensive electrical equipment must be added to perform control operations.

Pneumatic Load Cells: The advantages of pneumatic load cells are: compact units; inexpensive installation; no electricity required; readily adaptable to pneumatic control equipment; simple construction resulting in ease of maintenance; vibration is no serious problem; and motion of platform or weighing container is relatively small.

Disadvantages are: limited to single-cell application so that a large platform, hopper, or tank cannot be supported on several cells and the total forces added; a supply of compressed air is required; and speed of response is relatively slow.

Electronic Load Cells: The advantages of electronic load cells are: they are easily protected from corrosion; inexpensive installation; relatively easy remote indication and use of one remote unit to read from several scales; easily adapted to controls; and motion of platform or weighing container is very small.

Disadvantages are: relatively expensive, particularly small-capacity units; somewhat slower recovery from sudden changes in ambient temperature; load-cell mountings must be designed to minimize side forces; and electricity is required.

Nuclear Scales: Nuclear scales measure mass directly by absorption of alpha or beta rays.

The advantages of nuclear scales are: a simple installation; not affected by extraneous forces on container such as connections to tank or hopper and tension or stiffness of belt conveyor; no contact with material is required; no moving parts subject to wear or corrosion; and electrical output permits direct readout and adaptation to controls.

Disadvantages are: variations in geometry, as well as in the density of the material being measured, will affect accuracy; proper shielding and care are required to avoid exposure of personnel to radiation; frequent calibration is required to compensate for loss of source emission with time; and electricity is required.

A good source of information on weighing systems is presented by Colijn (1983). Norden (1993) discusses electronic weighing systems.

B23 LOADING AND UNLOADING OF RAILCARS AND HOPPER TRUCKS

B23.1 Types of Railcars and Hopper Trucks

Particulate solids are shipped in large quantities in railcars and hopper trucks. A number of different types of railroad cars are used for shipping solids such as:

- Standard box car
- Container car
- Air-activated gravity-discharge hopper car
- Closed multi-compartmented hopper car with integral vacuum unloading nozzles (at each hopper compartment)
- Pressurized tank car (compartmented)
- Standard covered hopper car
- Pressurized hopper car with aerated discharge outlet

Several different types of hopper trucks (trailers) are also used, such as:

- Air-activated gravity-discharge hopper van
- Pressurized container trailer
- Combination gravity and pneumatic discharge tank trailer
- Pressurized tank trailer with aerated slope sheets
- Self-loading and unloading hopper trailer
- Covered hopper trailer
- Standard van trailer

Kraus (1991) discusses these various types of railcars and hopper trucks and their design features and types of particulate solids that can be handled by them. Pictures are also presented of several of these railcar and hopper truck types.

B23.2 Railcar and Hopper Truck Loading

Railcars and hopper trucks are loaded from silos or gravity blenders which can be elevated or located at grade. If the silo or gravity blender is elevated the railcar or hopper truck is usually loaded by means of a loading spout (chute). If a railcar or hopper truck is compartmentalized, it may be necessary to move the railcar or truck to fill each compartment sequentially.

If a silo or gravity blender is located at grade, then a railcar or hopper truck is usually loaded by a pneumatic or mechanical conveyor. The vent air from the railcar or hopper truck is usually routed to a dust collection system to recover any entrained solids fines and contain emissions.

Cox (2000) discusses the configuration and operation of a bulk railcar loading station.

B23.3 Railcar and Hopper Truck Unloading

Railcars and hopper trucks are unloaded by several means, such as:

- Pneumatic conveying systems
- Powder pumps
- Pressurization of the railcar or hopper truck

Pneumatic conveyor unloading systems are very frequently used to unload railcars and hopper trucks. For some types of railcars a conveyor pickup device (also called an unloading pan or sled) is moved under the railcar and attached to the discharge nozzle by a suitable connecting device. Flexible hoses are attached to the unloading pan to supply conveying air and to connect the unloading pan to the conveying system piping. Then the solids are transferred from the railcar or hopper truck to a receiving vessel by dilute phase pneumatic conveying. The conveying system can be a vacuum or positive pressure system, depending on the type of railcar or hopper truck. In some types of railcars or hopper trucks air is used to fluidize the solids in the containers and pneumatically convey them to a receiving vessel via connecting air supply and material discharge hoses at the container discharge outlets. The air-activated, gravity-discharge hopper truck is equipped with a motor-driven blower to supply air for solids conveying through two blow-through type rotary airlock feeder valves (one under each discharge outlet, driven by a transmission shafting connected to an electric motor).

A powder pump also is often used to unload railcars and hopper trucks. It consists of a section with an impeller screw with gradually decreasing pitch which receives material from a hopper supplied by gravity, or by a mechanical or air-activated conveying system. This screw compacts the material as it is forced through a barrel housing and check valve into a mixing chamber. The mixing chamber is fitted with several nozzles that are supplied with high-pressure air from a compressor or plant air system, and directed toward the entrance to the conveying pipeline. The entering material from the screw drops directly into the stream of air in the mixing chamber and is conveyed through the line. Figure B-128 shows the components of a powder pump railcar unloading system.

Pressurized railcars and hopper trucks are unloaded by air or inert gas pressurization by connecting an air/inert gas supply hose and a material discharge hose to the respective connections on the railcar or hopper truck and transferring the solids to a receiving vessel.

Table B-10 (Kraus, 1991) lists the various types of railcars used and their capacities, means of discharging their contents, equipment required to unload their contents, and equipment required in the plant to convey the solids. Table B-11 (Kraus, 1991) is a similar table for hopper trucks.

B24 INSTRUMENTATION

This section discusses the various types of instrumentation used for the measurement and control of flow, level, pressure, and temperature in particulate solids processes and equipment.

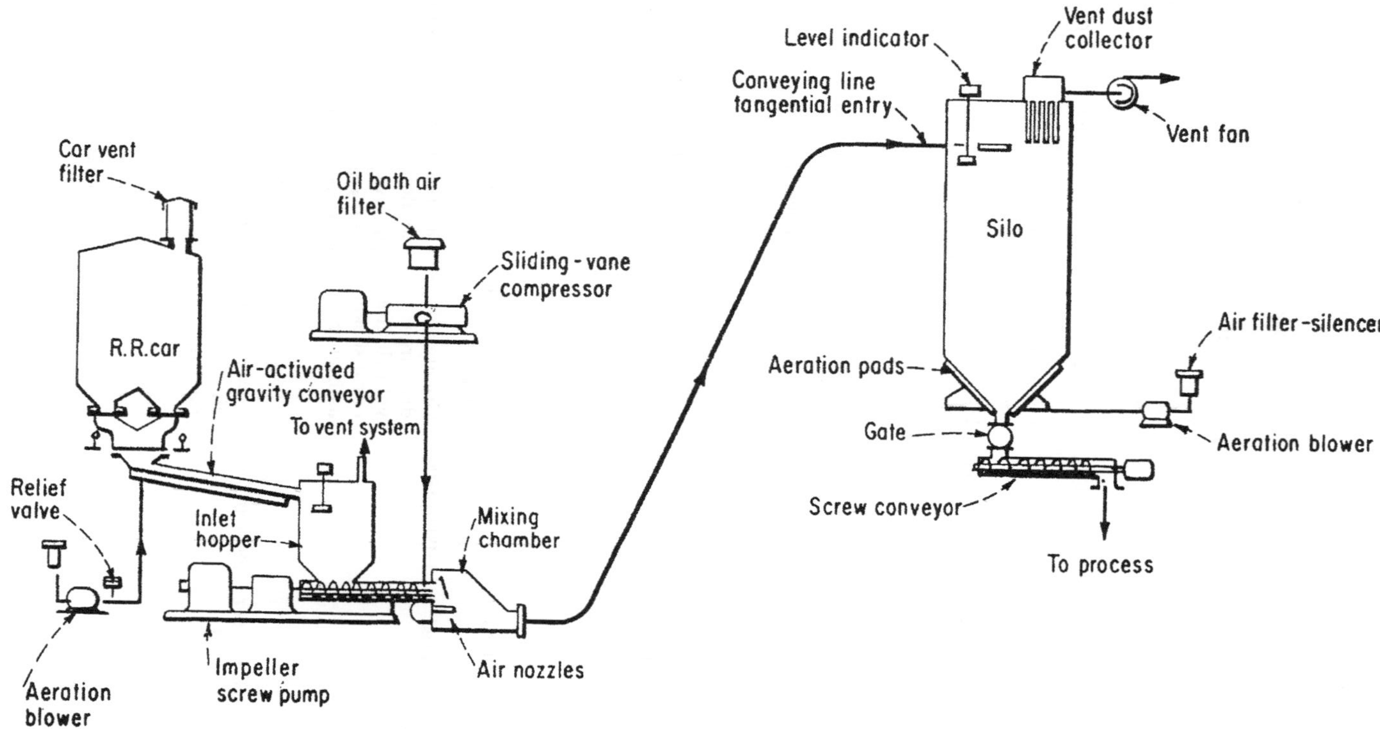

Figure B-128 Powder pump Railcar unloading system. (Source: Kraus, M. N., *Pneumatic Conveying Systems for Bulk Materials*, 3rd edition, 1991. Reprinted with permission of Prentice Hall PTR.)

TABLE B-10
Railcar Types and Unloading Equipment Used

Type of Transport	Volume (ft^3)	Means of Discharging Cargo	Equipment Required to Discharge Cargo	Equipment Required in Plant for Conveying
R.R. box car	3900 to 5800	Manually guided vacuum pickup nozzle	Vacuum unloading hose and pickup nozzle	Vacuum or V-P conveying system
Container car	1400 to 2400	High-pressure air fluidization of contents	Air supply hose and material discharge hose	Compressed air system
Air-activated gravity-discharge hopper car	2600 to 4566	Integral vacuum pickup nozzle Hose or chute to feeders of vacuum or pressure systems	Aeration air supply hose Material discharge hose or chute Discharge outlet adapter connection Car breather vent	Aeration blower system Vacuum, V-P or pressure conveying system
Closed compartmented hopper car with integral vacuum unloading nozzles (at each hopper)	2450 to 5700	Hose connection to vacuum system	Material discharge hose	Vacuum or V-P conveying system
Pressurized tank car (compartmented)	2400	High-pressure air fluidization of contents	Air supply hose and material discharge hose	Compressed air system
Standard covered hopper car	1900 to 4800	Gravity discharge with vbrator or car shaker assistance	Car shaker or vibrators Discharge chutes or vacuum pickup adapter connections	Vacuum, V-P, or pressure conveying system
Pressurized hopper car with aerated discharge outlet	2980 to 5700	Low-pressure air fluidization of contents (440 cfm at 15 psi)	Air supply hose (3-inch size) Material discharge hose(4-inch size; hoses may be made larger for high conveying rates)	Four-inch size conveying line to a vented bin (larger for high conveying rates) Aeration blower system

Source: Kraus, M. N., *Pneumatic Conveying Systems for Bulk Materials,* 3rd edition, 1991. Reprinted with permission of Prentice Hall PTR.

TABLE B-11
Hopper Truck Types and Unloading Equipment Used

Type of transport	Volume(ft^3)	Means of Discharging Cargo	Equipment Required to Discharge Cargo	Equipment Required in Plant for Conveying
Air-activated gravity-discharge hopper van	700 to 850	Integral aeration blower, conveying air blower, and pressure feeder at outlet connections	Material discharge hose	Conveying line to vented bin and a power supply receptacle
Pressurized container trailer	700 to 800	High-pressure air fluidization of contents (250 cfm at 25 psi)	Material discharge hose	Conveying line to vented bin
Pressurized tank trailer with aerated hopper	850 to 1950	Low-pressure air fluidization of contents (400–800 cfm at 15 psig)	Material discharge hose	Conveying line to vented bin
Self-loading and unloading hopper trailer	700 to 2100	Screw conveyor discharge into built-in pressure system	Material discharge hose	Conveying line to vented bin
Covered hopper trailer	700 to 800	Gravity discharge with vibrator assistance	Vibrators, discharge chutes, or vacuum pickup adapter connections	Vacuum, V-P, or pressure conveying systems
Standard van trailer	700 to 800	Bags or containers fork truck lifted to dumping station	Bag or container dump chute or hopper	Vacuum, V-P, or pressure conveying systems
Pressurized or vacuum tank for intermodal service	1500 to 1830	Low-pressure air fluidization of contents (fills via vacuum from railroad car, silo, or other source); 600 to 800 cfm at 15 psig or 13-in. Hg vacuum	Four- or 5-in.-diameter discharge hose or rein forced vacuum hose	Conveying line to vented bin

Source: Kraus, M. N., *Pneumatic Conveying Systems for Bulk Materials,* 3rd edition, 1991. Reprinted with permission of Prentice Hall PTR.

B24.1 Flow Instruments

The two most common types of flowmeters for particulate solids are: (1) the impulse type, and (2) the accelerator (or the Coriolis principle) type.

The impulse type is based on the principle that when a stream of solids impacts on a plate or a cylindrical surface at an angle, the resulting horizontal force relates to its mass flow rate. The force on the plate or cylindrical surface can be calculated by the use of impulse–momentum relations. In this type of flow meter solid particles are allowed to flow by gravity onto a calibrated spring-loaded resistance, the displacement of which, caused by the force of the falling particles, is a function of the flow rate of the solids. It is measured with a position transducer or transmitter. Figure B-129 is a schematic drawing of a cylindrical impulse-type solids flowmeter. These units are manufactured from steel or stainless steel with the sensing plate made out of stainless steel. They can handle free-flowing powders or granular and pelletized materials up to 0.5 inches in size. The smallest unit is claimed to have a range of 300 to 30,000 lb/hr, while the largest unit can handle flows up to 650,000 lb/hr. The standard units can be operated up to 140°F temperature, while special units are available for operation to up to 450°F. Metering accuracy is claimed to be 1% of full scale. Microprocessor-operated controls are available to integrate this flowmeter into batching or other automated material handling systems.

The accelerator (Coriolis principle) type flowmeter consists of a rotating measuring wheel with several guide vanes surrounding a central deflection cone. The wheel is mounted on a drive shaft, which extends upwards from the deflection cone. The wheel is inside a dust-tight enclosure with an off-center inlet above one side of the wheel and a central outlet below the wheel. The drive shaft is driven by an electric motor locate above and outside the

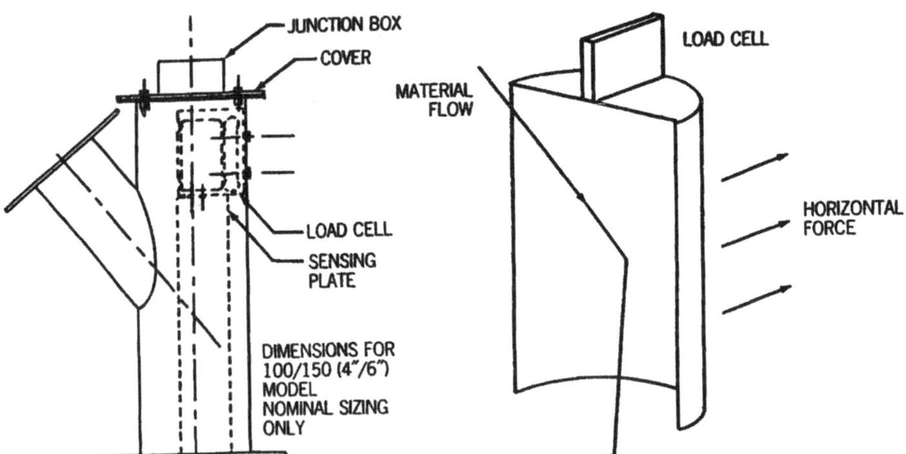

Figure B-129 Cylindrical impulse-type solids flowmeter.

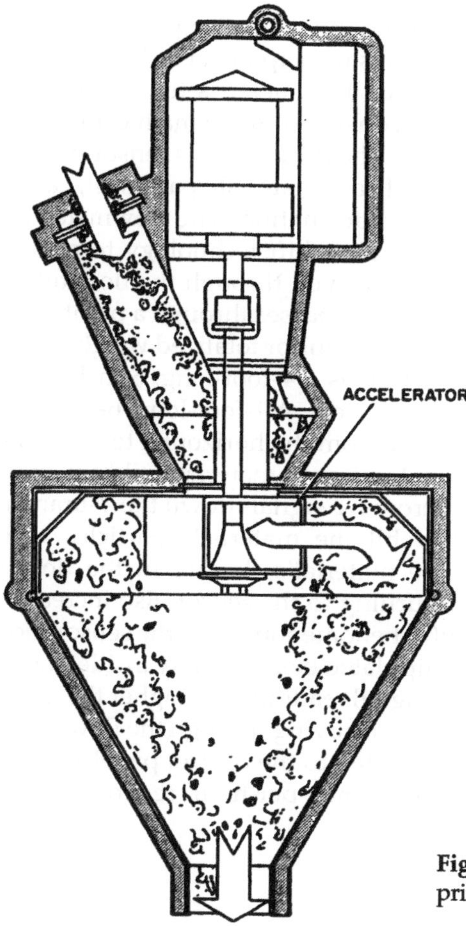

Figure B-130 Accelerator-type (coriolis principle) solids flowmeter.

enclosure. See Figure B-130 for a schematic of a accelerator-type flowmeter. In this type of solids flowmeter the solid stream enters the "accelerator" section (the rotating wheel) of the meter by gravity. As the solids enter the top of the rotating wheel section, the deflection cone deflects the particles outward in a radial direction. The Coriolis force acts in the tangential direction and produces a measurable reaction torque on the rotor. The change in torque is sensed by a torque-transducer. The amplified pneumatic signal is thus directly proportional to the mass flow of solids and can be used as the input to any pneumatic receiver instrument. Some units can handle flowrates up to 60 metric tons per hour with an accuracy of ±0.5% when the flow is over 2 metric tons per hour. It can handle particles with sizes up to 5 mm maximum, a maximum moisture content of 1%, and, and a bulk density of 18 to 188 lb/ft^3. They can handle free or moderate flowing, and fluidized solids, but not adhesive or sticky solids. Soderholm (1999) discusses the principles and application of an accelerator (Coriolis principle) mass flow meter.

Another, more recent, type of solids flowmeter is the RayMas™ unit (Anon 2001). The flowmeter assembly consists of three main components: a flow tube, a sensor, and a control panel. The flowmeters's sensor produces a continuous 4- to 20-milliamp signal as material passes it, and the flowmeter's control panel records the information. The flowmeter uses a continuous microwave signal and Doppler-shift technology to measure a material's flow rate and mass as it is conveyed through the flow tube. The flowmeter can calculate how much material passes through a point in a flow line over a given time in either an instantaneous or totalized weight-per-time measurement. The flowmeter's flow tube is a 4-foot-long, 6-inch-diameter section that can be installed in a piping system. A 1-foot-long, 6-inch-diameter Y-branch is positioned at a 55° angle from the horizontal to optimize the sensor's Doppler-shift sensitivity and to prevent material build-up on the Y-branch. The sensor transmits the microwave signals down the Y-branch and into the passing solids where they hit the material's particles and are reflected back to the sensor. As the material's density in the flow tube increases, the microwave signals returning to the sensor also increase, and vice versa, creating the Doppler-shift phenomenon. After the sensor receives the reflected signal, the signal is manipulated electronically inside the sensor and converted to a 4- to 20-milliamp signal readout. The higher the milliamp signal readout, the more material mass is moving through the flow tube. The milliamp signal readout is then transmitted to the control panel's electronic components, which convert it to a pounds-per-hour readout using the flowmeter's algorithm data.

B24.2 Level Instruments

Many techniques can be used for detection of level. The selection of a particular device depends on the application. Process variables such as pressure, temperature, and humidity, and product characteristics such as density, particle size and shape, particle size distribution, electrical conductivity, and moisture content all influence the choice of a level measurement system. Level measurement in a particulate solids application is more difficult than measurement of liquid level for the following reasons:

1. The surface is likely to be uneven.
2. The surface may contain local depressions and holes (rat holes).
3. The surface is not well-defined (fluffy powders).
4. Hydrostatic and buoyancy techniques cannot be used.

Level measurement systems for particulate solids may be employed to either give an indication of when a particular level has been reached (point or limit sensing), or to provide a continuous measurement of level.

Point (Limit Sensing) Devices

Point (limit sensing) devices are used to signal whether a silo or hopper is full or empty, and can be used as part of a control system or as back-up to a system which monitors the level continuously. The most widely used point level instruments are rotating paddles, vibration damping probes (tuning fork and single vibrating rod), capacitance probes, pressure switches, and radiation absorption instruments (optical, ultrasonic, microwave, or radiation). These are briefly discussed below.

The rotating paddle level sensor consists of a paddle that is slowly rotated at a fixed position inside a silo or hopper. When the paddle is covered, the resultant increase in torque is detected. Paddle sensors can be designed for a wide range of material characteristics; however, they do not tend to be used for light or very dense materials. Systems requiring large paddles can also be difficult to retrofit. The paddles are capable of accommodating a reasonable material adhesion, but as with other level devices, they should not be installed in line with the solids filling stream. If the paddle is to be used as a minimum-level switch (i.e., normally submerged), attention should be paid to the maximum loading on the device.

Vibration-damping level sensors employ a vibrating probe which is permanently located in the silo or hopper. The probe vibrates at its natural frequency in air, and upon contact with solids, this vibration is damped, and the change in amplitude detected. Two main types of sensors are available: the tuning fork type and the single vibrating rod type. The tuning fork type consists of two stainless steel tines that are electronically driven to oscillate in a manner similar to a tuning fork. A piezoelectric crystal attached to the tines is driven by an oscillator so that the tines vibrate at their natural frequency in air. A second piezoelectric crystal in contact with the tines produces a voltage signal due to the vibration. When solids come in contact with the tines the vibration is damped and the amplitude of the signal produced by the second crystal is reduced. The power used to maintain the oscillation of the tines is low and the vibration is damped by solids with a low density (i.e., less than approximately 30 kg/m^3). The unit should be installed in such a way that solids entering the silo or hopper do not fall onto it and the solids leaving the vessel are not trapped by the unit (i.e., the tines should be oriented to present a low cross section in the horizontal plane). These units are used in a wide range of applications, and if problems are encountered, they tend to be due to solids either adhering to or being trapped between the tines. The vibration of the tines does have a self-cleaning effect and this is helpful in situations where the moisture content is high and the solids have a tendency to stick. However, after submersion in some solids, the oscillation may not restart, with the result that the cleaning effect does not have a chance to operate. The single vibrating rod level sensor consists of a single rod driven by a piezoelectric crystal. The amplitude of the vibration is damped on contact with solids. This probe does not trap solids and the vibration again provides a self-cleaning effect. Both types of probes are suitable for high and low level detection, and should be mounted out of the direct flow of incoming solids. These probes are reliable, relatively inexpensive, and used extensively in many applications where level detection only is required. They have been applied for level detection of such solids as flour, sugar, cement, carbon black, animal feed, polystyrene chips, sawdust, and sand.

Capacitance probes operate on the principle that a change in capacitance between two electrodes occurs due to the presence of material with a dielectric constant different from air. One electrode is inserted in the vessel, while the other is provided by the wall of the vessel. The sensitivity of the device depends on the dielectric constant of the solids. Also, the capacitance of the system depends linearly on both the length of the probe and the dielectric constant, but does not depend significantly on vessel diameter, provided that the ratio of the vessel diameter to probe diameter is large (>100). The choice of electrode is determined by the application, and depends on the following:

1. Whether point level detection or measuring is required.
2. Properties of the solids such as dielectric constant, moisture content, electrical conductivity, adhesion propensity.
3. Process parameters such as pressure and temperature.
4. Vessel size and probe inlet location.

All electrodes must be insulated from the vessel. Fully insulated electrodes (coated with, for example. PTFE) are available for use with conducting powders or for solids where build-up is a problem.

Capacitance probes have no moving parts so wear is minimized, but it may be susceptible to material adhesion and the effects of moisture. They are easily adapted to high pressure or high temperature (up to 200°C) conditions. It is necessary to adjust their sensitivity at the installation site because the sensitivity depends on the shape of the vessel, and readjustment is necessary when the type of solids or moisture content of the solids changes. For point sensing the probe is often installed horizontally for low level sensing, and may be horizontally or vertically installed for high level sensing.

Pressure switches consist of a diaphragm installed in the side of a silo or hopper which is connected to a microswitch or a mercury switch. When the level of the solids rises to the level of the diaphragm, the pressure on the diaphragm increases and it deflects inward. The inward motion turns on a switch which can be interlocked to stop the filling of the vessel on high level or shut off a feeder at the outlet of the vessel on low level. After the solids level goes down, the diaphragm returns to its normal position and the switch goes off. This type of device has the advantage that it is non-invasive. However, the device may be susceptible to mechanical damage, particularly from high density solids. Diaphragms are made from synthetic rubber or metals such as stainless steel.

Radiation absorption sensors are devices that are based on the absorption of optical, ultrasonic, microwave, or gamma radiation. Optical and gamma devices consist of an emitter that is mounted on one side of a silo or hopper at a set-point height and a detector that is mounted on the other side. When the material fills to this level the radiation between emitter and detector is reduced. Optical radiation is seldom used in a particulate solids application due to the high sensitivity to airborne dust and particle adhering to either the sensor or the detector. Gamma ray systems do not require any equipment in contact with the solids being handled and can be installed without any modification to the tank as they are mounted on the outside of the vessel.

Ultrasonic level instruments can be used to detect point levels as well as continuously monitoring levels in a silo or hopper. They use an echo-sounding technique and measure the time it takes for a high-frequency acoustic pulse to return after reflection from the gas–solid interface. Ultrasound is sound whose frequency is above the limit of human hearing, that is, greater than 20,000 Hz. It can be produced by applying an oscillating voltage signal across a piezoelectric crystal. The applied voltage causes the crystal to deform and the deformation results in the production of ultrasound. The frequency of the ultrasound is dependent on the piezoelectric material. Because the piezoelectric effect is reversible, a crystal can be used as both a transmitter and a receiver. With ultrasonic devices care must be taken to ensure that:

1. The inflow of material does not cross the ultrasonic beam.
2. The time between ultrasonic pulses is sufficiently large to ensure that secondary echoes (for example, from the floor of the silo) do not cause spurious results. The minimum time between pulses is determined by the range of the instrument. Ranges from 1 meter to 60 meters are currently available.
3. The length of the ultrasonic pulse is sufficiently short so that it does not interfere with echoes from the solids. Typically, the time from the pulse being emitted until the system is ready to receive an echo is around 3 ms. During this time the pulse has traveled approximately 1 meter. No measurement can therefore be obtained from a surface nearer than 0.5 meters. This distance is known as the blocking distance of the sensor, and varies between 0.5 and 1 meters. This limitation can be overcome by mounting the sensor on an extension pipe above the vessel.

Ultrasonic systems must be compensated for temperature as the velocity of sound varies with the square root of absolute temperature. Pressure inside the vessel also affects the ability of ultrasonic devices as the vibration of the membrane may be dampened as the pressure increases. Process pressure limits of between 0.5 and 3 bar are typical. The energy reflected from a surface depends on the change in acoustic impedance between the solids on either side of the surface. In particulate solids the reflectivity will also depend on the structure of the interface, for example, if the material stored is a fluffy powder, there will be a low density gradient at the interface and the reflected pulse will be weak. This problem may sometimes be solved by using a unit with high acoustic power; however, the technique does rely on obtaining a detectable echo, and for some solids, this is not possible.

Continuous Level Measurement Devices

Continuous measurement of level in a silo or hopper can be achieved by mechanical, ultrasonic, capacitance, gamma ray, and microwave devices.

Mechanical devices are used extensively and a wide range of designs are marketed. In these systems, a weight attached to a cable is lowered under gravity. When the weight comes into contact with the surface of the solids, the tension in the cable is released, actuating a switch which reverses the motor and rewinds the cable. The length of the cable which has been unwound is a measure of the range to the material surface, and is of course, related to the level. The measurement can be either initiated manually on an on-off basis or the unit can be designed to operate continuously. The maximum measurement interval of this type of device is determined by the time taken to unwind and rewind the cable. The design and selection of the weight depends on the application. Parachute weights are used for very light, loose powders to prevent the weight penetrating the solids. Linen bag weights filled with the product can be used to avoid accidental contamination of the material in the vessel. The maximum measurement distances for these devices are typically about 75 meters, with quoted accuracies in the region of ±10 cm. These devices are simple, inexpensive, and particularly suited to measuring level in tall silos where short measurement intervals (short reading frequencies) are not required. However, they do contain mechanical components and regular maintenance is necessary.

Ultrasonic level devices are also used for continuous level measurement. The construction of the system and theory of operation is the same as for a point (limit sensing) capacitance system. The sensors are mounted from the top of the vessel and extend into the vessel to almost its entire length. Four different types of sensors are available: rod electrode, cable electrode with an insulated weight fixed to the bottom of the vessel, cable electrode with a gravity weight (not fixed), and wall-mounted electrode for nonconductive tanks. Continuous level measurement using a capacitance system is unsuitable if the solids are hygroscopic (e.g., lime) as the dielectric constant is highly dependent on moisture content.

Gamma radiation can also be used for continuous level measurement. Such a system has a single source of radiation and a radiation detector running the complete height of the vessel. Other configurations are possible using either multiple sources or multiple detectors. Care must be taken when using this technique to account for the half-life of the radioactive source, particularly if the device has been in long term use. The main advantage of this technique is that it is noninvasive. The main disadvantages are the high relative cost of these devices and the extra safety precautions necessary because of health hazards to operators.

Microwave (radar) devices have been used commercially to detect level in liquid storage tanks since the 1960s. Advances in recent years in hardware and analysis software have given radar an improved accuracy which has resulted in its increasing application in industry to measure the level in solids storage vessels.

There are two principle methods of using microwaves to detect the level of solids in a silo or hopper: the pulse-radar technique and the phase shift principle. In the pulse-radar technique a short burst of high frequency electromagnetic radiation is emitted, and the pulse is reflected at the surface of the solids and returns to a detector. The electromagnetic wave travels at the speed of light. If the time taken from the emission of the pulse to its detection can be measured, then knowing a velocity, a distance from the antenna to the solids surface can be determined. This can be used to assess the level in the vessel using simple geometry. The relatively short distance that the radar pulse has to travel, and hence, the very small time interval means that any instrument must have an extremely high resolution and a stable and accurate timer.

The phase-shift principle is employed by level measurement instruments. In this system a continuous electromagnetic wave is emitted, which is partially reflected from the solids surface. The phase relationship between the transmitted wave and the partially reflected wave will depend on the time delay between the two and, therefore, is dependent on the distance to the solids interface. This method requires a stable signal at a fixed frequency.

Modern level instruments use a combination of the two techniques in what is called synthesized pulse radar (SPR). This technique does not actually measure the time delay between pulses, but instead, a continuous signal is emitted and the phase change in the reflected signal monitored. The frequency of the emitted signal is swept, the phase shift "seen" at the detector from the reflected signal from the same surface will be dependent on the frequency. The information gathered on frequency and phase shift is analyzed using digital signal processing (DSP) to calculate a distance. The data processing can give a delay of a few seconds between information in and result out, but this is rarely a problem in level measurement. The advantages of these systems are that they are non-intrusive, have no moving parts, are relatively low cost and easy to install, and can operate in temperatures up to around 260°C (500°F). Problems can arise from signals which are reflected from geometrical features in the vessel other than the solids surface. Interference from other radio or microwave sources can result in false readings. One way to avoid many of the problems of false reflections is to employ a transmission line to "guide" the electromagnetic signal. A signal will travel down the cable and a portion of it will be reflected back from the point at which the cable is immersed in the solids. The measurement principle is phase tracking, much as described earlier.

The technique relies on the solids having a large enough effect on the local dielectric constant to produce a reflection, but systems have been reported to operate with solids bulk densities of the order of 10 lb/ft^3 and with particle sizes as large as 12.5 mm (½ inch). Moisture content variations or airborne dust have no effect on the measurement, and build-up of material on the cable can be allowed for by the DSP. These devices have much lower power requirements than conventional radar. They have found application in the plastics, food, brewing, chemicals, metal refining, and oil refining industries. Having a cable in a silo can, however, generate its own problems. For example, if the cable is hanging free, it can be moved off its true position during filling and discharging, or if anchored to the silo base, the loads on the cable which can occur particularly during silo filling can be considerable and must be allowed for when installing the cable.

Berrie (1996) and Knight and Pugh (1996) present information on various types of instruments for level measurement and control of particulate solids.

B24.3 Pressure Instruments

The types of pressure instruments used in particulate solids systems are similar to regular pressure instrumentation used for liquid or gas–vapor service, in other words, pressure gauges and differential pressure instruments. However, they have to be provided with special features for service with particulate solids (see Section 5.5.3). Pressure gauges are used for measuring pressure in a process vessel. Differential pressure is used usually to measure pressure drop, such as pressure drop across fixed and fluidized beds of solids.

B24.4 Temperature Instruments

Temperature measurement is required in the control of several types of particulate solids processing equipment (e.g., kilns, spray dryers, rotary dryers, etc.). Temperature sensors are classified in two main groups: thermal contact and radiation thermometers. The types used in particulate solids systems are identical to those used in liquid and gas–vapor systems.

Thermal Contact Temperature Sensors

The three most commonly used types of thermal contact temperature sensors are bimetallic thermometers, thermocouples, and resistance temperature devices (RTDs).

Bimetallic thermometers are composed of a composite material made up of strips of two or more metals fastened together. Because of the different rate of expansion of its components, the composite tends to change curvature when subjected to a change in temperature. With one end of a straight strip fixed, the other end deflects in proportion to the temperature change, the square of the length, and inversely as the thickness, throughout the linear portion of the deflection characteristic curve. Bimetallic thermometers are used at temperatures ranging from 580°C down to –180°C and lower. However, at low temperatures the rate of deflection drops off quite rapidly. Bimetallic thermometers do not have long-time stability at temperatures above 430°C.

Thermocouples are composed of two different types of metal wires connected at both ends, making a loop. One of the ends is called the hot junction (sensing point) and the other the cold junction (reference point). If there is a temperature difference between the two junctions, an electromotive force (emf) will be generated in the loop. which is a function of the temperature difference. The most commonly used thermocouples are the copper-constantan, iron–constantan, chromel–alumel, and platinum-rhodium. Table B-12 shows the composition, temperature range, emf generated, and accuracy of these thermocouples. Iron–constantan thermocouples are suitable for reducing atmospheres, chromel–alumel thermocouples for oxidizing atmospheres, and copper–constantan thermocouples for low-temperature atmospheres and also where the moisture content is high. Platinum–rhodium thermocouples are the most popular and accurate type, and can be used in either oxidizing or reducing atmospheres. Thermocouples are normally installed within a thermowell or protection tube to shield the sensor from corrosive action or mechanical damage, and to allow removal for maintenance. Consideration should be given to mechanical stresses on thermowells resulting from moving solids streams.

TABLE B-12
Types and Characteristics of Thermocouples

Material	Abbreviation	Temperature Range, °C	emf, μv/°C	Accuracy, °C
Copper–constantan	CC	–180–300	50	2–5
Iron–constantan	IC	0–600	60	3–10
Chromel-alumel	CA	0–1000	40	2–10
Platinum–10% rhodium	PR	100–1400	10	0.5–5

Resistance temperature devices (RTDs) depend on the inherent characteristics of materials to change in electrical resistance when they undergo a change in temperature. In essence, an RTD is an instrument for measuring electrical resistance that is calibrated in units of temperature instead of units of resistance (typically ohms). Industrial RTDs are usually constructed of platinum, copper, or nickel, and more recently of semiconducting materials such as thermistors (nonlinear temperature-dependent resistors). Platinum has a useful range of –200°C to 800°C; nickel from –80°C to 320°C; and copper from –100°C to 100°C. RTDs are being used more frequently nowadays because they are about ten times more accurate than thermocouples.

Radiation Temperature Sensors

These temperature sensors are non-contact type and measure temperature by means of thermal radiation. All solids emit thermal radiation energy depending on their temperature. The most common type is the radiation pyrometer. Its operating principle is based on detecting radiation energy emitted by a perfect emitter (black body) as given by Planck's law. Radiation pyrometers measure the total energy, and the radiation is gathered by lenses and focused on a detector such as a thermopile, silicon cell, thermistor bolometer, or a photocell. Thermopiles are composed of thermocouples connected in series so as to increase the sensitivity. Other types of pyrometers are the photoelectric pyrometer, disappearing filament pyrometer, and ratio (two color) pyrometers.

More information on temperature sensors is given by Liptak (1995).

REFERENCES

Alamzad. H. 2001. Prevent Premature Screen Breakage in Circular Vibratory Separators. *Chem. Eng. Prog.*, 97(5): 78–79 (May).

Albus, F. E. 1987. Fluid Energy Grinding or Jet Milling. *Powder and Bulk Engineering*, pp. 32–36, 41–47 (June).

Anon. 2001. Noncontact Solids Flow Meter Accurately Measures Material Flowrates. *Powder Handling & Processing*, 13: 64–70 (April).

Arnold. P.C., McLean, A. G., and Roberts, A. W. 1979. *Bulk Solids: Storage, Flow, and Handling*. TUNRA Ltd., University of Newcastle, New South Wales, Australia.

Ashworth, J. C. 1982. *The Scientific Approach to Solids-Drying Problems, Volume 2: Industrial Drying Practice*. Drying Research Limited, Wolverhampton, England, UK.

Austin, L. G. and Trass, O. 1997. Size Reduction of Solids- Crushing and Grinding Equipment. Chapter 12 in *Handbook of Powder Science & Technology*, 2nd ed., edited by M. E. Fayed, and L. Otten. New York: Chapman and Hall.

Bassett, R. 2001. What You Need to Know to Select an Automatic Sampler. *Powder & Bulk Engineering*, pp. 21–29 (March).

Bates, L. 2000. *Guide to the Specification, Design, and Use of Screw Feeders.* Bury St. Edmunds, Suffolk, UK: Professional Engineering Publishing.

Becker, S. 1999. Technology Developments in Bulk Bag Discharging. *Proc. 1999 Powder & Bulk Solids Conference,* pp. 9–17.

Berrie, P. G. 1996. Instrumentation for Level Measurement and Control. *Bulk Solids Handling,* 16(2): 265.

Bhati, M. V., and Cheremisinoff, P. N. 1977. Cyclones. Chapter 10, Part 1, *Air Pollution and Control Handbook,* edited by P. N. Cheremisinoff, and R. A. Young. New York: Marcel Dekker.

Brown, C. J., and Nielsen, J. (editors). 1996. *Silos: Fundamentals of Theory, Behavior and Design.* Stamford, CT: Thomson Learning.

Calvert, S. 1984. Particulate Control by Scrubbing. In *Handbook of Air Pollution Technology.* edited by S. Calvert and H. M. Englund. New York: John Wiley & Sons.

Calvert. S., Goldschmidt, J., Leith, D., and Mehta, D. 1972. *Wet Scrubber System Study,* Volume 1, *Scrubber Handbook.* Available from NTIS as PB-213016.

Capes, C. E. 1980. *Particle Size Enlargement.* Amsterdam and New York: Elsevier Science Publishing Company,

Carstensen, J. T. 1983. Tabletting and Pelletixation in the Pharmaceutical Industry. Section in *Handbook of Powder Science & Technology,* 1st ed., edited by M. E. Fayed and L. Otten. New York: Van Nostrand Reinhold.

Clement, S. and Purutyan, H. 2002. Narrowing Down Equipment Choices for Particle-Size Reduction. *Chem. Eng. Progr.,* 98(6), 50–54 (June).

Cole, G. 1990. *Pharmaceutical Production Facilities: Design and Applications.* New York: Ellis Horwood/John Wiley & Sons.

Colijn, H. 1983. *Weighing and Proportioning of Bulk Solids.* 2nd ed., Houston, TX: Gulf Publishing Company.

Colijn, H. 1985. *Mechanical Conveyors for Bulk Solids.* New York: Elsevier Science Publishing Company.

Cooksey, G. L. 1977. Rotary Valve Application Considerations. *Proc. Int. Powder & Bulk Solids Handling & Processing Conf.,* Rosemont, IL, May 10–12.

Cox, R. M. 2000. Configuring and Operating a Bulk Railcar Loading Station to Meet Your Needs. *Powder and Bulk Engineering,* pp. 23, 24, 26–28, 30, 32, 34, 35 (January).

Croom, M. L. 1995. *Filter Dust Collectors.* New York: McGraw-Hill.

Crouch, C. 2000. Safely Handling Your Powder with a Closed-Loop Pneumatic Conveying System. *Powder and Bulk Engineering,* pp. 43, 44, 46, 48, 50–52, 54, 56 (March).

Davidson, I. 1989a. Bulk Bag Design, Filling and Discharging. *Proc. 14th Annual Powder & Bulk Solids Conf.,* pp. 247–269.

Davidson, I. 1989b. Selecting and Using Bulk Bags—Part II. *Powder and Bulk Engineering,* pp. 19–27 (September).

DeCenso, A. J. 2001. Dry Screening of Granular Solids. *Chem. Engng.,* pp. 76-83 (April).

Fayed, M. E., and Skocir, T. S. 1997. *Mechanical Conveyors—Selection and Operation.* Lancaster, PA: Technomic Publishing Company.

Fruchtbaum, J. 1988. *Bulk Materials Handling Handbook.* New York: Van Nostrand Reinhold.

Gaylord, E. H., Jr. and Gaylord, C. N. 1984. *Design of Steel Bins for Storage of Bulk Solids.* Englewood Cliffs, NJ: Prentice-Hall, Inc.

Green, D. W., and Maloney, J. O. 1997. *Perry's Chemical Engineers' Handbook.* 7th ed., New York: McGraw-Hill.

Harnby, N., Edwards, M. F., and Nienow, A. W. *Mixing in the Process Industries.* Boston: Butterworths.

Heitbrink, W. and McKinney, W. M., Jr. 1987. Dust Control for Bag Opening Machines. *Powder and Bulk Engineering,* pp. 40–51 (May).

Heumann, W. L. 1997. *Industrial Air Pollution Control.* Chapter 8 on Cyclones. New York: McGraw-Hill.

Hicks, D. C., and Freese, H.L. 1989. Extrusion and Spheronizing Equipment, Chapter 4 in *Pharmaceutical Pelletization Technology,* edited by I. Ghelre-Sellassie. New York: Marcel Dekker, Inc.

Jenike, W. W. 1970. *Storage and Flow of Solids.* Bulletin No. 123. Salt Lake City: University of Utah Engineering Experiment Station.

Johanson, J. R. 1970. In-Line Blending. *Chem. Eng. Prog.,* 66(6): 50–55 (June).

Katz, J. 1989. *The Art of Electrostatic Precipitation.* Port Washington, NY: Scholium International Inc.

Kaye, B. H. 1996. *Powder Mixing.* New York: Chapman and Hall.

Kaye, B. H. 1997. Mixing of Powders. Chapter 11 in *Handbook of Powder Science & Technology,* 2nd ed., edited by M. E. Fayed and L. Otten. New York: Chapman and Hall.

Kennedy, M. G., and Cheremisinoff, P. M. 1977. Fabric Filter Dust Collectors, Chapter 13 in *Air Pollution Control and Design Handbook,* Part 1, edited by P. N. Cheremisinoff and R. A. Young. New York: Marcel Dekker, Inc.

Knight, E. A., and Pugh, J. R. 1996. Properly Select Level Measurement Devices for Bulk Solids. *Chem. Eng. Prog.,* 92(2): 50–55 (February 1996).

Leith, D., and Jones, D. L. 1997. Cyclones. Chapter 15 in *Handbook of Powder Science & Technology.* 2nd ed., edited by M. E. Fayed and L. Otten. New York: Chapman and Hall.

Liptak, B. G. 1995. *Instrument Engineers' Handbook—Process Measurement and Analysis Volume.* 3rd ed., Radnor, PA: Chilton Book Company.

Marcus, R. D., Leung, L. S., Klinzing, G. E., and Rizk, F. 1990. *Pneumatic Conveying of Solids.* New York: Chapman and Hall.

Masters, K. 1991. *Spray Drying Handbook.* Burnt Mill, Essex, England: Longman Scientific & Technical.

McKenna, J. D. and Turney, J. J. 1989. *Fabric Filter-Baghouses I.* ETS, Inc., Suite C-103, 3140 Chaparral Drive, SW, Roanoke, VA 24018.

Mead, W. J. 1964. *The Encyclopedia of Chemical Process Equipment.* Section on Mixing Equipment, pp. 641–662. New York: Reinhold Publishing Corporation.

Messman, H. C. and Tibbetts, T. E. (editors). 1977. *Elements of Briquetting and Agglomeration,* Las Cruces, NM: The Institute for Briquetting and Agglomeration.

Mills, D. 1990. *Pneumatic Conveying Design Guide.* Boston: Butterworths.

Moyers, C. G. 1997. Section 12 on Psychometry, Evaporation, Cooling, and Solids Drying in *Perry's Chemical Engineers' Handbook,* 7th ed., edited by D. W. Green and J. O. Maloney. New York: McGraw-Hill.

Mujumdar, A. S. (editor). 1995. *Handbook of Industrial Drying,* Volumes 1 and 2. 2nd ed., New York: Marcel Dekker, Inc.

Norden, K. E. 1993. *Electronic Weighing—Fundamentals and Applications.* Oxford, England and Boston: Butterworth-Heinemann.

Ogleby, S., Jr. and Nichols, G. 1969. *A Manual of Electrostatic Precipitator Technology.* Part I (PB 196380) and Part II (PB 196381). Available from the National Technical Information Service.

Parikh, D. M. and Parikh, C. K. (editors). 1997. *Handbook of Pharmaceutical Granulation Technology.* New York: Marcel Dekker, Inc.

Pietsch, W. 1997. Size Enlargement by Agglomeration. Chapter 6 in *Handbook of Powder Science & Technology,* 2nd ed., edited by M. E. Fayed and L. Otten. New York: Chapman and Hall.

Pietsch, W. 2002. *Agglomeration Processes: Phenomena, Technologies, Equipment.* New York: Wiley-VCH.

Pittman, E. D. 1985. Specifying Rotary Valves. *Chem. Engng.,* pp. 89-95 (July 8, 1985).

Prasher, C. L. 1987. *Crushing and Grinding Process Handbook.* New York: John Wiley & Sons.

Raymus, G. 1997. Handling of Bulk Solids and Packaging of Solids and Liquids. Section 21 in *Perry's Chemical Engineers' Handbook,* 7th ed., edited by D. W. Green and J. O. Maloney. New York: McGraw-Hill.

Ricklefs, R. D. 2000. Selecting a Sifter for Scalping, Removing Fines, or Grading. *Powder & Bulk Solids,* pp. 19–33 (December).

Ridgeway-Watt, R. 1988. *Tablet Machine Instrumentation in Pharmaceutics-Principles and Practice.* New York: Ellis Horwood/John Wiley & Sons.

Rotter, J. M. 2001. *Guide for Economic Design of Circular Metal Silos.* New York: Taylor & Francis.

Sherrington, P. J. and Oliver, R. 1981. *Granulation.* London: Heyden & Son Ltd.

Snow, R. H. 1997. Size Reduction and Size Enlargement, Section 20 in *Perry's Chemical Engineers' Handbook,* 7th ed., edited by D. W. Green and J. O. Maloney. New York: McGraw-Hill.

Soderholm, A. 1999. A Bulk Solid Mass Flow Meter Using the Coriolis Principle. *Powder Handling & Processing,* pp. 297–300 (July/September).

Stanley-Wood, N. G. 1983. *Enlargement and Compaction of Particulate Solids.* Boston: Butterworths.

Stein, M. R. 1990. Gravity Blenders: Storing and Blending in One Step. *Powder & Bulk Engineering,* 4(1): 32–36.

Stern, A. C. (editor). 1977. *Air Pollution,* Vol. IV, *Engineering Control of Air Pollution,* 3rd ed. New York: Academic Press, Inc.

Thomson, F. M. 1997. Storage and Flow of Particulate Solids. Chapter 8 in *Handbook of Powder Science & Technology,* 2nd ed., edited by M. E. Fayed and L. Otten. New York: Chapman and Hall.

Vetter, G. 1998. *Dosing Handbook.* Kidlington, Oxford, England: Elsevier Advanced Technology.

Weinekotter, R., and Gericke, H. 2000. *Mixing of Solids* (Powder Technology Series, No. 12). Norwell, MA:. Kluwer Academic Publishers.

White, H. G. 1963. *Industrial Electrostatic Precipitation.* Addison, MA: Addison-Wesley Publishing Company.

Williams-Gardner, A. 1971. *Industrial Drying.* Cleveland: CRC Press.

Wilms, H. 1992. Blending Silos: An Overview. *Powder Handling & Processing,* 4(3): 293ff.

Woodcock, C. R., and Mason, J. S. 1987. *Bulk Solids Handling.* New York: Chapman and Hall, pp. 31–32.

Zenz, F. A. 2001. Cyclone Design Tips. *Chem. Engng.*, pp. 60–64 (January).

Zoppa, J. 2001. Personal communication from Jack Zoppa, Vac-U-Max, Belleville, NJ to S. S. Grossel (September 5, 2001).

ACRONYMS AND ABBREVIATIONS

ACC	American Chemistry Council
ACGIH	American Conference of Governmental Industrial Hygienists
AIChE	American Institute of Chemical Engineers
AIHA	American Industrial Hygiene Association
AIT	Autoignition Temperature
ANSI	American National Standards Institute
APTAC	Automatic Pressure Tracking Adiabatic Calorimeter
API	American Petroleum Institute
ARC®	Accelerating Rate Calorimeter
ARSST	Advanced Reactive System Screening Tool
ASME	American Society of Mechanical Engineers
ASTM	American Society for Testing and Materials
AWS	American Welding Society
BAM	Bundesanstalt für Materialprufung
BIA	Berufsgenossenschaftliches Institut für Arbeitssicherheit
BMHB	British Materials Handling Board
BSI	British Standards Institute
CAAA	Clean Air Act Amendments (U.S.)
CAP	Community Advisory Panel
CART	Calculated Adiabatic Reaction Temperature
CAS	Chemical Abstract Services
CCPS	Center for Chemical Process Safety
CFD	Computational Fluid Dynamics
CFR	Code of Federal Regulations (U.S.)
CIRC	Chemical Incidents Report Center
CPI	Chemical Process Industries
CPSC	Consumer Product Safety Commission
CRHF	Chemical Reaction Hazards Forum
CSB	Chemical Safety and Hazard Investigation Board (U.S.)
DIERS	Design Institute for Emergency Relief Systems
DOT	Department of Transportation (U.S.)
DSC	Differential Scanning Calorimetry
DTA	Differential Thermal Analysis
EPA	Environmental Protection Agency (U.S.)
ESP	Electrostatic Precipitation
FDA	Food and Drug Administration (U.S.)
FFS	Fit-for-Service
FIBC	Flexible Intermediate Bulk Container
FMG	Factory Mutual Global
fpm	Feet per minute
GDC	General Duty Clause
GE	General Electric Company
GMP	Good Manufacturing Practice
GSA	General Service Administration (U.S.)
HAZOP	Hazard and Operability Analysis
HEPA	High Efficiency Particle Air
HRD	High Rate Discharge
HSE	Health and Safety Executive (UK)
IChemE	Institution of Chemical Engineers (UK)
ICI	Imperial Chemical Industries
IDLH	Immediately Dangerous to Life and Health

IEC International Electrotechnical Commission
IEST Institute of Environmental Sciences and Technology
ILO International Labor Organization
IR Infrared
IRI Industrial Risk Insurers
IRIS Integrated Risk Information System
ISA The Instrumentation, Systems, and Automation Society
ISO International Standards Organization
J Joule
LC_{50} Lethal Concentration-50%
LD_{50} Lethal Dose-50%
LEL Lower Explosive Limit
LEPC Local Emergency Planning Committee
LEV Local Exhaust Ventilation
LFL Lower Flammable Limit
LIW Loss in Weight
LOC Limiting Oxidant Concentration
LOTO Lockout/Tagout
MACT Maximum Available Control Technology
MAIT Minimum Autoignition Temperature
MARS Major Accident Reporting System
MEC Minimum Explosible Concentration
MI Mechanical Integrity
MIE Minimum Ignition Energy
MIT Minimum Ignition Temperature
mJ Millijoule
MOC Management of Change
ms Milliseconds
MSDS Material Safety Data Sheet
NAAQS National Ambient Air Quality Standard (U.S.)
NACE National Association of Corrosion Engineers
NEC National Electrical Code
NFPA National Fire Protection Association
NIOSH National Institute for Occupational Safety and Health
NIST National Institute of Standards and Technology
NOAA National Oceanic and Atmospheric Administration
NPCA National Paint and Coatings Association
OEL Occupational Exposure Limit
OSHA Occupational Safety and Health Administration (U.S.)
PBD Propagating Brush Discharge
PEL Permissible Exposure Limit
PHA Process Hazard Analysis
PPE Personal Protective Equipment
PR Pressure Ratio (Deflagration)
PS Pinch Style
PSM Process Safety Management
PVC Polyvinyl Chloride
RBI Risk-Based Inspection
REL Recommended Exposure Limit
RIBC Rigid Intermediate Bulk Container
RSIT Relative Self-Ignition Temperature
RSST Reactive System Screening Tool
RTECS Registry for Toxic Effects of Chemical Substances
SADT Self-Accelerating Decomposition Temperature
SEDEX Sensitive Detection for Exothermic Processes
SEPV Stepped End Pasted Valve
SOM Sewn Open Mouth
SIT Self-Ignition Temperature or Spontaneous Ignition Temperature
STEL Short Term Exposure Limit
TG Thermogravimetric Analysis
TLV® Threshold Limit Value
TSCA Toxic Substances and Control Act (U.S.)
UEL Upper Explosive Limit
UFL Upper Flammable Limit
UK United Kingdom
UL Underwriters Laboratories
UN United Nations
UV Ultraviolet
VDI Verein Deutscher Ingenieure
VSP Vent Sizing Package
VSP2 Vent Sizing Package, version 2
WEEL Workplace Environmental Exposure Limit
WHMIS Workplace Hazards Materials Information System
WHO World Health Organization

GLOSSARY

Acute Single, short-term exposure (less than 24 hr)

Aerodynamic Diameter The diameter of a sphere of the same particle density having the same terminal velocity in air or some other relevant fluid.

Aerodynamic Equivalent Diameter The diameter of a unit density sphere having the same settling velocity (due to gravity) as the particle of interest of whatever shape and density.

Angle of Repose Angle formed between horizontal plane and the slope line extending along the face of a particulate pile; used to characterize particulate fluidity

Antistatic Having the ability to dissipate charge at a sufficient rate to prevent hazards or nuisances under the condition of use.

Authority Having Jurisdiction The organization, office, or individual responsible for approving equipment, materials, an installation, or a procedure.

Autocatalysis The increase of the rate of reaction due to the catalyzing effect of the reaction products.

Autodecomposition The sustained decomposition of a substance without introduction of any other apparent ignition source besides thermal energy and without air or other oxidants present.

Autodecomposition temperature The minimum temperature for a specified test method, test apparatus (including material of construction and test volume) and initial pressure required to initiate self-sustained decomposition of a solid, liquid or gaseous substance without any other apparent source of ignition and without air or other oxidants present.

Autoignition Temperature (AIT) The minimum temperature required to initiate or cause self-sustained combustion, in air, with no other source of ignition.

Barrier/Isolator A sealed enclosure within which operations can be carried out without exposing the operators or the surrounding environment to contamination from the process materials inside it or vice versa.

Blanketing The technique of maintaining an atmosphere that is either inert or fuel-enriched in the vapor space of a container or vessel. Also called "Padding."

Bonding The process of connecting two or more conductive objects together by means of a conductor so that they are at the same potential as each other, but not necessarily at the same potential as the earth.

Breakdown Voltage/Strength The minimum voltage for spark breakdown to occur across a material of given thickness held between electrodes producing a uniform electric field under specified test conditions, expressed as volts per unit thickness.

Brush Discharge A higher energy form of corona discharge characterized by low frequency bursts or streamers. The simplest type forms between charged, isolated nonconductive surfaces and grounded conductors acting as electrodes.

Bulking Brush Discharge A partial surface discharge created during bulking of powder in containers, appearing as a luminous, branched channel flashing radially from the wall toward the center of the container. Its maximum effective energy with respect to dust ignition is believed to be 10–20 mJ (less than the MIE of Lycopodium).

Capacitance (C) The constant of proportionality between charge and potential difference for systems of conductive bodies. The capacitance, in Farads, is the charge in Coulombs which must be communicated to raise the potential difference by one volt.

Carcinogen A material that has been found by the International Agency for Research on Cancer or the National Toxicology Program to produce or potentially produce some type of cancer

Carr Indices Measures of particulate physical properties (such as bulk density) based on standard test methods.

Cascade Impactor Device used to collect and segregate particles by their aerodynamic diameters.

Charge (Q) An excess or deficit of electrons expressed in Coulombs. An electron carries an electric charge of -1.6×10^{-19} Coulombs.

Charge Density (Q_v) The quantity of charge divided by the total volume that contains this charge, expressed in Coulombs per cubic meter.

Charge Neutralizer A device that increases the conductivity of air so that charge drains away to ground.

Charge Relaxation The process by which separated charges recombine or excess charge is lost from a system.

Chemical Barrier *See* Suppressant Barrier

Chronic Long-term, usually multiple exposures

Cleanroom A room in which the concentration of airborne particles is controlled, and which is constructed and used in a manner to minimize the introduction, generation, and retention of particles inside the room and in which other relevant parameters (e.g., temperature, humidity, and pressure) are controlled as necessary.

Combustible Concentration Reduction The technique of maintaining the concentration of combustible material in a closed space below the lower flammable limit. This is also called "Air Dilution."

Combustible Dust Any finely divided solid material that is 420 microns or smaller in diameter (material passing through a U.S. No. 40 standard sieve) and presents a fire or explosion hazard when dispersed and ignited in air or other gaseous oxidizer.

Combustible Particulate Solid Any combustible solid material, composed of distinct particles or pieces, regardless of size, shape, or chemical composition.

Combustion A chemical process of oxidation that occurs at a rate fast enough to produce heat and usually light, in the form of either a glow or flames.

Compatibility The ability of materials to exist in contact without specified (usually hazardous) consequences under a defined scenario.

Conductive Having a conductivity greater than 10^4 picosiemens per meter (pS/m) or a resistivity less than 10^8 Ohm-meters (Ω-m).

Conductive Floor A floor having an average resistance between 2.5×10^3 Ohms as measured using specified electrodes placed a specified distance apart.

Conductive Hose A hose having an electrical resistance less than 10^3 Ohms per meter of hose length, based on measurement between the end connectors.

Conductivity (κ) The reciprocal of resistivity as Siemens per meter or more usually as picosiemens per meter, where 1 picosiemen (pS) = 1×10^{-12} Siemen.

Cone (or Conical Pile) Discharge See Bulking Brush Discharge.

Confidence Interval a range of values, intended to include the true LD_{50} with a specified degree of confidence

Containment (for Toxic Particulate Solids) The enclosure of process and/or control of the environment in which it is carried out in order to prevent the contamination of people by the materials used or vice versa.

Containment Strategy A generalized approach for limiting or eliminating exposure of personnel, products, or the environment to hazardous concentrations of substances.

Corona Discharge A self-sustained low energy electrical discharge with nonthermal ionization that takes place in the vicinity of an electrode of sufficiently low radius of curvature, in a medium whose presence is typically close to atmospheric.

Corrosive A chemical that causes visible destruction of, or irreversible alterations in, living tissue by chemical action at the site of contact.

Coulomb (C) The quantity of electricity on the positive plate of a capacitor of 1 Farad capacitance when the potential difference across the plate is 1 volt.

Critical Effect The first adverse effect, or its known precursor, that occurs as the dose rate (or concentration) increases.

Cube-Root (or Cubic) Law See K_{St} Value

Current (I) The rate of transfer of electricity normally expressed in Amperes (Coulombs per second).

Deflagration Propagation of a combustion wave through a fuel–oxidizer mixture at a rate that is less than the speed of sound in the unreacted medium and capable of producing a significant increase in pressure.

Deflagration Isolation A method employing equipment and procedures that interrupts the propagation of a deflagration flame front past a point (usually in a pipe).

Deflagration Pressure Containment The technique of specifying the design pressure of a vessel and its appurtenances so that they are capable of withstanding the maximum pressures resulting from an internal deflagration.

Deflagration Suppression The technique of detecting and arresting combustion in a confined space while the combustion is still in its incipient stage, thus preventing the development of pressures that could result in an explosion.

Deflagration Venting The reduction of pressure generated in a vessel by a deflagration by allowing the emergency flow of the vessel contents from the vessel by means of an opening in the vessel, thus avoiding the failure of the vessel by overpressure. The vent opening is usually closed by a pressure-relieving cover (e.g., rupture disk, explosion disk or hatch).

Denier Fiber weight per unit length (gram/9000-m length)

Detonation A release of energy caused by the propagation of a chemical reaction in which the reaction front advances into the unreacted substance at greater than sonic velocity in the unreacted material.

Dielectric Constant (ε_r) A dimensionless parameter expressing the ratio of the permittivity of a material to that of vacuum.

Dose The amount of test substance administered during testing, expressed in weight of test substance per unit weight of test animal (mg/kg)

D_{tex} Fiber weight per unit length (gram/10-km length)

Explosibility The ability of a dust to take part in a closed explosion when dispersed in air at a suitable concentration and in the presence of an effective ignition source.

Explosibility Index A numerical measure of the explosion hazard potential of a dust as determined by multiplying the dust Ignition Sensitivity Index by its Explosion Severity Index.

Explosion A rapid or sudden release of energy that causes a potentially damaging pressure increase.

Explosion-Pressure-Resistant Property of vessels and equipment designed to withstand the expected explosion pressure without becoming permanently deformed.

Explosion-Pressure-Shock-Resistant Property of vessels and equipment designed to withstand the expected explosion pressure without rupturing, but allowing permanent deformation.

Explosion Severity Index An overall measure of the potential hazard associated with a closed vessel explosion of a combustible dust as given by Equation [4.2] (page 156).

Fast-Acting Valve A valve that closes a path of deflagration propagation in a pipe or duct in response to upstream detection of a deflagration.

Flame A region in which chemical interaction between gases occurs, accompanied by the evolution of light and heat.

Flame Barrier A device that prevents transmission of a flame from a source to a receptor.

Flame Front That portion of the flame reaction zone moving into the unburned gas where the bulk of the reaction occurs and the medium reaches its ignition temperature.

Flame Front Diverter A device that opens in response to the pressure wave preceding the flame front of the deflagration, venting the pressure wave and flame front.

Flame Propagation The movement of a flame front in piping or equipment.

Flame Speed The speed of a flame front relative to a fixed reference point. Flame speed is dependent on turbulence, the equipment geometry, and the fundamental burning velocity.

Flammable Limits The minimum and maximum concentration of a particulate solid (dust) in a dust/gaseous oxidant mixture (usually expressed in percent by volume) defining the concentration range (flammable or explosive range) over which propagation of flame will occur on contact with an ignition source. *See also* Lower Flammable Limit and Upper Flammable Limit.

Flammable Range The range of concentrations between the lower and upper flammability limits.

Grounding (Earthing) Ideally, the process of connecting one or more conductive objects to ground so that each is at the same potential as the earth. By convention, the earth has zero potential. In practice, grounding is the process of providing a sufficiently small resistance to ground so that a static hazard cannot be created at the maximum credible charging current to a system.

Hazardous Particulate Solids In a broad sense, any particulate solids or mixture of particulate solids having properties capable of producing adverse effects to the health or safety of process equipment, human beings, or the environment. These can include such properties as combustibility, flammability, explosibility, toxicity, corrosiveness, etc.

Hot Work Any operation that uses flames or can produce sparks (e.g., welding).

Hybrid Mixture A mixture of a combustible gas with either a combustible dust or combustible mist.

Immediately Dangerous to Life and Health (IDLH) The concentration that poses a threat of either death, adverse health effects, or inability to escape danger without respiratory protective equipment.

Ignition Sensitivity Index An overall measure of the ignitability of a particulate material as determined from Equation [4.1] (page 156).

Inert Gas A nonflammable, nonreactive gas that can be used to render the combustible material in a system incapable of supporting combustion.

Inerting A technique by which a combustible mixture is rendered non-ignitable by addition of an inert gas or a noncombustible dust. *See* Oxidant Concentration Reduction.

Inhalation Reference Concentration The estimated continuous inhalation exposure concentration (mg/m^3) that is likely to be without appreciable risk of deleterious noncancer effects during a human lifetime.

Instability The degree of intrinsic susceptibility of a material to release energy through self-reaction (polymerizing, decomposing, or rearranging).

Interlock System A system that detects out-of-limits or abnormal conditions or improper sequences and either halts further action or starts corrective action.

Ionization The dissociation of air into ions by means of electrical energy.

Irritant A chemical, which is not corrosive, but which causes a reversible inflammatory effect on living tissue by chemical action at the site of contact.

Isolation A means of preventing certain stream properties (deflagration, mass flow, ignition capability) from being conveyed past a predefined point.

Isolation Room A room designed, constructed, and provided with control systems to prevent escape of health-hazardous particulate solids into the atmosphere or adjacent operating areas.

Karl Fischer Titration Chemical method for measuring particulate moisture content based on a reaction with an iodine bearing reagent

K_{St} Value The deflagration index of a dust cloud. It is a dust-specific measure of the explosibility, in units of bar-m/s, and calculated using Equation [6-10] (page 413). The equation is the so-called "cubic" or "cube root" law.

LD_{50} Median Lethal Dose A statistically derived single dose of a chemical that is expected to cause death in 50% of tested animals.

Limit Dose The upper limit dose used in testing (2000–5000 mg/kg)

Limiting Oxidant Concentration (LOC) Minimum oxygen concentration at which a suspended cloud of combustible particulate can be ignited

Lower Flammable Limit (LFL) That concentration of a combustible material in air below which ignition will not occur. It is often, interchangeably, called Lower Explosive Limit (LEL), and for dusts, the Minimum Explosible Concentration (MEC).

Lowest-Observed-Adverse-Effect-Level (LOAEL) The lowest exposure concentration at which there are statisitically and biologically significant increases in the frequency or severity of adverse effects between the exposed population and its appropriate control group, i.e., lowest observed adverse effect level.

Material Choke A mass of bulk solids or powders in a rotary valve or screw feeder that prevents a flame from being transmitted.

Maximum Explosion Pressure (P_{max}) The maximum pressure occurring in a closed vessel during the explosion of an explosible dust atmosphere determined under specific test conditions.

Maximum Rate of Explosion Pressure Rise [$(dp/dt)_{max}$] The maximum value of he pressure rise per unit time during explosions of all explosive atmospheres in the explosible range of a combustible particulate solid in a close vessel under specified test conditions.

Mechanical Integrity Program A program to ensure that process equipment and systems are and remain mechanically suitable for operation. It involves inspection, testing, upgrading and repairs of equipment, as well as written procedures to maintain on-going integrity of equipment

Minimum Explosible Concentration (MEC) The lowest concentration of a combustible dust in air, expressed in grams per cubic meter, that will propagate a flame. Also often called Lower Flammable Limit (LFL) or Lower Explosion Limit (LEL).

Minimum Ignition Energy (MIE) The minimum amount of energy released at a point in a combustible mixture that caused flame propagation away from the point, under specified test conditions. The lowest value of the minimum ignition energy is found at a certain optimum mixture. The lowest value is usually quoted as the minimum ignition energy.

Minimum Ignition Temperature (MIT) of a Dust Cloud The lowest temperature of a hot surface on which the most ignitable mixture of dust with air is ignited under specified test conditions.

Minimum Ignition Temperature (MIT) of a Dust Layer The lowest temperature of a hot surface at which ignition occurs in a dust layer under specified test conditions.

Mitigation Lessening the risk of an accident event sequence by acting on the source in a preventive way by reducing the likelihood of occurrence of the event, or in a protective way by reducing the magnitude of the event and/or the exposure of local persons or property.

Mutagenic Toxin A chemical that produces chromosomal damage.

Neurotoxin A chemical which produce their primary toxic effects on the nervous system

Noncombustible Material A material that, in the form in which it is used and under the conditions anticipated, will not ignite, support combustion, burn, or release flammable vapors when subjected to fire or heat

Nonconductive Possessing a conductivity less than 10^2 pS/m or a resistivity greater than 10^{10} Ω-m.

No-Observed-Adverse-Effect-Level (NOAEL) An exposure concentration at which there are no statistically and biologically significant increases in the frequency or severity of adverse effects between the exposed population and its appropriate control group; i.e. there are no adverse physiological effects.

Oxidant Any gaseous material that can react with a fuel (either gas, dust, or mist) to produce combustion. Oxygen in air is the most common oxidant.

Oxidant Concentration Reduction The technique of maintaining the concentration of the oxidant in a closed space below the concentration required for combustion to occur. This is commonly called "inerting."

Particulate Material Any solid material composed of distinct particles or pieces, regardless of size, shape, or chemical composition. Particulate materials include dusts, fibers, fines, chips, chunks, flakes, or mixtures of these.

Permissible Exposure Limit (PEL) The maximum permissible exposure limit for systemic workplace 8-hour time-weighted average exposures as established by OSHA

Phase Doppler Laser Diffraction Optical method for simultaneously measuring particles size and velocity distribution.

Plosive Density A chemical group contribution method for predicting the chemical instability explosion potential of a material.

Powder Generic term for subdivided solid material, comprising pellets, granules, and dust. Pellets have a diameter greater than 2 mm and typically above 3 mm, although they may contain granules and small quantities of dust. Granules (e.g., granulated sugar) have a diameter between 0.42 and 2 mm, although granular powders typically contain dust. Dusts have a diameter less than 0.42 mm and as low as 1 micron (μm). Suspended particles smaller than 1 μm are know as fumes. In European publications, pellets are typically described as "granules."

Predictive Maintenance Maintenance that is based on real-time condition-based data input to continuously adjust the profile of equipment performance expectations, and accordingly schedules inspections and repairs.

Preventive Maintenance Maintenance that seeks to reduce the frequency and severity of unplanned shutdowns by establishing a fixed schedule of routine inspection and repairs.

Probit Abbreviation for "Probability Integral Transformation" used for correlating dose–response probabilities.

Process Hazard Analysis (PHA) The identification of undesired events that lead to hazardous events, the analysis of the mechanisms by which these undesired events could occur, and usually the estimation of the consequences.

Propagating Brush Discharge An energetic discharge caused by the electrical breakdown of the dielectric layer in a capacitor and massive lateral surface discharge resulting in dissipation of most of the stored charge.

Purge Gas A gas that is continuously or intermittently added to a closed system to render the atmosphere in it noncombustible. The purge gas can be inert or combustible.

Purging The removal of oxygen from a closed system (equipment and piping) to below the LOC of the particulate solid by means of applying pressure, vacuum, and/or a purge gas.

Pyrophoric A chemical that will ignite spontaneously in air at a temperature of 130°F (54.4°C) or below.

Pyrophoric Particulate Solid A particulate solid that undergoes such vigorous oxidation or hydrolysis (often with evolution of highly flammable gases) when exposed to atmospheric oxygen or to water, that it rapidly ignites without an external source of ignition. This is a special case of spontaneous combustion.

Reduced Explosion Pressure (P_{red}) The maximum pressure developed in a vented enclosure during a vented deflagration, or during deflagration suppression.

Regional Deposited Dose The deposited dose (per unit surface area, mg/cm^2) of particles calculated for the respiratory tract region of interest.

Resistance The property of conductors, depending on material, dimensions, and temperature, which determines the current produced by a given difference of potential, expressed in ohms.

Resistivity The volume resistance of a sample of material having unit length and cross-sectional area. If the length is expressed in meters and the area in square meters, the resistivity is given in ohm-meters.

Reversal A situation in which a non-response occurs at one dose, and a positive response is observed at the next dose tested, or vice versa.

Runaway Reaction A reaction that is out of control because the rate of heat generation by an exothermic chemical reaction exceeds the rate of cooling available.

Sauter Mean Diameter The ratio of the cube of the volume mean diameter to the square of the surface mean diameter; roughly the ratio of the particle volume to its surface area.

Self-Accelerating Decomposition Temperature (SADT) The lowest temperature at which self-accelerating decomposition may occur with a substance in the packaging as used in transport.

Self-Heating of Powders *See* Spontaneous Combustion (Heating) of Powders.

Semiconductive Possessing a conductivity between 10^2 and 10^4 pS/m or a resistivity between 10^8 and 10^{10} Ω-m.

Sensitizer A chemical that causes a substantial proportion of exposed people or animals to develop an allergic reaction in normal tissue after repeated exposure to the chemical.

Short Term Exposure Limit (STEL) Is the limiting exposure concentration for exposure durations up to 15 minutes, as developed by the ACGIH.

Siemen Reciprocal ohm, formerly called a "mho." *See* Conductivity.

Smoldering Flameless, oxygen diffusion limited combustion within either a porous char-forming material or a particulate bed.

Smoldering Nest A small smoldering region within either a dust layer or a much larger quantity of particulate material.

Spark Discharge Transient discrete electric discharge which takes place between two conductors which are at different potentials, bridging the gap in he form of a single ionization channel.

Spark Extinguishing System An extinguishing system in which the radiant energy of a spark or an ember is detected and the spark or ember is quenched.

Specific Surface Area The ratio of the particle surface area to the particle mass.

Spinning Riffler A device for obtaining a representative sample of particulate material from a large bag or container. See Figure 4.3 (page 165).

Spontaneous Combustion (Heating) of Powders Ignition of powders in bulk caused by the rate of heat generation from oxidation and/or exothermic decomposition reactions of the powders being greater than the rate of heat loss to the surroundings.

Static Activation Pressure Pressure that actuated a vent closure when the pressure is increased slowly (with a rate of pressure rise less than 0.1 bar/min (0.15 psi/min).

Static Electricity The class of phenomena recognized by the presence of electrical charges, either stationary or moving, and the interaction of these charges, this interaction being solely by reason of the charges themselves and their position, and not by reason of their motion.

Stoichiometric Mixture A balanced mixture of fuel and oxidizer such that no excess of either remains after combustion.

Subchronic Exposure Multiple or continuous exposures occurring for approximately 10% of a species lifetime, usually over 3 months.

Suppressant A chemical agent used in a deflagration suppressant system to extinguish the deflagration.

Suppressant Barrier Isolation system using a suppressant.

Teratogenic Toxin A chemical that produces harmful effects on fetuses.

Threshold Limit Value (TLV) The maximum exposure concentration recommended by the American Conference of Government Industrial Hygienists (ACGIH) for long term exposures.

Threshold Limit Value (TLV®) An occupational exposure limit developed by the ACGIH which is defined as the maximum allowable concentration of an airborne contaminant that can be tolerated by man with no bodily discomfort or impairment of bodily functions, either immediately or after years of exposure. Three different categories of the TLV have been further defined as a time-weighted average (TLV-TWA), short-term exposure level (TLV-STEL), and a ceiling concentration (TLV-C). The TLV-TWA concentration limit is appropriate for a normal 8-hour workday and 40-hour workweek, and is an airborne level to which nearly all workers may be repeatedly exposed, day after day, without adverse effects. The TLV-STEL limit is the maximum concentration of an airborne contaminant which may be reached, but not exceeded, on up to four occasions during a day for a maximum of 15 minutes each time with each maximum exposure separated by at least one hour. The TLV-C limit is the atmospheric concentration of an airborne contaminant which may not be exceeded, even instantaneously.

Unstable material A material that, in the pure state or as commercially produced, will vigorously polymerize, decompose or condense, become self-reactive, or otherwise undergo a violent chemical change under conditions of shock, pressure, or temperature.

Upper Flammable Limit (UFL) That concentration of a combustible material in air above which ignition will not occur. It is often, interchangeably, called Upper Explosive Limit (UEL).

Vent Duct Piping attached to a vent on indoor equipment to direct the fireball and products of combustion to the outside of a building

Ventilation The process of supplying or removing an atmosphere to or from any space by natural or mechanical means.

Water-Compatible A material that is neither reactive with water nor incompatible with water and consequently can be extinguished with a water-based extinguishing system.

Water-Incompatible A material that does not chemically react with water but undergoes a change of phase or state upon mixture with water that renders it permanently changed or incompatible with the remainder of the process.

Water-Reactive A material that chemically reacts with water, producing some other compound that can represent a different set of fire protection concerns.

INDEX

D

I

J

K

L

M

N

O